Recommended by different Universities of West Bengal, Assam and other States of India

OUTLINES OF BIOLOGY

Meant for

PREMEDICAL, INTERMEDIATE, PRE-UNIVERSITY AND HIGHER SECONDARY COURSES

COVERING THE SYLLABUS OF THE BOARDS AND COUNCILS ALL OVER INDIA

K S Das MSc
Formerly Head of the Department of Botany
Surendranath College, Kolkata

AND

J B Mookerji MSc, MB
Formerly Head of the Department of Biology and Zoology
Surendranath College, Kolkata

New Central Book Agency (P) Ltd

LONDON

HYDERABAD ERNAKULAM BHUBANESWAR

DELHI KOLKATA PUNE GUWAHATI

REGD. OFFICE

8/1 Chintamoni Das Lane, Kolkata 700 009, India
ncbapvtltd@eth.net

OVERSEAS

NCBA (UK) Ltd, 149 Park Avenue North, Northampton, NN3 2HY, UK
ncbauk@yahoo.com

EXPORT

NCBA Exports Pvt. Ltd, 212 Shahpur Jat, New Delhi 110 049
ncbaexports@ncbaindia.com

BRANCHES

208B Peacock Lane, Shahpur Jat
New Delhi 110 049
ncbadel@ncbapvtltd.com

Shop Nos. 3 & 4, Vinayak Towers
681/B Budhwar Peth, Appa Balwant Chowk
Pune 411 002
ncbapune@ncbapvtltd.com

Shop No. 15, Bharati Towers
Forest Park, **Bhubaneswar** 751 001
ncbabub@ncbapvtltd.com

House No. 3-1-315, 1st Floor
Nimboliadda, Kachiguda, **Hyderabad** 500 027
ncbahydb@ncbapvtltd.com

GSS Shopping Complex, 1st Floor
Opposite College Post Office, Convent Road
Ernakulam 682 035
ncbaernk@ncbapvtltd.com

Radhamayee Complex (Second Floor)
Opp. SBI, Panbazar, Jaswanta Road
Guwahati 781 001
ncbaguwahati@ncbapvtltd.com

OUTLINES OF BIOLOGY • K S Das, J B Mookerji

First Published: July 1950
Ninth Edition: 1983
Reprinted: 1991, 1993, 1995, 1997, 1999, 2000,
2002, 2003, 2005, 2006, 2008, 2012

PUBLISHER

New Central Book Agency (P) Ltd
8/1 Chintamoni Das Lane, Kolkata 700 009

PRINTER

New Central Book Agency (P) Ltd
Web-Offset Division, Dhulagarh, Sankrail, Howrah

ISBN 978 81 7381 099 2

Price: ₹ 235.00

PREFACE TO THE EIGHTH EDITION

I have much pleasure in presenting the revised edition of 'Outlines of Biology'. Some new topics have been incorporated, and many chapters have been revised, often re-written, in the light of changed Syllabi prescribed by different Universities and Boards of Studies. Thus honest attempts have been made to make the volume thorough and uptodate, so that it can adequately meet the requirements of the students of different standards, for whom it is intended.

Let me express my sincere gratitude to Prof. Simananda Adhikari of Krishnanagar Govt. College for kindly going through the text and making very valuable contributions, particularly in the Zoology part. I am also thankful to many fellow teachers for their suggestions and keen interest in the improvement of the book.

K. S. DAS

PREFACE TO THE FIRST EDITION

'Outlines of Biology' is just an outcome of the long experience of the authors as teachers in one of the leading colleges of India and as examiners in more than one University. The growth of the 'science of life' has been very rapid and the students are getting increasingly attracted towards this fascinating and none-the-less important branch of natural science. The want of a standard textbook of Biology for the students of all the Universities in India and Pakistan is even now keenly felt. Though the authors do not claim this to be an ideal book, yet it is an humble attempt to remove that long-felt want.

The authors have tried their best to treat Biology as a single subject and avoid, as far as possible, the unfortunate tendency of splitting the parent science into its daughter components—Botany and Zoology. Such separation may possibly make the subject matter more straightforward for the students in the degree course, but at the same time this is sure to create confusion as regards the close relationship existing between the plants and the animals. With this end in view, the authors have presented a general account of life and of living things, the properties and structural details of matter in general and of living matter in particular, and the basic agreements and differences between the various living systems including a broad hint as to their mode of origin, in the introductory section. The activities connected with life are sought to be explained in terms of the physical science and the proximity between the various branches of the natural science is emphasised at every stage. The main subject matter is then treated in two parts—Botany, dealing with plants and Zoology, with animals. In the concluding chapters on evolution and heredity, the fundamental principles uniting the living things have been firmly established.

The authors have taken particular care in describing types available in this subcontinent and in citing examples from local flora and fauna. Diagrams have been prepared under our direct supervision and from nature as far as possible. It hardly needs mentioning that the syllabus of almost all the Universities of India and Pakistan have been covered in this volume and consequently a number of alternative types have been described.

The authors are quite conscious of their shortcomings. As the book had to be hurriedly seen through the press, some errors have unaviodably crept in. Criticism and suggestions from brother teachers will be highly appreciated.

We owe a deep debt of gratitude to Dr. I. Banerji, D.Sc., F.N.I., Prof. H. K. Mookerjee, D.Sc., D.I.C. and Dr. C. Datta, Ph.D., D.I.C., of the University College of Science, Calcutta, and to Prof. A. C. Dutta, M.Sc., of Cotton College, Gauhati, Assam, for valuable suggestions. We thank Professors A. K. Sinha, M.A., B.L., K. C. Banerji, M.A., A. N. Banerji, M.Sc., and R. C. Bhattacharyya, M.Sc., of Surendranath College for kindly going through portions of the manuscript. Lastly we express our hearty thanks to our young colleagues Sri S. K. Chaudhuri, M.Sc., and Sri K. P. Chakrabarti, M.Sc., for material help.

K. S. Das
J. B. Mookerji

CONTENTS

INTRODUCTION

PART I. BOTANY

MORPHOLOGY

ANATOMY

PHYSIOLOGY

ECOLOGY

SYSTEMATIC BOTANY

INTRODUCTION

BIOLOGICAL SCIENCE

Biology is the science of life. It is the study of life in man, in other animals as well as in plants. The word is derived from two Greek nouns, *bios* (meaning, life) and *logos* (meaning, a discourse). It was adopted as a scientific term by **Jean Lamarck** (1744–1829, Fig. 1) in the early part of the nineteenth century and soon spread into general use to denote the study of living organisms. Biology is a branch of the natural science. It is chiefly concerned with the structures and activities of plants and animals just as the various other branches of the natural science, namely astronomy, chemistry, physics and geology deal with the properties of lifeless matters.

Fig. 1. Jean Lamarck (1744–1829).

Biology, like physics and chemistry, is a basic science. It aims at explaining the activities of living organisms as well as various life phenomena in terms of chemistry and physics. It must, however, be admitted that life involves some factors outside the realms of physics and chemistry, and consequently all phenomena of life are not subject to the laws and theories of the inorganic sciences. Biology is concerned not merely with correlation between parts of the living body, but it also endeavours to find out the relationship between the organism as a whole and its environment. It enquires into the origin of life and explains the gradual evolution of the infinite variety of modern plants and animals. It focusses a strong searchlight through the dim vistas of the bygone ages to explore the *fossil remains* of prehistoric forms of life that are now extinct and lost to the world. It assigns a proper position to every kind of plant and animal by placing them into graded groups depending upon similarity and natural kinship.

LIFE

'Life' is abstract. It has, perhaps, no objective reality apart from the living body and, hence, evades definition. **Aristotle** (384–322 B.C.), the ancient Greek naturalist and father of modern biology (Fig. 2), recognised plants and animals as living objects, but failed to put forward a precise definition of life. Though life has not been accurately defined even today, yet the presence of life can easily be inferred from the activities of the organisms.

Life has been compared to the flame of a candle which is produced by the burning of the candle substance. Burning or oxidation is a chemical change that induces a rise of temperature in the surrounding invisible medium. Heated particles are thrown into violent vibrations which are manifested in the form of the flame emitting heat and light energies. It is to be noted that the flame has no separate existence apart from the candle, and that it persists as long as the burning or oxidation goes on. Life, similarly, finds expression in the various activities of plants and animals and it has no real existence outside the living body. The flame flickers out when the candle substance supplying energy to it is exhausted. Life ceases to exist when the store of energy in the living body is totally spent up.

Fig. 2. Aristotle (384–322 B. C.)

Life, then, involves a continuous expense of *energy* just comparable to the constant loss of heat and light by the candle flame. Energy is defined as the capacity for doing work. It exists in two principal forms—*potential* and *kinetic*. Energy capable of doing work, such as that stored in the living body or the candle substance, is known as potential energy. It is energy at rest. Kinetic energy is the dynamic variety. It is the energy which is manifesting itself, that is, which is actually at work. Life is a ceaseless conversion of the potential energy into the kinetic form.

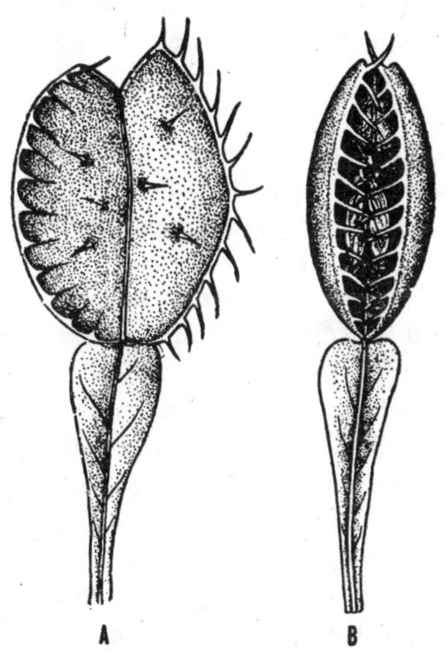

Fig. 3. A—a leaf of Venus's flytrap, an insectivorous plant. B—same with the insect in the fatal grip.

Activities of a living organism are constantly manifested in the conversion of the potential energy stored in its body into the kinetic variety. These find expression in movement, production of heat, light and electricity which are utilised by the organism for its own benefit. Every one is familiar with the *locomotion*, found in the animals. They swim actively, or fly easily or run about swiftly either in search of food or to avoid the enemy. Plants, too, are capable of producing motion. The lotus opens its petals at

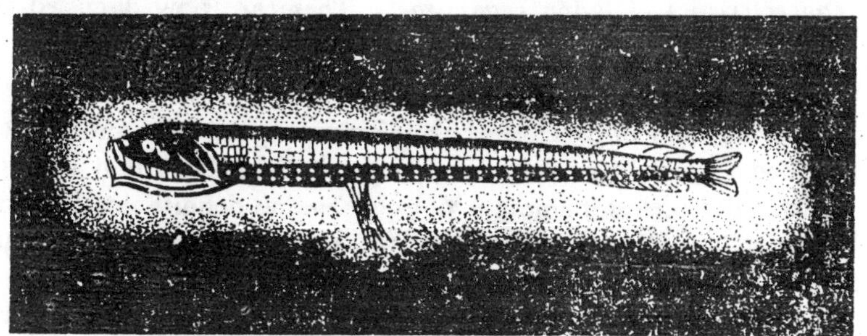
Fig. 4. A deep sea-fish glowing with double row of light organs on either side.

sunrise to close them up again at sunset. The leaves of the Venus's flytrap come together when touched by an insect to enclose the unfortunate creature in a fatal grip (Fig. 3). Heat energy is evolved from germinating seeds and from the body of a busy bee. The glow-worm emits light energy to find its way in the darkness of the night. The sea-fishes have ***phosphorescent patches*** along their sides glowing with light like the port-holes of an ocean-going liner (Fig. 4). They live at a depth of three to four thousand feet—a cheerless dark place where sun's rays fail to penetrate. The electric eels of the Amazon basin and the electric rays of the Indian ocean (Fig. 5) are capable of producing electricity in specialised *electric organs*. This is utilised by the fish to electrocute their enemies and stun their prey.

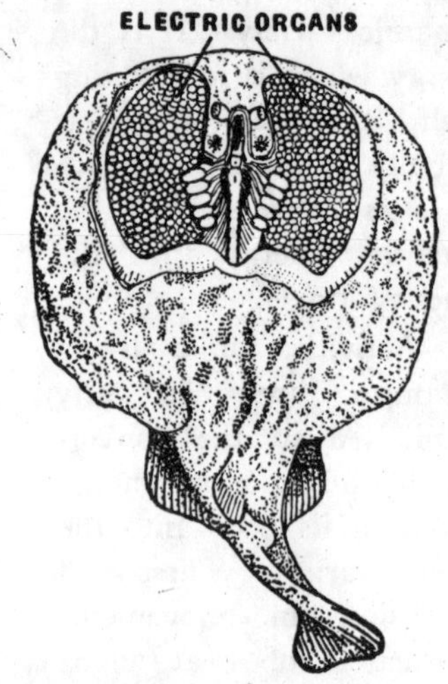

Fig. 5. The fish that produces electricity.

CHARACTERISTICS OF THE LIVING BODY

The concept of life is vague, but its manifestations in the living body are rather very distinct. It is better, therefore, that in all biological discussions the word 'life' be understood to mean the 'characteristics of living organisms'. These are now discussed in a simple manner point by point.

I. Size and Form of the Living Body

Every kind of organism has its own peculiar form and size of the body which may vary in different individuals within narrow limits. But lifeless inanimate objects have no definite size or form of the body. Water, for instance, may exist as a drop of rain or an ocean. It may be moulded into a mass of cloud or an iceberg or a small icicle. It is the same H_2O in all the cases, though varying in physical form and size to such a wide extent.

II. Composition of the Living Body

A living body is mainly composed of the living matter known as **protoplasm** (*protos*=first; *plasma*=mould or form). It is the essence of life. Protoplasm is a soft, semifluid substance resembling glue and consists of a mixture of complex organic compounds such as carbohydrates, proteins and fats along with inorganic salts in a watery medium. This living matter is never idle, because its ingredients are actively engaged in undergoing a series of chemical changes during the entire lifetime of the individual to which it belongs. An organism is said to be alive as long as its protoplasm remains active. Protoplasm is essentially the same in all living organisms whether it is a tiny microbe, a tree, a man or a huge elephant. The living parts of the body such as roots, stems, leaves of plants, and heart, lungs, liver of animals may differ widely in form and function, but these are all composed of the basic living material, the protoplasm, which is responsible for the vital activities that characterise life. It must, however, be admitted that some component parts of the living body such as the cork and wood of plants or the horns and hooves of cattle are dead substances produced by the activity of the protoplasm. These parts, though dead, are useful to their owners.

III. Organization of the Living Body

A living body is highly organized. It consists of closely correlated parts or organs such as the leaves of a plant or the lungs of an animal. Each organ is assigned to a particular vital function and

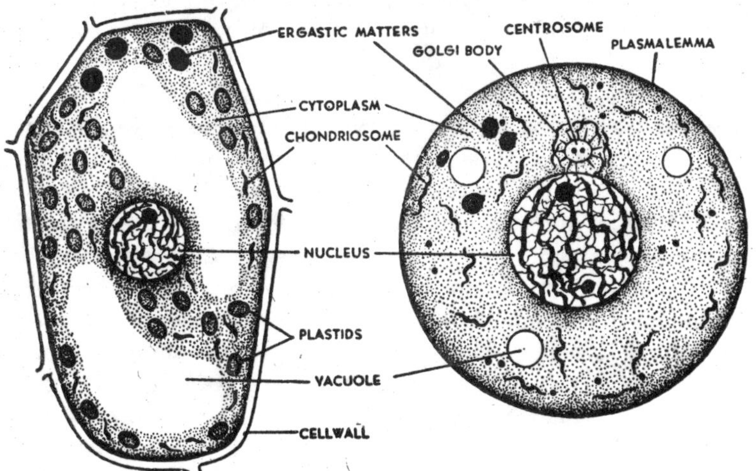

Fig. 6. Typical cells—plant cell (left) and animal cell (right).

there is a definite division of labour amongst the various organs in a living body. An organ is composed of small protoplasmic units known as the *cells* (Fig. 6). In fact, the entire protoplasm of a living individual is arranged and organized as small unit masses called cells. A cell, then, is an organized mass of protoplasm fit enough to carry on all the vital activities peculiar to life. It is not merely the structural unit comparable to the separate bricks in a building, but it is a functional or working unit as well. Whatever a plant or an animal does, is what its cells do. Groups of similar cells which perform a particular function are organized to form *tissues*, and likewise, collections of tissues constitute an *organ*. The organs are interrelated and act together in a spirit of co-operation. A living body is, therefore, a community of cells—a co-operative society—where the individual members (cells) as well as the organs (groups of cells and tissues) act in unison for the benefit of the entire organism. In the lowest form of life the entire protoplasm is organized into a single cell which by differentiation of its parts carry on the essential life functions.

IV. Metabolism

It has been pointed out that living protoplasm is constantly undergoing a series of chemical changes, and that cessation of protoplasmic activities results in disorganization and death. The sum total of these changes is collectively known as metabolism. The metabolic changes are of two kinds. A formative or constructive kind, in which new materials are added to the protoplasm, is known as *anabolism*. In this process the individual gains energy in potential form to be kept in reserve for future use. In the other kind, known as *katabolism*, the protoplasm is destroyed and lost. This is a means for conversion of the potential energy stored in the body of an individual into the kinetic form. If life is to remain in the protoplasm, both processes must continue side by side. By anabolism protoplasm is built up and potential energy is accumulated. Katabolism involves the conversion of the potential energy into the kinetic form. By katabolism the organic constituents of protoplasm are destroyed and reduced to inorganic materials. If anabolism exceeds katabolism, new materials accumulate in the cell and *growth* takes place. If, on the other hand, katabolism gains over anabolism, death is inevitable. This is life. In course of the life-processes new protoplasm is constantly built by synthesis of inorganic materials to replenish the wear and tear of the old organic matter which is thrown out and rejected.

It is very interesting to note that construction and destruction of protoplasm, a dual process of waste and repair, are going on simultaneously and side by side without rest.

The metabolic changes find expression in the various vital activities such as *nutrition*, *respiration*, *secretion*, and *excretion*. In fact, life may be defined as the external manifestation of the internal changes which are constantly occurring in the protoplasm of the organism. Metabolism is the means by which a living individual gains energy and spends the same to carry on its own day-to-day activities. The living machine is fuelled and cleansed automatically by its own metabolism.

(a) Nutrition : A living body requires energy to carry on its vital activities just as a machine requires fuel to run smoothly. The food is the fuel of the living body. The energy locked up in food molecules is utilised by plants and animals for this process of feeding, but in the strict sense it is a much more complicated affair and involves several stages running successively.

(i) Firstly, the food is taken within the body. This is *ingestion*. Plants ingest the raw materials of their food either in the form of a liquid or in the form of a gas. The liquid is absorbed through their roots and the gases enter mainly through pores present on the surface of their leaves. These simple inorganic raw materials are then converted within the plant body into complex organic food matters, the carbohydrates, proteins and fats, and remain stored up there for future use.

The animals cannot manufacture carbohydrates, proteins, and fats in their bodies. They get these food matters ready-made from the plants and ingest them in a solid or liquid state by their mouth.

(ii) The ingested food matters are not suitable for immediate utilisation. The complex substances are broken down into simpler forms and transformed into a soluble state by the process of *digestion*. Digestion of food is effected by certain chemical substances known as *enzymes* which are produced in the body cells of the organism for this purpose.

(iii) The digested food materials diffuse intimately from cell to cell. They are carried now to parts of the body where they are most needed. This is *absorption*.

(iv) The absorbed food is either utilised immediately for repair of the body cells and as a source of energy or they may be stored within a storage organ for future use. This is *assimilation* which is the ultimate goal of nutrition. It may be noted that in the course of assimilation, the food matters are ultimately transformed into the essential living substance, the protoplasm.

(v) All the ingested food may not be digested, absorbed, and assimilated completely. The residue is thrown out of the body by the process of *egestion.*

Nutrition, then, means energy-intake and storage in a potential form for future use. It is an anabolic process and is concerned with the building and repair of protoplasm.

(b) Respiration : This word implies much more than mechanical breathing, involving taking in of oxygen and giving out of carbon dioxide gas. It has been said that food provides the material from which energy can be extracted. Respiration is the process by which energy is released from food and other materials in the living body.

It is the motive force by which the potential energy stored in the living body is converted into an utilisable kinetic form. Respiration occurs spontaneously throughout the life period of all cells. Its complete cessation means death, and, therefore, it is one of the constant and invariable characteristics of living matter. Air or oxygen is usually required for respiration, but in exceptional cases it may proceed even in the absence of oxygen. In fact, free oxygen is injurious to certain microbes. In the majority of organisms respiration means combustion or oxidation of the complex food substances, particularly carbohydrates and fats. The result is the liberation of energy in the form of heat and production of carbon dioxide gas (CO_2) and water vapour (H_2O). The process of oxidation of glucose, which is a carbohydrate, may be represented by the following equation :

$$\underset{\text{(Glucose)}}{C_6H_{12}O_6} + \underset{\text{(Oxygen)}}{6O_2} = \underset{\text{(Carbon dioxide)}}{6CO_2} + \underset{\text{(Water)}}{6H_2O} + \text{Heat energy.}$$

In this respect, respiration is almost identical with the burning or oxidation of a lump of coal or a log of wood which also yields similar products along with release of heat and light energy. But respiration is a vital process. It takes place at a much lower temperature and is regulated by the vital activity of the protoplasm. The energy released by respiration is utilised by plants and animals to carry on their routine activities, whereas CO_2 and H_2O are breathed out.

Respiration, therefore, is essentially a destructive or katabolic **process.** It is a means for energy-release by breaking down the **organic** materials of the living body and changing them back into inorganic materials.

(c) Secretion and Excretion : The protoplasm produces certain new chemical substances during its anabolic and katabolic activities. Such substances, when useful to the organism, are called *secretions*. If, on the other hand, the new materials are useless or harmful to the organism, they are called waste products or *excretions*. Secretions and excretions are formed in specialised organs called *glands* within cavities called *alveoli*, and are finally released through tubes or *ducts* (Fig. 7).

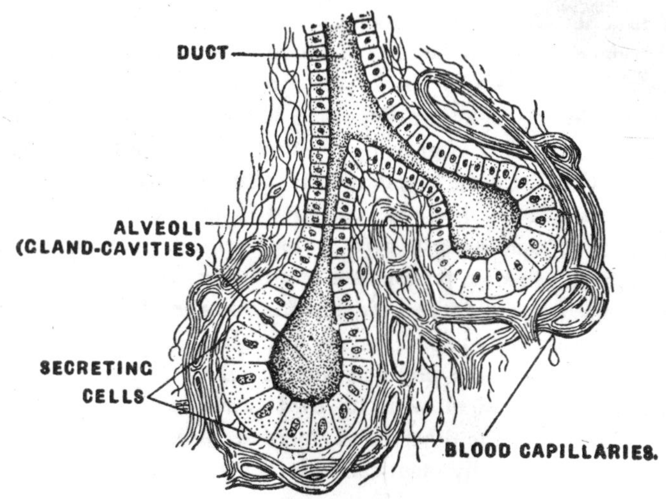

Fig. 7. Sectional view of an animal gland producing secretion.

Secretions often contain 'enzymes' or 'ferments', useful for digestion and other purposes. They may also contain chemical substances, called *hormones*, which have profound influence over the growth of the body as well as that of the mind. Enzymes and hormones are responsible for the chemical co-ordination of the living body, and control most of its metabolic activities.

The waste products or excretions are mostly formed during the combustion or oxidation of proteins, just as carbon dioxide and water are produced during oxidation of carbohydrates and fats. They are usually nitrogenous substances such as alkaloids, ammonia, urea, etc. As the organism has no use for the waste products, they are thrown out of the body along with other such substances.

Secretion and excretion, like respiration, are katabolic processes. The living body is automatically cleansed by excretion of waste products.

V. Growth

When anabolism exceeds katabolism, new organic substances are added between already existing particles of protoplasm, and the living body increases in bulk. This method of growth by wedging in of new substances between matters already in existence is known as growth by *intussusception*. A non-living body may grow in bulk by the mere deposition of identical matter on its outer surface. Thus a lump of sugar may be made to gain in size by keeping it suspended in a concentrated sugar solution when fresh particles of sugar are deposited on the outer surface of the original lump. This is growth by *accretion*, as opposed to growth by intussusception seen in plants and animals. It is to be noted further that a living body never grows in bulk at the expense of substances identical with its own protoplasm. A plant increases in size and builds its own protoplasm by utilising organic food matters manufactured out of carbon dioxide gas absorbed from the air and water with dissolved mineral matters from the soil. A man increases in bulk by taking plant products such as wheat and rice which are by no means identical with its own protoplasm. A non-living lump of sugar, on the other hand, increases in bulk at the expense of substance chemically identical with its own matter.

VI. Movement

Most of the living objects have the power of movement. Spontaneous movement of protoplasm is characteristic of plants as well as animals. This protoplasmic movement occurs within the cells and helps to distribute food and eliminate waste products. It can readily be observed in the cells of the water-plant *Vallisneria* (B. Patashaola). In a few of the lower forms of plants and animals are seen long whip-like protoplasmic structures called *flagella* or short hair-like processes called *cilia* which help the organisms to swim about in their watery habitat. The lowly *Amoeba* creeps about by means of *pseudopodia*, which are temporary outgrowths from the surface of its own protoplasm.

This inherent property of the protoplasm may be observed in a more prominent degree in majority of the higher animals which have developed specialised organs for locomotion such as fins for the purpose of swimming, wings for flying and legs for walking. Thus the living body is a perfectly automatic machine.

VII. Sensitiveness and adaptation to environment

Organisms are very sensitive and possess wonderful power of responding to external influences. Such influences are known as *stimuli*. A stimulus may be physical, such as heat or light, mechanical such as touch or contact, and even chemical, such as presence of acids or bases. A moth flies into a flame under the influence of light, and the leaves of the delicate sensitive plant shrink down modestly to touch (Fig. 8). The roots of a plant go deep into the

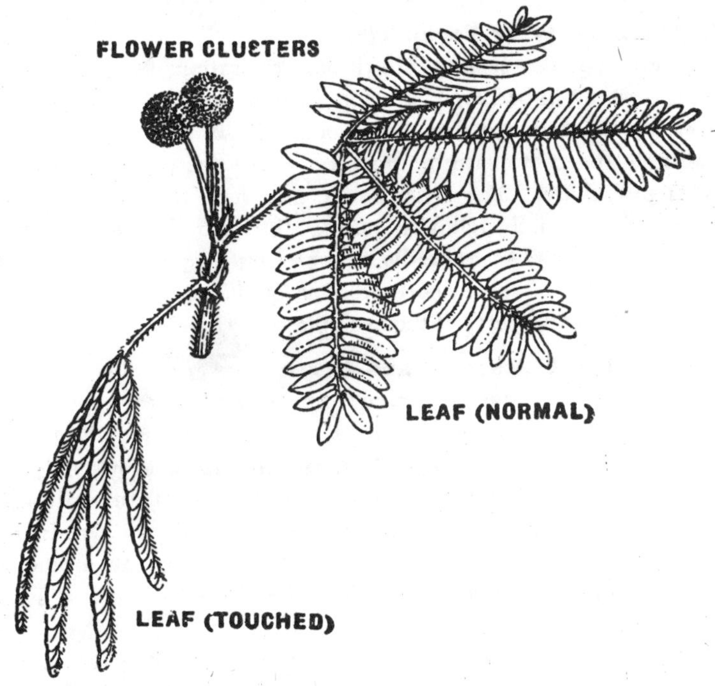

Fig. 8. The leaves of a sensitive plant shrinking modestly to touch.

soil under the influence of the force of gravity, and a muscle contracts when stimulated by an electric shock. The entire life of an individual is a series of reflexes or reactions between it and its surrounding environment. The organism responds to external stimuli with the definite purpose of maintaining its own life as well as for perpetuation of the whole race. Purposive behaviour is a feature of the living body.

VIII. Reproduction

Plants and animals have the power of producing young ones like themselves. This is reproduction. It is a unique process which maintains the continuity of life and occurs in all organisms usually after attaining maturity. The non-living can never reproduce its own kind.

There are various methods by which a living body reproduces its own kind. Whatever may be the method of reproduction, the young ones resemble the parents in the long run. So it is 'like from like'—referred to as *law of biogenesis* (*Omne vivum ex vivo*).

IX. Rhythmicity or Periodicity

Vital activities such as nutrition, respiration, secretion, etc., are all periodic in occurrence. A stage of intense activity is always followed by a distinct pause, however small this might be. In the meantime, the protoplasm gathers the vital force necessary for the next active phase. The heart, lungs, etc., have a rhythm of their own which is strictly followed, unless disturbed by some external factors. Growth and reproduction are also periodic in nature. Rhythm is one of the important signs of life and may be considered as a basic property of the protoplasm.

X. Life-cycle, Senescence and Death

An organism has a definite life-cycle. In the young or *trophic stage* it takes food and grows actively. When fully grown, it reproduces its own kind. In course of time it grows old and finally dies out. The trophic or growing stage of an individual always culminates in reproduction and is ultimately terminated by *senescence* and *death*. These four important events occur in the order mentioned, and constitute the *life-cycle* of every living organism.

Life is always limited in duration. The limit of life varies widely in different kinds of plants and animals. The tiny 'Dewali' insect completes its life-cycle and dies out in the course of a few hours, but the gigantic banyan tree lives for more than one hundred years. During the later part of life the vital activities decrease in intensity. The **metabolism** gradually slows down.

This sluggishness is the stage of senescence or old age which spares none. During old age the co-operation between the various cell-groups or organs gradually becomes feeble. Ultimately, death ends all. It is the gross failure of the co-operative society and is always preceded by the bankruptcy of the individual cells as well as the cell-communes.

LIVING AND NON-LIVING

As biology is mainly concerned with the study of living things, an attempt should now be made to distinguish them clearly from the lifeless inanimate objects. There is no difficulty in recognising a cocoanut palm or a guinea-pig as living objects and the rocks in the field or the sand grains in the sea-shore as non-living bodies. But, unfortunately, the difference between a living and a non-living system is not very sharp.

In recent years, biologists have discovered certain living units called *viruses*. They are so small that they cannot be recognised even with the help of an ordinary microscope. But they exhibit certain properties, for example, the ability to reproduce their kind, which are normally associated with life. The viruses can now be seen with the modern *electron microscope*. They resemble large molecules of matter to some extent and cause certain specific diseases such as 'small-pox' and 'measles' in animals. Plant diseases such as 'blue stem' in potatoes or 'curly top' in cabbages and other vegetables are also associated with virus infection.

Viruses are so small that they cannot be seen as individual particles under ordinary microscopes. That means most viruses are smaller than 210 millimicron ($m\mu$). Viruses range in size from about 17 to 450 millimicron.

The shape of all viruses are not same. They may be spherical (Polio virus), rod shaped (Tobacco mosaic virus or TMV), or tadpole like (Bacteriophage).

The structure of virus is rather simple. It has an outer coat of protein and an inner core of DNA or RNA. Most of the viruses that parasitise animal cells have a core of DNA. In some animal virus and in all viruses that parasitise the cells of higher plants the core is made up of RNA. The core in case of influenza and polio virus contains both DNA and RNA.

The viruses are completely inactive outside the host's cell. They require for their life processes the enzyme system present in the cells of their hosts. Once inside the host's cell the DNA or RNA, as the case may be, of the virus starts taking care of the host cell's biochemical activities. The host cell, instead of making its own characteristic nucleic acid and protein, begins to make different ones specifically suitable for the virus. Within the host cell the virus replicates producing hundreds of new viruses all exactly like the original virus that invaded the host cell (Fig. 8A).

Most plant viruses are spread by insects. The aphids and leaf-hoppers head the list. Plant viruses are transmitted by the mechanical rubbing of one infected leaf of a plant against another. In a few cases, seeds of plants are known to carry virus particles and thus to transmit the virus from one generation to the next.

Animal viruses are transmitted in a variety of ways. Viruses that cause pneumonia, the common cold, influenza, mumps, measles are expelled in droplets by coughing, sneezing or talking of the infected person and infect susceptible hosts. Some animal and human viruses like polio, infective hepatitis are excreted in feces. Flies carry the virus to food or drinking water and spread the disease.

All attempts to make the viruses to grow and reproduce outside the living cells have failed. Viruses are strict **parasities** and do not have any enzyme system of their own. The true nature of the viruses still remains a debatable issue. Are they primitive living particles that have not yet evolved the enzyme systems required for an independent existence? Or are they the descendents of some complex living organisms which have lost their capacity for an independent existence during the passage of time? The answer we may never know. However, it is suggested that the viruses most probably represent an intermediate stage between living and non-living forms of matter.

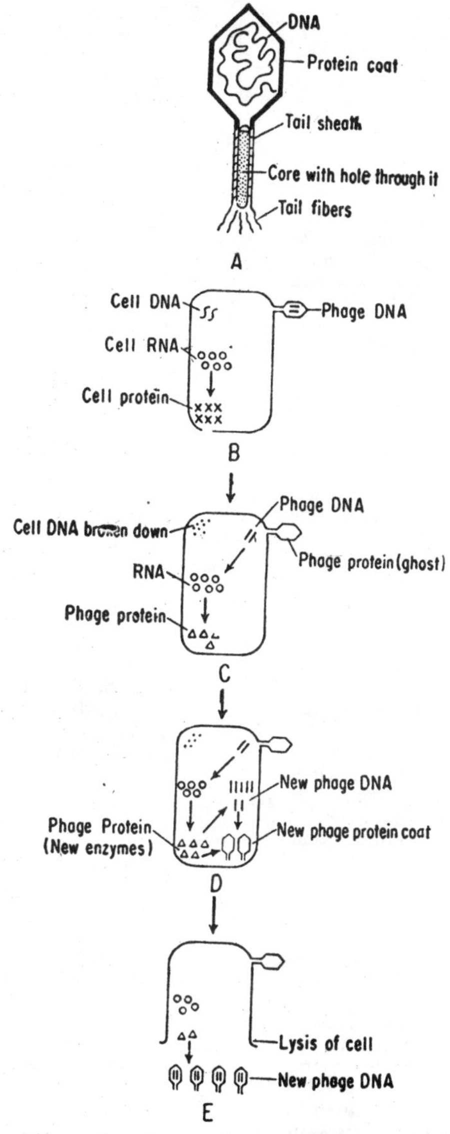

Fig. 8A. Life cycle of Bacteriophage showing morphology (A), attachment and injection (B), New phage formation (C-D), Lysis and emergence of new phage (E).

The principal points of difference between living and non-living objects are now recorded in a tabular form.

Living	Non-living
I. Each kind of plant or animal has a definite *form and size*, which may vary within narrow limits in different individuals of the same kind.	I. Non-living objects, such as masses of clouds or collections of water have neither a definite size nor any precise form of the body.
II. Living body is composed of protoplasm which is the *physical basis of life*. The protoplasm of a living individual is arranged in the form of one or more compartments —the *cells*—each of which is a structural as well as a functional unit of the living body.	II. Living protoplasm or *cells* are not found as component parts of the body.
III. A living body is well *organized*. It is composed of *cells*, *tissues*, and *organs* with division of labour, to carry on its routine vital activities.	III. No such *organization* exists.
IV. Metabolic changes such as *nutrition*, *respiration*, *secretion* of useful substances and *excretion* of waste products are going on constantly within the living body due to the vital activities of its own protoplasm. Life is the external manifestation of these internal protoplasmic changes.	IV. None of the metabolic changes may be detected in a non-living object which contains no protoplasm and therefore has no vital activity.
V. The living body is *automatic*. It is a *self-fuelling* and *self-cleansing machine*. Nutrition is the means for energy-intake of the living apparatus and respiration releases the energy to be utilised for its other activities. It cleanses itself by automatic excretion of its waste products.	V. A man-made machine is *never* strictly *automatic*. It has no innate power to manage its own affairs and requires to be cleansed and refuelled by man from time to time.

VI. Living body increases in bulk by *intussusception,,* that is, by wedging in of new particles in between already existing particles of protoplasm. While growing, the living body utilises substances other than its own protoplasm.	VI. *Growth* may occur occasionally by *accretion* or deposition of particles only on the outer surface of the body. Increase of bulk takes place at the expense of substances chemically identical to its own matter.
VII. The living body is *sensitive and* can *adapt* itself to its *environment* in an admirable manner. It *responds* to *stimuli* with some definite purpose.	VII. True *sensitiveness* is *absent;* voluntary power of adjusting to changes in the environment is nil. At least there is no *purposiveness* in their behaviour when stimulated.
VIII. A living body can *reproduce* its own kind and thus perpetuate its race.	VIII. There is *no power* to *reproduce* its own kind.
IX. A living body is *rhythmic.* There is a rhythm regulating all the vital activities. An intense activity by an organ is followed by a period of pause or rest.	IX. There is *no* definite *rhythm* and periodic activity is never met with as a rule.
X. A living body has a *life-cycle.* Each kind has a definite period of duration at the limit of which it tends to *grow old* and *dies.*	X. No *cyclical* phenomenon is observed. Period of duration is indefinite and there is neither *senescence* nor *death.*

PLANTS AND ANIMALS

Man has always been impressed by the vastness and variety of living things. Life manifests itself in infinite numbers and

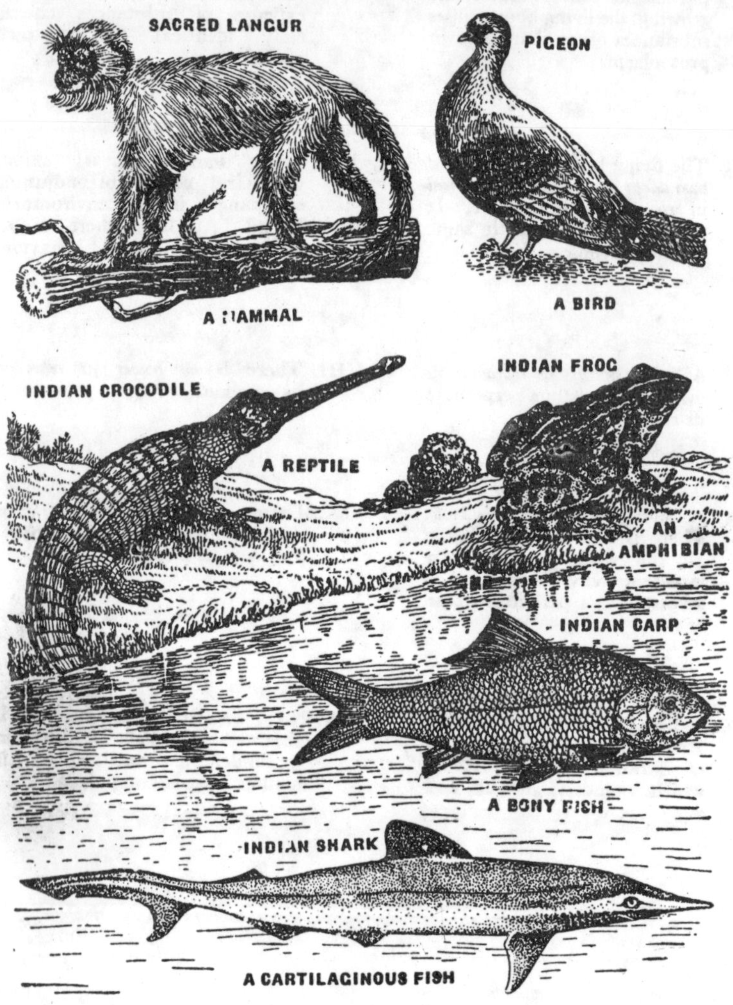

Fig. 9. Types of vertebrate animals.

diverse forms. Lcaving aside the less known and submicroscopic viruses, the plants and animals are the two important kinds of living things. Man himself is an animal, and so too are the hairy marnmals, feather-clad birds, scaly reptiles, warty toads and the finned fishes (Fig. 9). The oysters, butterflies, starfishes, earthworms, roundworms, tapeworms, corals, sponges and the micros-

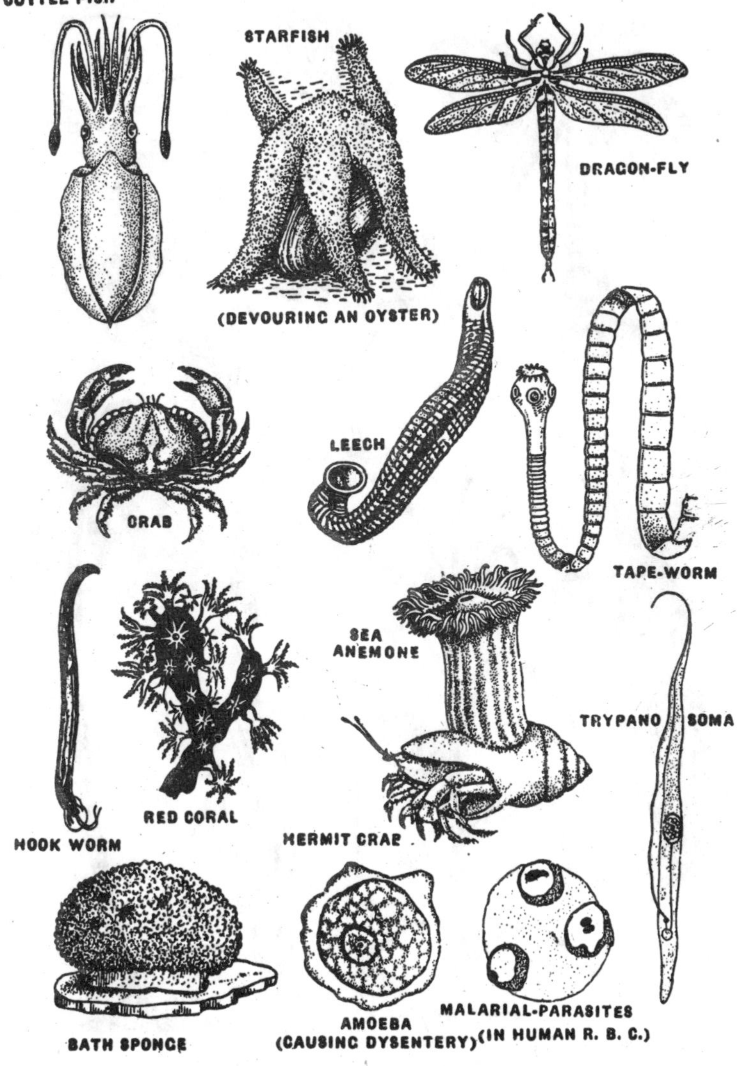

Fig. 10. Types of invertebrate animals.

copic creatures that cause malaria in man are also animals (Fig. 10).

Familiar plants are those that yield maize, pea, mustard, tube-rose, prickly pear, cocoanut palm and cycad (Fig. 11). Ferns,

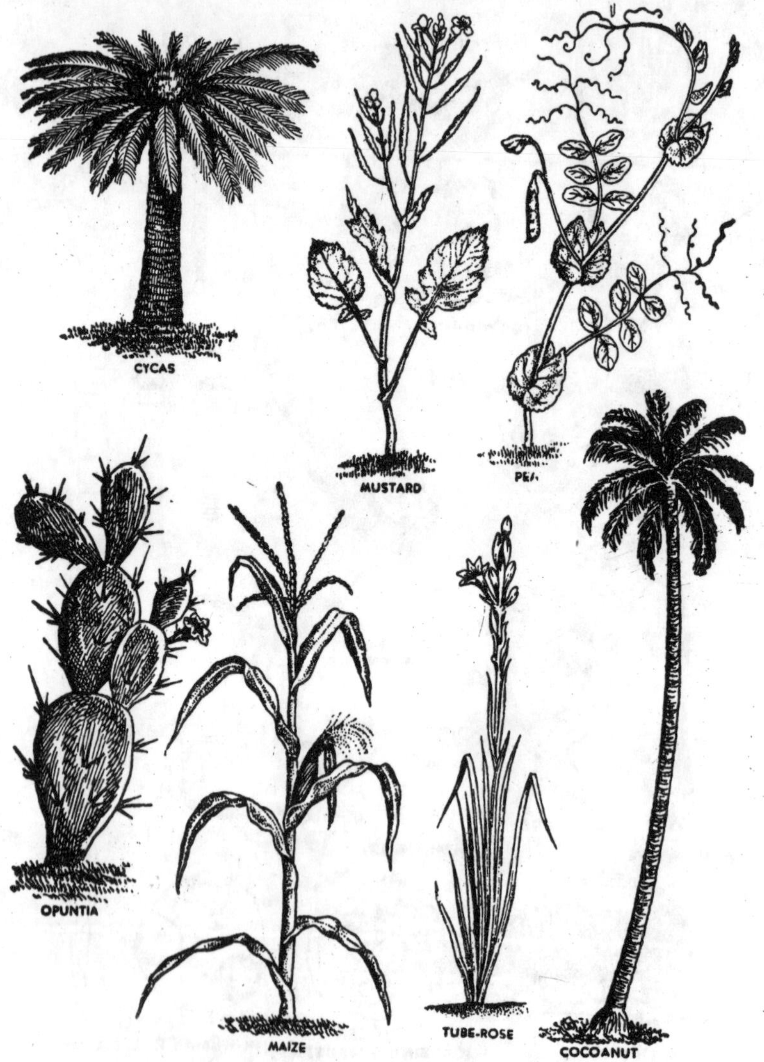

Fig. 11. Types of common plants (higher).

mosses, liver-worts, lichens, mushrooms, sea-weeds and pond-scums are also plants (Fig. 12).

The lowly microscopic forms such as green *Chlamydomonus*,

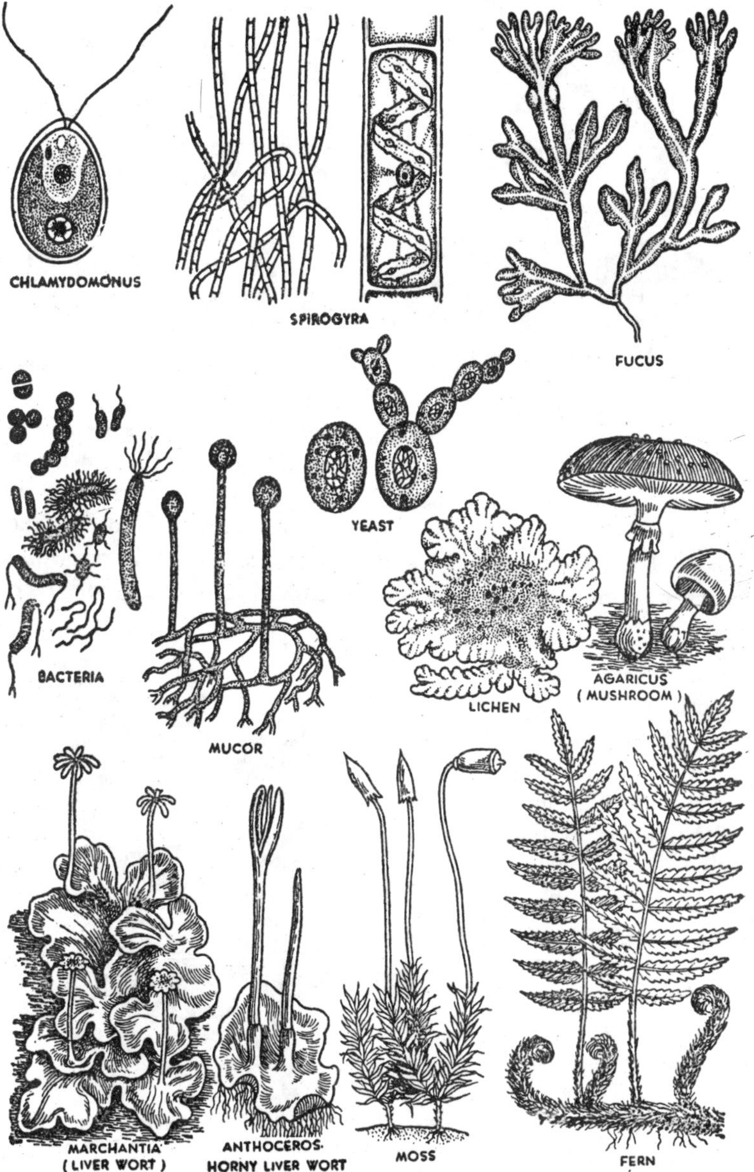

Fig. 12. Types of common plants (lower).

the colourless yeast, and the enormous assemblage of bacteria are plants too, which have interested man since **Leeuwenhoek** (1632–1723, Fig. 13) discovered them in the later part of the seventeenth century. Why should plants and animals assume so many forms? Why should there be such wide variations in size, colour, shape, internal structure and mode of life? The only answer to this question can be found in the adaptation of each kind to the particular environment in which it lives. Plants and animals are adapted perfectly to their own surroundings. This is effected by modification of forms and habit. They have to find their food and avoid their respective enemies. Failure to do this means extinction of the whole race.

Fig. 13. Antony Van Leeuwenhoek.

The Indian leaf-butterfly *Kallima* is eaten by birds. They avoid their enemy by imitating the form and pattern of a leaf while resting on the twig of a plant (Fig. 14). Such phenomenon is very common in the living world and is known as **mimicry.**

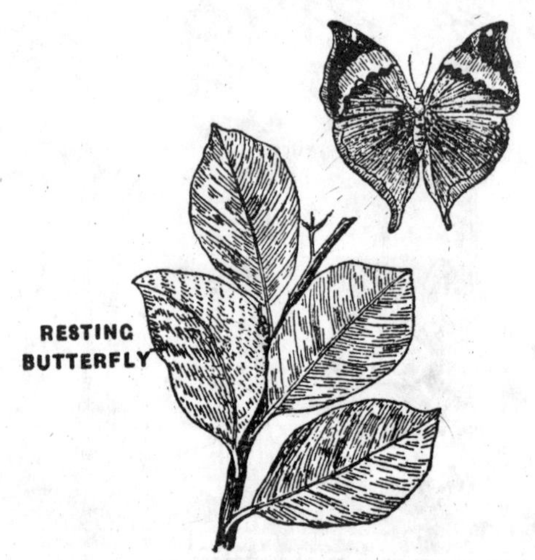

Fig. 14. The Indian leaf-butterfly protectively disguised amongst the leaves of a twig while resting.

Biology is a vast science. There has been a steady, yet rapid increase of knowledge about organisms during the last one hundred years. It is well nigh impossible today for a single individual to be acquainted with the diverse aspects of modern biology in minute details. For the sake of convenience, the science of life has been divided into **botany** and **zoology.** Botany (*botane*=a herb) deals with plants and zoology (*zoon*=an animal) is concerned with animals. These two main branches are further subdivided into a number of sub-sciences according to the nature and kind of plants and animals that are involved. Thus **bacteriology** (*bacterion*=a little rod) is a branch of botany related to the study of microbes and **entomology** (*entomon*=an insect) is a part of zoology devoted to the insects. Likewise, **anthropology** (*anthropos*=a man) is a special section of biological science which discusses the various races of mankind.

Higher plants are readily distinguished from higher animals but the distinction is not so sharply marked in some lowly organisms. For instance, the corkscrew-shaped microscopic Spirochaetes are claimed by the botanists as plants and by the zoologists as animals. The characteristic points of difference between the two groups are stated below in a tabular form:

Plants	Animals
1. The plant body usually bears branches. The palms, the *Cycas* and a few others are exceptions.	1. The animal body usually does not bear branches. The *Obelia* and a few plant-like hydroids are exceptions.
2. Majority of the plants, with the exception of fungi and bacteria, are green in colour due to the presence of chlorophyll. Plant cells, excepting those of the fungi and the bacteria, possess protoplasmic bodies called *plastids*. Of the three kinds of plastids, the chloroplasts contain the green pigment chlorophyll which occupies the most strategic position in the living world.	2. Majority of the animals, excepting lowly forms like *Euglena*, are not green, because they have no chlorophyll. Some animals, however, such as grasshoppers, may acquire chlorophyll from plants for the purpose of camouflage which ensures protection from enemies. Plastids, particularly chloroplasts, are usually absent in animal cells.
3. The protoplasm of the plant cell is usually surrounded by a rigid non-living cell wall composed of cellulose. This is an insoluble carbohydrate secreted by the protoplasm. In course of time the protoplasm itself may disappear and only the non-living cell wall is left behind. Such a cell is obviously dead.	3. A cell wall composed of cellulose is not found in the animal cell. As a rule, the animal cell is covered by a thin and elastic limiting membrane called *plasmalemma* which is composed of protoplasm. A non-living cell wall may be secreted by the protoplasm of the animal cell but this is always thin and is not composed of cellulose.

Plants	Animals
4. Vacuoles are present in plant cells. These are intracellular vesicles containing watery cell sap. Very often the vacuoles become so large that they occupy the major portion of the cell.	4. Vacuoles are scarce except in the protozoan cell body. When present they are much smaller in comparison to those found in plant cell.
5. A central body is rarely found in plant cells.	5. A star-shaped central body is usually present in animal cells. It lies near the nucleus and plays an important part during cell division.
6. Most plants, excepting a few lowly ones, are fixed to the soil by roots or root-like structures. They, therefore, have no power of locomotion.	6. Most animals have the power of locomotion. They can move from place to place in search of food. There are some, such as sponges, corals and sea-anemones, which are fixed and plant-like.
7. Plants can move parts of their body in response to external stimuli; but as they have no nervous system, the movement is slow in their case. Sensitive plant is an exception, because it responds quickly to touch.	7. Animals, as a rule, can move their body parts quickly in response to external stimuli, becauee majority of the animals possess a nervous system for producing nervous co-ordination.
8. Green plants absorb water with dissolved minerals from the soil and carbon dioxide gas from the air. With the help of chlorophyll, which they possess, they can synthesise complex carbohydrate food out of the two simple raw materials, carbon dioxide and water. This process is known as *photosynthesis*, because it always occurs in the presence of sunlight. In manufacturing proteins and fats, the carbohydrates form the ground substance. Thus the green plants can manufacture all the three important food substances, carbohydrates, proteins and fats. They, therefore, are said to be *autotrophic* or *holophytic* in their mode of nutrition. They can live in a world devoid of animals.	8. Animals, having no chlorophyll, cannot manufacture their food. They are directly or indirectly dependent on plants for the supply of proteins, fats and carbohydrates. So the animals are *heterotrophic* or *holozoic* in their mode of nutrition. They cannot continue to live in a world devoid of plants.
9. As already stated, plants absorb raw materials in liquid or gaseous forms. They cannot ingest solid food. The insectivorous plants are exceptions, because they capture insects and obtain a part of their proteins by digesting the latter.	9. Animals, excepting a few parasites such as tapeworms, ingest solid food. They cannot utilise gaseous raw materials like the plants. The green *Euglena* is an exception.

Plants	Animals
10. When required, plants digest their food bit by bit leaving no solid residue for egestion.	10. Animals cannot digest all the solids which they ingest. A residue is always left behind which is required to be egested.
11. Metabolism in plants is predominantly anabolic. They are constantly trying to build and store food substances. The result is that plants grow indefinitely till almost the last days of their lives.	11. Metabolism in animals is predominantly katabolic, because more active habit requires constant release of energy by the depletion of food. The result is that there is a definite growth period after which animals cease to grow.
12. Plants do not possess special excretory organs for eliminating their nitrogenous waste products. They utilise most of the nitrogenous substances for building proteins. The little that remains is transferred to barks, leaves, etc., and is thrown out that way.	12. Animals have special excretory devices for eliminating nitrogenous waste products. Substances like ammonia and urea are constantly formed in the animals body by the depletion of proteins. These substances are promptly thrown out by the excretory channels before they can do any harm.

MATTER

Biology seeks to explain the structure and properties of living matter in terms of physics and chemistry. It is, therefore, necessary to be acquainted with the fundamentals of the two sister sciences.

Matter is defined as anything which occupies space. It exists in three different states—**Solid, Liquid** and **Gas.** *Matter is indestructible.* This is the law of **conservation of the matter,** and is applicable to all substances whether living or non-living. Matter may, however, change from one state to another without suffering any loss of weight. Thus water may be frozen into a mass of ice or vaporised into steam.

Matter may exist in a number of different forms. An **element** in chemistry is a substance which cannot be split up into substances different from itself. Carbon, Oxygen, Nitrogen and Hydrogen are the common elements composing the living matter. Elements are often found in free state, but in nature, most matters exist in the form of a **mechanical mixture** or a **chemical compound.** Air is a mixture of nitrogen and oxygen. A **mixture** is made up of two or more substances each of which retains its own properties. The composition of a mixture is variable and the ingredients may occur in any proportion.

Elements may combine to form *compounds*. A **compound** is a substance composed of two or more elements combined in definite proportion. It is more or less *stable* but can be broken up into its component *elements* by ordinary chemical means. Thus, water, which can be decomposed into the elements hydrogen and oxygen, is a compound. It must be noted that the properties of a compound are entirely different from those of the elements composing it. Elements and compounds exist as *molecules*. The **molecule** is the smallest particle of a matter which retains all the properties of the original substance. If a molecule of a compound is decomposed the resulting parts no longer exhibit the characteristic chemical properties of the original molecule.

A molecule is, in turn, composed of *atoms*. An **atom** may be defined as the smallest particle of an element which takes part in chemical reaction. A molecule of helium is composed of only one atom, but a molecule of oxygen contains two atoms. According to modern idea, the atoms of all the different chemical elements are built up with only three kinds of particles called **protons** (+), **electrons** (−) and **neutrons** (±). The protons carry positive electric charge, the electrons are negatively charged and the neutrons have no charge. The nature of these particles are identical in all kinds of matter. It is the number and arrangement of these particles that determine the kind of element composed by them.

We shall close this section with a short discussion of the properties of matter in the **colloidal** state, because many of the substances of which the plants and animals are composed exist in the living cell as such. In fact, protoplasm owes most of its characteristic physico-chemical properties to the colloidal state of the substances composing it.

If a little sucrose (can sugar) or sodium chloride (common salt) is shaken up with water the resulting system will be a true **solution.** If, on the other hand, some fine river-bottom silt is stirred up with water in a test tube, a different sort of system will result. The silt particles do not dissolve but simply become *dispersed* throughout the watery medium. The suspended particles are large enough to be detected under a microscope. This is a **colloidal suspension.** It is not *stable*, because after a time the particles of silt gradually settle down and the system becomes separated into its original components in contrast to the *true solution*, which is absolutely stable. The particles of a colloidal solution are incapable of passing through the pores of a fine parchment membrane in contrast to those of the *true solution* which readily pass through it. This phenomenon, known as **dialysis,** was originally devised by

Graham to separate the **colloids** from the **crystalloids** yielding *true solution*, from which they can be recovered in the form of **crystals.** This division of matter into *colloids* and *crystalloids*, depending upon their ability to pass through the pores of a parchment membrane, is not accepted today, because most of the substances can, by suitable means, be brought into the colloidal state.

Thus sodium chloride, when shaken up with petroleum ether yields a colloidal solution although it forms a true solution when the dissolving medium is water. The size of the molecules in a colloidal solution is very large or they exist in a state of aggregation; that is why they cannot pass through the fine pores of a dialysing membrane.

A colloidal solution may either be a **suspensoid** (silt in water) or an **emulsoid** (starch or proteins in water).

The essential features of both forms are that they are systems consisting of two *phases*, the **dispersed phase** and the **continuous phase.** In a suspensoid the *continuous phase* is generally a liquid and the *dispersed phase* is a solid. But in an emulsoid both the *phases* are liquid. Milk and latex (milky juice exuding from the cut surface of certain plants like the India rubber plant) are the two natural *emulsions.*

Emulsions are much more stable than the **suspensoids,** because they are stabilised by the presence of a third component. These are the **emulsifiers.** If olive oil and water be vigorously shaken up in a test tube the resulting system will be an *unstable emulsion* but if a little *soap* or *gum* be added to the mixture previous to shaking, it will yield a *stable emulsion* from which the components do not easily separate out. Milk is an emulsion of fine droplets of fat in water stabilised by a protein which plays the role of an emulsifier. Butter, on the other hand, is an emulsion of fine drops of water dispersed through a medium of fat. Thus the continuous phase of milk has become the dispersed phase in the butter and *vice versa.*

It must be remembered that the colloidal solution is only a *state* and not a *form* of matter. Substances in colloid state do not constitute a definite *class* by themselves, since sodium chloride which is a definite crystalloid when the continuous phase is water, yields a colloidal solution when the continuous phase is changed into petroleum ether.

In **true suspensions** the components separate out only very slowly and the stability is ensured by the interaction of particles of the *dispersed phase*, which perform erratic, zigzag movement being bombarded from all sides by the particles of the *continuous phase*. This is known as the **Brownian movement.** It is one of the factors which stabilises a *true suspension.* It may be pointed out here that the continuous phase may not be always liquid. In some cases it may be solid, liquid or gas. The following examples will exhibit certain interesting colloidal combinations.

Common name	Dispersed phase	Continuous phase
Foam	Gas	Liquid
Fog	Liquid	Gas
Tobacco smoke or Soot	Solid	Gas
Ruby glass	Solid	Solid

Origin and Continuity of Life

The earlier Greek observers believed that living things originated spontaneously out of lifeless inorganic substances. Maggots were thought to be born of decaying carcasses and mice originating from river beds. Belief in *spontaneous generation* of the living things (ABIOGENESIS) prevailed throughout the *dark ages* of biology after the death of Aristotle in the fourth century B.C. It persisted even in the nineteenth century when eminent people firmly believed that at least the microbes are formed spontaneously. The first serious objection to the conception of "spontaneous generation" came from the Italian, Redi, who in 1860 proved that fly-maggots did not originate by **abiogenesis,** but from eggs of fly. During the later part of nineteenth century, Tyndall, in England, demonstrated that bacteria cannot generate by themselves in any infusion unless the latter is infected with their spores. **Pasteur** (Fig. 15) in France, established the principle

Fig. 15. LOUIS PASTEUR (1822–1895).

of sterilisation by heat or chemicals for guarding off and destroying the microbes. This is largely applied in modern surgery for preventing wound infection and also for protecting water, food and milk for public use from pollution by bacteria. In 1861 Pasteur confirmed the theory of 'Biogenesis' which states that new life comes only from pre-existing parent individuals (OMNE VIVUM EX VIVO). *Biogenesis* is effected by *reproduction* and is the only means for the continuity of life.

Biology in the Service of Mankind

The progress of human civilisation has, to a great extent, been moulded along the lines of development of the biological science. Man is a social creature. People live in towns and cities. This has created several health problems such as supply of pure water, disposal of sewage and prevention of contagious diseases, so that they may not break out in epidemic form. These problems are best solved by biology. Bacterial enemies of man have been discovered, and remedies against their spread and destruction are now possible by vaccination and other measures. No longer do plagues bring terror to a district. Penicillin, streptomycin, terramycin, chloromycetin, aureomycin, etc., are the magic bullets to shoot down the microbes, and all of them are the gifts of the biological science to mankind. Life-histories of animal parasites invading man, and of animals which quickly transmit them to a town-dwelling community by acting as carriers have been worked out and effective measures for their control have been planned. Moreover, an expert biological knowledge is essential for selecting the right kind of raw materials for the manufacture of drugs, because most of them are either vegetable or animal products.

The few facts that are mentioned here serve to illustrate how biology may be applied to mitigate human sufferings and increase the life span of man.

Modern man, unlike his primitive ancestors, does not depend upon fishing and hunting. He cultivates plants and rears fishes as well as cattle to get his food and other necessities of life. This has resulted in the development of agriculture, fishery and animal husbandry which are applied aspects of the basic biological science. Better and productive varieties of crop plants such as rice, wheat, jute, sugar-cane, cotton and pulses are now bred experimentally and distributed throughout the country. Disease-resistant grains are raised and seeds are vernalised by special treatments so as to yield the crop before the usual time of harvesting. Soil is saved from exhaustion by crop-rotation. Injurious insect pests are exterminated and plant diseases cured or prevented by adequate scientific measures. Domesticated animals, such as the poultry and the cattle, are improved with a view to increasing the egg or milk production by experimental breeding, and their diseases cured or controlled by preventive measures. Fishes are now reared in fisheries and the small fry distributed to fishermen with proper scientific instructions for growing them. Life-histories of the more important food-fishes are explored and suitable kinds are bred. Fishing industry has been improved by employment of adequate catching devices, such as traps and nets. Trawlers are used for fishing in the open sea, so as to increase the number of catches, and fish-oils are manufactured as side products. We are really proud of the fact that during last few years spectacular advances have been made in agriculture in our country. Employing scientific methods by way of selection of better seeds, adequate application of manures, and improved irrigation facilities agricultural practices have been practically revolutionised. Similar improvements have also been made in fisheries, silk-producing insects and diaries.

Silk industry has been greatly improved by sericulture. In this, better kinds of silk moths are experimentally bred with a view to increasing the rigidity, fineness or other qualities of the silk produced by them. Biology, therefore, has contributed largely towards obtaining better varieties and larger yields of food and other human necessaries.

Forests and wild game-animals are national resources from which man derives useful materials such as timber, fur and even food. Climate and crop position of the country depend, to a certain extent, upon the biological communities residing in perfect harmony amongst the forests. Proper and judicious conservation of the natural resources demands a fundamental knowledge of the principles of biology, so that they may be utilised at the right moment. Successful plant breeding, animal husbandry and conservation of forests are effective means for checking famine which is a scourage of the human society.

Man has three powerful adversaries to fight with. These are famine, disease and death. Biology is an effective weapon against each of these three human terrors.

Biology and Human Culture

Biological science should form an essential part of the education of every cultured person. It enables man to understand his very interesting and beautiful neighbours, the plants and the animals, which not only provide him with moments of thrill and enjoyment but also with food.

Robert Boyle, an eminent scientist of the seventeenth century and one of the founders of the Royal Society of London, said: "It is highly dishonourable for a reasonable soul to live in so divinely built a mansion as the human body in total ignorance of the very exquisite structure of the same". Biology initiates man to know the parts of his own body and the functions thereof, so that he may live more efficiently by the judicious use of his physical machinery.

Pope said that "the proper study of mankind is man". But no one can understand the nature of man without a foundation of the biological science for man's entire development including his intellect, his social customs and behaviours are biological phenomena. Biology is a necessary pre-requisite for the study of medicine, psychology and sociology. "Only a science which is directly related to life" said William James, one of the pioneers of modern psychology, "is really a science".

PART I
BOTANY

CHAPTER I

PLANTS : GENERAL CONSIDERATIONS

Botany (*botane*=herb) is that branch of biology which deals with the study of plants. So it is the science of plant life. As such it includes everything which has reference to plants. It considers the external appearance of the plant organs, their internal structure and organization, the different vital activities like nutrition, respiration, growth, reproduction carried on by plants, as well as a systematic classification based on resemblance and differences, their life histories, their adaptations to varying environmental conditions, relationship amongst themselves and with other living things, their distribution in the world in space and time and their economic values, particularly to mankind.

As Botany is a vast subject embracing everything having connection with plants, for the convenience of study it is usually divided into a few **branches** or subdivisions.

Morphology (*morphe*=form, *logos*=discourse) includes the study of the form and structure of the plant organs. The gross external features that can be examined with the naked eye, come under External Morphology. Internal Morphology or **Anatomy** (*ana*=asunder; *temnein*=to cut) is that branch which deals with the internal structure. These internal structures and other minute details cannot be observed without proper magnification. **Histology** (*histos*=cobweb) is the study of minute structures. It thus forms a part of anatomy. The use of the magnifying apparatus, the microscope is essential for anatomical studies. **Cytology** (*kytos*=cell) is the special study of the cell, the unit of structure and function. **Physiology** (*physis*=nature of life) deals with the different vital functions performed by the plants. It seeks to explain the vital processes like metabolism, growth, movement and reproduction in terms of physics and chemistry, as far as possible. **Taxonomy** (*taxis*=arrangement; *nomos*=law) or Systematic Botany is concerned with the classification of plants or division of the plant kingdom into smaller and smaller groups in a systematic manner, and with the naming of plants for identification. **Ecology** (*oikos*=house) considers the influence of environments (surroundings) on plants and plant groups and the various adaptations exhibited by plants according to the situations where they grow. **Plant geography** is the study of the distribution of plants in the different parts of the world and the factors responsible for it. **Paleo-botany** considers the plants that are extinct. This study is based on the fossil remains of the plants that existed in the dim past. **Phytopathology** or plant pathology is the study of the plant diseases with relevant suggestions about prevention and treatment. **Genetics** is a modern science dealing with studies of heredity in plants. **Economic botany** is the scientific knowledge of plants useful to mankind.

Importance of Plants and Scope of Botany

We are surrounded by a world of plants and have to depend on them for almost all our daily necessities. The principal items of our food are derived directly or indirectly from plants. The green plants have the unique power of synthesising complex organic food matters from simpler substances in the presence of sunlight. That process involves taking in of injurious carbon dioxide gas from the outside atmosphere and giving out of oxygen. In this way plants purify the air and thus make possible all life on earth. All the important food cereals like rice, wheat, maize, bazra, barley, oat are derived directly from the plants. Fats and oils derived from plants are used as food, and also in other industries like soap-manufacture, paints and varnishes, etc. Different spices, beverages like tea, coffee, cocoa, wines are also plant products. Most of the building materials like timber, wood and bamboos, are supplied by plants. Plant fibres are the chief materials for manufacture of clothings. Plants are the sources of a large number of medicines we use. The wonder drugs of the present age—penicillin, streptomycin, aureomycin, etc., are extracted from some of the lower forms of plants. Coal, which is an indispensable commodity for industry in the civilized world, is nothing but the remains of buried plants that existed long long before—in the past geological ages. Tar, petroleum, aniline dyes, saccharine, etc., are produced from coal. Various other essential substances like paper, rubber, resin and cork are also plant products. Further, the aesthetic value of the plants is by no means meagre. The world of plants has been the source of inspiration to many artists, poets, literatures and musicians from times immemorial.

Knowledge of Botany is an essential pre-requisite for the pursuit of many applied sciences like Agriculture, Horticulture, Sericulture, Animal Husbandry, Forestry, Pharmacology and Medicine.

The Plant Kingdom

Quite a large number of plants exists in the world of nature. They are not only numerous but varied. The smallest bacteria which are the causes of many diseases and which cannot be seen without the high magnification of a microscope, form one extreme and the huge forest trees form the other. Transitional forms are numerous. The problem is rather difficult and cannot be tackled unless some suitable and systematic plan of dividing the plant kingdom into smaller and smaller groups is devised. The classification that is followed in Botany is undoubtedly very systematic, as it is based on the relationships amongst the plant groups.

Plant kingdom is broadly divided into two groups, *viz.* (*i*) Flowerless plants or **Cryptogams,** and (*ii*) flowering plants or **Phanerogams** or **Spermatophytes.** Cryptogams never bear flowers and reproduce by some minute bodies called spores, or by simple fission. They are again divided into three groups:

(*a*) **Thallophyta.** A thallophyte is that plant where the plant body is an undifferentiated mass of cells (*thallus*), having nothing like stems, roots and leaves. The green thallophytes which usually grow in water are called **algae,** *e.g.* pondscum; and non-green thallophytes are the **fungi.** The common moulds growing on rotten organic substances and mushrooms, belong to this group.

(*b*) **Bryophyta.** This group includes plants like mosses with small stems and simple leaves but no roots. They grow in moist places often forming something like green velvety carpet. Some bryophytes have thalloid bodies.

(*c*) **Pteridophyta.** Here the plant body is well differentiated into roots, stems and leaves, and the organs have well-developed conducting strands. These are the highest cryptogams also known as vascular cryptogams. Ferns are the familiar examples.

Spermatophytes or Phanerogams. These are the higher plants. They bear flowers and reproduce by means of seeds. They are of two types: **Gymnosperms** (*gymnos*=naked; *sperma*=seed) and **Angiosperms** (*angeion*=a case). Gymnosperms are the naked-seeded plants where the seeds remain exposed; but the angiosperms are closed-seeded ones where seeds are covered, that is, remain inside the fruits. Almost all the flowering plants with which we are familiar belong to this group. Angiosperms are **dicotyledonous** or **monocotyledonous**, according to the number of cotyledons present in the embryo. Pea, mango, jack-fruit, etc., are dicotyledons, and rice, maize, palms, etc., are monocotyledons.

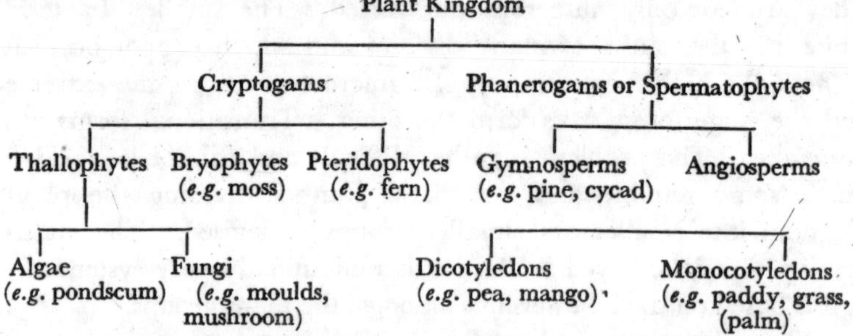

It may be of interest to note that more than 340,000 plants have been identified, described and named by the botanists. Approximately 110,000 of them are thallophytes, 20,000 bryophytes, 10,000 pteridophytes and nearly 200,000 are spermatophytes.

The above classification proves beyond doubt that in the plant kingdoms there is a gradual ascent in complexity of structure from the lowest thallophytes to the highest angiosperms. Gradual adaptability to terrestrial habital from aquatic is also a notable feature.

Apart from this systematic classification, the plants are also grouped on the basis of their mode of nutrition. It has already been mentioned that the green plants can manufacture complex organic food matters from the raw materials, water and carbon dioxide gas, in the presence of sunlight. So they are self-supporting and are called **autophytes.** Plants like *Cuscuta* or dodder (B. Swarnalata) do not manufacture food but grow on some other plants, referred to as 'hosts', and absorb sap from the host by sending out peculiar sucking roots (Fig. 44,2). They are called **parasites.** Parasites may be total or partial. *Cuscuta* and *Rafflesia* (Fig. 16), which produces the biggest flowers, a yard in diameter and nearly 25 lbs. in weight (native of Java and Sumatra), are total parasites. *Sapria Himalayana*, found in Eastern most part of India are just like *Rafflesia*, but smaller in size. *Loranthus* or mistletoe and sandal wood trees are known as partial parasites, because in addition to the sucking roots they also produce thick leaves for manufacture of food. Some orchids grow on the stout branches of trees and produce roots which freely hang in air. These roots are sometimes green in colour, and owing to presence of a tissue known as 'velamen' can absorb moisture from air. They are called **epiphytes** (Fig. 43). Fungi like moulds and mushrooms grow on dead rotten organic matters and draw nourishment from them. These plants are **saprophytes.** Roots of leguminous plants like pea and pulses are infected by some nitrogen-fixing bacteria which form swellings or nodules. These bacteria fix free nitrogen from the atmosphere and supply the same to the plant, and, in return, they get other food matters from the plant. Thus the bacteria and the plant lead a life of mutual friendship, referred to as *symbiosis*. They are called **symbionts.** The **insectivorous** or **carnivorous** plants are very interesting. Some of their organs are modified into structures suitable for attracting and catching insects. Those plants draw part of their nitrogenous food requirement from the bodies of the insects they catch. *Nepenthes* or pitcher plant is a small herbaceous plant in which the leaf-blade is modified into a hollow pitcher with interesting device for catching insect. *Drosera* or sundew is another small insectivorous plant. It has spatula-like leaves bearing glandular tentacles the tips of which glisten in sunshine like dew-drops. These sensitive tentacles actually are responsible for catching insects. *Utricularia* or bladderwort is a common aquatic plant with segmented leaves. Some of the leaflets are modified into swollen bladders for catching insects. An account of the insectivorous plants has been given in details in Chapter XIX.

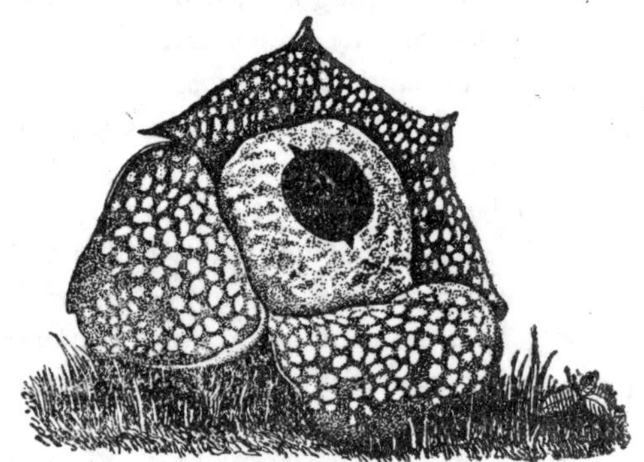

Fig. 16. *Rafflessia*, the biggest flower of the world—a root parasite.

According to *habitat* or situations where they grow, plants may be classified as follows:

Hydrophytes or water plants grow in water or places with plenty of water. They are normally delicate plants with poorly developed roots, segmented or ribbon-like leaves, soft and spongy stems with abundant air-spaces. *Vallisneria* (B. Pata shaola), water hyacinth, lotus, lily are the examples.

Mesophytes are common land plants growing in moderate

water supply. They stand midway between water plants and desert plants as regards structural peculiarities, possessing broad and thin leaves with stomata and well-developed roots with root-hairs.

Xerophytes are plants of deserts—very dry situations with extreme heat, dry air, intense light and high wind. They are characterised by very strongly formed root system; stunted aerial portion; rolled, folded or otherwise extremely reduced leaves with extra coating of wax on epidermis for preventing water loss, and arrangement for storage of water. Examples: Cacti, *Opuntia* (B. Fanimansha).

Halophytes grow in sea-shore or saline marshes where absorption of water becomes difficult due to preponderance of salt. A special type of vegetation of low forest called *mangrove* is found in Sundarbans. Stilt roots for extra support, special aerial roots for absorption of oxygen, cymose branching and peculiar germination called vivipary are characteristic features. Examples: *Heritiera* (B. Sundri), *Ceriops* (B. Goran).

According to the nature of stems plants can be put into a number of groups. Some of them have strong stems and so they can stand *erect.* The erect plants may again be unbranched ones, as in palms; with solid nodes and hollow internodes, as in bamboos; and also with a number of branches forming a spreading crown, as in banyan, mango.

Plants with weak stems cannot naturally maintain upright position. They either *creep* on the soil surface, as in Dub grass (B. Durba) and sweet potato (B. Ranga Aloo); or climb round a support by *twining* around it, as in *Clitoria* (B. Aparajita). Some of them have special sensitive organs for climbing called tendrils, as in Passion-flower (B. Jhumka Lata) and Gourd (B. Kumro).

Parts of an Angiospermic plant

Parts of a plant can be very conveniently observed in a common annual plant like sunflower (Fig. 17) or mustard (Fig. 18). The underground organs are the **roots,** by means of which the plant remains anchored to the soil. Moreover, the roots absorb water and mineral matters from the soil. The main root has many branches and sub-branches going in different directions in the soil to carry on the functions effectively. The aerial portion is called the **shoot.** The axis continuous to the root is the **stem** which bears **leaves** at the joints called nodes. The inner angle formed between stem and leaf is called the axil of the leaf. **Buds** are present at the axil. These buds, in course of time, develop into

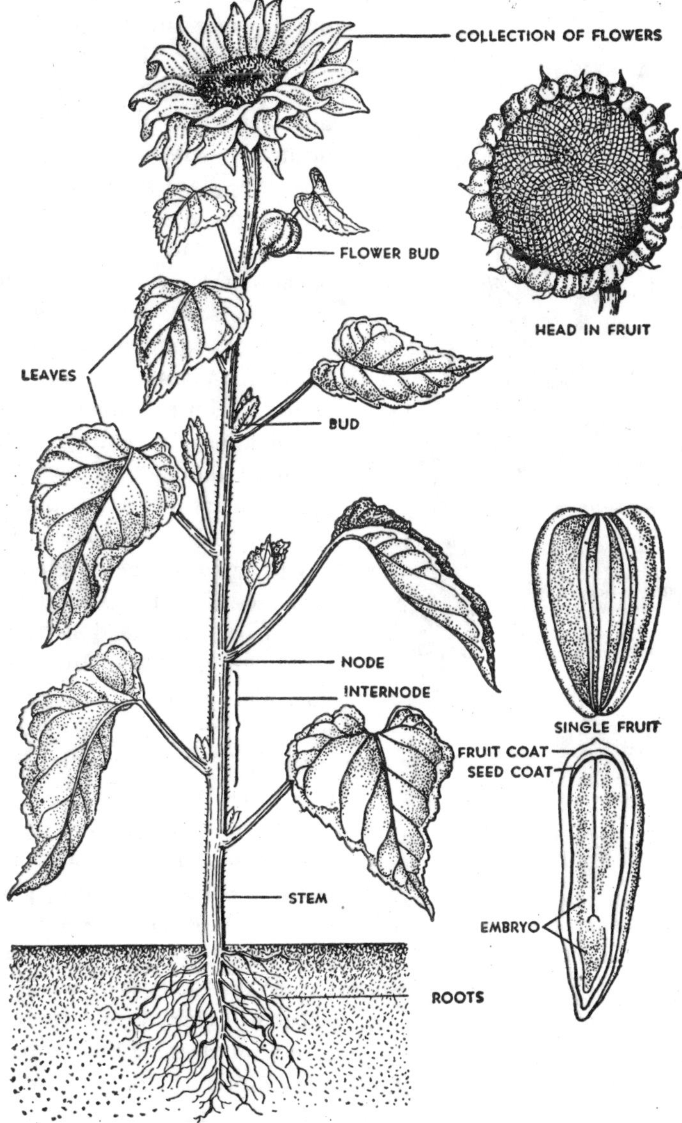

Fig. 17. A sunflower plant showing different parts.

branches. So the stem bears these organs and conducts water absorbed by the roots from the soil to the leaves. The leaves are the organs where complex organic food matters are manufactured out of water and carbon dioxide in sunlight. Prepared food matters are also conducted to different parts by stems and branches. After attaining maturity the plant **flowers.** Floral buds come out from the shoots and open out in time. Some part of the flower develops into **fruit** and the fruit contains **seeds.** Every seed contains a multicellular baby-plant, called **embryo,** which remains dormant. During germination this embryo wakes up from the state of dormancy and develops into a new plant.

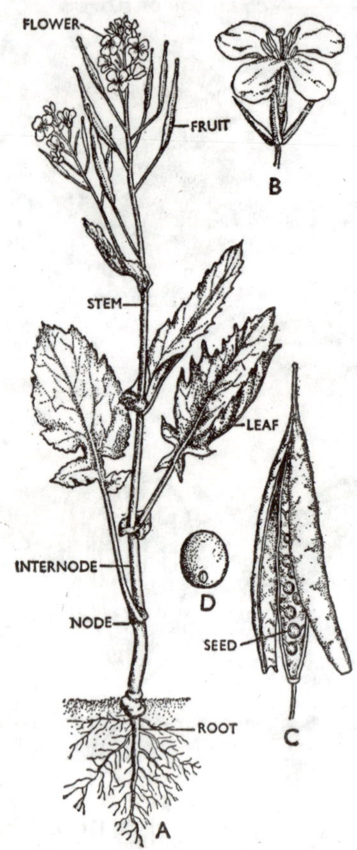

Fig. 18. A mustard plant showing different parts.

The plant as a living thing has to carry on the different life processes. The organs of the plant have particular functions to perform, but the life of the individual plant becomes possible only when all the organs work harmoniously. That shows a clear division of labour in plants. Roots, stems, leaves are the organs indispensable for the maintenance of the life of the individual plant. So, they are called **vegetative oogans** or **organs of vegetation.** But the flowers have nothing to do with the life of the plant. Even if a plant does not flower at all it will carry on its existence normally and will ultimately have natural death. But with the death of the plant its race will become extinct. Flowers are purely meant for multiplication or perpetuation of the race. They are called **reproductive organs** or **organs of reproduction.** Fruits and seeds mainly protect and help in dispersal of the new plant lying inside the seed.

MORPHOLOGY

CHAPTER II

THE SEED

The seed is usually covered by one or two coats. The most essential part of a seed is the **embryo** or the baby-plant which remains *dormant.* Dormancy means the temporary cessation of vital activities. During germination the embryo wakes up from the state of dormancy and takes to active life. The embryo has an **axis** (*tigellum*), the upper part of which is the **plumule** or the future shoot, and the lower part is the **radicle** or the future root. One or two **cotyledons** or seed leaves remains attached to the radicle-plumule axis. On the basis of the number of cotyledon the seeds are monocotyledonous (*mono*=one) or dicotyledonous (*di*=two). Every seed has stored food matters which are utilised by the growing embryo during germination. Food matters may remain stored up in the cotyledons of the embryo, or as a distinctly separate structure outside the embryo. In the latter case, the stored food is called **albumen** or **endosperm.** Seeds having albumen or endosperm are **albuminous** or **endospermic;** and if albumen is absent, the seeds are **exalbuminous** or **non-endospermic.**

Let us now examine a few common seeds. It is always convenient to get the seeds soaked in water and thus made soft before examination.

Dicotyledonous Seeds

Gram Seed. The gram seed is more or less rounded at one end and pointed at the other. It is covered by a brown seed coat called **testa;** the inner whitish coat is the **tegmen.** At the pointed end of the seed the testa bears a scar called **hilum.** It indicates the position of the attachment of the seed to the fruit wall. Near about the hilum there is a very minute opening called **micropyle** (*micro*=small; *pyle*=gate). If a soaked seed is pressed a droplet of water oozes out through this micropyle. The kernel is exposed when the seed coat is removed, which in the gram seed is nothing but the embryo. If the kernel is pressed the two fleshy cotyledons are found, which remain laterally hinged to the axis of the embryo at a joint called the **first node** or **nodal zone.** The lower part of the axis lying towards the pointed end is the **radicle** and the upper part lying between the two cotyledons is the **plumule.** The plumule has very minute leaves and looks like a feather. Food matters remain stored up in the fleshy cotyledons. So the gram seed is **dicotyledonous exalbuminous** (Fig. 19).

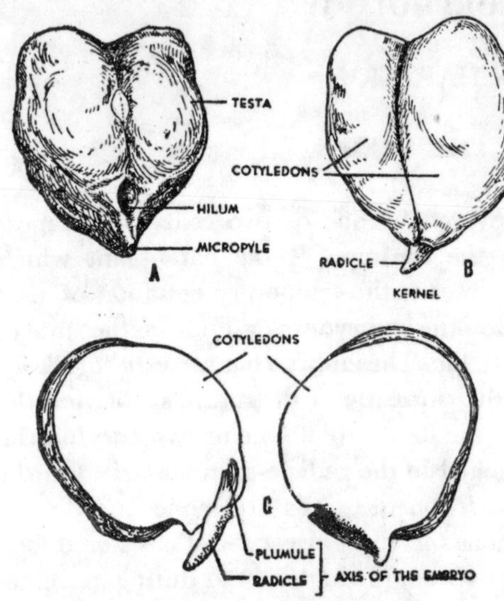

Fig. 19. Structure of Gram Seed. A—seed; B—kernel; C—same split open showing parts.

Pea Seed. Pea seed is more or less rounded in shape. It has only one coat, the pale white **testa.** On the testa there is the distinct **hilum** or the scar, and the **micropyle** is present near about the hilum. On removing the seed coat the kernel is exposed. It is made up of two fleshy **cotyledons** attached laterally to the axis of the embryo, the lower part of which is the **radicle** and the upper part, the plumule. So the kernel in pea corresponds to the **embryo.** Pea is also **dicotyledonous exalbuminous** (Fig. 20).

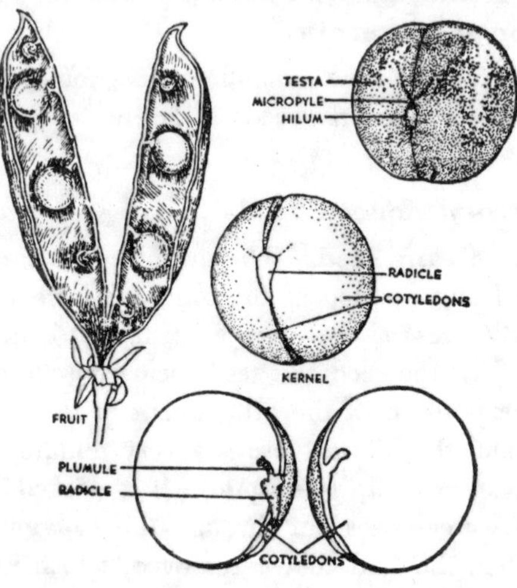

Fig. 20. Structure of Pea Seet.

Castor-oil Seed. It is more or less oblong in shape. The **testa** is hard, shell-like and has many sculpturings on the outer surface. At one end the testa has a spongy outgrowth called the **caruncle.** The hilum and micropyle occur at that end and usually remain concealed by the outgrowth. Next to the hard brittle testa there is a thin white papery coat. It is not really the second coat or tegmen but is known as **perisperm** (remnants of nucellus).

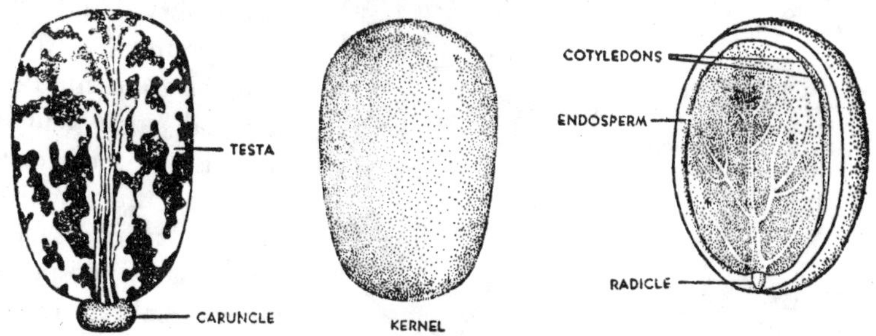

Fig. 21. Structure of Castor-oil Seed.

The kernel consists of **embryo** and **endosperm** or **albumen** surrounding the embryo. The embryo has small radicle and very minute plumule forming the axis and two thin leafy cotyledons placed one against the other. They have distinct veins which leave marking on the endosperm. Surrounding the embryo there is stored food matter, the endosperm. In a castor-oil seed the kernel is embryo *plus* albumen, and so it is a **dicotyledonous albuminous** seed (Fig. 21).

Dicotyledonous exalbuminous seeds ... pea, gram, pulses, beans, mango

Dicotyledonous albuminous seeds ... castor-oil seed, papaw, poppy

Monocotyledonous Seeds

Maize Grain. The maize grain is really a one-seeded fruit and not a seed, hence it is called a *grain.* The **seed coat** and the fruit wall or **pericarp** are inseparably united in the maize grain to form the outer pale yellowish coat. A whitish deltoid area is noticed on one side of the grain which marks the position of the embryo.

The embryo can be clearly seen if the grain is cut into two parts along the longer axis. It occupies only a very small portion of the grain, the remaining part being endosperm or stored food.

The embryo has an axis with radicle and plumule. The radicle is surrounded by a sheath called radicle sheath or **coleorhiza,** and the plumule has a similar sheath called plumule sheath or **coleoptile.** A shield-shaped body, called **scutellum,** lies between the axis of the embryo and the endosperm, thus dividing the grain into two unequal parts. Scutellum is the *single cotyledon* of the maize grain. It absorbs food from the endosperm for the embryo to be used during germination. Maize grain is **monocotyledonous albuminous** (Fig. 22).

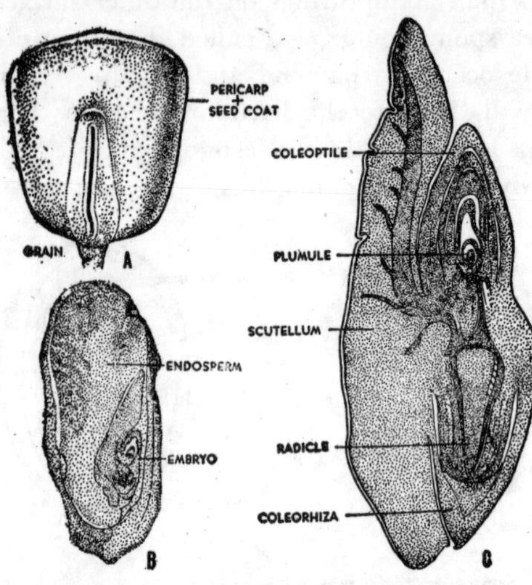

Fig. 22. Structure of Maize Grain. A—grain; B—same cut lengthwise; C—embryo enlarged.

Rice Grain. It is also a one-seeded fruit. The husk that encloses the grain is neither a part of the seed nor the fruit. It has four parts, two very small minute glumes, called *empty glumes*, at the base and two larger ones—the *flowering glume* and the *palea*, surrounding the grain.

The grain is exposed when the husk is removed. The reddish or whitish coat of the rice grain is the product of fusion of the pericarp and the seed coat. Here also the endosperm covers the major part of the grain, the minute embryo lying at one end beneath the palea. During polishing of rice the outer coat and the embryo are removed, leaving a small break at one end of the grain. Obviously what remains after polishing is nothing but endosperm.

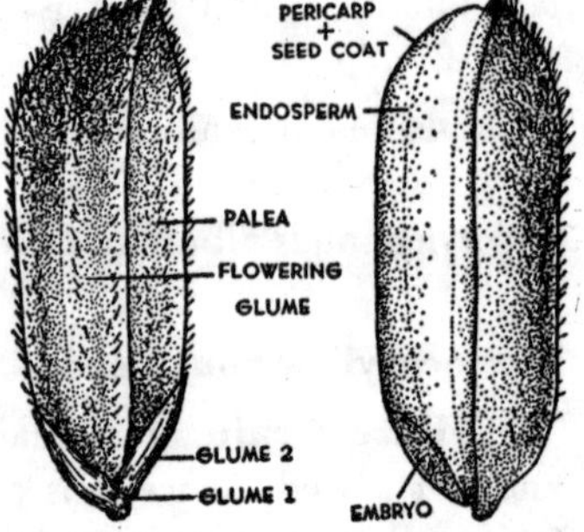

Fig. 23. Structure of Rice Grain.

The embryo is more or less like that of maize having radicle and plumule with coleorhiza and coleoptile respectively and the shield-shaped scutellum, the single cotyledon. The radicle is a bit curved in rice grain (Fig. 23).

So it is also **monocotyledonous albuminous.** Most of the monocotyledonous grains are albuminous with a few exceptions like orchids.

Dicotyledonous exalbuminous

- Gram Seed
  - Coats (testa and tegmen)
  - Kernel——Embryo ——
    - Axis ——
      - Plumule
      - Radicle
    - Cotyledons, 2 (fleshy)

- Pea Seed
  - Coat (testa)
  - Kernel——Embryo ——
    - Axis ——
      - Plumule
      - Radicle
    - Cotyledons, 2 (fleshy)

Dicotyledonous albuminous

- Castor-oil Seed
  - Coat (testa)
  - Kernel ——
    - Embryo ——
      - Axis ——
        - Plumule
        - Radicle
      - Cotyledons, 2 (leafy)
    - Endosperm

Monocotyledonous albuminous

- Maize Grain——
  - Coat (pericarp and seed coat)
  - Kernel ——
    - Embryo ——
      - Axis ——
        - Plumule with coleoptile
        - Radicle with coleorhiza
      - Cotyledon 1, (scutellum)
    - Endosperm (stored food)

- Rice-Grain
  - Grain——
    - Coat (pericarp and Seed coat)
    - Kernel ——
      - Embryo —
        - Axis ——
          - Plumule with coleoptile
          - Radicle with coleorhiza
        - Cotyledon 1, (scutellum)
      - Endosperm
  - Husk

GERMINATION OF SEEDS

Germination is the process by which the embryo wakes up from the state of dormancy and takes to active life. The process, in fact, covers all the changes from the earliest sprouting of the seed till it establishds itself as an independent plant. A mature seed is living. But the embryo remains dormant. The seed then contains only 10% to 15% of water, and this low water content is one of the factors responsible for dormancy. When certain external conditions are satisfied the dormant embryo begins to grow. Supply of **water,** favourable **temperature,** adequate supply of **oxygen** are main external conditions essential for germination, but, the seeds concerned must be *viable, i.e.* they must retain the power of germination. **Light** also has influence on germination of some seeds, but in most cases light retards germination at the early stages.

Water or Moisture. The supply of water is very important for germination. The seeds absorb water and swell up. This swelling is the earliest sign of germination. Seed coats are softened and rupture easily to make way for the radicle and plumule to come out. Due to absorption of water the dormant protoplasm is activated for carrying on the vital functions. It secretes *enzymes* or digestive juices, which convert the complex insoluble food matters stored up in the seed, either in the cotyledons or the endosperm, into their simpler soluble forms, so that the growing embryo may utilise them. Entrance of oxygen is also facilitated, and, as a result, the rate of respiration is accelerated.

Temperature. An optimum rate of temperature is essential for germination. Protoplasm cannot carry on vital activities at a very low or at a very high temperature. There is a minimum below which seeds will not germinate and a maximum above which germination will not take place. The suitable temperature, of course, varies with the seeds. Most of the seeds usually show signs of germination when temperature varies from 25°C. to 30°C.

Supply of Oxygen. The dry seeds respire very feebly; with germination the rate of respiration markedly increases. So a constant supply of oxygen becomes essential.

That these conditions are essential for germination can be shown by a simple experiment, called the *three bean experiment.* A glass beaker is half-filled with water. It is better to get water boiled beforehand to drive out the dissolved gases. A wooden block coated with paraffin is taken and three healthy bean seeds are fixed to the block by pins at regular intervals. Now the block is placed in the beaker in such a way as to keep one seed submerged in water, the second just at the water-level and the third above the water-level. The experiment thus set up is allowed to remain on the table for a few days. Water is to be poured from time to time to maintain the original level.

After a few days it is found that the first seed, *i.e.* one at the bottom, just shows signs of germination, the radicle comes out bursting the seed coat and there it stops. The second one (at the water-level) germinates nicely and the third one at the top remains unchanged (Fig. 24).

Fig. 24. Three bean experiment showing conditions essential for germination.

The conclusion that can be drawn is that the first one has germinated utilising traces of oxygen present in water, it has not grown further for want of oxygen, though it has plenty of water. The second seed has all conditions satisfied, so has germinated properly; and the third one above the water-level has shown no sign of germination, though it has adequate oxygen and suitable temperature.

In this experiment the influence of temperature on germination has only been indirectly proved. If a similar experiment is set up and the temperature is lowered by placing the beaker inside a bigger jar containing ice blocks, or the temperature is raised by passing steam through the bigger jar, it will be found that none of the seeds is showing any sign of germination. Thus it is proved by experiment that the conditions—supply of water, moderate temperature and a constant supply of oxygen, are essential for the germination of seeds.

Germination of Dicotyledonous Seeds

Gram Seeds. When the essential conditions are satisfied, the seed begins to germinate. At first it absorbs water and swells up. The testa ruptures near about the micropyle and the radicle is the first structure to come out of the seed. It goes downwards to produce the main root (primary root) of the plant from which subsequently come out many branch roots. In the mean time the stalks of the cotyledons elongate to make way for the plumule to come out, which actually goes upwards being pushed by the active **epicotyl.** The epicotyl is the region between the plumule and the point of attachment of the cotyledons, *i.e.* the nodal zone. The plumule gradually grows upwards to produce the green aerial portions of the plant. Food matters, stored up in the cotyledons, are used up by the growing embryo, so the cotyledons gradually shrivel up (Fig. 25).

It is to be noted that in the germination of gram seed the cotyledons remain inside the seed coat below the soil and the plumule goes upwards due to the active growth of the epicotyl. The position

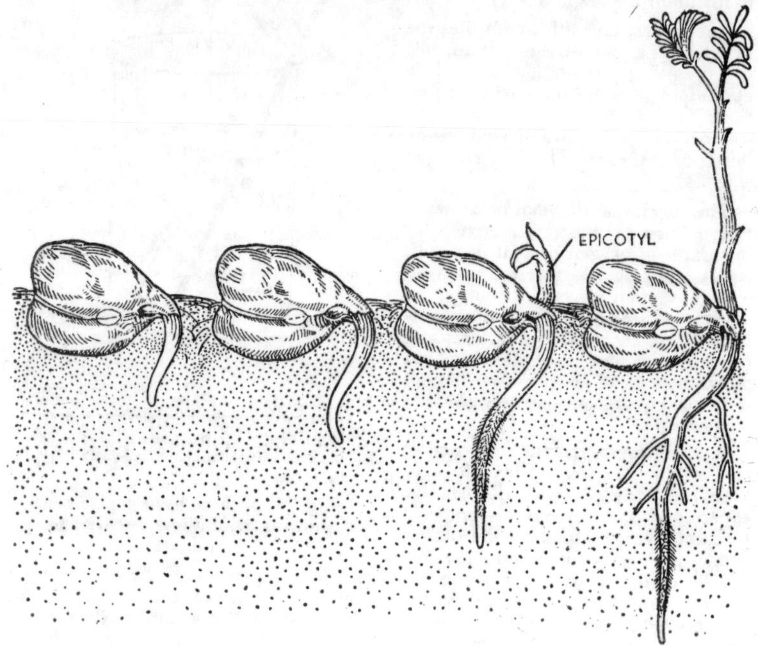

Fig. 25. Stages in the germination of Gram Seed (hypogeal).

of the seed is not disturbed. This type of germination is called **hypogeal** or **hypogeous** (*hypo*=below; *geo*=earth). Peas and many other exalbuminous seeds show this type of germination (Fig. 26).

Castor-oil Seed. In the castor-oil seed the seed swells up, the testa bursts near the spongy outgrowth, the caruncle, and the radicle comes out first. It goes downwards as usual to produce the primary root which later on bears many branches. Now the **hypocotyl,** the region between the radicle and the nodal zone, grows fast forming a curved arch or loop. This loop gradually straightens up, thus lifting the plumule and the cotyledons in the air above the soil level. Even then plumule and cotyledons remain surrounded by endosperm and remnants of test. The endosperm is gradually exhausted. The two cotyledons develop green colour and serve as the first pair of green

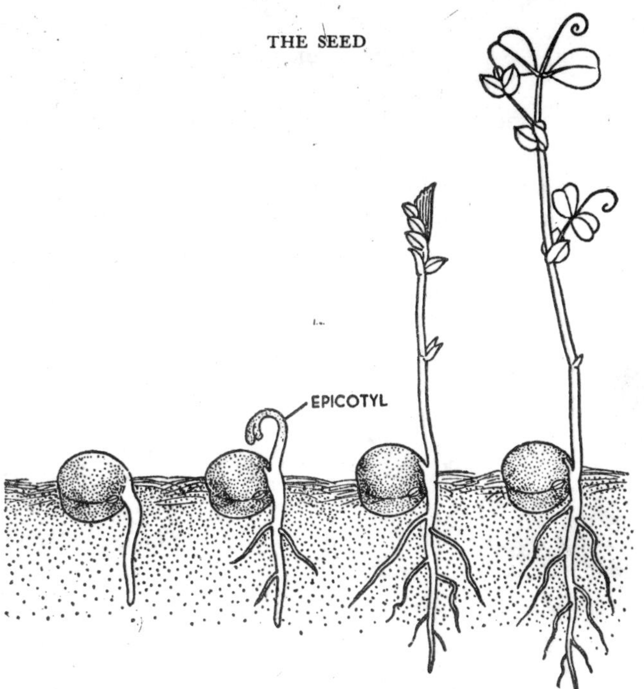

Fig. 26. Stages in the germination of Pea Seed (hypogeal).

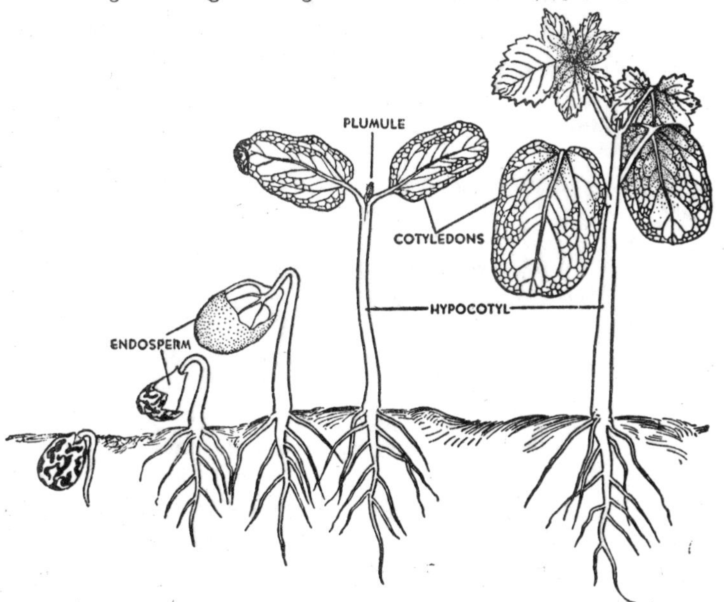

Fig. 27. Stages in the germination of Castor-oil Seed (epigeal).

leaves. Now plumule grows upwards to form the green aerial shoot.

So in the castor-oil seed the plumule goes upwards by the active growth and elongation of the hypocotyl which pushes the plumule, cotyledons, etc., above the soil level. Thus the position of the seed is definitely disturbed (Fig. 27). This is **epigeal** or **epigeous** germination (*epi*=upon, *geo*=earth). Gourd, tamarind also show this type of germination. In gourd the radicle usually develops into the root and the hypocotyl forms a loop which straightens up and pulls the rest above the surface of the soil. A peg-like projection develops at the base of the hypocotyl which pushes the seed coat, so that the cotyledons and plumule may easily come out of the seed. The cotyledons become green and large and behave like ordinary leaves [Fig. 28(A)]. Tamarind seed also shows same type of germination [Fig. 28 (B)]. Here the testa is very hard and the two cotyledons are quite large and thick. They become greenish in colour but do not serve as green leaves. The cotyledons gradually shrivel up and drop when the stored food matters are exhausted.

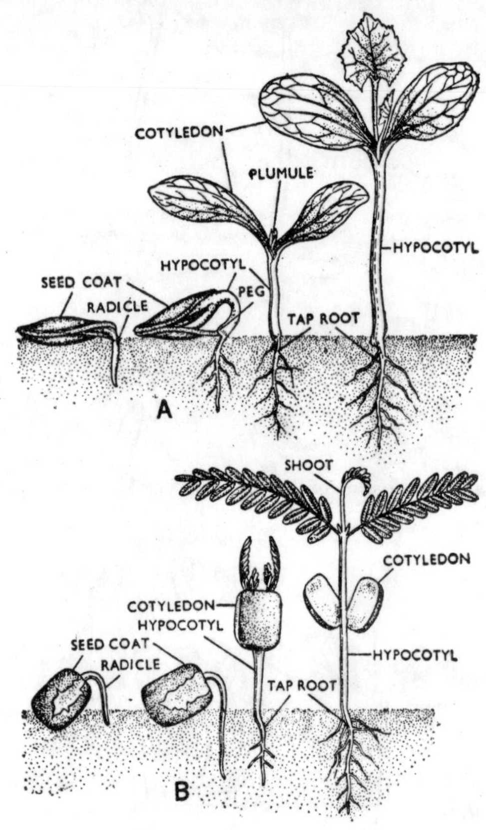

Fig. 28. Stages in epigeal germination in gourd (A) and tamarind (B).

Germination of Monocotyledonous Grains

Maize Grain. During germination of the maize grain the radicle at first comes out boring through the coleorhiza or radicle sheath and forms the primary root. This root does not persist but dies and is replaced by a tuft of adventitious fibrous roots from the base of the stem. In the mean time the plumule goes upwards with the plumule sheath or coleoptile. After growing to some extent the plumule pierces the sheath and grows up to produce the green aerial

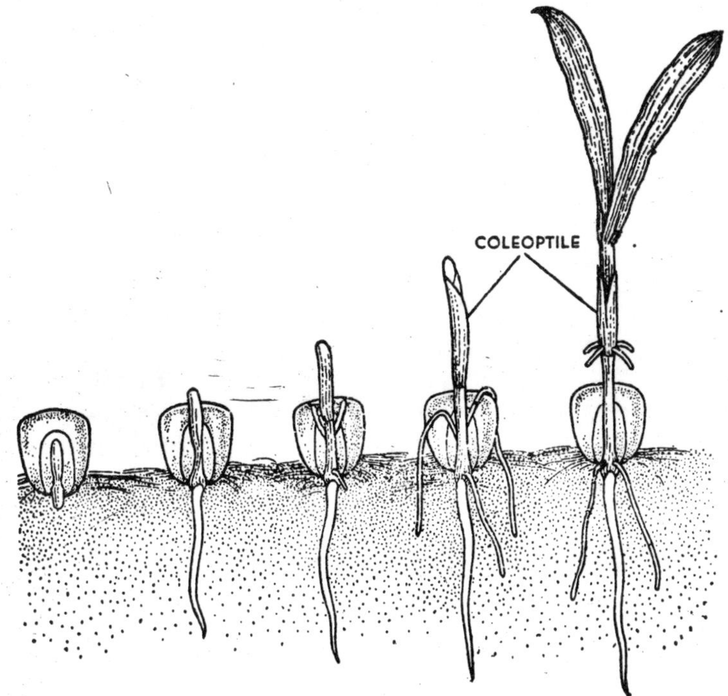

Fig. 29. Stages in the germination of Maize Grain (hypogeal).

organs. Scutellum or the single cotyledon continues to supply food matters to the growing embryo during germination. The mode of germination in maize grain is also **hypogeal** (Fig. 29). Rice (Fig. 30) and other grains also have this type of germination.

In cocoanut the innermost layer of fruit wall is the shell having three eye-like scars, and the embryo lies below one of the scars. The white kernel including the milk is the endosperm. During germination the lower end of the embryo—the cotyledon begins to grow as a spongy structure to absorb food materials stored in the endosperm. The upper end of the embryo grows through the eye carrying the

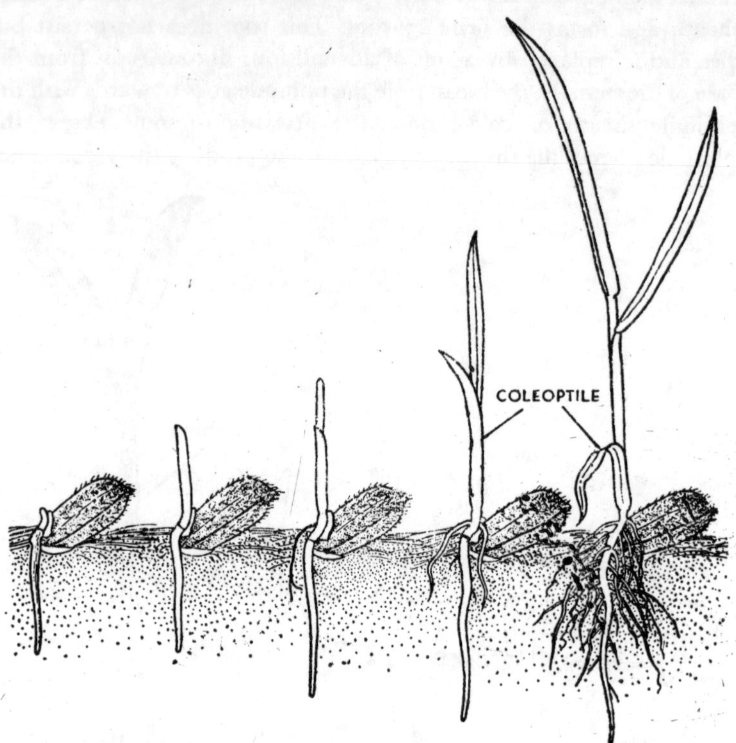

Fig. 30. Stages in the germination of Rice Grain (hypogeal).

radicle and the plumule; pierces through the fibrous coat and ultimately establishes itself.

Special type of Germination

Plants growing in saline marshes or near sea-shore show a peculiar type of germination known as **vivipary.** In that case, the seeds begin germination before their liberation from the fruits (Fig. 31). The radicle becomes elongated and considerably swollen. Then the seed gets detached from the parent plant and comes vertically downwards. The radicle pierces the muddy soil below and thus gets fixed. Lateral roots are soon formed for anchorage, and the plumule is kept above the surface of saline water. Examples: *Rhizophora, Ceriops.*

Dicotyledonous and Monocotyledonous Seeds and their modes of germination

Monocotyledonous Seeds	Dicotyledonous Seeds
1. One cotyledon in the embryo.	1. Two cotyledons.
2. Plumule is lateral and cotyledon is terminal.	2. Plumule is terminal, cotyledons are lateral.
3. Plumule and radicle are surrounded by coleoptile and coleorhiza respectively.	3. No sheath.
4. Mostly albuminous.	4. Albuminous or exalbuminous.
5. Germination—usually hypogeal.	5. Hypogeal or epigeal.
6. Plumule goes upwards with the plumule sheath.	6. Plumule is pushed upwards by the actively growing epicotyl or hypocotyl.
7. Primary root formed from radicle soon perishes and is replaced by a tuft of adventitious fibrous roots.	7. Radicle produces the primary root which persists and bears many lateral roots.

Functions of Cotyledons

Cotyledons or the seed leaves carry on many functions. In dicotyledonous exalbuminous seeds like pea and gram they serve as the store-houses of food. In albuminous seeds they absorb food matters from the endosperm and supply them to the growing embryo. Further, they protect the young plumule and serve as the first pair of green leaves in case of epigeal germination. In monocotyledons the cotyledon is the main absorbing organ which draws food from the endosperm for the embryo. In some cases it pushes the radicle and plumule out of the seed coat (*e.g.* Date-palm). In cocoanut the cotyledon serves as a sucker for absorption of food.

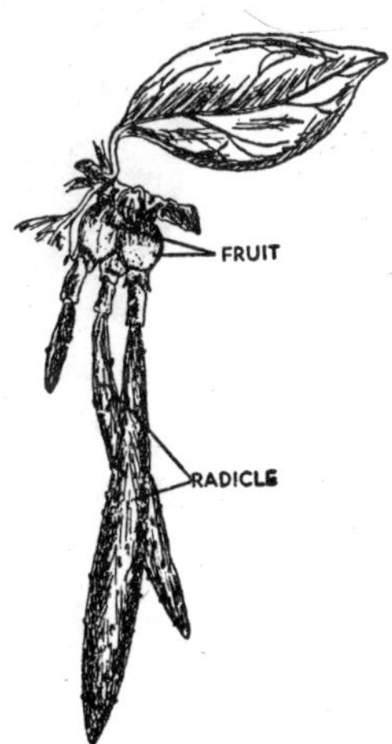

Fig. 31. Viviparous germination.

CHAPTER III

THE ROOT

The root is the descending axis of the plant and the downward prolongation of the radicle. It is usually non-green in colour and grows away from light. The root has no nodes and internodes. It bears only similar members, *i.e.* only roots. The main root developing from the radicle is called the **primary** root which, in turn, produces many **secondary** and **tertiary branches,** thus forming a system of roots going inside the soil in different directions. These branch roots develop from internal tissues of the primary root, so they are *endogenous* in origin. They are arranged in *acropetal* succession, *i.e.* older roots near the base and smaller and smaller ones towards the tip.

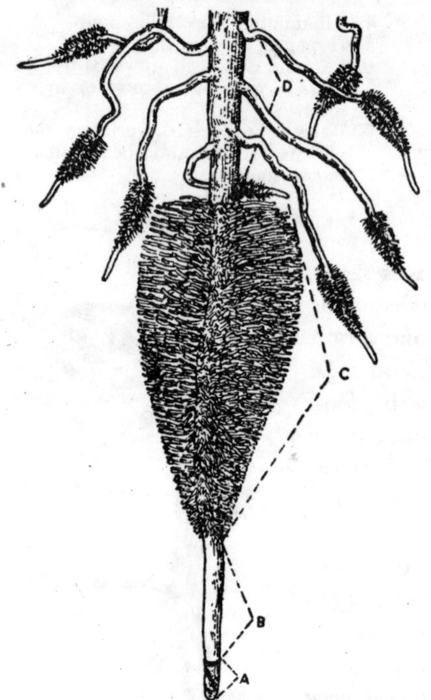

Fig. 32. Regions of a root. A—root cap region; B—region of active growth and elongation; C—root-hair region; D—permanent region.

Regions of the Root (Fig. 32). A root has usually four regions, in order, from the apex toward the base:

(i) **Root cap region.** The extreme tip of the root is usually protected by a cap or thimble-shaped body called the **root-cap.** When the root pushes its way through the soil, the tender and soft tip runs the risk of being injured due to friction with soil particles. The root-cap, serving as a buffer, protects the tip from that danger. Further, the cap secretes slimy mucilaginous matters which facilitate the course of the root through the soil. The root-cap is absent in many aquatic plants. In screwpine root-caps are multiple (Fig. 33).

(ii) **Region of active growth and elongation.** Next to the root-cap is the **region of active growth.** It is a very short region where cells divide actively. This merges into the region of elongation where the newly formed cells grow in length.

(iii) **Root-hair region.** This region is found next and here the outer cells of the roots grow out as unicellular hairs. The unicellular root-hairs are responsible for the absorption of water and mineral matters dissolved in water from the soil. By germinating mustard seeds on a piece of soaked blotting paper the dense root-hairs can be conveniently observed (Fig. 34). During the growth

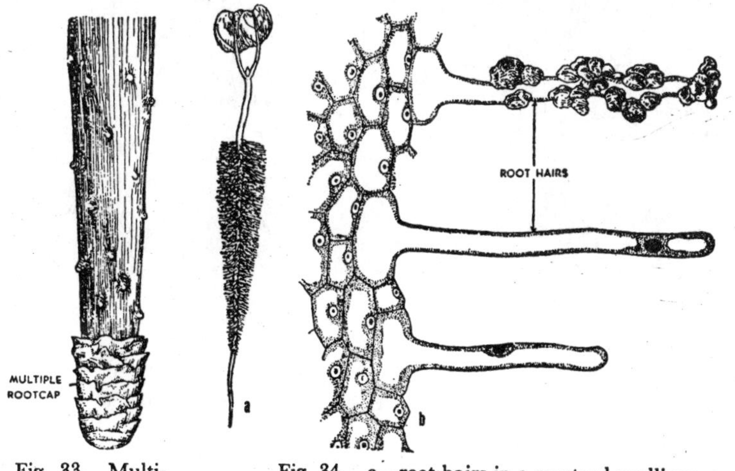

Fig. 33. Multiple root-cap of screwpine.

Fig. 34. a—root-hairs in a mustard seedling; b—some unicellular root-hairs (magnified) with adhering soil particles.

of the root old zone is exhausted and is replaced by new zones of root-hairs more and more towards the tip.

(iv) **Permanent region.** This is located higher up. Here secondary roots develop from the primary root. The branches also have four regions like the primary root.

Root Systems (Fig. 35)

In dicotyledonous plants the radicle produces the primary root which is quite prominent, that is, long and stout. This is called the **tap root.** The tap root bears many secondary and tertiary branches developing acropetally. All these branches can be easily distinguished from the main root. The whole system is called the **tap root system.** In monocotyledons like maize and rice the radicle produces the primary root which soon aborts and is replaced by a tuft of roots developing from the base of the stem. All

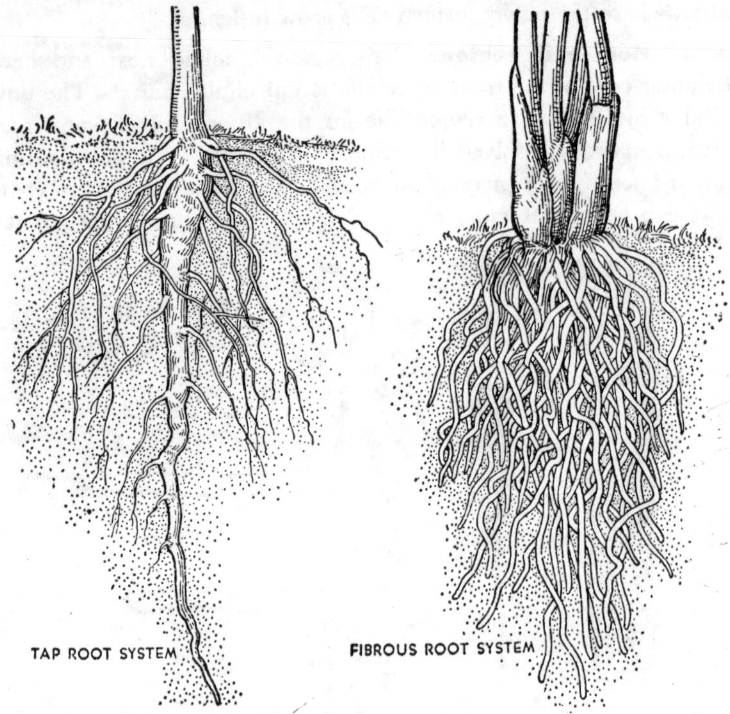

Fig. 35. Tap root system and fibrous root system.

of them are more or less similar and are called **fibrous roots** and the whole system is known as **fibrous root system.**

Normal and Adventitious Roots

Roots that develop from the radicle (primary root) or branches of the roots from that origin, are called **normal** or true roots; and roots growing from any position other than the normal point of origin, are called **adventitious** roots. So the tap root system, usually found in the dicotyledons, is normal and fibrous root system, common in the monocotyledons like grasses, is adventitious.

Normal Functions of the Root

Roots perform two normal functions. First, they fix the plants to the soil and thus secure anchorage. Secondly, roots are responsible for the absorption of water and dissolved mineral matters from the soil through the delicate root-hairs. Water is ultimately conducted to the leaves. The primary roots with their large number of secondary and tertiary branches in the dicotyledons, or the fibrous roots in the monocotyledons penetrate into the soil, spread out in different directions and thus effectively carry on the normal functions. In common land plants the root system, in fact, equals to or even exceeds the aerial portions (shoot system) both in length and volume. The fixation or anchorage is the *mechanical* function of the root; and absorption and conduction of water and dissolved mineral matters, are the chief *physiological* functions. Besides these, roots may perform many special functions which will be discussed in connection with the modified roots.

MODIFICATIONS OF ROOTS

Many roots undergo modifications for carrying on functions other than the normal ones. Some of the modified roots are normal, while many of them are adventitious.

Normal Roots Modified for Storage of food

Some roots become swollen and fleshy due to the storage of food matters and assume different shapes (Fig. 36) :

(a) **Fusiform.** Here the root with the hypocotyl becomes swollen. The swelling is maximum in the middle. It gradually tapers towards the ends, *i.e.* base and apex, *e.g.* radish. Practically the whole swollen portion in radish is the hypocotyl.

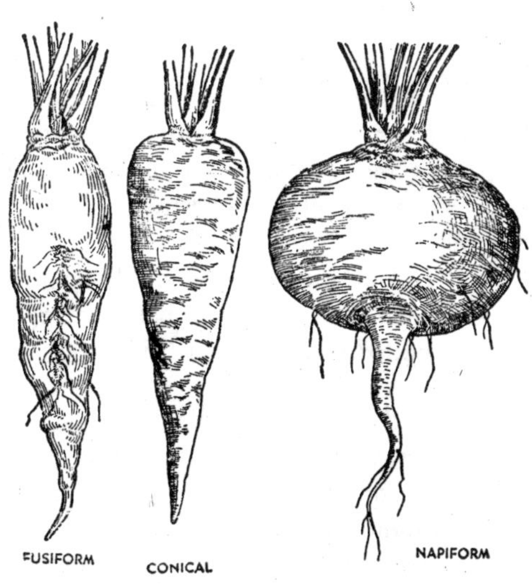

Fig. 36. Modifications of tap roots.

(b) **Conical**. Here also the root is fleshy; it is broadest at the base and gradually tapers towards the apex like a cone, *e.g.* carrot.

(c) **Napiform.** It is very much swollen at the base (with the hypocotyl), but abruptly tapers towards the apex, as in beet and turnip (Fig. 86).

Adventitious Roots Modified for Storage of food

(d) **Tuberous Root.** In plants like sweet potato which grow on the surface of the soil, some of the adventitious roots, developing from the stem, become swollen and fleshy due to the storage of food. They do not have any regular shape. These are called tuberous root (Fig. 37).

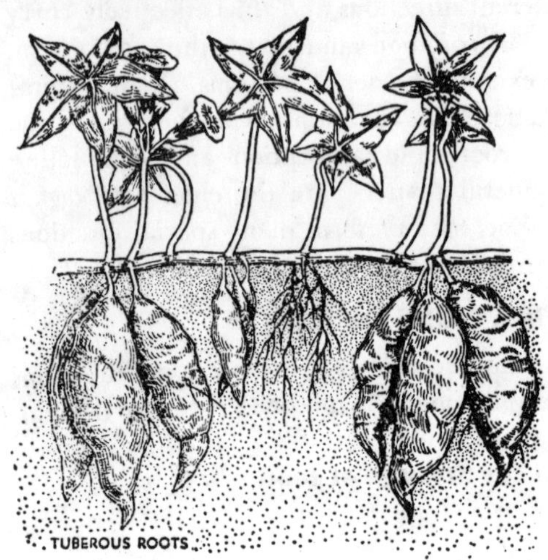

Fig. 37. Tuberous roots of sweet potato.

(e) In *Asparagus* (B. Satamuli), *Ruellia*, etc., quite a good number of tuberous roots are produced from the stem, forming a bundle or fascicle. They are called **fasciculated roots** (Fig. 38).

(f) In mango-ginger (B. Amada) the adventitious roots suddenly become swollen at the tips. They are called **nodulose** roots (Fig. 39).

(g) In many grasses the swellings are often found at frequent intervals giving the root a beaded appearance. These peculiar roots are called **moniliform** (Fig. 39).

Adventitious Roots Modified

1. **Prop roots.** Roots are normally underground organs. Some roots develop completely above ground and, so, are aerial. In the common banyan trees of our country roots are often found to hang freely in the air from the branches. In course of time they

reach the soil, become stouter and serve as so many extra supports or props. Thus they help the stem in bearing the weight of the heavy crown. These are called the **prop roots** (Fig.40).

2. **Stilt roots.** These are also supporting roots. In plants like screwpine (B. Keya) quite a good number of adventitious roots develop from the basal part of the stem and go downwards obliquely. These roots help the plant in maintaining the upright position. These are **stilt roots** (Fig. 41).

3. **Climbing roots.** Weak climbing plants like betel, vine, *Scindapsus* (B. Gaj-pipul), etc., produce some adventitious roots which cling to the supports and thus help the plants in climbing. They often secrete sticky juices or develop disc-like structures for that purpose (Fig. 42).

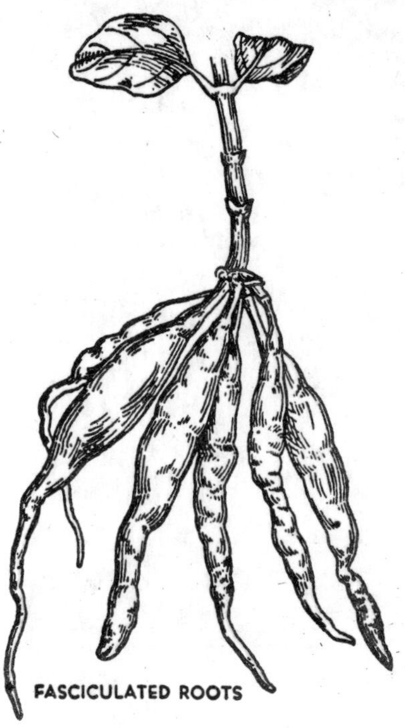

Fig. 38. Fasciculated roots.

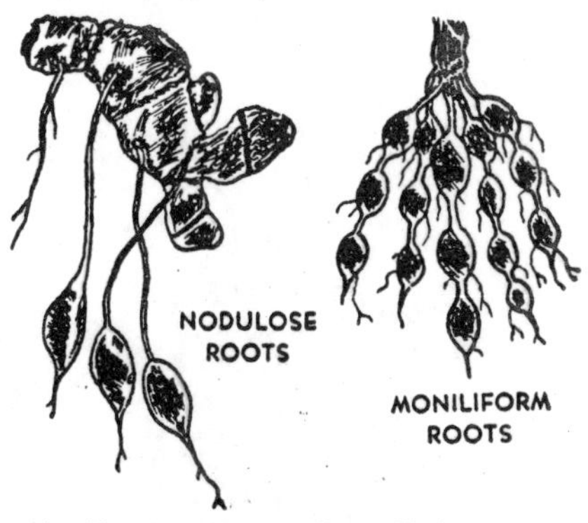

Fig. 39. Nodulose and moniliform roots.

4. **Epiphytic roots.** Many epiphytes, like orchids, produce long roots which freely hang in air. They have sponge-like tissues, called **velamen,** at the exterior by means of which they can absorb moisture from the air (Fig. 43).

5. **Sucking roots** or *haustoria.* Sucking roots or *haustoria* develop in the parasitic plants like *Cuscuta* or dodder (B. Swarnalata). These roots penetrate into the body of the host and draw nourishment from there without caring to

Fig. 40. Prop roots of banyan.

manufacture their own food (Fig. 44, 2 & 3).

6. **Breathing roots.** These roots are produced by the plants like *Heritiera* (B. Sundri), *Rhizophora* (B. Bora) growing in saline marshes or on the sea-shore where the soil is very poor

Fig. 41. Stilt roots of screwpine.

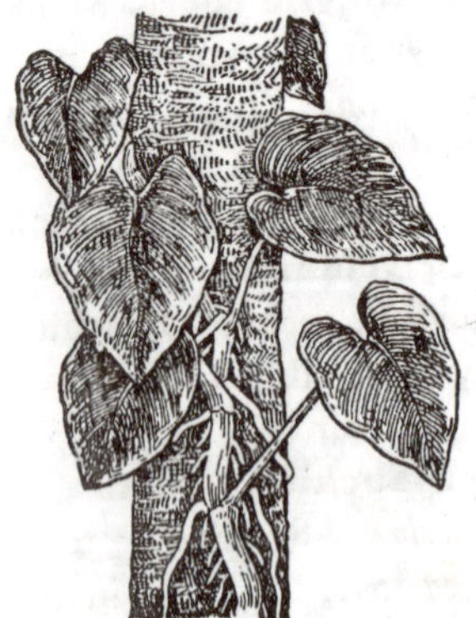

Fig. 42. Climbing roots of *Scindapsus*.

Fig. 43. Epiphytic roots of orchid.

in oxygen. Here some roots with pores at the tips come vertically above the surface of the soil and carry on the gaseous interchange with the outer atmosphere. They are also called *pneumatophores* (Fig. 44—1).

In an aquatic plant *Jussiaea* (B. Keshar, dam) some peculiar adventitious roots develop from the floating branches. They are light, soft and spongy due to the presence of air-spaces. Besides facilitating respiration, they give the plants buoyancy for floating on the surface of water (Fig. 45).

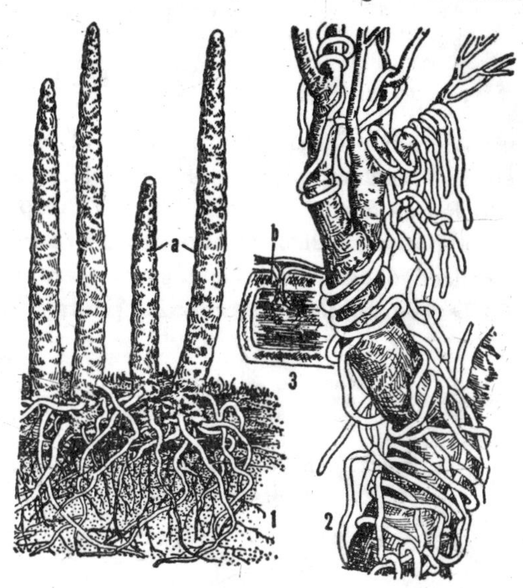

Fig. 44. 1. a— breathing roots; 2. *Cuscuta* growing on a host; 3. sucking root (b) penetrating into the host.

7. **Assimilatory roots.** Roots of plants like *Tinospora* (B. Gulancha) hang freely in the air and develop green colour. They can manufacture food mattters. The sub-merged roots of *Trapa natans* (water chest nut) coming out in pairs from the nodes are green and assimilatory in function.

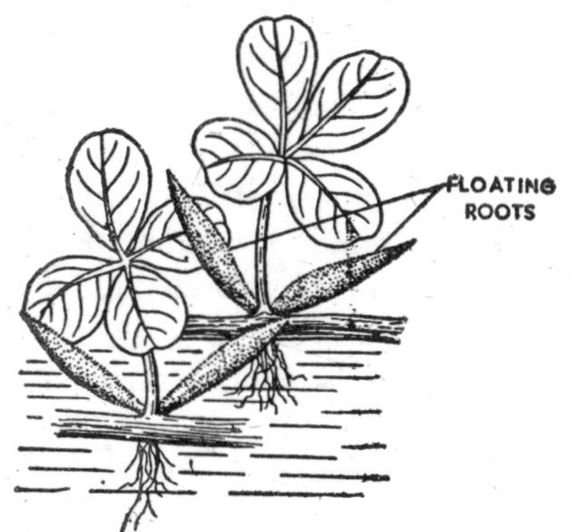

Fig. 45. Floating roots of *Jussiaea*.

Roots are also used for vegetative multiplication. Common vegetable plants like *Trichosanthes dioica* (B. Patol), sweet potato, etc., are propagated by means of roots. Many garden plants are also multiplied by root cuttings.

CHAPTER IV

THE STEM

Germination enables the dormant embryo to establish itself as an independent plant. The radicle produces the root system which ramifies in the soil, whereas the plumule forms the upper aerial portions called the **shoot** or the **shoot system.** The shoot is a collective term which includes the **stem,** the **buds** and **branches,** the **leaves** and the **reproductive parts** (Fig. 46). The axis between the two systems is continuous, as they develop from the radicle and plumule of the embryo.

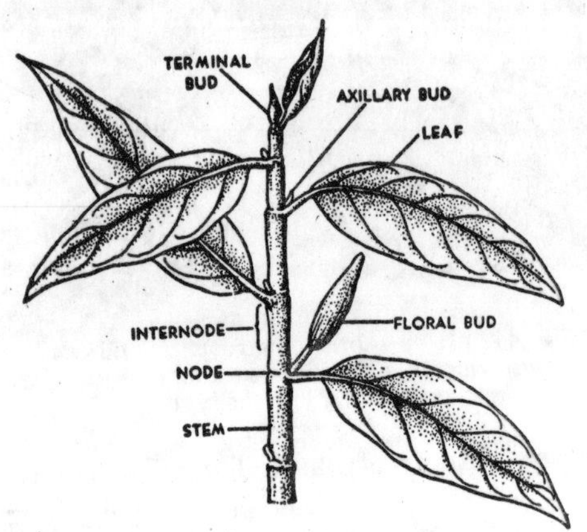

Fig. 46. Parts of shoot of *Michelia*.

The **stem** is the direct upward prolongation of the plumule and is the ascending axis of the plant. It is usually green in colour when young, and grows towards light. The stem has distinct **nodes** and **internodes,** and bears similar and dissimilar members, *i.e.* buds, branches, leaves, etc., and exposes them to proper amount of light and air. Branches are *exogenous* in origin. The terminal position of the stem is occupied by a bud called **apical** or **terminal** bud; and **lateral** or **axillary** buds are invariably present at the axils of the leaves. Many hairy outgrowths, usually multicellular (rarely unicellular), develop from the stem surface.

The morphological difference between the stem and the root is as follows:

The Stem	The Root
1. It develops from the plumule of the embryo.	1. It develops from the radicle of the embryo.
2. It is the ascending axis of the plant.	2. It is the descending axis.

The Stem	The Root
3. The stem is usually green in colour and grows towards light.	3. The root is non-green and goes away from light.
4. It has distinct nodes and internodes and bears similar and dissimilar members.	4. It has no nodes and internodes and bears only similar members.
5. Branches are exogenous in origin.	5. Branches are endogenous in origin.
6. At the apex there is a terminal bud.	6. Root-cap is present at the apex.
7. Multicellular or rarely unicellular outgrowths may develop anywhere on the stem surface.	7. Root-hairs are unicellular and they are the prolongations of the epidermal cells and not outgrowths. They are located at a particular zone called the root-hair zone.

THE BUD

The bud is a condensed rudimentary shoot. It is a compact body having an axis with a delicate growing point, nodes and very short unexpanded internodes and closely crowded young leaves (Fig. 47). The plumule is the first bud of the plant and the cabbage is a large bud. A good idea of a bud can be had by cutting the cabbage lengthwise when the growing point, short internodes and young leaves from nodes arranged acropetally, will be clearly seen.

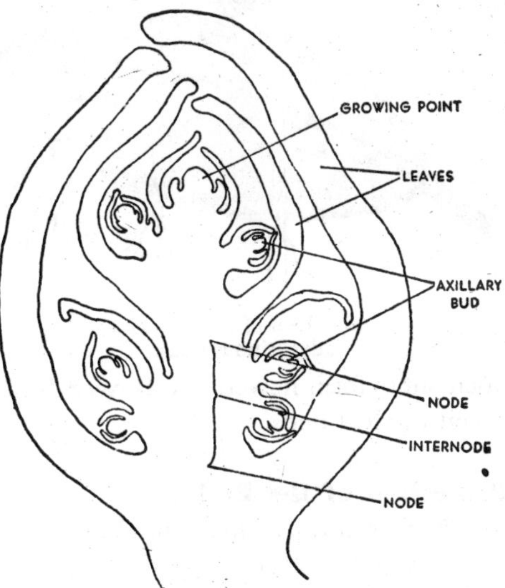

Fig. 47. Longitudinal section of shoot apex.

Buds are of two types, according to their position: (1) **Apical** or **terminal** buds are present at the apex of the stem and the branches. They are responsible for the growth in length of the axis. (2) **Axillary** or **lateral** buds originate from the axils of the leaves. Lateral buds usually grow into branches. Normally one bud develops from the axil of a leaf, but in some plants like *Duranta* (B. Mehdi), Rangoon creeper more than one bud may be present there. If each of the axillary buds develops into a branch then the plants would naturally be many-branched ones, what is not generally noticed in plants growing around us. It is due to the fact that many axillary buds remain dormant. Dormant buds are activated if the stem is injured. After the pruning of the branches quite a large number of new buds are found to come out. Some of them are activated dormant buds.

If buds arise from any position other than the normal ones, *i.e.* apex and axil of the leaf, they are called **adventitious.** Adventitious buds may develop from the stem, roots and even leaves. When the upper part of a stem is cut off, quite a good number of buds come out spontaneously from the injured part without any order. Some of them are dormant buds no doubt, but many of them are adventitious. Buds developing from the roots of sweet potato and *Dahlia* or from the leaf of *Bryophyllum* (B. Patharkuchi) are also adventitious ones (Fig. 48). Buds which develop into branches with leaves are the leaf buds, and those which open out to form flowers are **flower-buds.** Here we are discussing about the leaf buds.

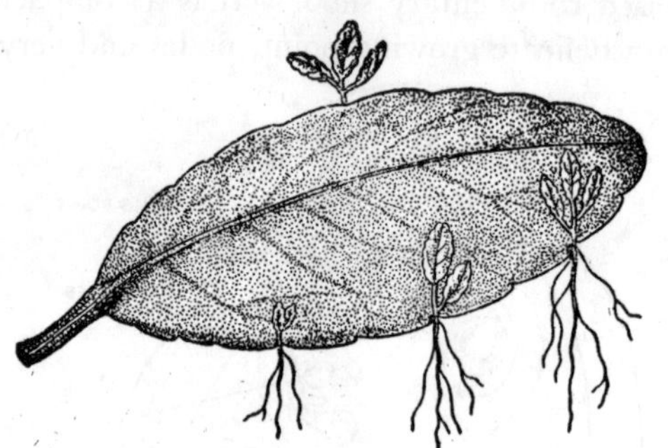

Fig. 48. Leaf of *Bryophyllum* with adventitious buds.

Protection of the Bud

Nature has provided suitable devices for the protection of the buds against external injuries and adverse climatic conditions. Many of them have scaly leaves, the so-called **bud scales,** to protect them, as in banyan, India rubber, etc. The majority of the buds of the tropical plants are naked, *i.e.* have no covering scales. Leaf-bases often act as sheaths to protect them. To prevent the loss of moisture many buds have dense hairy coating or secretions of wax and resinous matters, as in *Zizyphus* (B. Kul). The leaves of the bud may remain rolled or folded in order to secure protection from the sun and the rain. In cold countries buds formed in the late season remain dormant over the winter to open out in the next spring.

Modification of the Bud

Normally the axillary buds, if active, develop into branches. In Yam (B. Chupri aloo or Kham aloo) a more or less round fleshy body is found at the axil of the leaf (Fig. 49) which, in course of time, gets detached from the mother plant and develops into new

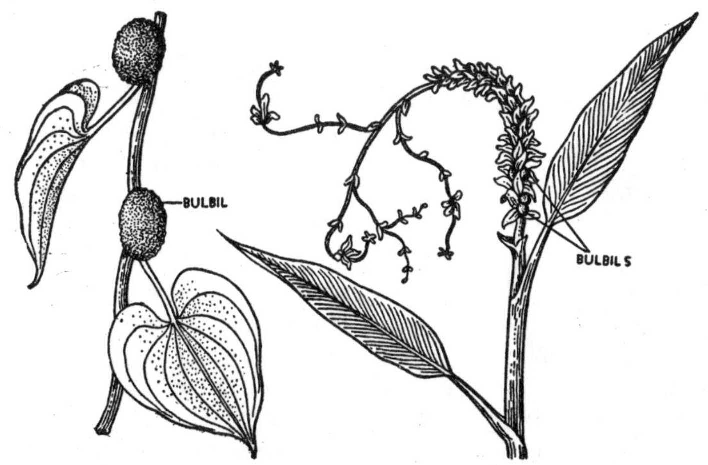

Fig. 49. Bulbils of Yam (left) and *Globba bulbifera* (right).

plant. This body is swollen due to accumulation of food matters. This is called **bulbil.** It is evident from the axillary position that the bulbil is a modification of the bud. In *Globba bulbifera* many bulbils are formed for reproductive purposes (Fig. 49).

BRANCHING

Branches are the similar members of the stems. It has been already mentioned that the axillary buds develop into branches by the elongation of the internodes accompanied by enlargement and expansion of the leaves. The mode of arrangement of the branches is called **branching.** There are two main types of branching : (i) **Lateral** and (ii) **Dichotomous.**

I. Lateral branching

In lateral branching the axillary buds, lying laterally or sideways, go on producing branches in acropetal order. Lateral branching, again, may be **racemose** or indefinite and **cymose** or definite.

Racemose. In the racemose type of branching the terminal bud continues to grow vigorously, producing a prominent primary axis, and the axillary buds develop into lateral branches which are, by all means, smaller than the main axis. Due to this type of branching the trees attain cone-shaped or pyramidal appearance (Fig. 51). It is also called monopodial (*mono*=one; *pod*=axis). Examples—*Polyalthia* (B. Debdaru), *Casuarina* (B. Jhau).

Cymose. In the cymose type of branching the growth of the primary axis is checked or arrested and a few branches or secondary axes grow vigorously for some time. They also meet the same fate and new branches exceed them in vigour; so the process goes on. As a result the whole tree takes up a dome-shaped appearance. All the common spreading trees have this type of branching (Fig. 51). Examples—mango, banyan.

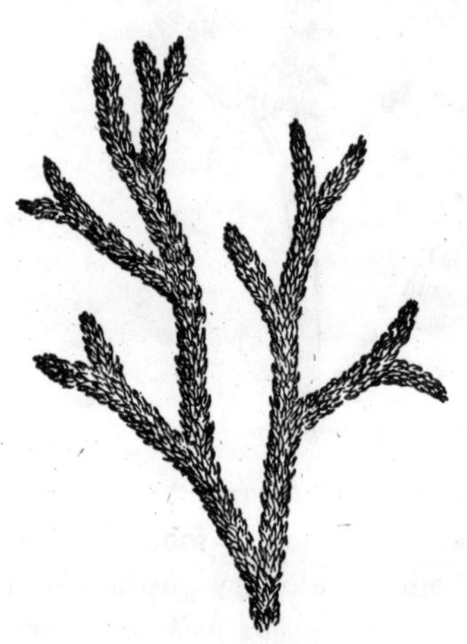

Fig. 50. Part of *Lycopodium* plant showing dichotomous branching.

II. Dichotomous branching

It is common only in lower plants (cryptogams) where the growing point bifurcates or divides into two equally vigorous parts. The branches behave in the same manner (Fig. 50). Example—*Lycopodium.*

KINDS OF STEMS

Stems show wide diversity in their forms. Normally they are aerial and cylindrical bodies. Triangular stems in sedges (B. Mutha ghass), quadrangular in the sacred tulsi (*Ocimum*), ribbed or with distinct ridges and furrows in *Cucurbita* (B. Gourd) and hollow jointed stems in the bamboos and grasses are also not uncommon. Most of the plants have erect stems, *i.e.* the stems are strong enough to maintain upright position. They attain different shapes according to the nature of branching (Fig. 51).

They are called (i) **excurrent,** when the plant attains pyramidal or cone-like form due to racemose type of branching, as in *Polyalthia* (B. Debdaru), *Casuarina* (B. Jhau) and (ii) **deliquescent,** when they have spreading crowns due to cymose type of branching, *e.g.* banyan, mango. Monocotyledonous plants like palmyra palms, cocoatnut palms have tall woody unbranched stems with a crown of leaves at the apex. They are called **caudex.** Unbranched jointed stems of sugar-canes, grasses are the **culms.** Many plants, on the other hand, have stems too **weak** to stand erect. These weak-stemmed plants either lie prostrate and thus trail or creep horizontally on the surface of the soil or climb round a support, having special devices for the purpose.

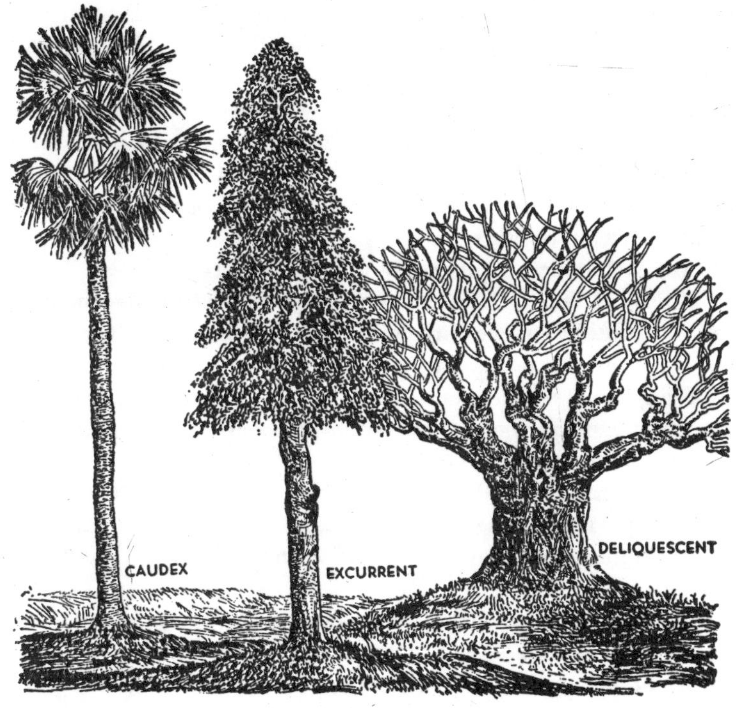

Fig. 51. Erect stems.

According to the nature of the stem the plants are often classified into the following groups :

1. **Herbs.** These are the small plants with soft stems having scanty woody tissues. On the basis of duration of life, herbs may

be **annuals, biennials** and **perennials.** If they grow for only one season and complete life cycle during that period, they are called annuals, *e.g.* mustard, Balsam (B. Dopati), paddy, jute. Those which complete their life cycle in two seasons are called biennials. In the first season the biennials manufacture and store sufficient food and in the next season they bear reproductive organs, *e.g.* English radish, carrot. It is to be noted that these plants are biennials in cold countries but in our country they grow as annuals. If the plants continue for more than two seasons, they are called perennials. Banana, ginger, *Canna* (B. Sarbajaya), are perennial herbs.

2. **Shrubs.** These are woody plants which produce so many branches that the main stem often cannot be distinguished. Shrubs are perennial plants, *e.g. Hibiscus* (B. Java), pomegranate, *Duranta* (B. Mehdi).

3. **Trees.** These are tall woody plants with distinct pillar-like main axis or *trunk*, *e.g.* mango, banyan, palms. Trees which shed their leaves at a particular season of the year are called *deciduous;* whereas those which have no definite period for shedding leaves are called *evergreen.*

Weak-stemmed plants are of the following types :

4. **Trailers.** These are the weak-stemmed plants which just trail horizontally on the surface of the soil. Some of them do not bear roots at the nodes and are called **procumbent,** *e.g. Basella* (B. Puin); but sweet potato, *Cynodon* (B. Durboghass) and many other grasses bear roots at the nodes to fix themselves to the soil. They are called **creepers.**

5. **Climbers.** These are the weak plants which grow up-right only with the help of some kind of support. They are classified into following types according to their modes of climbing :

(*a*) *Root climbers.* Betel, vine, *Scindapsus* (B. Gaj-pipul), etc., climb with the help of adventitious roots from the stems which, in fact, cling to the support on which the plant climbs (Fig. 42).

(*b*) *Hook climbers and scramblers.* Many weak plants produce hard, pointed hook-like bodies for climbing. *Artabotrys* (B. Kan-thali champa) bears hooks on the floral axis (Fig. 52, 3). Pointed prickles are present in many wild roses and canes for the same purpose.

(*c*) *Tendril climbers.* Tendrils are the slender leafless bodies remarkably sensitive to contact. As soon as they come in contact

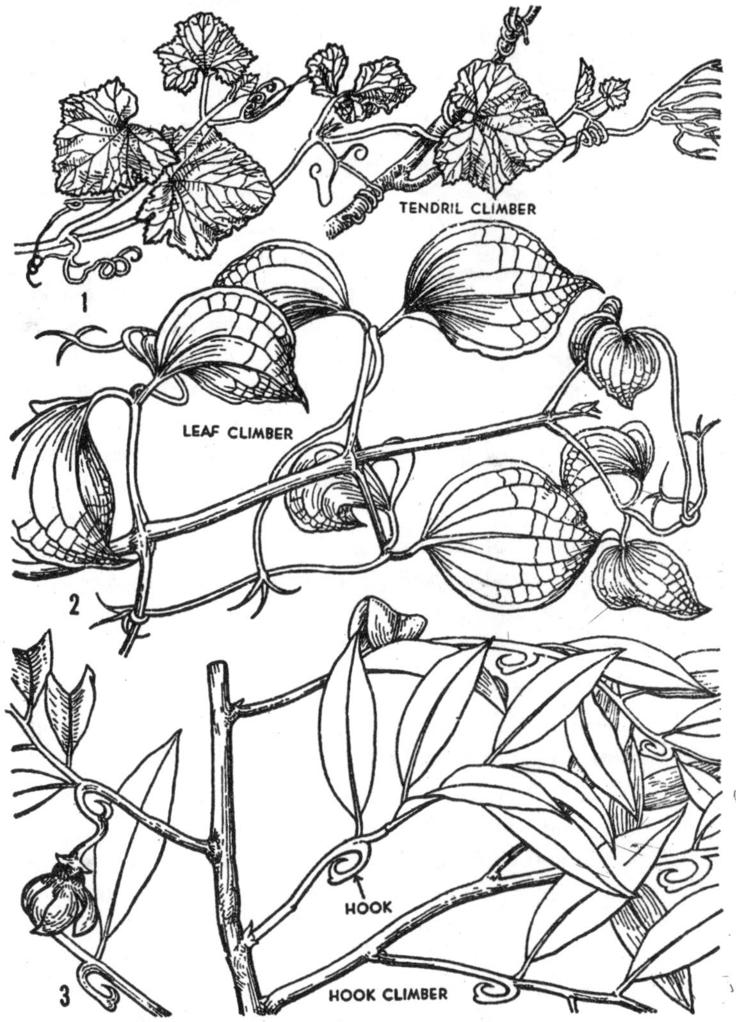

Fig. 52. Climbers. 1—tendril climber; 2—leaf climber; 3—hook climber.

with any support they quickly coil round it and thus help the plant in climbing. Plant-parts like stems (Fig. 52, 1), leaves and even floral axis are modified into tendrils. Examples—*Cucurbita* (B. Gourd), pea, *Smilax* (B. Kumarika), *Antigonon.*

(*d*) *Leaf climbers.* Parts of the leaves of some plants also coil round the support. The leaf stalks of *Clematis* (Fig. 52, 2) and *Nepenthes* (Pitcher plant, Fig. 177, 1), leaf tip of *Gloriosa* (B. Ulat chandal, Fig. 68, 6), are good examples.

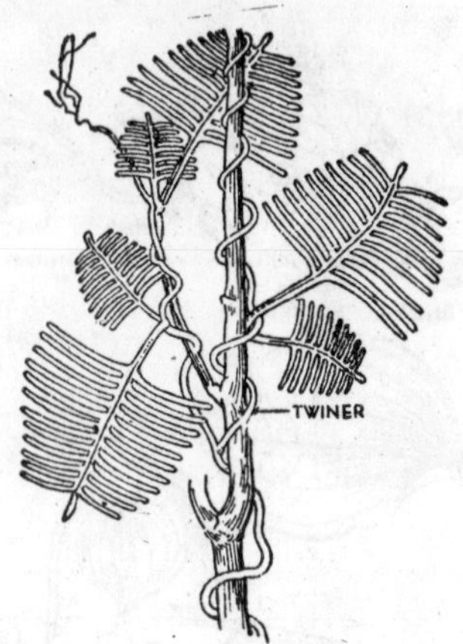

Fig. 53. Twiner of *Ipomoea.*

(*e*) *Stem climbers* or *Twiners.* Some plants twine round the support by their weak slender stems and branches, having no specialised structures for the purpose. They may twine clockwise as in *Dolichos lablab* (B. Sheem), or may twine anti-clockwise as in *Clitoria* (B. Aparajita) and *Ipomoea quamoclit* (B. Tarulata, Fig. 53).

(*f*) *Liane* or *Woody climbers.* These are fast growing woody plants which twine round the stems and branches of tall trees and thus go up in search of light and air. Examples—*Quisqualis* (Rangoon creeper), *Hiptage madhablata* (B. Madhabilata).

Normal Functions of the Stem

1. The stem bears the leaves and branches and exposes them to proper amount of light and air.

2. It conducts water with dissolved mineral matters absorbed by the roots from the soil to the green leaves where food matters are usually synthesised. It also conducts the elaborated food matters to the different part of the plant. So the stem really connects the absorbing organs (roots) with the green leaves—the specialised region for the manufacture of food. This conduction of water and food becomes possible due to the presence of continuous tube-like bodies—the conducting strands, inside the roots, stems and leaves.

3. It bears the reproductive organs, flowers, etc., which are essential for the perpetuation of the race. Besides these, many stems serve as organs for storage of food and water and also for vegetative propagation.

MODIFICATIONS OF STEMS

Stems and branches of some plants often undergo modifications for performing special functions. Modified stems are usually of three types—(I) **Subaerial,** (II) **Underground** or Subterranean and (III) **Aerial** or Metamorphosed stems.

I. Subaerial Modifications

Slender branches develop from the axils of the leaves of some plants which usually run horizontally on the surface of the soil. Most of them are subaerial and some may be partly subterranean. Their main function is **vegetative multiplication.** As a matter of fact, subaerial modified stems are instrumental in very rapid multiplication of certain plants.

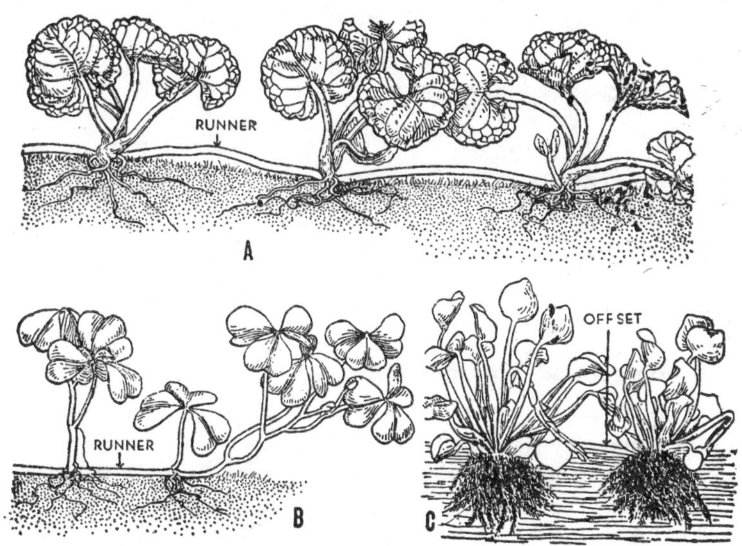

Fig. 54. Subaerial modifications of stems. A—runner of *Hydrocotyle;* B—runner of *Oxalis;* C—offset of water hyacinth.

(A) **Runner** or **stolon.** It is a slender branch which develops from the axil of a leaf and grows prostrate on the surface of the soil. It has nodes and long internodes. Roots develop at the nodes to fix the runner to the soil, and the buds at the axils of the scale leaves grow into aerial shoots which, again, produce runner to behave in the same manner. Thus quite a good number of runners are formed from the mother plant (Fig. 54, A & B). Each of them can lead independent existence if they get severed from the mother plant. Examples—*Oxalis* (B. Amrul). *Hydrocotyle* (B. Thulkuri).

(B) **Offset.** It is a short condensed runner. It also develops from the axil of a lower leaf of the mother plant as a short slender branch which produces roots and a rosette of leaves from the tip (Fig. 54, C), *e.g.* water hyacinth (B. Kachuri pana), *Pistia* (B. Bara pana).

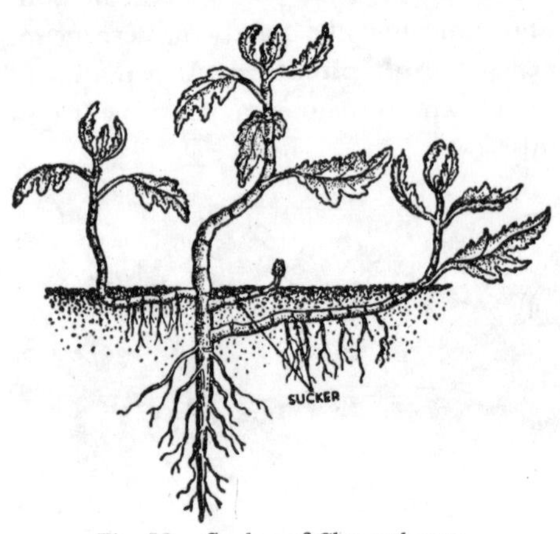

Fig. 55. Sucker of *Chrysanthemum*.

(C) **Sucker.** It is a branch which grows to certain extent beneath the surface of the soil, then comes upwards and produces the green aerial portions (Fig. 55). Adventitious roots arise, as usual, from the subterranean portion. Examples — *Chrysanthemum* (B. Chandramallika), Mint (B. Pudina).

II. Underground Modifications

Unlike normal stems, they grow beneath the surface of the soil and are non-green in colour, so they are often confused with roots. Presence of nodes and internodes, scale leaves at the nodes, axillary and terminal buds are distinct morphological characters to show that they are stems and not roots. Underground modified stems are usually fleshy due to storage of food matters. Their main functions are (*i*) **perennation** or tiding over the unfavourable period of the year, (*ii*) **storage of food matters,** and, lastly, (*iii*) **vegetative multiplication.** There are four types of underground modified stems :

1. **Rhizome** (Figs. 56 & 57, F). It is a fleshy modified stem which grows horizontally beneath the surface of the soil. It has distinct nodes and internodes and minute scale leaves at the nodes.

Lateral buds come out as usual from the axils of **scale leaves** and so, it becomes branched. The terminal bud produces the green aerial shoot in favourable season which dies after flowering, leaving a scar on the rhizome. Rhizome bears many adventitious roots. Examples—ginger, turmeric, mango ginger, fern, etc. In *Alocasia* (B. Mankachu) the rhizome is vertical and is called **root stock** (Fig. 57, G).

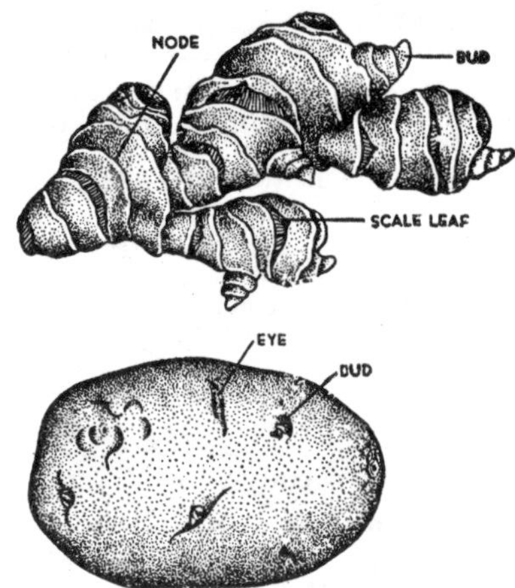

Fig. 56. Underground modification of stem—rhizome (above) and tuber (below).

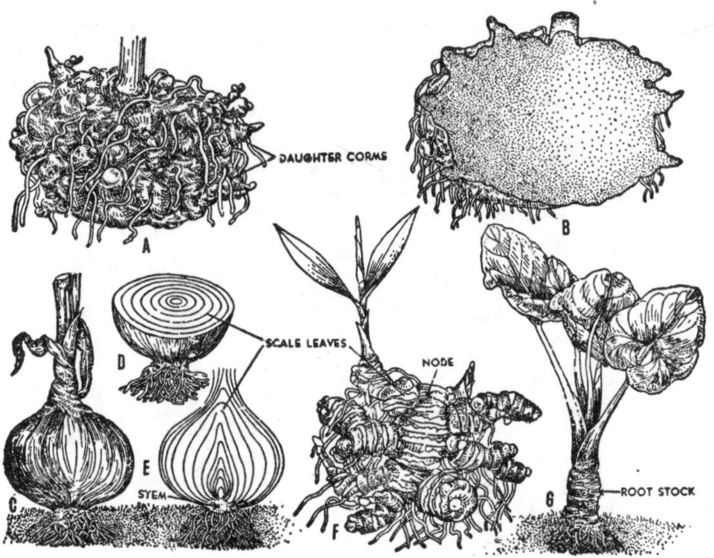

Fig. 57. Underground modifications of stems. A—corm of *Amorphophallus;* B—same cut lengthwise; C—bulb of onion; D—cross-section showing scale leaves; E—longitudinal section showing flattened disc; F—rhizome of ginger; G—root stock of *Alocasia.*

2. **Tuber** (Figs. 56 & 58). In potato plant slender branches develop from the underground part of the stem and grow more or less horizontally beneath the surface of the soil. The growth of

the branch is arrested and the tips begin to enlarge due to accumulation of food matters. These swollen ends of the underground branches are the **tubers.** They have so many buds, commonly called **eyes,** situated at slightly depressed pits at the axils of scale leaves. Adventitious roots are usually absent, *e.g.* potato.

Students often commit mistake about the morphology of potato and sweet potato. Potato is a modified stem and sweet potato is a modified adventitious root. Condensed nodes and internodes, scale leaves and **eyes** are clear characters to establish that potato is a modified stem.

It is interesting to note that the potato plant bears three types of stem,—first, green aerial stems; second, underground stem growing beneath surface of the soil which may be called a stolon; and third, the swollen tip—the tuber.

3. **Corm** (Fig. 57, A & B). It is a very fleshy stout modified stem. It is usually vertical and produces green aerial portion from the terminal bud. Scale leaves are present and daughter corms are formed from the axils of the scale leaves. Corm is really a much condensed rhizome. Adventitious roots develop from the body of the corm. Example—*Amorphophallus* (B. Ol).

Fig. 58. Tubers of potato.

4. **Bulb** (Fig. 57, C, D & E). It is a short underground stem. Here the stem is reduced to a flattened disc with compressed nodes and internodes. Fleshy scale leaves develop from the disc, practically enveloping it. The scale leaves may bear axillary buds

which, in course of time produce daughter bulbs. The terminal bud grows into the aerial shoot during favourable season. A cluster of adventitious roots are given off from the disc. The scale leaves

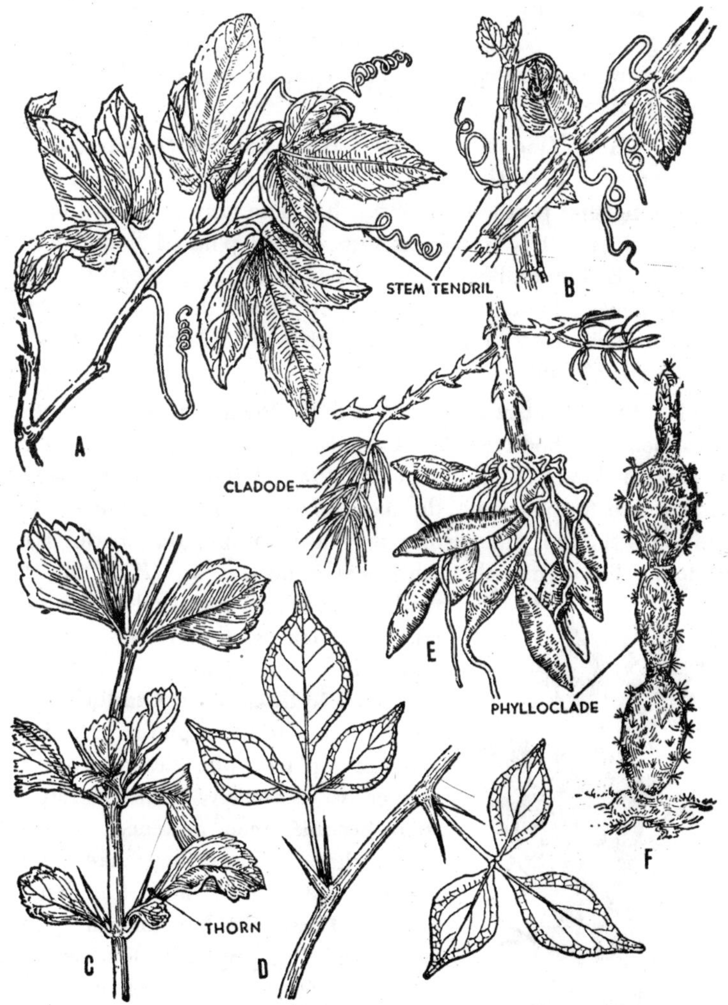

Fig. 59. Metamorphosed stems. A—tendril of passion flower; B—tendril of *Vitis;* C—thorn of *Duranta;* D—pointed structure of wood apple; E—cladode of *Asparagus;* F—phylloclade of *Opuntia.*

are fleshy due to the storage of food matters. Bulbs which are covered by some brown papery scales are called **tunicated bulbs,** *e.g.* onion. In lily the scale leaves overlap each other and have no papery covering or tunic. They are known as **scaly bulbs.**

III. Aerial Modifications or Metamorphosed Stems

In some plants the stems undergo extreme modifications resulting in the change of structures associated with change of functions. They are called metamorphosed stems. The following are the types :

1. **Thorn** (Fig. 59, C & D). It is a hard sharp-pointed structure used as a defensive organ. It develops from the axil of the leaf, the normal position of a branch. The growth of that branch is arrested and it is converted into a hard pointed body for self-defence. Besides axillary origin, it has other morphological characters of stems. Thorn often bears leaves from the nodes, and even flowers. They may be simple or branched. Common example—*Duranta* (B. Mehdi).

2. **Stem tendrils** (Fig. 59, A & B). Weak plants often produce coiled sensitive bodies, called tendrils, for the purpose of climbing. In passion-flower (B. Jhumkolata), *Vitis* (B. Harjora), etc., similar tendrils are formed from the axils of leaves or from the terminal position. So morphologically they are nothing but modifications of stem. Sometimes they also bear scale leaves. (Here the stem tendrils are being discussed; other plant parts like leaves, flower axis may also be modified into tendrils.)

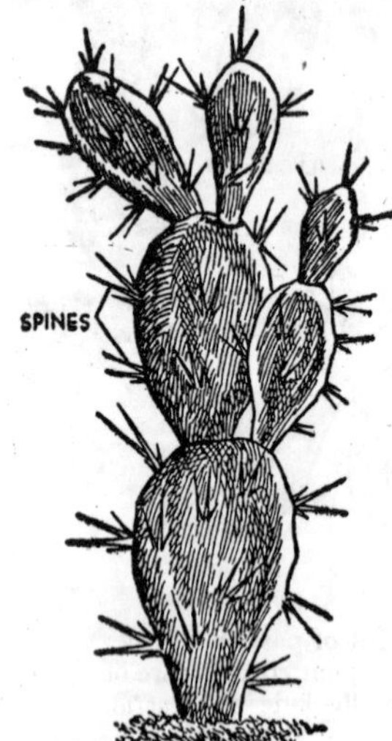

Fig. 60. Phylloclade of *Opuntia*.

3. **Phylloclade** or **Cladode** (Figs. 59, E, F & 60). In plants growing in very dry places like deserts, the leaf surface is extremely reduced and the stems become green, flattened or rounded in shape. Thus they take up the general appearance and functions of leaves. These peculiar metamorphosed stems are called **phylloclades.** Examples—Cacti, *Opuntia* (B. Fanimanasha). Phylloclade of one internode is called cladode, *e.g. Asparagus* (B. Satamuli). Phylloclades have nodes and internodes, minute scale leaves or spiny modifications of leaves and flowers.

CHAPTER V

THE LEAF

The leaves are the most conspicuous vegetative organs of the plants. They are lateral dissimilar outgrowths or appendages of the stems or branches. Leaves are usually green, flat, expanded organs of limited growth. They develop from the nodes having invariably buds at their axils, and remain arranged in acropetal order. Leaves originate as exogenous outgrowths from the growing point of the stem. These protuberances are called **leaf-primordia,** which, by continued growth, develop into the mature green leaves called foliage leaves (Fig. 47).

Functions of the Leaf

Leaves are of paramount importance as they are mainly responsible for the *manufacture of food.* Photosynthesis, as the process is called, is the main function of the leaf. Owing to the presence of **chlorophyll,** the green colouring pigment, the leaves can manufacture complex organic food matters (like sugar and starch), out of water and carbon dioxide gas absorbed from the soil and air respectively, with the aid of sunlight. Leaves also carry on **respiration,** an energy releasing process, which involves intake of oxygen and outgo of almost equal volume of carbon dioxide. All living organs, of course, have this function. The third function of the leaf is **transpiration** or giving out of excess of water as water vapour. Plants usually absorb water from the soil much in excess of their need. They get rid of the surplus water by transpiration. Besides these normal functions, leaves protect the axillary buds, often store up water and food matters and perform other special functions as well.

Parts of a Typical Leaf (Fig. 61)

A typical or an ideal leaf has usually three parts :

(*i*) **Leaf base,** by means of which the leaf remains attached to the stem or branch.

(*ii*) **Leaf stalk** or **petiole,** the cylindrical stalk which connects the leaf base with the flat blade.

(*iii*) **Leaf blade** or **lamina,** the green flat expanded part of the leaf. It goes without saying that the blade is the most important part.

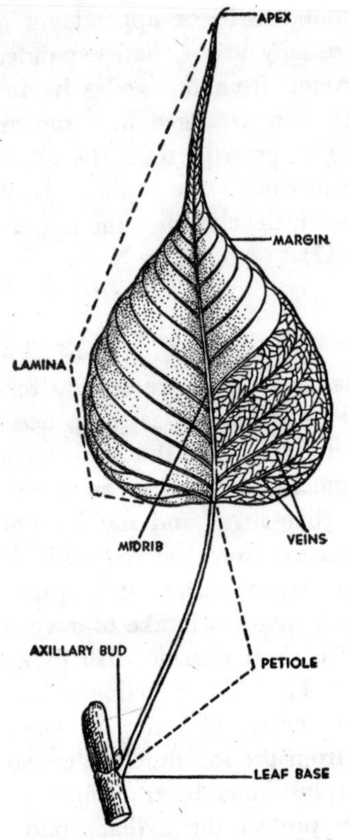

Fig. 61. Parts of a typical leaf.

The blade has usually a prominent rib running up to the tip. It is the **midrib.** The midrib has many branches and sub-branches distributed in the lamina. They are called **veins.** Veins really form the skeleton of the leaf on which softer materials remain inserted, and they are the channels for conduction of water and food. The outer edge of the leaf forms the **margin,** and the extreme tip, the **apex.**

THE LEAF-BASE

It is by no means a conspicuous part of the leaf. Some leaves like those of the pea, sensitive plant, mango have swollen leaf-bases. Such a leaf-base is called **pulvinus** (Fig. 62, a). In many monocotyledons and a few dicotyledons the leaf-base grows into a sheath partly or wholly surrounding the stem. They are sheathing leaf-bases, *e.g.* palms, grasses, banana (Fig. 62, b), coriander.

The Stipules

The leaf-base often bears a pair of lateral appendages or outgrowths called **stipules.** If stipules are present, the leaves are called **stipulate,** *e.g.* rose, China-rose; and if **stipules** are wanting, they are called **ex-stipulate,** *e.g.* mango, guava. They normally protect the buds.

The stipules are of the following kinds (Fig. 63):

1. **Free lateral.** These are two small green stipules from the two sides of the leaf-base which remain quite free from each other, *e.g.* China-rose (Fig. 63, 1), cotton.

2. **Adnate.** Adnate stipules are found in roses. The stipules unite with the petiole up to a certain distance and form two wings on the two sides (Fig. 63, 2).

3. **Interpetiolar.** The stipules of two opposite leaves unite by their outer margins and form a structure lying in between the petioles of the two leaves, *e.g.* *Ixora* (B. Rangan, Fig. 63, 3), *Anthocephalus* (B. Kadamba).

4. **Ochreate** or **Ochrea.** Here the two stipules unite to form a hollow tube surrounding a part of internode above, *e.g.* *Polygonum* (B. Panimarich, Fig. 63, 5), *Rumex* (B. Ban palang).

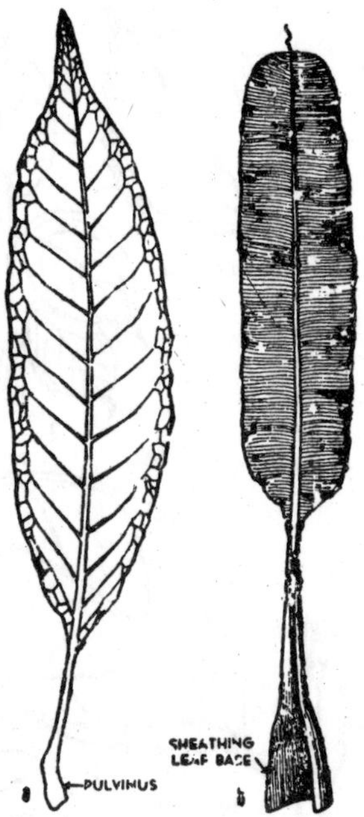

Fig. 62. a—leaf of mango with pulvinus; b—leaf of banana with sheathing leaf-base.

5. **Bud Scales.** These are rather prominent stipules which protect the terminal bud when young. With the gradual opening of the bud the pale membranous stipules fall off, *e.g.* banyan, India-rubber, Jack fruit plant (Fig. 63, 4).

Modifications of Stipules

Some stipules are modified for special functions. They are:

I. **Foliaceous.** These are large flat leaf-like stipules as found in the pea (Fig. 63, 6). Here the terminal leaflets of the compound leaf are modified into tendrils and the pair of foliaceous stipules take up the functions of leaves. In some varieties of wild pea the whole leaf is modified into a tendril.

II. **Tendriller.** These stipules are found in *Smilax* (B.

Kumarika) where a pair of tendrils develop from the leaf-base; obviously they are meant for climbing (Fig. 63, 8).

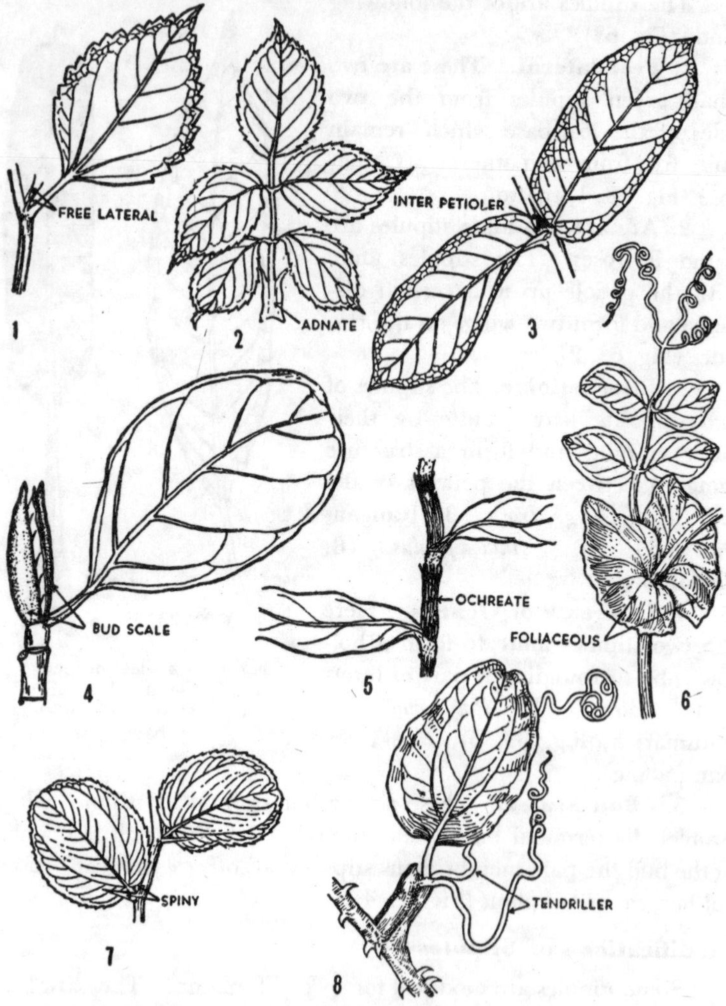

Fig. 63. Kinds of stipules. 1—free-lateral of China-rose; 2—adnate of rose; 3—interpetiolar of *Ixora;* 4—bud scales of jack fruit plant; 5—ochreate of *Polygonum;* 6—foliaceous of pea; 7—spinous of Indian plum; 8—tendriller of *Smilax.*

III. **Spinous.** In *Acacia* (B. Babla) or *Zizyphus* (B. Kul) a pair of pointed outgrowths are present at the leaf-base (Fig. 63, 7). These modified stipules are meant for defensive purpose.

THE PETIOLE

The petiole is the cylindrical stalk connecting the lamina with the base. It exposes the leaf lamina to proper light and air, and also conducts water from the stem to the leaf blade and elaborated food matters from the leaf to the stem.

If the petiole is present, the leaf is called **petiolate,** *e.g.* pipul, mango; and if the petiole is wanting, the leaf is called **sessile,** *e.g. Calotropis* (B. Akanda), *Gloriosa* (B. Ulat chandal). Normally the petiole lies in the same plane with the lamina, but in some leaves like those of the lotus, castor, etc., the petiole is attached at the back more or less at right angles to the blade. They are called **peltate** leaves.

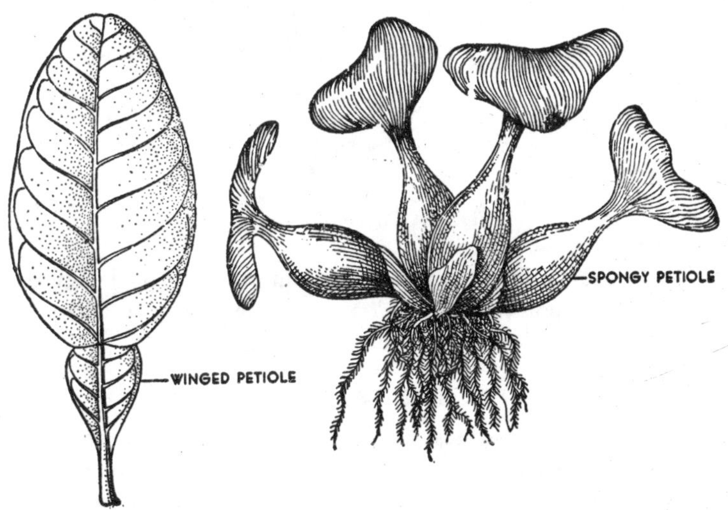

Fig. 64. Winged-petiole of lemon (left) and spongy petiole of water hyacinth (right).

Usually the petiole is a cylindrical stalk. It is grooved in banana, and swollen and spongy in water hyacinth due to the presence of abundant air chambers (Fig. 64). In lemon leaves it forms a pair of flat wings on the two sides and is called **winged petiole** (Fig. 64). Leaves of Australian *Acacia* (B. Akasmoni) have peculiar petioles. They become vertically flattened and lamina-like in appearance. As a matter of fact, the leaves of these plants fall off soon and their functions are taken up by flat lamina-like petioles. These modified petioles are called **phyllodes** (Fig. 65).

Petioles of plants like *Clematis*, *Nepenthes* or pitcher plant are modified for climbing.

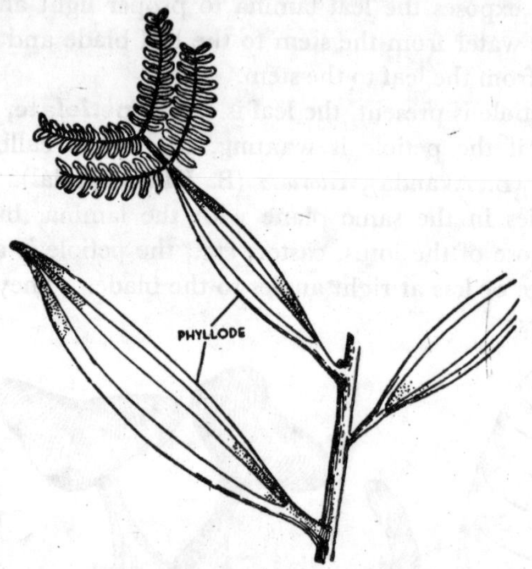

Fig. 65. Phyllode of *Acacia*.

In grasses a membranous hairy outgrowth, called **ligule,** is present at the point of the union of the sheathing leaf-base and the blade. These leaves are known as **ligulate.**

THE LAMINA

The lamina is the most conspicuous and important part of the leaf. It is characterised by green colour, thinness and flatness. The structure is so nicely constructed that the maximum number of chlorophyll containing cells may be exposed to light and air, and the different functions involving gaseous exchange between the internal tissues and the outer atmosphere, may be conveniently carried on.

Lamina should be discussed with reference to the form or shape, the margin, the apex and the veins.

Forms of Lamina

Leaves show wide variations as regards their forms or shapes. There are the long needle-like leaves of the pines and *Casuarina* (B. Jhau), and in the other extreme there are the round disc-like leaves of the lotus and *Victoria regia.* Transitional forms are numerous. An account of the more common forms of lamina is given here (Fig. 66):

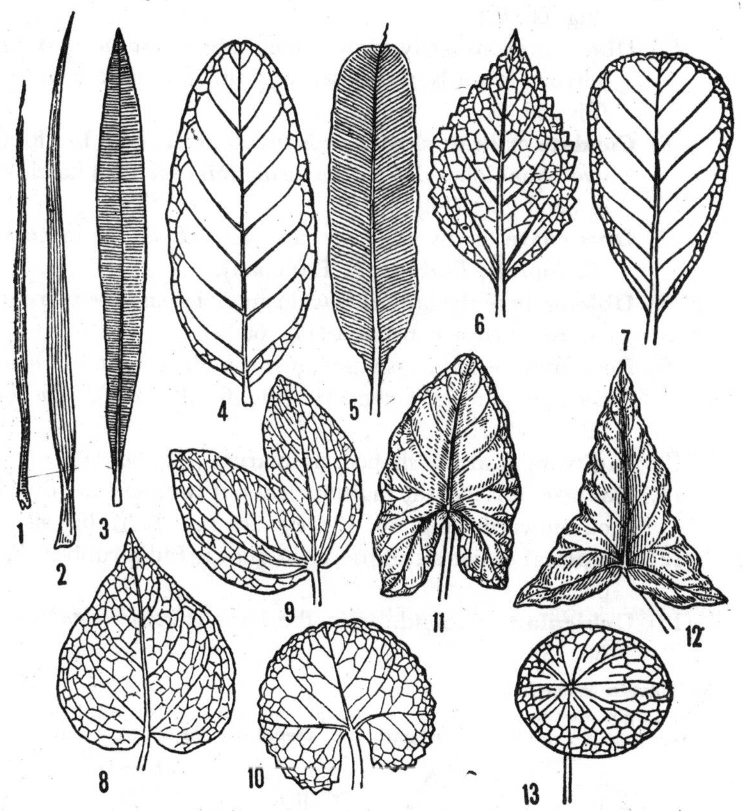

Fig. 66. 1—acicular; 2—linear; 3—lanceolate; 4—elliptical; 5—oblong; 6—ovate; 7—obovate; 8—cordate; 9—obcordate; 10—reniform; 11—sagittate; 12—hastate; 13—orbicular.

(1) **Acicular,** the long needle-like leaves, as in the pine (Fig. 66, 1).

(2) **Linear,** the lamina here is quite long and narrow; the two margins run more or less parallel to each other but meet at the apex, *e.g.* grass (Fig. 66, 2).

(3) **Lanceolate,** the length of lamina is much greater than its breadth and it tapers towards both ends more or less like a lance, *e.g. Polyalthia* (B. Debdaru), custard apple (Fig. 66, 3).

(4) **Ovate** or egg-shaped, when it is broad at the base but gradually becomes narrow towards the apex, *e.g.* banyan (Fig. 66, 6).

(5) **Obovate** or inversely ovate, which is broad at the apex and narrow towards the base, *e.g.* almond, jack fruit leaf (Fig. 66, 7).

(6) **Cordate** or heart-shaped, it is broad at the base but forms two equal lobes due to the formation of a deep notch, as in betel vine (Fig. 66, 8).

(7) **Obcordate** or inversely cordate (Fig. 66, 9), as in *Oxalis* (B. Amrul), *Bauhinia* (B. Kanchan).

(8) **Oblong,** here the lamina is wide and long and the two ends are rounded, *e.g.* banana (Fig. 66, 5).

(9) **Reniform** or kidney-shaped, it is rounded above but has a deep notch towards the base (Fig. 66, 10), *e.g. Hydrocotyle* (B. Thulkuri).

(10) **Sagittate** or arrow-shaped, as in arum (Fig. 66, 11).

(11) **Hastate,** it is like the sagittate leaf but two lobes are directed outwards (Fig. 66, 12), *e.g. Ipomoea* (B. Kalmi shak).

(12) **Elliptical** or like an ellipse, as in guava, India rubber (Fig. 66, 13).

(13) **Orbicular** or round, as in the lotus, *Victoria regia* (Fig. 66, 13).

The Margin of the Leaf (Fig. 67)

Margin may be (a) **entire,** *i.e.* even and smooth, *e.g.* banyan, banana; (b) **wavy,** as in *Polyalthia* (B. Debdaru); (c) **serrate** or like the teeth of a saw,

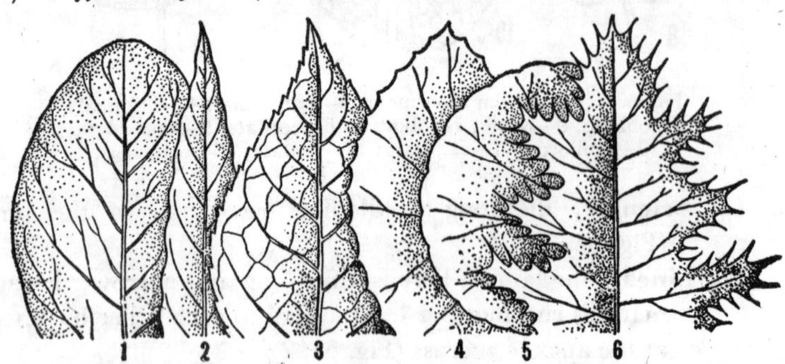

Fig. 67. Margin of lamina. 1—entire; 2—wavy; 3—serrate; 4—dentate; 5—crenate; 6—spiny

e.g. China-rose; (d) **dentate,** where the teeth are directed outwards, *e.g.* water-lily; (e) **crenate,** having rounded teeth, as in *Hydrocotyle* (B. Thulkuri); and (f) **spiny,** having spiny projections, as in Mexican poppy (B. Shialkantha).

The Apex of the Leaf (Fig. 68)

The apex may be (a) **acute,** as in mango, China-rose; (b) **acuminate,** long tapering apex, as in pipul; (c) **obtuse,** as in banyan; (d) **mucronate,** apex ending in an abrupt sharp point, *e.g. Vinca* (B. Nayantara); (c) **cuspidate,** long sharply pointed apex, *e.g.* date palm, pine-apple; and (f) **tendrillar,** as in *Gloriosa* (B. Ulat chandal).

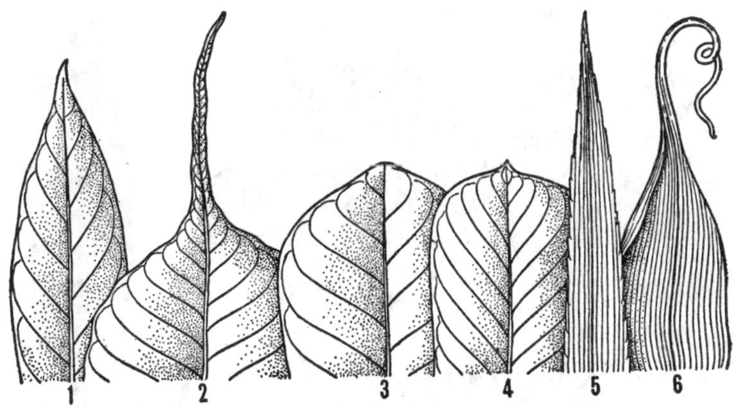

Fig. 68. Apex of the lamina. 1—acute; 2—acuminate; 3—obtuse; 4—mucronate; 5—cuspidate; 6—tendrillar.

Venation (Fig. 69)

Veins are the conducting strands distributed in the leaf lamina. They are solely responsible for the conduction of water and prepared food materials. Moreover, they form the skeleton of the leaf on which softer materials remain inserted. Veins can be conveniently observed by holding the leaf against light. The mode of arrangement of veins is called **venation.** There are two principal types of venation: (i) **Reticulate** or netted venation and (ii) **Parallel** venation.

Reticulate venation. The midrib or a few prominent ribs produce many branches and subbranches which ramify in the leaf lamina in such a manner as to form a network. This is reticulate venation (*reticulum*=net). Some of the veinlets are so small that they are not visible to the naked eye.

Reticulate venation may be **unicostate** or **multicostate,** according to the number of prominent ribs. If there is only one prominent rib, *i.e.* the midrib producing branches, as usual, it is **unicostate** or **pinnate** (*uni*=one; *costa*=rib), *e.g.* pipul, mango.

But if more than one equally prominent ribs develop from the base, the venation is **multicostate** or **palmate.** Multicostate, again, may be **convergent** or **divergent.** If the prominent veins converge (*i.e.* meet or tend to meet) towards the apex, it is called

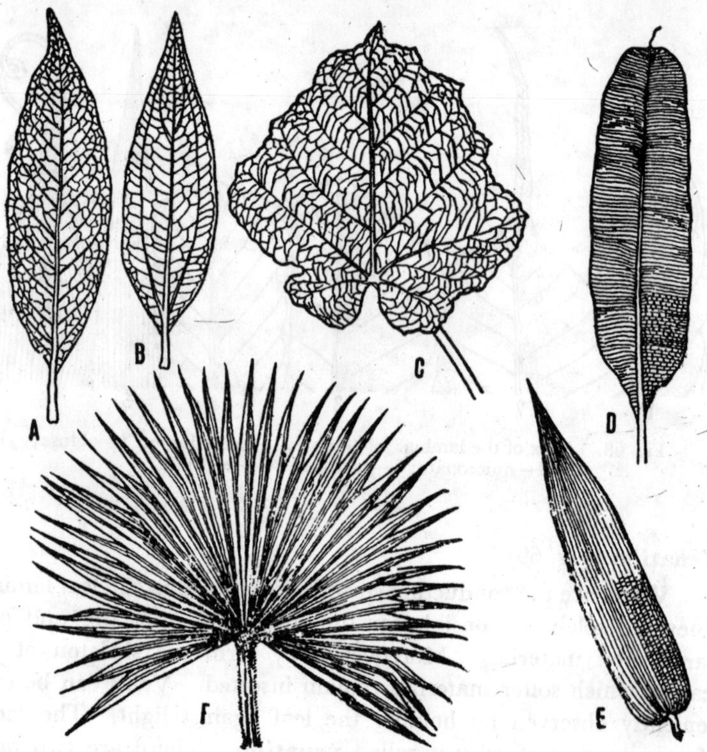

Fig. 69. Types of Venation.
A—reticulate unicostate; B—reticulate multicostate convergent; C—reticulate multicostate divergent; D—parallel unicostate; E—parallel multicostate convergent; F—parallel multicostate divergent.

multicostate convergent, as in *Zizyphus* (B. Kul), Cinnamon leaf. But if the veins go away from one another or diverge, they are called **multicostate divergent,** as in castor-oil leaf, *Cucurbita.* Dicotyledonous leaves have usually reticulate type of venation.

Parallel venation. This type of venation is usually found in the monocotyledons. Here the veins run more or less parallel to one another without forming a network. Parallel venation may also be **unicostate** or **multicostate.** Banana leaf is a good example of **parallel unicostate** venation. In bamboos and grasses a good number of veins arise from the base and meet at the apex. So in these cases venation is **parallel multicostate convergent.** The palm leaves have many prominent veins but they diverge, showing **parallel multicostate divergent** venation.

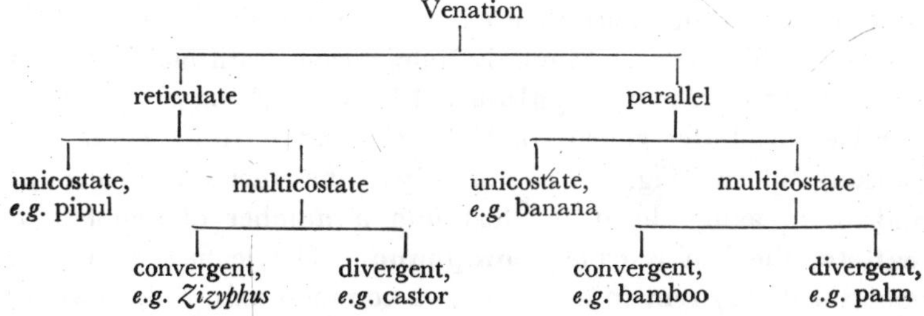

SIMPLE AND COMPOUND LEAVES

A leaf is called **simple** when it has a single blade with entire

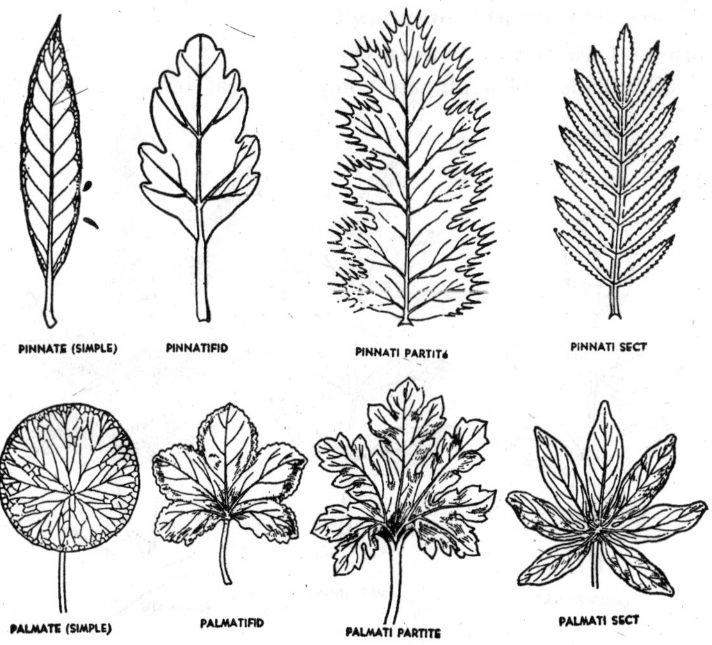

Fig. 70. Incisions of the lamina—pinnate and palmate types.

or incised margin. It is simple so long it presents a single appearance. According to the degrees of incisions, the suffixes 'fid', 'partite' and 'sect' are used. A simple leaf with pinnate or unicostate venation is called **pinnati-fid,** when it has incisions less than halfway towards the midrib, as in *Chrysanthemum* (B. Chandramallika); it is **pinnati-partite,** if the incisions extend beyond halfway, *e.g.* *Argemone* (B. Shialkanta); and it is **pinnati-sect,** when the incisions almost reach the midrib, as in marigold, *Ipomoea* (B. Tarulata). Similarly simple leaves with multicostate or palmate venation may be **palmati-fid,** *e.g.* lady's finger; **palmati-partite,** *e.g.* bitter gourd (B. Uchche); and **palmati-sect,** as in morning glory (Fig. 70). But when the incisions reach the midrib, breaking down the leaf into a number of segments or **leaflets,** the leaf is called **compound.** The leaflets of a compound leaf may remain attached to a common axis, called **rachis,** like the pinnae of a feather; or may be jointed or *articulated* to a common point on an axis like the outstretched palm. The former types are called **pinnate compound** leaves; and the latter are known as **palmate compound** leaves.

Pinnate Compound Leaves (Fig. 71)

The leaflets of a pinnate compound leaf remain attached to

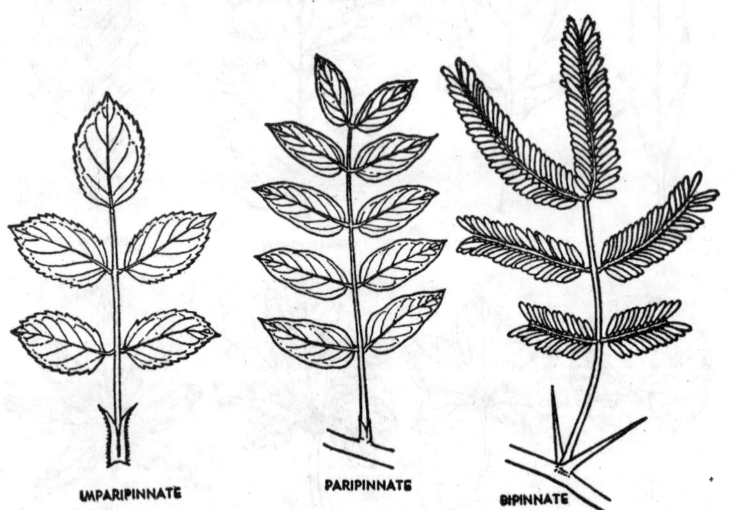

Fig. 71. Pinnate Compound Leaves.

an axis called **rachis.** If they are arranged in pairs, the rachis ends abruptly and the number is even, the leaf is called

paripinnate as in tamarind, *Cassia* (B. Kalkasunde), but if a terminal leaflet is present on the rachis, naturally making the number odd, the leaf is called **imparipinnate,** as in rose, *Clitoria* (B. Aparajita). Pinnate compound leaves may be **unipinnate** or once pinnate, **bipinnate** or twice pinnate, **tripinnate** or thrice pinnate, and so on. Paripinnate leaves of tamarind and imparipinnate leaves of roses are unipinnate, as they have leaflets attached directly to the rachis. In bipinnate leaves the rachis is branched and the leaflets are arranged pinnately on the secondary axes of the rachis, as in the sensitive plant (B. Lajjabati). The **tripinnate** leaves have the third series of branches bearing leaflets, *e.g.* horse-radish. When the leaves are pinnate more than thrice they are called **decompound,** *e.g.* carrot.

Palmately Compound Leaves (Fig. 72)

Here the leaflets are jointed or articulated to a common point on the axis. According to the number of leaflets present, **palmate compound** leaves may be (i) **unifoliate,** with only one leaflet, as in lemon; (ii) **bifoliate,** with two leaflets, *e.g. Balanites* (B. Hinghan); (iii) **trifoliate,** with three leaflets, as in wood apple, *Oxalis* (B. Amrul); (iv) **quadrifoliate** with four leaflets, as in *Marsilea* (B. Sushni); and (v) **digitate,** with five or more leaflets, as in silk-cotton.

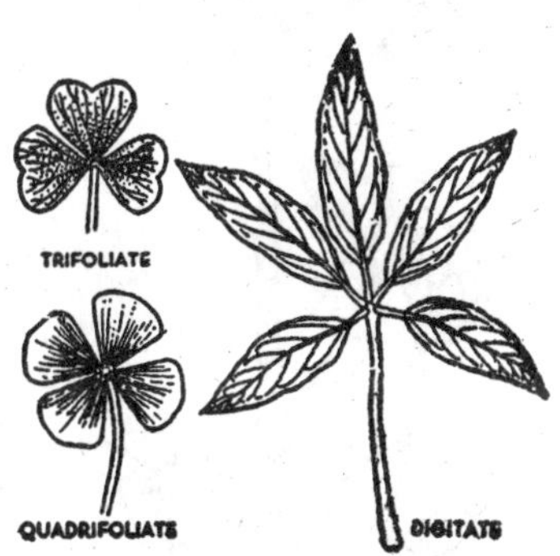

Fig. 72. Palmate Compound Leaves.

Difference between a pinnate compound leaf and a short branch

A pinnately compound leaf with many leaflets often closely resembles a branch with simple leaves. The following characters would show the difference between the two :

(1) A compound leaf has no terminal bud which the branch always has.

(2) A compound leaf develops from the node of the stem or the branch, bearing buds at their axils; whereas the short branch is axillary in position.

(3) If stipules are present they are found at the base of the

rachis of a compound leaf; whereas in a short branch the individual leaves bear the stipules.

(4) Buds are never present at the axils of the leaflets of a compound leaf, but the simple leaves of a short branch invariably bear buds at the axils.

PHYLLOTAXY

The leaves are not haphazardly arranged on the stems and branches but they come out in a definite order. This arrangement

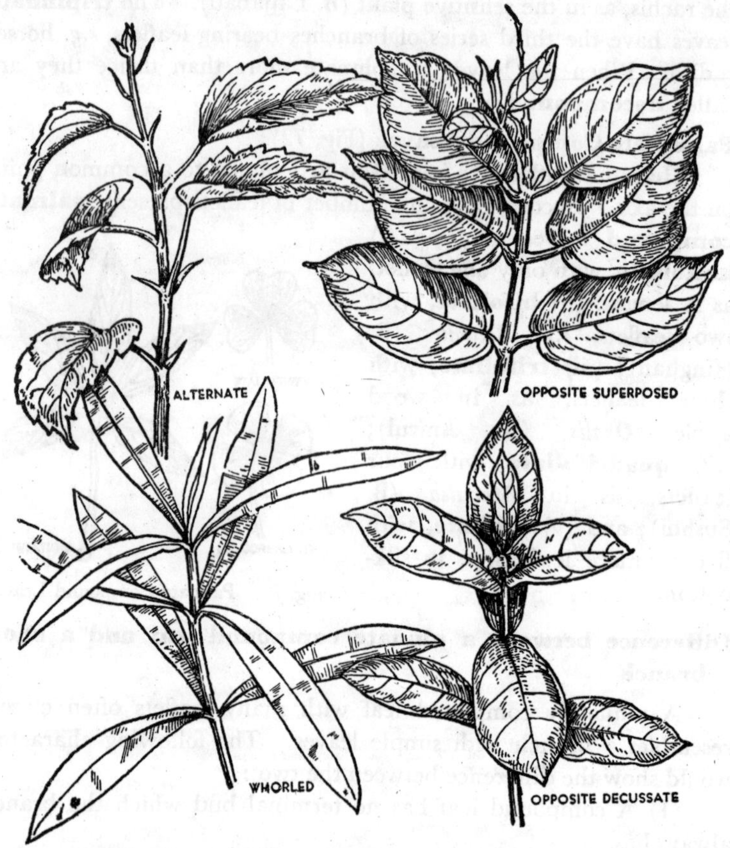

Fig. 73. Kinds of Phyllotaxy.

of leaves on the stems and branches is known as **phyllotaxy,** meaning leaf order. The main object of phyllotaxy is to expose the leaves to proper illumination for different vital functions, avoiding shading as far as possible.

There are three principal types of arrangement (Fig. 73).

(1) **Alternate,** when a single leaf arises from a node and naturally the leaves are alternately placed. This arrangement is also called *spiral,* because if an imaginary line is passed through the bases of the leaves in the order of their development, it would form a spiral round the stem.

(2) **Opposite,** when two leaves develop from a node opposite to each other. Opposite arrangement may be (i) **superposed,** when one pair of leaves stands just above the lower pair. Here leaves are arranged in two vertical rows and the internodes are long to avoid shading as far as possible, *e.g.* guava, Rangoon creeper; and (ii) **decussate,** when one pair of leaves is at right angles to the next upper or lower pair, forming four vertical rows of leaves on the stem or branch, *e.g. Ocimum* (B. Tulsi), *Ixora* (B. Rangan), *Calotropis* (B. Akanda).

(3) **Whorled** or **Verticillate,** when more than two leaves arise from a node forming a whorl, *e.g. Nerium* (B. Karabi), *Alstonia* (B. Chatim).

Leaf Mosaic

Plants growing in shady places with poor light exhibit a peculiar sort of arrangement of leaves. Here all the leaves come up and arrange themselves side by side with a view to utilising the maximum amount of light. They form something like a mosaic (Fig. 74), hence called **leaf mosaic,** *e.g. Oxalis* (B. Amrul), Garden Nasturtium.

Fig. 74. Leaf Mosaic.

Heterophylly

Leaves of different forms are sometimes found on the same plant. This condition is known as **heterophylly** (*heteros*=different; *phylla*=leaves). Heterophylly is particularly noticed in aquatic plants. *Sagittaria* or arrow-head, a common water-plant, has two kinds of leaves; the submerged ones are ribbon-shaped, whereas the aerial ones are sagittate (arrow-shaped). Many other aquatic plants bear finely dissected submerged leaves and flat expanded floating or aerial leaves. Heterophyllous condition in aquatic plants is an adaptation to two different environmental conditions. Even land plants often bear different types of leaves. Leaves vary considerably in shape in plants like *Pterospermum* (B. Kanak champa), *Artocarpus chaplasa* (B. Chaplas), coriander, etc. The first formed leaves in wood apple, *Mormodica* (B. Uchche), etc., are quite different from later-formed ones.

Kinds of Leaves

Besides the green foliage leaves which have been just discussed, there are other types of leaves as well. **Cotyledons** of the embryo are the seed leaves, and so are the first leaves of the plants. The **scale leaves,** as found in modified stems, are small membranous bodies. **Bracts** are specialised leaves often brightly coloured, bearing floral buds at the axils. The beautifully coloured petals and other floral parts are also specialised leaves meant for reproduction.

Some Descriptive Terms

The green leaves having two distinct surfaces are called **dorsiventral.** The upper or ventral surface which is directly illuminated by sunlight is usually dark-green in colour, and the lower or dorsal surface is of pale-green colour, *e.g.* mango, banana. Monocotyledonous leaves usually grow erect where both the surfaces are equally exposed to light. They are called **isobilateral** (*iso*= same; *bi*=two; *lateral*=sideway) leaves, *e.g.* paddy, grass. Leaves of onion are cylindrical bodies, round in cross-section. They are **centric** leaves. Common foliage leaves arising from aerial organs are called **cauline** (*caulis*=a stem). In radish, pine-apple, etc., leaves develop from very short underground stem, as if they have come out from the roots. They are called **radical** (*radix*=root) leaves.

Duration of Leaves

Many leaves fall off very early; they are called **caducous;** leaves which fall off at the end of the season are called **deciduous,** and those which last for more than a season and have no fixed time for shedding are called **persistent.**

MODIFICATIONS OF LEAVES

Leaves are often modified for carrying on special functions. An account of the modified leaves is given below :

1. **Leaf tendrils.** The whole leaf or more commonly the parts of the leaves are modified into slender sensitive tendrils for climbing. The whole lamina in *Lathyrus* or wild pea, the terminal leaflets in pea, the apex in *Gloriosa* (B. Ulat chandal), the petiole in *Clematis* and the stipules in *Smilax* (B. Kumarika) are modified into tendrils.

2. **Spines.** The lamina or a part of it may be modified into hard sharp-pointed structure, called **spine,** for the purpose of

self-defence. In common *Opuntia* (B. Fani manasha) the stem becomes green and flattened (phylloclade) and the leaves are converted into spines. Similarly the apex of date-palm, the margin of *Argemone* (B. Shialkanta), stipules in *Acacia* (B. Babla) are modified into spines. (It should be noted that thorns and spines carry on the same function; but thorns are modified stems and spines are modified leaves. The curved sharp-pointed outgrowths of roses are the **prickles.**)

3. **Storage leaves.** Many leaves become fleshy and succulent due to the storage of water and food. Here the leaves are partly modified. Leaves of *Agave*, *Aloe*, *Portulaca* (B. Nune-shak), the scaly leaves of onion are familiar examples.

4. **Insect-catching leaves.** The insectivorous plants have peculiar leaves nicely adapted for catching insects (Fig. 176).

(a) **Nepenthes** or Pitcher plant. Here the leaf-base is modified into a flat lamina-like body, the petiole into a slender coiled tendril for climbing and the lamina into the pitcher proper. The pitcher has a coloured hood for attracting the insects.

(b) **Utricularia** or bladder-wort. The bladder-worts (B. Jhanji) have much dissected compound leaves. Some of the leaflets are modified into bladders having special devices for catching insects. Each bladder has a valve-like door which opens only inwards. The small aquatic insects push in the valve and are caught. In course of time they die and the bodies are digested.

(c) **Drosera** or sundew. It is another insectivorous plant where the spatula-shaped leaves bear many tentacles, the tips of which glisten in the sun and look like dew drops. Small insects are attracted, come and rest on the leaf only to find the tentacles bend and imprison them. By the secretion of enzymes the insect body is decomposed. After completing digestion the tentacles resume their original position and get ready for another prey.

5. Leaves of *Bryophyllum* (B. Pathar kuchi), *Begonia* bear epiphyllous buds and thus help in vegetative multiplication.

Homologous and Analogous Organs

The organs which are morphologically similar are **homologous,** irrespective of their structures and functions. Thus thorns of *Duranta*, tendrils of passion-flower, phylloclades of *Opuntia*, rhizome of ginger and ordinary branches are homologous organs, though they carry on diverse functions.

But the organs having the same physiological functions regardless of their morphological origin are **analogous.** So, the tuberous root of sweet potato and the stem tuber of potato are analogous. Similarly all the tendrils like those of pea (modified leaves), of passion-flower (modified stems), of *Antigonon* (modified inflorescence axis) are analogous organs, though they are morphologically different.

Defensive Mechanism in Plants

In spite of the fact that plants are our mute and innocent neighbours, they have quite a good number of enemies. It is known to us that animals cannot manufacture their food. Many animals, particularly the herbivorous ones, considerably injure the plants during their search for food. As most of the plants remain rooted to the soil, they cannot run away from the enemies. Nevertheless, they have their own mechanism for self-defence, often very interesting ones, a brief account of which is given below :

1. **Armature.** Very often plants are provided with hard sharp-pointed bodies which serve as weapons for self-defence. They may be morphologically different, but all of them carry on the same function. *Thorns* are the pointed organs, as found in *Duranta* (B. Mehdi). These are metamorphosed stems. Leaves or parts of leaves are also modified into sharp-pointed bodies, as in date-palm and Cactus. They are called *spines*. Roses and canes usually bear curved outgrowths on their bodies for the same purpose. They are *prickles.*

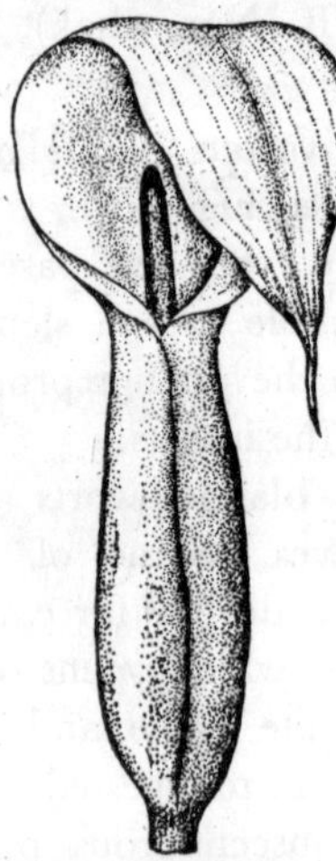

Fig. 75.
Spadix of *Arisaema.*

2. Formation of dense hairs on the surface of the organs, as a check against attacking animals is quite common in plants. Glandular hairs of *Jatropa* (B. Bheranda), *Plumbago* (B. Cnita), etc., secrete sticky substances which adhere to the mouths of the grazing animals and make them uncomfortable. The stinging hairs of nettles and *Mucuna* (B. Alkusi) are nice examples of defensive mechanism in plants. These hairs are long and have sharp brittle tips. As soon as they come in contact with animals, the tips break and make pricks or punctures on the skin of the animals. Immediately a fluid is secreted by the gland lying at the basal region, which is injected, so to say, on the animal and creates the irritation.

3. During metabolic activities various waste products are formed in the plants. Alkaloids found in *Cinchona*, Nux-vomica, poppy, tobacco, the milky juice, latex, present in many plants; different types of crystals, particularly the needle-shaped crystals of calcium oxalate occurring in different varieties of aroids, are the common waste products which become helpful in warding off the animals. Besides these, many plants have bitter and disagreeable taste, *e.g.* bitter gourd, and repulsive odour, as in *Typhonium* (B. Ghet Kachu), and are thus avoided by the animals.

4. Some plants escape the attacks of the animals by imitating some dangerous animals like the venomous snakes. A common Aroid known as *Arisaema* or snake plant, possesses a greenish purple spathe which expands over the spadix in such a manner that it looks like the hood of a cobra (Fig. 75). Such phenomenon is known as **mimicry.**

5. Plants like mango often give shelter to various types of ants and other stinging insects which act as an army, and protect the plant whenever necessary. Such an adaptation is called **myrmecophily.**

CHAPTER VI

THE INFLORESCENCE

In the previous chapters the important vegetative organs like roots, stems and leaves have been discussed. They are essential for the existence of the individual plant, which bears flowers after attaining maturity. Flowers are specialised organs meant only for reproduction and have nothing to do with the maintenance of the plant itself. This is further evident from the fact that in many plants flowers come as the fore-runners of death.

Flowers may arise singly, as in China-rose, or more commonly, in groups at the axils of special leaves or at the terminal position. A branch or a branch system bearing a cluster of flowers is called an **inflorescence.**

The stalk of a single flower is called **peduncle.** In case of an inflorescence, the branched floral axis is called **rachis.** The branches are the **pedicels.** Flowers having pedicels are called **pedicellate** and those without pedicels or stalks are **sessile.** The inflorescence axis developing from an underground stem is known as **scape,** as in onion, tube-rose.

Bracts

Flower buds, like the leaf buds, develop either at the apex of the stem or at the axils of special leaves called *bracts.* The bracts are generally small green bodies, though in some cases they become brightly coloured. Flowers having bracts are called *bracteate* and those without bracts are called *ebracteate.* Bracts may be: (i) leaf-like or *foliaceous,* as in *Adhatoda* (B. Basak); (ii) coloured and boat-shaped, called *spathe,* surrounding the whole or major part of the inflorescence, as in banana, arum; (iii) *petaloid* or brightly coloured like petals of the flowers, as in Poinsettia (B. Lalpata); (iv) small membranous *glumes,* as in paddy, and grass; (v) *scaly,* as in the florets of sunflower; (vi) *involucre,* forming a whorl of bracts surrounding the inflorescence, as in sunflower, marigold, etc.; and (vii) *epicalyx,* a whorl of green bracts placed beneath the calyx, as in China-rose.

Kinds of Inflorescence (Fig. 76)

Inflorescence may be of two distinct types, *viz.* **racemose** or indefinite and **cymose** or definite. Racemose inflorescence is one in which the main axis continues to grow giving off flowers laterally. The flowers develop in *acropetal* order from the base upwards and the mode of opening of the flowers is *centripetal* (*i.e.* from circumference towards the centre). It is also called indefinite, because the main axis does not end in a flower, but continues indefinitely.

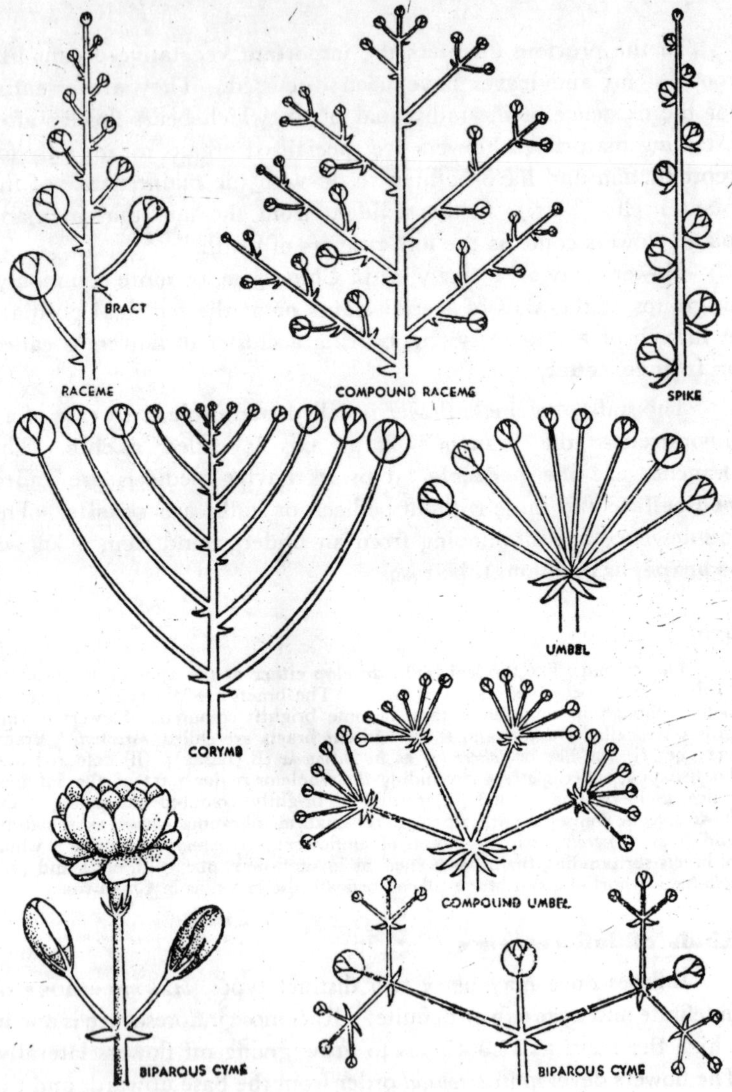

Fig. 76. Kinds of inflorescence (diagrammatic).

Racemose inflorescences are of the following types :

1. **Raceme.** It is a racemose inflorescence in which the

main axis goes on producing lateral stalked or pedicellate flowers in acropetal order. Example—*Crotolaria* (B. Atasi, Fig. 77, A), mustard. A compound raceme, in which the rachis is branched and branches bear the pedicellate flowers in racemose fashion is called **panicle,** *e.g. Peltophorum.*

Fig. 77. A—raceme of *Crotolaria;* B—spike of tube-rose.

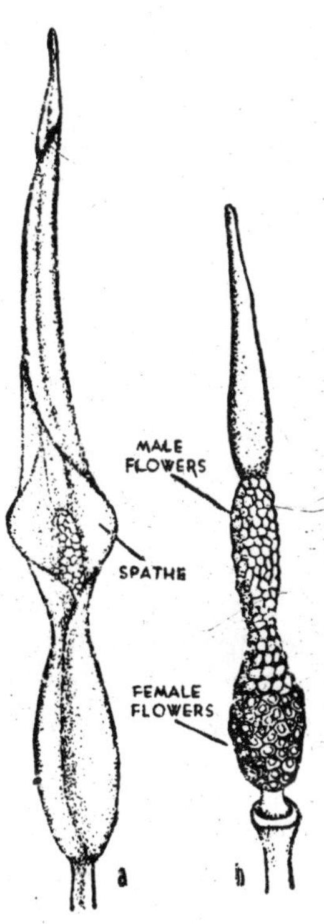

Fig. 78. Spadix of arum. a—with spathe; b—spathe removed.

2. **Spike.** It is just like a raceme, but the flowers are sessile, *e.g. Achyranthes* (B. Apang), tube-rose (Fig. 77, B).

3. **Spadix.** It is a kind of spike having a fleshy axis with sessile flowers, which remains enclosed in the coloured boat-shaped bract, the spathe. Familiar examples are arum (Fig. 78), banana. Palms have branched spadix.

4. **Catkin.** It is a spike with long pendulous axis bearing only unisexual flowers, *e.g.* Mulberry (B. Toonth).

5. **Corymb.** It is a racemose inflorescence in which the main axis is comparatively short and the older flowers have longer stalks than the younger ones, so that all of them are raised more or less to the same level. Example—*Iberis* (Candytuft, Fig. 79, A).

6. **Umbel.** Here the main axis is very short and a number of flowers with pedicels of same length develop from a point giving the inflorescence an umbrella-like appearance, e.g. *Hydrocotyle* (B. Thulkuri). Bracts often develop at the base of the pedicels. Coriander (B. Dhoney, Fig. 79, B). Anise (B. Mouri) are good examples of **compound** umbels.

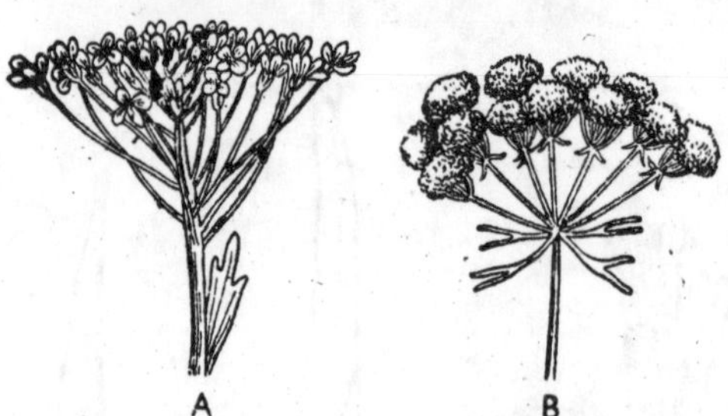

Fig. 79. A—corymb of *Iberis* (Candytuft); B—compound umbel of coriander.

7. **Capitulum.** In this case the rachis is very much shortened to form a flat convex body called **receptacle.** Many sessile flowers are present on the receptacle in centripetal order. The florets are usually of two types, **ray florets** at the circumference and the **disc florets** at the centre; the florets have small scaly bracts and the whole inflorescence is surrounded by an involucre of bracts. Though it looks like a single flower, it is really an inflorescence. Familiar examples—sunflower (Figs. 80 & 17), marigold, *Dahlia*, and other members of sunflower family.

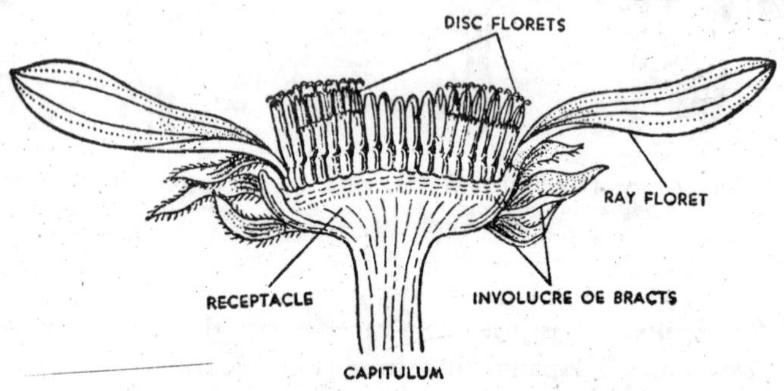

Fig. 80. Capitulum—section of sunflower.

Cymose Inflorescence

In cymose inflorescence the main axis is terminated by development of a flower. Lateral axes, one or more bearing younger flowers, originate from below the terminal one. The arrangement is *basipetal*, *i.e.* oldest one at the tip and the younger ones away from it. Thus the mode of opening of flowers here is *centrifugal* (from centre towards circumference). According to the branches formed, cymose inflorescence may be (i) **uniparous,** (ii) **biparous** and (iii) **multiparous.**

In uniparous or monochasial cyme only one lateral branch is produced below on the same side of the branch bearing it, as in *Hamelia*, or on opposite sides in a zigzag fashion, as in *Heliotropium* (B. Hatisoor). In biparous or dichasial cyme two lateral branches are formed regularly. Jasmine, *Nyctanthes* (B. Sheuli), etc., are the common examples. In multiparous cyme or polychasium more than two branches are formed, *e.g. Calotropis* (B. Akanda).

(The multiparous cyme apparently looks like an umbel. A careful examination of the position of flowers, old and young, will reveal the difference between them.)

Mixed Inflorescence

Usually plants have either racemose or cymose type of inflorescence; but some plants have mixed inflorescence (partly racemose and partly cymose) in which the branches are arranged in racemose manner and the flowers in cymose fashion, *e.g. Melia* (B. Neem), *Clerodendron* (B. Bhat), mango and some grasses.

Special Types of Inflorescence

Hypanthodium. It is a special type of inflorescence where the rachis becomes fleshy and hollow cup-shaped with a small opening at the tip, guarded by scales. Many small unisexual flowers, male and female, are present on the inner surface of the hollow receptacle. Examples —figs (Fig. 81), banyan, pipul.

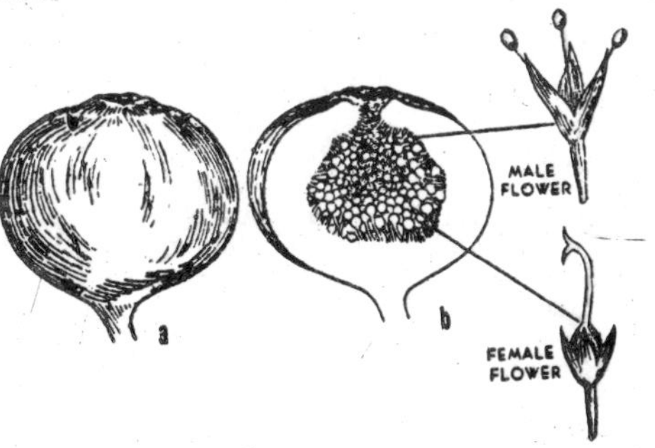

Fig. 81. a—hypanthodium of fig; b—same cut lengthwise showing male and female flowers.

Verticillaster. It is a special condensed cymose inflorescence occurring at the axils of opposite decussate leaves (Fig. 82). It forms a cluster of flowers surrounding the stem, *e.g. Leonurus* (B. Raktodrona).

Cyathium. It is another special inflorescence with very much reduced flowers. The whole inflorescence has an involucre of bracts, often with glands, which encloses a very much reduced female flower on a long stalk at the centre surrounded by some equally reduced male flowers with shorter stalks. Example—*Euphorbia* or Spurges (Fig. 83).

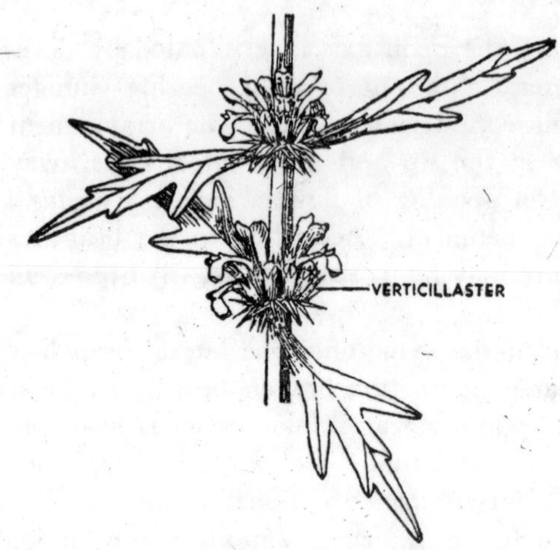

Fig. 82. Verticillaster of *Leonurus*.

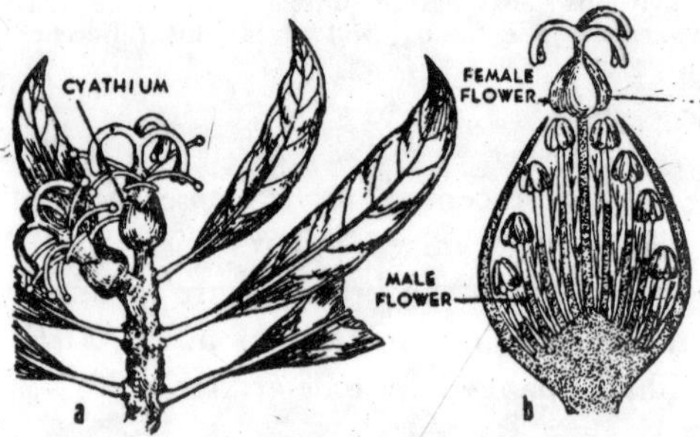

Fig. 83. a—cyathium of Poinsettia; b—same cut lengthwise.

Kinds of Inflorescence

Racemose	Cymose	Mixed	Special
1. Raceme, *e.g. Crotolaria*	(1) Monochasium, *e.g. Hamelia*	*e.g.*, *Melia*	(1) Hypanthodium, *e.g.* fig
2. Spike, *e.g.* tube-rose	(2) Dichasium, *e.g.* jasmine		(2) Verticillaster, *e.g. Leonurus*
3. Spadix, *e.g.* arum	(3) Polychasium, *e.g. Calotropis*		(3) Cyathium, *e.g. Euphorbia*
4. Catkin, *e.g.* mulberry			
5. Corymb, *e.g.* candytuft			
6. Umbel; *e.g.* coriander			
7. Capitulum, *e.g.* sunflower			

CHAPTER VII

THE FLOWER

The flower is the most conspicuous and attractive organ of the plant. It is solely concerned with reproduction, ultimately resulting in the formation of fruits and seeds. Flower is equivalent to a shoot, having an axis with a number of specialised leaves meant for the purpose of reproduction.

A typical flower has four sets of floral leaves arranged on a short axis, called **thalamus,** in a definite order (Fig. 84). They are **calyx, corolla, androecium** and **gynoecium.**

I. **Calyx** is the outermost whorl with the individual parts, *sepals*. They are usually green inconspicuous bodies. Sepals protect the essential organs in the bud stage.

II. **Corolla** is the second set consisting of *petals*. Petals are usually brightly coloured or sweetly scented and are thus most conspicuous and showy parts of the flower. Besides protection of the essential organs at the bud stage, they attract insect visitors for bringing about pollination.

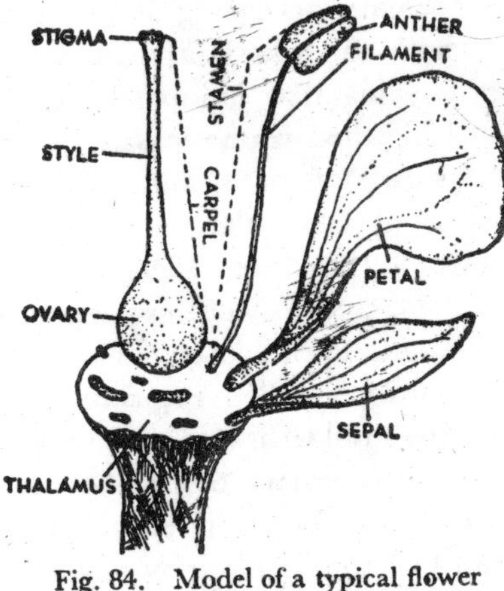

Fig. 84. Model of a typical flower showing one member of each whorl.

III. **Androecium** is the third whorl composed of *stamens*. Stamens, also called *microsporophylls*, are the male organs of the flowers. Each has usually a slender **stalk** called **filament** surmounted by a swollen body known as **anther.** The anther has generally two lobes, each usually possesses two sac-like bodies called pollen-sacs. Many small unicellular **pollen grains** or *microspores* are present inside the pollen-sacs. Pollen grains bear the male gametes or reproductive units.

IV. **Gynoecium** or **pistil** is the fourth and the innermost set.

It is also a collective term, the individual parts are *carpels* or *megasporophylls*. Each carpel has usually a swollen basal **ovary,** a long slender continuation of ovary, called **style,** and the terminal receptive portion known as **stigma.** Egg-shaped bodies, called **ovules,** are present inside the ovary. These ovules, really the immature or potential seeds, contain the female gametes.

Of the four sets of floral leaves the last two, androecium and gynoecium, are directly concerned with reproduction. They are known as **reproductive** whorls or **essential** whorls. The outer two sets, calyx and corolla, do not take any active part in reproduction, but just help in the process. So they are called **accessory** or helping whorls. Calyx and corolla are often collectively referred to as **perianth.** This term applies particularly to the monocotyledonous flowers where sepals and petals cannot be distinguished.

A flower having all the four floral sets is said to be **complete,** but if one or more sets are absent, it is **incomplete.** When both stamens and carpels are present in the same flower, it is **hermaphrodite** or **bisexual;** if only one of them is present, the flower is **unisexual.** Unisexual flower may be **staminate** or male, or **pistillate** or female, according as the stamens or carpels are present in it. If both the whorls are absent the flower is **neuter.** It goes without saying that unisexual and neuter flowers are incomplete. A plant is called **monoecious** (*mono*=one, *oikos*=house) when it bears both staminate and pistillate flowers, *e.g. Cucurbita* (B. Gourd); but when the plants bear either staminate or pistillate flowers only, they are called **dioecious,** *e.g.* palms. Some plants like mango, mangosteen, etc., bear all types of flowers—bisexual, unisexual and even neuter. They are called **polygamous.**

A flower is called **regular** when the individual parts of the whorls are similar in size and shape, as in China-rose; but if the parts are dissimilar, the flower is irregular, *e.g.* pea, *Crotolaria* (B. Atasi). A flower is termed **actinomorphic,** when it can be cut into two equal halves by any number of planes passing through the centre. Regular flowers are usually actinomorphic. It is **zygomorphic,** when it can be cut into two equal halves by only one plane through the centre, *e.g.* pea. When even that is not possible, a flower is said to be **asymmetrical** or having no symmetry, *e.g. Canna* (B. Sarbajaya).

If the number of members in each whorl is three, or multiple of three the flower is called **trimerous.** Similarly they may be

tetramerous (in fours) or **pentamerous** (in fives). Usually monocotyledons have trimerous and dicotyledons have tetramerous or pentamerous flowers.

THALAMUS

The **thalamus** is the short abbreviated axis bearing the four sets of floral leaves. It is the swollen end of the peduncle or pedicel with four nodes and very much compressed internodes. The floral leaves remain inserted on the nodes in whorls or spirally. The axis-nature of the thalamus becomes quite evident in some flowers in which it (thalamus) is considerably long and the internodes are distinct. The internode between the calyx and corolla is known as **anthophore,** that between corolla and androecium is **androphore,** and the one between androecium and gynoecium is **gynophore.** Androphore is distinct in passion-flower and gynophore in *Pterospermum* (B. Kanak champa). In *Gynandropsis* (B. Hurhure) both androphore and gynophore develop to form what is known as **gynandrophore** (Fig. 85).

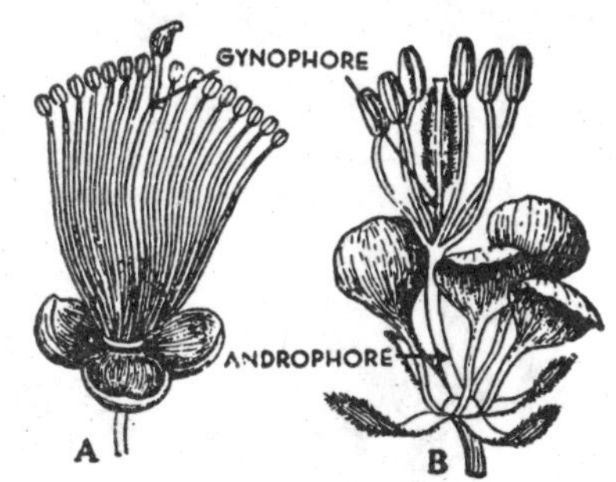

Fig. 85. A—flower of *Capparis* showing gynophore; B—flower of *Gynandropsis* showing androphore and gynophore.

The thalamus may be a bit elongated, as in *Michelia* (B. Champaka), convex in flowers like China-rose, *Datura*, cup-shaped in roses, and it may have spongy flat top, as in lotus.

According to the nature of the thalamus and mode of insertion of the first three floral sets with respect to the gynoecium, flowers may be **hypogynous, perigynous** and **epigynous** (Fig. 86).

Hypogynous flower is that in which the thalamus is convex and dome-shaped, and the pistil occupies the top position of the thalamus, other whorls remaining inserted *below* the pistil one after another in normal order. The ovary here is called *superior*. Examples—China-rose, *Datura*.

Perigynous flower has concave or cup-shaped thalamus forming rims or flanges, on which stamens, petals and sepals remain inserted. The pistil occupies the central position and the other whorls are placed *around* it. The ovary is said to be *half-inferior*. Examples—pea, roses.

Epigynous flower is that in which the thalamus is hollow and cup-shaped, and completely fuses with the lower part of the

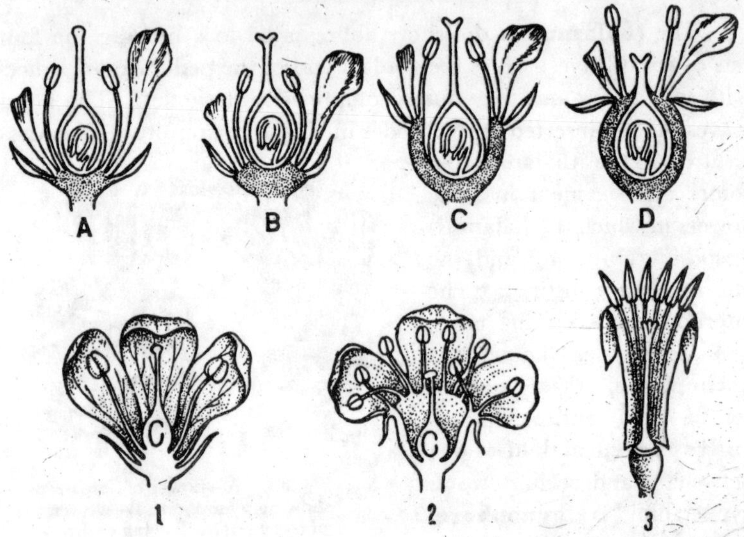

Fig. 86. Insertion of floral leaves on the thalamus. A, B & 1—hypogynous; C & 2—perigynous; D & 3—epigynous.

pistil (ovary). The stamens, petals and sepals are placed *above* the pistil. The ovary in epigynous flower is *inferior*. Examples—Sunflower, *Cucurbita* (B. Gourd), tube-rose.

Flower is equivalent to a shoot

It has been stated that the flower is equivalent to a shoot. The following facts may be cited in support of the above statement.

The floral buds, like the leaf-buds, develop either at the terminal or axillary positions. The floral members remain on the thalamus in whorls or in spirals resembling the phyllotaxy of the leaves. The arrangement of the floral parts in respect to each other (aestivation) is also similar to the arrangement of the leaves in the bud stage. Modifications of floral buds into bulbils (buds modified into swollen bodies for vegetative multiplication) is found in plants like *Aloe*, pine-apple.

Normally the thalamus is a short abbreviated axis with suppressed nodes and internodes. The axis-nature of the thalamus is evident in the flowers like *Pterospermum* (B. Kanak champa), passion-flower, *Gynandropsis* (B. Hurhure), etc., where the internodes between the whorls are quite distinct. A peculiar feature, as found in some roses and pears, is that the thalamus instead of terminating at the flower, continues or prolongs through it and bears normal leaves. Such condition, referred to as 'monstrous development', further proves the axis-nature of the thalamus.

The leaf nature of the members of accessory whorls—sepals and petals, is rather clear, in view of the fact that they are usually expanded bodies resembling the foliage leaves in structure and nature of venation. In *Mussaenda* (Fig. 88) one of the sepals spreads out like a leaf. Main difficulty arises in case of the essential members—stamens and carpels. The gradual transition from one member to another—sepals to petals, petals to stamens, is distinct in water-lily (Fig. 87). In many cultivated flowers the stamens are found to be converted into petals, what is commonly called 'doubling of flowers'. Rose, China-rose, *Hibiscus mutabilis* (B. Sthal padma) are the examples. The stamens and style become petaloid in *Canna*, a common garden flower. All these facts go to show that the stamens and carpels are also morphologically equivalent to the leaves.

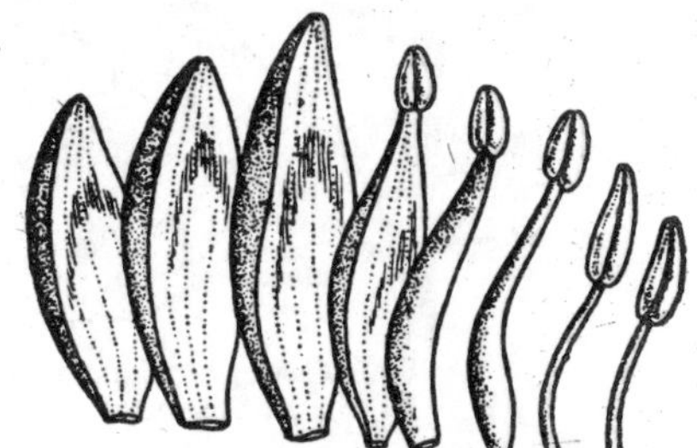

Fig. 87. Gradual transition of floral leaves in water-lily.

I. CALYX

The **calyx** is the outermost accessory set of floral leaves, consisting of *sepals*. Sepals are usually green inconspicuous bodies. They protect the essential whorls against mechanical injuries and desiccation in the bud stage. So, their function is mainly protective. Sepals of some plants become coloured like the petals and are called petaloid, *e.g. Caesalpinea* (B. Krishnachura). In *Mussaenda* one of the sepals is transformed into a flat leaf-like structure (Fig. 88). In the sunflower family the sepals are modified into hairy bodies called *pappus* which help in the dispersal of fruits (Fig. 113, 3).

If the sepals are free the calyx is called **polysepalous,** and if they are united the calvx is **gamosepalous.** The gamosepalous calyx is more or less like a tube in which the number of sepals may be easily determined by counting the teeth or lobes.

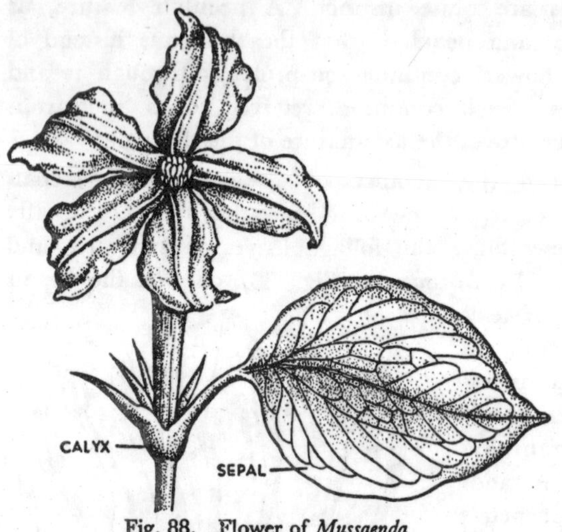

Fig. 88. Flower of *Mussaenda* with a leaf-like sepal.

In many cases sepals fall off before the flowers fully open out. They are called **caducous** sepals, *e.g.* Poppy; while in others they fall off with other parts after fertilisation. These are **deciduous** sepals. In some flowers sepals persist even up to the fruit stage and are called **persistent,** *e.g.* brinjal, tomato. Sometimes they not only persist but grow to form a part of the fruit, as in *Dillenia* (B. Chalte), *Physalis* (B. Tepari). These are called **accrescent** sepals.

II. COROLLA

It is the second accessory whorl of the flower, the individual parts of which are called *petals.* They make the flowers showy and attractive, as the petals are brightly coloured and often sweetly scented. The function of corolla is twofold : firstly, the protection of the essential whorls in the bud stage ; and secondly, the attraction of insects for pollination. The petals of *Polyalthia* (B. Debdaru), custard apple, etc., are called **sepaloid,** as they are green and sepal-like in appearance. If the petals are free, the corolla is called **polypetalous;** and if they are united, it is **gamopetalous.** The free petals often have two parts—a narrow stalk called **claw** and the upper flat blade known as **limb.** Petals having claws are said to be **clawed,** as in pink, radish, mustard (Fig. 89, 1).

The corolla is regular, if all the petals are similar in size and shape; if they are dissimilar, it is irregular.

Forms of Corolla

Different forms of corolla are found in flowers. They may be free or united, regular or irregular. It is, by no means, easy to

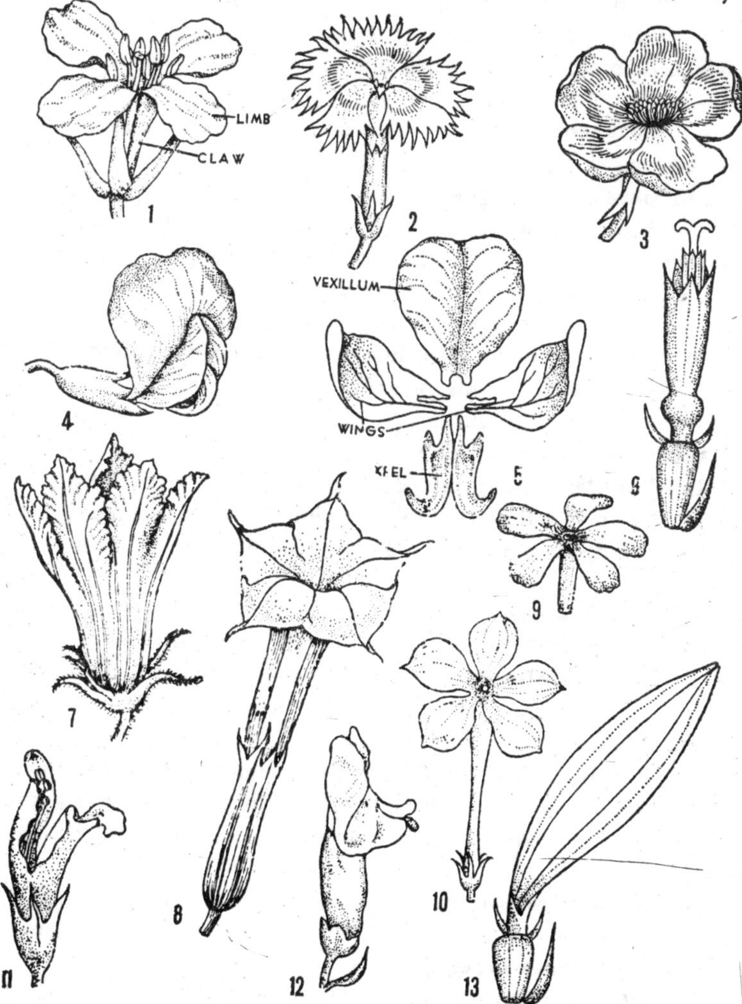

Fig. 89. Forms of corolla.
1—cruciform; 2—caryophyllaceous; 3—rosaceous; 4—papilionaceous; 5—same showing parts; 6—tubular; 7—campanulate; 8—infundibuliform; 9—rotate; 10—hypocrateriform; 11—bilabiate; 12—personate; 13—ligulate.

give a precise account of all the forms found in nature. The following are the principal and more common forms (Fig. 89):

A. **Polypetalous Regular**

1. **Cruciform.** It consists of four free clawed petals arranged in the form of a cross (Fig. 89, 1). Examples—mustard, radish. It is characteristic of the mustard family (*Cruciferae*).

2. **Caryophyllaceous.** It has five clawed petals, the limbs of which lie more or less at right angles to the claws, *e.g.* pink (Fig. 89, 2).

3. **Rosaceous.** It is composed of five petals without claws, *e.g.* roses, tea (Fig. 89, 3).

B. **Polypetalous Irregular**

1. **Papilionaceous.** It looks more or less like a butterfly (*papilion*=butterfly). It consists of five free petals. The largest and the outermost one is called **standard** or **vexillum;** two lateral ones are known as **wings** or **alae;** and the innermost two, called **keel** or **carina,** unite to form at boat-shaped body. Examples—pea (Fig. 89, 4 & 5), beans, *Sesbania* (B. Bak). It is characteristic of the pea family (*Papilionaceae*).

C. **Gamopetalous Regular**

1. **Tubular.** Here the petals unite to form a tube-like body having lobes or teeth at the tip, by counting which the number of petals is determined, *e.g.* disc florets of sunflower, marigold (Fig. 89, 6).

2. **Campanulate.** Five petals unite to form a bell-shaped corolla, as in gourd (Fig. 89, 7), *Physalis* (B. Tepari).

3. **Infundibuliform.** It consists of five petals which unite to form a funnel-shaped corolla gradually spreading towards the tip, *e.g. Datura* (Fig. 89, 8), *Thivetia* (B. Kalke).

4. **Rotate.** Here the tube is short and the spreading limbs are arranged more or less at right angles to the tube, giving the corolla wheel-shaped appearance (Fig. 89, 9), *e.g.* brinjal, *Nyctanthes* (B. Shieuli).

5. **Hypocrateriform.** It is like the rotate one but with a comparatively long tube (Fig. 89, 10), *e.g. Vinca* (B. Nayantara).

D. **Gamopetalous Irregular**

1. **Bilabiate** or two-lipped. Here the petals by union form two parts with an open wide mouth between them. The lower part is made of three petals and the upper of two (Fig. 89, 11), *e.g. Ocimum* (B. Tulsi), *Leonurus* (B. Drone).

2. **Personate.** It is like the bilabiate one but the mouth is closed, as the two lips are placed very near each other, *e.g.* snapdragon (Fig. 89, 12), *Lindenbergia* (B. Basanti).

3. **Ligulate** or strap-shaped. The five petals are united in such a way as to form a short tube at the base, but they spread out above like a strap, *e.g.* ray-florets of sunflower, marigold (Fig. 81, 13).

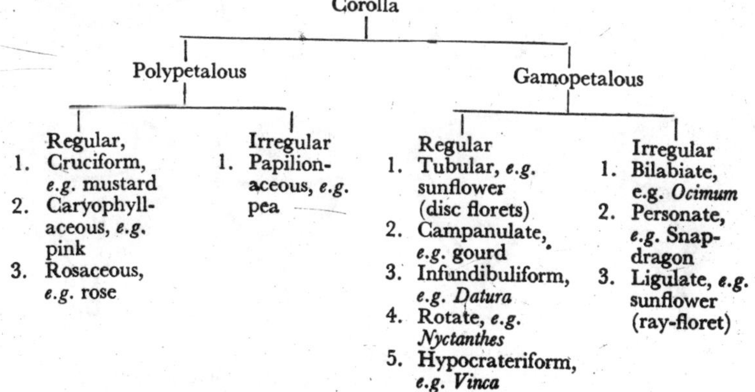

Appendages of the Corolla

The corolla is often provided with spurs or long projections, as in Balsam (B. Dopati), and outgrowths, called **corona,** which enhance the beauty of the flowers. Corona is present in passion-flower (B. Jhumko) as a beautiful whorl of outgrowth between the corolla and the androecium. Nectaries, the honey-secreting glands, are sometimes present at the base of the petals. These are natural extra-devices for attracting insects for pollination.

AESTIVATION

Aestivation is the mode of arrangement of sepals and petals in the bud stage in respect to one another (Fig. 90). It is of the following types :

(i) **Valvate,** when the members either touch or lie side by side very closely but do not overlap, as in *Calotropis* (B. Akanda), *Artabotrys* (B. Kantali champa).

(ii) **Twisted** or *contorted*, when the members are arranged in such a way that each of them is overlapped at one margin and overlaps the next member by the other margin, as in China-rose.

(iii) **Imbricate.** It is modification of twisted arrangement where one member is completely **external,** *i.e.* overlaps by both the margins, one member is completely **internal,** *i.e.* is overlapped at both the margins and the rest remain arranged as in twisted aestivation, *e.g. Cassia* (B. Kalkasunde).

(iv) **Vexillary.** Here out of five petals the largest is the outermost. It includes two laterals which, in their turn, enclose the innermost two. It is characteristic of papilionaceous corolla like pea, *Sesbania* (B. Bak).

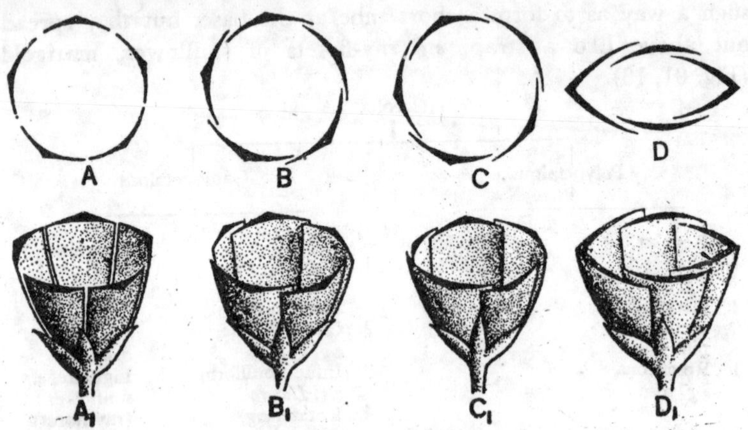

Fig. 90. Types of aestivation.
A & A_1—valvate; B & B_1—twisted; C & C_1—imbricate; D & D_1—vexillary.

Perianth

The accessory whorls calyx and corolla are often referred to as **perianth.** In monocotyledons the perianth is not differentiated into two sets, as in tube-rose, *Crinum* (B. Sukhadarsan), etc. If the lobes are free, the perianth is **polyphyllous** and, if united, it is **gamophyllous.** In grasses the perianth is represented by a pair of small scales called **lodicules.**

III. ANDROECIUM

Androecium is the third set and the first essential whorl of the flower. It is composed of *stamens* or *microsporophylls* which are male organs. A stamen has usually two parts, *viz.* a slender stalk called **filament** and a swollen body at the tip of the filament known as **anther.** The anther has usually two lobes on the two sides; each lobe, in its turn, has two bag-like bodies, called **pollen-sacs** or *microsporangia,* which contain innumerable small unicellular **pollen grains** or *microspores.* In China-rose one pollen-sac is present in each anther lobe. The prolongation of the filament which holds together or connects the two anther lobes is called **connective.** Vascular bundle runs through the connective to convey food to the growing anther.

All stamens do not always bear fertile anthers. In *Cassia sophera* (B. Kalkasunde) four out of ten stamens are sterile. They are called **staminodes.** In *Canna* (B. Sarbajaya) staminodes are coloured and petal-like in appearance. Outgrowths on the stamens, called staminal corona, are found in *Calotropis* (B. Akanda).

Attachment of the Filaments to the Anthers

The stamens are of the following types according to the modes of fixation of anthers to the filaments :—

(i) **Basifixed** or innate, when the filament is attached just to the base of the anther, as in mustard (Fig. 91).

(ii) **Adnate,** where due to the presence of a well-marked connective the filament seems to run throughout the whole length of the anther, *e.g. Michelia* (B. Champaka).

(iii) **Dorsifixed,** when the filament is attached to the back of the anther, as in passion-flower, *Sesbania* (B. Bak, Fig. 99).

(iv) **Versatile,** when the filament is attached at the back of the anther by a sharp point, so that the anther may freely swing in air, *e.g.* grasses, paddy.

Dehiscence of the Anthers

With the maturity of the anthers the two pollen-sacs coalesce due to the fact that the wall between them breaks up. Pollen grains then remain scattered in the cavity. The anther dehisces or bursts to liberate the pollen grains, so that they may be carried by different agencies to the stigmas of flowers. Dehiscence takes place mostly by longitudinal splitting, as in *Datura*, by pores formed at the apices of anthers, as in brinjal, or by a number of valves (valvular), as in cinnamon.

Relative Length of the Stamens

The stamens of a flower may be of equal or unequal length. *Ipomoea* (B. Kalmi) has five stamens, all of which are of unequal length. Flowers of mustard family have usually six stamens in two whorls. Two of the outer whorl are short and four of the inner whorl have long filaments. They are called **tetradynamous** stamens (Fig. 91). *Leonurus* (B. Drone), *Ocimum* (B. Tulsi), etc., have four stamens, of which two are long and two short. They are called **didynamous** stamens.

TETRADYNAMOUS STAMENS

Fig. 91. Tetradynamous stamens of mustard.

Union of the Stamens

Stamens may unite amongst themselves or with other whorls. The union of similar members is called *cohesion;* and that of dissimilar members is spoken of as *adhesion.*

Cohesion of the Stamens. As the stamens have two main parts —filaments and anthers, two conditions in cohesion are possible, *viz.* (i) stamens may unite by filaments, anthers remaining free; or (ii) they may be united by anthers, filaments remaining free (Fig. 92). The union by filaments results in the formation of bundle or **adelphy.** If one such bundle is formed by union of filaments, the stamens are called **monadelphous,** as in China-rose, cotton. If two bundles are formed, they are **diadelphous,** as in pea, *Sesbania* (B. Bak); and when more than two bundles are formed, they are called **polyadelphous,** as in silk-cotton. On the other hand, when they unite by the anthers, filaments remaining free, the stamens are called **syngenesious,** as in gourd, sunflower.

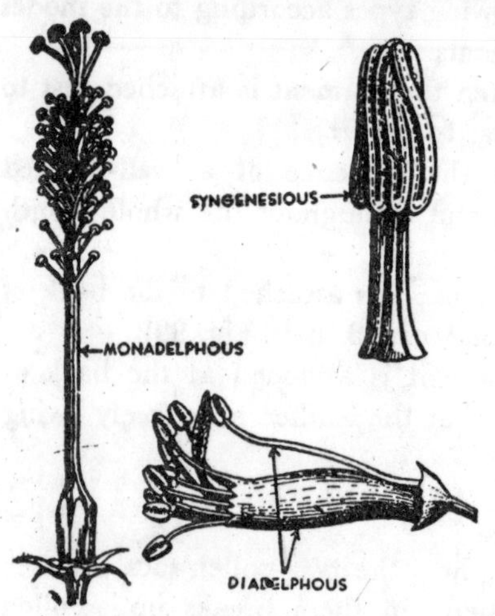

Fig. 92. Cohesion of stamens.

Adhesion of the Stamens. The stamens may unite with the next lower whorl, *i.e.* corolla or perianth, or with the next upper whorl, the pistil. When they unite with petals, stamens are called **epipetalous** (Fig. 101). Almost all the flowers with gamopetalous corolla have epipetalous stamens. If they adhere to perianth lobes, stamens are called **epiphyllous,** as in tube-rose (Fig. 102). The stamens are called **gynandrous,** when they are attached to the pistil, *e.g. Calotropis* (B. Akanda).

Pollen grains

The pollen-sacs or *microporangia* of the anthers contain some special cells, called pollen-mother cells or *microspore-mother cells.* Each of them undergoes *reduction* division to produce a tetrad of

pollen grains or *microspores.* These are small dusty unicellular bodies usually round or oval in shape. A pollen grain has protoplasm with a single nucleus surrounded by two coats. The inner coat, known as **intine,** is thin, delicate and made of cellulose; and the outer coat, called **exine,** is thick, cutinised and often provided with warty or spiny projections. Some unthickened areas or weak spots, called **germ pores,** are present here and there on the exine. The uninucleate pollen grains generally become binucleate before liberation from the sacs (Fig. 103).

In some flowers the pollen grains, instead of remaining loose, are aggregated together into masses called **pollinia,** as in *Calotropis* (B. Akanda), orchids.

IV. GYNOECIUM OR PISTIL

Gynoecium or **pistil** is the last and innermost set of floral leaves occupying the central position on the thalamus. It is composed of *carpels* or *megasporophylls* which are the female organs of the flowers. A carpel has usually three parts : (i) the basal swollen portion is the **ovary,** (ii) the slender cylindrical region is the **style** which usually ends in the receptive spot called (iii) **stigma.** The ovary contains one or more egg-shaped bodies known as **ovules,** which are nothing but immature seeds.

Pistil may be made of one, two, three or many carpels. Accordingly they are monocarpellary, bicarpellary, tricarpellary or polycarpellary. The monocarpellary pistil is called *simple,* while others are termed as *compound.* Foliar nature of a carpel is evident in a simple pistil which may be compared to a simple leaf. If the leaf is folded along the mid-rib, the two margins meet forming a hollow cavity, which resembles a simple pistil. The line where the two margins meet is known as *ventral* suture and that corresponding to the mid-rib is the *dorsal* suture. Along the ventral suture soft cushion-like bodies or ridges, called **placentae,** are formed to which the ovules remain attached.

The compound pistil is called **apocarpous,** when the carpels are free. Here all the carpels remain inserted on the thalamus, each having its own ovary, style and stigma, as in *Michelia* (B. Champa), *Artabotrys* (B. Kantali champa). But if the carpels are united, the pistil is known as **syncarpous** where the common

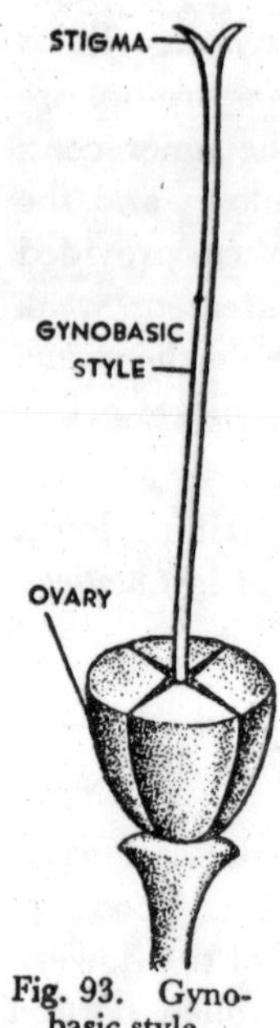

Fig. 93. Gynobasic style.

ovary and style are usually formed, carpels usually remaining free only in the region of stigmas, *e.g.* China-rose (Fig. 98), cotton, *Datura.*

The style normally is the slender cylindrical continuation of the ovary. In *Ocimum* (B. Tulsi) family the ovary is four-lobed and the style originates from the base of the ovary or from the thalamus. This type of style is called **gynobasic** (Fig. 93). Usually styles are deciduous, but silky persistent ones are found in maize (Fig. 221).

The stigma is the terminal receptive spot. It may be flat and sticky due to secretion of stigmatic fluid, and feathery in case of monocotyledons. These are devices for facilitating reception of pollen grains.

PLACENTATION

The ovary may be unilocular (*uni*=one, *loculus*=chamber) or multilocular (*multi*=many, *loculus*=chamber). The ovules or *megasporangia* remain attached to the cushion-like parenchymatous protuberances of the ovary, called **placenta,** and the mode of arrangement of the placentae is known as **placentation.**

Types of Placentation (Fig. 94)

1. **Marginal.** Marginal placentation is characteristic of the pea family, where the pistil is simple, and placenta bearing the ovules lies at the ventral suture.

2. **Parietal.** Here the pistil is polycarpellary and the placentae are distributed at the regions where the margins of the carpels meet. The ovary is unilocular and the placentae groups correspond to the number of carpels present, *e.g.* papaw, gourd.

3. **Axile.** In axile placentation also the pistil is polycarpellary. The margins of the carpels, instead of meeting, bend downwards to form an axis at the centre. Placentae bearing ovules are distributed around the axis. So the ovary becomes multilocular, the

number of loculi being equal to the number of carpels present, *e.g.* orange, China-rose, *Datura.*

4. **Free Central.** Here the placentae are present around the axis, as in axile placentation, but due to the absorption of the partition walls the multilocular ovary is rendered unilocular. Sometimes remnants of the partition walls may be noticed. Example—pink.

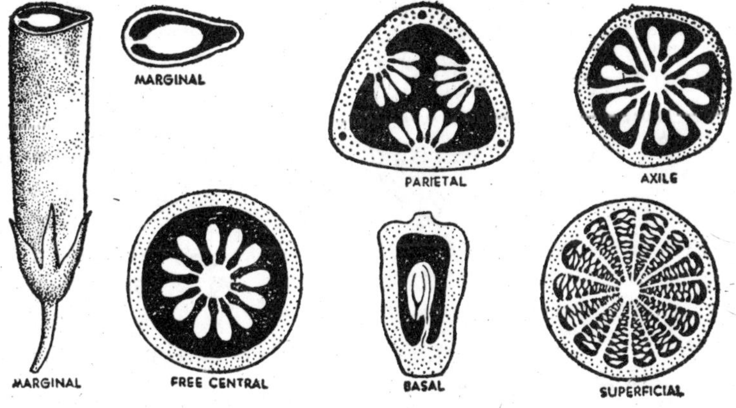

Fig. 94. Types of placentation.

5. **Superficial.** It is also like axile having many chambers but placentae with ovules are present all over the inner surface of the partition walls, as in water-lily.

6. **Basal.** Here the ovary is unilocular, and the placenta bearing a single ovule develops directly on the thalamus at the base of the ovary, *e.g.* Sunflower, *Dahlia.*

THE OVULE

The **ovule** or *megasporangium* develops into the seed after fertilisation. Each ovule is an egg-shaped body, having a stalk called **funiculus,** by means of which it remains attached to the placenta. The main soft body of the ovule is known as **nucellus** which remains surrounded by two coats called **integuments.** The integuments, outer and inner, almost invest the ovule, only leaving a small opening, known as **micropyle,** at the tip. In some ovules only one integument is present. The broad basal region, where the integuments meet the nucellus, is known as **chalaza.** An oval body, called **embryo-sac,** is embedded in the nucellus towards the micropylar end. The embryo-sac has usually eight nuclei. Three of them aggregate at the micropylar end forming the **egg-apparatus.** The middle one of the three is the **egg** or *female gamete*, and the other two are called **help cells** or **synergids.** Three collecting at the chalazal end are known as **antipodal cells;** and the remaining two, called polar nuclei, either lie side by side or fuse up at the central region, forming what is known as **definitive nucleus** (Fig. 95). The embryo-sac is the most important part of the ovule.

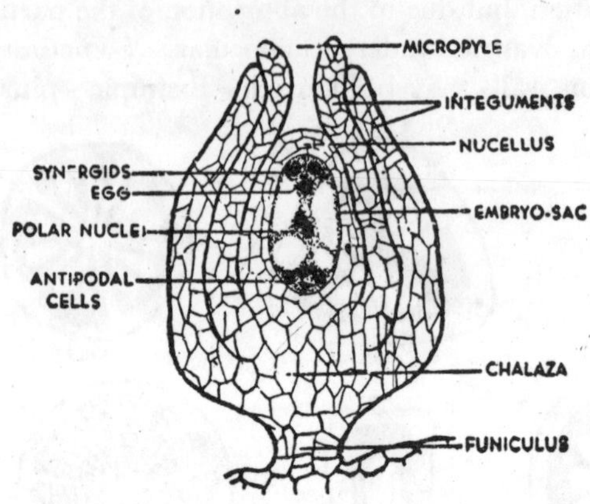

Fig. 95. An ovule in longitudinal section.

Formation of the Ovule and the Embryo-sac

The ovule originates as a small rounded protuberance of soft tissue (nucellus) on the placenta. Gradually collar-like outgrowths or integuments begin to develop which cover the nucellus, leaving an opening, the micropyle, at the tip. Before complete formation of the integuments a hypodermal cell at the apex of the nucellus becomes conspicuous. This is the *megaspore-mother cell.* It undergoes *reduction* division to produce four cells in a row, what is known as **linear tetrad.** Three of them generally disintegrate and the innermost one survives. The nucleus of the functional megaspore now divides by *mitosis* into two, which move to the two ends of the megaspore. Each of them again divides twice. So that four nuclei are present at each end. Three of them at the micropylar end form the egg-apparatus with the central prominent egg and the two synergids. Three aggregated at the chalazal end constitute the antipodal cells. The remaining two, often called polar nuclei, move to the centre of the megaspore to produce the definitive nucleus (Fig. 96).

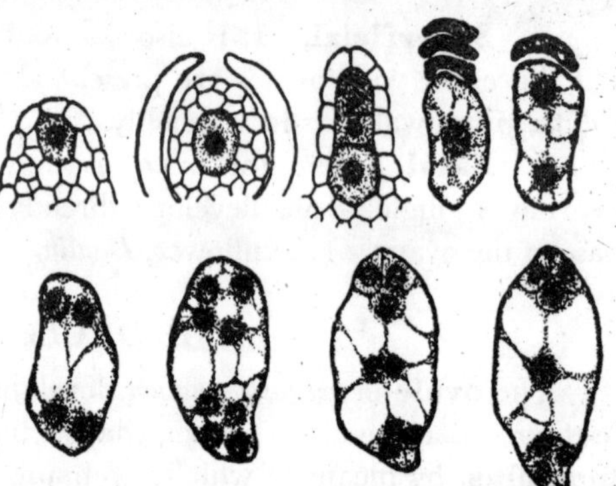

Fig. 96. Stages in the development of the embryo-sac.

This is an outline of the process of normal type of development of the embryo-sac in the angiosperms.

Forms of Ovules

Ovules are of the following types (Fig. 97):

(1) **Orthotropous** or straight, when the ovule is erect with the funiculus, micropyle and chalaza in the same straight line. Here micropyle lies towards the apex and chalaza towards the base.

(2) **Anatropous** or inverted, when the ovule is inverted due to overgrowth of the funiculus, so that micropyle lies at the base near the funiculus and chalaza at the other end. This type is more common.

(3) **Amphitropous** or transverse, where the main body of the ovule is placed more or less at right angles to the funiculus, *e.g. Lemna* (B. Kshudi-pana).

(4) **Campylotropous** or curved, where the main body of the ovule is curved or bent becoming horse-shoe-shaped and is arranged more or less at right angles to the funiculus, *e.g.* mustard.

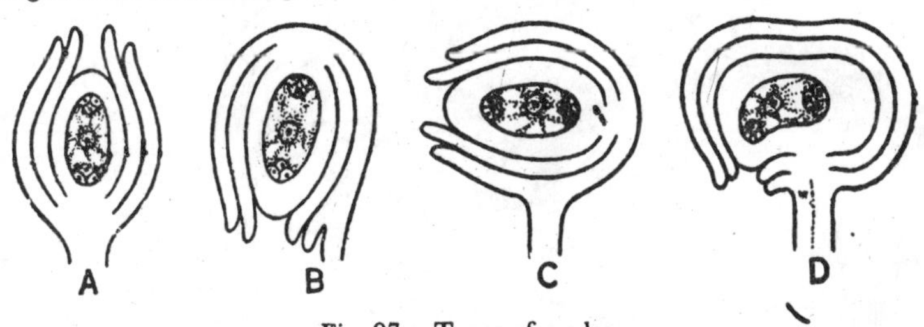

Fig. 97. Types of ovules.
A—orthotropous; B—anatropous; C—amphitropous;
D—campylotropous.

Floral diagram

The parts present in a flower and their arrangement with reference to one another may be represented by diagrams called floral diagrams. They have a number of concentric rings corresponding to the floral sets present, where united and free parts can be shown. The position of the floral axis is represented by a 'dot'. The best idea about a floral diagram may be had by cutting a floral bud crosswise.

Floral formulae

Floral formula represents the parts of a flower with their respective position. Some signs and symbols are used here. K stands for calyx, C for corolla, P for perianth, A for androecium and G for gynoecium. Union is shown by putting the letters within brackets, and adhesion, by lines above the whorls concerned. Members of the same set present in two whorls are represented by putting '+' (plus) sign between them. ∞ stands for many members in a whorl. The superior or inferior ovary is indicated by putting horizontal lines below or above the number representing carpels. Formulae of a few common flowers are given below:

Mustard —K_{2+2}, C_4, A_{2+4}, $G_{(\underline{2})}$.

Pea —$K_{(5)}$, C_5, $A_{(9+1)}$, $G_{\underline{1}}$.

China-rose —$K_{(\underline{5})}$, $\overline{C_5 \quad A_\infty}$ $G_{(\underline{5})}$

Datura —$K_{(5)}$ $\overline{C_{(5)} \quad A_5}$ $G_{(\underline{2})}$

Tube-rose —$\overline{P_{(3+3)} \quad A_{3+3}}$ $G_{(\overline{3})}$

DESCRIPTION OF A FEW COMMON FLOWERS

1. China-rose (*Hibiscus rosa-sinensis*) (Fig. 98)

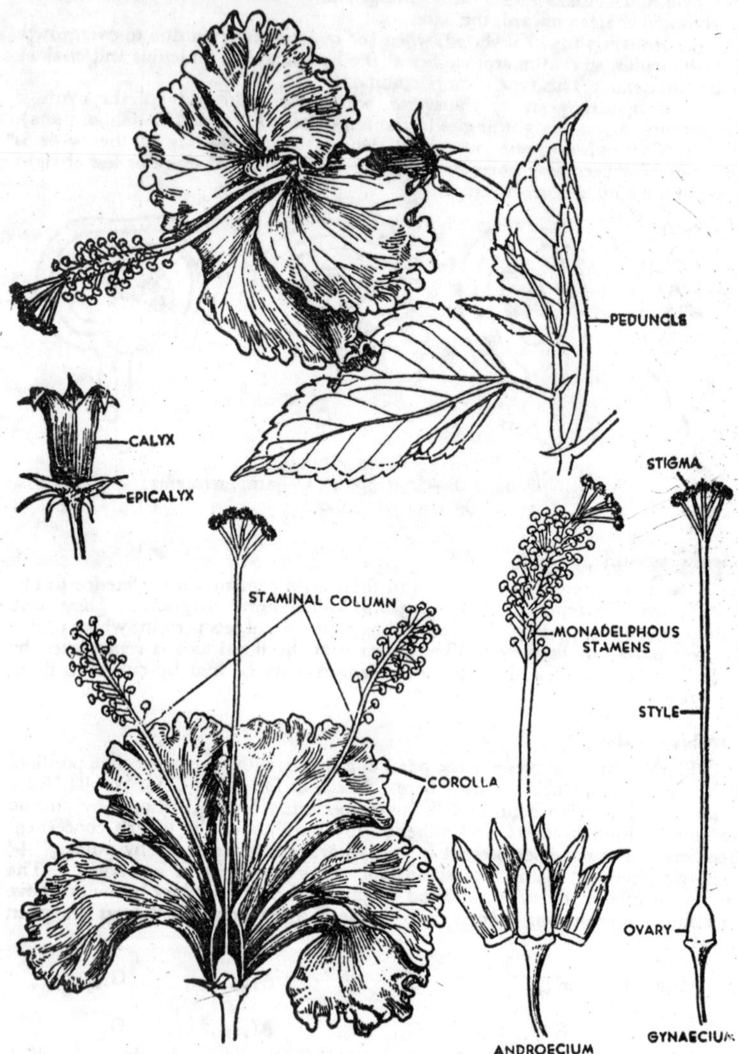

Fig. 98. Flower of China-rose and its different parts.

Flowers—solitary, axillary, pedunculate, complete, regular, hermaphrodite, hypogynous. Bracts 5 or more forming a whorl called epicalyx.

Calyx—sepals 5, gamosepalous, tubular.

Corolla—regular, showy, petals 5, slightly united at the base with the staminal column; aestivation—twisted.

Androecium—stamens many, monadelpous—united by filaments to form a hollow staminal column; anthers—free, reniform, one-celled.

Gynoecium—carpels 5, syncarpous; style passes through the staminal column; stigma 5, free; ovary—superior, 5-chambered with many ovules; placentation—axile.

2. *Sesbania grandiflora* (B. Bakful) (Fig. 99)

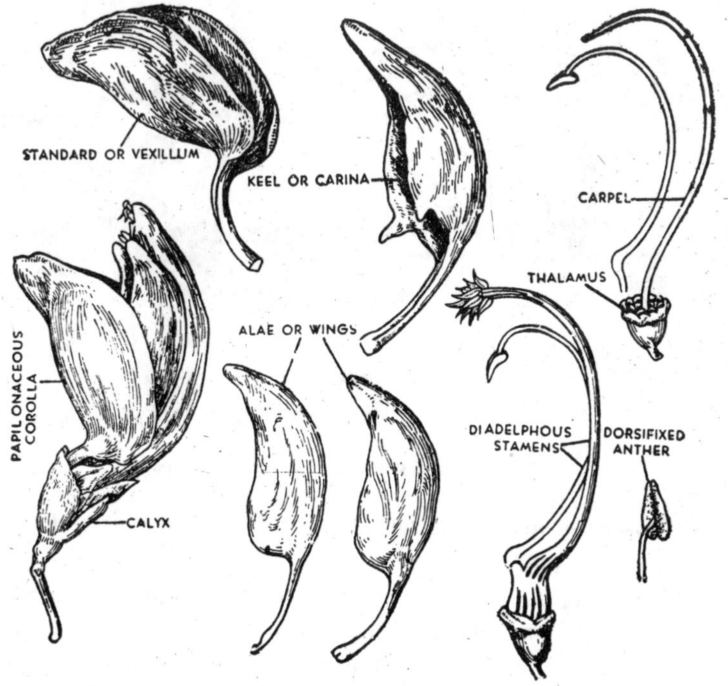

Fig. 99. Flower of *Sesbania* and its different parts.

Flowers—in axillary racemes, pedicellate, complete, irregular, zygomorphic, hermaphrodite, slightly perigynous.

Calyx—sepals 5, gamosepalous, 2-lipped or 5-toothed.

Corolla—petals 5, free, papilionaceous—the outermost and the largest one is standard or vexillum, the two lateral petals are wings or alae and the innermost two, keel or carina, unite to form a boat-shaped body; aestivation—vexillary.

Androecium—stamens 10; 9 united by filaments to form a bundle, one remaining free—diadelphous; anthers—free, dorsifixed.

Gynoecium—carpel 1, elongated; ovary—superior, one-chambered with many ovules; placentation—marginal.

3. **Gourd** (*Cucurbita maxima*) (Fig. 100)

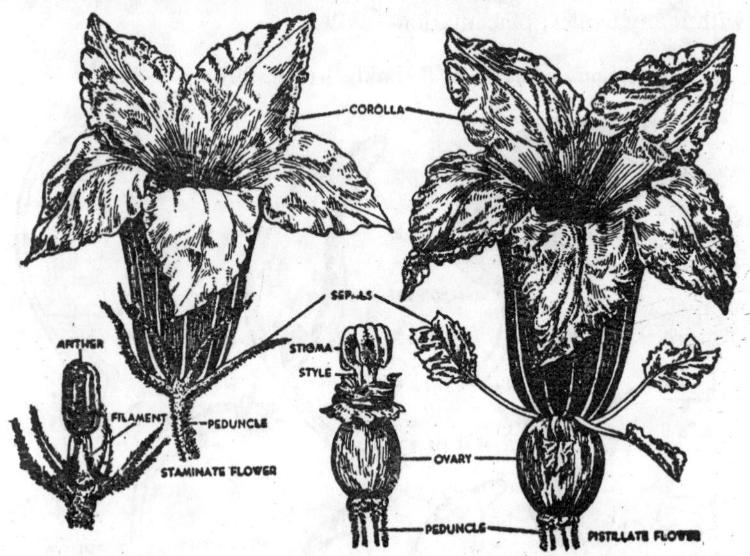

Fig. 100. Flowers of gourd and their different parts.

Flowers—solitary, pedunculate, regular, incomplete, unisexual (monoecious), epigynous.

Calyx—sepals 5, united at the base, lobes linear in male flowers and leafy in female flowers.

Corolla—petals 5, gamopetalous, campanulate or bell-shaped.

Male flowers

Androecium—stamens 3, united by anthers, filaments remaining free—syngenesious; anther lobes twisted in form of inverted 's' (sinuous).

Female flowers

Gynoecium—carpels 3, syncarpous; ovary—inferior, one-chambered; ovules—many; placentation—parietal; style short; stigma 3, each forked or bi-fid.

4. *Datura fastuosa* (B. Dhutra) (Fig. 101)

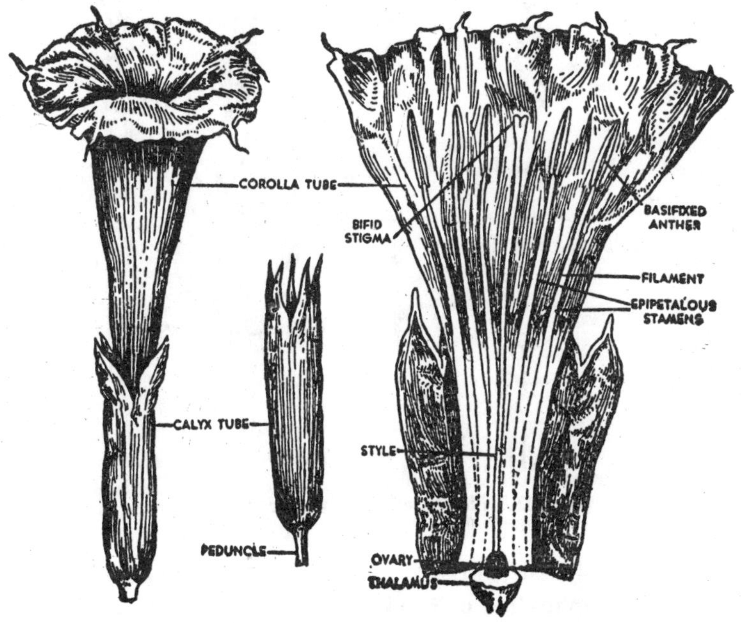

Fig. 101. Flower of *Datura* and its different parts.

Flowers—solitary, pedunculate, complete, regular, hermaphrodite, hypogynous.

Calyx—sepals 5, gamosepalous, tubular.

Corolla—petals 5, gamopetalous, infundibuliform or funnel-shaped.

Androecium—stamens 5, united with the corolla tube—epipetalous; anthers—basifixed.

Gynoecium—carpels 2, syncarpous; ovary—superior, two-chambered but four-chambered below by false partition wall; ovules—many; placentation—axile; style—1, long; stigma—bifid.

5. **Tube-rose** (*Polianthes tuberosa*); B. Rajanigandha (Fig. 102).

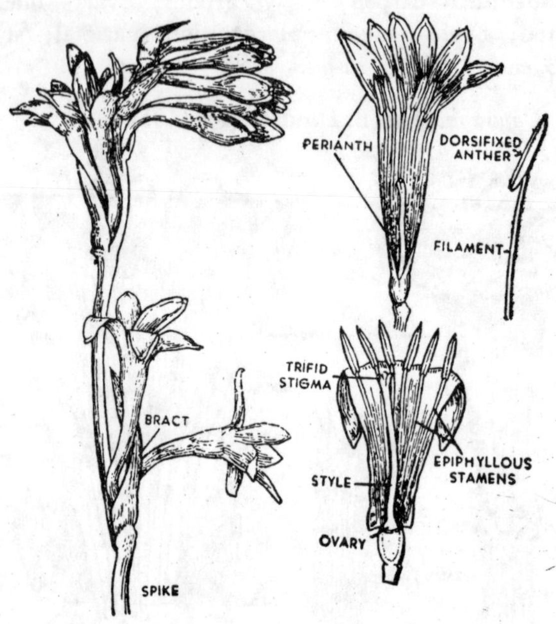

Fig. 102. Flowers of tube-rose and their parts.

Flowers—sessile, bracteate, arranged on a scape (spike), regular, hermaphrodite, epigynous.

Perianth—lobes 6, gamophyllous, in 2 whorls.

Androecium—stamens 6, attached to the perianth lobes—epiphyllous; anthers—dorsifixed.

Gynoecium—carpels 3, syncarpous; ovary—inferior, 3-chambered; ovules—many; placentation—axile; style—long; stigma—trifid.

CHAPTER VIII

POLLINATION AND FERTILISATION

It has been mentioned in the previous chapter that flowers are essentially meant for reproduction through seeds, and that the two reproductive organs, stamens and carpels, contain male and female gametes inside pollen grains and ovules respectively. For fertilisation, which involves a union of the two dissimilar gametes, the pollen grains must be carried to the stigma of the carpels. This process of transference of the pollen grains from the anthers to the stigma is known as **pollination.**

Pollination is of two types: **self-pollination** or *autogamy* and **cross-pollination** or *allogamy.* If pollen grains are transferred from the anthers to the stigma of the same flower or a flower of the same plant, the process is called self-pollination; but if pollen grains are carried to the stigma of a flower of the same species growing on a different plant, it is cross-pollination.

All unisexual flowers are cross-pollinated and even the bisexual flowers which, in fact, form the vast majority, have special contrivances for favouring cross-pollination. Cross-pollination is secured through many agencies like insects, wind, water current and animals. The insect-pollinated flowers are called **entomophilous,** (*entomon*=insect), wind-pollinated ones are **anemophilous,** (*anemos* =wind), water-pollinated flowers are **hydrophilous,** bird-pollinated flowers are **ornithophilous** (*ornitho*=bird), and animal-pollinated ones are known as **zoophilous** flowers. The majority of the plants growing around us bear insect-pollinated or wind-pollinated flowers.

The entomophilous flowers have many insect visitors like bees, butterflies, moths, etc. These flowers are usually very showy, having bright colours. Many of them, particularly the nocturnal flowers like *Nyctanthes* (B. Shieuli), jasmine lack bright colour but have very fragrant odour. Flowers of *Typhonium* (B. Ghet Kachu) have very disagreeable odour. Small flowers are aggregated together into inflorescence only to make themselves attractive to their visitors. Less attractive flowers often have corona for enhancing their beauty. Nectaries and honey glands are quite common.

Insects, during their visits to the flowers in search of honey, carry the pollen grains from flowers to flowers. The pollen grains are usually sticky and provided with spiny and warty projections, so that they can adhere conveniently to the insect-bodies. The stigmas are also sticky due to secretion of stigmatic fluid which facilitates the reception of the pollen grains. Examples—China-rose, lotus, roses, jasmine, tube-rose, *Cestrum* (B. Hasnahana).

The anemophilous flowers are small insignificant ones, neither brightly coloured nor sweetly scented. The stamens have usually versatile anthers which swing freely in the air and liberate pollen grains to be carried away easily even by the gentle breeze. Pollen grains are small, light and dusty. They are abundantly produced, as there is always the risk of a huge wastage. The stigmas are generally branched or feathery which expose maximum surface for receiving the light smooth pollen grains from air.

Hydrophilous flowers are present only in the aquatic plants like *Hydrilla*, *Vallisneria* (B. Pata shaola), *Ottelia*, etc. Mode of pollination in *Vallisneria* is very interesting. It bears unisexual flowers. The pistillate flowers have long stalks to keep them at the surface of water, whereas the staminate ones have very short stalks and so remain under water. Before pollination the male flowers get detached from the mother plant and float on the water. By water current they are carried to the proximity of the female flowers where the anthers burst to set the pollen grains free. Thus pollination takes place, after which the stalks of female flowers coil to bring the ovary under water.

Bird-pollinated or ornithophilous flowers are large, brightly coloured and showy. A large number of them usually remain aggregated together and they possess profuse quantities of honey and pollen grains. This type of flower is common in forests where birds of different kinds are present. Silk cotton is a good example.

Zoophilous flowers are those where animals like snails, birds, etc., serve as agents for pollination. Arum, *Anthocephalus* (B. Kadamba), silk cotton, *Butea* (B. Palas) are the familiar examples.

Advantages and Disadvantages of Cross-pollination

Cross-pollination is more advantageous than self-pollination, though the latter is a surer process. The progeny of cross-pollination is healthier and better suited for the struggle for existence, as they inherit paternal and maternal characters from two different plants. They give better yield and many new varieties are produced. The only disadvantage of cross-pollination is that the process is dependent on so many external agencies.

Adaptations for Cross-pollination

Nature has some contrivances for preventing self-pollination and favouring cross-pollination.

1. **Unisexuality** of flowers. In plants bearing unisexual flowers self-pollination is practically out of question and cross-pollination is inevitable.

2. **Self-sterility.** In some plants pollen grains have no effect and are, thus, sterile on the stigma of the same flower, *e.g.* Orchids.

3. **Dichogamy.** In some bisexual flowers stamens and carpels do not mature at the same time. Two conditions are thus possible, *viz.* (*i*) *protandry*, when stamens mature much earlier than the carpels, as in coriander, *Carum* (B. Joan) and (*ii*) *protogyny*, when carpels ripen before the anthers mature, as in *Michelia* (B. Champa). In any case self-pollination is avoided.

4. **Heterostyly.** Self-pollination is rendered impossible in some flowers with peculiar forms. They bear long stamens with short styles, or short stamens with long styles. Examples—*Oxalis* (B. Amrul), *Linum* (B. Linseed).

5. **Heikogamy.** Bisexual flowers with stamens and carpels maturing at the same time often prevent self-pollination by placing barriers between the two, *e.g. Aristolochia* (B. Ishermul), *Calotropis* (B. Akanda).

The only adaptation for favouring self-pollination is noticed in **cleistogamous** flowers which remain under the soil and thus ever-closed, *e.g. Commelina* (B. Kanshira).

FERTILISATION

Fertilisation is the process of the union of two dissimilar gametes, male and female. Pollen grains are carried to the stigma by pollination. It is a necessary preliminary to the more important process, fertilisation. A general account of the process of fertilisation in higher plants is given below:

The pollen grains rest on the stigma of the carpels. Uninucleate pollen grains generally become binucleate before liberation from the anthers. Of the two, the larger one is the **tube nucleus** and the smaller one is the **generative nucleus.** The stigmatic fluid stimulates germination of the pollen grain, as a result of which the exine is stretched and intine with the protoplasmic contents comes out through a weak spot, **germ pore,** to form a tube, called the **pollen tube.** The tube nucleus is located at the tip of the pollen tube and is followed by the generative nucleus. Now the pollen tube elongates and pushes its way through the style ultimately reaching the ovary. During the growth of the pollen tube, the tube nucleus gradually disintegrates and the generative nucleus divides into two nuclei, called **sperm nuclei.** They are the male *gametes* (Fig. 103). The pollen tube moves on through the micropyle to the nucellus and embryo-sac when its tip swells and bursts liberating the two male gametes (Fig. 104). One of the synergids is usually crushed during the process. One male gamete fuses with the egg, *i.e.* female gamete, and that results in actual *fertilisation.* The fertilised egg soon secretes a wall around itself and forms the *zygote,* called **oospore,** which in course of time, develops into the **embryo.**

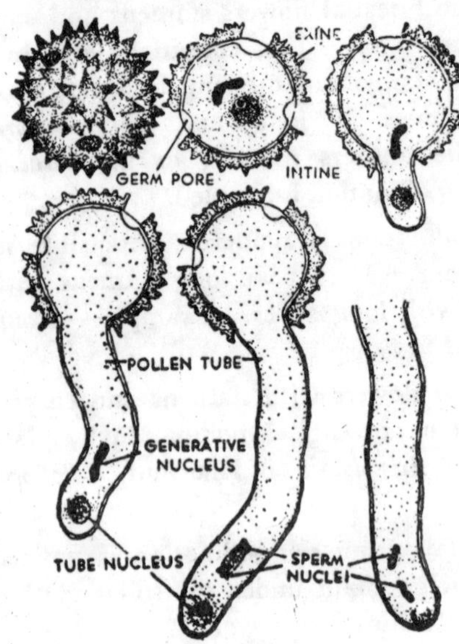

Fig. 103. Stages in the germination of the pollen grain and development of the pollen tube.

The second male gamete moves on to the definitive nucleus located at the centre of the embryo-sac and fuses with it, forming the **endosperm nucleus.** Though the second method is subsidiary to fertilisation it is nevertheless important, because the endosperm nucleus ultimately produces the nutritive tissue **endosperm** by method of free cell formation. This behaviour of the male gametes is called *double fertilisation.*

Fertilisation is a very important process which restores the original '2*n*' or *diploid* number of chromosomes. The pollen grains are formed from pollen-mother cells by *meiosis* or reduction division. Hence they have *haploid* or '*n*' chromosomes. Megaspores also have '*n*' chromosomes. Only when the two gametes meet the number becomes '2*n*' or diploid, and that is first noticed in the oospore. The endosperm has '3*n*' chromosomes, as it is formed

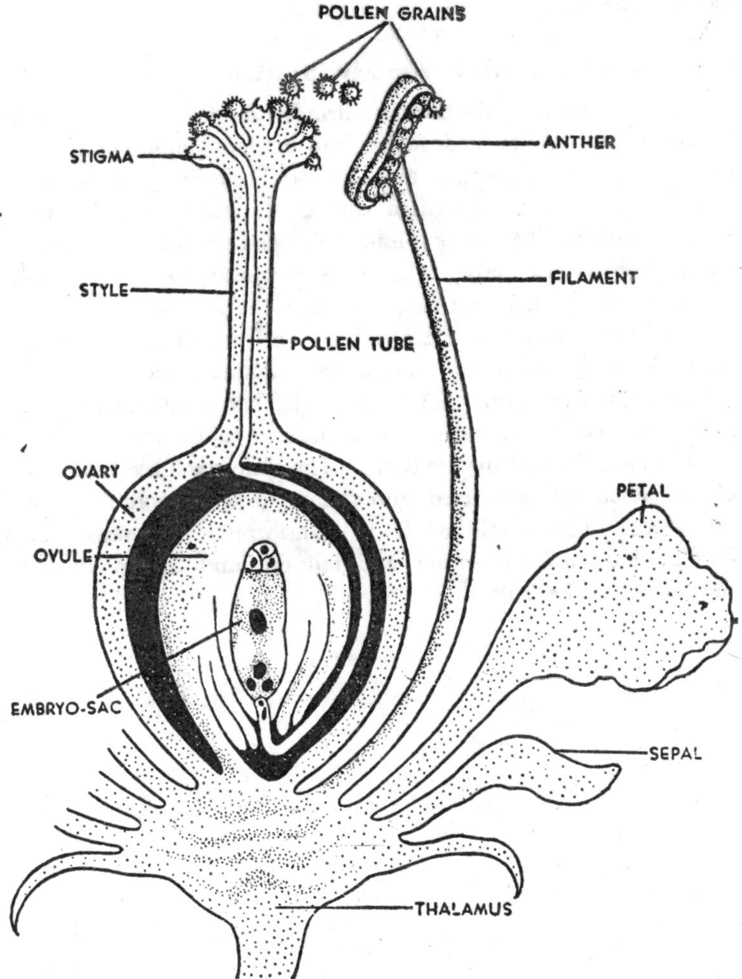

Fig. 104. Longitudinal section of a flower showing process of fertilisation (diagrammatic).

by 'triple nuclear fusion', since the endosperm nucleus is the product of fusion of 3'*n*' nuclei—two already present in the definitive nucleus and the third coming from the male gamete.

Normally pollen tube reaches the embryo-sac through the micropyle, whatever may be the position of the ovule. This method is called **porogamy.** In some plants like *Casuarina* (B. Jhau), the pollen tube moves through the base piercing the chalaza. This is **chalazogamy.**

Changes in the Ovule due to Fertilisation

The stimulus of fertilisation brings about a series of changes in the ovule. Oospore develops into embryo, endosperm nucleus into endosperm. Synergids and antipodal cells generally disappear. Nucellus is usually exhausted during the formation of embryo. In some seeds like castor, four O'clock remnants of nucellus peisist even in the seed. This is called **perisperm.** Endosperm is also used up in many seeds, so that the kernel corresponds to the embryo; they are exalbuminous or non-endospermic seeds. In some seeds, however, endosperm remains present. They are albuminous or endospermic seeds. The two integuments develop into two seed coats, testa and tegmen. Often outgrowths like caruncle are formed on the testa. The simple soluble food matters of the ovule are converted into their complex insoluble forms in the seeds. The whole ovule is transformed into seed. Calyx, corolla, etc., usually wither and fall off, and the ovary usually develops into the fruit.

CHAPTER IX

THE FRUIT

As a result of fertilisation the ovary is transformed into fruit. So a fruit may be said to be a normally *ripened* or *mature* ovary. In some flowers the stimulus of fertilisation extends to other parts as well which grow and are associated with the ovary in the formation of the fruit, *e.g. Physalis* (B. Tepari), *Dillenia* (B. Chalte). The former types are **true** fruits and the latter are **false** or **spurious** fruits. Very rarely fruits are formed without fertilisation. They are called *parthenocarpic* fruits, *e.g.* banana.

The fruit wall is known as **pericarp.** It may be dry and membranous, or fleshy and succulent. Pericarp of fleshy fruit is often differentiated into three regions—(*i*) **exocarp** or **epicarp** or the outermost skin, (*ii*) **mesocarp,** the central juicy or fibrous portion and (*iii*) **endocarp,** the innermost portion which is often stony, as in mango [Fig. 110 (B)].

Fruits are broadly classified into three groups:

A. **Simple** fruits are those which develop from monocarpellary or polycarpellary syncarpous pistil. So there is only one ovary in the flower which will develop into one fruit, *e.g.* pea, gourd, mango.

B. **Aggregate** fruits are those which develop from polycarpellary apocarpous pistil. As the carpels are free, each of them has its ovary to form a fruit. Thus a group of fruits is formed from a single flower, *e.g. Michelia* (B. Champa), *Artabotrys* (B. Kantali-champa).

C. **Multiple** or **Composite** fruit, where the inflorescence develops into a fruit. Thus one fruit is formed from many flowers, *e.g.* pine-apple, jack fruit, fig.

A. Simple Fruits

Simple fruits may be dry or fleshy, according to the nature of the pericarp. Some dry fruits burst automatically to liberate the seeds. They are called dehiscent; others which do not dehisce are *indehiscent fruits.*

Dry indehiscent fruits are usually one-seeded. Following are the types:

(*a*) **Achene** is a dry indehiscent one-seeded fruit from a superior simple pistil where the seed coat is separable from the pericarp, *e.g. Naravelia* (B. Chagalbati, Fig. 113, 4).

(*b*) **Caryopsis** is just like an achene but the seed coat and pericarp are inseparably united, *e.g.* maize (Fig. 22), grasses.

(*c*) **Cypsela** is a one-seeded fruit from an inferior bicarpellary pistil, as in sunflower (Fig. 113, 3).

(*d*) **Nut** is a one-seeded fruit from a superior bi- or polycarpellary pistil with tough and woody pericarp, *e.g.* chestnut.

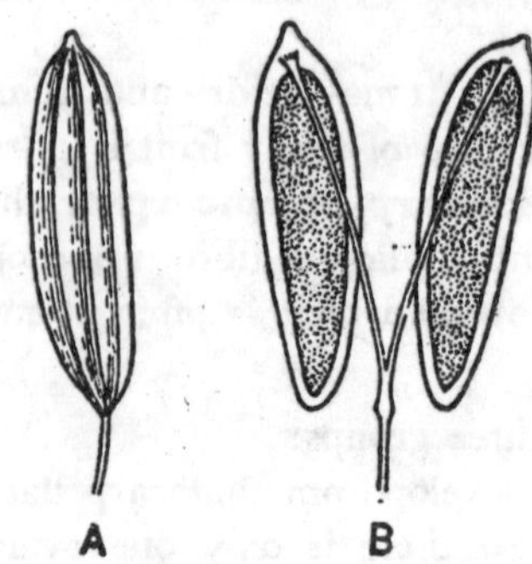

Fig. 105. Schizocarpic fruit of anise.

(*e*) **Samara** is the same type of fruit with a pair of winged outgrowths, *e.g. Hiptage* (B. Madhabilata), *Dipterocarpus* (B. Garjan, Fig. 113, 5).

(*f*) **Schizocarpic** fruits are dry indehiscent ones from inferior bicarpellary pistil. After ripening the two one-seeded parts separate and remain attached to an axis, *e.g.* coriander, anise (B. Mouri, Fig. 105).

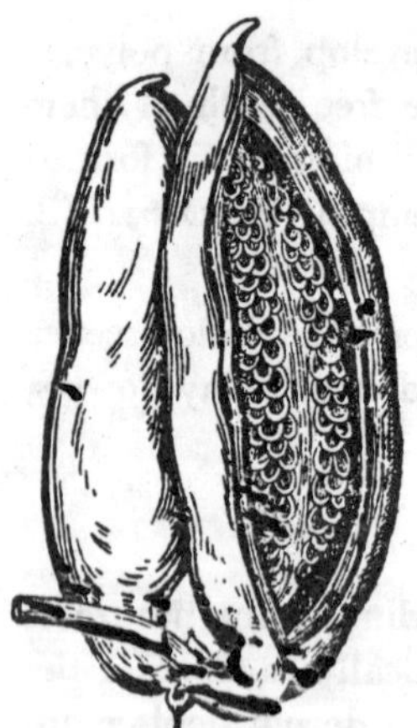
Fig. 106. Follicle of *Calotropis*.

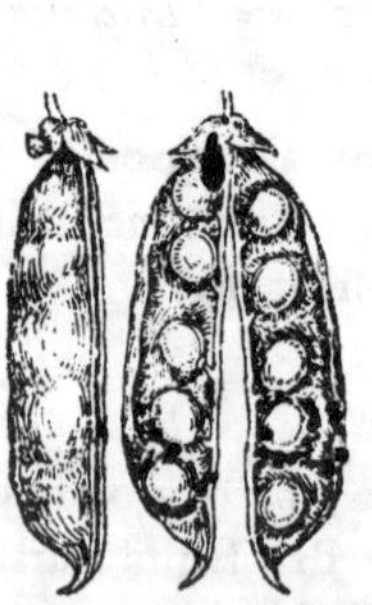
Fig. 107. Legume of pea.

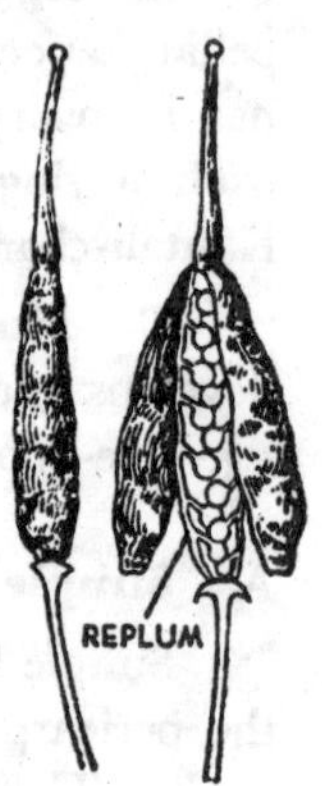

Fig. 108. Siliqua of mustard.

Dry dehiscent fruits:

(*a*) **Follicle** is a dry dehiscent fruit from a monocarpellary pistil which dehisces by one suture only, along the ventral suture, *e.g. Calotropis* (B. Akanda, Fig. 106).

(*b*) **Legume** or **Pod** is a many-seeded fruit from a monocarpellary pistil which dehisces by both the sutures, ventrally and dorsally, *e.g.* pea (Fig. 107), *Crotolaria* (B. Atasi). Legume is often constricted into one-seeded parts. It is called **lomentum,** *e.g. Acacia* (B. Babla).

(*c*) **Siliqua** is a dry dehiscent many-seeded fruit from a superior bicarpellary pistil which dehisces from the base upwards. The seeds remain attached to a false partition wall called *replum, e.g.* radish, mustard (Fig. 108).

(*d*) **Capsule** is a dry many-chambered fruit from a polycarpellary pistil which dehisces by splitting into a number of parts, *e.g. Datura*, cotton, lady's finger (Fig. 109).

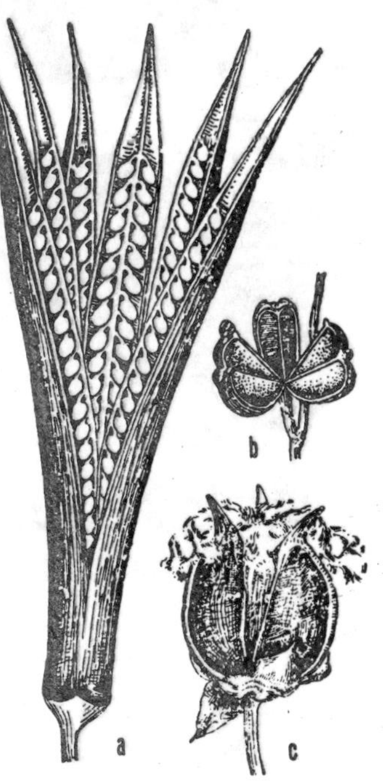

Fig. 109. Capsules of (a) Lady's finger; (b) Castor; (c) Cotton.

Fleshy fruits (Fig. 110).

Fleshy fruits have succulent or juicy pericarp. They are usually indehiscent. Seeds are set free after decomposition of the pulp.

(*a*) **Drupe** is a fleshy fruit with one or more seeds from a monocarpellary or polycarpellary syncarpous pistil where the pericarp is differentiated into outer epicarp, juicy mesocarp and stony endocarp, *e.g.* mango [Fig. 110 (A & B)], Indian plum, etc. Cocoanut is a fibrous drupe in which mesocarp is not juicy but fibrous.

(*b*) **Berry** or **Bacca** is a fleshy fruit, usually many-seeded, from a simple or compound pistil where stony endocarp is lacking and seeds remain freely in the pulp, *e.g.* tomato [Fig. 110 (C & D)], brinjal, banana.

(*c*) **Pepo** is a fleshy fruit like a berry from an inferior syncarpous pistil with parietal placentation. Seeds remain attached in series to the placentae. Pepo has usually a tough rind, *e.g.* gourd [Fig. 110 (E & F)].

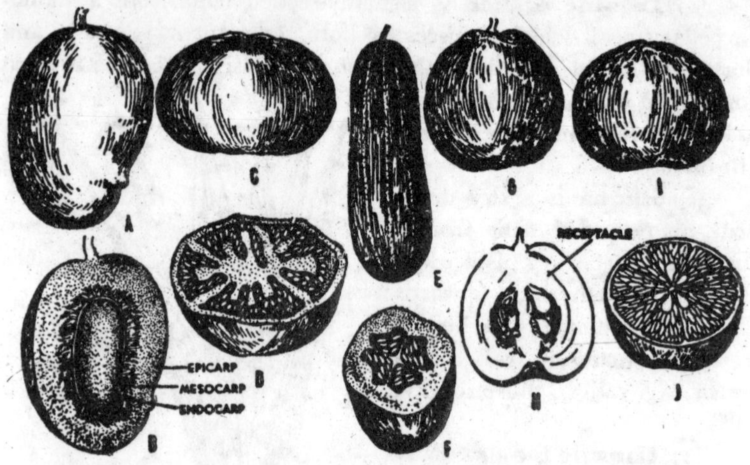

Fig. 110. Fleshy fruits.
A & B—drupe of mango; C & D—berry of tomato; E & F—pepo of cucumber; G & H—pome of apple; I & J—hesperidium of orange.

Fig. 111. Aggregate of berries of *Artabotrys.*

Fig. 112. Multiple fruit.
a—sorosis of pine-apple;
b—same cut lengthwise.

(*d*) **Pome** is a false fleshy fruit from cyncarpous pistil surrounded by thalamus, which, in fact, forms the main edible part of the fruit, *e.g.* apple [Fig. 110 (G & H)], pear.

(*e*) **Hesperidium** is a many-chambered fruit from a syncarpous pistil where epicarp and mesocarp unite to form the separable rind. Many juicy hairs develop from the pericarp, which constitute the edible part of the fruit, *e.g.* orange [Fig. 110 (I & J)], lemons.

B. Aggregate Fruits

It has been mentioned that the aggregate fruits develop from polycarpellary apocarpous pistil. The aggregate or group of fruits from a single flower is known as an *etaerio*. They may be etaerio of achenes, as in *Naravelia* (B. Chagalbati), etaerio of berries, as in *Artabotrys* (B. Kantali champa, Fig. 111).

C. Multiple Fruits

Due to the stimulus of fertilisation often the whole of inflorescence develops into a single fruit. These types of fruits are multiple.

(*a*) **Sorosis** is a multiple fruit from an inflorescence spadix with fleshy axis, *e.g.* pine-apple (Fig. 112), jack fruit.

(*b*) **Syconus** develops from the special inflorescence, hypanthodium with fleshy hollow receptacle, *e.g.* fig.

Classification of Fruits

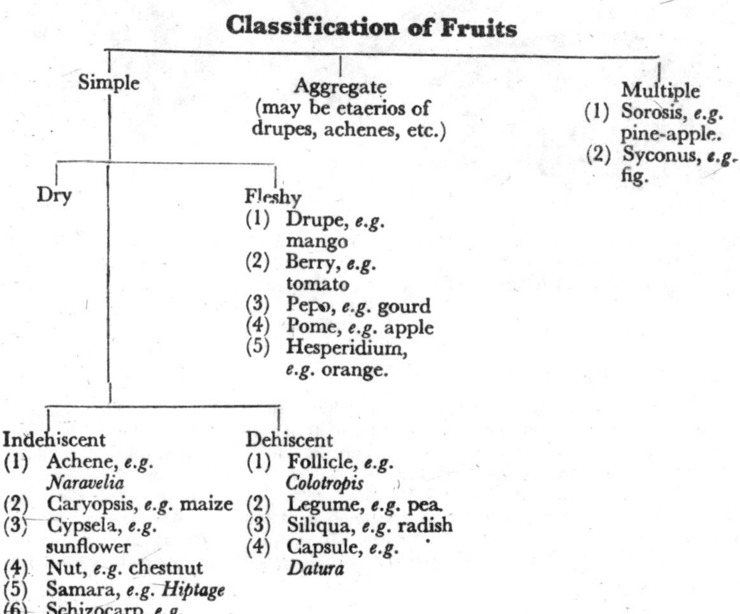

CHAPTER X

DISPERSAL OF FRUITS AND SEEDS

Fruits and seeds are usually formed in abundance, so that they may be distributed over a wide area to ensure the multiplication of plants. If all the seeds and fruits fall near about the mother plant and germinate there, the new plants cannot have healthy growth. They will naturally be so overcrowded that the nutrient materials of the soil will be quite inadequate and the other conditions will be unfavourable for growth. The result will be a keen struggle for existence; only the stronger ones will survive and the rest will perish. To ensure proper continuation of the race, plants disperse seeds and fruits far and wide. They are produced in large numbers just for covering the risk of wastage.

The dispersal takes place through several agencies like wind, animals and, water; some of the devices are really interesting.

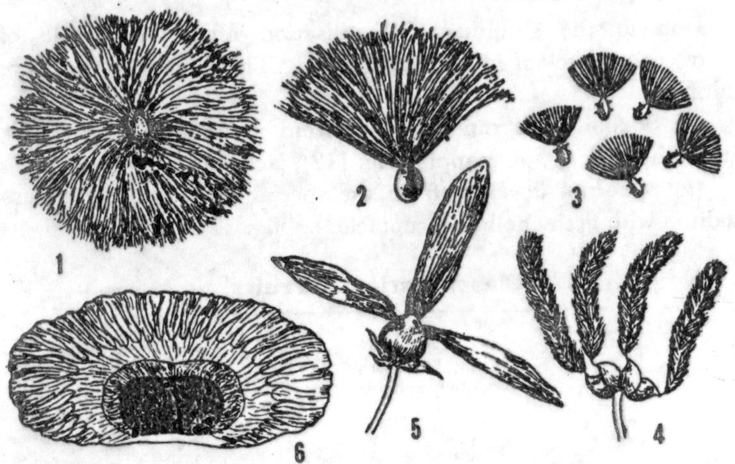

Fig. 113. Seeds and fruits with hairy and winged outgrowths. 1—seed of cotton; 2—seed of *Calotropis;* 3—fruits (cypsela) of sunflower family; 4—achenes of *Naravelia;* 5—samara of *Hiptage;* 6—winged seed of *Oroxylon.*

Dispersal by Wind (Fig. 113)

Some fruits and seeds are so small and light that they may be easily carried by wind. Many of them develop crowns of hairy outgrowths and winged expansions. often acting like parachutes, which help them in distribution. Hairy outgrowths are present on the seeds of cotton, *Calotropis* (B. Akanda), *Nerium* (B. Karabi). Many small fruits of sunflower family have modified hairy sepals. Persistent hairy styles are found in *Naravelia* (B. Chagalbati). Winged expansions on the seeds are common in *Moringa* (B. Sajina), mahogany. Fruits of *Hiptage* (B. Madhabilata). *Dipterocarpus* (B. Garjan) have winged outgrowths for the same purpose.

Dispersal by Animals

Many fruits and seeds are provided with spiny projections or sticky glands to adhere to the animal bodies, and are thus scattered. *Andropogon* (B. Chore Kanta), *Achyranthes* (B. Apang) have stiff hairs on the pericarp; curved hooks and barbs are present in *Martynia* (B. Bagnak, Fig. 114); *Xanthium* (B. Okra), *Plumbago* (B. Chita) have glands by which they stick to the animal bodies.

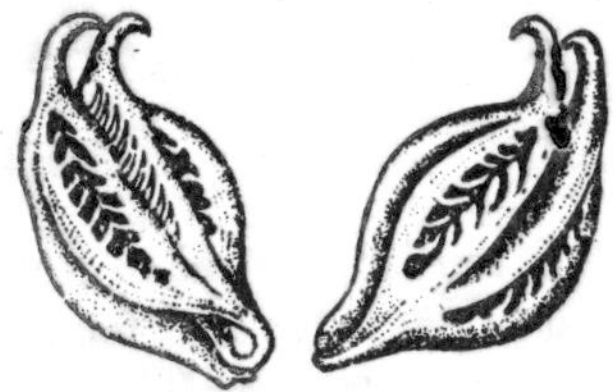

Fig. 114. Fruits of *Martynia* with curved hooks.

Fleshy fruits like tomato, figs, develop beautiful colours to attract animals like birds, squirrels and bats. The small seeds are carried by those animals from place to place. Some of the fruits are eaten up by animals and seeds remain uninjured even when they pass through their alimentary canals. The excreta of the animals rather forms a more congenial soil for the germination of the seeds. That explains the occurrence of the seedlings on the roof-tops and cornices of big buildings. Ants, mice and other rodents, store small fruits and thus become responsible for their dispersal. Even human beings are instrumental in the dispersal of many seeds and fruits of economic interest and ornamental value. Plants like water hyacinth, papaw, pine-apple which grow abundantly all over India today have come from foreign countries through human agency.

Dispersal by Explosive Mechanism

Certain fruits burst with a bit of force to scatter the seeds away from the mother plant. Familiar examples are Balsam (B. Dopati), *Oxalis* (B. Amrul), castor. Fruits of *Ruellia* (B. Chatpati), *Andrographis* (B. Kalomegh), burst suddenly when they come into contact with moisture. Legumes of *Clitoria* (B. Aparajita), dehisce by both the sutures and the two halves twist just to scatter the seeds.

Dispersal by Water

Aquatic plants and plants growing on river banks and sea-shore have fruits and seeds which are dispersed through water. They have usually fibrous tissue for floating on water surface, and protective devices so that the embryo may not be damaged. Fruits of cocoanut, *Nipa* (B. Golpata) are common examples. Lotus fruits remain embedded in the spongy thalamus. The seeds of water-lily contain air-spaces in the testa for ready dispersal by water.

ANATOMY

CHAPTER XI

THE CELL

If a thin slice is taken from any part of a plant and observed under a microscope, it is found to be composed of many chambers or compartments resembling the chambers of a honeycomb. Each of these chambers is called a **cell.** A plant cell has usually an outer firm non-living boundary called the **cell wall** and a cavity which contains granular jelly-like **protoplasm.** The organised mass of protoplasm of each cell is called **protoplast.** Protoplasm, as we know, is the essence of life. It has been very aptly described by the famous nineteenth century biologist Thomas Henry Huxley as the 'physical basis of life'. Due to the presence of the distinct non-living wall the plant cells appear to be separate from one another. But this separation is by no means complete. Protoplasm is a continuous living system. Through the small pores on the

MATTHIAS JACOB SCHLEIDEN (1804–81) THEODORE SCHWANN (1810-82)

Fig. 115.

cell wall fine fibres of protoplasm, called **plasmodesmata,** pass to the contiguous cells, establishing the organic continuity of protoplasm.

A cell is defined as the *unit of structure and function.* It is a unit of structure more or less in the same sense as a piece of brick is a unit in a building. Plant-body is built up structurally of cells. It is called a unit of function, because a cell contains a unit mass of protoplasm which is solely responsible for all the vital activities.

In 1665 an Englishman Robert Hooke observed a thin slice of bottle cork under a microscope improved by himself. He noted many chambers and suggested the term 'cell', on the basis of their resemblance with the so-called cells or chambers of a bee-hive. As bottle cork is a dead thing Hooke certainly noticed only the cell wall and not protoplasm. Italian Professor Marcello Malphigi (1628–94) and English physician Nehemiah Grew (1641–72) also carried on studies on cellular structures. A theory, known as 'cell theory', was postulated by Matthias Jacob Schleiden (1804–81), a German botanist, and Theodor Schwann (1810–82), a German zoologist (Fig. 115), in 1838–39, who claimed that all living things, plants and animals, are cellular in nature, *i.e.* they are essentially made up of cells. The announcement of cell theory was an incident of paramount importance in the history of our knowledge about cell. Early workers concentrated their attention practically on the distinct cell wall, possibly because the transparent protoplasm escaped their notice. A Frenchman Felix Dujardin appreciated the importance of cell contents to which he applied the term 'sarcode'. Hugo von Mohl, a German botanist, was, however, the first man to recognise the importance of the living contents in the middle of the nineteenth century, and he gave it the name protoplasm. Thomas Henry

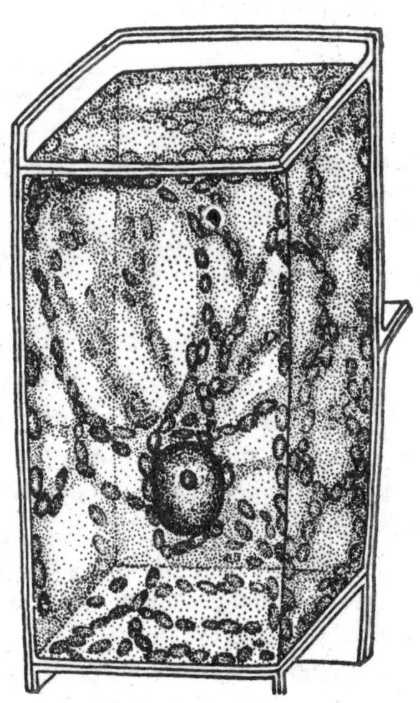

Fig. 116. A plant cell in perspective.

Huxley in 1868 described it as 'the physical basis of life'. Hanstein in 1880 proposed the term protoplast for a cell.

Though even today we use the term cell as proposed by Hooke, present-day conception of a cell is that it is essentially a mass of protoplasm (protoplast) with or without the wall. In contradistinction to animal cells, the plant cells are usually walled. Even in plant kingdom 'naked' cells, *i.e.* cells without walls are not altogether absent. Reproductive cells are usually naked. A cell is called *dead*, when it has lost its protoplasm like the cells of bottle-cork noted by Hooke.

Cells are so numerous and diverse that it is rather difficult to give a precise account of their shape and size. All shapes between round or circular on one hand and elongate needle-like on the other, are possible. The majority of them, however, are more or less isodiametric--round, elliptical, square, pentagonal, hexagonal, etc. (Figs. 116, 117). The average diameter of these cells ranges

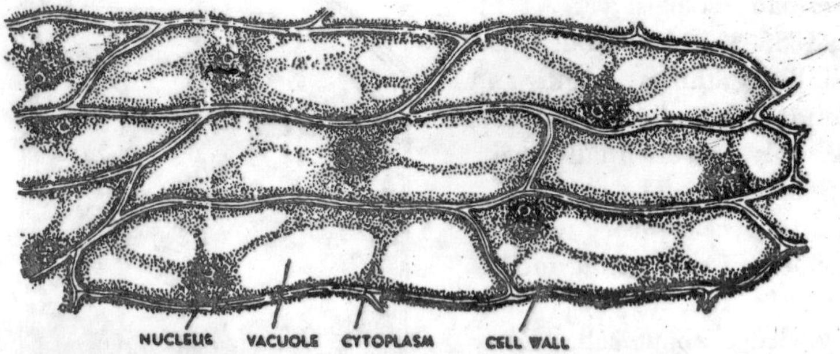

Fig. 117. A few cells from the epidermis of onion scale.

between 0.01 mm. to 0.1 mm., though there are cells much smaller and much larger in size; and average length of elongate cells ranges between 8 to 10 mm.

Some lower organisms are made up of single cells. They are called *unicellular*. But the vast majority of the plants have bodies built of innumerable cells. They are *multicellular* organisms.

A plant cell is considered separable into two distinct parts, *viz.* (*i*) the **protoplast** or the organised mass of protoplasm, and (*ii*) the surrounding non-living **cell wall.**

THE PROTOPLAST

The organised mass of protoplasm present in a cell is called protoplast. Protoplasm is really the essence of life. It has very aptly been described by Huxley as the physical basis of life. The manifestations of life, such as metabolism, irritability, growth and reproduction are all due to mysterious activities of protoplasm.

Physically protoplasm is a colourless, transparent jelly-like fluid, always saturated in a high percentage of water. Under the microscope it looks granular, due to the fact that a large number of granules of varying shape and size remains suspended in the watery background. The amount of water may be as high as 98%. With the decrease of water protoplasmic activities also become more and more feeble ; ultimately it comes to a standstill. Protoplasm contracts on application of heat, strong chemicals, electric shock, etc., and may again expand on removal of those external influences. This power of contraction and expansion, referred to as 'contractility', is an inherent property of protoplasm. It is a *colloidal* system.

Chemical composition of protoplasm is very illusive. It cannot be analysed in the living condition. During the process of analysis it is killed. The problem becomes still more complicated, because protoplasm always undergoes chemical changes. It is not a chemical compound; it can only be called a wonderful *organization* in which various substances of diverse chemical nature remain dispersed in watery medium and constantly undergo chemical changes. It is, however, mainly proteinaceous in nature. Proteins are composed of carbon, hydrogen, oxygen, nitrogen, usually sulphur and often phosphorus. Besides proteins, carbohydrates, fats and many inorganic salts enter into the composition of protoplasm.

The protoplast usually possesses a highly specialised, dense, round or elliptical body called the **nucleus**; the remaining granular portion is termed **cytoplasm.**

Cytoplasm

The extremely delicate outermost layer of cytoplasm lining the cell wall is known as **plasma membrane** (ectoplasm). It is of ultra-microscopic fineness and is *semi-permeable* in nature. This layer, which may be compared to the surface film of a drop of water, is of utmost physiological importance, because it controls the entrance and exit of fluids to and from the cells. The remaining granular portion of cytoplasm is called **endoplasm.** In young cells cytoplasm almost fills up the volume, but with the growth of the cells some distinct cavities, called **vacuoles,** are noticed in the cytoplasm. Even young cells have vacuoles, but they are very small. In adult cells a number of vacuoles may be present or they may coalesce to form a large central vacuole, pushing cytoplasm with the nucleus towards the cell wall (Fig. 118). In that case the

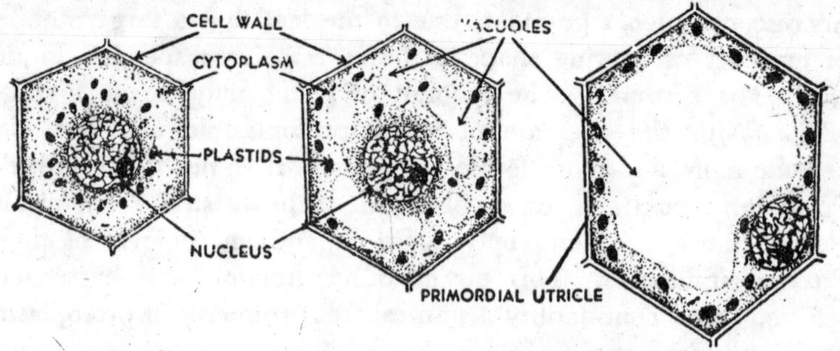

Fig. 118. Typical plants cells in different stages of development.

lining protoplasm is called *primordial utricle.* *Vacuoles* are not empty but contain watery fluid with various matters in solution. This fluid is called **cell sap.** Thus vacuoles often serve as store-house of reserve foods and different by-products of metabolic changes of protoplasm. Many vacuoles possess water-soluble pigments or colouring matters called *anthocyanin.* They give red, blue and purple colourations to many young leaves, flowers and fruits. The thin layer of cytoplasm lining a vacuole is called **tonoplasm.**

Nucleus

The **nucleus** is the most highly organised part of the protoplast. It controls or regulates all the activities and is thus of utmost importance. The nucleus is universally present in all plant cells excepting some lower plants where true well-formed nuclei may be absent, but they have corresponding nuclear materials. Nuclei are never formed *de novo* or afresh, but always originate from the pre-existing ones.

They are usually round or elliptical in shape. The shape of the cell has influence on the shape of the nucleus to some extent, so that in long cells the nuclei are also correspondingly elongated. The nucleus is quite prominent in young cells where it usually occupies the central position; but in an adult cell it is not so conspicuous, because the nucleus does not correspondingly increase with the growth of the cell. In an adult vacuolated cell it may be present in the centre or pushed towards the cell wall with primordial utricle. Sometimes nucleus remains suspended by delicate cytoplasmic threads in the midst of a large central vacuole (Fig. 119). Whatever may be the position, the nucleus always remains associated with cytoplasm. It has a special protein, *nuclein*, in its composition which is rich in phosphorus.

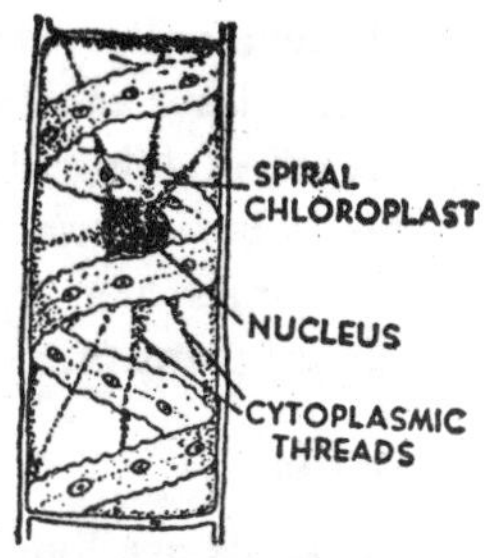

Fig. 119. A cell of *Spirogyra*, the nucleus is suspended by cytoplasmic threads.

Usually the cells of the higher plants are uninucleate, *i.e.* have only one nucleus in each cell. The presence of more than one nucleus or multinucleate condition, as it is called, is common in some lower plants. The size of the nucleus is quite variable. The average diameter, however, ranges between 10 to 15 micra. (1 micron, written as μ, is the unit of micro-measurement. It is equal to 1/25000th part of 1 mm. or roughly 1/1000th part of an inch).

The nucleus has an outer delicate membrane, known as **nuclear membrane,** which demarcates it from the surrounding cytoplasm. The main-body of the nucleus is filled up with a dense fluid called **nuclear sap** or *karyolymph.* Many crooked threads, called **chromonemata,** remain suspended in the karyolymph. Chromonemata are made of **chromatin** matters which have a strong affinity for taking up stains. Thus they are quite distinct in fixed preparations. They often anastomose to form a network referred to as **chromatin reticulum.** Chromonemata ultimately are transformed into the most important bodies, called **chromosomes,** which play a prominent role in cell division. They are mainly composed of a complex chemical deoxyribose nucleic acid (DNA) and a protein, and are chiefly concerned with heredity. It also encloses one or more highly stainable refractive bodies called **nucleoli** or plasmosomes (Fig. 120), which are largely the storage places of another nucleic acid—ribonucleic acid (RNA).

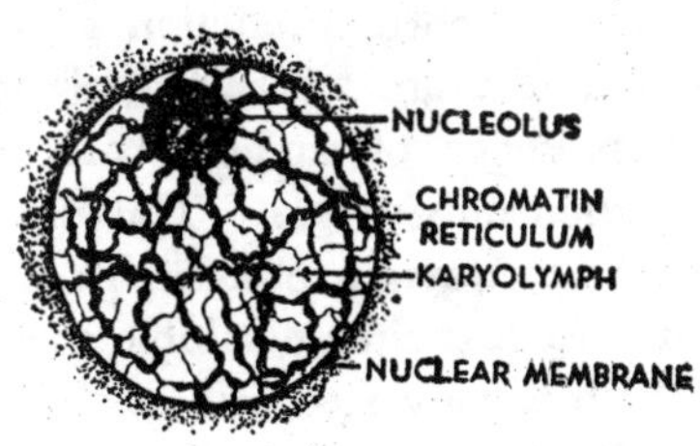

Fig. 120. Structure of a nucleus.

RNA also present in cytoplasm is concerned with protein synthesis.

Functionally, the nucleus is of the highest importance. It is the controlling centre of all vital activities and is, therefore, called the brain of the cell. It plays a prominent part in cell division which is indispensable for growth as well as reproduction. The chromosomes are regarded as the bearers of heredity characters which are transmitted from the parents to the offspring.

Plastids

Plastids are living inclusions of cytoplasm. They have distinct protoplasmic basis. In size they are much smaller than the nucleus, and a fairly good number of them occur in a cell. They are universally present in all plant groups with the exception of some lower plants like fungi and bacteria. Plastids are not formed anew but originate from pre-existing ones. The colourless ground substance of the plastid is called **stroma.** Plastids are the colour-bearers of the plants. According to the colours they bear, plastids are of three types, *viz.* (*i*) **chloroplasts** or green plastids, (*ii*) **chromoplasts** or plastids coloured other than green, and (*iii*) **leucoplasts** or colourless plastids.

Chloroplasts or the green plastids (Fig. 122) are present in the aerial parts of the plants which are exposed to sunlight. The green colour is due to the pigment, *chlorophyll.* Chloroplasts are usually small bodies, round, elliptical or disc-like in shape ranging between 4 μm and 6 μm in length. The peculiar star-shaped, spiral (Fig. 119) or reticulate plastids are found in some algae. Chloroplasts carry on the most important function, *photosynthesis* or manufacture of carbohydrate food in the presence of sunlight, out of water and carbon dioxide gas. Thus they occupy the most strategic position in the living world. At first simple soluble sugar is formed which is converted into insoluble starch grains (assimilatory starch), only to be reconverted into sugar, with nightfall.

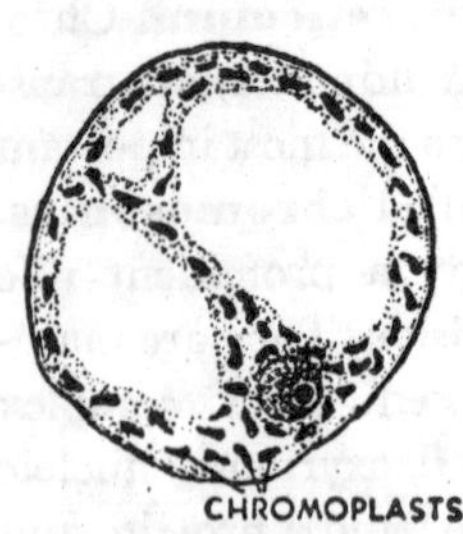

Fig. 121. A cell from flesh of tomato, showing chromoplasts.

Chromoplasts (Fig. 121) are coloured other than green. The colour, ranging between red and yellow, is imparted by the pigments, *xanthophyll* and *carotin.* Chromoplasts are irregular in shape—rod-like, angular, forked, etc. The function of chromoplasts in flowers and fruits is colour manifestation for attracting insects and animals for pollination and dispersal of fruits, but their function in some underground organs is rather obscure.

Leucoplasts or the colourless plastids are present in the organs which are not exposed to sunlight. They are small, often irregular in shape. Some of the leucoplasts, called **amyloplasts,** are concerned with the storage of food. They convert the simple soluble sugar manufactured by chloroplasts into their complex insoluble forms, starch grains (reserve starch), for future use.

Plastids are present even in the embryonic cells, when they are too small to be seen under the microscope. They are called *plastid-primordia* or **preplastids,** which multiply and mature with the growth of the cell to develop into different types of plastids. These are the precursors of all the plastids. Thus all the plastids are essentially same, as they have the same origin. This is quite evident from the fact that one type of plastid is readily converted into another. The potato tuber has leucoplasts which are often converted into chloroplasts when exposed to sunlight. Some fruits nicely show conversion of plastids. When very young they have leucoplasts, with growth they become green possessing chloroplasts which are converted into chromoplasts with the ripening of the fruit.

Chondriosomes *or* Mitochondria

These are small bodies universally present in the cytoplasm of plant cells. They are smaller than plastids and occur as rod-lets or granules. The function of chondriosomes was not clearly known for a long time. Now it has been definitely established that they are responsible for cellular oxidation and liberation of energy; so they are aptly called 'power-house' of the cell.

Movement of Protoplasm

Protoplasm is always in a state of motion. Naked masses of protoplasm show movements of two types, *viz.* (*a*) **amoeboid** movement, in which the protoplast moves slowly by alternate contraction and expansion like the small unicellular animal, amoeba; (*b*) **ciliary** movement, in which naked masses of protoplasts are provided with whip-like projections, called **cilia,** by means of which they can freely move, *e.g.* zoospores of algae, male gametes of fern, moss.

The cells of higher plants with distinct cell wall often exhibit

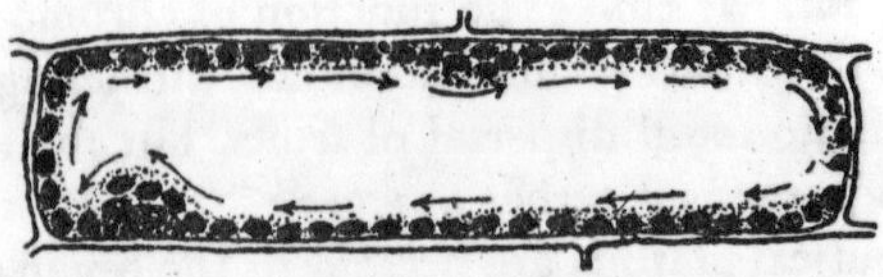

Fig. 122. Streaming movement (rotation) of protoplasm in a cell of *Vallisneria* leaf.

movement of protoplasm. Ribbon-like leaves of *Vallisneria* (B. Patashaola) have elongated cells with large central vacuoles. Protoplasm lining the cell wall shows nice streaming movement along the cell wall in a definite direction. This is called **rotation** (Fig. 122). **Circulation** is another kind of streaming movement in cells with more than one vacuole where the movement is irregular, *i.e.* not in a definite direction. Circulatory movement is found in staminal hairs of *Rhoeo* (Fig. 123).

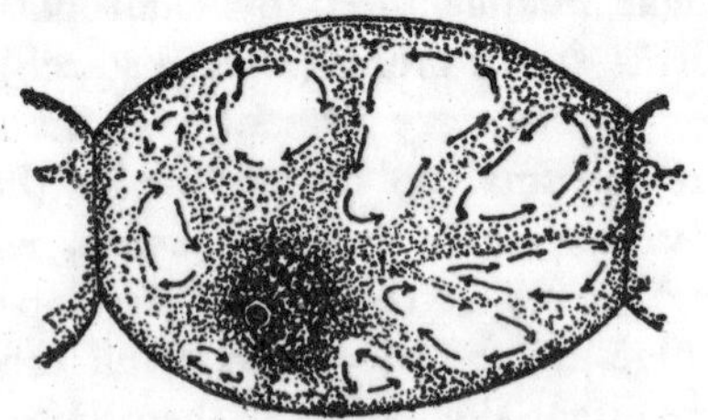

Fig. 123. Streaming movement (circulation) of protoplasm in a cell of staminal hair of *Rhoeo*.

ERGASTIC SUBSTANCES

Non-protoplasmic materials present in the cells are commonly called **ergastic substances.** These products of metabolic changes of protoplasm may be soluble or insoluble in water. They may occur in the cell sap of the vacuoles or in the protoplasm itself. On the basis of their utility to the plants, ergastic substances are classified into the following three groups: (I) **Reserve** materials, (II) **Secretory** materials, and (III) **Excretory** materials.

I. Reserve Materials

Reserve materials are those non-living inclusions which are directly concerned with the nutrition of plants, *i.e.* they serve as plant food matters. They are again of three types, *viz.* (A) **Carbohydrates,** (B) **Nitrogenous matters,** and (C) **Fats** and **Oils.**

(A) **Carbohydrates.** These are composed of carbon, hydrogen and oxygen, where the latter two occur in the same proportion as in water. On heating carbohydrates get charred, leaving the black mass, carbon. The following are the main carbohydrate food matters of plants :

1. **Sugars.** Sugars are the simplest soluble carbohydrates of plants. **Glucose** or grape sugar ($C_6H_{12}O_6$) **is** manufactured by

chloroplasts in the presence of sunlight. **Fructose** or fruit sugar with the same formula occurs in many fruits. **Sucrose** or cane sugar ($C_{12}H_{22}O_{11}$) is abundantly present in sugar-canes and beets. This is our table sugar.

2. **Inulin.** Inulin, $(C_6H_{10}O_5)_n$, is a soluble carbohydrate present in root tubers of *Dahlia* and of some plants of that family. When the root tubers are steeped in alcohol or glycerine for a few days, the soluble inulin precipitates out in the form of beautiful fan-shaped crystals (Fig. 124).

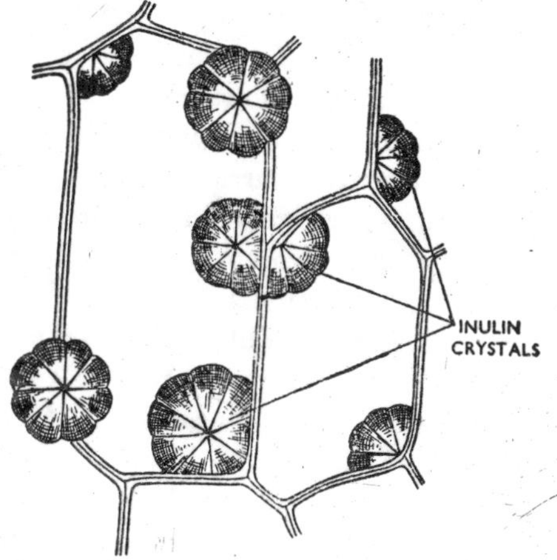

Fig. 124. Inulin crystals in the cells of root tuber of *Dahlia*.

3. **Starch Grains** (Fig. 125). Starch grains are the complex plant food and are universally found in all plant groups with the exception of fungi and bacteria. Starch grains formed by the chloroplasts are called **assimilatory** starch, which are converted into sugar soon. **Reserve** starch grains are produced by the amyloplasts out of simple sugar. When we speak about starch, we usually mean the reserve starch grains. They are abundantly present in the cotyledons, endosperm, roots, underground stems, etc. Starch grains are of varying shapes. Every grain has a shiny point, called the **hilum,** which is the centre of formation. Around the hilum starchy matters are deposited layer after layer, giving the grain a

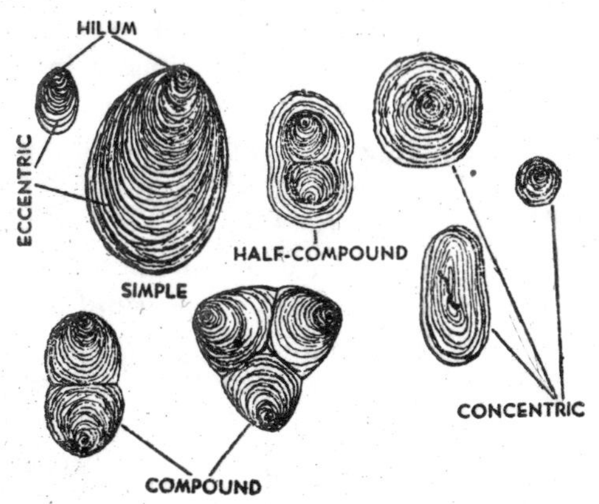

Fig. 125. Different types of starch grains.

stratified appearance. The layers are known as **lines of stratifications.** In the starch grains of potato tuber the hilum is located at one end of the grain due to unequal deposition. They are called **eccentric** grains; whereas those in the cotyledons of pea are known as **concentric,** as they have lines of stratifications around the hilum. Starch grains are **simple,** when they have one hilum with lines of stratifications. More than one simple grain may be adpressed together to form a **compound** grain. An intermediate form is noticed where the grain has two hila, their own stratified lines, but they are ultimately surrounded by common lines of stratifications. Such grains are called **semi-compound** or **half-compound.**

As the plants cannot take solid food, starch grains are converted into sugar before assimilation. Starch grains have a very characteristic test, *viz.* they turn blue when treated with iodine solution. The formula of starch grain is $(C_6H_{10}O_5)_n$, where value of '*n*' is not known.

4. **Glycogen.** Glycogen, $(C_6H_{10}O_5)_n$, is another insoluble carbohydrate like starch present usually in fungi. As glycogen occurs abundantly in animal bodies, it is also called *animal starch.*

(B) **Nitrogenous reserve materials.** These are very complex chemically. They have nitrogen, usually sulphur and often phosphorus in addition to carbon, hydrogen and oxygen. They are mainly of two types, *viz.* complex insoluble **proteids** and their simple soluble forms, **amino acids.**

Proteids may be present in many parts of the plants. They are usually insoluble in water but dissolve readily in strong acids and alkalies. A common form of proteid, called **aleurone grain,** is found in the endosperm of castor-oil seed. Each aleurone grain is more or less round in shape which encloses a large crystalline body called **crystalloid** and a small rounded one celled **globoid** (Fig. 126). Of the two, the crystalloid only is the nitrogenous matter, the globoid chemically being a double phosphate of calcium and magnesium. The occurrence of crystalloid and globoid is not always constant in aleurone grains. Proteid grains

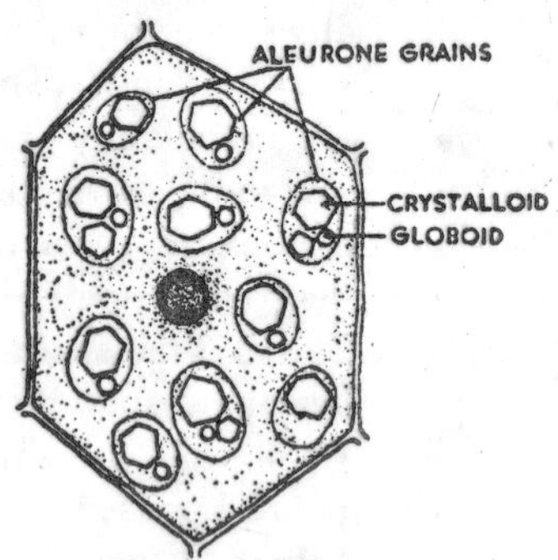

Fig. 126. Aleurone grains in a cell from endosperm of castor-oil seed.

are converted into simple soluble amino-acids for assimilation, and as such they travel to the different parts of the plants. That proteid grains are very complex chemically, is evident from the formula of a common proteid, known as *gliadin*, present in wheat. It is $C_{685}H_{1068}N_{198}O_{211}S_{6}$.

(C) **Fats and Oils.** These are also important energy-giving reserve materials of plants. Like carbohydrates they are composed of carbon, hydrogen and oxygen, but the latter two do not occur in the same proportion as in water, there being more carbon and hydrogen and less oxygen. Fats, which are liquids at ordinary temperature, are called *oils*. They are abundantly present in many seeds. The formula of a common vegetable fat, *palmitin*, is $C_{51}H_{98}O_{6}$. Fats are always lighter than water and leave greasy spots on paper. They are insoluble in water but dissolve in ether and chloroform. The fat of castor-oil seed is sparingly soluble in alcohol. When treated with caustic soda or caustic potash fats saponify, *i.e.* they form glycerine and soap, the latter is nothing but a sodium or potassium salt of a fatty acid. Fats are formed from fatty acids and glycerine. In plants they usually occur as oils, and are extracted by pressure.

II. **Secretory Materials**

These are not food matters but are indirectly useful to the plants. Colouring matters or pigments like chlorophyll, xanthophyll, carotin, etc., are certainly not plant food matters, but chlorophyll, as we know, is indispensable for manufacture of food. Similarly other pigments also have positive utility.

(1) **Colouring matters.** The green pigment chlorophyll is certainly indispensable for manufacture of food but it itself is not a food. Other pigments are also indirectly useful for the plants.

(2) **Enzymes.** These are the digestive agents secreted by protoplasm. They have the wonderful power of converting complex insoluble food matters into their simple soluble forms, but they themselves remain unaffected during the process. Thus enzymes are the organic *catalysts*. They convert starch into sugar, proteids into amino-compounds, and fats into fatty acids and glycerin. Enzymes are *specific* in action, thus the enzymes responsible for the breaking down of starch into sugar have no action on proteids or fats. Enzyme re-actions are sometimes *reversible* *i.e.* the enzymes having help in converting starch into sugar are also instrumental in the formation of starch out of sugar.

III. Excretory Materials or Waste Products

These are absolutely useless for the plants. Plants have no distinct excretory systems as animals have, but they too have devices for getting rid of waste products. There are a fairly good number of excretory products of which just a few will be discussed here.

(1) **Alkaloids.** These are complex nitrogenous matters which occur in plants. They are bitter in taste and some of them are extremely poisonous. Alkaloids are the active principles of many herbal medicines. They are usually extracted by dissolving them in alcohol. Quinine present in the bark of *Cinchona*, morphine in poppy, caffeine in coffee, nicotine in tobacco are some common alkaloids.

(2) **Tannins.** These are bitter substances present in the cell sap of many plants and also in the cell wall of dead tissues like the bark of the trees. The fruits of myrobalans (B. Hartaki, Amlaki, Bahera) contain enough of tannins. In banana, guava, mango, etc., tannins disappear with the ripening of the fruit. Tannins turn black with iron salts, and are so used for manufacture of ink. They are also used for tanning leather.

(3) **Latex.** It is the milky fluid (often watery, as in banana) found in many plants. Latex is an emulsion of various substances in a watery matrix. It is the chief commercial source of rubber. Latex of papaw contains the enzymes—papain, which is helpful in the digestion of protein food.

(4) **Essential oils.** These are the volatile oils which occur in many plants. The fragrant odour of flowers like rose, jasmine, lotus, is due to presence of esential oils in special glands. These oils, unlike fixed oils, are extracted by distillation.

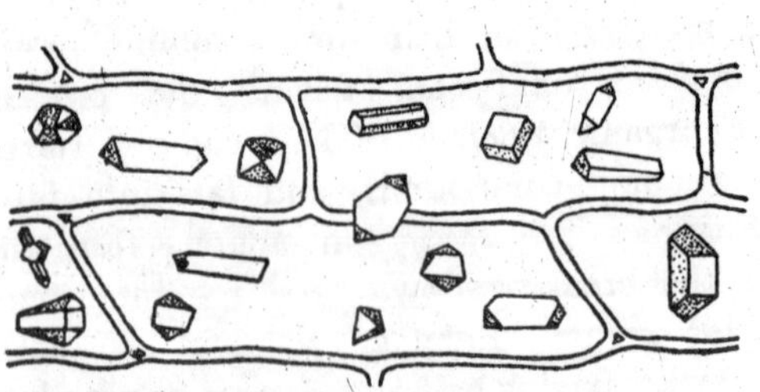

Fig. 127. Solitary crystals of calcium oxalate from dry scale of onion.

(5) **Mineral crystals.** Crystals of various forms are present in the cells. They may occur singly or a large

number of them may remain conglomerated together attaining peculiar shapes.

Calcium oxalate crystals are abundant in the plants, particularly in the underground organs. Solitary crystals of calcium oxalate may be rod-like, cubical, pristmatic, octahedral, etc. They are common in the dry scales of onion (Fig. 127). **Raphides** are the crystal-bundles which look like a bunch of needles inside a sac. **Sphaeraphides** are beautiful crystal-aggregates which have star-like appearance. Raphides and sphaeraphides are present in *Pistia* (Fig. 128), arum etc.

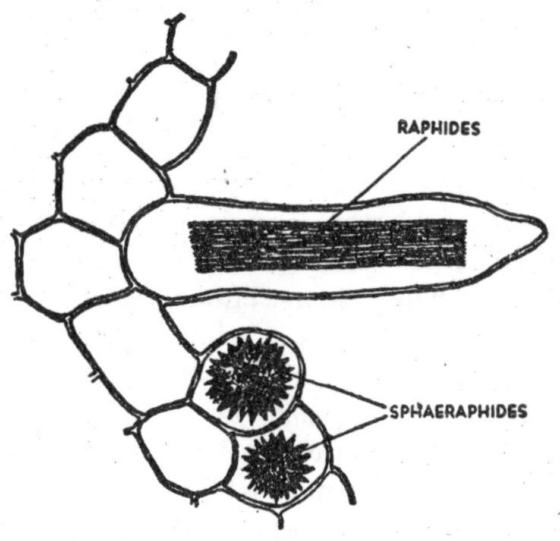

Fig. 128. Raphides and sphaeraphides in a section of the petiole of *Pistia*.

Calcium carbonate crystals are often aggregated together on the epidermis of leaves of banyan, India-rubber. Here cells of innermost layer of epidermis often enlarge to accommodate crystals of calcium carbonate deposited on a peg-like projection of the cell wall. This crystal-aggregate, called **cystolith,** looks like a bunch of grapes (Fig. 129). Cystoliths of irregular shape are present in the leaves of *Momordica* (B. Uchche).

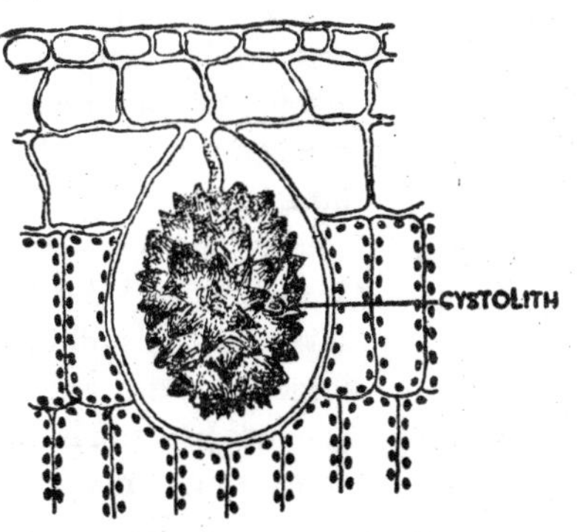

Fig. 129. Cystolith in a section of leaf of India-rubber.

(6) **Organic acids.** They are present in the cell sap of many plants, particularly the unripe fruits which taste sour. Common organic acids are tartaric acid in tamarind, oxalic acid in *Oxalis*, citric acid in citrus fruits (lemons).

THE CELL WALL

Most of the plant cells have distinct rigid walls. Cell wall is regarded as a secretory product of the protoplast. Naked protoplast always runs the risk of injuries from external influences. For self-protection a thin delicate layer is formed, which is the primary cell wall. Besides protection, the cell wall delimits the protoplasts, and gives definite shape, requisite strength and rigidity to the cell.

The cell wall is made up primarily of an insoluble carbohydrate, cellulose, with the formula $(C_6H_{10}O_5)_n$, where the value of '*n*' is not known. The thin delicate primary wall has *pectin* substances which ultimately change into calcium pectate. New layers deposited by the protoplast on the primary wall are mainly of cellulose. The primary walls of the living cells are continuous, but for some minute breaks through which fine cytoplasmic fibrils, plasmodesmata (page 106), pass from cell to cell, connecting thereby the protoplasts of the adjacent cells (Fig. 130).

During the growth of the cell wall generally the primary wall gets stretched and new cell wall materials are deposited on the inner-side of the primary wall in layers one after another. So the wall becomes considerably thick and stratified in appearance. This is called growth by **apposition.** Growth of the cell wall in surface area may be due to intercalation of new materials in between the parts of the original wall. This type is known as growth by **intussusception.** The primary cell wall, lying between the secondary thickening of the adjacent cells, is clearly noticed and known as **middle lamella.** This middle lamella, composed of calcium pectate, is the cementing material which binds the cells in a multicellular structure (Fig. 132).

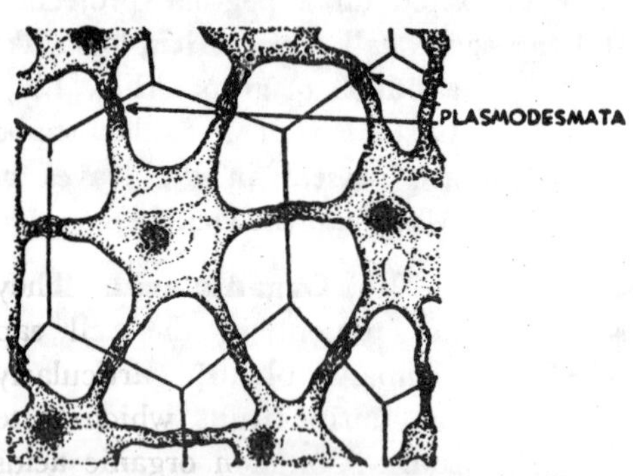

Fig. 130. Plasmodesmata in the cells from the endosperm of date seed.

Secondary walls made of cellulose or modifications of cellulose

are deposited either uniformly all over the primary wall or may be localised here and there. Due to localised thickenings the following patterns are formed, particularly in the water-conducting elements of plants which look very interesting under the microscope (Fig. 131).

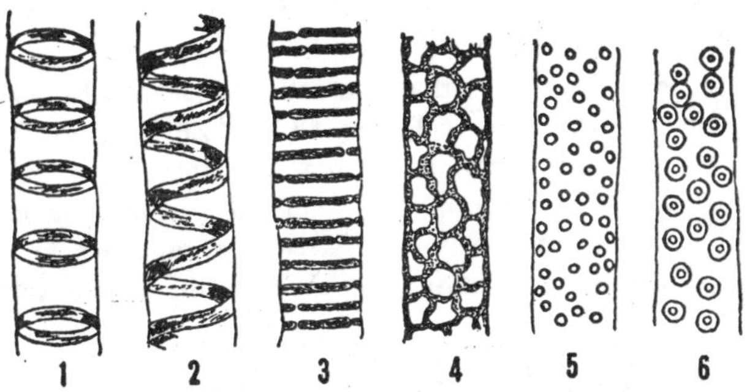

Fig. 131. Different types of thickening of the cell wall. 1—annular; 2—spiral; 3—scalariform; 4—reticulate; 5—pitted (simple); 6—pitted (bordered).

(1) **Annular** or ring-like, when secondary matters are deposited in the form of rings on the inner side of the original wall. (2) **Spiral,** when they are present in the form of spiral bands, (3) **Scalariform** or ladder-like, when secondary matters are deposited in the form of bars, one above another, so that they look like so many rungs of a ladder. (4) **Reticulate** or in form of a network, and (5) **Pitted,** when the cell wall is almost uniformly thickened leaving some small unthickened areas, called 'pits', here and there. Pits are passages for facilitating movement of fluids, and so they are usually formed in pairs on the two adjacent cells lying side by side. After the formation of pits the primary wall lying between them is called the *closing membrane*. Pits are of two types: **simple** pits and **bordered** pits. Simple pits have equal deposition and look like so many holes on the wall in surface view. Bordered pits are those

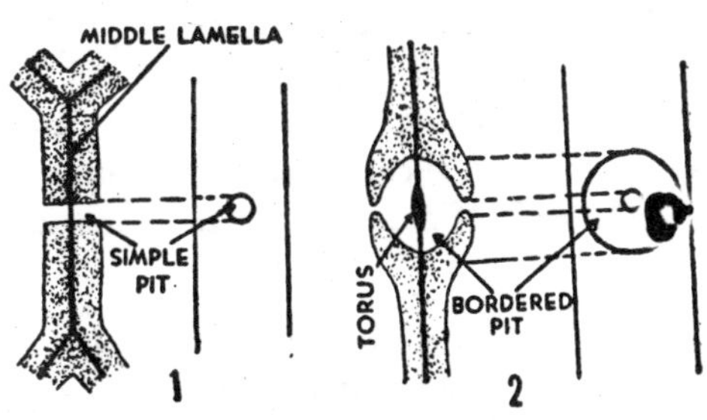

Fig. 132. Thickening of the cell wall. 1—sectional and surface view of simple pit; 2—same of bordered pit.

where the secondary depositions are unequal and bulge out from both sides, forming dome-shaped bodies. Each is represented by two circles, smaller circle to indicate the opening at the mouth and the larger (which, in fact, is the border), indicates the cavity inside. The closing membrane in a bordered pit often swells up at the middle forming what is known as **torus.** Torus can change its position and thus regulate diffusion of fluids (Fig. 132).

It has been mentioned that the cell wall is mainly composed of insoluble carbohydrate, cellulose. Cellulose is thin, delicate, to some extent elastic and permeable to water. It is of much commercial value as linen, paper, artificial silk like rayon, etc., are manufactured from cellulose. It turns blue when treated with chlor-zinc-iodine solution and dissolves in strong sulphuric acid.

According to the necessities of plants, cellulose often undergoes modifications when it is either transformed into other substances or those matters are deposited on existing cellulose walls.

Lignification. Lignin is profusely deposited on the walls of the cells, specially of the woody parts of plants. During lignification of the wall protoplasm usually vanishes and the cells become dead. Lignin gives the necessary strength and rigidity to the plant organs to stand against external injuries. Lignified walls turn bright yellow when treated with acid aniline sulphate solution.

Cutinisation and Suberisation. Cutin and suberin are fatty substances deposited on the wall to make them impermeable to water. Thus they serve as water-proofs. Outer walls of the epidermal cells are cutinised for checking evaporation of water. The layer formed by deposition of cutin is called **cuticle.** Suberin is present in the cork cells. The skin of potato tuber has suberised walls.

Mucilaginous change. Some seeds like *Plantago* (B. Ishab-gul) have mucilaginous hairs on the walls which swell up quickly by absorbing water and can retain that water. Mucilage is also present in plants like China-rose, lady's finger, Indian aloe.

Mineralisation. Mineral matters, like silicon particles, oxalates and carbonates of calcium, are frequently deposited on the walls. Leaves of grasses and bamboos have rough surface due to strong impregnation of silica. Calcium carbonate crystals in form of cystolith are present in the leaves of india rubber and banyan (Fig. 129). Calcium oxalate crystals are found in many plants (Figs. 127, 128).

Intercellular Spaces. Young cells usually remain compactly arranged, so that spaces in between them are absent. During the growth of the cells some re-adjustments occur, and as a result, small spaces arise at the corners of the cells. These are called **intercellular spaces** (Fig. 136). Though usually small, often some of them become quite big. They are known as cavities. Intercellular spaces may originate by two methods. In some cases during the growth the common walls between the adjacent cells split up and spaces appear. These are said to be formed *schizogenously* (*schizo*= split; *genesis*=beginning). Resin ducts of pine and secretory ducts present in the members of sunflower family are examples of schizogenic spaces. The second type of intercellular spaces arises due to the complete dissolution of a number of cells. These are *lysigenous* (*lysis*=loosening; *genesis*=beginning) spaces. Large air-spaces of the aquatic plants and some monocotyledonous roots are the examples.

CELL DIVISION

The formation of new cells from the pre-existing ones and their enlargement are important factors in the growth of the plants. A plant, in fact, starts its life as a single cell. In course of time the cells undergo repeated division to produce many new cells, as a result of which they grow and develop into mature plants.

An account of the processes of cell division is given below:

I. Somatic Mitosis or Karyokinesis

The somatic cells or the body cells divide by a much complicated process where the nucleus plays the most important role. The nucleus divides first into two exactly equal daughter nuclei, and that process is followed by cytoplasmic division to make the cell division complete. Nuclear division is called **mitosis** or karyokinesis and cytoplasmic division is known as **cytokinesis.** Vegetative parts of the plant grow by this method of cell division.

Mitosis (Fig. 133) is a much complicated and continuous process in which the nucleus undergoes a series of changes to divide ultimately into two daughter nuclei. For the convenience of study, the whole process has been divided into four stages or phases. This

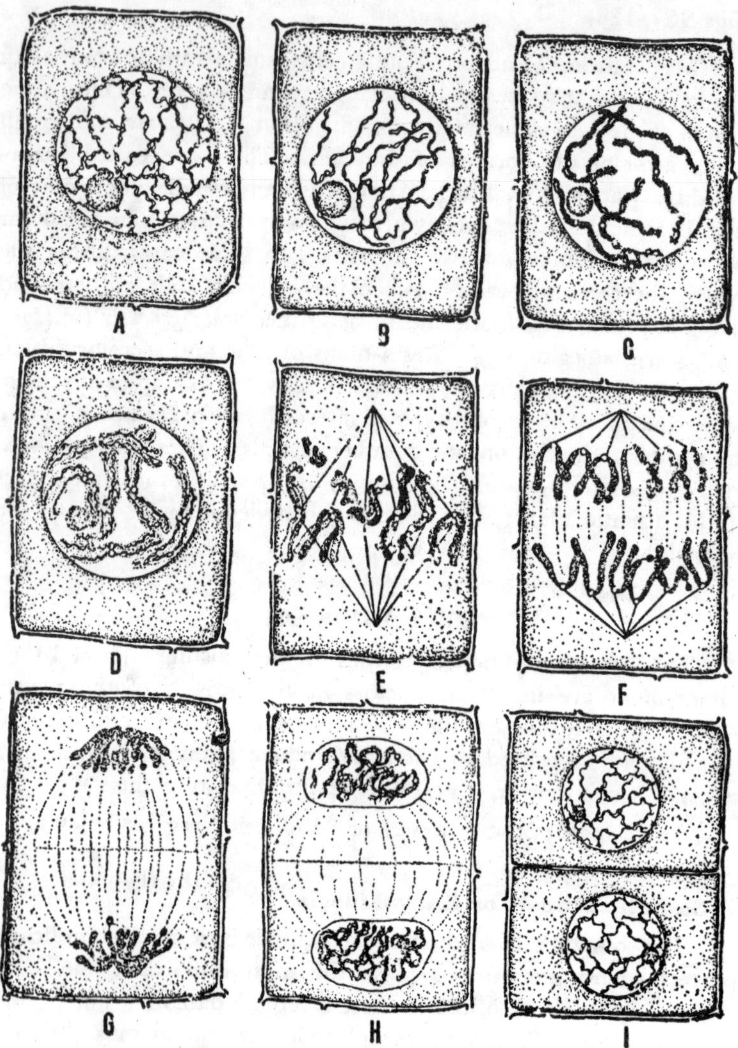

Fig. 133. Successive stages in Somatic Mitosis.
A—resting nucleus; B to D—prophase; E—metaphase; F—anaphase; G, H—telephase; I—cell division complete.

division, though purely man-made, is recognised in the whole biological world.

1. **Prophase** (the early phase). It begins with the earliest recognisable changes in the 'metabolic' or resting nucleus. The nucleus slightly enlarges and the crooked chromonemata, which form the reticulum, separate and become quite distinct. They undergo progressive shortening and thickening, and form thread-like bodies, called **chromosomes.** The number of chromosomes is always constant for a particular species of plant. Deeply stainable matter, called 'matrix', appears in which chromonemata remain embedded. Each chromosome undergoes longitudinal splitting into two equal and identical halves, called **chromatids,** which may often remain coiled or twisted round each other. In the mean time the nuclear membrane and the nucleolus gradually disappear, and some fibrils make their appearance in the nuclear field.

2. **Metaphase.** It is a short phase when the nuclear membrane and the nucleolus completely disappear, and by further extension of the fibrils a *bipolar spindle* is formed. The two ends of the spindle are the poles and the central portion is called the equator. The chromosomes arrange themselves at the equator, just preparatory to migrating towards the poles. Each of the chromosomes has a spindle-fibre attachment region, called **centromere,** to which a fibre remains attached (traction fibre). Other fibres of the spindle (supporting fibres) extend from pole to pole.

3. **Anaphase.** It is the moving phase when the chromatids are repelled from the equator and move towards the poles. As the attachment regions move first, the chromosomes appear U- or L-shaped, often with unequal arms. One set of chromosomes moves towards one pole and the other set towards the opposite pole. The cause of the movement is still uncertain and has aroused a good deal of controversy. It is likely that a tractive force is exercised by the spindle fibres.

4. **Telophase** or reconstruction phase. This is the last phase when chromosomes have reached the poles and are crowded together. Their individuality is lost. Matrix disappears and the crooked chromonemata form the reticulum. Nuclear membrane and nucleolus re-appear, and thus two daughter nuclei, with the same number of chromosomes, are reconstructed.

Cytokinesis

During telophase the fibres of the spindle expand outwards almost touching the lateral walls. Protoplasmic materials accumulate at the equatorial region in form of small droplets which ultimately coalesce to form a plate called **cell plate.** The cell plate by chemical and physical changes is transformed into the intercellular substance, middle lamella, on which cellulose particles are deposited by the protoplasts to form the primary walls.

Somatic mitosis is the most important method of cell division which takes place in all the vegetative parts of the plants. It is usually confined to some regions called *meristems*, like root-tips and stem-tips. This is *equational* division, because the distribution of chromosomes in the two daughter nuclei are exactly equal, both qualitatively and quantitatively. The daughter nuclei are exactly alike the mother nucleus, as they have the same number of chromosomes. The number of chromosomes is constant for a species. A cell may divide a million times, but the method is such that the number will remain same. The number of chromosome in pea is 14, in onion 16, in tobacco 48, in maize 20.

II. Meiosis

Meiosis or *reduction* division (Fig. 134) is a very complicated method of cell division which is restricted only to the *reproductive* cells. This method involves two divisions, of which the first division is reductional and the second one is mitotic or equational. The process is being described rather briefly. Like mitosis it also passes through four phases—prophase, metaphase, anaphase, and finally, telophase.

During prophase slender thread-like bodies, the chromosomes, are formed as usual. Two chromosomes approach each other and become very intimately associated. They may remain coiled or twisted around each other, but do not actually fuse. This selective pairing is called **synapsis,** which is a fundamental feature of meiosis. Taking arbitrarily, if there are six chromosomes, they are arranged in three pairs. The chromosomes forming the pair are called **homologous** ones, and the pairs are known as **bivalents.** Soon, while in contact, each chromosome undergoes longitudinal splitting and become double. In metaphase the split chromosomes move to the equator of the spindle. Since each member of the homologous pair has split longitudinally into two halves the chromosomes at the equator appear as **tetrads,** consisting of four chromatids. During the next stage, anaphase, the original chromosomes forming a pair, each consisting of two chromatids—**dyads,** are separated and move towards the poles. This is separation or **disjunction** of the two members which formed the homologous pair, since the two chromatids of each dyad constitute an original chromosome. In the telophase two daughter nuclei with reduced or halved number of chromosomes are reconstructed. This number is designated

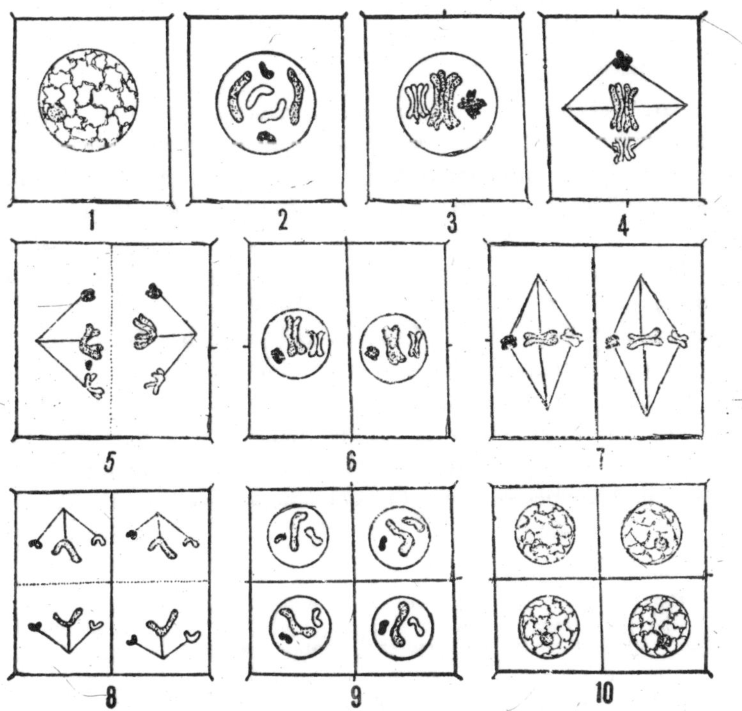

Fig. 134. Successive stages in meiosis (diagrammatic).
1—resting nucleus; 2—prophase of first division; 3—same, pairing of homologous chromosomes; 4—metaphase of first division; 5—anaphase of first division; 6—two nuclei with *haploid* chromosomes; 7 to 9—stages in second division (mitotic); 10—resting nuclei in four daughter cells (spores).

as *haploid* or '*n*' number; so the original number is *diploid* or '2*n*'. This division is immediately followed by a second division, which is mitotic or equational. Thus four nuclei are formed at the close of meiosis, each nucleus receiving one of the four chromatids of each tetrad. Now cytokinesis occurs, resulting in the formation of four cells.

Meiosis or reduction division takes place somewhere in the life cycle of all plants having sexual method of reproduction. The *gametes*, male and female, have reduced or *n* chromosomes. '2*n*' number is restored in the *zygote*, the product of fusion of the two gametes. Thus the number of chromosomes remains constant in a species.

III. Amitosis

Amitosis or direct nuclear division takes place only in lower organisms. Here the nucleus simply undergoes constriction and ultimately breaks up into two parts, which are in all probability unequal. The process is very rare and simple, as there is no spindle, no chromosomes and their manoeuvres and no wall formation.

Budding is a peculiar amitotic process found in some species of the unicellular organisms, yeasts. Here a small bud is formed from the mother cell which enlarges and cytoplasm flows from the mother cell to the bud. In the mean time the nucleus divides by constriction into two parts, one of which remains in the mother cell and the other migrates to the bud. The bud is gradually abstricted. It may be at once detached from the mother cell or may remain connected, often forming chains of cells (Fig. 194).

Compound Microscope

Compound microscope is a very useful optical instrument. It gives a high magnification necessary for working out the details of minute organisms and also for studying the internal structures of plants and animals.

The instrument rests on a horse-shoe-shaped *base* (foot) with a short *pillar*. There may occur an *inclination joint* at the upper end of the pillar where the main body of the apparatus remains screwed. It bears a *stage* (platform) with a central hole and a pair of clips. Below the stage lies a sub-stage consisting of a *condenser* and an *iris-diaphragm*. Light is focussed from a *mirror* attached to the pillar. Condenser is there to concentrate light and the diaphragm for regulating it. The body of the microscope has an *arm*, often bent, with an attached *body tube* having another tube inside known as *draw tube* which is graduated in millimetres. At the upper end of the body tube occurs the *eyepiece*. Two *objectives*—low-power and high-power—remain fitted on a revolving *nosepiece* at the lower end of the body tube. Two *adjustment screws*—coarse and fine—work on rack and pinion system for bringing the object under proper focus.

The material to be seen under the microscope should be sufficiently thin so that light may pass through it. It should be mounted in a liquid medium on a clean glass slide and covered with a cover-slip. Now it has to be placed on the stage taking care that the material lies at the centre of the hole and clipped. Light should be focussed properly and the body tube with lower-power objective is to be brought down with the help of coarse adjustment screw to such a level that the distance between the slide and the objective is roughly 1/4 inch. Having put the eye on the eyepiece the objective is to be raised slowly till a clear vision is achieved. For still higher magnification low-power objective is to be removed and the high-power objective brought in and only the fine adjustment screw is to be used for adjustment.

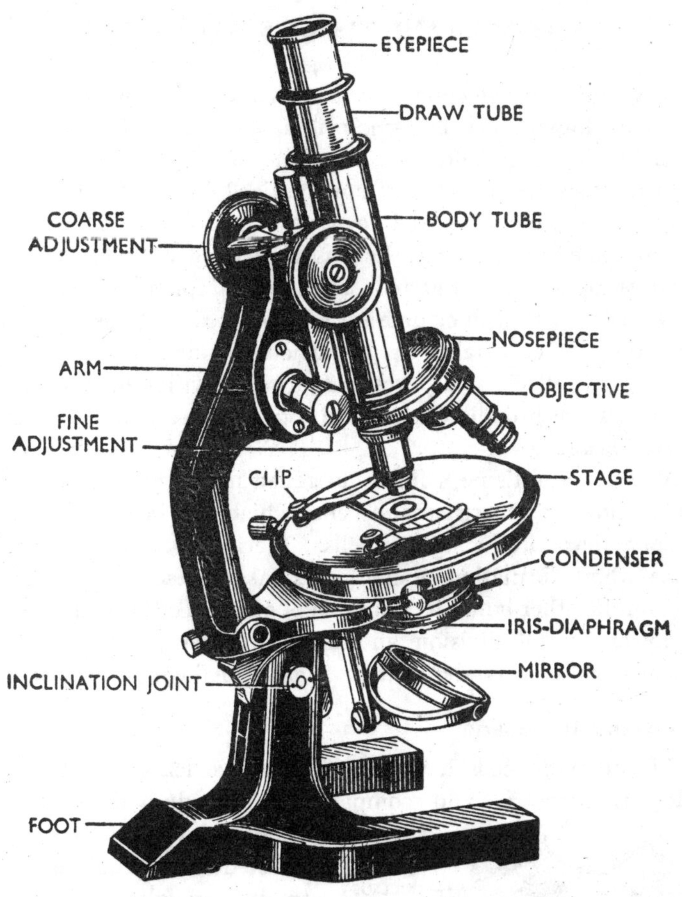

A compound microscope.

CHAPTER XII

TISSUES AND TISSUE SYSTEMS

A **tissue** is a collection of cells which are more or less similar in size and shape, have the same origin, are governed by the same laws of development, and have the same function.

Plants may be unicellular and multicellular. The unicellular plant may be well compared to a very unfortunate man left all alone in some island, where he has to depend on himself for all the necessities of life. But in a multicellular plant, as in human society, there is a distinct division of labour, in response to which every group of cells takes up a particular function. Tissues are the groups of cells formed in response to this division of labour.

Tissues are broadly classified into two groups : **meristematic** and **permanent.**

Meristematic tissue, commonly called meristem, is a collection of cells known as meristematic cells, which are in a state of cell division and are producing new cells. Meristems are thus organised regions where formation of new cells takes place. A permanent tissue, on the other hand, is composed of mature cells which have lost the power of cell division and have attained permanent form. Permanent cells always derive their origin from the meristematic cells.

Meristematic Tissue

Meristematic cells have some distinctive features. They are usually isodiametric and compactly arranged without leaving intercellular spaces. Cells have dense cytoplasm, very small vacuoles and quite prominent nuclei. Plastids are in the proplastid stage and ergastic matters are usually absent. Walls are thin, homogeneous and made of cellulose (Fig. 135). These are the characters of truly meristematic cells; departures are of course noticed in some cases, as in cambium of the vascular bundles.

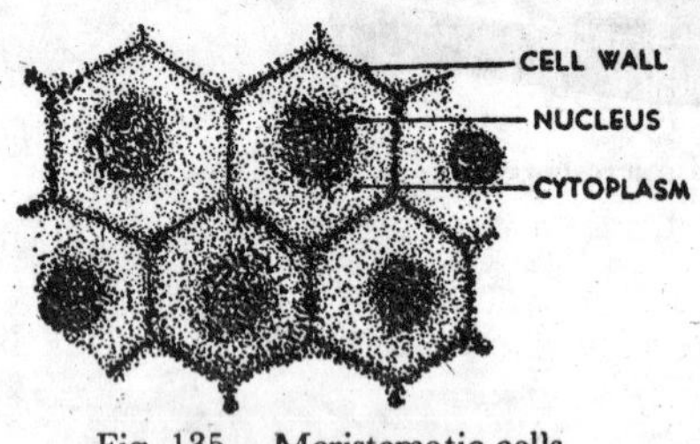

Fig. 135. Meristematic cells.

Meristems may be classified on the basis of origin and position. According to origin, they may be **promeristem, primary** meristem and **secondary** meristem. Those which occur at the very embryonic stage, what, in fact, constitutes the foundations of the organs, are called **promeristems. Primary** meristems build up the fundamental parts of the plants. They usually derive their origin from promeristem and retain the meristematic nature. Some permanent cells regain the power of cell division; they are called **secondary** meristems, *e.g.* cork cambium and interfascicular cambium.

According to position, meristems may be **apical, intercalary** and **lateral. Apical** meristems are located at the apices of the axis, *i.e.* stem-tip and root-tip, referred to as *growing point.* Apical meristems are responsible for growth in length of the axis. In plants like grasses, *Equisetum*, some meristematic cells are left behind during the growth of the apical meristem, and are intercalated between regions of permanent cells. They are called **intercalary** meristems, which, of course, gradually lose meristematic nature and become permanent. **Lateral** meristems, as the name implies, are strips of meristematic cells placed sideways or laterally on the axis, *i.e.* parallel with the sides of the organ in which they occur, *e.g.* cambium and cork-cambium. Growth in thickness, as in dicotyledonous stems, is entirely due to the activities of lateral meristems.

The primary meristem at the apical region is usually differentiated into three zones, *viz.* **dermatogen, periblem** and **plerome.** The dermatogen is the outermost one-layered zone passing right over the apex. It forms the epidermis of the adult region. The periblem is one-layered at the tip, but many-layered below. It gives rise to the cortex. The plerome is the innermost zone which produces the central cylinder or *stele.* Some cells of the plerome show tendency of elongation. They are called *procambial strands*, which form the vascular bundles.

Permanent Tissues

Permanent tissues are **simple** or **complex,** according to the nature of component cells. A **simple** permanent tissue is composed of only one type of cells and is thus homogeneous; whereas the **complex** permanent tissue is heterogeneous, consisting of more than one type of cell elements.

Simple permanent tissues are of following types :

Parenchyma. Parenchymatous tissue (Fig. 136) is usually made of isodiametric cells with intercellular spaces. The cells are living and contain vacuolated protoplast. The cell wall is thin, homogeneous and made up of cellulose. Only the parenchyma cells of epidermis or outermost skin of aerial organs have cuticularised outer walls. Parenchyma is by far the most abundant type of tissue distributed in all the organs of the plants, underground and aerial. Their main function is the manufacture and storage of food matters, which is referred to as *vital* function. Besides, the parenchyma cells may serve as storehouse of water and various waste products. They also help in the conduction of water and food matters in solution. The parenchyma cells of the epidermis are protective in function. In the aquatic plants the parenchyma cells often assume stellate (star-like) or armed appearance due to presence of abundant air chambers. These cells, also called *aerenchyma*, give buoyancy to the plants.

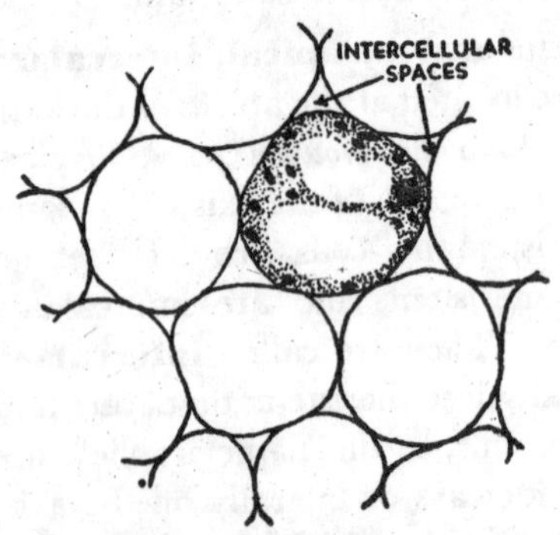

Fig. 136. Parenchyma.

Collenchyma. Collenchymatous tissue (Fig. 137) is composed of somewhat elongate cells with peculiar thickenings, confined to the corners of the cells. They often remain interlocked. The cells have protoplast. Chloroplasts are normally absent, but may be present in some cells. They usually do not manufacture food.

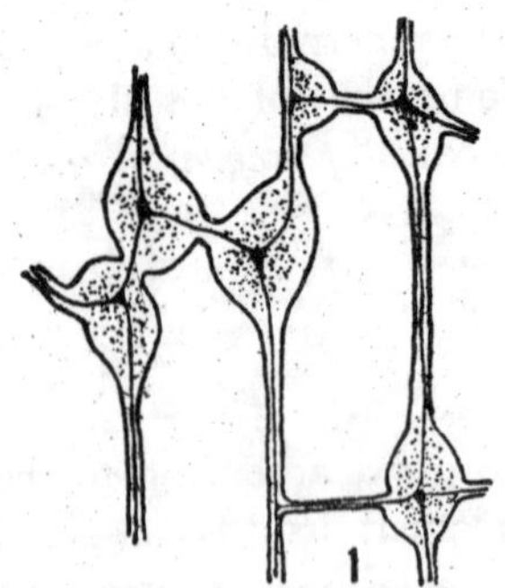

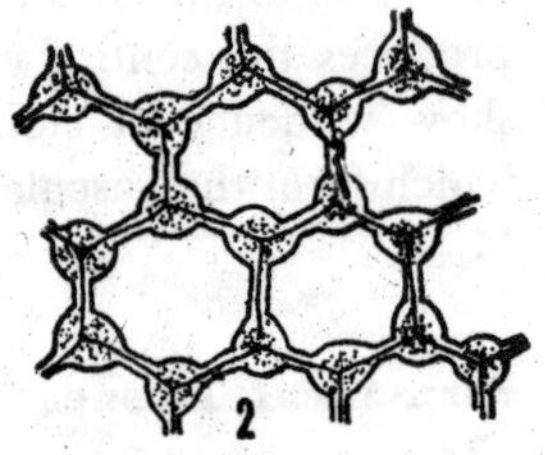

Fig. 137. Collenchyma.
1—longitudinal section.
2—cross-section.

Collenchyma cells look pentagonal or hexagonal in cross-section. The soft and plastic walls are made up of cellulose and pectin. Extra-deposition of cellulose on walls abutting on intercellular spaces gives them characteristic thickened corners. Collenchyma occurs chiefly in the superficial regions of stems (beneath the epidermis) and midribs of leaves. They are abundant in the rapidly elongating organs like leaf-stalk and floral stalk. Though soft, it is a *mechanical* or strength-giving tissue, particularly meant for the support of growing organs. It is called a temporary supporting tissue.

Sclerenchyma. Sclerenchymatous tissue (Fig. 138) consists of long needle-like cells with pointed ends. As the cells are much longer than their breadth they are also called *fibres*. They gradually lose protoplasm and become dead. In cross-section they look typically angular. The walls are hard and lignified. Simple pits are present. Sometimes lignification becomes so much pronounced that the central cavity (lumen) is almost blocked. They usually occur in groups or patches with peculiarly interlocked or dovetailed ends. They are present in both aerial and underground organs. Sclerenchyma is the most important *mechanical* or strength-giving tissue of the plant.

Sclereids or stone cells (Fig. 138) are modified sclerenchyma. They are not elongated but isodiametric or irregular in shape. The gritty texture of fruits like pears and guava, is due to the presence of sclereids. There are some peculiar types of sclereids occurring in many leaves. Ones projecting into the intercellular cavities in the stem and leaf of water-lily are hair-like in appearance; and so are also called 'internal hairs'.

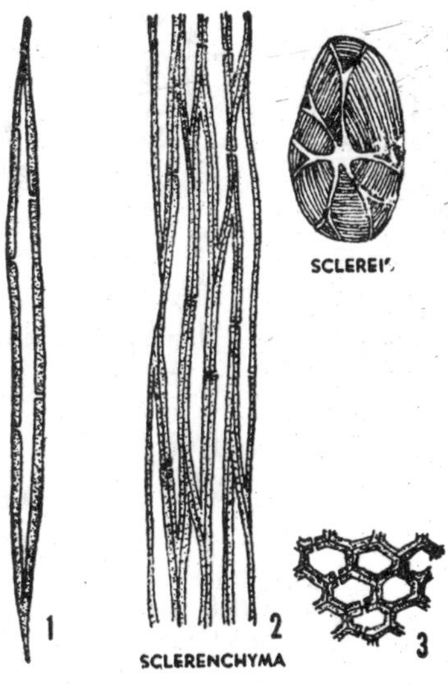

Fig. 138. Sclerenchyma.

1—a single cell; 2—a few cells in longitudinal view; 3—cross-sectional view; and a single sclereid.

Complex Tissue

A complex tissue is heterogeneous, as it consists of more than one type of cell elements. The two main complex tissues are **xylem** and **phloem,** present in the vascular bundles or conducting strands of higher plants. Vascular system is continuous in the plant body, and is instrumental in the conduction of water and soluble food materials.

Xylem. Xylem is the hard woody part of the vascular bundle which is concerned with the conduction of water. It also gives sufficient mechanical strength to the plant members. The following elements form this complex tissue: (*i*) tracheids, (*ii*) tracheae or vessels, (*iii*) some parenchyma cells called xylem parenchyma or wood parenchyma, (*iv*) sclerenchymatous fibres called xylem fibres.

Tracheids (Fig. 139) are elongated cells with fairly large cavities and chisel-like tapering ends. They extend parallel to the long axis of the organs. Tracheids are typically angular in cross-sections. Protoplasts disappear; so tracheids are dead cells. The cell wall is hard, lignified with different types of localised thickenings, of which bordered pits are more common. They are the primitive conducting elements, forming very important elements of xylem of the gymnosperms, but may also occur in the angiosperms. Tracheids are primarily meant for the conduction of water. Secondarily, they can give mechanical support with the aid of hard lignified walls.

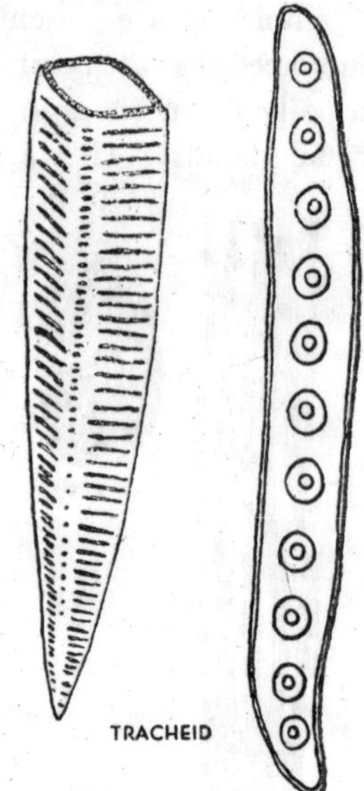

Fig. 139. Tracheids, scalariform (left) and bordered pitted (right).

Tracheae or **vessels** (Fig. 131) are long tube-like bodies occurring parallel to the long axis of the organs. A trachea is formed from a row of cylindrical cells, attached end to end, by the total dissolution of the partition walls (Fig. 140). So they are long tube-like bodies with fairly large cavities. Tracheae are more or less circular in cross-section. The protoplast vanishes with the formation of the tube. The cell wall is hard and lignified with different types of localised thickenings from annular to pitted. Tracheae are the most important elements of xylem in the angiosperms, and are thus the advanced conducting elements. Their main function is conduction of water, as they are quite well-suited for the purpose. They can give mechanical strength as well.

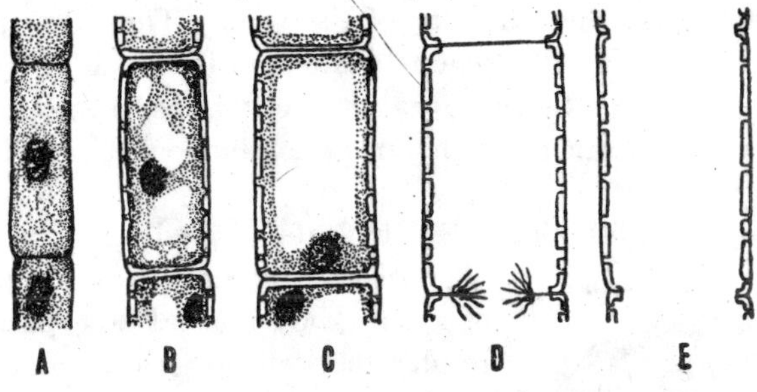

Fig. 140. Formation of a vessel—stages.

It should be noted that a tracheid is a *single* cell, but a trachea or vessel is a tube-like body formed from a *row* of cells.

The first formed xylem vessels have smaller cavities and usually annular and spiral thickenings. They are known as **protoxylem** vessels; but lately formed ones, called **metaxylem** vessels, have wider cavities and reticulate, scalariform or pitted thickenings (Fig. 147).

Phloem. The complex tissue phloem is composed of some thin-walled elements. It is responsible for the conduction of elaborated food materials. The elements of phloem are mainly—(*i*) sieve tubes, (*ii*) companion cells, and (*iii*) phloem parenchyma. Often some fibres, called phloem fibres, may remain associated with them.

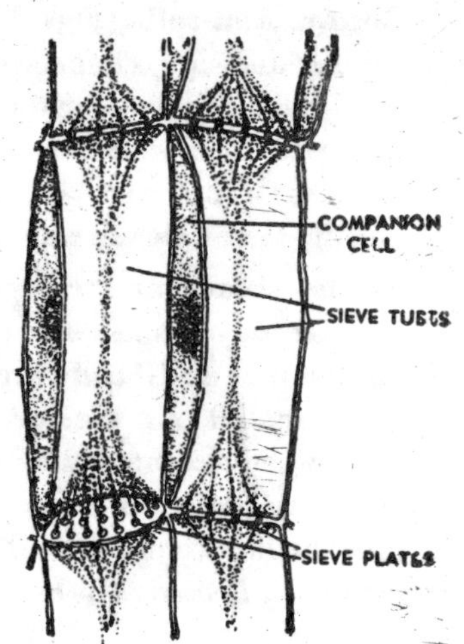

Fig. 141. Sieve tubes and companion cells in longitudinal view.

Sieve tube (Figs. 141 & 142) is a long tube-like body formed by a series of cells where the partition walls are partially absorbed in a sieve-like manner. The perforated partition walls are called the **sieve plates** through which cytoplasm of adjoining cells are in communication. Sieve plates may be horizontal or obliquely inclined. Unlike vessel, a sieve tube is living, having lining cytoplasm with a large central vacuole. The nucleus disintegrates with the maturity of the tube. The cell sap of the vacuoles is rich in nitrogenous matters. The wall is thin and made of cellulose. Sieve tubes are very important elements of phloem, and perform the function of conduction of prepared food materials.

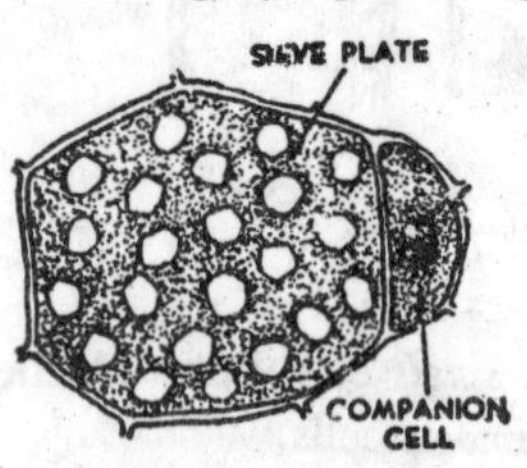

Fig. 142. Tranverse section of sieve tube, showing sieve plate and companion cells.

A highly glistening carbohydrate, called callose, is often deposited on the sieve plate in the form of a pad known as **callous pad.** As a result, the cell to cell communication is temporarily or permanently cut down. The callous pad formed with the approach of winter usually dissolves in spring to re-establish the communication. This is the seasonal callus. Permanent or definitive callus may also be formed in the functionless sieve tubes.

Companion cells (Figs. 141 & 142) are closely associated with sieve tubes and are connected with them by pores. They are elongated cells with dense cytoplasm and prominent nuclei.

Elongate cells with sieve-like perforate walls are present in gymnosperms. They are called **sieve cells.** These are primitive, whereas the sieve-tubes are advanced elements.

Besides simple and complex tissues, there are special ones, called **secretory** tissues, concerned with the secretory or excretory materials of plants. **Glands** are the common secretory tissues which store up essential oils, etc. Faint spots on the leaf of lemon and petals of sweetly scented flowers are the glands. Similarly **ducts** with resin are present in some plants. **Laticiferous** ducts are long tube-like bodies containing the milky fluid latex. They occur abundantly in *Plumeria* (B. Katgolap), *Calotropis* (B. Akanda).

TISSUE SYSTEMS

The tissues, which are adapted for the performance of the same function, together form a tissue system. There are three tissue systems in plants, *viz.* (*i*) epidermal tissue system, (*ii*) ground tissue system, and (*iii*) vascular tissue system.

I. Epidermal Tissue System

This tissue system consists of the outermost skin or **epidermis** of the plant organs. Epidermis is generally uniseriate or single-layered and protective in function. Many-layered epidermis, called *multiple* epidermis, is present in the leaves of banyan and roots of orchids. Epidermal cells are of modified parenchymatous type. In surface view, they look irregular in shape. They are compactly arranged without leaving intercellular spaces. The cells are living with peripheral cytoplasm, round central vacuole, nucleus and minute leucoplasts. Chloroplasts are present only in the epidermal cells of aquatic and shade plants and guard cells of the stomata. In aerial organs the outer wall is cuticularised for checking the loss of water. Besides that, deposition of waxy or resinous matters may take place on the surface. Many outgrowths, mostly multicellular and rarely unicellular, may develop on the epidermis of stems and leaves. The epidermis of root has no cutinised outer wall and bears unicellular hairs at the root-hair zone. Root-hairs are, of course, prolongations of the cells themselves. The epidermis of root is also called **epiblema** or **piliferous layer** (Fig. 153).

Epidermis is a continuous layer but for some minute openings, called **stomata** (singular, stoma) present in the aerial parts of the plants. Every stoma has a small opening called **stomatal aperture** and two crescent-shaped cells known as **guard cells** (Fig. 143). Guard cells have dense cytoplasm, nucleus, chloroplastids and even starch grains. Semilunar protoplasmic guard cells can be very easily distinguished from vacuolated irregular epidermal cells. Stomata may be conveniently observed in surface view by peeling off the epidermis of the leaf. Stoma is formed from a mother cell which splits lengthwise into two parts leaving a small

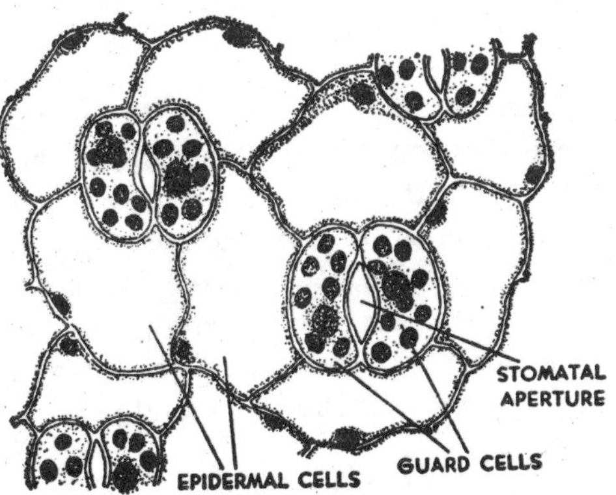

Fig. 143. Stomata in surface view, showing details and irregular epidermal cells.

slit between them. The slit develops into stomatal aperture and the two cells into two guard cells. The aperture opens into a cavity, called **sub-stomatal** chamber, which is in communication with the intercellular space system of the internal cells (Fig. 144). The guard cells regulate the opening and closing of the stomata which usually remain open in day time and close up with the nightfall. In day time the guard cells manufacture sugar. The concentration of sap increases and water is absorbed from the neighbouring epidermal cells by osmosis. The guard cells thus become *turgid* or full of water, then they bulge out in outward direction making the aperture wide open. At night there is no manufacture of sugar, concentration of sap decreases. The guard cells lose water and become *flaccid*. Then they lie side by side, so that the aperture is closed.

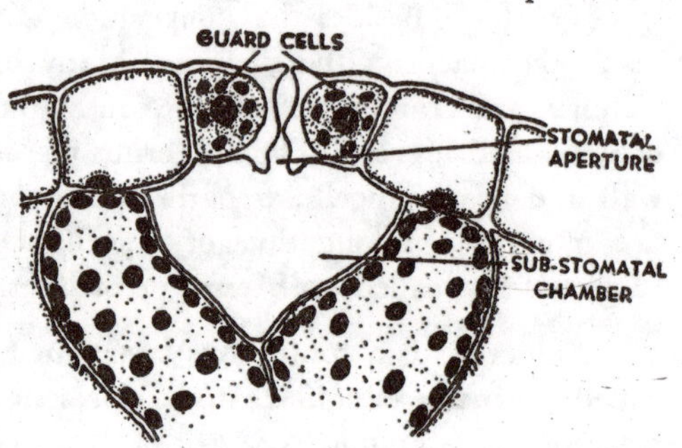

Fig. 144. Stoma in sectional view.

Stomata are present in green aerial organs of the plants, particularly in the green leaves. Dorsiventral leaves have stomata on the lower surface, isobilateral and centric leaves have evenly distributed stomata; and in floating leaves they are on the upper surface. In an individual leaf there are more numerous stomata near the apex, whereas minimum number is found near the base, the middle portion having a distribution which is an average of the base and the apex. Stomata usually occur in parallel rows in the leaves of monocotyledons, and lie scattered in the reticulately-veined leaves of the dicotyledons. Their number is fairly large which may range between a few thousand to over a hundred thousand per cm.

Stomata carry on the gaseous interchange between the internal tissues and the outer atmosphere. Normally the guard cells remain in the same level with adjacent epidermal cells. But in plants growing in dry situation the stomata are located at the bottom of a depression. They are known as *sunken* stomata, which can considerably reduce transpiration.

II. Ground Tissue System

This tissue system is heterogeneous consisting of different types of tissues, of which parenchyma is most abundant. It extends from beneath the epidermis right up to the centre, thus covering the major portion of the organs. In stems and roots ground tissue system is differentiated into two zones—**extra-stelar** or tissues lying outside the stele and **intra-stelar** or those present inside the stele. **Stele** is the central cylinder or column of the organ which includes the vascular bundles. Extra-stelar ground tissue, known as **cortex,** is often differentiated into three zones, *viz.* **hypodermis, parenchyma** and **starch sheath** or **endodermis;** and intra-stelar ground tissues into **pericycle, medullary rays** and **medulla** or **pith.** In leaf the ground tissue goes by the name of **mesophyll** tissue, usually differentiated into palisade and spongy cells. The ground tissues will be discussed in detail along with primary structures of the organs.

III. Vascular Tissue System

It consists of the **vascular bundles** or conducting strands present in the different plant organs. This tissue system is continuous in roots, stems and leaves. A vascular bundle is composed of two complex tissues—xylem and phloem, also called *hadrome* and *leptome* respectively. In stems of dicotyledons and gymnosperms a strip of lateral meristem, called **cambium,** is present between xylem and phloem, and this is responsible for growth in thickness of the stems. If cambium is present, the bundle is called *open;* if absent, the bundle is said to be *closed.*

According to the position of xylem and phloem, the vascular bundles are of the following types (Fig. 145):

(*a*) **Radial.** Here xylem and phloem occur as separate patches and remain arranged on alternate radii. Radial vascular bundles found only in roots.

(*b*) **Collateral.** In this type xylem and phleom together form a bundle and thus lie on the same radius, xylem is internal, *i.e.* towards centre, and phloem is external. Collateral bundle may be open, as in many dicotyledonous stems, or closed, as in monocotyledonous stems.

(*c*) **Bicollateral.** It is a collateral bundle where xylem is in the centre, two strips of cambium, outer and inner, and two patches of phloem, outer and inner, occur on the two sides of xylem. Thus the sequence is outer phloem, outer cambium, xylem, inner cambium and inner phloem. This type of bundle is present in *Cucurbita* family.

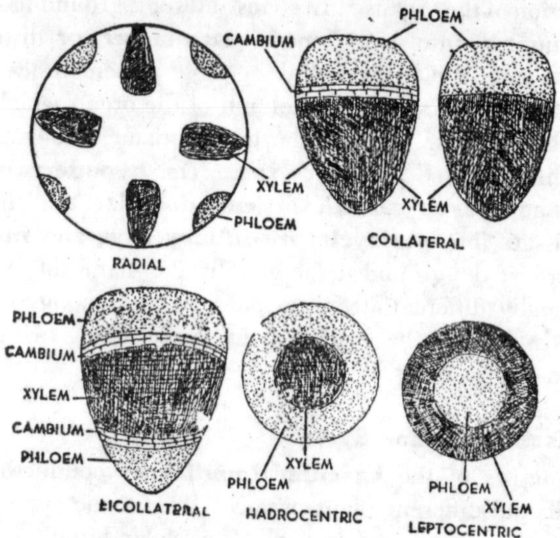

Fig. 145. Types of vascular bundles.

(*d*) **Concentric.** Here one type of tissue surrounds the other. If xylem is in the centre surrounded by phloem, the bundle is called **hadrocentric** or *amphicribral*, as in fern; and when phloem is surrounded by xylem, the bundle is **leptocentric** or *amphivasal*, *e.g. Dracaena*. Concentric bundles are always closed.

CHAPTER XIII

THE PRIMARY BODY

The embryo develops into a plant with the root-stem axis and a number of appendages. Apical meristems present at the root-tip and stem-tip divide and produce new cells. Permanent tissues produced by the apical meristems are called **primary permanent** tissues, and they together form the **primary body** of the plant.

INTERNAL STRUCTURE OF STEM

Dicotyledonous stem

There are many dicotyledons growing in varying situations. Two typical dicotyledonous plants have been selected for the study of internal structure of stem. They are sunflower and *Cucurbita*.

Young Sunflower stem (Figs. 146 & 147)

If a thin and uniform transverse section is taken from a young sunflower stem and observed under the microscope, the following plan in the arrangement of tissues will be noticed:

I. **Epidermis.** It is the single-layered outermost skin composed of tabular parenchyma cells which are compactly arranged without having intercellular spaces. Outer walls are cuticularised. Stomata may be present here and there on the epidermis. Many multicellular hairy outgrowths are noticed.

II. **Cortex.** It is the extra-stelar ground tissue lying internal to epidermis. Cortex is differentiated here into three zones:

(*a*) **Hypodermis.** It is composed of a few layers of collenchymatous cells with thickened corners forming a band just beneath the epidermis. They are meant for mechanical support.

(*b*) **Parenchyma.** Next to collenchymatous hypodermis many layers of parenchyma cells with intercellular spaces are found. They often enclose a number of glands.

(*c*) **Starch-sheath or endodermis.** It is the last layer of cortex. It is one-layered with barrel-shaped closely fitted cells forming a wavy band, delimiting the central cylinder or stele.

These cells contain abundant starch grains, hence called starch sheath.

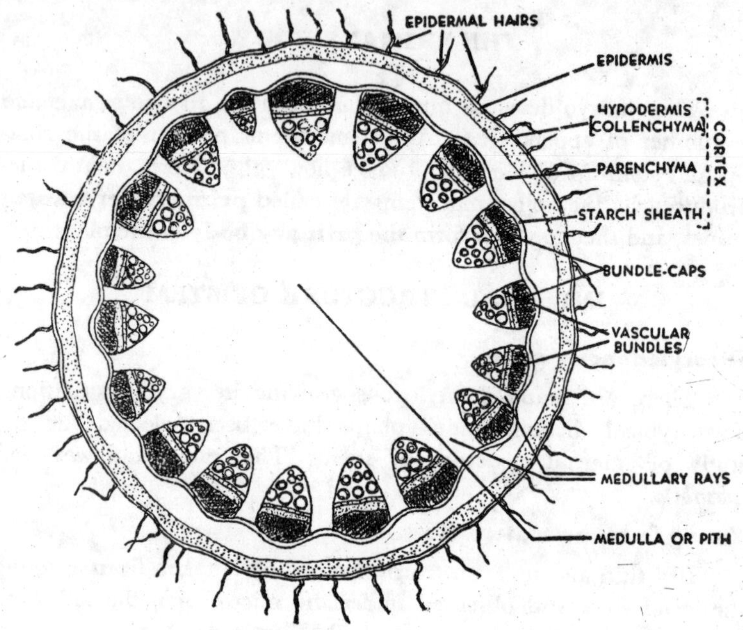

Fig. 146. Transverse section of young sunflower stem showing the plan of arrangement of tissues (diagrammatic).

III. The **Central Cylinder or Stele.** It encloses many vascular bundles arranged in form of a distinct ring and the intra-stelar ground tissues.

(*a*) **Vascular bundles.** These are placed more towards the epidermis than towards the centre. Bundles are *collateral* and *open.* **Xylem** with its component parts, vessels, tracheids, wood-fibres and wood parenchyma, lies towards the centre. Protoxylem vessels with smaller cavities and annular or spiral thickenings are always placed towards centre, and metaxylem vessels, with wider cavities and reticulate, scalariform or pitted thickening, towards, the epidermis. This arrangement, typical of stems, is called **endarch.** **Phloem** with sieve tubes, companion cells and phloem parenchyma lies outside or towards the circumference. A strip of lateral meristem, called **cambium,** is present between xylem and phloem.

(*b*) **Intra-stelar** ground tissues. These are differentiated into three zones: (*i*) **Pericycle.** Next to endodermis there is a many-layered region, called **pericycle,** lying outside the bundles. Pericycle in the sunflower stem is made up of two types of cells. Against

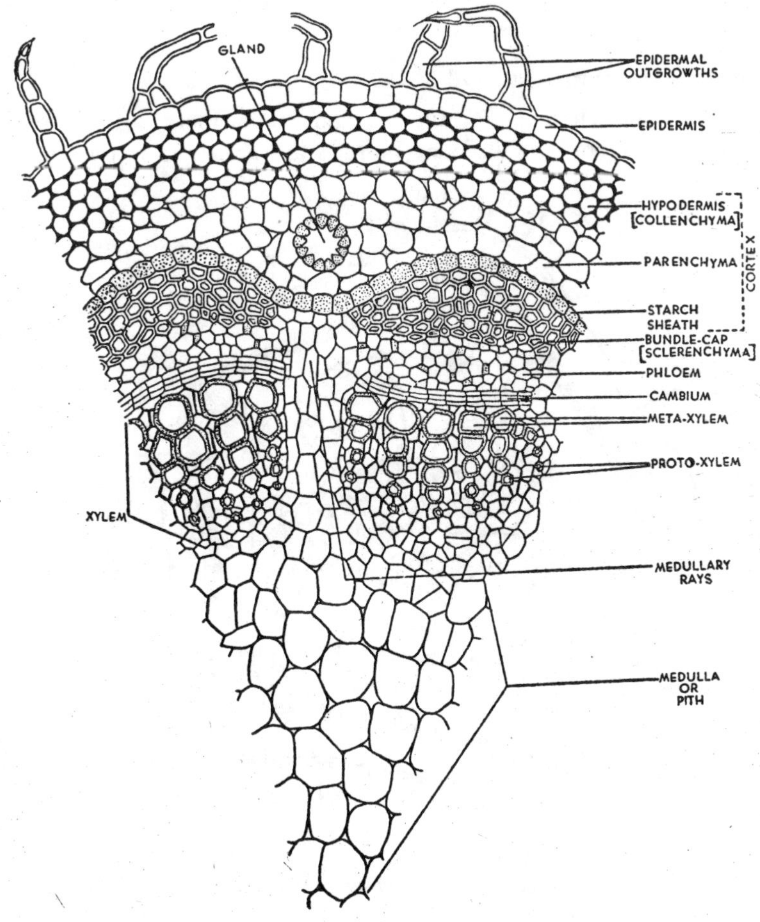

Fig. 147. Portion of transverse section of a young sunflower stem showing the plan of arrangement of tissues.

every vascular bundle there is a patch of sclerenchyma forming something like a cap on the bundle. These patches are called bundle caps or **hard bast.** In between the two bundles, pericycle is parenchymatous. Thus the pericycle is heterogeneous.

(*ii*) **Pith or Medulla.** It is the large central portion of the stem occupied by thin-walled colourless parenchyma cells with lots of intercellular spaces.

(*iii*) **Medullary rays.** Parenchymatous cells present in between every two vascular bundles look like rays radiating from the pith; they are called *medullary rays*.

Cucurbita stem (Figs. 148 & 149)

Cucurbita stem has wavy margin, thus there are distinct ridges and furrows. Vascular bundles are arranged in two rings,

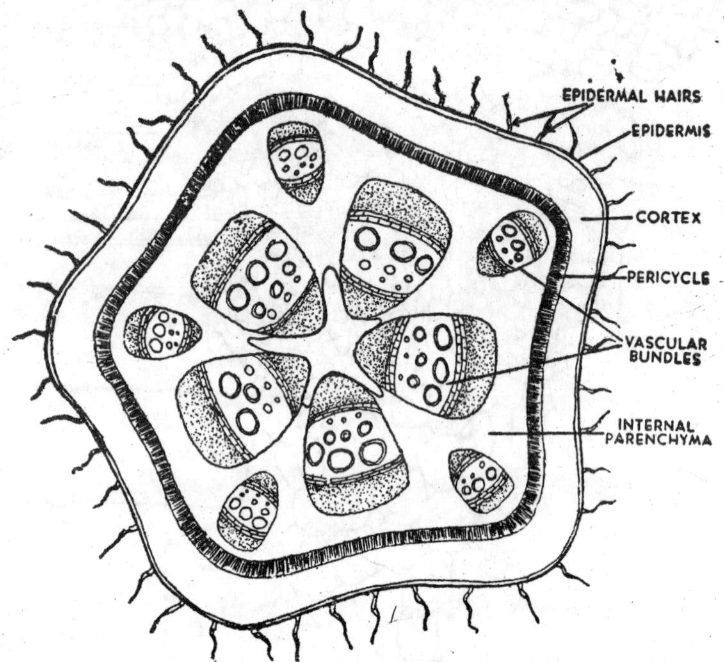

Fig. 148. Transverse section of young *Cucurbita* stem showing the plan of arrangement of tissues (diagrammatic).

one ring of smaller bundles against the ridges and one ring of larger bundles against the furrows. The central hollow cavity is formed by disintegration of pith. The following plan of arrangement of tissues is noticed here :

I. **Epidermis.** It is single-layered with cuticularised outer walls and many multicellular hairs.

II. **Cortex.** It is differentiated into collenchymatous hypodermis, a small parenchymatous region, and single-layered starch sheath with tabular cells attached end to end.

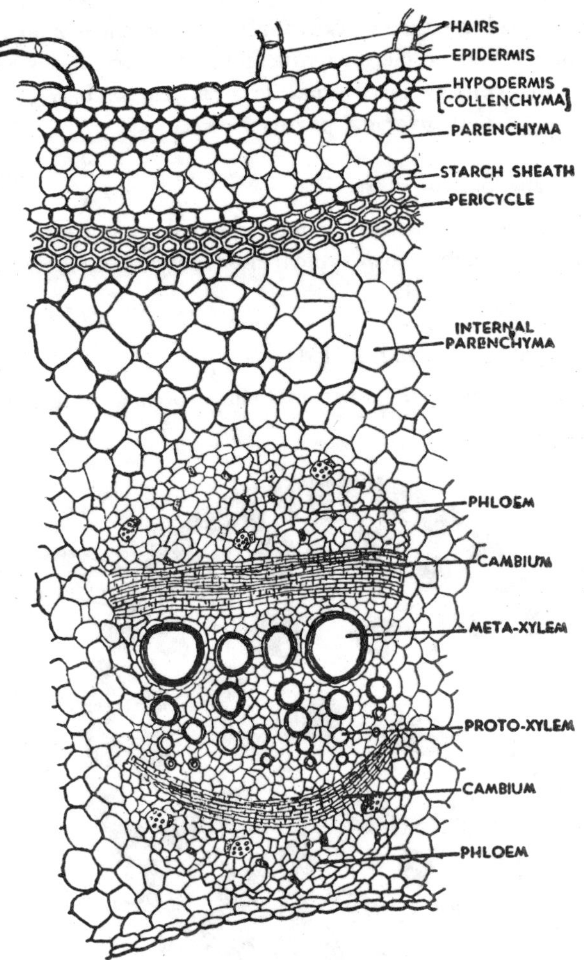

Fig. 149. Portion of transverse section of *Cucurbita* stem.

III. **The Stele.** It encloses vascular bundles and intra-stelar ground tissues.

Vascular bundles are *bicollateral.* Xylem is in the centre, two strips of cambium on the outer and inner sides of xylem and two patches of phloem on the two sides of cambium. Xylem and phloem have the same elements as in sunflower. Sieve tubes and companion cells can be clearly observed here :

Intra-stelar tissues. **Pericycle** is homogeneous, as it is made up of a few layers of sclerenchyma forming a continuous band. All parenchyma cells lying internal to pericycle and in between the bundles are called **internal parenchyma.**

Monocotyledonous stem

The internal structure of monocotyledonous stem will be discussed with reference to a typical monocotyledon, the maize stem, and *Canna* scape.

Maize stem (Figs. 150 & 151)

I. **Epidermis.** It is as usual single-layered with cuticularised outer walls.

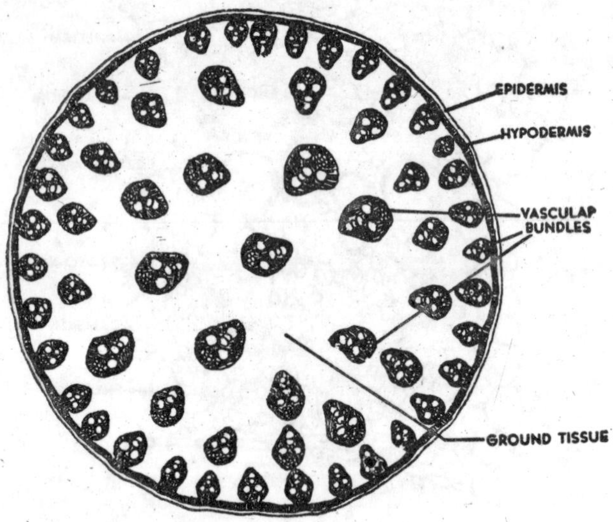

Fig. 150. Transverse section of a maize stem showing the plan of arrangement of tissues (diagrammatic).

II. **Ground tissues.** Next to epidermis there are two or three layers of sclerenchyma forming the hypodermis. Ground tissues internal to hypodermis are **all** parenchymatous in nature showing no differentiation into cortex, starch sheath, etc. They are simply called ground tissue.

III. **Vascular** bundles are numerous and scattered in the ground tissue, more crowded towards the periphery than towards the centre. The bundles are *collateral closed.* That is why monocotyledons usually do not grow in thickness. **Xylem** has the usual elements. Two metaxylem vessels lie outwards and two protoxylem vessels towards the centre. Their arrangement is more or less like the letter Y. Beneath the lowest protoxylem there is a cavity called *protoxylem cavity.* Phloem is much smaller and is composed of only sieve tubes and companion cells, phloem parenchyma being absent. The whole bundle remains surrounded by sclerenchymatous tissue forming what is called the **bundle sheath.**

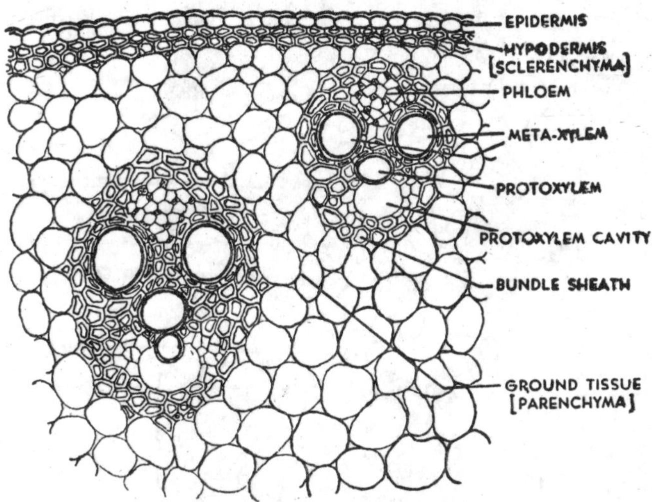

Fig. 151. Portion of transverse section of maize stem.

Flowering stem or the scape of Canna (Fig. 152)

It also shows typical monocotyledonous structure, though a bit different from the maize stem.

I. **Epidermis** is the single-layered protective zone with cuticularised outer walls.

II. **Ground tissues.** Next to epidermis there are a few layers of parenchyma cells forming the cortex which is immediately followed by a band of chloroplast-containing cells called *chlorophyllous tissues*. Patches of sclerenchyma remain internally attached to chlorophyllous tissue here and there. The remaining portion is parenchymatous ground tissue enclosing the scattered vascular bundles.

III. **Vascular** bundles are *collateral closed*. Each of them has small amount of xylem and phloem, with only sieve tubes and companion cells. Sclerenchyma cells do not form a sheath surrounding the whole bundle, as in the maize stem, but remain in two patches on the outer and inner sides of the bundle. The patch on the outer side is larger forming something like a cap and that on the inner side is much smaller.

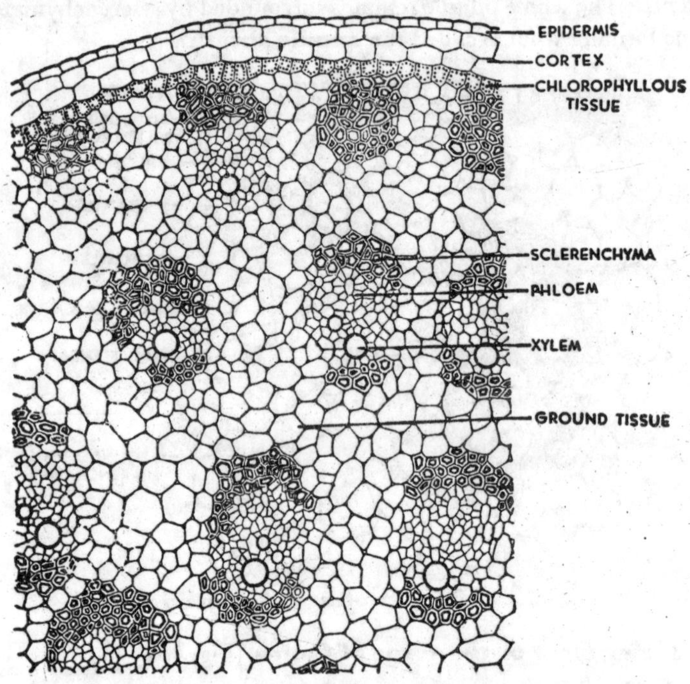

Fig. 152. Portion of transverse section of *Canna* scape.

It has been noticed that all these stems have protective epidermis, ground tissues and conducting tissue with xylem and phloem. Xylem is typically endarch in the stems. There are, however, some points of difference in the structures of dicotyledonous and monocotyledonous stems which are given below :

Dicotyledonous stem	Monocotyledonous stem
1. Ground tissue is well-differentiated into extra-stelar and intra-stelar regions. Extra-stelar ground tissue is differentiated into three zones:—collenchymatous hypodermis, parenchyma and starch sheath; and intra-stelar ground tissue consists of pericycle—homogeneous or heterogeneous, large medulla and medullary rays.	1. Ground tissue is not well-differentiated into distinct regions. Hypoderms is sclerenchymatous in the maize stem, the rest of ground tissue is usually parenchymatous in nature.
2. Vascular bundles are arranged in distinct rings—one or two. Bundles are collateral or bi-collateral but always open. So dicotyledonous stems increase in thickness.	2. Vascular bundles are numerous and irregularly distributed in the ground tissue. Bundles are collateral but closed, and so monocotyledonous stems do not usually grow in thickness.
3. Xylem is larger, with a few rows of vessels.	3. Xylem is smaller with only a few vessels.
4. Phloem is composed of sieve tubes, companion cells and phloem parenchyma.	4. Phloem has only sieve tubes and companion cells and hence, is much smaller.
5. No sclerenchymatous sheath is present. Pericyclic fibres are found as bundle caps or continuous band.	5. Sclerenchymatous bundle-sheath is present wholly or partly surrounding the bundle.

INTERNAL STRUCTURE OF ROOTS

Dicotyledonous root (Fig 153)

A thin section of a dicotyledonous root shows the following tissues:

I. **Epidermis.** It is single-layered and composed of thin-walled cells. The outer walls of epidermal cells are not cutinised. Many epidermal cells prolong to form long hairy bodies, the typical unicellular hairs of roots. Epidermis of root is also called **epiblema** or piliferous layer (*pilus*=hair; *ferous*=bearing).

II. **Cortex.** It is quite large and extensive in roots. Cortex is made of thin-walled living parenchymatous cells with leucoplasts, which convert sugar into starch grains. The last layer of cortex is **endodermis.** It is of universal occurrence in roots. Endodermis is composed of one layer of barrel-shaped cells which are closely arranged without having intercellular spaces. The endodermal cells have thickened radial walls, which are called *Casparian strips*, after the name of Caspary, the gentleman who first noted them.

III. **Stele** or central cylinder. Next to endodermis there is a single-layered **pericycle** made up of thin-walled parenchyma cells. Pericycle is the seat of the origin of lateral roots.

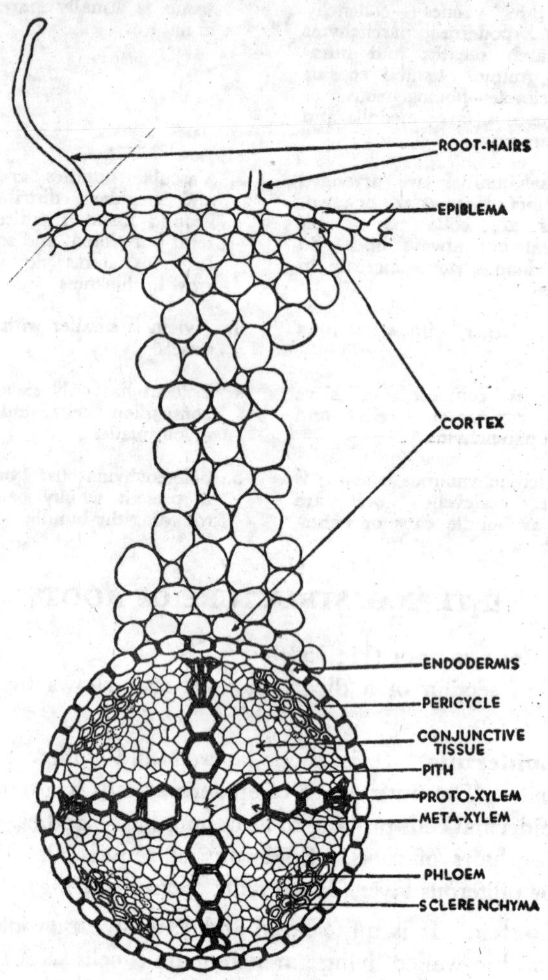

Fig. 153. Portion of transverse section of young dicotyledonous root (gram) showing the plan of arrangement of tissues.

Vascular bundles are typically *radial* in roots. Xylem and phloem form separate patches and are intervened by non-conducting cells. In dicotyledonous roots the number of bundles is limited.

Xylem has protoxylem towards circumference abutting on peri-

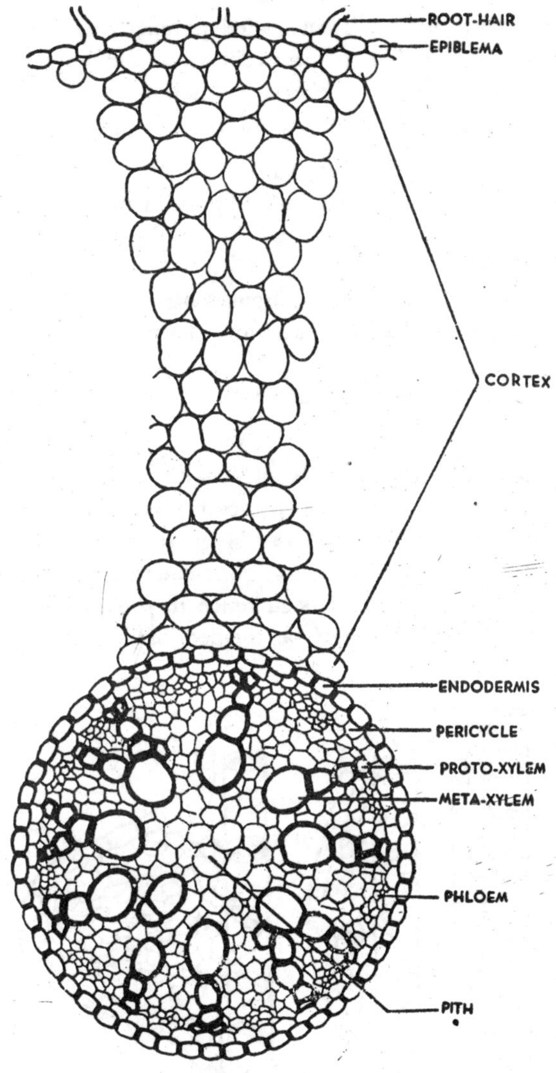

Fig. 154. Portion of transverse section of a monocotyledonous root (arum) showing the plan of arrangement of tissues.

cycle and metaxylem towards centre. This is called *exarch* arrangement (cf., endarch arrangement of stems). Phloem with sieve tubes, etc., form patches arranged alternately with xylem. A small patch of sclerenchyma cells is present outside every group of phloem.

Conjunctive tissue. Thin-walled parenchymatous cells lying in between xylem and phloem groups constitute the conjunctive tissue.

Pith. At the centre there is a small parenchymatous pith. It may be even absent in dicotyledonous roots.

Monocotyledonous root (Fig. 154)

A monocotyledonous root has almost identical structure and plan of arrangement of tissues, excepting the following points of difference :

(1) Bundles in monocotyledonous roots are numerous and not limited, as in dicotyledonous roots; and the metaxylem cavities are wider.

(2) A large pith is always present.

Exodermis

The roots have the piliferous layer with root-hairs. When the root-hairs are exhausted a new protective band is formed in the outermost portion of cortex. A few layers of cortex undergo modifications to form the said protective zone called **exodermis.** Exodermis is almost of universal occurrence in the monocotyledonous roots. In some cases it may be formed before the epidermis is decayed.

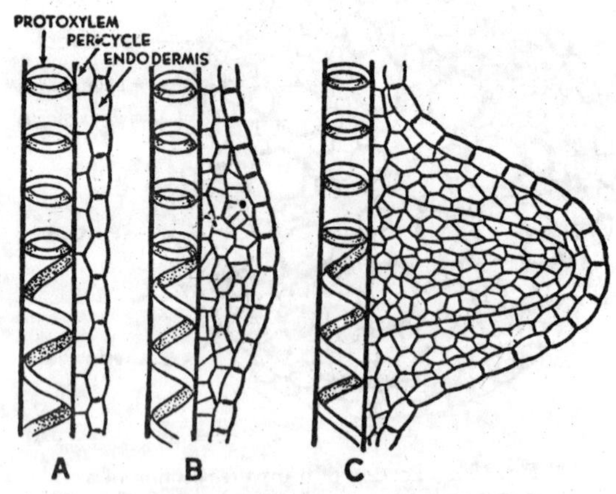

Fig. 155. Origin of lateral root—stages.

Origin of lateral roots

Lateral or branch roots originate from the pericycle, a layer lying inside the stele. So they are **endogenous** in origin. A few cells of the pericycle against the protoxylem become meristematic, and then by division produce new cells. These cells by further division give rise to the lateral roots which come out by piercing the endodermis and all the tissues in front of it (Fig. 155).

Root-Stem Transition

It has been mentioned that the vascular bundles are continuous in roots and stems. But roots have radial bundles with exarch xylem and stems usually have collateral bundles with endarch xylem. In that case continuity of bundles in the root-stem axis becomes impossible, unless some orientations occur somewhere on the axis. The region where changes really take place is known as *transition region.* It is usually the hypocotyl where radial bundles of the roots with exarch xylem are gradually changed into collateral bundles of stems with endarch xylem. There are different ways of transition. The basic point is that the xylem swings or rotates by 180° and joins on the phloem. Thus radial bundles become collateral and exarch xylem becomes endarch.

STRUCTURE OF LEAF

Dorsiventral leaf (Fig. 156)

Leaves having two surfaces, *dorsal* and *ventral*, are called dorsiventral. The upper surface of these leaves is directly illuminated by the sun's rays. A thin section of such a leaf would show the following structure :

I. **Epidermis.** There are two epidermis, upper and lower. The epidermis is usually one-layered with a row of tabular cells, the outer walls of which are properly cuticularised. Stomata are present on the lower epidermis.

II. **Ground tissues.** The ground tissue of leaf is called **mesophyll** tissue. It is differentiated into two types of cells. Towards

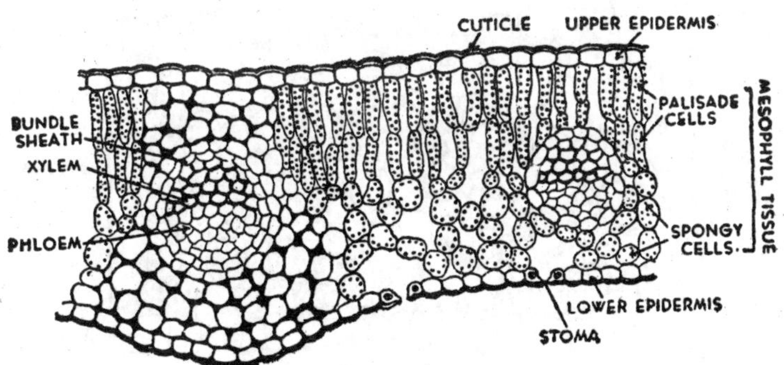

Fig. 156. Section of a dorsiventral leaf showing the plan of arrangement of tissues.

upper epidermis two or three layers of elongated columnar cells with abundant chloroplasts remain rather compactly arranged more or less at right angles to the epidermis. They are called **palisade** parenchyma. Towards the lower epidermis the cells are irregular or isodiametric in shape, and are very loosely arranged with profuse intercellular spaces. They are **spongy** parenchyma. Next to each stoma on the lower epidermis there is a cavity, called sub-stomatal chamber, which is in communication with the intercellular spaces of the internal tissues.

III. **Vascular** tissues. Veins of the leaves are the vascular bundles. The size of the bundle depends on the position one chooses to take in making the section. A section through the midrib shows a vascular bundle with xylem and phloem, **xylem** with tracheids, vessels, etc., on the outer side and **phloem** with sieve tubes, etc., on the inner side. Patches of collenchyma may be present below the upper and lower epidermis for mechanical support. But in a section passing through a smaller vein the bundle remains surrounded by a row of colourless parenchyma cells forming what is known as *bundle sheath* or *border parenchyma.*

Isobilateral leaf (Fig. 157)

Both the surfaces of isobilateral leaves are evenly illuminated and thus they show less differentiation of tissues. Stomata are distributed both on the upper and the lower epidermis. Mesophyll

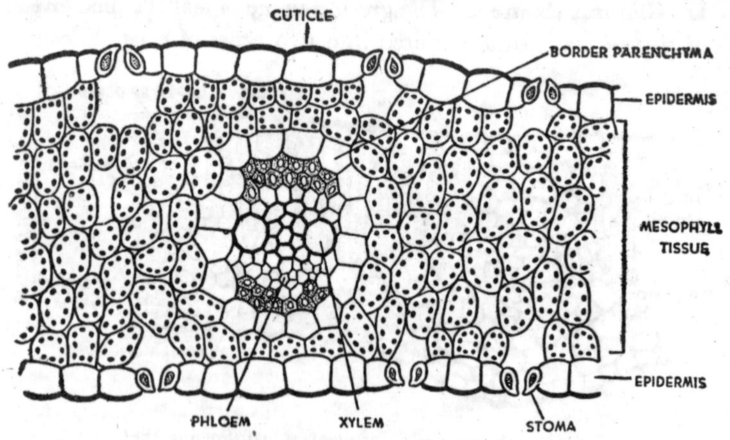

Fig. 157. Section of an isobilateral leaf.

is not differentiated into palisade and spongy cells. Patches of sclerenchyma occur here and there for mechanical support.

CHAPTER XIV

SECONDARY GROWTH IN THICKNESS

It has been mentioned that the primary structure of the plant is formed by the apical meristem. Majority of the dicotyledons show further increase in thickness referred to as the **secondary growth** in thickness. The secondary growth is possible due to the formation of new tissues, called the **secondary** tissues, entirely by the activity of the *lateral meristems*, like **cambium** and **cork cambium.**

Secondary Growth in Dicotyledonous Stem (Fig. 158)

A. In the intra-stelar region

The secondary growth always begins in the intra-stelar region. The meristematic cells of the cambium of the vascular bundles, called **fascicular** cambium, begin to divide and produce new cells on the outer and inner sides. Almost simultaneously a few parenchymatous cells of the medullary rays become meristematic in a line with the fascicular cambium of the vascular bundle and join with the latter to form a complete cambium ring. Newly formed strips are known as **inter-fascicular cambium.** A cambium cell divides into two cells, one of which remains meristematic, and the other one is modified either into a xylem element or a phloem element. That is how the cambium perpetuates itself. The newly formed cells on the inner side are modified into **secondary xylem** elements, and those on the outer side into **secondary phloem** elements. Thus the primary xylem and phloem are gradually pushed apart from each other. The cambium cylinder always produces more xylem than phloem; so the ring, with all the tissues in front of it, is pushed more and more towards the periphery.

Secondary xylem. It has the same elements as in primary xylem, *viz.* vessels, tracheids, fibres and parenchyma. The formation of secondary xylem is not uniform throughout the year. Particularly in the places where seasonal variations are acute, some concentric rings are found in the hard woody part. They are called **annual rings.** Each annual ring represents a year's addition of secondary xylem, so by counting them the age of the plant may be determined. Spring is the favourable season for plant growth, when a large number of new buds come out and unfold. The secondary xylem formed during spring is made up mainly of vessels with wide cavities. This wood is called **spring wood** or early wood. In autumn the rate of assimilation decreases, and, thus, the wood formed is mainly composed of hard lignified elements with smaller cavities. This wood is called **autumn wood** or late wood. Winter is a period of rest when cambial activities come to a standstill, which is followed by spring and autumn of next year. Consequently a sharp line of demarcation exists between small-celled

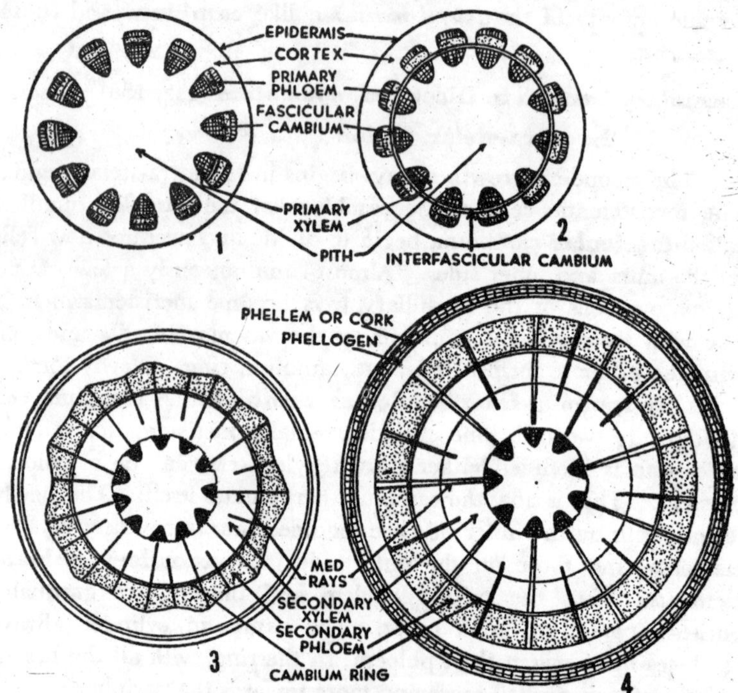

Fig. 158. Diagrams of transverse sections of dicotyledonous stems showing stages in secondary growth in thickness up to two years.

autumn wood of previous year and large-celled spring wood of current year. Spring and autumn wood together form an annual ring.

After considerable growth of the trees the central portion of the wood becomes very hard, durable and usually dark-coloured due to deposition of resin, tannins, etc. This hard central wood is called **heart wood** or *duramen*. Heart wood does not conduct water but only serves as mechanical column. The outer light-coloured wood which, in fact, conducts water is called **sap wood** or *alburnum*. Though heart wood is more valuable from the commercial point of view, but biologically sap wood is more important.

Secondary phloem. It is made up of sieve tubes, companion cells, phloem parenchyma and fibres. Phloem shows no corresponding annual rings.

Secondary rays. Some cambium cells produce thin-walled parenchyma which run as strips or channels in radial direction between the external and internal regions. They, in fact, belong to secondary phloem and secondary xylem, and are known as secondary rays or vascular rays.

B. In the extra-stelar region

Due to the formation of secondary tissues in the intra-stelar region considerable pressure is exerted on the epidermis which becomes stretched and tends to rupture. To prevent that a new meristem, called cork-cambium or **phellogen,** is produced on the superficial layers of cortex. Cork-cambium cells divide as usual forming parenchymatous **phelloderm** or secondary cortex on the inner side and cork cells or **phellem** on the outer side. Cork cells are brick-shaped ones which lose protoplasm and their walls are suberised. These suberised cork cells serve as secondary protective tissues. But as they are impervious to water, the cells lying outside the cork do not get any supply of food. They ultimately die and dry up to form what is known as **bark.** So all the dead tissues lying outside the active cork-cambium constitute the bark. New phellogen may be formed in the deep-seated layers of cortex which also behaves in the same manner and produces quite thick bark. A continuous bark is called **ring bark;** but if the bark is formed in patches, as in guava, it is called **scale bark.**

Lenticel

After the formation of the secondary tissues the stomata present in the epidermis can no longer function. For carrying on gaseous interchange with the outer atmosphere new aerating pores, called **lenticels,** are formed. Lenticel originates just beneath a stoma. The cork-cambium there, instead of producing suberised cork cells, gives rise to some loosely arranged parenchyma cells, called **complementary** cells, which rupture the epidermis and look like small swellings on the stem. The opening with the complementary cells is called the **lenticel** (Fig. 159).

Secondary growth in thickness in Dicotyledonous Roots (Fig. 160)

Dicotyledonous roots have limited number of radially arranged vascular bundles. With the initiation of secondary growth some parenchymatous conjunctive tissue cells become meristematic

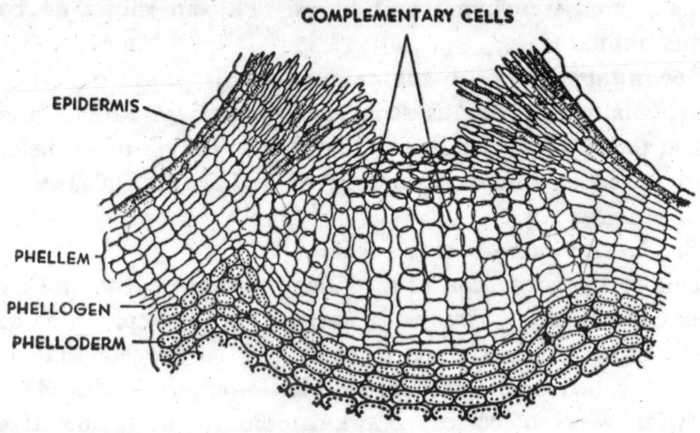

Fig. 159. Transverse section through a lenticel.

beneath the phloem groups, thus forming strips of meristems equal to the number of phloem groups present. The cells of one-layered pericycle now divide. The meristems by further extension reach the pericycle abutting the protoxylem and form a complete wavy band of cambium passing internal to phloem and external to xylem. The first formed meristem is much more active than the lately formed ones and produces more xylem on the inner side. As a result the wavy cambium ring is rendered circular.

Now the cambial cells divide to produce secondary tissues in the same manner as in dicotyledonous stems. Cambium against every protoxylem group gives rise to parenchyma cells which run as longitudinal strands and are called main medullary rays.

As a result of addition of new tissues in the intra-stelar region a pressure is exerted on the epidermis. The pericycle as a whole becomes meristematic and constitutes the **phellogen** or cork-cambium. Cork and bark are produced in the same manner as in stem. Lenticels may also be formed.

Fall of Leaves

Deciduous trees usually shed leaves with the approach of winter, when the leaves usually turn yellow due to breaking down of chlorophyll into colourless substances and increase in the concentration of carotinoid pigments. The immediate structural reason for the fall of a leaf is the formation of a separation-layer, called **absciss layer,** at the base of the petiole surrounding the bundles

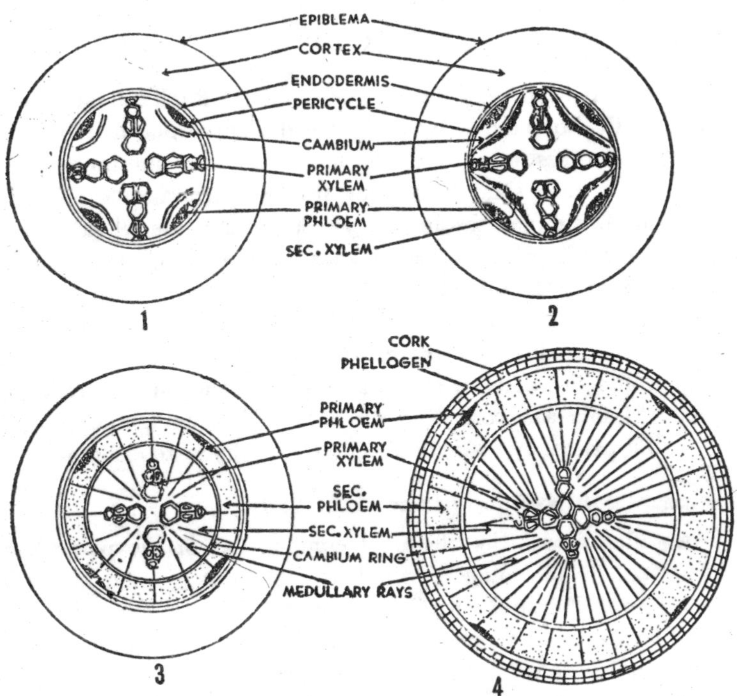

Fig. 160. Diagrams of transverse section of dicotyledonous root showing stages in secondary growth in thickness.

(Fig. 161). The parenchymatous cells there undergo chemical changes often dissolving the cementing materials and thus form the absciss layer. Now the leaf remains attached only by the bundle. By its own weight or due to wind it breaks and the leaf falls. The wound is healed up by the formation of cork, leaving a mark there called leaf-scar.

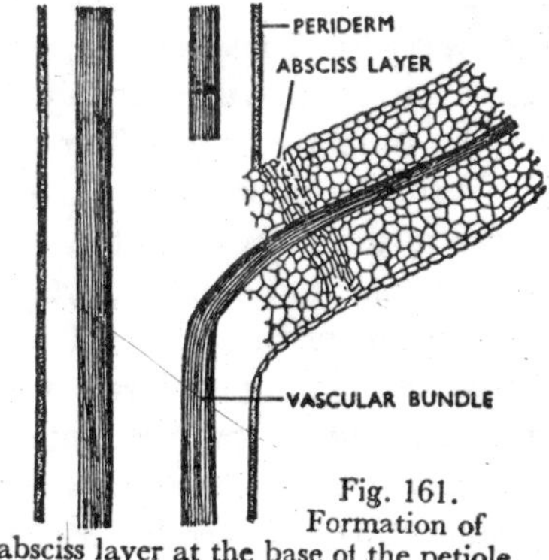

Fig. 161. Formation of absciss layer at the base of the petiole.

PHYSIOLOGY

CHAPTER XV

NUTRITION

Plant body is made of material units, cells. Each cell has protoplasm usually surrounded by a rigid wall. Protoplasm undergoes metabolic changes and produces many ergastic substances. As plants cannot take solid food, they manufacture complex organic matters out of simple substances. We know that the whole animal kingdom, including human beings, has to depend on plants for the supply of food. Chemically, protoplasm is a very complex matter having proteids, carbohydrates, fats, and other inorganic substances in composition. Proteids are made of carbon, hydrogen, oxygen, nitrogen, and in addition, often sulphur and phosphorus; carbohydrates and fats are composed of carbon, hydrogen and oxygen. Cell wall is primarily made of insoluble carbohydrate, cellulose.

Chemical composition of the plant body

An idea about the chemical composition of the plant body can be had by the following method. A plant is dried at a high temperature. Naturally it will lose water. The percentage of water in plants is very high, often as much as 95%. The proportion of water can be determined by proper weighing. The dried plant gets charred forming a black mass of charcoal. That is mainly carbon which forms almost half the dry weight of the plant. Now it is burnt. The combustible matters like proteins, carbohydrates and fats are converted into carbon dioxide, sulphur dioxide, ammonia, water vapour, and other gases and thus escape, which may be properly collected and composition noted. The white incombustible residue left behind is **ash,** which is nothing but the oxides of the metals constituting the plant body. The percentage of ash varies; but on the average, it forms nearly 5 per cent. of the total weight of the plant. On proper analysis ash is found to be composed of quite a large number of elements, many of them, of course, being present only in minute traces. The following elements are, however, always present in ash: potassium (K), calcium (Ca), magnesium (Mg), iron (Fe), sodium (Na), sulphur (S), phosphorus (P), chlorine (Cl), and silicon (Si). Elements like manganese (Mn), boron (B), zinc (Zn), copper (Cu), and molybdenum (Mo) may be present in slight traces.

Elements essential for healthy growth of a plant

It is found that thirteen elements are constantly present in the plant body, as revealed by ash analysis. The mere presence in the composition does not necessarily mean that they are all essential for the growth and nutrition of plants. What elements are essential and what are not, can be determined by water-culture experiment. As plants absorb most of the elements from soil, a standard solution is prepared with those elements, and plants are grown there. By conducting a series of experiments (to be discussed presently) it has been found that ten elements are very important for the nutrition of plants and the remaining three, referred to as non-essential elements, are not directly concerned with nutrition. It has been found that minute traces of the elements like boron, zinc, copper, manganese, and molybdenum are equally necessary for the normal development of the plants. Thus essential elements are put into two groups.

Essential elements:

(1) Major elements are macro-elements—C, H, O, N, S, P, K, Mg, Ca, Fe.

(2) Trace elements or micro-elements—B, Zn, Cu, Mn, Mo.

That the above elements are essential for the healthy growth of a plant can be proved by **water-culture experiment** (Fig. 162). It is undoubtedly a laborious one. A few wide mouthed jars are taken, thoroughly washed and rinsed in nitric acid. Seeds are germinated on sterilised saw dust. A few healthy seedlings of more or less same size are selected for experiment. Jars are labelled as 1, 2, 3, 4 and so on. The first jar 1 is filled with normal culture solution, *i.e.* a solution with all the elements essential for growth of plants. Culture solution should not be alkaline. Culture solutions of various compositions are used. **Knop's normal culture solution** is a quite suitable one.

Knop's normal culture solution

Potassium nitrate, KNO_3—1·0 g.
Potassium phosphate KH_2PO_4—1·0 g.
Magnesium sulphate $MgSO_4$—1·0 g.

Calcium nitrate, $Ca(NO_3)_2$—3·0 g.
Ferric chloride, $FeCl_3$—a few drops.
Water—6 litres.

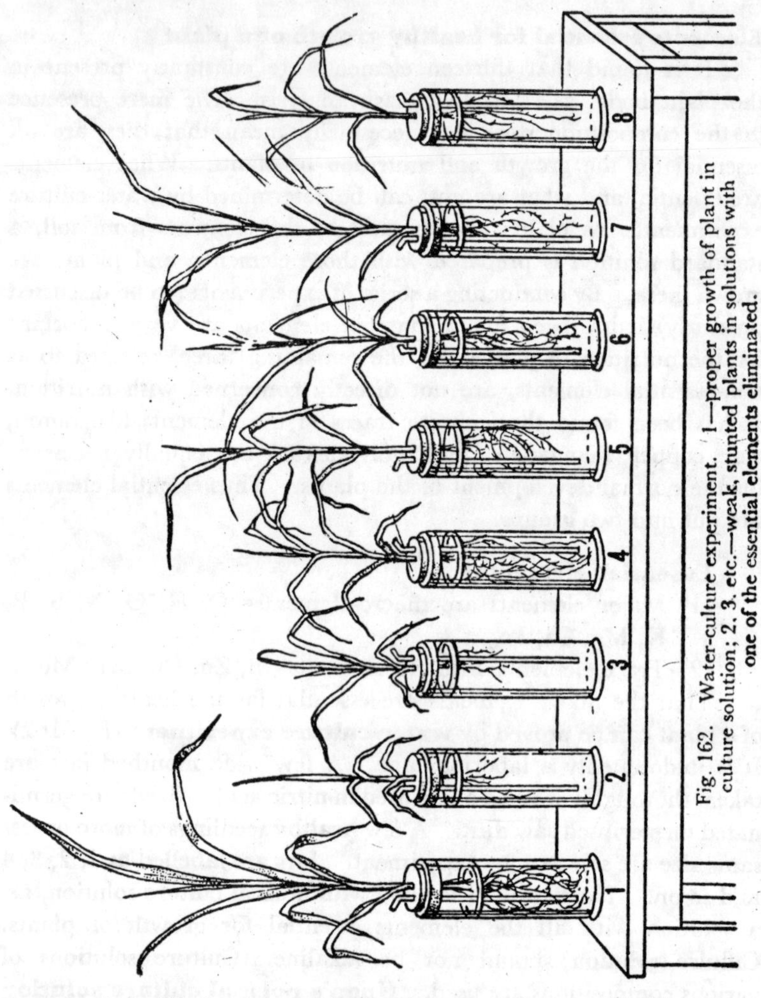

Fig. 162. Water-culture experiment. 1—proper growth of plant in culture solution; 2, 3, etc.—weak, stunted plants in solutions with one of the essential elements eliminated.

Jar No. 1 is filled with normal culture solution; a healthy seedling is selected and the roots after proper washing are inserted in the jar through the cork fitted at the mouth. A bent glass tube is put for aeration of roots and the jar is covered by black paper to cut off light. Now a series of jars are similarly filled with plants but one element, in turn, is eliminated from the solution poured in every jar; thus jar 2 has no potassium salt, jar 3, no calcium salt and so on. Thus the experiment is set up.

After a few days it is observed that only the first one is having normal healthy growth and the rest are all weak, deficient, and stunted. Plant in jar 2 without potassium salt is weak and bears unhealthy leaves; in jar 3 without calcium salt the plant has poorly developed roots and spotted deformed leaves; leaves are yellow (chlorotic) in that without iron salts, and so on. That shows beyond doubt that the elements used in water culture solution and hydrogen, oxygen and carbon are absolutely necessary for the healthy growth of the plant. The essential elements are absorbed either as salts and compounds from the soil or as gas from the air by the two main absorbing organs, roots and leaves. Recent researches have, however, shown that minute traces of manganese, boron, zinc, copper and molybdenum are also essential for normal growth. Unlike the essential elements, they do not enter into the composition of plant body but they are absolutely essential as *catalysts* or *growth regulators*. Deficiency symptoms of these trace elements are not always observed for they are generally present in the soil or in the water culture where even the very pure analar reagents have traces of these elements. The requirements of these trace elements are fantastically small, not more than 1–3 parts per 100 million parts.

Carbon is absorbed from the air as carbon dioxide gas. Though the percentage of this gas in atmosphere is 0·03, that is the only source of carbon for the plants which really forms nearly half of the dry weight of the plant and is the primary constituent of protoplasm, cell wall, food matters, etc.

Hydrogen and **oxygen** are taken mainly as water from the soil. Free oxygen is also absorbed from air during respiration. They enter into the composition of cell wall, reserve materials, and the protoplasm itself. Water is indispensable for protoplasmic activities and also for translocation of dissolved food materials.

Nitrogen forms an important element of protein and a considerable part of protoplasm. Though atmosphere has 78% of molecular nitrogen, it is not available to plants as such. It enters all visible plants and animals and comes out again without taking part in the life processes. Nitrogen mainly comes from soil as nitrates and ammonia salts. The soil bacteria play important roles in converting nitrogenous matters into nitrates for ready absorption. Atmospheric nitrogen can be fixed by some free nitrogen-fixing bacteria and also by some bacteria which inhabit the roots of leguminous plants forming characteristic nodules. The legume bacteria lead a life of mutual friendship (symbiosis) with the plants. The amount of nitrogen fixed by the bacteria is not very high—about 10% of the total nitrogen fixed in the soil. Certain blue green algae can also fix nitrogen, which has considerable economic importance in increasing fertility of the rice fields, particularly in the tropics. During thunder storm free nitrogen becomes available to some extent. The nitrogen of the air combines with oxygen to form nitric oxide—$N_2+O_2=2NO$. This nitric oxide unites with O_4 of the air to form nitrogen peroxide—$2NO+O_2=2NO_2$. The nitrogen peroxide dissolves in falling rain water and forms nitrous acid and nitric acid—$2NO_2+H_2O=HNO_2+HNO_3$, and reach the soil, where they combine with metals like calcium and potassium to form nitrites and nitrates. The nitrates are directly absorbed by the plants, whereas the nitrites are further oxidised to nitrates. Insectivorous plants get part of their nitrogen requirement from the bodies of the insects they capture.

Nitrogen occurs in the soil in organic form as complex proteins, and in inorganic form as nitrates and nitrites, and also ammonia and ammonium ions. The proteins of plant and animal bodies are broken down to simpler substances by the action of *decomposers*, the soil micro-organisms and fungi. Ammonia is such a simple substance which readily dissolves in water. It reacts chemically with H-ions to form Ammonia ions NH_4^+. Only some plants can utilise nitrogen in this form for building up complex proteins. In case of most of the plants a group of micro-organism called *nitrifying bacteria* of the soil are responsible for converting the ammonium compounds into nitrates. The process *nitrification*, as it is called, occurs in two stages—first, by the action of *nitrite-bacteria* ammonium compounds are oxidised into nitrites (NO_2^-), and then nitrites are acted upon by the *nitrate-bacteria* to be finally oxidised into nitrates (NO_3^-). Nitrates thus formed are readily absorbed by the plants and utilised. Nitrifying bacteria work only under *aerobic* conditions, *i.e.* when oxygen is available. Under *anaerobic* condition another group of micro-organism, the *denitrifying bacteria* change the remaining nitrates into free nitrogen which gradually escapes to the atmosphere.

The whole picture, referred to as *nitrogen cycle* is represented by the following schematic diagram:

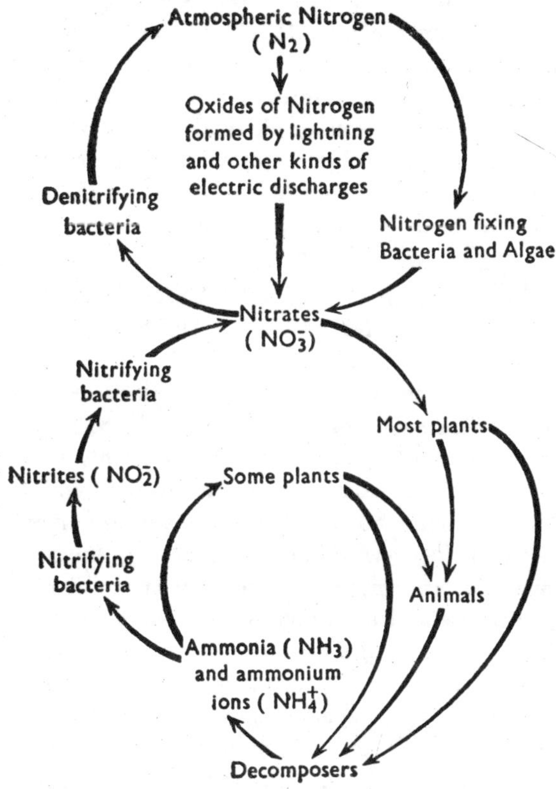

Sulphur is absorbed as sulphate from the soil. It is present in proteins and protoplasm.

Phosphorus is also obtained as phosphates from the soil. It is richly present in nucleo-proteins and has direct influence on cell-division and growth.

Potassium is absorbed as salts from the soil. It is important for the formation of healthy vegetative organs and maturation of fruits and seeds. Potassium is present abundantly in the meristematic region. It helps in the synthesis of reserve materials—carbohydrates, proteins, and fats.

Magnesium is also taken as salts. It is present in chlorophyll and has influence on the formation of proteins.

Calcium is essential for healthy growth. In absence of calcium, plants get diseased. It is abundantly present in the middle lamella of the wall. Calcium neutralises organic acids like oxalic acid forming calcium oxalate. It is also absorbed as salts.

Iron. Iron does not actually enter into the composition of chlorophyll, but minute quantities are essential for the formation of chlorophyll. Plants become chlorotic in absence of iron. Perhaps iron acts as a catalyst for the formation of complex chlorophyll structure.

Sodium, chlorine, and silicon are present in the ash. But they are not essential for the normal growth of the plants.

ABSORPTION OF WATER AND DISSOLVED MINERAL MATTERS

Most of the essential elements are obtained as salts and compounds in very dilute solutions from the soil which, besides providing substratum to hordes of plants, serves as the only source of water and dissolved mineral matters for the growth and nutrition of plants. Soil is composed of particles of varying sizes and shapes of disintegrated rocks with air-spaces among them. It includes gravel, sand, silt, and minute particles of clay. Dead bodies and excreta of animals and the different plant parts like leaves, twigs, etc., decompose due to the action of innumerable micro-organisms present in soil, forming usually black humus which increases the porosity, water-holding capacity, and fertility of the soil. On the basis of the nature of the predominating particles, soil may be sandy, clay, loam, etc. Rains and snow are the chief source of water for the soil. After rainfall some water evaporates and some percolates down through the soil particles under the influence of gravity. During the passage of water downwards much of it is retained by the soil particles in the form of films round them and also filling the spaces in between them. This is called **capillary** water. The films of capillary water of adjacently lying particles coalesce to form a continuous system which is really the main source of water for plants. Even when soil loses capillary water each particle firmly holds a very thin film of water, called **hygroscopic** water, which can be driven out only by the application of strong heat. Plants cannot absorb this water. The spaces between the particles which do not hold water remain filled with air. This air is necessary for respiration and growth of the roots.

In higher plants water and dissolved mineral matters are absorbed through root-hairs which are cylindrical prolongations of the epidermal cells, localised at particular zones on the roots. A large number of root-hairs are formed, and when the old ones perish, they are replaced by new root-hairs during the elongation of the roots. Long root-hairs make their way through the interspaces of the soil particles and come in contact with capillary water. Each hair has a large central vacuole with concentrated cell-sap and thin layer of protoplasm lining the cellulose cell wall. Nucleus moves on to the tip of the hair.

Osmosis

Water is actually obtained by **osmosis.** Osmosis is a physical phenomenon where diffusion takes place when two liquids of different concentration are separated by a semi-permeable membrane, and it continues till the two liquids attain same concentration.

If water and strong sugar solution are separated by a permeable membrane, water will flow more rapidly towards sugar solution and sugar solution will move towards water till an equilibrium is secured. By the term membrane is meant a thin sheet or layer of plant, animal or of artificial origin. *Permeable* membrane is one through which a liquid with dissolved matters may diffuse safely, *e.g.* cellulose cell wall. There are membranes, natural or artificial, which allow only the solvent, *i.e.* water to pass and not the solute. They are called *semi-permeable* or selectively permeable, *e.g.* fish bladder, parchment, etc. An *impermeable* membrane is one which allows neither the solvent nor the solute to pass through it, *e.g.* suberised walls. If plasma membrane were perfectly semi-permeable, no absorption of dissolved substances from the soil by osmosis would have been possible. But it becomes possible because this layer allows certain substances to

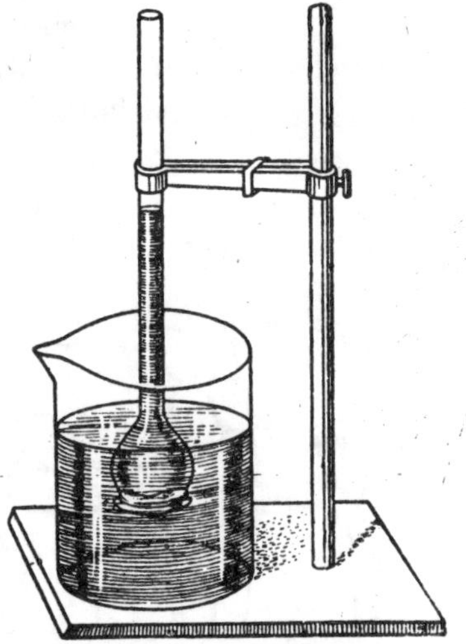

Fig. 163. Experiment on osmosis.

pass through it under one set of circumstances, while under another set their entry is restricted; so this is **selective absorption.** Complexity of the structure of the plasma membrane and its behaviour under different circumstances are thought to be responsible for this permeability. Flow of weaker solution towards the more concentrated one, called **endosmosis,** is always more rapid than the flow of stronger solution to weaker one, referred to as **exosmosis.** Due to endosmosis a pressure, called **osmotic pressure,** is set up.

Experiment 2. Osmosis (Fig. 163). A thistle funnel is taken and the wide end of the funnel is tied with fish bladder which is a semi-permeable membrane. The funnel is filled with strong sugar solution and the level in the narrow tube is noted. Now it is put in a beaker with water, so that the wide end remains dipped in water. Water-level in the narrow tube almost immediately begins to rise showing endosmosis.

Absorption of water takes place through the root-hairs which reach the capillary water of the soil particles. The cellulose cell wall is the permeable membrane between the external weak capillary water and internal concentrated cell sap of the vacuoles having various organic and inorganic matters, particularly sugar. The lining plasma membrane serves as the semi-permeable membrane and exerts selective influence. Osmosis is set up and outer weak solution flows into the root-hair. Semi-permeable plasma membrane regulates entrance of fluids and prevents exosmosis to some extent.

Turgidity

As a result of osmosis the root-hairs, which are really epidermal cells, become full of water and exert hydrostatic pressure on

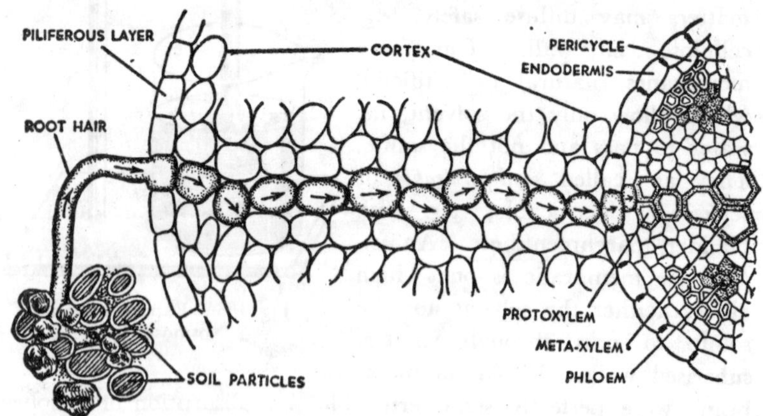

Fig. 164. A section of root showing flow of water to the xylem vessels.

the cell wall. The latter gets stretched and, due to elastic nature, tends to regain original size exerting pressure, called *wall pressure*, again on the contents. This tense distended condition is known as **turgidity**, and such a cell is called **turgid.** The hydrostatic pressure exerted by the contents on the wall, referred to as **turgor pressure,** tends to force water out of the cell. It is to be noted that wall pressure which is inwardly directed is always equal to the turgor pressure which acts in the opposite direction. The turgor pressure is released as water is forced into the adjacent cortical cell from the root-hair. The cell, in turn, passes water to the next one and so on, till water reaches the vascular elements (Fig. 164). Thus the movement of water inwards is secured by cell-to-cell osmosis. Turgidity is important for this flow of water, for growth and various movements.

Experiment 3. Cell-to-cell osmosis (Fig. 165). Cell-to-cell osmosis may be demonstrated by a very simple experiment. The skin is removed from a large potato tuber and slices are taken off from all sides for giving it the shape of a square block. Now central portion is scooped off, so that it looks like a cup. Strong sugar solution is put in the cup and the level is marked by inserting a pin. The whole potato block is put in a dish coloured with a drop or two of eosin. After some time it is noted that level of water has risen in the cup and also that water has become coloured. Sugar has slightly flown to the water in the dish.

Importance of water in plant life

Life originated in water. Lowest plants like the green algae live in water. Even to the higher plants, which through series of changes and adjustments have adapted themselves to terrestrial habits, water is of paramount importance.

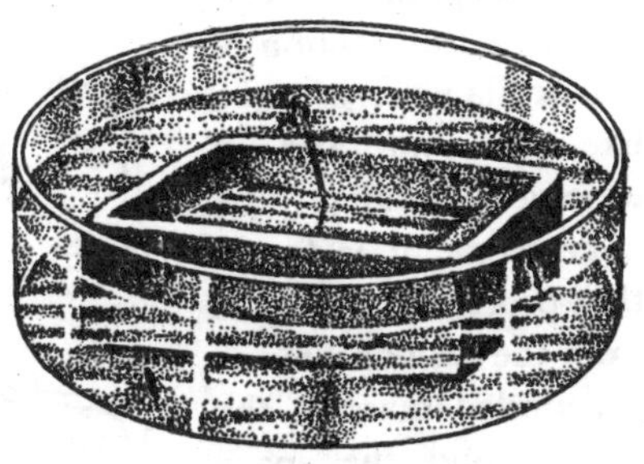

Fig. 165. Cell-to-cell osmosis—potato-cup experiment.

Protoplasm, the physical basis of life, remains saturated in a high percentage of water. In fact, functioning of the mysterious fluid protoplasm depends on the large amount of water it contains. In the active and young vigorously growing cells water content is very high, approaching 95%. In the hard woody portions the amount of water is nearly 50%. Dormancy of the seeds is due to low water content, varying from 10% to 15%. Protoplasm becomes less and less active if the normal amount of water is removed, until a point is reached when death ensues.

A certain amount of water in the vacuoles of the cells is necessary to maintain the normal condition of the turgor, what is essential for the growth of the cells and also for the **erectness** of the plant.

Water is essential for continual translocation from one part of the plant to another part. As far as it is known, all foods and dissolved materials translocated in a plant move from cell to cell in watery solution.

Many of the various organic compounds are formed by the combination of water with certain materials that enter the plant from the soil and the air.

A supply of water in the soil is also needed to replace the water lost by transpiration.

The process of absorption of inorganic salts by the root-hairs is very complex and little understood. Probably salts come in by diffusion through the plasma membrane independently of the movement of water.

Plasmolysis

If a few drops of strong sugar solution or 10% potassium nitrate is put on a thin section of leaf or filaments of algae like *Spirogyra* and is observed under the microscope, it is found that protoplasm retreats from the wall and forms a rounded mass at the centre, and the space between the wall and mass of protoplasm is filled with external solution (Fig. 166). This shrinkage of protoplasm is called **plasmolysis.** It is obviously due to the fact that sugar or KNO_3 solution being more concentrated has drawn water from

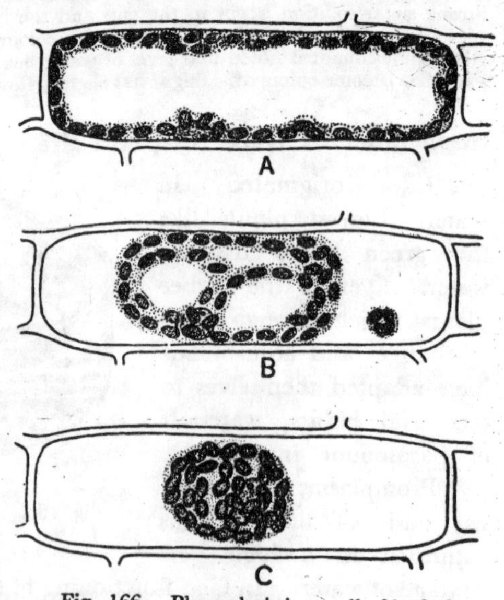

Fig. 166. Plasmolysis in a cell of leaf of *Vallisneria*, A—normal; B & C—stages of plasmolysis.

cell-sap by osmosis. Rounded mass of protoplasm expands to get back to original position as soon as the external influence is removed. Plasmolysis proves the permeability of the cell wall.

Root Pressure

When root-hairs absorb water by endosmosis the turgor pressure forces water to the next layer of cortical cells. They, in turn, send water to the next layer, and the process goes on. The cells become alternately turgid and flaccid, as a result of which something like an intermittent pumping is set up. Ultimately water is forced to the xylem vessels through their variously thickened lignified walls. The pressure, which is responsible for forcing water to the vessels and its conduction upwards, is known as **root pressure.** The magnitude of root pressure is not great, rarely exceeding more than 1–2 atmosphere. That water is forced upwards by root pressure can be demonstrated by the following experiment:

Experiment 4. Root Pressure (Fig. 167). A healthy potted plant is selected and the stem is cut across 5·0–7·5 cms or 2–3 inches above the soil level. A T-shaped tube is fixed to the stem with rubber tubing, and a manometer filled with water and mercury is connected with the tube by a cork, as shown in the figure. All connections are made air-light with molten paraffin. Now the level of mercury in the manometer is noted. After some time there is a distinct change in the mercury level which rises in the long arm. The rise of mercury column is due to collection of water exuded from the cut-stem in the T-tube.

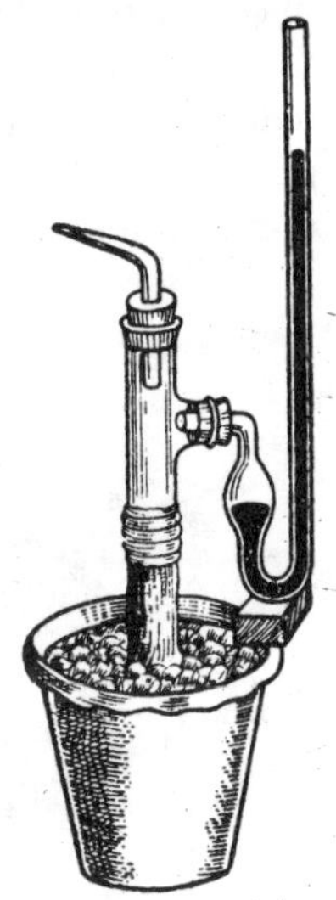

Fig. 167. Experiment on Root Pressure.

TRANSPIRATION

Plants usually absorb water much in excess of their need through the innumerable root-hairs. Some amount of water is utilised for metabolic activities and the surplus goes out as water vapour. This process of getting rid of excess of water in the form of water vapour, is known as **transpiration.** It may take place through the exposed aërial parts of the plants. But as structure and position of leaves are very much conducive to the process, they are the main transpiring organs of plants. It should be noted that transpiration is not simple evaporation, but a physiological process where the outgo of water is *regulated by protoplasm.* A twig detached from a plant would evaporate much more water than one attached to it.

Maximum transpiration takes place through the stomata present on the leaves. This is called **stomatal transpiration.** In day time the two guard cells manufacture sugar and thus the sap is concentrated. By osmosis they absorb water from the adjoining cells and become turgid, when they stand away from each other making the stomatal aperture wide open. The spongy cells evaporate water into the intercellular spaces which accumulates at the sub-stomatal chamber and ultimately diffuses through the stomatal opening, provided, of course, the outside air is dry. At night the turgor pressure of the guard cells decreases and they become flaccid, naturally closing the aperture. The epidermal cells have cuticle on the outer walls which is almost impervious to water. In plants of moist and shady places cuticle is very thin, and feeble transpiration becomes possible through it. This is **cuticular transpiration.** It is about 10% of the entire transpiration through the leaf, the remainder being stomatal transpiration. In the still air, however, the rates of cuticular transpiration in many places have been found to be as high as 30% of the total transpiration through leaves. Transpiration may also take place through the lenticels often referred to as **lenticular** transpiration. The amount of water lost by transpiration is surprisingly high. For example, a

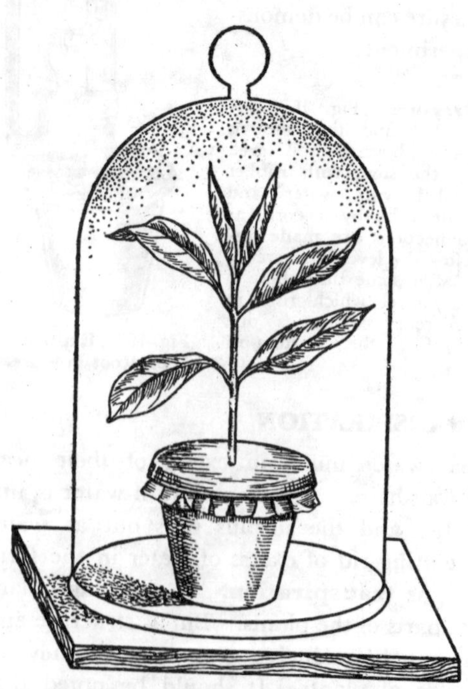

Fig. 168. Experiment on Transpiration.

maize plant during a growing season may lose by transpiration as much as 246 litres or 54 gallons of water. Due to transpiration from the leaf surface, a pull or suction force is produced, as a result of which water is drawn upwards. This is termed as **transpiration current.**

That plants lose water by transpiration can be shown by a simple experiment.

Experiment 5. Transpiration (Fig. 168). A healthy potted plant is selected. The soil of the pot is carefully covered by a sheet of rubber to prevent liberation of moisture. A coating of molten paraffin will also serve the purpose. Now the pot with the plant is covered by a bell-jar. After a couple of hours small drops of water like dew drops will be found on the inner surface of the bell-jar. Water vapour given off by transpiration is converted into drops coming in contact with the cool surface of the bell-jar.

The dorsiventral leaves, which form the majority, have stomata on the lower surface, through which evaporation takes place. Here is another experiment to demonstrate transpiration through lower surface of leaves.

Experiment 6. Transpiration from lower surface of a dorsiventral leaf. A piece of filter paper is dipped in 5% cobalt chloride solution which is pink in colour. The filter paper naturally turns pink. Now it is carefully dried over a flame when it turns blue. Thus cobalt chloride paper is pink when moist and blue when dry. The paper is cut into two pieces and placed on two sides of a dorsiventral leaf. They are covered with glass slides or mica sheets and vaseline is carefully put on the margins to prevent contact with moisture. They are fixed with a clip, as shown in the figure (Fig. 169). After some time it is noticed that the paper on the lower side turns pink much more rapidly than one on the upper surface. Thus the experiment shows vigorous transpiration from the lower surface of a dorsiventral leaf where the stomata are present.

Fig. 169. Experiment showing more rapid Transpiration from the lower surface of a dorsiventral leaf.

It is expected that water lost by transpiration is compensated by absorption from the soil. Though the relation between transpiration and absorption depends on many factors, normally amount of water lost might be equal to the amount absorbed. If more water is lost than absorbed the plant may wilt temporarily, reviving where there is again a supply of water. Permanent wilting due to continued absence of water will inevitably bring death to that tissue.

Experiment 7. Transpiration and absorption. (Fig. 170). A wide-mouthed glass bottle with a graduated side tube is taken. Through a split cork at the mouth the roots of a plant are put inside the bottle which is filled with water. A few drops of olive oil are poured on the water surface at the side tube to prevent evaporation. All the connections are made air-tight. The bottle with the plant is weighed and the weight noted. After a few hours the water-level in the side tube falls suggesting the amount of water absorbed by the roots. Now the bottle is again weighed. The difference in weight gives the amount of water transpired. Thus the relation between transpiration and absorption is worked out

It has been mentioned that transpiration exerts a suction force to draw water upwards. This can be demonstrated by an experiment.

Experiment 8. Suction due to transpiration. A healthy twig of a plant is cut under water. It is fitted with a cork to one end of a glass tube. The tube is completely filled with water and is placed on a trough of mercury with the help of a clamp as shown in the figure (Fig. 171). All the connections are made air-tight. After some time it is noted that mercury has risen in the tube, suggesting loss of water by transpiration and the suction force to draw mercury in the tube.

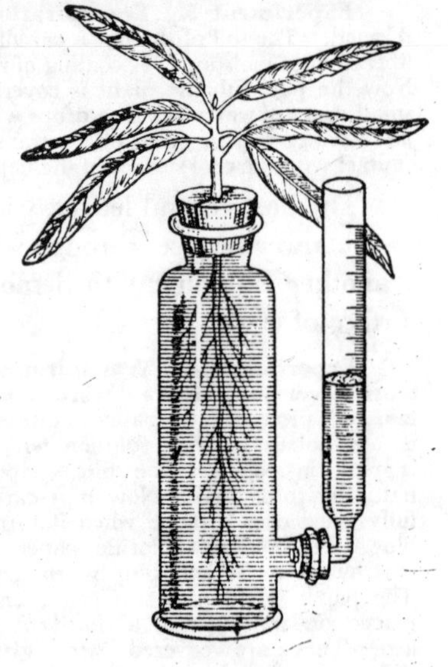

Fig. 170. Experiment on relation between Transpiration and Absorption.

The rate of transpiration may be measured quantitatively with the help of an apparatus called **potometer.**

Experiment 9. Rate of transpiration. A Ganong's Potometer (Fig. 172) is taken. It is a twice-bent glass tube with graduations on the horizontal portion, the upper end has wide mouth to hold a twig fitted with a cork and the lower end is dipped in water in a beaker or specimen tube. A water reservoir is fitted to the horizontal tube by a stop-cock.

The apparatus is filled with water. A twig is cut under water and inserted quickly to the wide mouth of the potometer through a cork. The narrow end is lifted for a moment to allow an air-bubble to enter into it and again dipped in water. All connections are made air-tight with paraffin. It is found that the bubble moves along the tube due to suction. By taking reading from the graduated tube rate of transpiration per unit time may be determined. The bubble may be pushed back by opening the stop-cock of the water reservoir and the experiment may be again started.

Factors affecting Transpiration

The following external factors influence the rate of transpiration in plants.

1. **Humidity of air.** As transpiration involves diffusion of water vapour from regions of high concentration (intercellular

spaces of leaves) to that of low concentration (outside air), humidity or amount of water vapour in the atmosphere, surrounding the plant has considerable influence on the process. On damp foggy days the rate of transpiration decreases as the outer air remains saturated with water vapour. The less moisture there is in air, the greater will be the rate of transpiration.

2. **Light** or **Illumination.** It has marked effect on transpiration. The opening and closing of the stomata, through which by far the maximum amount of water is lost, depend on light. Moreover, due to absorption of radiant energy and its transformation into heat, temperature of the leaf is raised bringing about an increase in transpiration rates.

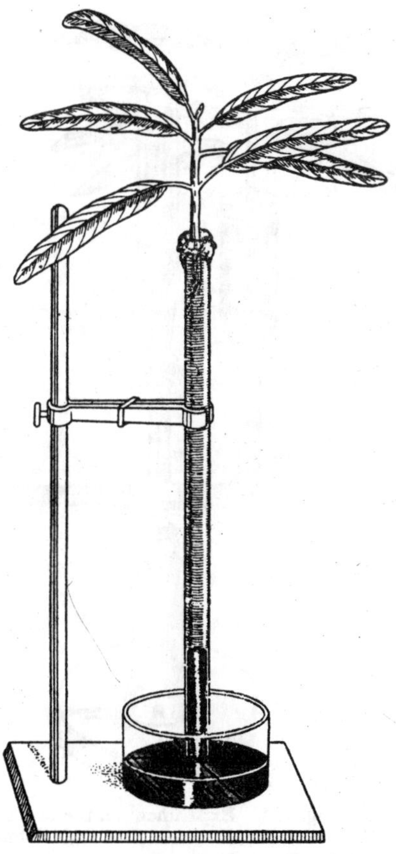

Fig. 171. Experiment on Suction due to Transpiration.

3. **Temperature.** It increases the rate of transpiration as it hastens transformation of water into water vapour.

4. **Wind.** By wind or air current water vapour given off during transpiration is removed; thus saturation of the surrounding air is avoided which otherwise would retard the rate of transpiration. Winds of high velocity, however, retard transpiration, because the stomata close up due to high winds. Moreover, winds of high velocity bring about a reduction in temperature which undoubtedly affects transpiration.

5. **Atmospheric Pressure.** Atmospheric pressure affects

transpiration, because it has influence on the rate of vaporisation. When atmospheric pressure is high, the rate of transpiration is low. However, in any given locality variations in the atmospheric pressure are too slight to have any considerable effect upon transpiration rates. Plants growing in high altitudes have distinctly lower atmospheric pressures, and as such, those plants have significantly

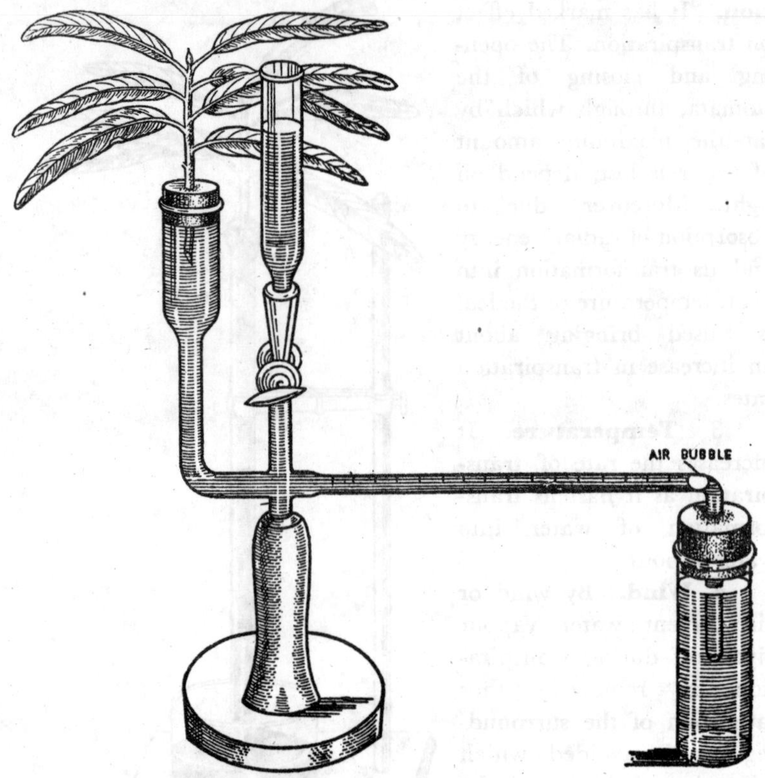

Fig. 172. Experiment on the rate of transpiration current with Ganong's Potometer.

high rates of transpiration, if other environmental factors are not limiting.

6. **Soil factors.** As all necessary water is absorbed from the soil, factors like water content, composition, temperature, concentration of soil solution, etc., indirectly influence the rate of transpiration.

Regulation of Transpiration

Plants maintain wonderful balance between the intake and outgo of water. The aquatic plants are so adapted that they can transpire any amount of water, because the loss may be readily made good by absorption. Land plants regulate transpiration having cutinised epidermal cells, deposition of waxy and resinous matters, etc., and also by the suberised corky bark in the old stems. These are all contrivances for reducing transpiration. Plants growing in very dry situations like deserts (xerophytes) have their transpiration rates reduced to the minimum not because of any inherent peculiarities of their structure and physiological behaviour but due to the only fact that the water supply of the soil in which they grow is so low that very little absorption of water can occur. Their leaf surfaces are extremely reduced. They are rolled or folded, have very thickly developed cuticle on epidermis together with dense hairy outgrowths and coating of waxy matters. Stomata usually remain sunken. Leaves are often modified into spines as in *Opuntia*. It should be mentioned that stomatal regulation of transpiration is only significant, when the aperture is almost closed. Once the stomata are even half open, there is no question of regulation, but environmental factors such as humidity, temperature, etc., are much more important.

Significance of Transpiration

Transpiration is regarded as the natural accompaniment of the physiological processes involving gaseous exchange between the internal tissues and the outer atmosphere. It is unavoidable because, when the stomata will open for gaseous exchange they must allow water vapour to pass out. Transpiration is positively dangerous for the plants having limited water supply. For ordinary plants transpiration must, however, be considered as a necessary evil, because the metabolic processes taking place in plants are certainly intimately related with transpiration. It has been conclusively proved that there is no relationship between the rate of transpiration and absorption of nutrient matters. The whole structure of the leaf was not possibly meant to check transpiration. A thin flat surface provided with large intercellular spaces ending in numerous openings (stomata) was never intended for checking transpiration but rather for facilitating gaseous exchange. Transpiration, of course, facilitates the mass movement of inorganic salts through the conducting channels. It is considered to have a cooling effect on plants, as considerable amount of heat is used in converting liquid into vapour, but the difference in temperature brought about is certainly not great to have any significant effect on the transpiration rates.

Guttation

Generally water goes out as vapour. In some plants water in liquid form is given off as small drops from the margins and apices of leaves. This is called **guttation** or exudation of water. It takes place when conditions are not favourable for transpiration. Guttation occurs particularly on moist nights following hot days. Small drops of water on the grasses of the lawn, often mistaken for dew drops, come out by guttation.

ASCENT OF SAP

It has been discussed that plants lose enormous amounts of water by transpiration. Water is absorbed from the soil and is conducted upwards through roots and stems ultimately to reach the leaves. This becomes possible due to the presence of continuous conducting channels in the organs. Vessels and tracheids of the xylem are the main paths for ascent of water with dissolved matters, what is termed *sap*. The fact can be easily demonstrated by placing a twig in water coloured with a harmless dye like eosin. Within a short time the veins become coloured, and sections also show colour in the xylem vessels and tracheids. A simple experiment of this nature with an inflorescence of lovely tube-rose is really very interesting. The inflorescence stalk is put in water coloured with eosin. It is noticed after a few hours that the beautiful white flowers are turning reddish in colour. The standard 'ringing experiment' has been cited as evidence that ascent of sap takes place through xylem elements. It is this: in a plant the soft outer portions up to cambium are removed in the form of a ring (girdled) but the xylem is left intact. Leaves above the ring do not show sign of wilting. Girdling, of course, turns fatal in the long run, because removal of phloem interrupts downward movement of food.

The factors responsible for the ascent of sap, particularly up to the top of all trees like *Eucalyptus*, pines, etc., have been subjects of keen interest and extensive investigations since a long time; but no entirely satisfactory explanation has been offered. However, the probable theories governing the ascent of sap are given below:

Root Pressure. The root pressure forces water up to the xylem vessels becomes evident from characteristic 'bleeding' from decapitated stumps of small plants. But this force is not sufficient for lifting water in the tall trees. Moreover, root pressure definitely decreases when the rate of transpiration is vigorous. Even negative pressure is registered in the stems. Root pressure might be all right for small herbs and shrubs in the tropical countries; but it has never been found to be of sufficient magnitude in the European countries.

Capillarity. As the vessels and tracheids are capillary tubes, water is expected to rise by capillarity. The smaller the diameters of the capillary tubes, the higher the rise of water—is the general rule. So for ascent of sap it would be expected that the tall trees should have narrow tubes. But just the reverse is true. Particularly in spring when the rate of assimilation is vigorous and there is great need of water, the vessels formed have wide diameters (spring wood). So capillarity as a force responsible for the ascent of sap, cannot be accepted.

Imbibitional Force. It was thought to be the cause of movement of water through xylem by German physiologist, Sachs. The forces of imbibition are certainly great and may seem adequate for carrying water to required distance, but the imbibition rates are extremely slow. And for imbibition to be effective, the cell wall must carry water. But it has been clearly demonstrated that water moves up through the lumen of the xylem vessels.

Indian physiologist, Acharyya Jagadish Chandra Bose, believed that the living cells of the cortex exert osmotic forces, as a result of which sap moves upwards by a sort of pumping. Bose claimed to have proved conclusively that pulsatory activities brought about by alternate contraction and expansion of the living cells, thought to be the cells of the innermost layer of cortex, maintain the ascent of sap in plants. According to Bose, xylem vessels merely act as reservoirs of water. The pulsatory activities have been compared to heart beats of animals by which water is pumped up to next higher cells. Though the experimental evidence as yet available does not support Bose's conclusions, still the possibility that the living cells adjoining the tracheids and vessels may influence the ascent of sap through the stem cannot be altogether dismissed.

Cohesion Theory. Due to evaporation from the leaf surface a tremendous suction force or pull is produced which works on the top of water column present in the xylem. Water particles cohere in the xylem vessels and tracheids, thus forming a continuous column from roots to the leaves. This property makes it possible for it to be pulled to the top of even the tallest tree by the force resulting from transpiration. The xylem vessels actually function merely as tubes through which water is drawn. The column of water resists breaking because of cohesion between water molecules. Water is pulled up from the roots to the leaves very much as a rope is pulled through a hard rigid tube. As a result of the pull, sap is lifted upwards even in tall trees. This theory, which explains the ascent of sap purely on a physical basis, is the most widely accepted one, in spite of its certain weaknesses. Undoubtedly this is the only mechanism adequate for the actual water conduction rates through the plants.

METABOLISM

In the introductory chapter it has been mentioned that the mysterious fluid protoplasm always undergoes changes involving series of chemical reactions. The constructive or *anabolic* changes lead to the synthesis of complex organic food matters as well as protoplasm out of simpler raw materials. These complex matters are essential for nutrition, assimilation and growth. Destructive phase or *katabolism* accomplishes the breaking down of complex organic matters into simpler elements, thereby releasing considerable energy stored up during anabolic phase. Thus there is going on simultaneously a continued dual process of building up and breaking down of waste and repair. The sum-total of the changes is referred to as **metabolism** which is a characteristic feature of the living things.

Plants are called **autotrophic** and **heterotrophic,** according to their modes of nutrition. The green plants can manufacture their own food and are thus autotrophic or self-supporting. We know that the food matters of the plants are carbohydrates, proteins and fats. The manufacture of carbohydrate food is a very important constructive process of the plants. Proteins and fats are made out of carbohydrates through very complex chemical processes.

PHOTOSYNTHESIS

Photosynthesis is the process of manufacture of carbohydrate food matters like sugars and starch by the chloroplasts of the cells out of water and carbon dioxide gas in the presence of sunlight. As absorption of radiant energy supplied by sunlight is an indispensable factor, the process is called **photosynthesis** or carbon assimilation (*photo* =light, *synthesis*=construction).

It is a monopoly of the green plants. Though all the green aerial portions are capable of constructing food, the flat green leaves are best suited for the purpose. Their fineness, thinness and the natural exposure of their maximum surface to light and air as well as their internal structure are very much conducive to the process of photosynthesis. The chloroplasts present abundantly in the mesophyll cells are instrumental to the absorption of radiant energy from the sunlight. It is transformed into chemical energy for bringing about a series of chemical reactions which culminate in the formation of sugar and starch with accumulated potential energy, capable of being oxidised during respiration.

Photosynthesis, as performed by green plants, is a very slow process utilising about 0·5 to 3% of the total radiant energy falling upon the leaf surface. But this is the only mechanism of keeping the solar energy stored up for all out work on the surface of the earth. If the green plants had not solved this little problem of converting solar energy into chemical energy of the carbohydrates, the continued existence of life would have been impossible on earth. Theoretically there seems to be no difficulty for man, the most highly developed animal, to solve this little technical problem of converting the solar energy into some other potential form before the solar energy is converted into heat and lost beyond recovery. But man has not solved this problem yet, *i.e.* artificial photosynthesis has not yet been achieved. We still cannot prepare one molecule of sugar from its constituents, CO_2 and H_2O, in spite of all our technical and scientific advance and attainments. The day artificial photosynthesis will be possible, mankind will discard the green plants and will have at their disposal an unlimited and inexhaustible source of energy.

The Process

Water absorbed from the soil is conducted upwards through the xylem vessels to reach the mesophyll cells of the leaf. Stomata open due to the turgor pressure of the guard cells and carbon dioxide gas comes in. It dissolves in imbibed water saturating the cell wall and diffuses in solution to the chloroplasts of the mesophyll cells. The chemical reactions taking place in photosynthesis are not clearly understood. With the aid of radiant energy, six molecules of carbon dioxide unite with six molecules of water to produce one molecule of simple sugar, glucose, and six molecules of oxygen as end product.

During the process energy is stored up in sugar molecule in potential form. Oxygen usually goes out through the stomata; some may as well be used for respiration. It has been calculated that the volume of oxygen liberated is precisely equal to the volume of carbon dioxide absorbed.

$$6CO_2+6H_2O+\text{light energy (674 cal.)} = C_6H_{12}O_6 + 6O_2.$$

The process is not actually so simple. For a long time it was taken for granted without any conclusive experimental evidence that an intermediate product, formaldehyde (CH_2O), is formed. It was thought that as formaldehyde is very unstable and poisonous, six molecules are quickly condensed into a molecule of glucose. This process of condensation is referred to as **polymerisation.** The accumulation of sugar results in too much concentration of sap which retards photosynthesis. So sugar is converted by chloroplasts into insoluble starch grains by **condensation.** A number of molecules of glucose condenses to form one molecule of starch by elimination of one molecule of water for each molecule of glucose used. These starch grains, known as assimilatory starch, are very temporary bodies, as they are converted into sugar with nightfall so that the starch-free leaves can again carry on photosynthesis next morning. Carbohydrates are assimilated as sugar and travel to different parts probably through the sieve tubes in the phloem. Surplus sugar is converted into reserve starch and other complex carbohydrates in the storage regions. The probable reactions have been represented thus:

$$CO_2+H_2O=CH_2O\text{ (formaldehyde)}+O_2.$$
$$6CH_2O=C_6H_{12}O_6\text{(sugar)}.$$
$$(C_6H_{12}O_6)_n-(H_2O)_n=(C_6H_{10}O_5)_n\text{(starch)}.$$

But it has been conclusively demonstrated recently that formaldehyde is never formed as an intermediate product in photosynthesis. The most probable intermediate in photosynthetic process is a phosphate compound, actually a phospho-glyceric acid.

Though sugar is the first-formed carbohydrate of photosynthesis, insoluble starch is the first visible product. It can be demonstrated in the leaves with characteristic iodine reaction. A leaf collected in the evening is decolourised by boiling for a few minutes in alcohol over a water bath. If the leaf is now treated with iodine solution, it turns blue indicating the presence of starch grains; but a leaf collected early in the morning will not give iodine reaction, because it is starch-free.

The whole idea of photosynthesis has been that CO_2 enters the plant and breaks down into C and O_2. Carbon was correctly thought to be **acquisition,** (which is retained) and O_2 came out. So the ratio O_2/CO_2 was unity. So far as the origin of oxygen in photosynthesis is concerned, it has been taken for granted that it came from CO_2, after carbon has been retained in plants. Several modern investigators have proved conclusively in 1941 that all the oxygen evolved in photosynthesis came not from CO_2 but from H_2O, and that photosynthesis is a purely *oxidation-reduction* process between CO_2 and H_2O—CO_2 being reduced, *i.e.* hydrated by the reductant H_2O, which is oxidised to oxygen by the removal of H_2. Removal of H_2 is oxidation by all the sense of the term. Then the equation of photosynthesis, according to recent cla ification, should be either

$$CO_2+2H_2O=CH_2O+H_2O+O_2, \text{ or } CO_2+4H_2O=CH_2O+3H_2O+O_2.$$

All the oxygen evolved is derived from water and not from CO_2. Recent evidences have conclusively demonstrated that formaldehyde is never formed as an intermediate product in photosynthesis.

The point is again verified in the photosynthesis is pigmented sulphur bacteria which live in an atmosphere of sulphuretted hydrogen. These bacteria use H_2S in place of H_2O and make carbohydrate food from CO_2 and H_2S in presence of sunlight—$CO_2+2H_2S=CH_2O+H_2O+2S$. The reduction of CO_2 here leads to the production of sulphur and not 1 part of sulphur and 2 parts of oxygen or of SO_2, which really proves that O_2 in photosynthesis of green plants comes from H_2O and not from CO_2. The pigmented sulphur bacteria and the world of green plants may be thus called *photo-autotrophic* (independent in presence of light) in contrast to the other groups of non-pigmented sulphur bacteria which also utilise sulphur compounds like H_2S. These bacteria which have been termed *chemautotrophic* (independent when there is supply to chemical energy) derive their energy from oxidation of inorganic compounds like H_2S. The reactions are exothermal and the energy thus liberated can be used up by these bacteria for the synthesis of carbohydrate food. The chemautotrophic bacteria thus can synthesise carbohydrates without chlorophyll and in complete absence of light.

The abolition of formaldehyde hypothesis and the conclusive evidence about origi.. of *O_2 from H_2O* and *not from CO_2*, have for the first time since the discovery of photosynthesis was announced, made the picture clear about the most fundamental biochemical reaction in the living cells. The chemical mechanism of photosynthesis, as conceived today based on undeniable evidences and shorn of its formaldehyde-beclouded atmosphere, stands on solid foundations and can be summarised as follows:

It is an oxidation-reduction reaction between CO_2 and H_2O. In the first part of the reaction, photo-chemical dehydrogenation of water takes place at the expense of absorbed light energy by chlorophyll. Water loses its hydrogen, *i.e.* it is oxidised, and the liberated hydrogen forms a hydrogen donator with any compound which would take up hydrogen easily (A).

In the second reaction the hydrogen donator, *i.e.* a compound formed in the first reaction, which would readily give up hydrogen, would reduce CO_2. The second reaction might take place in dark—a chemical reaction, but the for mation of the hydrogen donator must be stimulated by light as dehydrogenation

of water is a photo-chemical reaction. It must be evident, here, that the present conception envisages that evolution of O_2 and the reduction of CO_2 take place in two separate steps, as can be seen from the following summary equations:

$$2A + 2H_2O \xrightarrow[\text{light energy}]{\text{chlorophyll}} 2AH_2 + O_2 \text{ (oxygen evolved).}$$

$$2AH_2 + CO_2 \xrightarrow{\text{chemical reaction}} CH_2O + 2A + H_2O.$$

A=any compound which would readily take up hydrogen (hydrogen acceptor). CH_2O here does *not* mean formaldehyde, but the *simplest* sugar.

That oxygen is given out during photosynthesis can be demonstrated by the following experiment.

Experiment 10. Photosynthesis: Evolution of oxygen (Fig. 173). A few submerged plants like *Elodea, Hydrilla* are taken in a beaker half-filled with water and are covered by an inverted funnel. A test-tube filled with water is placed on the stem of the funnel. Now the beaker is exposed to sunlight for some time. It is found that bubbles of gas evolve and collect on the end of the test-tube displacing water. By standard tests it can be determined that the collected gas is oxygen. The aquatic plants draw their CO_2 supply from the large amount dissolved in water. If distilled water is used instead of tap water, there will be no evolution of O_2, owing to lack of CO_2. If the tap water is previously boiled all the dissolved CO_2 will be driven off, and as a result there will be no evolution of O_2 and no photosynthesis.

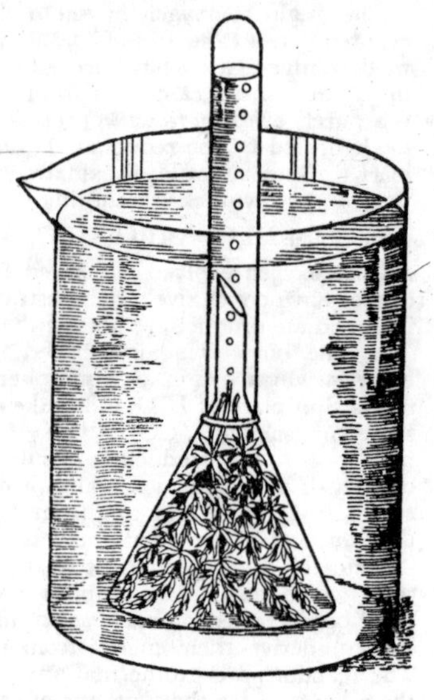

Fig. 173. Evolution of oxygen bubbles during Photosynthesis.

Factors affecting Photosynthesis

Photosynthesis depends upon some external and internal factors. As a matter of fact, the factors jointly influence the rate of carbohydrate formation in plants. The external factors are carbon dioxide content of the atmosphere, supply of water, light and temperature. As all these external factors act simultaneously on the photosynthetic process, the factor which is in the relative minimum will act as the *limiting* factor, *i.e.* will control the rate of the process. The important internal factor influencing the process is chlorophyll, the green-colouring pigment of the chloroplasts.

Carbon dioxide. This gas is present in air in only 0.03%, *i.e.* only 3 parts in 10,000. In industrial areas and above soil surface the percentage may be a bit higher. But it is the only source of carbon which, as we have seen, forms nearly half of the dry weight of plants. Many plants can utilise higher percentage of carbon dioxide. In fact, encouraging results were obtained by the application of extra carbon dioxide to crop plants. Carbon dioxide absorbed from air is replaced by the respiration of plants and animals, decomposition of organic matters, etc. The rate of photosynthesis increases when the concentration of carbon dioxide in air reaches near about 1%. As the percentage is very low this factor possibly decides the rate of photosynthesis under normal conditions.

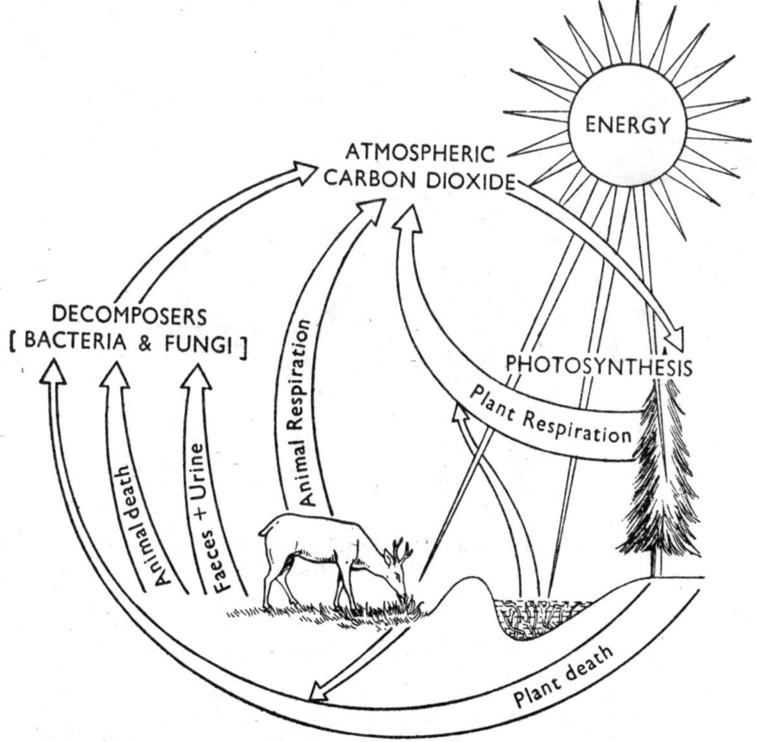

Fig. 174. Carbon cycle.

It may appear that continued intake of CO_2 from the atmosphere for photosynthesis by the green plants, both aerial and aquatic, results in deficiency of that gas in the air. But there are various ways by which CO_2 comes back to air to make up the deficiency.

During respiration, a process going on day and night, the plants and animals give out CO_2 and take in O_2. Plant and animal bodies after death and their refuses are gradually decomposed and broken down to simpler elements by a group of organisms—fungi and bacteria. These are referred to as decomposers and by their action ultimately CO_2 is released to the air. Moreover during combustion of coal and wood CO_2 is given out. Thus the total volume of this gas remains constant in the air. It may be represented by a cycle (Fig. 174) —something like a wheel having no beginning or end, known as **Carbon Cycle.**

That photosynthesis is not possible without the supply of carbon dioxide can be shown by the following experiment:

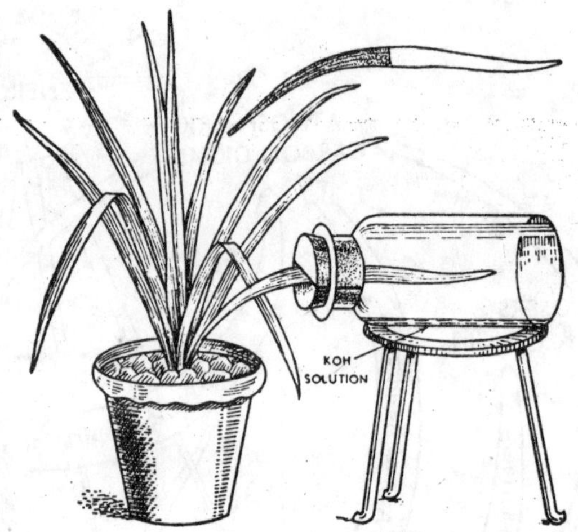

Fig. 175. Moll's experiment on Photosynthesis.

Experiment 11. To show that photosynthesis is not possible without CO_2, Moll's experiment (Fig. 175). A wide-mouthed bottle with a split cork at the mouth is placed sideways. A bit of caustic potash solution is put inside the bottle. Now a potted plant with long leaves is selected and terminal half of a leaf is put inside the bottle through the cork and the basal half remains outside. Even a leaf detached from the plant will serve the purpose, but the petiole in that case should be put in water. The bottle is made air-tight by putting vaseline on the cork. The bottle with the leaf is exposed to light till evening when the leaf is taken out, decolourised and subjected to iodine test. Half of the leaf lying outside the bottle gives positve reaction; whereas the half shows no starch formation, evidently due to the fact that carbon dioxide in the bottle has been absorbed by caustic potash. ɪt goes without saying that the experiment should be set up early in the morning with a starch-free leaf.

Water-supply. As it is another raw material for the manufacture of carbohydrates, water-supply has definite influence on the process. Low water-supply would certainly decrease the rate of photosynthesis.

Light. The enormous quantity of energy essential for the formation of carbohydrates comes from sunlight. The radiant energy absorbed is not, of course, lost but remains stored up in the product of photosynthesis as potential energy. Of the seven different rays of the sun, light red and blue rays are mainly absorbed by the chloroplasts. In higher plants chlorophyll is not formed in the absence of light when the plants get etiolated, though in many lower plants light is not essential for chlorophyll formation. That photosynthesis will not take place in the absence of light can be demonstrated by the following experiment:

Experiment 12. To prove that light is essential for photosynthesis (Fig. 176). A potted plant is kept in a dark room for two or three days, so that all the leaves become starch-free. To be doubly sure, one leaf is plucked and tested with iodine. Now portions of one or two leaves of the plant are covered carefully with black paper on the upper and lower sides with the help of a clip. The whole plant is exposed to sunlight. In the evening those leaves are collected, decolourised and tested with iodine. The exposed parts of the leaves turn blue, but the covered parts do not give iodine reaction, suggesting thereby that starch is not formed in the absence of light.

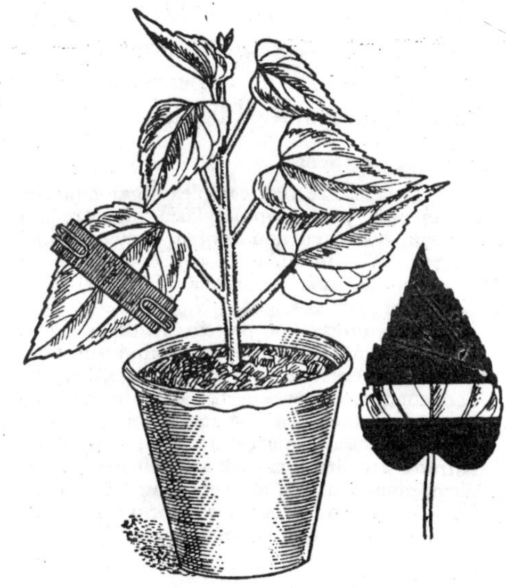

Fig. 176. Experiment to prove that light is essential for photosynthesis.

Temperature. Though photosynthesis takes place even in very low temperature, yet an optimum temperature near about 32°C is most suitable for the process. This factor, however, varies with the plants and climatic conditions.

Chlorophyll. The most important internal condition for photosynthesis is the presence of green pigment, chlorophyll, in the plastids. Photosynthesis is possible only because the chloroplasts can absorb radiant energy from sunlight. Chlorophyll is a complex organic matter. It is really a mixture of two pigments—chlorophyll a, which is blue-green in colour and chlorophyll b, which is yellowish-green. In higher plants the ratio of chlorophyll a and b is remarkably constant, varying from 3–3, 5 to 1. Two yellow pigments, carotin and xanthophyll, accompany chlorophyll. The yellow pigments are collectively called *carotenoids*. They are invariably present whenever there is chlorophyll, but they can also occur independently, as in many flowers and fruits. Chlorophyll can be extracted by boiling in alcohol. Chlorophyll also dissolves in acetone, chloroform, benzene, etc. Anethyl alcoholic solution of chlorophyll, if shaken with benzene, separates into two layers—the upper benzene layer contains chlorophyll a, chlorophyll b and carotin, while the lower alcohol layer contains xanthophyll. The complete separation of the four pigments involves still further elaborate methods. Chlorophyll in alcoholic solution exhibits fluorescence, *i.e.* it appears brownish-red by reflected light and green by transmitted light.

Protein Synthesis

Proteins are very complex organic matters found in the plants. They form an integral part of protoplasm and are also essential for nutrition. Besides carbon, hydrogen and oxygen, they contain nitrogen, often sulphur and phosphorus. The formulae of a few common proteins would show how complex they are. *Gliadin* (a protein of wheat)—$C_{685}H_{1068}N_{196}O_{211}S_5$; *Zein* a common protein of maize—$C_{726}H_{1101}N_{184}O_{208}S_3$.

The process of protein synthesis is complicated. Carbohydrates formed by photosynthesis serve as ground substance. Nitrogen, sulphur and phosphorus are derived from soil as salts, *i.e.* as nitrates, sulphates and phosphates and conducted upwards through the xylem vessels. Proteins are mainly constructed in the leaves having abundant sugar. This process is independent of light. Necessary energy is obtained by oxidation of sugar during respiration. Nitrates are first reduced into nitrites. Then amino acids, the simple soluble nitrogenous matters, are synthesised out of nitrites and sugars. Finally, amino acids are linked up together to form complex protein molecules by enzyme action. Thus proteins are chains of amino-acids.

Fat Synthesis

Fats are very important energy-giving food. They also make up a part of protoplasm. They are composed of carbon, hydrogen and oxygen like the caroohydrates, but the relative hydrogen-oxygen proportion is not maintined, the fats containing relatively little oxygen. *Olein*—a fat present in olive oil has the formula $C_{37}H_{104}O_6$. Fats are abundantly present in storage regions.

The process of fat synthesis is equally complex. Light and chlorophyll have no influence on this process. Fats are manufactured out of simpler matters, glycerine and fatty acids. Glycerine is possibly formed from glucose and fatty acids by fermentation of sugars. So here also the products of photosynthesis are the ground materials. By the action of enzyme, *lipase*, glycerine and fatty acids are condensed to form fats.

SPECIAL MODES OF NUTRITION

We have dealt with the normal methods of nutrition in the green plants, which can manufacture their own food. These self-supporting plants are called **autotrophic** or autophytes. Some plants, however, are devoid of chlorophyll and thus cannot synthesise their own food. They have to depend on external sources of food. These are **heterotrophic** plants or heterophytes.

Parasites

Plants, which grow on other living organisms and draw nourishment therefrom, are called **parasites,** and the living organisms attacked are referred to as the **hosts.** Many fungi lead parasitic existence on plant hosts and often cause enormous damage to the crop plants. They send out slender thread-like filaments into the body of the host for absorbing food matters. Phanerogamic parasites may be devoid of chlorophyll and are thus entirely dependent on the host. They are known as total parasites. *Cuscuta* or dodder (B. Swarnalata), *Rafflesia*, etc., are examples of total parasites. *Cuscuta* has weak stems which twine round the host. It produces special sucking roots or *haustoria*, which penetrate into the tissues of the host. These roots absorb food matters, water and salts from vascular bundles of the host plant (Fig. 44, 2). *Orobanche* is another total parasite growing on the roots of the plants of the potato family (*Solanaceae*), particularly tobacco. *Rafflesia*, yielding the largest flowers in the world, is a remarkable root-parasite (Fig. 16). It is a native of Java and Sumatra. Partial parasites have normal green leaves for manufacturing food, but having no connection with the soil they draw the entire supply of water and dissolved salts from the host. *Loranthus*, commonly growing on the stems of mango and other trees, is a partial parasite. Sandal wood plants attach themselves to the roots of other trees. So they are also partial parasites.

Saprophytes

Saprophytes are the non-green plants which derive food directly from decaying organic matters. Many fungi like moulds and mushrooms growing on rich humus or decaying organic matters are saprophytes. They usually send out small slender filaments, called *hyphae*, into the substratum and make use of the dead organic matters as food by enzyme action. A few flowering plants like *Monotropa* entirely lack in chlorophyll and draw nourishment from the partially decayed organic matters of the soil, like the saprophytic fungi.

Symbionts

Symbiosis is an association of two different kinds of organisms leading a life of mutual friendship. In connection with nitrogen-assimilation it has been mentioned that some bacteria can fix free nitrogen from the air. The bacteria attack the roots of leguminous plants and give rise to characteristic swellings or nodules by stimulating cell division in the cortical region. They convert nitrogen into nitrogenous compounds within the cells which partly become available to the plants. In exchange, the bacteria draw carbohydrate food matters from the plant. The bacteria alone cannot fix nitrogen, nor can the host leguminous plant. Only when there is an intimate association between the bacteria and the root-hair host cells nitrogen fixation can occur. Thus it is a case of symbiosis, what may also be termed reciprocal parasitism. These plants are called **symbionts.**

Some forest trees have roots encased by many soil fungi. This fungus-root association is called **mycorrhiza.** The fungus-infected roots lack in root-hairs and the fungal hyphae take up the function of the absorption of water and salts from the soil.

It has been observed that the root nodules of the legumes are sometimes pink or reddish in colour. Only recently the pink colour has been definitely identified to be *haemoglobin* with the characteristic absorption bands. The presence of haemoglobin in nature has been considered as general only in the vertebrate animals. This view may have to be seriously modified by the discoveries that the pigment exists in a ciliate *Parmoecium* and in the root nodules of the legumes. Neither the plant cells (host) nor the bacteria grown separately synthesise haemoglobin. There is thus no doubt that haemoglobin itself or the process connected with its synthesis may be directly linked with nitrogen fixation. The bacteria, after they have formed nodules by stimulating growth and multiplication of cells, supply the host plant with one or more factors essential for the synthesis of haemoglobin.

Insectivorous or Carnivorous plants

These are green plants capable of manufacturing carbohydrates. As the name suggests, the plants have peculiar devices for catching insects which supply them with a part of nitrogenous food requirements. Insect bodies are digested by secretion of enzymes and ultimately absorbed. Some of the common insectivorous plants will be described here (Figs. 177 & 178).

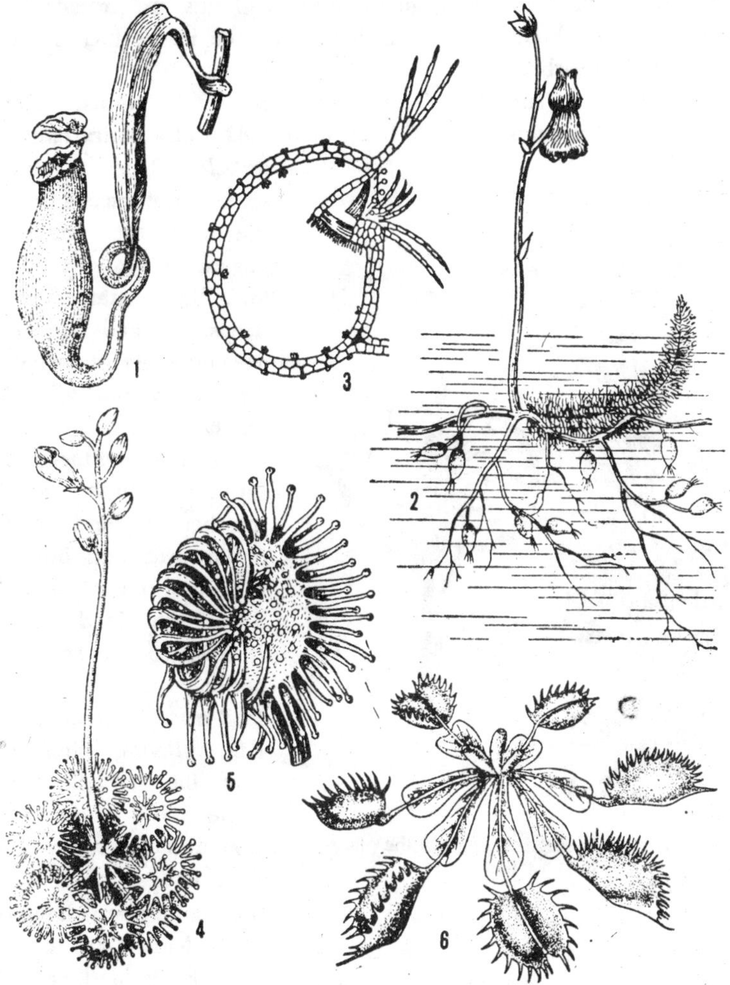

Fig. 177. Insectivorous plants. 1—*Nepenthes;* 2—*Utricularia,* with bladders magnified; 3—a bladder of *Utricularia* much enlarged; 4—*Drosera;* 5—a leaf of *Drosera* much enlarged; 6—*Dionaea.*

Nepenthes or pitcher plant. These are herbaceous plants. The leaf blade is modified into the hollow pitcher having a coloured hood at the top for attracting the insects. Many hairs remain pointed downwards inside the body of the pitcher. Small insects slip down in the cavity and get drowned in water which partly fill the pitcher. A large number of glands are present on the inner wall which secrete enzymes, and the insect bodies are digested and absorbed.

Drosera or sundew. These are also small plants. The spathula-like leaves of *Drosera* bear many glandular hairs called *tentacles*. Due to the secretion of viscous fluid the tips of the tentacles glisten in sunlight and look like so many dew-drops. Small insects are naturally attracted. As soon as one alights on the leaf some movement of the tentacles is initiated. Quite a good number of them bend downwards and imprison the poor insect. Then the dead body is digested by enzyme action and nitrogenous food matters are absorbed. Having finished that the tentacles resume their original position and get ready for next prey.

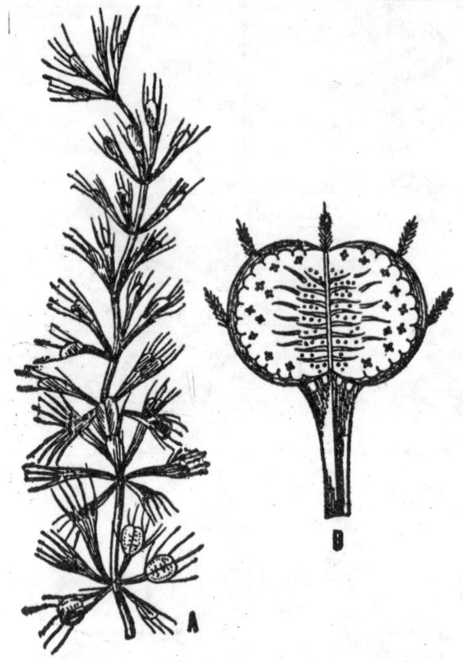

Fig. 178. *Aldrovanda:* A—part of the plant. B—One leaf open.

Utricularia or bladderwort. These are aquatic plants having much segmented compound leaves. Some of the leaflets are modified into swollen bladders for catching insects. Each bladder has a valve-like door which opens inwards, so that when the insects push in, they are caught. Glands present on the inner surface of the bladder secrete enzymes to digest the insect bodies.

Dionaea or Venus's fly-trap. It is a small insectivorous plant native of America. Here the leaves have three very sensitive hairs on each half, and a good number of teeth-like projections occur on the margin. When an insect alights on the blade, the two halves close quickly, as if they are hinged to the mid-rib and the teeth-like projections get interlocked (Fig. 3). Thus the insect is caught and the body is digested by usual enzyme action.

Aldrovanda. It is a root-less free-floating aquatic plant (Fig. 178) commonly found in the saline marshes near Calcutta and in the jheels of East Bengal. Here also the leaves have many sensitive hairs on the upper surface and teeth-like projections on the margin. The two halves of the leaves fold along the mid-rib and entrap the insects, the mechanism being same as in Venus's fly-trap.

TRANSLOCATION AND STORAGE OF FOOD

Water with dissolved mineral matters is absorbed and conducted upwards through the xylem elements to the leaves. Leaves are, in fact, the organs where complex organic food matters are mostly synthesised. Food matters are consumed particularly by the actively growing regions in simple soluble forms. As plants manufacture in excess of immediate need, the surplus food moves to the storage regions, to be stored up for future use. When necessary the stored food becomes available for nutrition. As the seat of manufacture is quite away from growing region and storage region, there must be arrangement for translocation of food. This transport takes place through the complex tissue, phloem, the sieve tubes taking the most active part.

For storage, simple soluble food matters are converted into their complex insoluble forms and they remain as such. The parenchymatous cells of pith, pith rays and cortex store food matters. Many plant organs are specially adapted for the purpose. Modified roots like radish, carrot, sweet potato, ginger, potato and other modified stems, fleshy leaves, bulbils, inflorescence of cauliflower store food matters. The seeds invariably have accumulated food in cotyledons or endosperm, what is converted into simple forms to be utilised by the growing embryo during germination.

DIGESTION AND ASSIMILATION OF FOOD

Food matters mostly remain stored up in the plants in complex insoluble forms. For utilisation and translocation they are changed into simple soluble matters. **Digestion is the** process of converting complex insoluble food matters **into their soluble** and diffusible forms, so that they may be properly utilised. Thus starch is converted into sugar, proteins into amino acids and fats into fatty acids and glycerine. Digestion may be *intracellular* or *extracellular*. In the former type, what is, in fact, common in plants, digestion takes place inside the cell. In extracellular digestion food matters are digested outside the cell and the products are used by the cell. The insectivorous plants and some fungi have extracellular digestion.

Digestion is really a process of hydrolysis, a chemical process, when complex matter is rendered simple by addition of water. It is mainly accomplished by the digestive juices, **enzymes,** which are complex organic colloidal substances secreted by protoplasm. They have the power of greatly accelerating the rate of breaking down complex food matters into their simple form without themselves being used up during the process. Thus they are organic *catalytic* agents. Action of enzyme is *specific*, *i.e.* each enzyme acts on one specific group of substance. Enzymes catalysing starch into sugar have no action on proteins or fats. Enzyme reactions are sometimes *reversible*. Those which catalyse breaking down of starch into sugar are also instrumental in synthesis of starch out of sugar. A few common enzymes are: **diastase,** which converts starch ultimately into sugar; **protease,** which breaks down proteins into amino acids; and **lipase,** which catalyses fats into fatty acids and glycerine.

The digested foods are mostly transformed into protoplasm. This process involving an actual incorporation of non-living matters into the living substance is known as **assimilation.** It is the final stage to find the ultimate goal of constructive metabolism. The process of actual transformation is not clearly known. Assimilation results in production of more protoplasm that is needed for growth and repair.

CHAPTER XVI

RESPIRATION

All living things need a constant supply of energy for their various activities of life. Sunlight is the ultimate source of energy. It has been mentioned that the solar energy is absorbed and stored up in carbohydrates during photosynthesis. In respiration carbohydrate is broken down, as a result the potential energy is transformed into kinetic form. Though it is a destructive or katabolic process, yet respiration is extremely beneficial, because it releases the necessary energy for performing the life functions.

Plants take in free oxygen from the air which enters through the stomata and lenticels and diffuses through the continuous intercellular space system to reach the living cells. It oxidises sugar, particularly glucose, into carbon dioxide and water vapour, thereby liberating considerable energy. The reaction may be represented thus:

$$C_6H_{12}O_6 + 6O_2 = 6CO_2 + 6H_2O + \text{energy (674 cal.)}$$

Though sugar is the main oxidisable material used during the process, other materials and in extreme cases, even protoplasm may be broken down for the liberation of energy. Carbon dioxide and water vapour go out through the stomata and lenticels. At night when stomata remain closed gaseous exchange takes place through the lenticels. Thus respiration is essentially a process of biological oxidation in which oxygen absorbed from the air breaks down the food (sugar) into carbon dioxide and water, thereby releasing the stored energy. It should be noted that respiration is not mechanical breathing involving simple inhalation of oxygen and exhalation of carbon dioxide, but a chemical process meant for the liberation of energy for the maintenance of life.

Respiration goes on day and night in all the living cells irrespective of their position in the plant organs; but the rate is vigorous in growing regions like root-tip, stem-tip, flower-buds and germinating seeds. During day time when the rate of photosynthesis is higher than the rate of respiration and when both the processes are going on simultaneously, all the oxygen liberated in photosynthesis is not consumed in respiration, whereas all the CO_2

liberated in respiration is completely consumed in photosynthesis and some more CO_2 must enter the plant for the higher rate of photosynthesis. As a result, during day time, the exchange of gases between the plant and the atmosphere is that O_2 is liberated and CO_2 is consumed. Thus the effects of respiration are completely masked during day time. During night, however, there is undoubtedly no photosynthesis, and only exchange of gases between the plant and the atmosphere is the effect of the process of respiration—O_2 consumed and CO_2 liberated.

Respiration in plants can be proved by the following simple experiment:

Experiment 13. Respiration (Fig. 179). A few germinating seeds or flower-buds are put in a flask with a cork at the mouth. Care must be taken to remove all the green parts from the flower-buds, otherwise the CO_2 liberated will be at once utilised in photosynthesis. A glass tube is fitted through the cork. Now one or two sticks of caustic potash are introduced into the flask. Two cotton plugs are put to keep the flower-buds and the caustic potash sticks in position. The flask with the tube is inverted over a trough of mercury, thus cutting off connection with external air. It is fixed in position with clamp and stand.

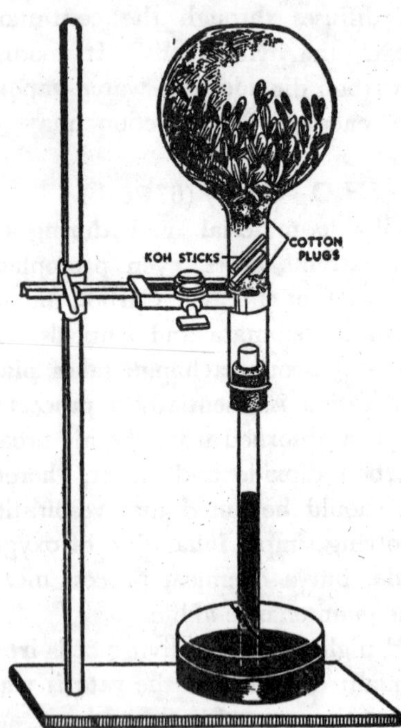

Fig. 179. Experiment on Aerobic Respiration.

After a few hours it is noted that the mercury has risen in the tube covering nearly one-fifth of the total volume. The volume of mercury risen can at its maximum be one-fifth of the entire volume of the flask, for O_2 in the atmosphere is only about 20%. It is due to the fact that carbon dioxide given out during respiration has been absorbed by caustic potash producing partial vaccum. Oxygen present in the flask has been utilised for respiration. The gas that remains in the flask is nitrogen. We know that nearly one-fifth of the volume of air is oxygen. So this experiment, besides showing intake of oxygen and giving out of carbon dioxide, also proves that volume of oxygen taken in is almost equal to the volume of carbon dioxide given out.

Photosynthesis and Respiration compared

Photosynthesis	Respiration
1. Photosynthesis is a constructive (anabolic) process by which food matters are synthesised, and so the plants gain in dry weight.	1. Respiration is a destructive (katabolic) process by which food matters are broken down and plants lose in weight.
2. Photosynthesis is confined to the chlorophyll-containing cells and takes place only in presence of light.	2. All the living cells respire day and night independent of chlorophyll and light.
3. Solar energy is absorbed and stored in sugar molecule as potential energy.	3. Energy is released—potential energy is transformed into kinetic energy.
4. Carbon dioxide and water are used, and carbohydrates and oxygen are the end products.	4. Oxygen is taken in to oxidise sugars, and carbon dioxide and water are the end products.
5. Photosynthesis is represented by the formula $6CO_2+6H_2O+\text{energy}=C_6H_{12}O_6+6O_2$.	5. Respiration is written thus: $C_6H_{12}O_6+6O_2 = 6CO_2+6H_2O+\text{energy}$.

Aerobic and Anaerobic Respiration

So long there is a normal supply of oxygen, respiration goes on uninterruptedly in all the living cells giving out water and carbon dioxide as end products. This is normal or **aerobic** respiration. If the supply of oxygen is cut off, plants would ultimately perish by suffocation, though death does not come immediately. Respiration continues for a short period. This is **anaerobic** or **intramolecular** respiration. Anaerobic respiration can only continue for a short time in higher plants. Ultimately the plants die, if they are deprived of CO_2 for longer time, as the energy released in the process is not sufficient for the continuance of the life processes for considerable length of time. Moreover, the accumulation of alcohol, one of the end products of the process, might reach toxic concentration causing permanent harm to the tissues. Anaerobic respiration ceases when the concentration of alcohol produced reaches about 15%. There is really a close similarity between aerobic and anaerobic types of respiration. It is known for certain now that in plants during first stage of respiration, either aerobic or anaerobic, no oxygen is needed. *Zymase*, the enzyme (organic catalyst secreted by living cells) extracted from yeast and which has also been found in all other green plants, is the agent responsible for the first part of respiration—which is known as *anaerobic cleavage* or breakdown. The sugars like glucose are converted by zymase into some other intermediate products. In presence of oxygen, *i.e.* in aerobic respiration, these intermediate products are further broken down and completely oxidised into CO_2 and H_2O; whereas in absence of oxygen, *i.e.* in anaerobic respiration, the intermediate products, under further action of zymase, are converted into C_2H_5OH (alcohol) and carbon dioxide, according to the following scheme:

CO_2+H_2O

Starch $\xrightarrow{\text{Hydrolysis}}$ Sugars ($C_6H_{12}O_6$) $\xrightarrow{\text{Zymase}}$ (Anaerobic cleavage)

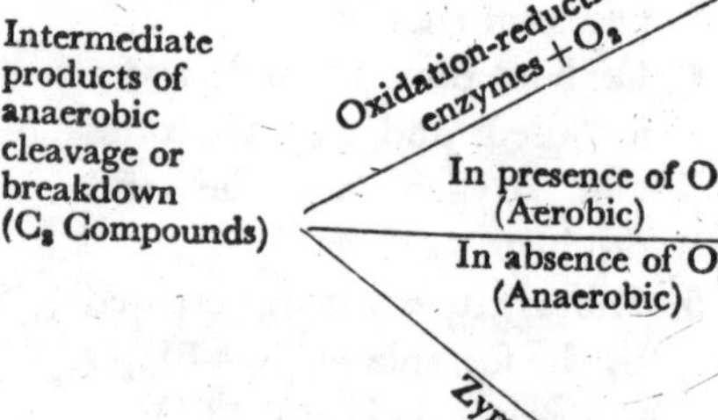

$C_2H_5OH+CO_2$

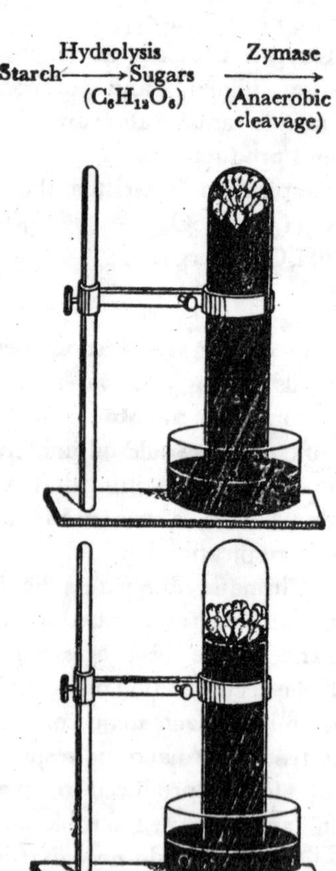

Fig. 180. Experiment on Anaerobic Respiration.

The amount of energy released in anaerobic respiration is much less in comparison to that liberated during aerobic process.

Here is an experiment to demonstrate anaerobic respiration:

Experiment 14. Anaerobic Respiration (Fig. 180). A test-tube is filled with mercury. A few properly soaked seeds are taken and after removing the seed-coats they are introduced one by one with the help of a bent forcep in the test-tube. The seed-coats are removed in order to facilitate the exchange of gases. The test-tube is placed in position with the help of clamp. After a day it is noticed that mercury column has come down and the closed end of the tube is filled with a gas. Now a bit of caustic potash is introduced in the same manner which absorbs the gas and mercury column again rises to fill up the tube, thus proving that the gas evolved is carbon dioxide.

Alcoholic Fermentation

Many micro-organisms like fungi and bacteria derive energy by anaerobic process, though it is rather uncommon in higher plants. This process is also called **alcoholic fermentation.** Yeast is the most common unicellular organism which excites alcoholic fermentation by secretion of an enzyme, zymase, breaking down sugar into alcohol and carbon dioxide with liberation of energy. $C_6H_{12}O_6$+Zymase=$2C_2H_5OH$+$2CO_2$+Energy (25 cal.)+Zymase. There is thus no difference between alcoholic fermentation in yeast and the anaerobic respiration which takes place in higher green plants in absence of oxygen. In both the processes sugar like glucose is the main respiratory substrate and both are catalysed by enzyme, zymase. The end products are also the same. In anaerobic respiration the reaction takes place inside the living cells, whereas alcoholic fermentation implies the fermentation outside the living cells, *i.e.* when the solution of sugar is catalysed by the enzyme, zymase. There are other types of fermentation where, instead of alcohol, organic acids such as lactic, butyric, etc., are formed. Yeasts and many other groups of bacteria have only anaerobic method of respiration and they thrive best in an oxygen-free atmosphere. They are called *anaerobes*, in contrast to the higher green plants and other bacteria which may be termed *aerobes*.

CHAPTER XVII

GROWTH AND MOVEMENT

Growth is one of the fundamental features of the living organisms. Non-living objects like rocks and crystals may also grow by accretion or deposition of new materials, but growth in living things is something distinctly different. It is really a complex phenomenon regulated by protoplasm. During anabolic activities complex food matters are synthesised and ultimately assimilated. As a result, an increase in size occurs and new tissues are formed. In katabolic activities complex matters are broken down for the liberation of energy, which is essential for life processes. The two processes go on simultaneously, but anabolism gains over katabolism and the ultimate result is growth. Growth may be defined as a permanent increase in mass or volume, accompanied by irreversible change of form or structure. Usually growth is attended by increase in weight. But in germinating seeds or sprouting potato tuber there is a loss of weight during growth, obviously due to utilisation of stored food matters.

In the lower plants, like unicellular organisms, growth is simple enlargement of the cell; but in the vast majority of plants, besides enlargement, new cells are continually formed which enlarge and mature in due course. The process of growth consists of three phases; (*a*) **Phase of formation,** involving formation of new cells. In plants it is localised in the meristematic regions. The apical meristems of stem-tip and root-tip are primarily responsible for the growth in length of the axis; and lateral meristems, cambium and cork-cambium, are instrumental to the secondary increase in thickness in the stems of dicotyledons and gymnosperms and roots of dicotyledons. (*b*) **Phase of elongation** lies next to the phase of formation. Here the cells enlarge, protoplasm gets more and more vacuolated ultimately lining the cell wall. This is the region of active growth. (*c*) **Phase of maturation** is the final phase where the cells assume permanent shape and form.

The rate of growth is imperceptibly slow in plants and so can be hardly detected and measured without the help of suitable instrument. Certain plant parts like tendrils of *Cucurbita*, shoot of bamboos, etc., however, exhibit rapid growth. **Auxonometer** is a common instrument used for measuring growth by which the actual growth in length may be magnified many times and conveniently recorded. **Lever auxonometer** or **arc-indicator** has a pointer fixed to a wheel and a graduated arc as shown in figure 181.

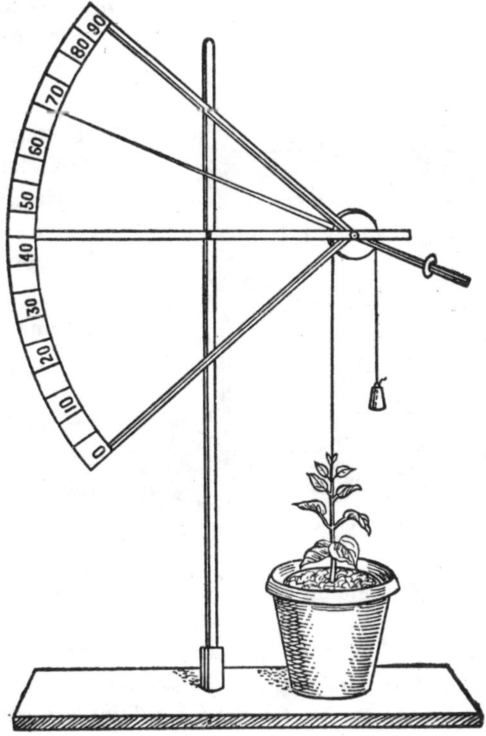

Fig. 181. Arc-indicator for measuring growth.

A cord attached to the growing tip of a potted plant is passed over the wheel and allowed to remain suspended at the other end by a weight. With the growth of the tip the indicator moves and the increase is recorded on the graduated arc, of course, on a magnified scale. From the record thus obtained the actual increase may be calculated.

That maximum growth takes place in the region behind the root-tip may be shown by a simple experiment.

Experiment 15. Region of maximum growth (Fig. 182). A few seeds are germinated. The radicle of a seedling is marked carefully with water-proof Indian ink at regular intervals, say of 1 mm., with a space-marker. The seedling is allowed to grow normally. It can be well done by placing the seedling in a bottle making proper arrangement for aeration. After a couple of days it is found that the lines are widest at the region next to the **root-tip.**

Factors affecting Growth

The rate of growth is influenced by the simultaneous operation of a number of external and internal factors. Some factors affecting growth are discussed below:

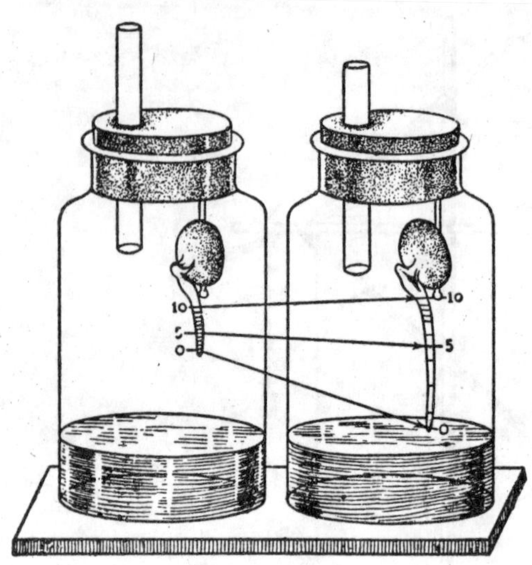

Fig. 182. Experiment to show the region of maximum growth.

Supply of nutritive materials. Growth, as we know, is the result of the predominance of the constructive phases of metabolism over the destructive ones. For growth, plant organs always need sufficient supply of food which is digested and assimilated to form new protoplasm and other substances like cellulose to construct the skeleton of the plants. Like animals, plants also require a 'balanced ration', *i.e.* food matters in proper proportion for their activities and growth.

Moisture or water. Turgidity of the cells is a pre-requisite for growth, for the maintenance of which an ample supply of water is necessary. Water is also essential for carrying food matters to the growing regions. Protoplasm itself remains saturated in a high percentage of water. Generally plants of habitats with abundant water supply have luxuriant growth, as found in the tropical forests.

Soil. The nature of the soil, and particularly the nutrient matters present in the soil, profoundly influence the growth of plants, as we have seen in water culture experiments.

Oxygen. A supply of oxygen is necessary for respiration, by which process energy for the vital activities is released.

Temperature. It has a marked effect on the rate of growth. A suitable temperature, which varies with climate and plants, is most conducive to growth. There are three cardinal points, *viz.* minimum, below which growth will not take place; optimum, which is most favourable and gives the highest rate of growth; and maximum, above which growth will not be possible. The optimum temperature, on the average, is near about 30°C. Within proper limits the rate of growth is roughly doubled with the rise of temperature by 10°C.

Light. Though at the initial stage light, instead of accelerating the process, retards growth, yet it plays a very important role in the growth of the plants. It is the ultimate source of energy that is stored up during photosynthesis in the carbohydrates without which growth would not be possible. Besides, it has influence on the formation of chlorophyll and on the opening of the stomata. Some lower plants may carry on throughout their life in darkness, but higher plants get etiolated in the absence of light. The organs become elongated and soft, are deprived of chlorophyll and thus grow or become yellowish in colour. These plants ultimately perish for want of food. Only plants with sufficient stored up food matters can continue to live in darkness for a long time.

The duration of light has a positive influence on growth and particularly on flowering and fruiting. The response of plants to the daily period of illumination is called **photoperiodism.** It has attracted the attention of many investigators in recent years who have shown experimentally that the rate of growth and time of flowering could be profoundly affected by the length of day-light. It may be of much economic importance, in view of the fact that by artificially changing the length of day-light the vegetative growth and flowering may be considerably controlled.

Growth Hormones. These are organic growth-stimulating substances secreted by the plants. Much have been heard about hormones secreted by ductless glands in the animal bodies. Even plants secrete hormones which regulate growth, regeneration and the healing of wounds. Like vitamins very minute quantities of hormones are needed for proper action. Plant hormones go by the name **auxins.**

Grand Period of Growth

The rate of growth is not uniform in plants. Plant cells or organs or even the whole plant show variations in the rate of growth. It starts slowly, the rate is gradually accelerated till it reaches the maximum, then falls slowly and ultimately comes to a standstill. The whole period from the initiation to the stoppage is referred to as the **grand period of growth.** It may be easily determined by measuring any growing plant organ at regular intervals.

The rate also varies with factors influencing growth. It is more rapid at night than in the day, obviously due to inhibiting influence of light. This is called diurnal variation of growth. There are seasonal variations as well. For many plants winter is very unfavourable for growth, whereas spring is the favourable period.

Vernalisation

Vernalisation is the process by which seeds are subjected to treatments before sowing, to hasten flowering in the plants they would give rise to. The famous Russian botanist Lysenko suggested this pre-treatment method. The principle of vernalisation, to put very briefly, is as follows: Seeds are soaked as usual. When germination has commenced moisture is controlled to arrest further growth. Now the seeds are subjected to treatments like exposure to light and darkness, high or low temperature. Seeds sown after these treatments would bring about earliness in flowering.

As vernalisation has immense possibilities in agriculture, it has created enough of sensation in recent years. Soviet Russia is the only country where it has been extensibly practised in fields with good results. Vernalisation is receiving the attention of investigators all over the world. In our country some scientists are engaged in research on vernalisation of economic plants.

According to some, vernalisation is an old practice under a new name. A process similar to vernalisation was recorded in an American reference book on agriculture some time in the middle of the last century. Indian farmers also subject the seeds to pre-treatment (by cutting off light) before sowing, with good results without really knowing anything about the principle of vernalisation.

MOVEMENT

Movement is another characteristic feature of the living things. Since most of the plants remain rooted to the soil they are unable to move from place to place like the animal. The capacity of movements of plants is often consequently doubted. Many lower forms of plants like unicellular organisms have the distinct power of locomotion. The higher plants show movements by changing position of the organs in various ways. Opening and closing of the flowers, behaviour of the sensitive tendrils while coiling round a support, drooping of the leaflets of the sensitive plants when touched, movements of aerial stems and underground roots in response to light and force of gravity are some glaring instances of plant movements. There is no denying of the fact, however, that the rate of movement in higher plants is usually very slow.

Movements may be of two classes:

(A) **Movement of Locomotion,** where the entire organism may move from place to place.

(B) **Movement of Curvature,** where the organs of the stationary plants change positions and thus curve in different ways.

Both the kinds of movements may be (*i*) **autonomic** or spontaneous, *i.e.* without any external influence, or (*ii*) **induced,** *i.e.* due to some kind of external influence or stimulus. The capacity of responding to stimulus is **irritability** which, we know, is an inherent property of protoplasm. Irritable movements help the plant organs to adjust themselves favourably to the environmental conditions, so that the vital activities may be most advantageously carried on.

A. Movements of Locomotion

(*i*) **Autonomic.** Amoeboid and ciliary movements of naked masses of protoplasm and streaming (rotatory and circulatory) movements of protoplasm in walled cells are instances of autonomic movement of locomotion. This type of movement of many algae may be seen under the microscope.

(*ii*) **Induced.** Movement of the entire organism may be influenced by the external stimuli like light, chemical substances, etc. These are called **taxisms** or tactic movements. **Phototaxis** is the induced movement of locomotion, where the stimulus is light.

Many algae show phototaxis by moving towards weak light and away from strong light. **Chemotactic** movement in response to stimulus of chemical substances is exhibited by the male gametes of fern and moss. Fern spermatozoids rush towards the egg being attracted by malic acid and salts of malic acid secreted by the archegonium. Similarly sucrose secreted by the moss archegonia attracts the sperms. Many bacteria move towards the source of oxygen.

B. Movement of Curvature

(*i*) **Autonomic.** In fixed plants autonomic movements of organs are associated with growth or variations in the turgidity of the cells. Growth movements are clear in twiners and leaves. A stem twines round the support only on account of unequal growth on two sides of the organ. The same is the case with rolled or folded leaves, which remain rolled due to rapid growth on the under surface and open out due to the rapid growth of the upper surface. Autonomic movement due to variations of the turgidity of the cells is noticed in *Desmodium gyrans* or Indian telegraph plant, where the two lateral leaflets move up and down from morning till evening, but the terminal one remains fixed (Fig. 183).

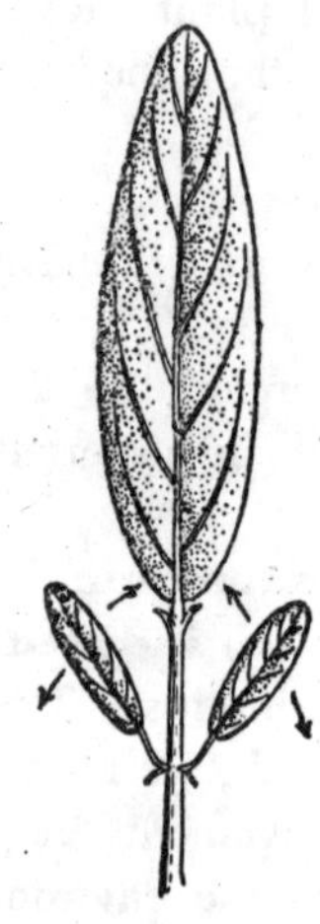

Fig. 183. Movement of variation in Indian telegraph plant.

(*ii*) **Induced.** Induced movements of curvature are caused by external stimuli like light, force of gravity, moisture, etc. These movements are again of two kinds, *viz.* **tropisms** or **tropic movements,** where the direction of stimulus determines the direction of movement; and **nasties** or **nastic movements,** where the direction of movement is independent of the direction of stimulus.

Tropic movements are usually of the following types:

1. **Phototropism.** Phototropism or heliotropism is the movement initiated by the stimulation of light. Stems always grow towards light. So they are **positively phototropic.** The roots go away from light and are **negatively phototropic.** Leaves orient themselves at right angles to the source of light just for absorbing the maximum solar energy required for photosynthesis. They are transversely phototropic or **diaphototropic.** When potted plants

are placed on windows, the stem is always found to go outwards, *i.e.* towards the source of light. Phototropic curvatures may be shown by growing a mustard seedling in culture solution in a jar (Fig. 184).

2. **Geotropism.** The capacity of roots and stems to respond to the force of gravity is called **geotropism.** Roots always go downwards showing **positive geotropism** and stems grow upwards and are thus **negatively geotropic.** The lateral roots and branches are **transversely geotropic.** If a seed is germinated and the axis is placed horizontally on a tray in a dark room (for eliminating the stimulus of light), in a day or two it will be found that the root is bending downwards and the stem upwards showing positive and negative geotropism of root and stem respectively. Curvature usually takes place in the region of elongation behind the tip.

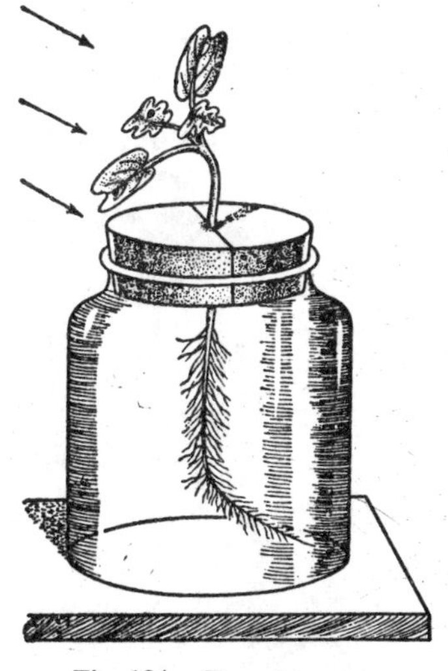

Fig. 184. Experiment on Phototropism.

3. **Hydrotropism.** It is the induced movement caused by the stimulus, moisture. Roots always grow towards the source of water and are thus **positively hydrotropic.** An interesting example of negative hydrotropism can be observed in the pneumatophores of mangroves where the roots are simultaneously negatively hydrotropic and negatively geotropic.

4. **Haptotropism.** This is the movement of curvature in response to contact with foreign bodies. Tendrils quickly coiling round a support, nicely exhibit haptotropic movement.

5. **Chemotropism.** This movement in response to chemical stimulus is noticed in the leaves of *Drosera*, where the nitrogenous matters induce the bending down of the tentacles. The movement of the pollen-tube through the style towards the embryo-sac is also an instance of chemotropism.

Nastic movements

Nastic movements are independent of the direction of stimulus, *i.e.* the stimulus acts from all directions and the direction of the response bears no relation whatsoever with any particular direction of the stimulus. Movements exhibited at the opening and closing of flowers in response to the variations of light intensity and temperature or both, are called **nyctinastic** movements. Flowers like lotus, lily, sunflowers open in the morning and close in the evening; similarly nocturnal flowers like jasmine, *Cestrum* (B. Hasnahana) open in the evening. Leaves of many plants,

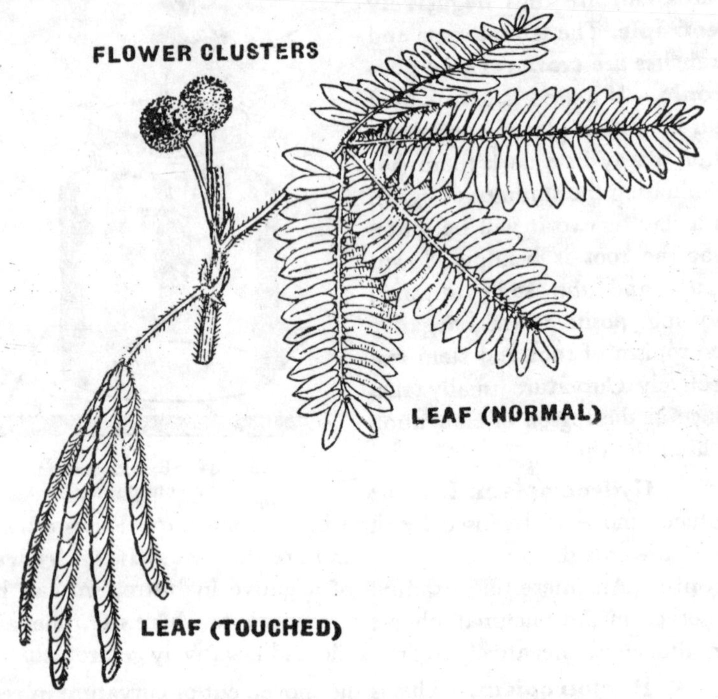

Fig. 185. Seismonastic movement in *Mimosa pudica*.

particularly of family Leguminosae, show sleep movements resulting in the folding up of leaflets with the fall of light and temperature.

A movement as a result of shock stimulus is called **seismonasty.** An unusual but interesting type of seismonastic movement is exhibited by the leaves of sensitive plant, *Mimosa pudica* (B. Lajjabati). Leaflets of this plant close up on touch (Fig. 185). Though not directive, intensity of stimulus markedly affects the degree of movement. On slight shock one or two pairs of leaflets collapse but stronger stimulation results in closure of many leaflets and the ultimate drooping of the leaf. Rapid conduction of the stimulus to all the aerial leaves is a striking feature of this movement in sensitive plant. Movements are due to the changes in the turgour of the cells of the swollen pulvinus (leaf-base). Here the stimulus is perceived by the leaf, whereas the action takes place at a distance from it; there is apparently a sensory organ which receives the stimulus, and the motor organ is the swollen pulvinus which brings about the movement. An analogy is apparent here with the nervous mechanism in animals.

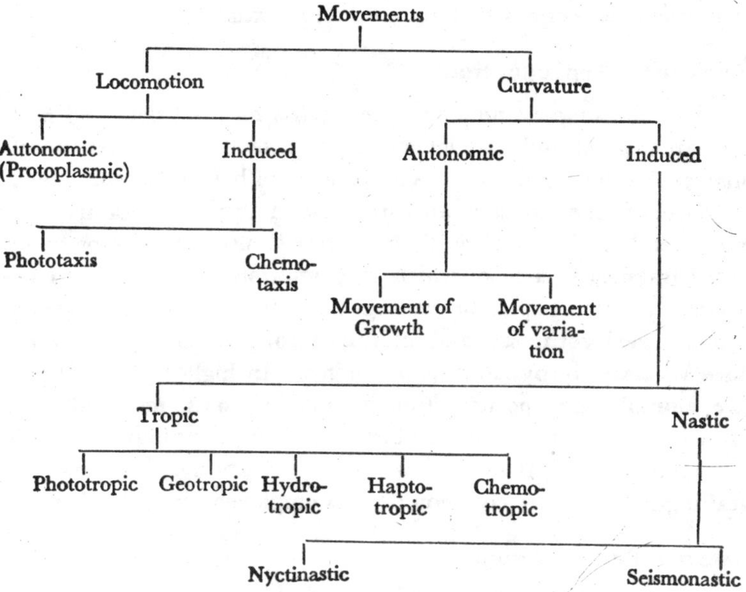

CHAPTER XVIII

REPRODUCTION

Only living things have the capacity of producing young ones like themselves according to the *law of biogenesis*. This is **reproduction.** As the duration of life is limited, reproduction ensures continuation or perpetuation of the race. Plants reproduce by three methods: **vegetative, asexual** and **sexual.**

Vegetative Reproduction

It is the simplest method, where parts may get detached from the parent plant and lead independent existence. In lower unicellular plants like yeast, outgrowths in form of buds come out which, in course of time, are separated from the mother cell. In filamentous algae like *Spirogyra* the filament breaks up into one- or more-celled fragments, each of which may multiply and carry on independently. Bryophytes like mosses and liverworts produce special bodies, called **gemmae,** and multiply through them. Ferns reproduce vegetatively by rhizomatous stems. In higher plants vegetative multiplication occurs through modified roots, modified stems like runners, rhizomes, etc., bulbils and adventitious buds. Artificially multiplication is secured by cuttings, layerings, buddings and graftings. These are common devices employed by the florists.

Asexual Reproduction

Here special minute, usually unicellular bodies, called **spores,** are formed which directly can produce new individuals. Spores may be naked or walled. Naked motile ciliated spores of algae are called **zoospores.** The spores of bryophytes, pteridophytes and higher plants are non-motile and have cell wall. The spore-containing sac is called **sporangium.**

Sexual Reproduction

This method essentially involves the fusion of two usually naked sexual or reproductive units called **gametes.** Individually gametes are of no use for reproduction, but by fusion they give rise to the **zygote,** which alone can produce a new plant. When the two gametes are morphologically similar, they are called **isogametes.** Product of fusion of two isogametes is **zygospore** and the process is known as **conjugation.** Dissimilar gametes are called **heterogametes.** Obviously one of them is male and the other is female. Here zygote is called **oospore** and the process is **fertilisation.** Of the two heterogametes, the male, called **antherozoid** or **spermatozoid,** is, as a rule, much smaller than the female gamete and is motile and active. The female gamete is much larger, always stationary or sessile and passive. It is called **egg, ovum** or **oosphere.** The male gamete-containing sac is known as **antheridium** and the female gamete-containing one is **oogonium** or **archegonium.**

Parthenogenesis

Though sexual reproduction essentially demands union of two gametes, in some lower plants one of the gametes (usually the female one) behaves like a zygote without uniting with the corresponding one. This process is called **parthenogenesis,** and the spore concerned is known as **azygospore** or **parthenospore.**

Apogamy and Apospory

Normally the plants having sexual method of reproduction show an alternation of two generations—*gametophytes* with haploid chromosomes, and *sporophytes* with diploid chromosomes. But there are cases in which one generation may be wholly or partly eliminated. If the eliminated part be gametophyte, the phenomenon is called **apogamy;** and if that is sporophyte, the phenomenon is known as **apospory.**

In many ferns buds may vegetatively develop on the prothallus which grow into sporophytic plants. Thus formation of sporophyte has taken place without the act of fertilisation, which is not certainly normal. In some seed-plants certain cells of the nucellus, which belongs to the sporophyte, may develop into embryos, thus completely eliminating the whole gametophytic generation. These are cases of apogamy.

Apospory is quite common in some ferns in which prothallus, the gametophyte, may develop from the sporangia, and even from the leaf. The prothallus so formed bears sex-organs like a normal one, though it has not developed from a spore.

There are ferns and seed-plants which exhibit both apogamy and apospory.

ECOLOGY

CHAPTER XIX

PLANTS IN RELATION TO ENVIRONMENT

In the preceding chapters reference has been frequently made to external factors influencing characters, structures and different physiological activities of the plants. These external factors collectively constitute **environment.** The branch of botany which seeks to investigate the relation of the plants to their environment is known as **ecology.** Ecology, in fact, includes studies on morphological forms, anatomical structures and physiological activities. Plants adjust themselves marvellously to the varying environmental conditions. This power of adjustment is referred to as **adaptation.**

Factors making up environment may be divided into:—(*i*) **climatic,** which act through atmosphere, (*ii*) **edaphic,** which operate through soil; and (*iii*) **biotic,** which come from living things, plants and animals.

Climatic Factors

Important climatic factors affecting plants are temperature, light, humidity, precipitation and wind. All the factors are interdependent. Further, climate markedly affects soil and biotic factors. In a fast country like ours climate varies from tropical to temperate.

We have seen that all physiological processes go on most actively at optimum **temperature** which is widely variable with plants and climatic conditions. There are the cardinal points, minimum and maximum, below and above which activities are not possible. **Light** is the ultimate source of radiant energy indispensable for photosynthesis. It has positive influence on transpiration, growth, development and movement. Water vapour present in air is termed atmospheric **humidity.** It precipitates mainly in form of rain. A part of rain water percolates down the soil and becomes available to the plants. Rainfall has marked effect on vegetation. In our country rainfall varies from about 7·5 cms or 3 inches a year in the arid regions of Rajputana to nearly 1200 cms in Cherrapunji in Meghalaya, the region of the highest rainfall in the world. In India rains are associated with monsoon winds. **Wind** profoundly affects the rate of transpiration, dispersal of fruits and seeds and shape and branching of trees. Plants adjust themselves to stand against high wind prevailing in situations like sea-shore and deserts.

Edaphic Factors

Physical nature and chemical properties of soil, and particularly amount of water present in the soil available to the plants are important edaphic factors.

Biotic Factors

These include influence of living things on plants. Men, grazing animals, birds, earthworms, and even fungi and bacteria present in soil variously influenced the vegetation.

Ecological Classification of Plants

As water is of the greatest importance to the plants, it is customary to classify the plants into the following ecological groups on the basis of their water relation.

1. **Hydrophytes** or aquatic plants.
2. **Mesophytes,** the common land plants, flourishing best in moderate water-supply.
3. **Xerophytes,** plants of very dry regions like deserts.
4. **Halophytes,** plants growing in sea-shore or saline marshes where absorption of water becomes difficult.

Hydrophytes

These are plants which grow in situations with high percentage of water. They may be free-floating, submerged or amphibious.

Roots and analogous organs are very poorly developed as the plants can absorb water by entire surface. Stems are usually soft and spongy. Leaves are generally dissected, ribbon-like in submerged plants with very scanty cuticularisation. Stomata and palisade cells are lacking. Conducting and mechanical tissues are very poorly developed. Air-spaces are abundant for giving buoyancy to the plants. Many aquatic plants often show heterophylly. Vegetative multiplication far exceeds sexual reproduction. Fruits and seeds are often provided with air chambers for facilitating dispersal by water.

Amphibious plants have poorly developed root system and ample air-spaces like typical hydrophytes; but they more or less resemble the mesophytes in characters of leaves and arrangement of conducting tissues. Common plants—*Vallisneria* (B. Pata shaola), *Hydrilla*, *Utricularia* (B. Jhanji), water-hyacinth, *Pistia* (B. Barapana), lotus, lily, water-plantain, arrowhead.

Mesophytes

These are the common land plants growing in moderate water-supply. So far as structural peculiarities are concerned, they are intermediate between hydrophytes and xerophytes. Leaves are usually broad and thin and possess well-developed stomata. Production of root-hairs is a distinctive character.

Xerophytes

These are plants growing in very dry situations like deserts. They have adaptations to stand against extreme heat, dry air, intense light and high wind.

Root system is very strongly developed, both in length and volume. Aerial portions are stunted in growth. Transpiring surfaces are very much reduced for minimising the loss of water. Leaves are often modified into spines and scales; they may be thin, rolled or folded. Strong development of cuticle with extra-coating of wax, resinous matters, sunken stomata, formation of dense hairs, etc., are distinctive characters. Succulence for storage of water is very common. Common plants—Cacti, *Opuntia* (B. Fanimanasha), spurges, *Ravenala* (traveller's tree, B. Pantha padab).

Halophytes

These are plants growing in saline marshes or on sea-shore where absorption of water becomes difficult due to the preponderance of salt. This type of soil is called *physiologically dry*.

Plants of saline marshes form a special type of vegetation of low forest, called **mangrove,** as found in Sundarbans. The plants produce stilt roots for extra support and special aerating roots or *pneumatophores* for absorbing oxygen from air. Succulence is common in some plants. They usually show cymose type of branching. A special mode of germination, called **vivipary,** is exhibited by the mangrove plants (Fig. 37). Common plants—*Rhizophora*, *Heritiera* (B. Sundri), *Ceriops* (B. Goran).

CHAPTER XX

GENERAL PRINCIPLES OF CLASSIFICATION

Diverse forms of plants, ranging from the smallest bacteria to the gigantic forest trees, exist on the earth. They are not only numerous but varied in nature. It has been mentioned in the introductory chapter that more than 340,000 plants have been identified, described and named by the botanists. The number is increasing, as newer ones are being discovered. Quite a large number of them possibly still remain unknown. The main object of systematic botany is to bring about an orderly system out of chaos by classifying the plants into groups, on the basis of their relationship.

Even the early Greek naturalists devised classifications. They used as their bases arbitrary characters like utility, habit, etc. So their systems were *artificial*. The present system of classification began with Carolus Linnaeus, the famous Swedish botanist of the 18th century, who was referred to as the 'father of classification'. He classified plants into groups on the basis of reproductive or sexual characters, particularly on the characters of the stamens. This classification was also artificial, as it was based on one character. It can be compared to the arranging of books in a library in alphabetic order of the names of authors, disregarding altogether the subject matters of the books. In the nineteenth century George Bentham and Joseph Hooker introduced a natural system based on a number of characters and attempted to trace the 'blood relationship' amongst the plants through common descent. Though other systems have since then been suggested, Bentham and Hooker's classification is widely accepted in our country and is followed in the herbaria.

According to this system, the plant kingdom is divided broadly into two sub-kingdoms: **Cryptogamia** or flowerless plants and **Phanerogamia** or **Spermatophyta,** or the flowering plants. Flowerless plants reproduce by simple fission or by minute, usually unicellular spores; and the flowering ones reproduce by seeds containing multicellular embryo. Cryptogamia, again, are divided into three classes, *viz.* **thallophyta, bryophyta** and **pteridophyta.** Thallophyta include the lowest plants. The plant body of a thallophyte is an undifferentiated mass of cells called *thallus*. The green aquatic thallophytes like pond-scum, sea-weeds are called **algae;** and the non-green ones like moulds and mushrooms are **fungi.** Bryophytes like mosses may have small stems and simple leaves but not true roots. Hairy bodies called *rhizoids*, carry on the functions of roots. They always lack in vascular tissues, and usually live in moist situations. Some bryophytes have thalloid bodies with rhizoids to fix them on the substrata. Pteridophytes are the highest cryptogams. They are primarily land plants with the plant bodies well-differentiated into roots, stems, and leaves which flourish well in moist and shady places. They possess distinct vascular bundles. Ferns belong to this group. Phanerogams or Spermatophytes are the highest plants which bear flowers and reproduce by seeds. They possess complex vegetative organs and very well-developed vascular tissues. Spermatophyte are divided into **Gymnospermia** or naked-seeded plants (*e.g.* pine, cycad, etc.) and **Angiospermia** or covered-seeded plants. In gymnosperms ovules remain exposed and are directly fertilised; whereas in angiosperms ovules remain within the ovary and which, as we know, with the stimulus of fertilisation develops into the fruit containing seeds. Angiosperms are **dicotyledonous** (*e.g.* pea, mango) and **monocotyledonous** (*e.g.* maize, rice), according to the number of cotyledons in the embryo. They are, again, divided into a number of sub-classes. An outline of this classification has been given in the introductory chapter.

Darwin's doctrine of evolution, which is certainly the most epoch-making event in the history of biological studies, maintains that the higher forms have been derived from the simple ones that lived in the past. That doctrine prepared the way to natural classification based on relationship through common descent. The above system shows, beyond doubt, a gradual ascent in the complexity of structures from the lowest to the highest plants indicating, at the same time, gradual adaptation from aquatic to terrestrial or land habit.

Units of Classification

The **species** is the unit of classification. All individual plants, which resemble one another in important vegetative and reproductive characters and differ visibly from other plants, constitute a species. All the plants of a species are thus supposed to have descended from the same ancestor. Let us take an example. All banyan trees have similar characters, no matter where they grow. So they form a species, say banyan species. Similarly pipul forms a species; fig forms a third species. Plants belonging to the same species may show variations in minor characters due to environmental conditions and hybridisation. They are but *varieties*.

Banyan, pipul and fig plants bear the same type of reproductive organs. They form a **genus** (pl. genera). Thus a genus is a collection of species having similar reproductive characters. Allied genera constitute a **family** or **natural order;** natural orders are again grouped into **cohort,** cohorts into **series;** and so on, till plant kingdom is reached.

Nomenclature

Plants identified and described must be named. Though almost all the plants have common names but they do not serve our purpose. For proper co-ordination amongst workers names should be internationally acceptable. **Linnaeus** introduced 'binomial system' of nomenclature which has been followed all over the world. Every name has two parts, the generic part and the specific part. The generic part of the name is written in capital and the specific part in small initial letter. Both the names as a rule are indicated in italics and the name of the author in Roman script. Banyan is *Ficus benghalensis*, pipul is *Ficus religiosa* and fig is *Ficus hispida*. So all the three belong to the same genus *Ficus* but form different species. The abbreviated name of the author or the person who first published the name of the plant with a suitable description, is put after the specific name. Pea is *Pisum sativum* Linn. Here Linn indicates that Linnaeus was the author

CRYPTOGAMS

CHAPTER XXI

THALLOPHYTA

This group includes the lowest and the most primitive plants. The plant body of every thallophyte is an undifferentiated mass of cells, called *thallus*, showing no division of labour. Organs like root, stem and leaves are never found in this group. Some thallophytic plants are small unicellular organisms; while many of them are multicellular. They are divided into two groups: **algae,** the green thallophytes, and **fungi,** the non-green ones. Points of difference between the two groups are shown below :—

1. Algae contain chlorophyll and, so are green in colour.	1. Fungi are devoid of chlorophyll and, hence, are non-green.
2. Algae can manufacture their own food. They are *autophytes* or autotrophic plants.	2. Fungi cannot manufacture food but depend on other sources for the supply of food. They are *heterophytes* or heterotrophic plants. Fungi may be *parasitic* on living organisms, both plants and animals, or *saprophytic* on dead rotten organic matters.
3. Algae can grow only in light, as that factor is indispensable for food-synthesis.	3. Fungi can grow both in light and darkness.
4. The plant bodies of algae are composed of true parenchymatous cells.	4. Bodies of fungi are usually made of fine threads, called *hyphae*, which usually form an interwoven mass. A collection of hyphae is called *mycelium*. The interwoven mycelia often look like parenchyma cells and, hence, called pseudo-parenchyma or plectenchyma.
5. The cell wall in algae is usually made up of cellulose.	5. Here the wall is made of fungus-cellulose or chitin.
6. Starch grains are usually the reserve materials in algae.	6. In fungi reserve materials are mainly glycogen and oil.

Thallophytic plants reproduce by three methods. First, very often the plant body is broken up into pieces or fragments; thus the number is increased by simple fragmentation, a **vegetative** method. Secondly, minute **asexual** spores are formed for reproduction. In some algae the spores are provided with ship-like projections, called *cilia*, which impart distinct power of locomotion. Those spores are called **zoospores.** Lastly, **sexual** reproduction takes place by union of sexual cells or *gametes* which may be similar or dissimilar. Gametes are naked protoplasmic bodies. In this group the **gametangia** or gamete-containing sacs are usually very simple and unicellular. **Parthenogenesis** is also common in thallophytes.

ALGAE

OSCILLATORIA

Oscillatoria is a very common blue-green alga growing abundantly in muddy places, ditches, drains and similar damp situations forming colonies. It belongs to a very primitive group of algae, there the cells are prokaryotes, and a blue pigment, *phycocyanin*, is present besides the photosynthetic ones diffused in the protoplast.

The plant body is an unbranched filament, with a row of cells attached end on end. The terminal cell is rounded in shape. A distinct swaying or oscillating movement is exhibited by the filament, what can be clearly seen under the microscope—and hence the name *Oscillatoria*. Each filament has a gelatinous sheath as envelope.

The cells are square in shape and shorter than length. Protoplast is differentiated into two distinct portions—the peripheral coloured portion, called *chromoplasm*, and the colourless central portion named *central body*. The coloured portion is made of chlorophyll and associated pigments in form of granules with diffused blue pigment phycocyanin. Absence of plastids is a notable feature. Various reserve materials—usually glycogen (not starch), protein and fat globules occur in the chromoplasm.

The colourless central body is regarded as the 'incipient' or degenerate nucleus without a nuclear membrane and nucleolus. Only chromatin materials are present. It has no power of meiotic cell division. The two portions of protoplast may be considered corresponding to the cytoplasm and nucleus of normal cells.

Reproduction takes place by vegetative method only. Some dead cells, called *separation discs*, occur here and there along which the filament breaks up into small pieces or fragments, called **hormogonia.**

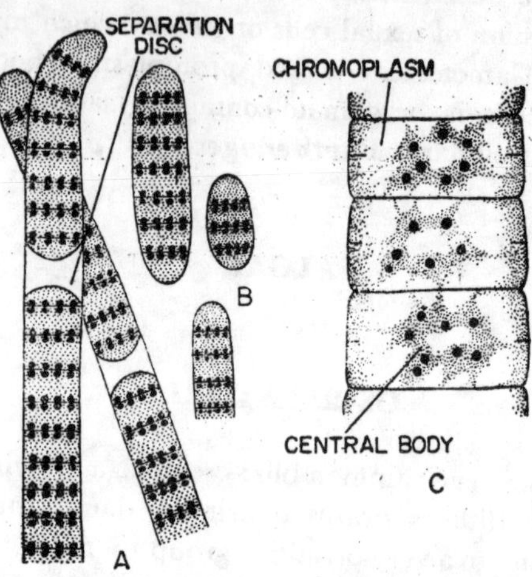

FIG. 185A. *Oscillation*. A—filaments, B—hormogonia, C—cells magnified to show nature of protoplast.

Each of them may be made of one or a few cells, and may grow into a new filament by cell-division in one plane only. The new filaments also exhibit swaying movement.

NOSTOC

Nostoc is a blue-green alga very commonly growing in masses or colonies embedded in a thick mucilagenous matrix, in ponds, ditches and similar damp places. Some of them may also occur endophytically in the intercellular cavities of plants like *Anthoceros* and *Lemna*.

The plant body is an unbranched filament made of rounded cells, attached end on end, attaining a distinct beaded appearance. A pretty good number of them, each with its own gelatinous sheath, remain embedded in a mass of mucilage, looking like a number of beaded chains interwoven together.

The cells are prokaryotic like those of *Oscillatoria*. The protoplast is differentiated into peripheral coloured portion, chromoplasm, with green chlorophyll and blue pigment phycocyanin in form of granules, and reserve food globules; and the colourless central portion, the *central body*, representing the 'incipient' nucleus which has no nuclear membrane and nucleolus. Here and there on the filament some vegetative cells enlarge, become colourless which may occur either terminally or in intercalary position. These cells are known as *heterocysts*. Each of them has two pores at the poles, through which cytoplasmic strands pass to adjacent vegetative cells. The poles are ultimately closed by button-like thickening of the wall forming the so-called *polar nodules*. In case of a terminal heterocyst there is a single pole and polar nodule.

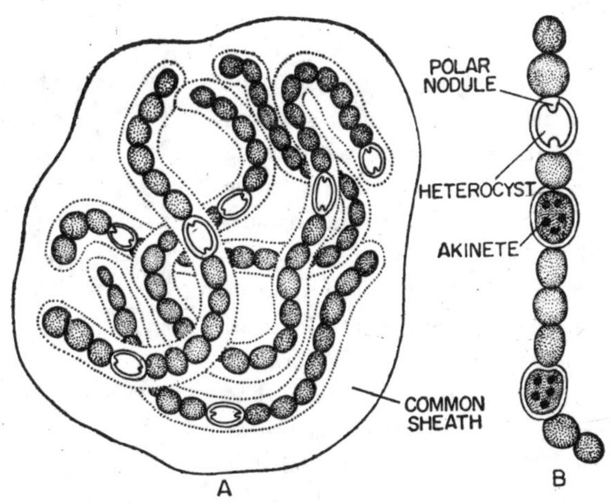

FIG. 185B. *Nostoc*. A—filaments embedded in common sheath. B—portion of filament enlarged.

Reproduction. *Nostoc* reproduces vegetatively by *fragmentation*. The filament breaks up into pieces—the *hormogonia*, at the point of heterocysts. The fragments grow in one plane and remain in a tangled mass in the mucilagenous matrix.

Some vegetative cells singly or in a row often enlarge and become full of stored food to serve as resting spores. These bodies, called *akinetes*, are formed mainly for tiding over unfavourable periods. In course of time they germinate and give rise to new filaments.

SPIROGYRA

Spirogyra is a very common free-floating fresh-water alga. It occurs in stagnant water in ponds and ditches and also in springs and slow-running streams. A large number of them usually grow in tangled masses. It is commonly called *pond-scum*.

Plant body

The plant body is a long filament of cylindrical cells attached end to end, hardly showing any difference between base and apex. Colourless *holdfasts* or attaching organs are rarely formed in those growing in running water. The cell wall is composed of cellulose and pectin, the latter dissolves in water to produce a gelatinous sheath surrounding the filament which, in fact, gives *Spirogyra* the characteristic slimy nature. Every cell is long and cylindrical with lining cytoplasm and large central vacuole (Fig. 186). The nucleus remains suspended in the vacuole by fine cytoplasmic threads. One or more chloroplasts remain arranged in the form of spiral bands with smooth or toothed margins. The name *Spirogyra* has been derived from the peculiar spiral chloroplasts. Some shiny refractive bodies, called **pyrenoides,** are present here and there on the spiral bands. These nodule-like bodies are stored food matters, each having a protein-centre surrounded by starchy layers. The filaments grow in length by cell-division.

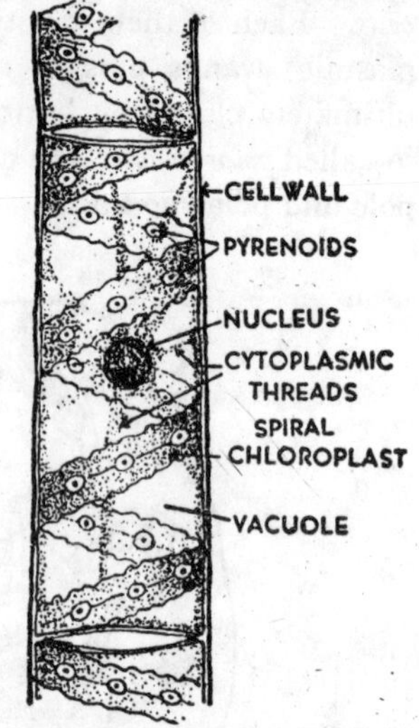

Fig. 186. A portion of filament of *Spirogyra*.

Reproduction

Reproduction takes place by vegetative and sexual methods. *Spirogyra* reproduces vegetatively by simple **fragmentation,** when the filament breaks up into one- or more-celled pieces which grow and multiply by cell-division. The splitting of the filament into pieces may be accidental or due to softening of the middle lamella between two cells.

Sexual reproduction takes place by **conjugation,** an isogamous method, which involves union of similar gametes. Conjugation may occur between cells of two different filaments, or rather rarely, between the adjacent cells of the same filament. In the former case, two filaments of *Spirogyra* approach each other and lie side by side touching at a number of points. They are surrounded by a common sheath. Short protuberances come out at the points of contact which elongate, naturally pushing the two filaments away from each other. On dissolution of the end walls, pathways or tubes are established connecting the two cells of different filaments. These are called **conjugation tubes.** In the mean time, protoplast of every cell loses water, contracts and forms rounded

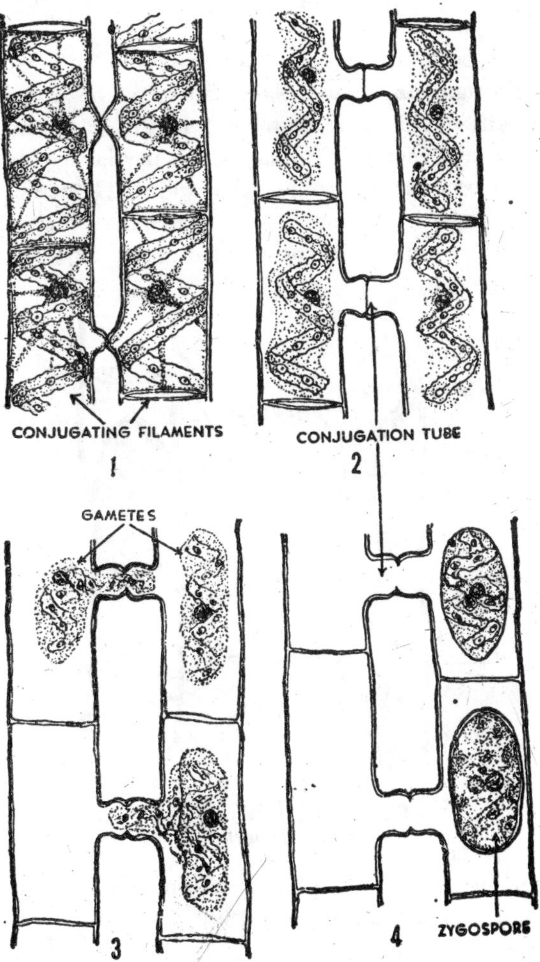

Fig. 187. Successive stages in scalariform conjugation of *Spirogyra*.

mass at the centre. Contracted mass of protoplast of every cell function as the *gamete*, and so every cell is a *gametangium*. The gametes of one filament move through the conjugation tubes in amoeboid fashion to unite with the gametes of the corresponding filament. By union of gametes *zygotes*, called **zygospores,** are formed. It is often found that all the cells of one filament are empty and all the cells of the corresponding filament contain zygotes. This type of conjugation is called **scalariform conjugation** (Fig. 187).

In some species of *Spirogyra* another type of conjugation, called **lateral conjugation,** takes place between the gametes of the adjoining cells of the same filament. In this case also there is a formation of outgrowth, accompanied by partial dissolution of the side wall, to produce the conjugation tube. Gametes are formed in the same fashion. A gamete of a cell moves through the conjugation tube to unite with the gamete of the adjacent cell. After

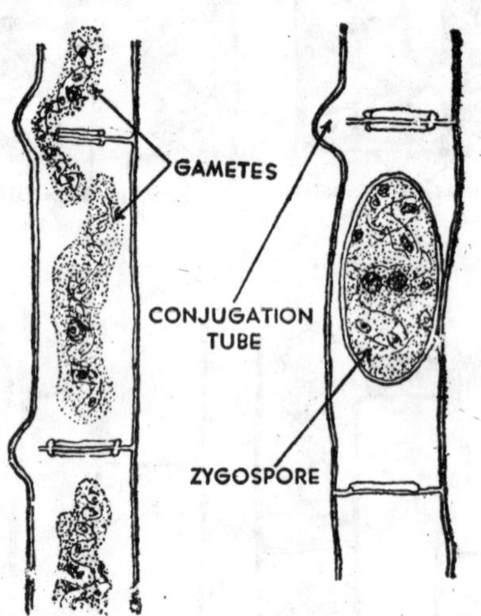

Fig. 188. Lateral conjugation in *Spirogyra*.

conjugation, it is often found that alternate cells contain zygotes and alternate cells are empty (Fig. 188).

Though *Spirogyra* produces similar gametes or *isogametes*, there is differentiation of sex. The moving gametes are male and stationary ones are female. Thus the two filaments involved in scalariform conjugation are unisexual, and the one in lateral conjugation is bisexual.

The zygospore formed by union of two gametes, soon secretes a tough wall around itself. The wall, mainly made of cellulose, has three coats, of which the second one contains chitin. Nuclear fusion is considerably delayed, and chloroplasts contributed by the male gamete are supposed to disintegrate, and those of the female gamete persist in the zygote. It is liberated by rupture of the wall of the cell and falls to the bottom of the tank or ditch wherever the filaments may grow. After a period of rest, usually covering the unfavourable season, the zygospore germinates to produce a new filament. The *diploid* zygote nucleus undergoes *reduction* division to produce four *haploid* nuclei; three of them disintegrate and

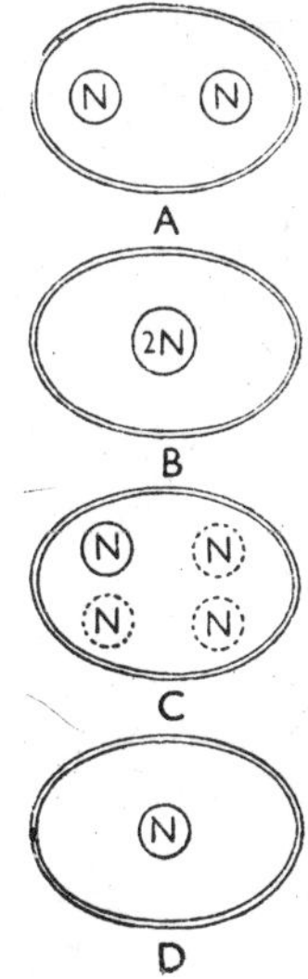

Fig. 189. Nuclear changes in the zygote of *Spirogyra* (diagrammatic).

Fig. 190. Germinating zygospore.

only one survives (Fig. 189). Now the outer wall bursts and the innermost wall with protoplasmic contents comes out in the form of a tube. The cells divide and ultimately give rise to a new *Spirogyra* filament (Fig. 190).

Parthenogenesis

Often conjugation fails in *Spirogyra*, when one of the gametes surrounds itself by thick wall and behaves like a zygote, without actual union with the corresponding one. Such a spore is called **azygospore,** and the process is parthenogenesis.

ULOTHRIX

Ulothrix is a common green filamentous alga growing

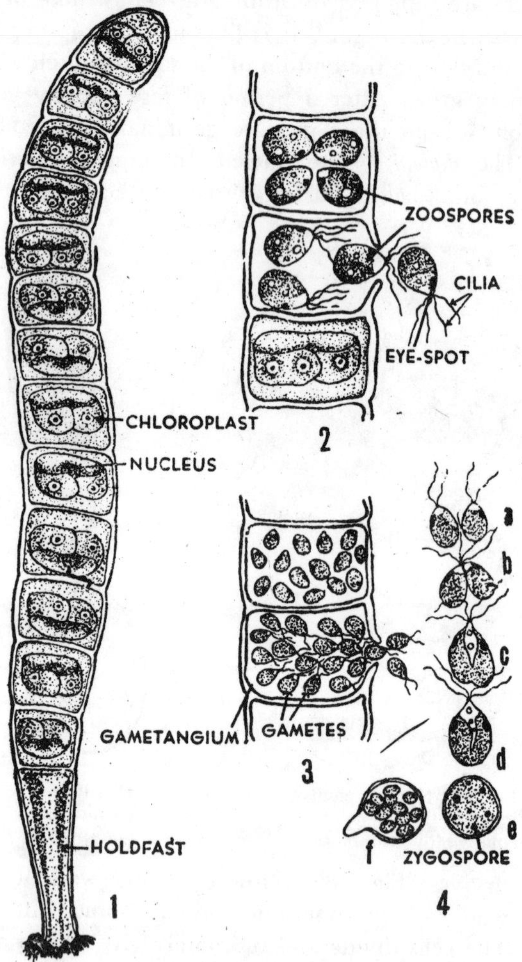

Fig. 191. *Ulothrix:* 1—the plant; 2—portion of filament showing formation and liberation of zoospores; 3—gametangium and liberation of gametes; 4—(a to d) union of gametes; e—formation of zygospore; f—same producing zoospores.

abundantly in fresh-water of ponds, reservoirs and slow-running streams. A few species are marine.

Plant body

The thalloid plant body is a long unbranched filament composed of short cylindrical cells attached end to end. Only the colourless basal cell elongates and serves as the *holdfast* to attach the filament to convenient substratum. Every short cell has cytoplasm, a single nucleus and chloroplast in the form of a parietal band with many pyrenoids (Fig. 191, 1). The filaments grow in length by cell division.

Reproduction

Reproduction takes place by all the three methods—vegetative, asexual and sexual.

Vegetative reproduction occurs by simple fragmentation, when the filament breaks up into fragments, each of which grows in length by cell division and carries on normal activities.

Asexual reproduction takes place by the formation of ciliated spores called **zoospores.** The protoplast of any cell, excepting the basal holdfast, may produce one or more naked uninucleate pear-shaped bodies; each of them develops into a zoospore. As many as 32 zoospores may be formed in a cell. The zoospore is provided with four cilia, and has a conspicuous *eye-spot* near the cilia (Fig. 191, 2). The lateral wall of the cell breaks forming an opening or pore, through which the zoospores come out. They swim f.eely for some time, a day or more, and ultimately settle down by getting themselves attached to any object by the ciliated end. Cilia are withdrawn and cell wall is formed around each of them. Now they germinate and by cell division develop into new filaments.

Sexual reproduction takes place by conjugation usually at the end of the growing season, when conditions are less favourable. Gametes are formed in a large number and liberated just like the zoospores. They are biciliate and smaller in size. Coming out of the cells, gametes produced by different plants unite in pairs to form quadriciliate zygote or **zygospore** (Fig. 191, 3 & 4). Cilia are withdrawn and a thick wall is formed around the *zygote.* After a period of rest, the *diploid* zygote nucleus undergoes reduction division and ultimately four or more zoospores are produced. Each of them develops into a new plant.

FUNGI

MUCOR

Mucor, also called mould, is a very common saprophytic fungus growing abundantly on decayed organic matters, particularly on those rich in carbohydrates—starch and sugar. Soft white cottony patches of *Mucor* are frequently found on rotten bread, vegetables and dung.

Plant body

The plant body is a copiously branched **mycelium,** which is a collection of slender non-septate threads called **hyphae** (sing. hypha). The wall, made of fungus-cellulose, encloses cytoplasm with many vacuoles and innumerable nuclei (Fig. 192). So they are *coenocytic*. Glycogen and oil globules are present to serve as reserve food. The hyphae become thinner and thinner, the more they penetrate into the substratum for absorbing nourishment. Non-septate mycelium becomes septate on attaining old age and during reproduction.

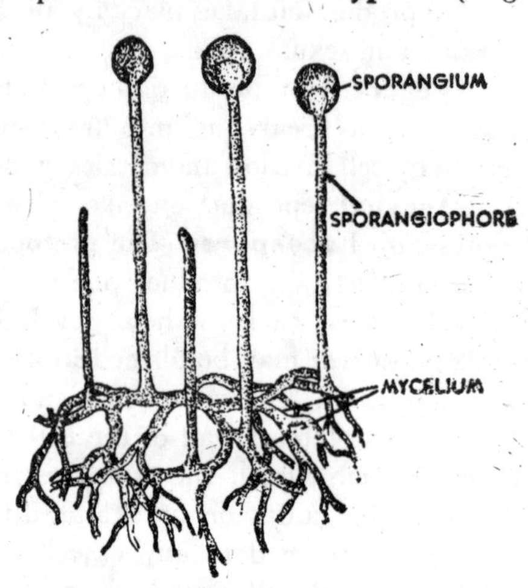

Fig. 192. Mycelium of *Mucor* with sporangiophores and sporangia.

Reproduction

Mucor is reproduced by asexual and sexual methods. During **asexual** reproduction a number of stout aerial hyphae shoot up from the superficial mycelium. After growing to a certain extent, the tip of each of them swells, and some protoplasm with reserve food flows to the enlargement from the adjoining region. Protoplasm collects more densely towards periphery, the central portion remaining comparatively thin and vacuolate. A good number of vacuoles arrange themselves between the outer denser protoplasm, called **sporoplasm,** and the central thinner protoplasm, known as **columella-plasm.** The flattened vacuoles coalesce, and as a result, a distinct cleft is formed separating the two regions; now a wall is constructed along the cleft delimiting the central sterile dome-shaped region, called **columella,** which projects into the enlargement. By progressive cleavage or furrowing the sporoplasm now breaks up into a good number of angular masses, each with cytoplasm and many nuclei. They round off, tough black walls are secreted and finally become **spores.** *Mucor* spores are also called **gonidia** (sing. **gonidium**). The enlargement containing the spores is the **sporangium** or **gonidangium,** and the hypha bearing the sporangium at the tip is called **sporangiophore** or **gonidangiophore** (Fig. 193). The outer wall of the sporangium dissolves in water and the spores are liberated. Often a number of spores remain held in suspension. On suitable medium each spore germinates forming one or more germ tube which gives rise to new mycelium.

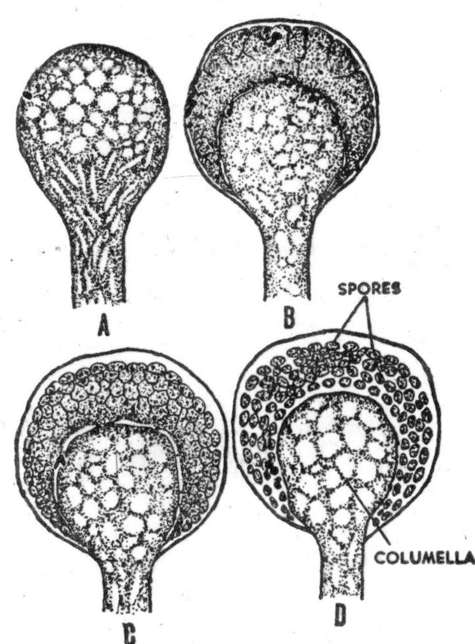

Fig. 193. Stages in the development of sporangium, columella and spores of *Mucor*.

Sexual reproduction in *Mucor* takes place by conjugation. Uusually hyphae of two sexually different strains, designated as + strain and − strain, send out club-shaped branches called **progametangia,** which touch at the tips. Terminal portion of each branch swells and is cut off by a transverse septum. The compartments thus formed, function as gametangia, and their protoplasmic contents as gametes. The gametes here are multinucleate and hence called **coenogametes.** The remaining portion of hyphal branch is knwon as **suspensor.** On dissolution of the wall between them, the gametes fuse, cytoplasm with cytoplasm and nuclei with nuclei in pairs. The nuclei which do not fuse in pairs ultimately disintegrate. The zygote (**zygospore**) thus formed enlarges and secretes a tough wall around it which is often warty or spiny. After a period of rest it germinates, when the outer wall bursts and the inner wall with the protoplasmic contents comes out as an unbranched tube (Fig. 194). This is called **promycelium.** Tip o

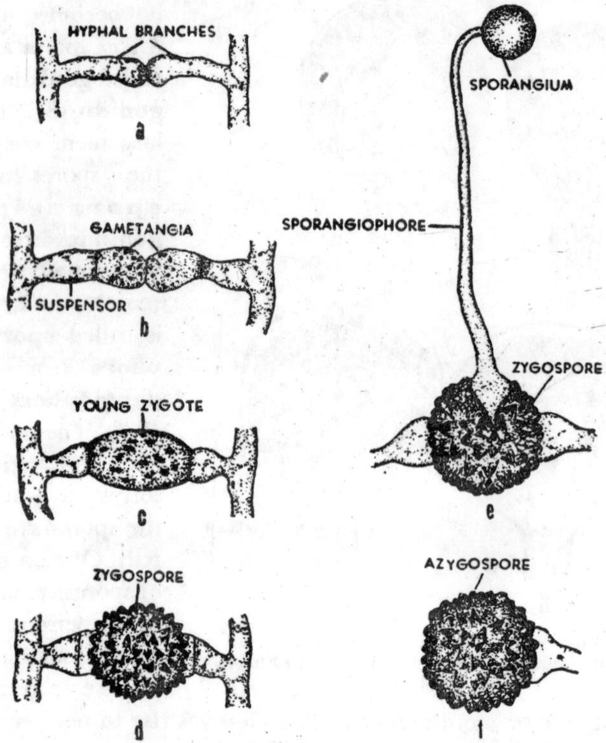

Fig. 194. *Mucor* : a-d—stages in the formation of Zygospore; e—zygospore germinating; f—an Azygospore.

promycelium enlarges and produces spores, each of which can give rise to a new mycelium. The spores produced in the sporangium resulting from the germination of a zygospore are either all + or all −, never both.

Though *Mucor* produces isogametes, it shows distinct differentiation of sex. American botanist Blakslee showed in 1904 that zygote formation in *Mucor* is possible only when gametes produced by two different strains meet. He termed them as + and – strain, rather than male and female for the two mycelial types. Morphologically, there is not much difference between the two, only + mycelium grows a bit more vigorously than its – counterpart. Such species are called *heterothallic*.

Parthenogenesis is not uncommon in this fungus. One of the gametes may behave like a zygospore without actual pairing. This is called **azygospore** or parthenospore.

Yeast condition or Torula stage of Mucor

If a portion of non-septate mycelium is put in sugar solution, it readily develops septa and finally breaks up into many one-celled parts called *oidia*. Like yeast, oidia multiply by budding and can also excite alcoholic fermentation in sugar solution. This is yeast condition or torula stage of *Mucor*.

YEAST (SACCHAROMYCES)

Yeast is a common saprophytic fungus growing in sugary substances. They are abundantly present in grape juice, vineyards, nectaries of flowers and sugary exudates of plants like date-palm, juice and palmyra-palm juice.

Plant body

The plant body is very simple. Yeasts are unicellular organisms. Each cell is elliptical or round in shape having a distinct cell wall. There is granular cytoplasm, a single nucleus and granules of glycogen and protein and oil globules as reserve food. It was believed that the nucleus is a degenerate one due to occurrence of a vacuole, but now it has been found to be a true nucleus.

Reproduction

Under favourable conditions, *i.e.*, when there is sufficient food, yeast cells reproduce **vegetatively** by budding. This is the most common method of reproduction in this fungus. A bud or protuberance arises at one end of the cell and gradually enlarges. The nucleus divides into two by mitosis. One nucleus with some cytoplasm and food flow from the mother cell to the bud. By constriction the bud is separated from the mother cell. By this process a large number of buds are produced which may often remain in the form of short chains (Fig. 196). The idea that the nuclei divide directly by amitosis during the process of budding has been found to be incorrect.

A few of them are called *fission yeasts*, in which the protoplast of the mother cell is separated into two parts by the formation of a septum, rather than by constriction as in budding.

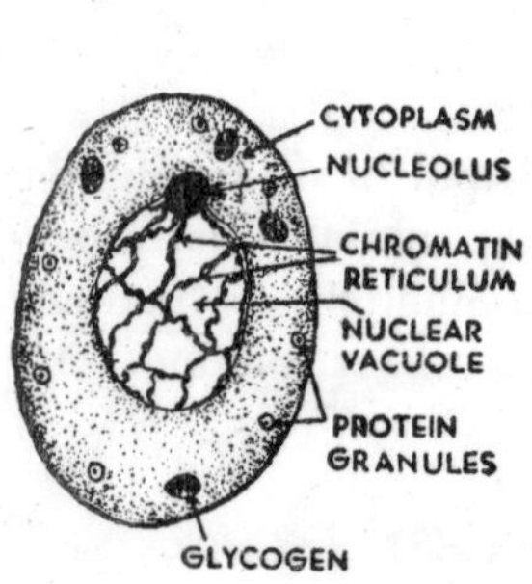

Fig. 195. One yeast cell (magnified).

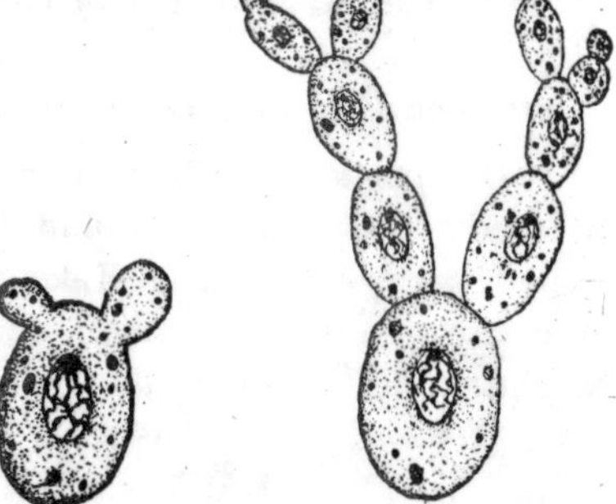

Fig. 196. Stages of budding in yeast.

When food supply is exhausted yeast cells produce spores. Individual cells enlarge and the nuclei divide once, twice, or thrice, forming 2,4, or 8 nuclei in each cell. Cytoplasm collects round each nucleus and ultimately resistant spores are formed. The spores are called **ascospores** and the mother cell (sporangium) is knwon as **ascus**. The ascus wall breaks to liberate the ascospores. Each of them goes on reproducing by budding in suitable medium. This process, described as **asexual,** is really **parthenogenetic** (Fig. 197).

Sexual reproduction takes place in some species of yeast by conjugation. Two yeast cells approach each other and touch, where short protuberances are formed. Dissolution of the wall results in the formation of a short conjugation tube connecting the two cells. The two nuclei move to the tube which broadens considerably and, as a result, the whole thing takes up more or less barrel-shaped appearance. The nuclei fuse in the tube forming zygote (*diploid*) nucleus. The zygote nucleus usually divides thrice, of which the first division is reduction division. Cytoplasm collects round each nucleus and usually eight spores are delimited. The spores are the **ascospores** and the barrel-shaped sporangium is the **ascus** (Fig. 198). Ascospores come out of the ascus and reproduce by budding is suitable medium.

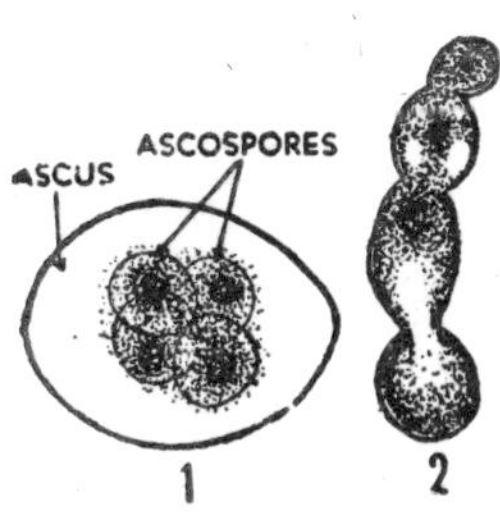

Fig. 197. 1—Formation of Ascospores; 2—Germination of Ascospore and formation of new cells by budding.

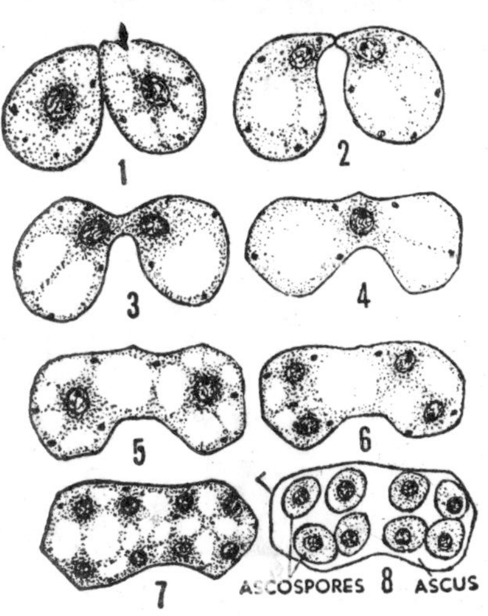

Fig. 198. Stages in the conjugation of yeast and formation of Ascospores.

Alcoholic Fermentation

Yeast has the property of setting up alcoholic fermentation in sugary solution. We know that alcoholic fermentation is an energy-releasing process brought about by micro-organisms under anaerobic conditions. Yeast cells secrete an enzyme, *zymase*, which decomposes sugar into alcohol and carbon dioxide with liberation of energy. CO_2 comes out as bubbles forming froth. It may be represented thus:

$$\underset{\text{(Sugar)}}{C_6H_{12}O_6} + \underset{\text{(Enzyme)}}{\text{Zymase}} \rightarrow \underset{\text{(Alcohol)}}{2C_2H_5OH} + 2CO_2 + \text{Zymase} + \text{Energy (25 cal.)}$$

For this particular property yeasts are used commercially in breweries for the manufacture of alcoholic beverages like wines, beers, etc. The same principle applies in the preparation of indigenous liqour, toddy, from sugary exudates. They are also used in bakeries or 'raising' of breads. Yeast cells excite alcoholic fermentation in bread paste (dough), and carbon dioxide bubbles, while escaping on application of heat, raise the bread.

Yeasts are rich in vitamin B complex. So they have nutritional value as well.

PENICILLIUM

Penicillium is a common saprophytic fungus growing on decayed organic matters like bread, jam, jelly, vegetables and fruits and even on damp shoes and leather. It is known as green or blue mold. The spores of this fungus are abundantly present everywhere and often cause considerable damage to fruits and vegetables. Some species are also used in industries. Sir Alexander Flemming isolated the wonder

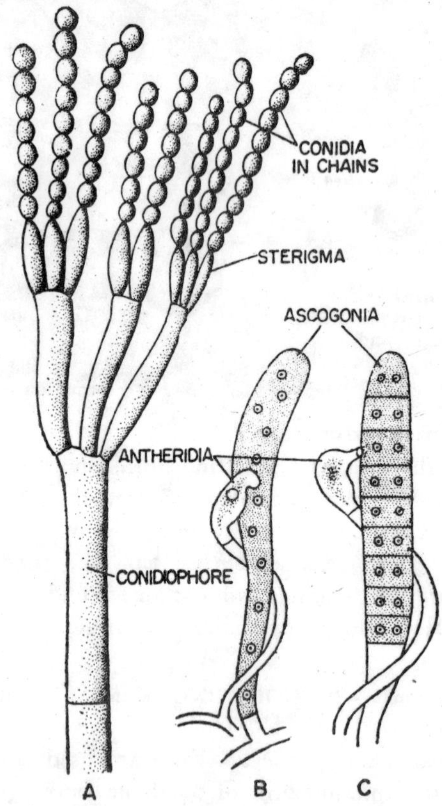

Fig. 198A. *Penicillium.* A—conidiophore with branches, sterigmata and conidia in chains. B & C—antheridia and ascogonia.

drug, *penicillin* from *Penicillium notatum* in 1929. The antibiotic penicillin had revolutionised medical science and proved to be a real boon to humanity.

The mycelium is composed of much branched septate hyphae

occurring in tangled masses. They ramify extensively on the substratum and many of them penetrate into it to serve as rhizoids. The hyphal cells are multi-nucleate.

Reproduction. This fungus reproduces mainly by asexual method, through the spores called *conidia*, which are formed in very large number. Sexual method of reproduction has also been reported in some species, though the stages are not very clearly known.

Asexual. Some stout hyphae come out erect from the mycelium and function as **conidiophores.** Smaller branches develop from the tip of the conidiophore, which again divide to form a row of closely-packed branches called **sterigmata.** Unbranched chains of asexual spores—**conidia,** are cut off from the tip of the sterigmata in basipetal order (Fig. 198A).

The terminal portion of conidiophores with branches and chains of conidia together looks like a broom and is called *penicillus*—meaning broom. The conidia are globose, ovoid or elliptical with smooth or spiny surface and usually green in colour. They are uninucleate at the early stage but may become multinucleate in some cases. The conidia are easily dispersed by wind, and germinate on a suitable substratum. A germ tube is first formed which ultimately develops into a new mycelium.

Sexual. Sexual reproduction, though not clearly known, is *oogamous.* A hypha comes out erect, enlarges, becomes club-shaped and develops into the **ascogonium.** The nucleus divides again and again to form 32-64 nuclei dispersed in the cytoplasm of the ascogonium. In the meantime another branch comes out of a neighbouring hypha which twines round the ascogonium. The terminal part of that branch is cut off by a septum, swells and forms the **antheridium.** It comes in contact with the ascogonium where a pore is formed for movement of the protoplast of the antheridium to the ascogonium. Though gametic union has not been established it may be assumed that the process takes place. By formation of septa the multinucleate ascogonium gives rise to a row of binucleate cells. Many hyphae with binucleate cells now develop from these cells—which are called *ascogeneous hyphae.* They become septate, each cell having two nuclei, and the terminal cell develops into an **ascus.** The two nuclei of the ascus fuse to form the zygote nucleus, which then divides thrice, the first division being reductional, and ultimately produces eight **ascospores.** Meanwhile sterile vegetative hyphae send out many branches around the sex organs forming a closed fruit body—the **ascocarp.** This cover made of hyphal cells is known as *cleitothecium,* the inner layer of which is nuritive in function. With maturity of the ascospores the asci dissolve leaving them free and scattered in the cleistothecium. The periderm now decays liberating the ascospores which are easily blown off by wind.

BACTERIA (SCHIZOMYCETES)

Bacteria are the smallest living things which are visible only under very high magnification of a microscope. Many of them are positively harmful to plants and animals, as they live parasitically

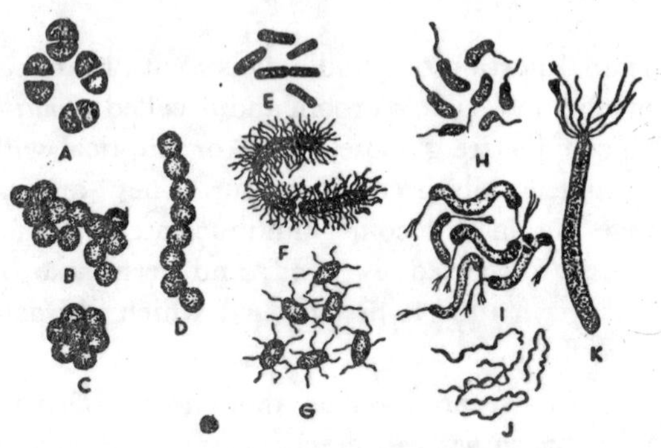

Fig. 199. Bacteria; A-D—Coccus forms; E-G—Bacillus forms; H-K—Spirillum forms.

on the host, and cause diseases. Some bacteria, on the other hand, are beneficial in different ways. They exist in all types of situations and can withstand extreme heat or cold and other unfavourable conditions. Bacteria are abundantly present in the air we breathe, in the food we take, in the water we drink, in the soil we inhabit, in different objects we touch and even inside our bodies.

Bacteria were first noted and described by a Dutch merchant Leeuwenhoek in the seventeenth century. In the nineteenth century famous French scientist **Louis Pasteur** asserted that sterilised fruit juice remained free from bacterial contamination. He really dealt the death blow to the theory of spontaneous generation. **Robert Koch** in the same century discovered that bacteria are the causes of cattle-disease anthrax and tuberculosis and Asiatic Cholera. In fact, Pasteur and Koch are the builders of a new branch of biology, called **bacteriology,** which has advanced by rapid strides in recent years and revolutionised medical science.

Bacteria form a group of extremely small unicellular plants. Each has protoplasm which a surrounding limiting membrane which contains no cellulose. They may remain united together by forming filaments or irregular masses. Well-formed nuclei are always absent but chromatin granules remain distributed, which serve as corresponding nuclear material. Plastids and chlorophyll are absent, though some pigments may often be noticed in them. Occasionally a large number of bacteria remain embedded in a gelatinous sheath forming a colony. They vary in size, smallest ones may even be less than 1 micron in length. According to their forms, bacteria are of three principal types (Fig. 199), *viz.* spherical or **coccus** (pl. cocci) forms, rod-shaped or **bacillus** (pl. bacilli) forms and spiral or **spirillum** (pl. spirilla) forms. Some of them are non-motile, while others are provided with cilia.

Reproduction

Bacteria reproduce very rapidly by fission or fragmentation. They are constricted in the middle and ultimately break up into pieces of equal size.

The rate of multiplication is surprisingly high. Prolific reproduction is checked by external factors like scarcity of food. Under unfavourable conditions bacteria produce spores. Usually one spore is produced in each cell, where protoplasm contracts and is surrounded by a tough resistant wall. Spores can stand very high temperatures, poisonous substances and other extremely unfavourable conditions. Spore-formation is regarded as a device for tiding over the unfavourable period. They are present everywhere. Under favourable conditions, each spore germinates to produce a bacterium.

Bacteria usually cannot manufacture their food and are perforce dependent on other sources. Many of them live parasitically on plants and animals including human beings. Parasitic bacteria are the causes of many diseases like cholera, tuberculosis, pneumonia and typhoid fever. They cause disease by direct attack upon the host tissue or by the liberation of toxins or by both. Disease-causing bacteria are called **pathogenic.**

Importance of Bacteria

Saprophytic bacteria grow on dead rotten organic matters. They are not harmful but rather beneficial. In the soil they work as scavengers. Bacteria bring about decay of plant and animal bodies into simpler materials like water, carbon dioxide, nitrates, sulphates, phosphates, etc., which may again be used by the green plants. Different types of bacteria work out this decay step by step by secretion of enzymes. Thus they prevent accumulation of organic matters and the essential elements are set free, which otherwise would have been permanently 'locked up' in dead bodies.

Bacteria considerably increase the fertility of the soil. Though nearly 80% of air is nitrogen, most plants are unable to use it, and we know that this essential element is mainly absorbed as nitrates from the soil. Common free-living bacteria *Azotobacter* and *Clostridium* can fix free nitrogen from air and convert it to organic nitrogenous compounds which are ultimately set free and broken down to nitrates. Necessary energy is obtained for the purpose by oxidation of carbon compounds of the soil. Nodule-forming bacteria, *Pseudomonas radicicola* infect roots of leguminous plants and lead symbiotic life (p. 190). They also fix nitrogen. Nitrifying bacteria, *Nitrosomonas* and *Nitrobacter* can oxidise ammonia and nitrites respectively, derived from plant and animal remains of the soil, into nitrates, so that they may be absorbed by the roots of green plants. Some denitrifying bacteria of soil have the power of reducing nitrates into free nitrogen which is lost from the soil and passes into air. They live under anaerobic conditions.

Besides these beneficial activities of the bacteria inhabiting the soil, there are many others which are also important for different industrial processes. Sugar of milk is changed by some bacteria into lactic acid which brings about coagulation of milk-protein (casein). This is really the curdling of milk. Bacterial actions are advantageously utilised in the manufacture of cheese, cream and butter, in the curing of tobacco leaves, tanning of hides to leather, retting of fibres of jute and flax, manufacture of vinegar and alcohol and various other industrial processes.

VIRUS

Viruses are the smallest and the most primitive living organisms. In fact they are so small that they can pass through the best porcelain filter and cannot be seen even under the most powerful light microscope. Thay they can multiply with amazing rapidity and create positive disease symptoms in living bodies cannot be questioned. A virus disease of plant—TMV or tobacco mosaic virus was reported in the last century. A large numbers of diseases due to virus infection has been definitely known now. Influenza, common cold, mumps, chicken pox etc. are common virus diseases. Cattle and plants also suffer from different types of virus infections.

Investigations with the aid of electron microscope have made the picture clear. Viruses multiply very rapidly inside the living body forming innumerable ones exactly similar to themselves, but outside the host they are perfectly inactive. So they are strict parasites. They have no cellular structure nor they show any sign of respiration and metabolism. For these reasons they were believed to be something intermediate between non-living matter and living organisms.

The virus particles usually range between 10-20 mμ in length, spherical to rod-shaped in appearance. Each has a core of nucleic acid—DNA, often RNA or both surrounded by a protein sheath as a protective film. They are complex organisms having genetic mechanism.

Bacterial viruses, commonly known as *bacteriophage* are tadpole shaped and attack the bacteria.

LICHEN

Lichens (Fig. 200) constitute an extraordinary group of mostly grayish-green plants widely distributed all over the world. They grow on rocks, tree-trunks, dead wood and on ground, and can withstand extreme temperature. A lichen is a peculiar combination of an alga and a fungus, the two live deriving mutual benefit. Major part of a lichen is usually composed of fungal hyphae enclosing some algal cells irregularly or in a definite layer. Alga manufactures carbohydrate food which becomes available to the fungus, and the fungus, in return, absorbs and retains water and thus keeps the algal cells moist. That makes life of lichens possible in very dry situations where individually none of them can thrive. Thus lichen is an excellent example of *symbiosis* or reciprocal parasitism.

According to forms, lichens are of three principal types, *viz.* (i) **crustose** forms, which grow on rocks and barks as incrustations,

(ii) **foliose** forms, with flat leaf-like thalli, and (iii) **fructicose** forms, which hang from trees or grow erect on the ground.

Reproduction

Lichens multiply vegetatively when small fragments break away from the parent plant and carried by different agencies. Each piece can produce a new thallus. Often special reproductive bodies, called **soredia,** are formed as bud-like outgrowths on the thallus. A soredium consists of a few algal cells surrounded by fungal hyphae. They get detached from the plant, are carried away by wind and give rise to new thalli. Lichens also reproduce by spores called **pycnidiospores.** The spores germinate to produce fungal hyphae which develop into lichen, after coming in contact with a suitable alga.

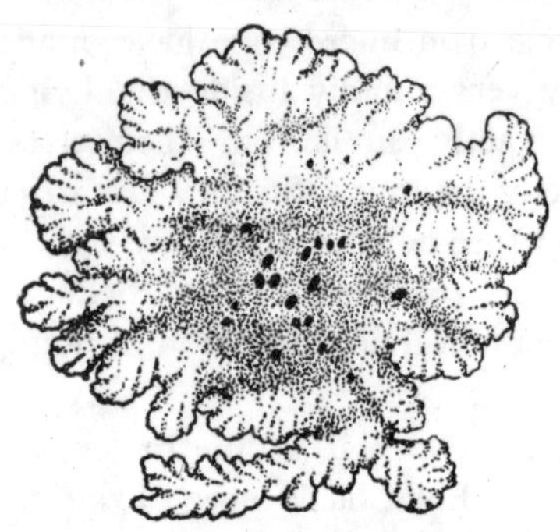

Fig. 200. A foliose lichen.

Economic importance

Lichens growing in extremely cold regions serve as main food for animals like reindeers. Some of them contain pigments. Common litmus used as an indicator of acidity and alkalinity, is extracted from lichen. A few of them are also used in perfume industry. Lichens bring about disintegration of rocks on which they grow, and thus prepare suitable ground for growth of other plants.

CHAPTER XXII

BRYOPHYTA

This group stands between thallophyta and pteridophyta. Bryophytes are small plants usually growing in moist situations. The plant body is thalloid or differentiated into short stem and simple leaves, but true roots are absent. Unicellular or multicellular hairs, called rhizoids, carry on the functions of roots. The organs do not possess characteristic vascular elements. Gametangia, male and female, called antheridia and archegonia respectively, are multicellular and have outer jackets of sterile cells. Alternation of generations is quite well-established.

MOSS

Moss is a very common representative of this group. Mosses grow abundantly in groups or patches on damp wall, rocks, barks of trees, etc., particularly during rainy season, often forming dense green velvety mats. These are small plants, roughly 2·54 cms or one inch in height.

Plant body

The plant body is very simple. It has a small branched stem with spirally arranged simple leaves (Fig. 201). Multicellular hairy outgrowths, called **rhizoids,** develop from the base of the stem which carry on the functions of roots, true roots being altogether absent. They have no vascular tissues; some elongated cells at the centre of stem and midribs of leaves are solely concerned with conduction of water and food.

Reproduction

The plant bears gametangia at the apices of the stem and branches. So they are called **gametophores** or gamete-bearers. A longitudinal section of the apex reveals that **antheridia** and **archegonia** remain intermixed with many sterile hairy processes called **paraphyses** (Fig. 203). Mosses may be monoecious or dioecious.

The leaves of a male shoot often spread out in the form of petals of a flower and hence commonly called 'moss flower', though technically, it has nothing to do with a flower. Each antheridium is a club-shaped body with a short stalk. It is normally green in colour but turns orange-red with maturity. Antheridium contains many spermatozoid-mother cells in the central region, which produce a fairly large number of sperm cells. Every sperm cell develops into a biciliate spermatozoid or antherozoid (male gamete). They

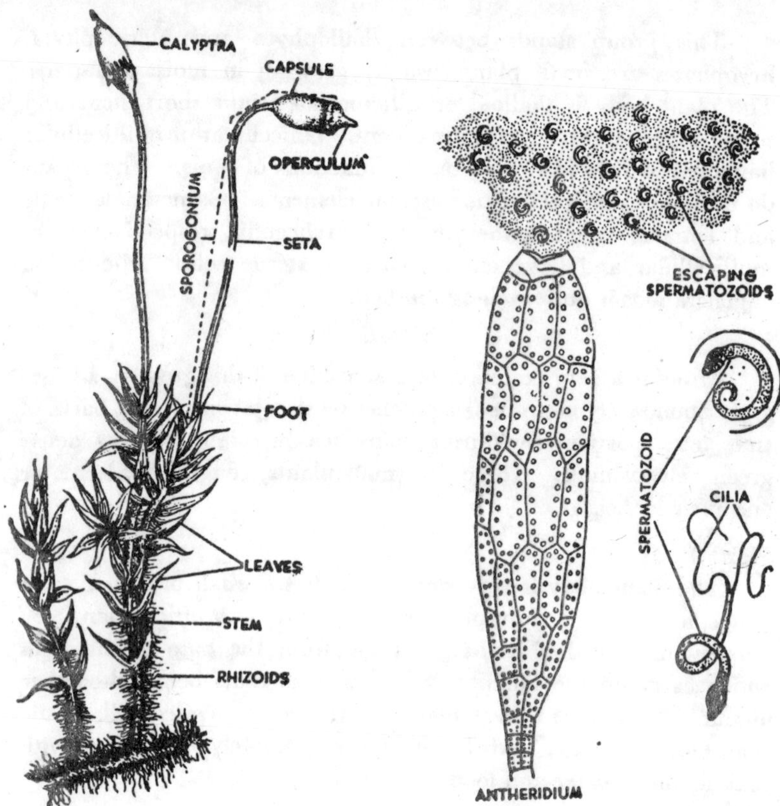

Fig. 201. Moss plant.

Fig. 202. Mature antheridium liberating spermatozoids.

remain within the antheridium till moisture is available. When the mature antheridium gets moisture, the mucilaginous contents surrounding the sperm cells absorb water and swell up. A pressure is set up, as a result of which the sperm cells are liberated in a mass through an apical opening. Outer mucilaginous wall of each sperm cell dissolves in water, and a naked biciliate spermatozoid is set free (Fig. 202). It moves about in surrounding film of water.

The female shoot similarly bears archegonia with paraphyses (Fig. 204, left). Each archegonium is a distinct flask-shaped body with a massive swollen lower portion called **venter** and a long projecting **neck.** The venter contains two cells, *viz.* the lower and the larger one is the female gamete called **egg** or **ovum,** and the

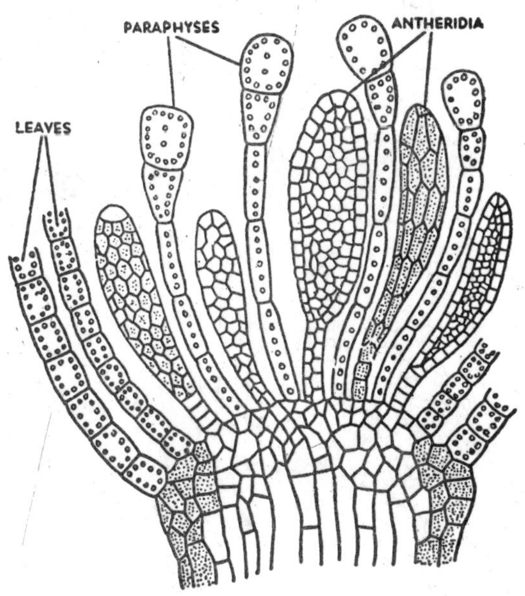

Fig. 203. Longitudinal section of apex of moss stem bearing antheridia and paraphyses.

other one is known as **ventral canal cell.** The neck contains a few cells called **neck canal cells** (Fig. 204, right). With maturity, the tip of archegonium opens and all cells excepting the egg disintegrate forming mucilaginous mass. Thus a thorough passage is established to the only surviving cell, egg. Further, archegonium secretes chemical substance, sucrose, to attract the male gametes. A large number of them rush towards the egg; one of them passes down the canal and fuses with the egg. This fusion is **fertilisation,** and the zygote formed is the **oospore,** which soon secretes a wall around itself. Oospore begins to divide inside the venter of the archegonium and forms the embryo-sporophyte, which by further division and elaboration gives rise to the complex but important structure called **sporogonium** of moss. The sporogonium has three parts—(i) lower **foot,** by means of which it remains attached to and draw water from the plant, (ii) a long stalk, called **seta,** which conducts water absorbed by foot, and (iii) the swollen terminal pear-shaped portion called **capsule** which is destined to produce the asexual spores (Fig. 201). It goes without saying that capsule is the most important part of the sporogonium. Capsule remains partly

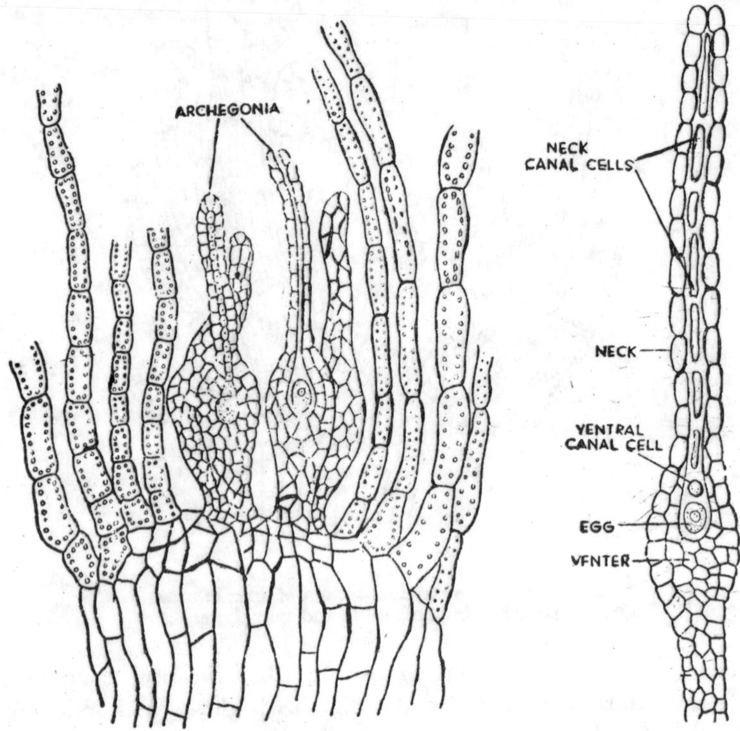

Fig. 204. Left—Longitudinal section of apex bearing archegonia; right—a mature archegonium.

covered by a delicate and temporary cap or hood called **calyptra,** which is soon blown off by wind. During the formation of sporogonium the seta elongates rapidly, and as a result, the upper portion or archegonium ruptures and is carried upwards by the elongating seta. Thus the calyptra is nothing but the remnants of the archegonial wall.

The capsule is a complicated multicellular body, having arrangements for formation and liberation of spores. A longitudinal section of the capsule would reveal the following structures (Fig. 205).

The basal solid portion is called **apophysis.** Capsule has an outer epidermis with stomata. Next to epidermis there is a zone of parenchyma cells with chloroplasts called subepidermis. Internal to that, there are many air cavities traversed by delicate strands

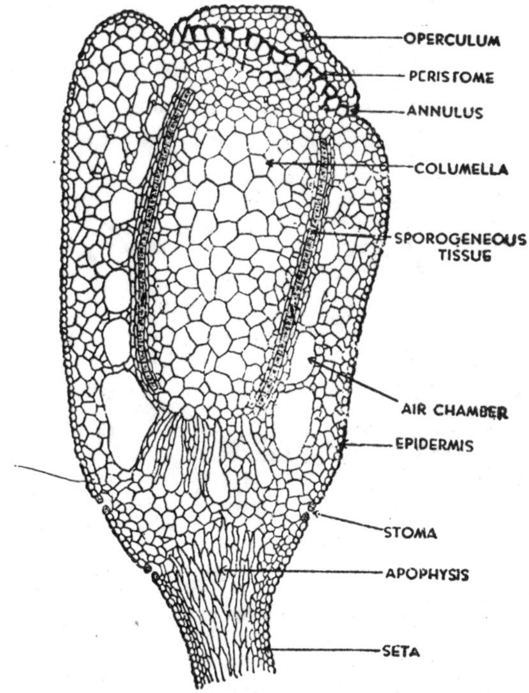

Fig. 205. Longitudinal section through the capsule of moss.

of cells. Two layers of compact protoplasmic cells are found next towards right and left, surrounding the central sterile parenchymatous portion called **columella.** The protoplasmic cells constitute the **sporogenous** tissue or **spore-sac.** Every cell of spore-sac functions as a **spore-mother cell.** It undergoes reduction division and gives rise to a tetrad of **spores.**

A multicellular lid, called **operculum,** is present at the mouth of the capsule. At the region where operculum meets the main body of the capsule, there is a row of large cells, called **annulus,** with cuticularised walls. The last layer of operculum is known as **peristome.** It consists of cells with teeth-like projections which respond to atmospheric changes in humidity. When atmosphere is moist, they fit in tightly with the internal cells; and when air is dry, they withdraw. After the formation of spores in the capsule, annulus breaks which is followed by the withdrawal of peristome teeth. As a result, the whole operculum is blown off, making a clean passage for the liberation of the spores.

Spore, as we know, is a unicellular body with two coats, intine and exine. When conditions are satisfied, spore germinates to form a green thalloid filamentous body called **protonema** (Fig. 206).

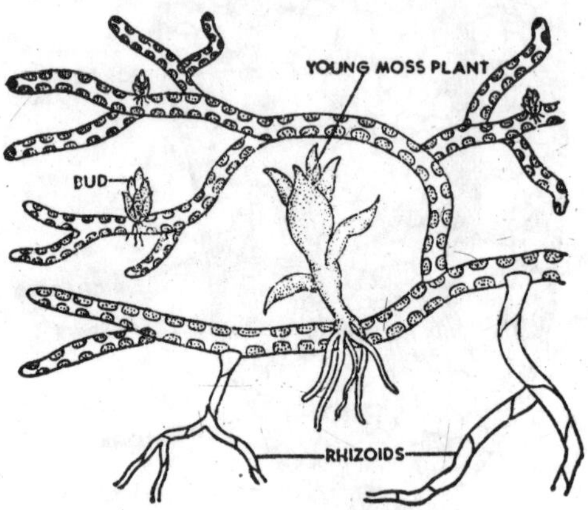

Fig. 206. Protonema of moss.

The protonema bears many lateral buds, each of which develops into a new moss plant, thus completing the life-history.

Alternation of Generations

Moss plant reproduces by two methods, sexual and asexual. Plant itself bears the sexual units, sperms and egg. This part is called gamete-bearing part of **gametophyte.** When the two gametes fuse, zygote is formed, which, as we have seen, develops into the sporogonium. Sporogonium in its turn bears asexual spores. That part is spore-bearing generation or **sporophyte.** The two generations regularly alternate in the life-history. This is the phenomenon of **alternation of generations.** Two generations differ in form and structure, in methods of reproduction and also in number of chromosomes in the nuclei of the cells. The gametophyte always contains reduced or *haploid* number of chromosomes, represented as 'n' number; whereas the sporophyte has double or *diploid* number of chromosomes represented as '2n' number. Two cardinal points in the life-history determine the limits of gametophyte and sporophyte. They are (i) *reduction division* in the spore-mother cells with subsequent formation of spores, which marks the beginning of gametophyte, and (ii) *fertilisation* involving union of two gametes

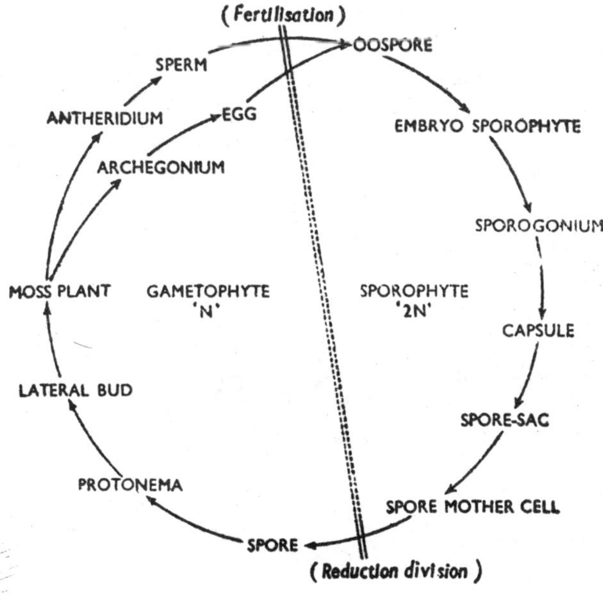

Fig. 207. Life-cycle of moss.

which restores the '2n' number ('n' being contributed by each gamete), and initiates the sporophyte (Fig. 207).

In moss gametophyte is the more prominent generation represented by the plant itself and the protonema; and sporophyte represented by sporogonium is partially dependent or *semi-parasitic* on the gametophyte. Calyptra, the temporary hood on the capsule, the most important part of sporophyte, really belongs to gametophyte. Another point to be noted is that in the same life-history moist condition is necessary for liberation of spermatozoids and dry air for dispersal of spores.

CHAPTER XXIII

PTERIDOPHYTA

This group includes the most highly developed cryptogams, like ferns and horse-tails, distributed all over the world. Like higher plants they have plant bodies well-differentiated into stems, roots, and leaves, and the organs possess prominent vascular bundles. So pteridophytes are also called vascular cryptogams. Alternation of generation is quite distinct in this group. Unlike bryophytes, the sporophyte here is the most prominent generation and the gametophyte is much smaller, which may be independent or dependent of the sporophyte.

FERN

Ferns are widely distributed all over the world. They may grow in any type of situation ranging from xerophytic to aquatic. The vast majority of them, however, flourish very well in moist and shady places.

Plant body

Most of the ferns are small perennial herbs. Tree ferns with long pillar-like stem and a crown of leaves at the top are found only in tropical countries. The plant body is differentiated into stems,

Fig. 208. A fern plant with underground stem, adventitious roots, adult leaf and a young leaf showing circinate vernation.

roots and leaves with well-developed vascular bundles. The stem is usually a creeping underground rhizome bearing many adventitious roots It remains covered by a large number of persistent leaf-bases. Pinnately compound leaves, often called **fronds,** spring up from the rhizome. The leaves, when young, remain coiled inwards, thus exhibiting *circinate* vernation (Fig. 208). The underground rhizome, young leaves and rachis of adult leaves remain covered by brown hairy scales called **ramenta** (singular—ramentum). This is a safeguard against drought. Circinate arrangement of young leaves and presence of ramenta are characteristic features of ferns.

Reproduction

Unlike moss, the fern plant is the spore-bearing generation or **sporophyte.** The leaflets bear some brown kidney-shaped dots, called **sori** (sing. sorus), on the under surface (Fig. 209). They are usually present directly on the veins or vein-endings.

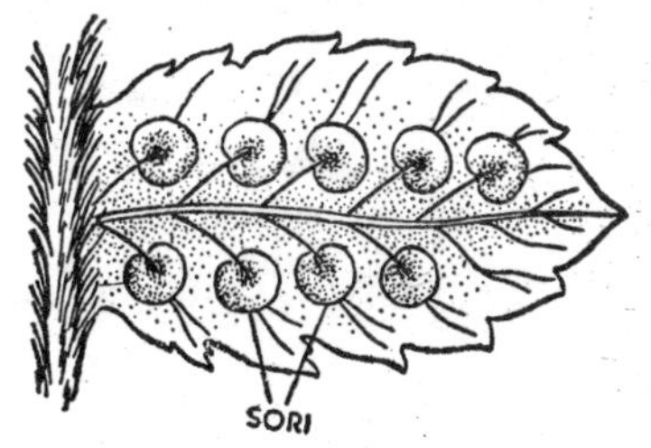

Fig. 209. A leaflet of fern bearing kidney-shaped sori.

A section of a leaflet passing through a sorus shows that it (sorus) is a collection of **sporangia,** which develop from the soft parenchymatous cushion-like outgrowth called **placenta.** The sporangia, when young, remain protected by a curved outgrowth of the leaf-tissue, called **indusium** (Fig. 210). Each sporangium has two parts, *viz.* a multicellular stalk surmounted by the swollen body called the capsule. The capsule has the outer wall made up of one layer of cells. It encloses a definite number of large cells called *spore-mother cells.* The number is usually sixteen; in some cases it may be twelve. Every spore-mother cell undergoes *reduction* division to produce a tetrad of spores. Reduction division with subsequent formation of spores, marks the beginning of gametophytic generation. Along the margin of the capsule wall there is a row of cells with inner and radial walls cuticularised. This row covering the major part of the capsule is called **annulus.** A few thin-walled cells forming the weak spot, so to say, on the capsule wall constitute, what is known as **stomium** (Fig. 211). After formation of spores the annulus straightens up in response to

atmospheric changes in humidity, and the capsule bursts at the

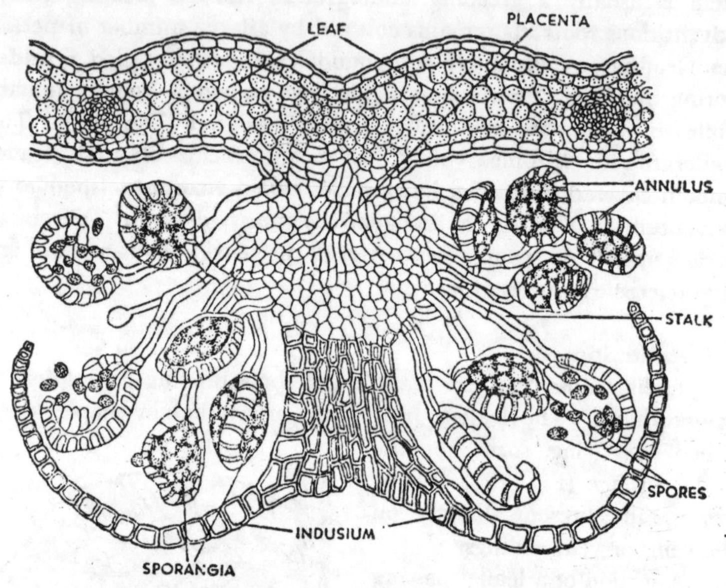

Fig. 210. Cross-section of a fern leaflet through a sorus.

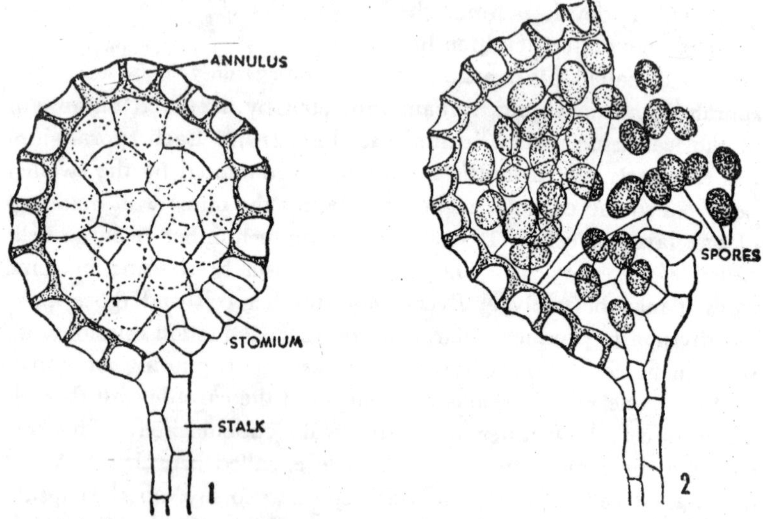

Fig. 211. Fern; 1—a sporangium; 2—bursting of the capsule at stomium and liberation of spores.

stomium liberating the spores, which are easily carried away by wind.

Spore is a unicellular body with two coats, outer thick exine and thin inner coat, intine. Under suitable conditions spore germinates. Exine bursts and intine comes out in form of tube which, by repeated cell division, gives rise to a green flat heart-shaped thalloid body called **prothallus.** It develops unicellular hairs,

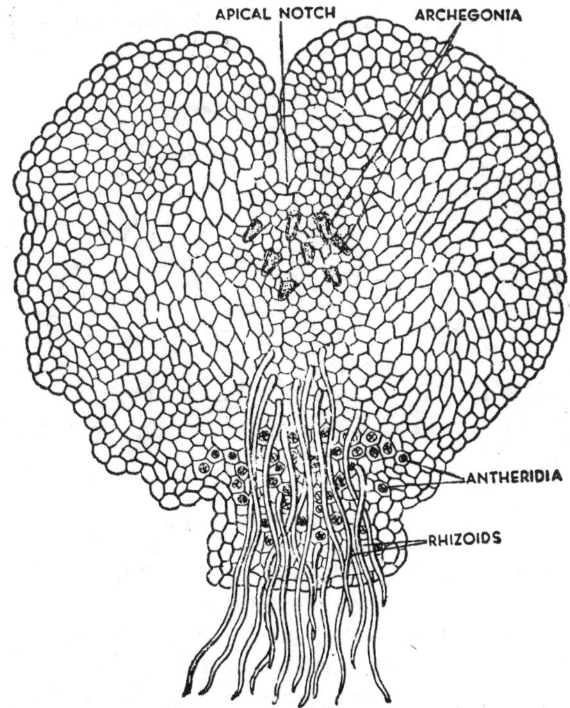

Fig. 212. Mature prothallus (gametophyte) of fern viewed from below.

called **rhizoids,** from the under surface for fixation and absorption of water and mineral matters. Thus the small prothallus, barely 1·0 cm in size, is an independent body. Prothallus bears the sex organs, and so represents the **gametophyte** of fern. Antheridia are located at the basal part amongst the rhizoids and archegonia near the apical notch (Fig. 212).

An antheridium is a spherical structure with an outer jacket of sterile cells. It encloses a good number of spermatozoid-mother cells; each of them produces a spermatozoid or antherozoid (male gamete). In presence of water sperms, surrounded by remnants of the mother cells, are liberated from the antheridium through an apical opening. The spermatozoids are naked spirally twisted multiciliate bodies (Fig. 213, left). They swim freely in surrounding film of water.

Archegonium is a flask-shaped body with the basal swollen portion, **venter,** and the short slightly curved **neck.** Venter remains embedded in the body of the prothallus, only neck projects out. Venter contains the female gamete, egg or ovum, and a ventral

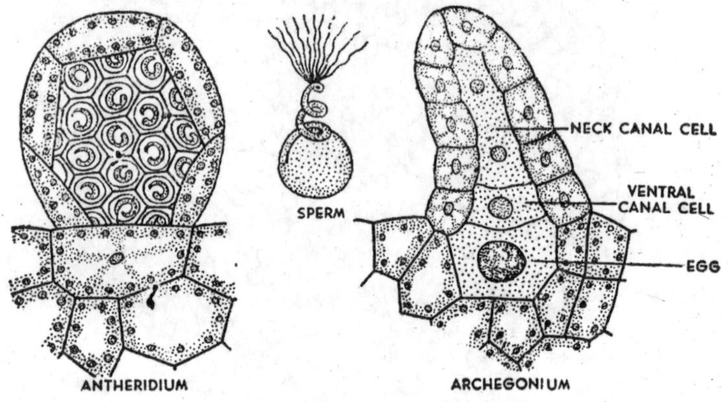

Fig. 213. Antheridium and archegonium of fern.

canal cell; and the neck contains a binucleate neck canal cell (Fig. 213, right). At maturity of the archegonium, all the cells excepting the egg disorganise, an opening is formed at the tip, thus establishing a thorough passage to the egg. Besides mucilage, archegonia secrete chemicals, malic acid and salts of malic acid, which attract the sperms. Quite a good number of them move towards the archegonium. One passes down the canal and fertilizes the egg. The **oospore,** thus formed, soon secretes a wall around itself. By repeated cell division oospore gives rise to the multicellular **embryo,** which develops into a new fern plant, thus completing the life-history.

Alternation of Generations

The above life-history shows that it has two distinct parts, the plant itself reproducing asexually by spores and the prothallus reproducing sexually by gametes. **Sporophyte** and **gametophyte,** as they are called, regularly alternate in the life-history, establishing 'alternation of generations'. Sporophyte, the more prominent generation, contains *diploid* or '2n' number of chromosomes in the nuclei of the cells. *Fertilization* is the starting point of this generation and continues up to spore-mother cells. As soon as there is

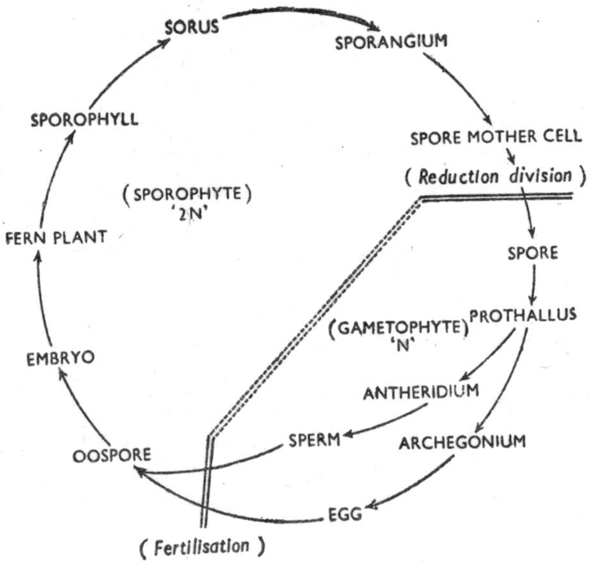

Fig. 214. Life-cycle of fern.

reduction division of the spore-mother cell, number of chromosomes is reduced to half and gametophyte begins. So spore is the starting point of gametophyte, which continues right up to male and female gametes before their union. The gametophyte, though much smaller and inconspicuous, is independent of the sporophyte. The two cardinal points, fertilization and reduction division, determine the limits of the two generations (Fig. 214).

CHAPTER XXIV

SPERMATOPHYTA (PHANEROGAMS)

This group occupies the highest position in the plant kingdom. In fact, the vast majority of plants with which we are familiar, belong to this group. Development of flower and formation of seeds through fertilization are the outstanding features of spermatophytes. The sporophyte here is still more elaborate, complex and the most prominent generation, and gametophyte is extremely reduced and dependent on the sporophyte. Spermatophytes are divided into two groups—Gymnosperms and Angiosperms.

GYMNOSPERMS

Gymnosperms are the naked-seeded plants. They formed dense vegetation in the past geological ages, but many of them are now extinct. Gymnosperms really constitute a group intermediate between angiosperms and pteridophytes. Flowers are very simple, devoid of accessory whorls. *Microsporophylls* (stamens) and *megasporophylls* (carpels) are often aggregated together in cones. Ovules present on the surface of the megasporophyll are directly pollinated by pollen grains. There is nothing like ovary, style and stigma; and naturally there is no fruit. Seeds remain naked. Unlike angiosperms, endosperm is formed before fertilization.

CYCAS (Cycad)

Cycas is a very ancient gymnosperm, now to be found only in the tropical countries. They are cultivated in gardens for ornamental purpose. Some of them are common sago-palms.

The plant looks more or less like palms or tree ferns. The stem is short, erect, stout and usually unbranched. It is covered by leaf-bases, and bears a crown of conspicuous pinnately compound leaves at the summit, arranged spirally alternating with brown scale-leaves (Fig. 215). Like ferns, they also exhibit circinate vernation. Leaflets are thick, leathery and provided with thickly cutinised epidermis and sunken stomata. *Cycas* has tap root system. Some roots often develop characteristic swellings due to the attack of certain algae and may help in nitrogen fixation.

Reproduction

Cycas is dioecious, *i.e.* microsporophylls and megasporophylls are borne by different plants. A large number of microsporophylls remain aggregated together around an axis forming a compact ovoid cone or **strobilus** (Fig. 216, A). Each microsporophyll is broadened above gradually becoming narrow below. Apical and basal portions are sterile, and the rest bear innumerable sporangia on the lower surface grouped in sori, a sorus consisting of 3 to 5 sporangia (Fig. 216, B & C). Each *microspore-mother cell* produces a tetrad of *microspores* or pollen grains by reduction division. Microspore begins germination before liberation from the sporangium. A small cell, called **prothallial cell,** is cut off at one end, which solely represents the male **gametophyte.** The rest develops into a small **generative** cell and a large **tube** cell. In such three-celled condition the spore is liberated.

Fig. 215. *Cycas* plant.

The megasporophylls do not form cone but constitute a body more or less leaf-like in appearance (Fig. 217). It is non-green and covered by dense brown hairs. Ovules, usually varying from 2 to 10, are present on the lower margin of each megasporophyll. Ovules are quite large. Each of them mainly consists of nucellus invested by a single thick integument, which is fused with the nucellus almost throughout the entire surface, only leaving a small region near the micropyle. Here, in fact, a small chamber, called

pollen chamber, is formed for collecting the pollen grains. The *megaspore-mother cell* undergoes reduction division to form a *linear tetrad* of *megaspores*, of which the lowest one survives and the rest degenerate. By free cell formation a tissue, called **endosperm** or female prothallus, is formed, which represents the female gametophyte. Towards the micropylar region a few (3 to 6) archegonia develop, each having a short neck and venter. The latter contains the egg or female gamete.

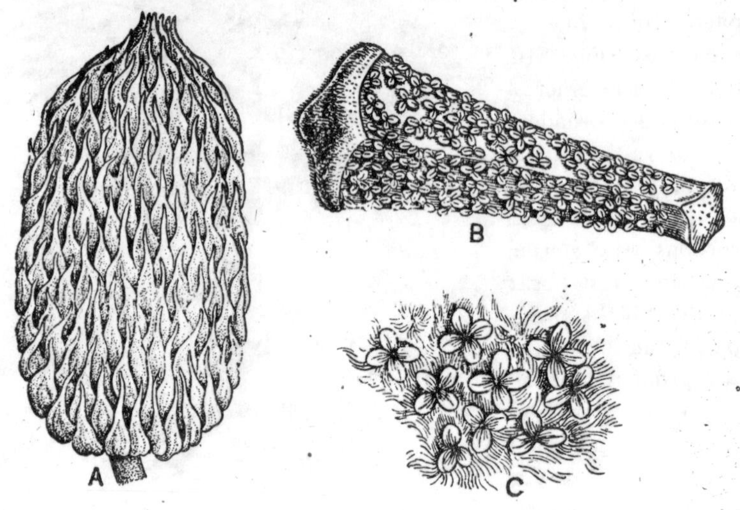

Fig. 216. *Cycas:* A—male strobilus; B—microsporophyll with sori on the lower surface; C—microsporangia (magnified).

Pollination and Fertilization

Pollen grains carried by wind are caught in the mucilaginous secretion of the ovules and ultimately lodged in the pollen chamber. Now tube cell begins to elongate forming the pollen tube which goes through the nucellus. The generative cell divides into a sterile stalk cell and a body cell. The body cell gives rise to a pair of quite large naked multiciliate **spermatozoids** which are set free when the end of the pollen tube bursts. They move to the archegonia, and one of them fuses with the egg, thus accomplishing fertilization.

The resulting zygote (oospore) soon secretes a wall round itself. It undergoes repeated cell division (by free-cell formation) and produces a multinucleated body called **pro-embryo.** Some of the cells from the upper region elongate to form a long slender much-coiled body, known as **suspensor,** which pushes the basal part of the pro-embryo deep into the female prothallus (Fig. 218). Those basal cells develop into embryo with a central axis and two cotyledons. The whole structure is ultimately transformed into a seed, which consists of embryo, endosperm and a seed-coat developed from the integument. After attaining maturity, they get detached and may germinate under suitable conditions to produce new plants. The mode of germination is hypogeous.

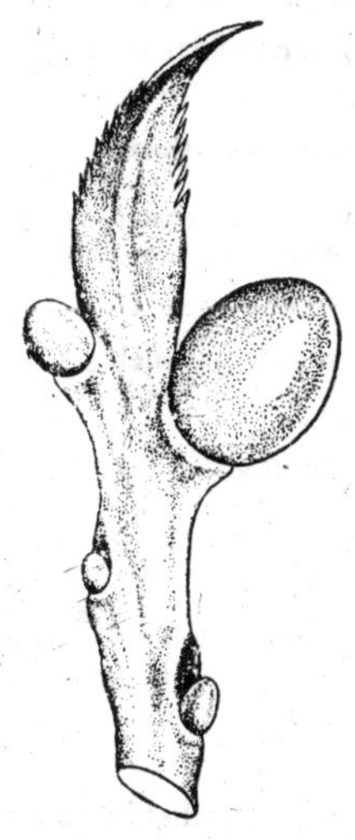

Fig. 217. *Cycas:* megasporophyll with ovules.

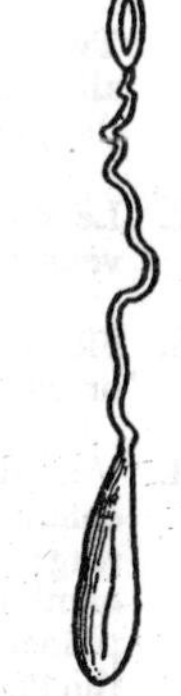

Fig. 218. *Cycas:* embryo with long-coiled suspensor.

Alternation of Generation

Cycas shows distinct alternation of two generations, sporophyte represented by the plant, and gametophyte by the male and female prothallus (endosperm). Sporophyte is most dominant, and gametophytes are extremely reduced and dependent on the sporophyte.

ANGIOSPERMS

Angiosperms are the most highly developed and conspicuous plants, capable of growing in all types of situation. They show wide variations as regards form and structure, ranging from very small aquatic duckweeds to the gigantic forest trees. Angiosperms are characterised by the development of distinct flowers with accessory and essential whorls. One or more carpels unite by margins to form basal swollen ovary which encloses the ovules. Pollen grains are received by the stigma and not directly by ovule, as in gymnosperms. After fertilization, the ovary develops into fruit and ovules into seeds. Thus angiosperms are 'closed-seeded' plants, the seeds remaining inside the fruit. Archegonia are not found in this group, and endosperm is formed after fertilization.

The plant is the dominant sporophyte and gametophytes are very small and reduced.

Two subdivisions of angiosperms are dicotyledons and monocotyledons, depending mainly on the number of cotyledons in the embryo. Though much has been already said about them, the main points of difference between the two groups are given below.

Dicotyledons	Monocotyledons
1. Two cotyledons are hinged laterally to the axis of the embryo; the plumule is terminal.	1. Only one terminal shield-shaped cotyledon is present; the plumule is lateral.
2. Leaves have usually reticulate venation.	2. Venation is usually parallel.
3. Flowers are generally pentamerous or tetramerous.	3. Flowers are trimerous.
4. Vascular bundles of stems are collateral, open and arranged in ring; due to the presence of cambium between xylem and phloem, secondary growth in thickness is possible.	4. Vascular bundles are numerous and scattered irregularly in the ground tissue. They are collateral, closed, thus secondary growth is normally absent.
Roots have limited number of radial vascular bundles.	Many vascular bundles in mono cotyledonous roots.

FAMILIES OF DICOTYLEDONS

SUBCLASS I—POLYPETALAE

1. Family—Cruciferae (Mustard family)

Common plants—Mustard (Fig. 219), radish, turnip, cauliflower, cabbage.

Plants—annual herbs, often with pungent juice.

Leaves—radical or cauline, simple, alternate, exstipulate, often irregularly lobed.

Inflorescence—raceme.

Flowers—pedicellate, ebracteate, complete, hermaphrodite, regular, hypogynous.

Calyx—sepals 4, in 2 whorls, free.

Corolla—petals 4, free, clawed, cruciform.

Androecium—stamens, 2+4, two in outer whorl with short filaments and 4 in inner whorl with long filaments, tetradynamous, free.

Gynoecium—carpels 2, syncarpous. Ovary—superior, one-chambered but rendered two-chambered by the formation of a false septum called replum. Ovules—many. Placentation—parietal.

Fruit—a siliqua (Fig. 108).

Seed—exalbuminous.

2. Family—Malvaceae (Cotton family)

Common plants—Cotton, China-rose (Fig. 98), Lady's finger (B. Dhenros), Indian mallow (B. Petari), Madras hemp, holly hock.

Plants—herbs, shrubs or trees, often copiously mucilaginous.

Leaves—simple, petiolate, alternate, stipulate (free-lateral), venation—reticulate multicostate.

Flowers—solitary, pedunculate, bracteate (a whorl of bracts, epicalyx) complete, hermaphrodite, regular, hypogynous.

Calyx—sepals 5, gamosepalous.

Corolla—petals 5, united with the staminal column at the base, aestivation—twisted.

Androecium—stamens many, monadelphous (united by filaments to form a bundle, anthers remaining free), epipetalous. Anthers —reniform, one-celled.

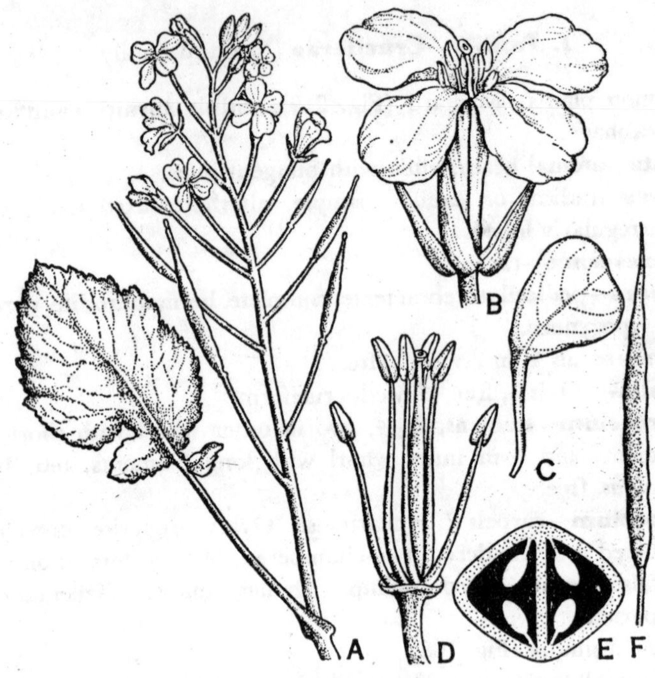

Fig. 219. Mustard (Cruciferae).
A—part of the plant with flowers and fruits; B—a flower with cruciform corolla; C—a clawed petal; D—tetradynamous stamens; E—section of ovary showing parietal placentation; F—a fruit (siliqua).

Gynoecium—carpels 3 to many, syncarpous. Ovary—superior, many-chambered, number of chambers being equal to the number of carpels present. Placentation—axile. Ovules—one to many in each chamber. Style—one, long, passing through the staminal column. Stigma—equal to the number of carpels.

Fruit—capsule.

3. Family—Leguminosae (Pea family)

Plants—Herbs, shrubs or trees, often climbers. Roots form nodules due to bacterial infection.

Leaves—usually compound, stipulate.

Flowers—solitary or in raceme. complete, bisexual, regular or irregular, slightly perigynous or hypogynous.

Calyx—sepals 5 or 4, united.

Corolla—petals 5 or rarely 4, free or united.

Androecium—stamens 10 or less, free or united.

Gynoecium—carpel 1, with marginal placentation, ovary—superior.

Fruit—usually a legume.

This is a large family. It has been divided into three subfamilies on the basis of the character of corolla and androecium.

Subfamily—Papilionaceae

Common plants—Pea, pulses, *Sesbania* (B. Bak) [Fig. 99], groundnut, sissoo.

Leaves—usually, imparipinnate, stipulate.

Flowers—solitary or in raceme. Complete, irregular, bisexual, slightly perigynous.

Corolla—petals 5, free, papilionaceous, aestivation—vexillary.

Androecium—stamens—10—(9)+1, diadelphous or monadelphous.

Description of a pea plant (*Pisum sativum* Linn)

Pea plant (Fig. 220) is a common annual herb cultivated during the winter for the seeds. It is a weak plant and climbs with the help of tendrils. The roots are infected by nitrogen fixing bacteria and they form characteristic nodules. Leaves are pinnately compound (imparipinnate), where the terminal leaflets are modified into tendrils. The leaf-base is swollen, forming pulvinus. A pair of foliaceous stipules is present. Flowers are lateral, solitary or in racemes. They are complete, irregular, zygomorphic, bisexual and slightly perigynous. Calyx is composed of five united sepals. Corolla is papilionaceous, made up of five free petals. The largest and the outermost one is the standard or vexillum, two lateral petals are wings or alae, and the innermost two, called keel or carina, unite to form a boat-shaped body. Aestivation is vexillary. Androecium consists of ten stamens, nine united to form a bundle, one remaining free, diadelphous. Gynoecium is monocarpellary, one-chambered with ovules in two series. Placentation is marginal. Ovary is elongated and superior. Fruit is a legume dehiscing by both the sutures. Seeds are exalbuminous and germination is hypogeal. The plant is the sporophyte; the gametophytes, represented by pollen tube and embryo-sac, are extremely reduced and dependent on the sporophyte.

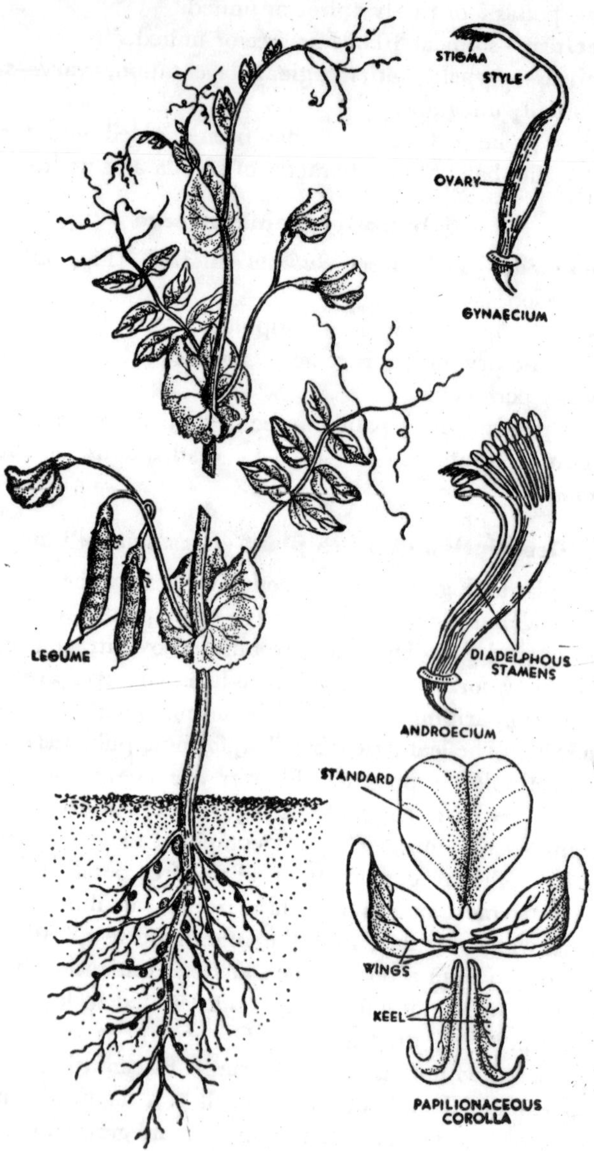

Fig. 220. Pea plant and dissected parts of a flower.

Subfamily—Caesalpinieae

Common plants—Tamarind, *Saraca* (B. Asoke), *Cassia* (B. Kalkasunde), Peacock flower (B. Krishnachura).

Inflorescence—racemose.

Flowers—complete, bisexual, slightly irregular, hypogynous.

Corolla—petals 5, free, aestivation—imbricate.

Androecium—stamens 10, free, or less by abortion.

Subfamily—Mimoseae

Common plants—Sensitive plant, *Catechu* (B. Khayer), *Acacia* (B. Babla), rain tree, *Albizzia* (B. Siris).

Leaves—usually bipinnate.

Inflorescence—a globose head or spike.

Flowers—complete, bisexual, regular, hypogynous, usually tetramerous, aestivation—valvate.

Androecium—stamens many, free.

4. Family—Cucurbitaceae (Gourd family)

Common plants—Gourd (Fig. 100), water-melon, cucumber.

Plants—climbing herbs with simple or branched extra-axillary tendrils, usually monoecious, rarely dioecious.

Leaves—simple, petiolate, exstipulate, with reticulate multicostate venation.

Flowers—usually solitary, incomplete, regular, unisexual, epigynous.

Male flower

Calyx—sepals 5, united.

Corolla—petals 5, usually united, bell-shaped.

Androecium—stamens usually 3, often 5, united by anthers (syngenesious), anthers sinuous.

Female flower

Calyx and **corolla** as in male flowers.

Gynoecium—carpels 3, syncarpous, with parietal placentation, usually one-chambered—rarely spuriously three-chambered, stigma—3, each with two branches.

Fruit—usually pepo.

Seed—exalbuminous.

SUBCLASS II—GAMOPETALAE

5. Family—Compositae (Sunflower family)

Common plants—Sunflower (Figs. 17, 80), marigold, *Dahlia*, *Zinnia*, *Chrysanthemum*, saff flower.

Plants—usually herbs.

Leaves—simple, usually alternate, exstipulate, with reticulate venation.

Inflorescence—capitulum with involucre of bracts. The central flowers with tubular corolla are called **disc florets** and peripheral ones with ligulate corolla are **ray florets** (Fig. 8).

Disc florets—complete, regular, bisexual, bracteate, epigynous (Fig. 89, 6).

Calyx—modified into hairy pappus or absent.

Corolla—petals 5, united, tubular.

Androecium—stamens 5, united by anthers (syngenesious), epipetalous.

Gynoecium—carpels 2, syncarpous. Ovary—inferior, one-chambered with a single ovule on basal placenta, stigma—bifid.

Ray florets—incomplete, unisexual (female) or neuter, irregular, epigynous (Fig. 89, 13).

Corolla—petals 5, united, ligulate, irregular.

Gynoecium—as in disc florets.

Fruit—cypsela.

Seed—exalbuminous.

6. Family—Solanaceae (Potato family)

Common plants—*Datura* (Fig. 101), tobacco, potato, brinjal, tomato, chillies.

Plants—herbs or shrubs.

Leaves—simple, petiolate, alternate, exstipulate.

Inflorescence—cymose.

Flowers—complete, regular, bisexual, ebracteate, hypogynous.

Calyx—sepals 5, united, persistent, often accresent.

Corolla—petals 5, united, different forms.

Androecium—stamens 5, epipetalous, alternating with corolla lobes.

Gynoecium—carpels 2, united, obliquely placed. Ovary superior, two-chambered, often rendered four-chambered by the formation of false partition walls. Ovules—many. Placentation—axile, stigma—bifid.

Fruit—usually berry, often capsule.

Seed—exalbuminous.

SUBCLASS III—INCOMPLETAE

7. Family—Euphorbiaceae (Spurge family)

Common plants—Castor-oil, spurges, croton, emblic myrobalan (B. Amlaki).

Plants—herbs, shrubs, trees, often with milky or watery latex.

Leaves—usually simple, alternate, opposite or whorled, stipulate.

Inflorescence—racemose, cymose or special inflorescence cyathium with extremely reduced flowers (Fig. 83).

Flowers—incomplete, regular, bracteate, unisexual, monoecious or dioecious, hypogynous.

Perianth—simple or absent.

Androecium—(in male flowers only)—stamens 1 or many.

Gynoecium—(only in female flowers)—carpels 3, syncarpous, ovary—superior, 3-chambered with one to two ovules in each chamber, style 3, stigma 3, each bifid.

Fruit—capsule.

Seed—usually albuminous with spongy outgrowth.

FAMILIES OF MONOCOTYLEDONS

SUBCLASS I—PETALOIDEAE

8. Family—Liliaceae (Onion family)

Common plants—Onion, garlic, *Asparagus* (B. Satamuli), *Smilax* (B. Kumarika).

Plants—mostly annual or perennial herbs with different types of stems (bulb, rhizome, cladode, etc.).

Leaves—simple, radical or cauline or both.

Inflorescence—usually racemose.

Flowers—regular, bracteate, usually bisexual, hypogynous.
Perianth—members 6, in two whorls, free or more commonly united (gamophyllous), petaloid.
Androecium—stamens 6 in two whorls, free, epiphyllous.
Gynoecium—carpels 3, syncarpous, ovary—superior, 3-chambered, ovules—many, placentation—axile, stigma—3.
Fruit—berry or capsule.
Seed—albuminous.

SUBCLASS II—SPADICIFLORAE

9. Family—Palmaceae (Palm family)

Common plants—Cocoanut, palmyra palm, date-palm, betel-nut, cane.
Leaves—large, compound with sheathing leaf-base.
Inflorescence—spadix, often quite large with boat-shaped bracts—spathes.
Flowers—small, bisexual or unisexual, hypogynous.
Perianth—members 6, in two whorls.
Androecium—stamens usually 6, often 3.
Gynoecium—carpels 3, united, ovary—superior, 3-chambered or 1-chambered.
Fruit—drupe or berry.
Seed—albuminous.

SUBCLASS III—GLUMIFLORAE

10. Family—Graminaceae (Grass family)

Common plants—Rice, maize, wheat, barley, oat, bamboo, sugar-cane, grasses.
Plants—annual and perennial herbs, rarely woody, as in bamboos.
Stem—cylindrical, usually jointed at nodes and hollow at internodes (solid stems in maize and sugar-cane).
Leaves—simple, alternate, 2-ranked with sheathing leaf-base, a membranous ligule is present at the junction of leaf-base and blade, linear with parallel venation.

Inflorescence—spike or panicle of spikelets, spadix in female inflorescence of maize. Each spikelet is one or more flowered and is enclosed by glumes (bracts) arranged alternately, first and second glumes are small and empty; the third one called flowering glume, bears a sessile flower at its axil and the fourth is 2-nerved palea.

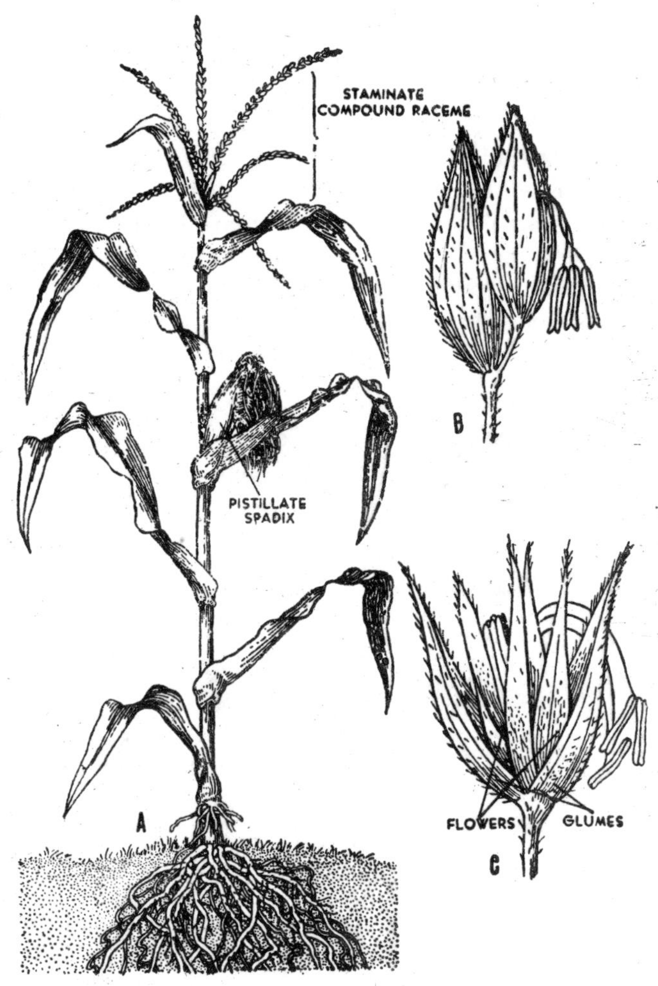

Fig. 221. Maize plant.
A—a plant; B—paired spikelets; C—spikelet opened.

Flowers—usually bisexual, often unisexual, monoecious (*e.g.* maize), hypogynous.

Perianth—represented by a pair of scales called lodicules, rarely lodicules absent or more than two.

Androecium—stamens 3, sometimes 6 (*e.g.* bamboos) with long versatile anthers.

Gynoecium—carpel 1, ovary superior, one-chambered with one ovule. Stigma—2, feathery.

Fruit—caryopsis.

Seed—albuminous with single cotyledon, scutellum.

Description of maize plant (*Zea mays* Linn)

Maize or Indian corn (Fig. 221) is a stout annual plant cultivated for the grains during the rainy season. It forms a staple food in some parts of India. Roots are of fibrous adventitious type. Primary root aborts after germination, and is replaced by fibrous adventitious ones from the base of stem. Stems are stout, solid with

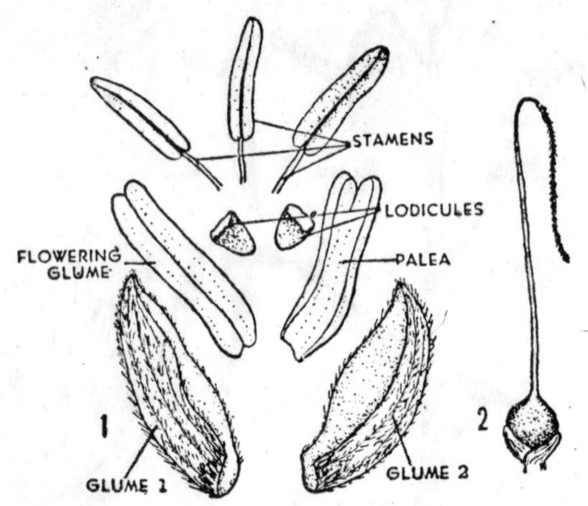

Fig. 222. Maize: 1—dissected male flower; 2—carpel of female flower.

distinct nodes and internodes. Leaves are simple with entire margin, quite large, linear in shape and arranged alternately. The arrangement is distichous or 2-ranked. They have sheathing leaf-base surrounding a part of internode above. Membranous ligule is present at the junction of base and blade. Venation is parallel. All the organs have rough surface due to impregnation of silica. Inflorescences are of two types, *viz.* terminal staminate compound raceme or panicle and axillary pistillate spadix.

Male inflorescence consists of many spikelets. The spikelets occur in pairs—the lower one is sessile and the upper one is stalked. Each spikelet is two-flowered. They have four glumes. The first and the second glumes located at the base, are sterile, and the third one, called flowering glume, and the last one, the two-nerved palea, enclose a flower. Flowers are obviously unisexual. Perianth is represented by a pair of fleshy scale-like bodies called **lodicules.** Androecium is composed of three free stamens with long prominent linear anthers (Fig. 222, 1).

Female inflorescence is a spadix arising from the axil of a lower leaf. A large number of spikelets are arranged closely on the fleshy axis, forming what is known as cob. It remains surrounded by a few large hyaline bracts called **spathes.** Each spikelet, like the male ones, is two-flowered and has protective glumes. Of the two, usually the upper flower is fertile and the lower one aborts. Lodicules are absent. Gynoecium is monocarpellary. The ovary is one-chambered, superior, ovoid in shape with a single anatropous ovule. The styles are long and silky. Persistent styles hang out in tufts from the apex of the cob. Stigma is long and feathery (Fig. 222, 2).

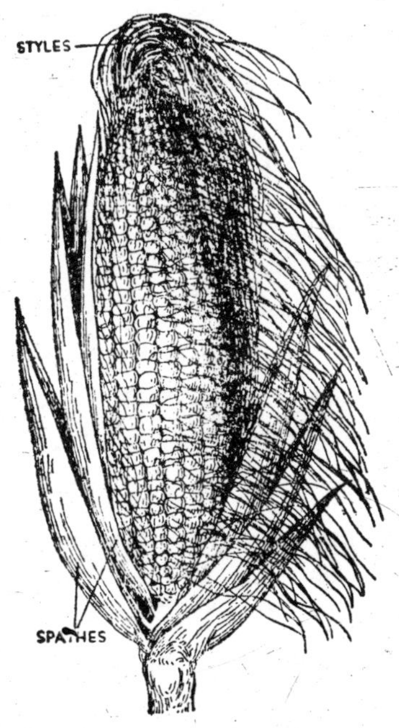

Fig. 223. Cob (pistillate inflorescence) of maize.

Flowers are wind-pollinated. Small dusty pollen grains are easily caught by the feathery stigma of the carpel. Fertilization takes place by normal process. Fruit is a caryopsis with a single closely fitted seed where fruit wall and seed-coat are inseparably united. A large number of fruits, called grains, remain densely crowded on the spongy axis (Fig. 223). Seeds are albuminous. The single cotyledon, scutellum, serves as the absorbing organ. Germination is hypogeal.

The plant itself is the most prominent generation, the sporophyte, with *diploid* (2N) chromosomes in the nuclei of the cells, and the gametophytes with haploid (N) chromosomes are represented by embryo-sac and pollen tube. They are very small, reduced and dependent on the sporophyte. Gametophytic generation begins with *reduction division* in microspore-mother cell and megaspore-mother cells with subsequent formation of microspores and megaspores respectively. With fertilization '2*n*' number is restored and sporophyte begins.

PART II
ZOOLOGY

CHAPTER I

ANIMALS : GENERAL CONSIDERATION

Zoology is the biology of animals. This means that Zoology is the science of animal life. All the facts, theories and laws related to animals are included within the scope of the Zoological science. But before attempting the study of any individual animal it is necessary to consider the following subjects: (1) a brief sketch of the conditions under which the animals live, that is **animal habitats;** (2) a rapid survey of the entire *animal kingdom* to learn the **variety of animal life** and (3) their **classification** into main groups; (4) a general account of the structure of the animal body in relation to function, that is, **physical organization of the animal body.**

Animal Habitats

Animals are found almost everywhere on the earth. The fathomless depth of the ocean, the highest peak of the mountain, the freezing cold of the polar region and the scorching heat of the desert, each has its own peculiar **fauna** or stock of animals. Most of the animals are **free-living.** They inhabit the salt water of the sea or the fresh water of the river. They may live *on* or *in* the soil and some may take to the air. Lastly, many animals reside *on* or *within* the bodies of other animals, forming **animal associations.** The particular locality in which any one kind of animal lives is known as its *habitat.*

Different animal habitats may be broadly divided into two groups—**Aquatic** and **Terrestrial.**

Aquatic Habitats

The animals which live in water are called aquatic animals. According to the nature of the water aquatic animals may be *marine* or *fresh-water.*

Marine animals. About three-fourths of the earth's surface is covered by the oceans. The salt water serves as the home for the **marine animals** who can survive neither in fresh water nor on land. The sea beach is occupied by the **littoral animals,** some of which live on sand, others in mud, and still others remain attached to submerged rocks or to sea weeds. The depth of the ocean is inhabited by the **abyssal** or **deep-sea animals** which are collectively known as the *benthos*. The open ocean is thickly populated by the surface-dwelling **pelagic animals** which either swim freely or float passively and are drifted from place to place by the high waves. The pelagic fauna is thus subdivided into *nekton*, which includes the active swimmers, and *plankton*, which includes the smaller passive forms that are at the mercy of the currents. Marine animals are more or less restricted to their own respective zone—either the sea beach, or the abyssal depth of the bottom, or the surface of the open sea.

Fresh-water animals. Fresh-water animals are found in ponds, pools, rivers, lakes and swamps. Some prefer to live in *stagnant water* but others choose the *running stream*. Each of these habitats may contain a number of communities, such as those occupying (1) the beach, (2) the open water, and (3) the bottom.

TERRESTRIAL HABITATS. Terrestrial animals are those who live on the land. They may be simple surface dwellers or they may burrow beneath the soil and thus become **subterrestrial.** The **surface communities** may choose to live on rocks, or plains, or desert or damp forest. The **aerial animals** spend a part of their time in the air, but they depend on the surface for rest. A number of animals are **arboreal,** homing amongst the branches of the tree.

Animal associations. Some animals habitually live *upon* or *within* other animals. There are *four* main types of such animal associations: (1) *commensalism*, (2) *mutualism*, (3) *symbiosis*, and (4) *parasitism*. These four categories are absolutely man-made, which are not discontinuous in nature but are diverse aspects of the same general laws.

Commensalism indicates an association between two individuals which live together with no metabolic interdependence and the relationship is not necessarily continuous for the life of the partners. Commensals may be *external* as well as *internal*. The best example of commensalism is the suckerfish Remora which develops a sucker for attaching to sharks. The Remora is thus carried from place to place and protected from enemies. They never injure their host and feed upon scraps that are rejected by the shark.

Mutualism involves an association between two animals in which there is a metabolic dependence, but it is not obligatory for existence. Association of hermit crab and a sea-anemone may be taken as example of mutualism.

Symbiosis indicates an association between two individuals in which there is metabolic dependence beneficial to both and both the partners are mutually benefited. In extreme cases, the two partners cannot live without each other. One of the best example of symbiosis is the association between termites and a kind of small flagellates which live within their digestive canal. The termites feed on wood but are unable to digest the same; the flagellates help them by digesting the wood and thus save them from actual starvation. In return, the flagellates are safely housed, protected from enemies and supplied with food by their host.

Parasitism is an association between two individuals in which one is metabolically dependent on other and is benefited at its expense; the one benefited is the parasite. The host is always the losing partner. The protozoa causing malaria is a parasite; the host in this case is man. The parasites may live externally (ectoparasite) or internally (endoparasite).

Variety of Animal life

Animals are numerous. There are different kinds of animals varying from one another in *size*, *form*, and mode *of life*.

Size of animals varies widely. The smallest animal is a parasite living within the red blood cells of a cattle. It measures about 1 *micron*, that is 1/1000 millimeter in diameter. The largest animal is the whale which measures 250 cms in length and weighs about 150 tonnes. Each kind of animal exists in groups of similar individuals, namely, dogs, sparrows, frogs, cockroaches, etc. Each kind of animal is known as a **species.** Members belonging to the same species, though varying somewhat among themselves, have common features which set them apart from all other kinds. Moreover, the various kinds themselves exist naturally in larger groups. Thus, the dogs and jackals are quite different from the wolves and from each other, and yet they all constitute a large group of more or less similar animals. Again, these larger groups are capable of being themselves grouped into larger and larger division. For example, dogs, jackals and wolves are placed along with cats, tigers and lions in a big subdivision, because they all possess certain common features. Evidently, the animals can be grouped on the basis of their similarities and differences. There are about *one million* kinds or **species** of animals in the **animal kingdom.** Zoologists have arranged them into *ten* major groups or **phyla.** The members belonging to the same **phylum** are united by certain common characteristics. Besides these, there are a few less important phyla which have been omitted from the following list.

I. **PHYLUM PROTOZOA** (approximately 30,000 known species): UNICELLULAR ANIMALS LIKE AMOEBA, PARAMOECIUM, MONOCYSTIS AND MALARIA PARASITE.

Protozoa are microscopic in size. Each individual consists of only one cell which has to carry on all the vital activities. They are abundantly found in water containing decaying organic matter. Some, such as the dysentery amoeba and the malaria parasite, live within other animals. Still others live in damp soil, or in fresh water, or in the sea.

The single-celled condition is an important feature which sets the protozoa apart from all other animals. These unicellular creatures have therefore been placed in the **subkingdom protozoa,** which includes only one phylum, the protozoa. The remaining phyla of animals, all of which are many-celled, comprise the **subkingdom metazoa.**

II. **PHYLUM PORIFERA** (approximately 5000 known species): SPONGES.

These are pore-bearing sedentary animals found mostly in the sea. A few species occur in the fresh water but none on the land. The sponges, like plants, are attached to a substratum. The outer surface of the sponge is perforated by numerous pores and the body wall is supported by a framework which is composed of lime, or of silica or of an organic substance called **spongin.**

III. **PHYLUM CNIDARIA** (approximately 10,000 known species): HYDRA, JELLY-FISHES, SEA-ANEMONES AND CORALS.

Most of the cnidaria are marine but Hydra is found in fresh water. Some, such as the corals and sea-anemones, are attached to a substratum; others are slow moving or adapted for drifting in the water. All are radially symmetrical. This means that the animal is the same all round, and has no right or left side. It is symmetrical around a median vertical axis, and can be divided into similar halves by a number of vertical planes. Body wall is composed of two layers; it encloses a central digestive cavity which communicates with the exterior by only one opening, the mouth. Thus, the cnidarian body is essentially a two-layered hollow sac opening by the mouth; the sac may be tubular, as in hydra, or saucer-shaped, as in jelly fish. There are movable arm-like structures near the mouth, called **tentacles,** which carry peculiar **stinging cells** for stunning the prey.

IV. **PHYLUM CTENOPHORA** (approximately 80 species): BEROE, HORMIPHORA, PLEUROBRACHIA.

The phylum derives its name from two greek words—*Ktenos*= comb, *phoros*=bearing. Ctenophores are all marine. They have biradially symmetrical bodies. They possess eight meridionally placed ciliated plates. They resemble the cnidarians on many counts but differ from them in not having the nematocysts. Their ecto-mesoderm is gelatinous and bear mesenchymal muscle cells. They possess a specialised aboral sense organ and the tentacles bear adhesive cells. All are planktonic.

V. **PHYLUM PLATYHELMINTHES** (approximately 6500 known species): FLAT-WORMS, FLUKES AND TAPE-WORMS.

These are flat, unsegmented, worm-like creatures with soft and bilaterally symmetrical body. In a bilaterally symmetrical animal there is a right side and a left side, a fore end and a hind end, a dorsal or back surface and a ventral or front surface. There is only one plane of symmetry by which the body can be divided into two equal halves. Leaf-like liver-flukes and ribbon-like tape-worms are parasites but there are several free-living species, marine as well as fresh-water. Digestive canal is incomplete, with only one opening, the mouth; there is no anus. Excretion of waste products is effected by peculiar **flame cells.**

VI. **PHYLUM NEMATHELMINTHES** (approximately 10,000 known species): ROUND-WORMS.

These are cylindrical, unsegmented, worm-like animals with soft, bilaterally symmetrical body, tapering at both the ends. Digestive canal is complete, with two openings, a mouth in front and an anus behind; it is a straight tube running through the body from end to end. Most of the group are aquatic. A few inhabits damp soil. Others, such as hook-worms, thread-worms and filaria worms are parasites of man and cattle.

VII. **PHYLUM ANNELIDA** (approximately 7500 known species): Earth-worms, Leeches and Sand-worms.

These are true worms with soft, elongated, bilaterally symmetrical body, divided into a series of ring-like segments or **metameres.** The annelids are, therefore, known as the segmented worms. The annelidan body is built on the tube-within-a-tube plan. The outer tube represents the body wall and the inner tube represents the digestive canal. The two tubes are separated from one another by a space called body cavity or **coelom.** Most of the annelids, such as the sand-worms, are marine; others, like the leeches, are fresh-water; but the earth-worm is subterrestrial.

VIII. **PHYLUM ARTHROPODA** (approximately 750,000 known species): Prawns, Crabs, Cockroaches, Centipedes, Millipedes, Scorpions, and Spiders.

Arthropods are bilaterally symmetrical, segmented animals with soft parts of the body protected by a hard *chitinous* external skeleton. Each segment of the body bears paired legs or **appendages** which are jointed. This phylum is the largest of the animal phyla and includes nearly three-fourths of all the known species of animals.

IX. **PHYLUM MOLLUSCA** (approximately 90,000 known species): Clams, Oysters, Snails, Cuttle-fishes and Octopus.

Molluscs are unsegmented and without appendages. The soft parts of the body are enclosed in a hard *calcareous shell*, as in snails and oysters. A fleshy muscular **foot** for locomotion is often present. Many of the molluscs are marine, some are fresh-water, and a few like the garden snails are terrestrial.

X. **PHYLUM ECHINODERMATA** (approximately 6,000 known species): Starfishes, Sea-urchins, Sea-cucumbers and Sea-lilies.

Echinoderms are characterised by spiny skin. All are marine, inhabiting the shore and bottom of the sea. A few such as the sea-lilies are attached; but the majority are free to move about. Locomotion is very sluggish and effected by peculiar structures called **tube-feet.** This is the only phylum possessing a *water-vascular system*. The body is radially symmetrical and star-like as in starfishes, brittle-stars and basket-stars.

XI. **PHYLUM CHORDATA** (approximately 100,000 known species): BALANOGLOSSUS, ASCIDIANS, AMPHIOXUS AND VERTEBRATES.

The chordates possess a stiff supporting rod, called **notochord.** Leaving aside a few lower forms, such as balanoglossus, ascidians and amphioxus, all chordates are vertebrates.

Vertebrates possess the backbone which forms the supporting skeleton for the long axis of the body. Vertebrate body is bilaterally symmetrical and is typically composed of head, trunk and tail. There are two pairs of appendages, either in the form of paired fins or limbs, or wings. They comprise the highest animals and include man. Vertebrates are divided into the following classes: (1) the **cyclostomata** including lampreys and hag fishes which are round-mouthed and without a lower jaw; (2) the **chondrichthyes** or cartilaginous fishes such as sharks and electric rays; (3) the **osteicthyes** or body fishes like Bhetki and Rohu; (4) the **amphibians** such as toads, frogs and salamanders with moist, naked skin; (5) the **reptiles** including snakes, lizards, tortoises and crocodiles with scales on their outer surface; (6) the **aves** or birds with feathers and wings for flight; (7) the **mammals** including duck-billed mole, kangaroo, guinea-pig and man, with hairy skin and with young ones fed by the mother with her own breast-milk.

Classification and Nomenclature

It is a matter of common observation that individuals related to one another tend to be similar and the degree of the similarity depends upon the degree of the kinship. Animals which are like enough to be included in the same species owe their likeness to descent from a common ancestor. In classifying animals, zoologists are guided by *two* important **principles:** (1) animals, possessing similar structures and therefore descended from a common ancestor, are placed in the same natural group; (2) different groups of like animals agreeing with one another in some important feature and therefore related to one another, are included in a bigger group. In other words, classification is based on similarity of structure and descent from common ancestry.

Probably the first attempt to classify animals on a scientific principle was made by **Aristotle** (384–322 B.C.). He used structural similarity as the basis of classification; but some of his animal groups were **artificial** that is, the members of the group were not united by a common ancestry. Aristotle's work was greatly perfected by **Carolus Linnaeus** (1707–1778). This Swedish biologist is referred to as the father of classification, because he devised the system that has been developed into the current **Linnaean system** of classification.

Fig. 1. **Carolus Linnaeus** (1707–1778)

In classifying animals into **natural** groups Linnaeus recognised certain grades or degrees of resemblance, which are spoken of as species, genera, families, orders, classes and phyla.

Species. Each kind of animal roughly corresponds to a *species* (plural, *species*). For example, man and dog each represents a species. A species may be defined as a group of individuals which resemble one another very closely and can mate with each other successfully for reproducing their own kind; interbreeding is seldom possible among animals belonging to different species.

Genus. Two or more species agreeing in many important features are included in a *genus* (plural, *genera*). For example, the common Indian toad appears to be closely related to the American toad and the two species resemble one another; hence they are included in the same genus.

Higher Categories. Genera having common characters constitute a **family** and likewise similar *families* are placed in an **order,** orders into a **class,** and classes into a **phylum.** The phylum (*phylon*=a race) is the largest subdivision of the **animal kingdom.** There may be intermediate groupings in all these divisions, such as **subphyla, super-classes, subclasses, suborders,** etc. The terms family, order and class, are of course used in a special sense quite different from their ordinary meaning.

Phylogenetic tree. Relationship of the various animal groups may be represented by a branching tree-like diagram. Each leaf

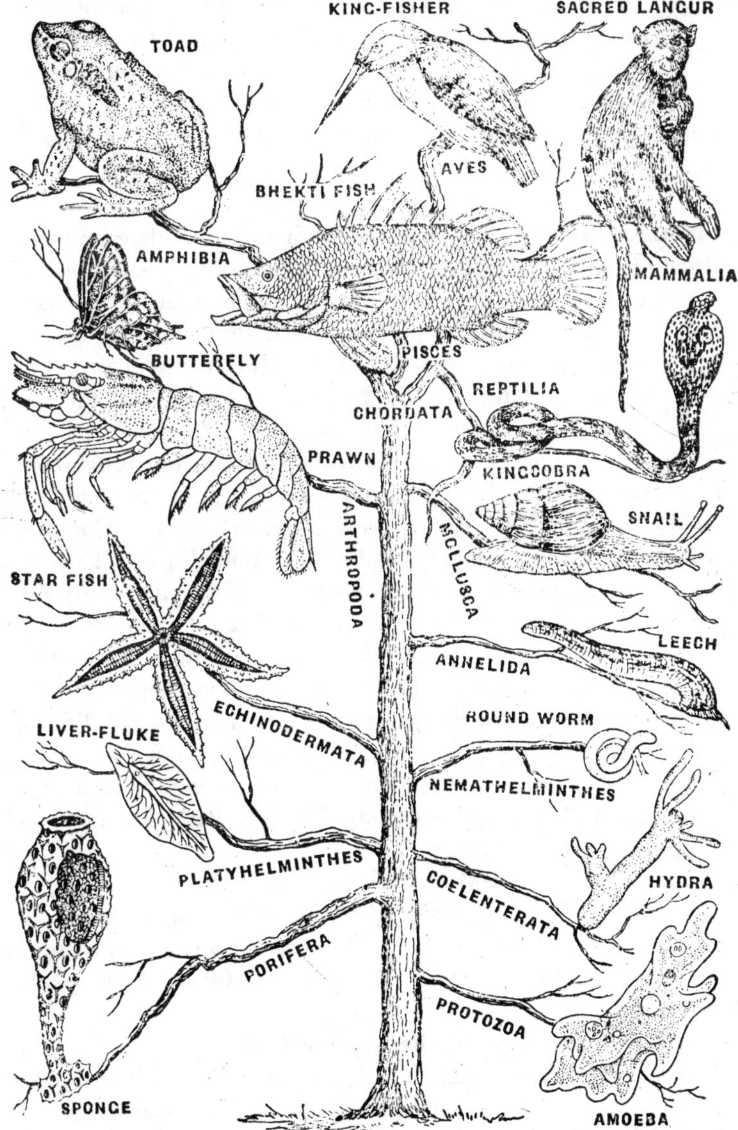

Fig. 2. 'Phylogenetic tree' representing the animal kingdom. The lowest phylum is placed at the bottom and the highest at the top.

on the tree represents a **species** of animal. The fine shoot bearing the leaves constitutes a **genus** and the little twig that carries the shoots comprises a **family.** The small branch bearing the twigs represents an **order** and the big branch carrying the smaller ones is equivalent to a **class.** The ten main branches of the phylogenetic tree bearing the larger branches compose the major divisions or **phyla** of the animal kingdom. The lowest and the most primitive group is placed near the bottom of the tree, whereas the highest group occupies the topmost position.

Nomenclature. The common names of animals vary widely in different provinces of the same country and more than one common name is often applied to the same animal even in the same language. Linnaeus advocated the use of Latin for naming animals and thus prevented confusion. The **scientific name** of an animal is composed of two Latin words; the first word stands for the *genus* and the second for the *species.* The system of using the generic and specific names is known as the **binomial system of nomenclature.** The generic name is always written with a capital letter but the following specific name begins with a small letter. Moreover, scientific names must either be *italicised* or underlined. Thus the common Indian toad is *Bufo melanostictus.* Here the word *Bufo* represents the genus and the word *melanostictus*, the species. The American toad is *Bufo boreus;* it belongs to the same genus *Bufo*, but is of a different species. The common Indian frog is known as *Rana tigrina;* it belongs to the genus *Rana* which is entirely different from the genus *Bufo.* The European bull frog, *Rana temporaria,* is of the same genus as the Indian frog, but belongs to a different species. Similarly, the American bull frog is *Rana catesbeiana* and the leopard frog is *Rana pipiens.*

The toad *family* **Bufonidae** containing *B. melanostictus, B. boreus*, etc., and the frog *family* **Ranidae** containing *R. tigrina*, R. *temporaria*, etc., are included in the *order* **Anura.** The anura along with other orders form the *class* **Amphibia** of the *phylum* **Chordata.**

Physical Organization of the Animal body

The animal body is composed of one or more cells. The simplest animals are the unicellular protozoa. In such a primitive group, a single cell forms the entire organism, and the living matter of the cell has to carry on all the vital activities, such as nutrition, respiration, excretion, growth and reproduction. The higher animals are the multicellular metazoa. All metazoa are composed of a large number of cells. In such a creature, there is marvellous unity and division of labour among the various cells which form the body. Metazoan cells differ in shape and size according to the special nature of the function which they are to perform. Some of the cells carry on respiration, others nutrition and still others reproduction. Thus, there is a **physiological division of labour.**

Tissues. In higher metazoa, cells are aggregated to form tissues. A **tissue** may be defined as a collection of similar cells adapted to carry out the same general function. Four principal types of tissues are recognised in all higher animals. These are (1) *epithelial tissue*, (2) *connective* or *supporting tissue*, (3) *muscular tissue*, and (4) *nervous tissue*.

Epithelial tissues consist of cells which cover surfaces and line up cavities. Epithelial cells are usually arranged in such a way as to form a *membrane*. They may also be modified to form solid or hollow structures, called *glands*, which either secrete useful substances or excrete waste products.

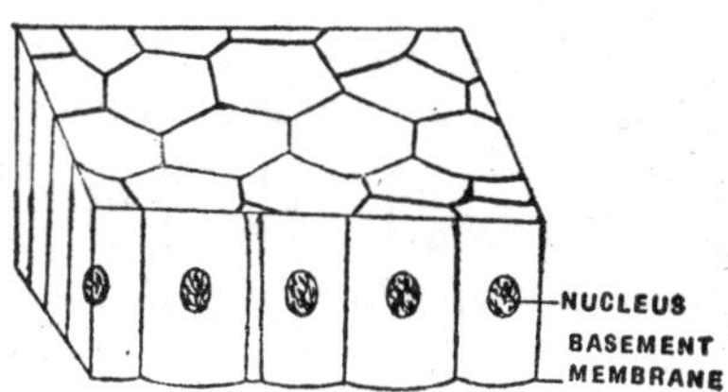

Fig. 3. Epithelial tissue showing cells.

Connective tissues include bones, cartilages, other supporting materials and blood. It is composed of only a few cells and fibres. The main bulk of the tissue is a non-living intercellular substance called *matrix* which is secreted by the cells. This ground substance hardens up in bones and cartilages.

Muscular tissues form the muscles of the body. A muscle is composed of muscle cells which are adapted to contraction and relaxation, thereby causing movement and locomotion.

Nervous tissue is composed of nerve cells and nerve fibres. It is capable of transmitting impulses from one part of the body to another.

Thus with the increase of structural complexity, the body cells become more and more specialised and there is a greater tendency to develop tissues. Further consideration and full description of the tissues will be undertaken in Chapter XVIII.

Organs. A single tissue does not usually work alone. The efficiency is greatly increased when two or more tissues function together. An **organ** may be defined as an association of tissues for performing a special function. The lung, for example, is an organ for carrying out respiration; it is composed mainly of epithelial and connective tissues. Similarly, the human hand is an organ for manipulating and grasping objects; it consists of all the four kinds of tissues.

Organ systems. An isolated organ cannot work satisfactorily. Several organs act together in the metazoan body for effecting a particular important function. Such a group of organs is known as an *organ system.*

The body of a fish or a frog is comparable to a huge factory with several separate departments. Each department is entrusted with a special kind of job and the various departments act in co-operation with one another. The organ systems, just mentioned, are the various departments of the animal machinery. Each system performs its own part and the benefit is shared by the others.

The organ systems which are found in the higher metazoa are: (1) *integumentary*, (2) *skeletal*, (3) *muscular*, (4) *alimentary*, (5) *circulatory*, (6) *respiratory*, (7) *excretory*, (8) *reproductive*, (9) *nervous*, and (10) *endocrine*. Figure 4 illustrates the organ systems of the common toad.

1. **Integumentary system** includes the **skin** and its various outgrowths, such as hairs, nails, scales and feathers. The principal function of this system is *protective*. But the skin is a 'jack of all trades' and often helps in the performance of other functions, such as *respiration*, *excretion* and *regulation of body temperature*.

2. **Skeletal system** consists of the main supporting framework of the body. It is composed of **bones, cartilages, ligaments** and other *hard structures*. A skeletal framework lying outside the body in animals like crabs and snails is known as the **exoskeleton.** It protects soft parts from external injury. In vertebrates, there is an **endoskeleton** inside the body which functions as the main prop for support and protects internal organs. Moreover, the shape of the animal depends, to some extent, on the shape of its skeleton.

3. **Muscular system** is composed of **muscles** attached to bones, and also of muscles arranged on the walls of organs such as

heart, digestive canal, etc. The muscles are the motive power in the physical machinery of an animal and serve to produce move-

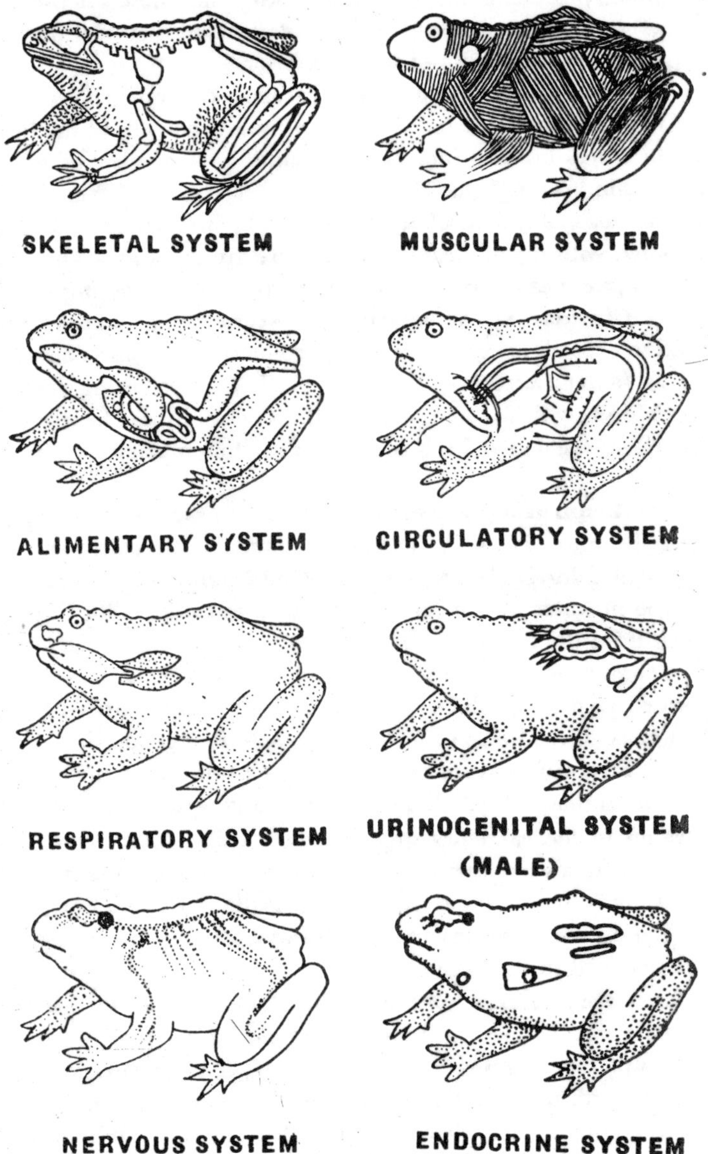

Fig. 4. Organ systems of toad.

ment and locomotion. The *skeletal muscles* are said to be **voluntary** because they are contracted and relaxed according to the wish of the animal possessing them. *Heart muscles* and those on the wall of other soft organs are said to be **involuntary** because they act independently of the wish of their possessor.

4. **Alimentary** or **digestive system** is concerned with taking in of food, preparing it for absorption by all the cells of the body, and then egesting the indigestible residue. It consists of a long tube running through the body from end to end, and composed of **mouth, pharynx, oesophagus, stomach, intestines** and **anus.** Associated with the alimentary tube, there are digestive glands such as **liver, pancreas,** etc., which secrete *enzymes* for digesting food.

5. **Circulatory** or **vascular system** brings about transportation of food, oxygen, and waste products. It consists of organs such as **heart, arteries, veins,** etc. The transporting medium is the **blood,** which is pumped into the arteries by the heart. The veins bring the blood back to the heart, thus making a complete circuit.

6. **Respiratory system** constitutes the oxygen-intake apparatus of the animal body. Terrestrial animals inhale oxygen into organs called **lungs:** but most of the aquatic animals breathe oxygen dissolved in water by special organs known as **gills.** Blood absorbs oxygen from the lungs or gills and transports the gas to the tissues, where combustion and release of energy occurs and carbon dioxide is formed as a by-product. This carbon dioxide is carried by the blood back to the lungs or gills, from where it is expelled.

7. **Excretory system** removes waste products of metabolism from the animal body. Blood carries away waste products from all tissues to the excretory organs, which collect and throw them out. In animals like earthworms and leeches, the process is accomplished by a set of tubes called **nephridia.** It vertebrates, the principal excretory organs are the **kidneys** which expel nitrogenous waste products in the form of urine.

8. **Reproductive system** essentially consists of the **gonads.** The male gonads are the **testes** (singular, *testis*) and the female gonads are the **ovaries** (singular, *ovary*). The gonads are often associated with the excretory organs and when this happens, the two separate systems are described together as the **urinogenital system.**

The gonads by manufacturing and releasing *germ cells* (**spermatozoa** in males and **ova** or **eggs** in females) are responsible for the production of young individuals of the same species.

9. **Nervous system** enables the different parts of the body to work together, so that the organism can function as a whole. It is composed of organs such as **brain, spinal cord, nerves,** etc. Associated with the nervous system there are **receptor organs,** such as *eyes, ears, nose,* etc., each of which is expert in the perception of a special kind of sensation.

The nerves are of two kinds: (1) **afferent** or **sensory** and (2) **efferent** or **motor.** An afferent or sensory nerve connects a receptor organ with either the brain or the spinal cord. A motor nerve runs the opposite way; it connects the brain or cord with an **effector organ,** that is, either a muscle or a gland.

The manner in which **nervous co-ordination** occurs is best explained by a simple example. The smell or sight of food is perceived by the brain via sensory nerves. The brain promptly decides the steps that are to be taken and sends down its orders along the motor nerves. The hands are directed to grasp the food and put it into the mouth; the teeth, tongue and pharynx are then prompted to chew and swallow the food step by step in a co-ordinated manner. Lastly, the digestive glands are ordered by the brain to secrete enzymes for digesting the food that has been swallowed.

10. **Endocrine system** consists of a number of *ductless glands,* such as **thyroid, pituitary, adrenal,** etc., which release chemical substances called **hormones.** The hormones are washed away by the blood and carried to every part of the body. By the combined effort of these chemical substances the animal body regulates its growth and metabolism.

Thus a higher animal possesses two co-ordinating systems: (1) the nervous system for producing **nervous co-ordination,** and (2) the endocrine system for effecting **chemical co-ordination.**

The organ systems as listed above, are best studied in a familiar animal. The common toad, *Bufo melanostictus,* is very suitable for this purpose. It is abundantly found in all parts of Indian subcontinent, and can, therefore, be procured easily for laboratory work almost throughout the year. The toad is a typical vertebrate. The animal is large enough for a beginner to dissect easily. It has, therefore, become a convention to begin zoology with the study of toad, which, in fact, is called the grammar of Zoology. The anatomical peculiarities of the frog are somewhat similar and hence frogs are good substitutes in places where they are easily available.

CHAPTER II

THE COMMON TOAD

(*Bufo melanostictus*)

Habitat

Toads are noted for their ugly look, noxious skin secretions and peculiar squatting posture. They are usually found in dark corners of the garden, in shady ditches and swamps, and concealed under cover of damp rubbish heaps. These animals are nocturnal and often appear just at dusk, hopping along the roadside and feeding upon insects, caterpillars, earthworms and garden snails. During cold weather, toads bury themselves underground and lie hidden in a dormant condition, that is, they **hibernate** (*hiberna*= winter quarter). With the increasing warmth of the spring they wake up from their winter sleep and resume normal activities.

The adult toad habitually lives on land; but when frightened it never hesitates to jump into a pond and stay away from the shore as long as it likes. Moreover, they must come to a pond for laying their eggs. The young ones are the fish-like tadpoles which hatch out of the eggs and grow in water for about twelve weeks. At the end of this period, the aquatic tadpoles are quickly changed into the terrestrial toads. The young toads now emerge from the water and come to live on land. The toad, therefore, is **amphibious** (*amphi*=both; *bios*=life), because it is quite at home on the land as well as in the water.

External Features

The body of a toad is divisible into *head*, *trunk* and *limbs*. Head and trunk are not separated by a flexible neck; this condition is very suitable for aquatic (*aqua*=water) life. The adult has no tail.

The toad is a **bilaterally symmetrical** animal; that is, if the animal is cut through the middle of its length, the parts on the right half will be similar to those on the left half.

For the purpose of description, the back of the animal is spoken of as the *dorsal* and the underneath as the *ventral surface;* the part of the body which is foremost during locomotion is the *anterior end* and the rear is the *posterior end*. The part of a limb which lies nearest to the trunk is *proximal* and the opposite is *distal*. Each side of the body is said to be *lateral* and structures on the middle line of the body are *medial*.

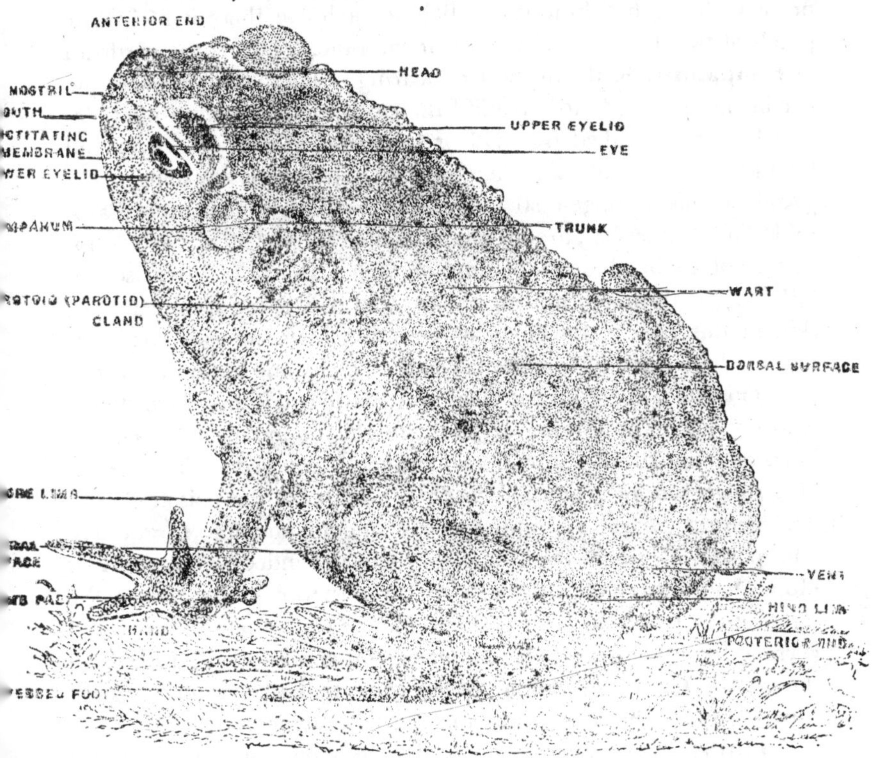

Fig. 5. The common toad (*Bufo melanostictus*).

Head. The head is nearly semicircular in outline with the bluntly rounded **snout** in front. The **mouth** is a wide gaping slit which extends from side to side along the anterior end of the snout; it is meant for the ingestion of food. On the dorsal surface of the head, there are two small openings near the anterior end. These are the nostrils or **external nares** which lead into the nasal passages and serve as the inlet or outlet of air. On either side of the head, there is a large protruding eye which can be closed by means of eyelids. Each eye is provided with three eyelids: (1) a thick and prominent upper eyelid; (2) a rudimentary lower eyelid; and (3) a transparent third eyelid or **nictitating membrane** which is attached below to the lower eyelid and can be stretched up to cover the eyeball. There are no eyelashes. The bulging eyes enable the animal to see all round, and the nictitating membrane protects the eyeball from dirt. Behind each eye there is a circular patch of tightly stretched greyish membrane. This is the eardrum or **tympanum** for the purpose of hearing. There is no pinna of the ear in the toad. A little behind the eardrum, there is a round elevated structure called **parotid** or **parotoid gland.** These extend behind into the trunk and when the toad apprehends any danger, the glands secrete a pungent sticky juice for repelling the enemy. Parotoid glands, therefore, are organs of offence and defence. The ventral surface of the head forms the floor of the mouth cavity. Its rise and fall along with breathing can be observed in a living specimen. The skin in this part of the body is loose and may, in male toads, be inflated by a dark-coloured vocal sac. The female has no vocal sac.

Trunk. The trunk is short and flattened from above downwards. There is nothing to indicate the boundary between the head and the trunk. The dorsal surface is covered with innumerable rough **warts** which indicate the position of the poison glands. The ventral surface is comparatively smooth and free from warts. The posterior end of the trunk is broadly rounded. The **vent** or **cloacal aperture** lies dorsally at the posterior end, between the bases of the two hind legs; it serves for the exit of faeces, urine and germ cells. A toad, when disturbed, often passes water through the vent. There is a widely spread rumour that handling of toads causes warts on human hands; this is a false superstition.

Limbs. Attached to the trunk are a pair of *forelimbs* and a pair of *hindlimbs.* The hindlimbs are kept in a folded state when the animal is at rest.

The **forelimbs** are short and divisible into segments by movable joints. The proximal part, by which it joins the trunk, is known as the arm or *brachium.* To this is joined the forearm or *antebrachium.* Distal to the forearm is an insignificant wrist or *carpus* followed by a hand or *manus.* There are four digits or fingers corresponding to those on our own hand; but the thumb is absent. The male toad develops a cushion-like swelling, called *thumb pad*, at the bases of the index and middle fingers.

The **hindlimbs** are longer and stronger than the forelimbs. Each hindlimb consists of a proximal segment called thigh or *femur* which is followed by the shank or *crus*. Distal to the shank is a long ankle or *tarsus*, a foot or *pes*, and five slender digits or toes of which the fourth toe is the longest. The toes are united by a thin fold of skin called *web*.

The limbs are very suitable for the purpose of hopping on land as well as for swimming in water. A toad, when disturbed, jerks off its strong hindlimbs, leaps into the air, and finally lands on the ground at a considerable distance. This peculiar method of locomotion is known as hopping. When taking the hop, the long ankles and feet are pressed against the ground and used as levers. The animal swims by alternately bending and straightening its powerful hindlimbs, the webbed feet being used as paddles.

Integument

The skin is rough but moist. It does not bear scales, feathers, hairs or claws and is, therefore, said to be naked. Toads are almost of the same colour as the earth; the dorsal surface of the animal is greyish-black and the ventral surface is yellowish-grey. They are thus very difficult to locate when not moving. Toads have the power of changing the colour of their skin, so that they may match with the background and thus become invisible to the enemy. This phenomenon is known as **protective colouration.** The tree frog *Hyla* is better than toad in this respect and can change its colour very quickly.

The skin forms a cover for the entire body. It protects the underlying soft parts from injury and excludes disease germs. It serves as an additional respiratory organ by absorbing oxygen and giving out carbon dioxide. Toads never drink water; they freely absorb water by their skin.

The skin is composed of two distinct layers: (1) an outer *epidermis*, and (2) an inner *dermis*. The **epidermis** consists of several layers of epithelial cells. The superficial epidermal cells are thin, scale-like and devoid of nuclei. They form a horny layer called **stratum corneum** which is shed from time to time as one piece. This process is known as **ecdysis** or moulting. The deeper epidermal cells are tall and polygonal. They form the **stratum germinativum** which produces new cells to replace the old horny layer.

The **dermis** is much thicker than the epidermis. It is composed of glands, pigment cells, involuntary muscle cells, blood vessels and nerves, all enclosed and held together by connective tissue. Scattered in the superficial part of the dermis are spider-like pigment cells or **chromatophores** which can change the colour of the animal by simply shifting their position. The chromatophores are either black or yellow, but never green. The **glands** produce the rough warts on the outer surface. There are two kinds of glands: (1) *poison glands* and (2) *mucous glands*. They manufacture useful fluids or *secretions* which pass out by narrow ducts and moisten the surface of the skin. The **poison glands** are larger and more numerous. They secrete a thick whitish fluid with burning taste which repels the enemy. The **mucous glands** secrete a thin watery fluid which keeps the skin moist and shiny.

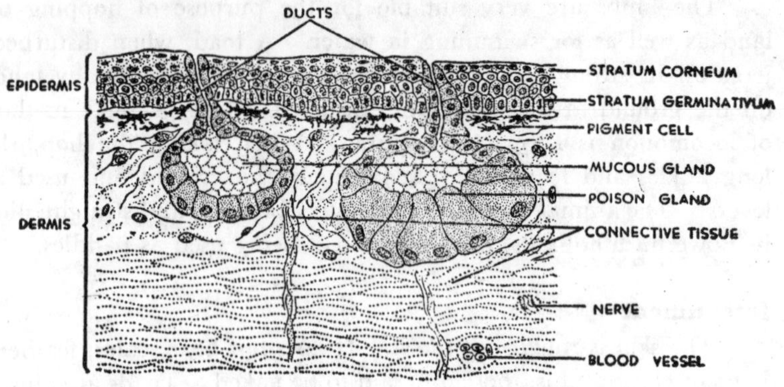

Fig. 6. Internal structure of toad's skin.

Body wall and Body cavity

The toad's body is essentially a double tube. The outer tube is formed by the body wall and the inner tube by the digestive canal. The two tubes are separated from one another by the body cavity or coelom.

The body wall is composed of the following structures: (1) *skin*, (2) *subcutaneous connective tissue*, (3) *muscle layers*, and (4) *peritoneum*. The skin is loosely attached to the underlying muscles by a thin layer of yellowish white connective tissue. There are large spaces enclosed between the skin and the muscular layer. As these are filled with a colourless fluid called *lymph*, they are known as the *subcutaneous lymph spaces*. The peritoneum lies beneath the muscular layers. It is a thin glistening membrane which forms an internal lining for the whole of the body cavity or coelom. Most of the internal organs of toad lie in the coelom and are exposed by cutting down the different layers of the body wall through an incision in the midventral line. The peritoneum not only lines the body cavity but it also encircles the digestive tube and other organs, suspending them to the body wall by fan-shaped folds called *mesentery*. The relative position of the various internal organs or *viscera* is shown in Fig. 7.

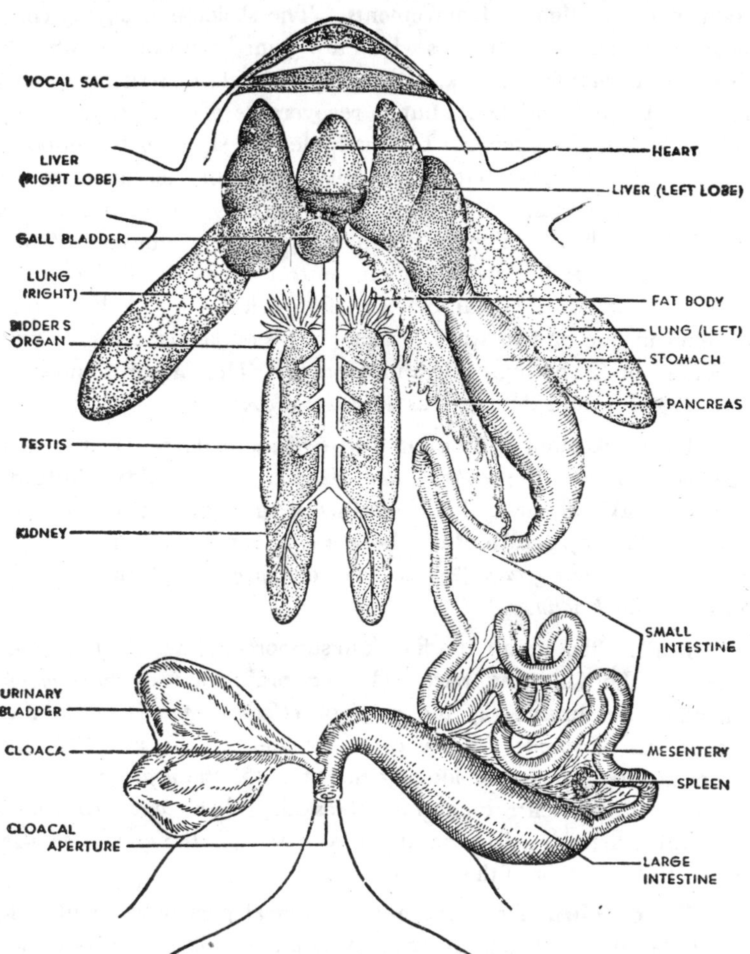

Fig. 7. General viscera of a male toad.

The integumentary system, including the skin and its various functions, has already been discussed. We shall now proceed to study the other organ systems of the toad.

Skeletal System

The skeleton lies internally and is, therefore, known as the *endoskeleton*. It supports the soft parts, protects important organs from injury and serves for the attachment of muscles which are used in locomotion and movements. The skeleton is chiefly composed of bones and cartilages which are joined with one another to form an internal framework. A *cartilage* is an elastic structure which may be twisted and bent, but it recovers its original shape when released from the stretch. In a tadpole, the skeleton is composed entirely of cartilages; but in the adult, most of the cartilages are replaced by bones. A bone is hardened by the deposition of minerals, such as calcium phosphate, in its ground substance. Many bones are hollow and their central cavity is filled with fat and *bone-marrow* in which blood cells are formed. A bone may develop in one or two ways: either (1) by replacing a pre-existing cartilage, or (2) by replacing a membrane. The former is known as a *cartilage bone* and the latter as a *membrane bone.*

Toad's skeleton is divisible into two main parts: (1) the *axial skeleton* and (2) the *appendicular skeleton.* The **axial skeleton** forms the axis of the body; it consists of the *skull* and the *vertebral column.* The **appendicular skeleton** is composed of: (1) the *pectoral* or *shoulder girdle,* (2) the *pelvic* or *hip girdle,* (3) the *forelimbs,* and (4) the *hindlimbs.*

Skull. The broad and flat skull supports the head. It is composed of the following parts: (1) the brain case or *cranium* for enclosing the brain; (2) the *nasal* or *olfactory capsules* containing the organs of smell; (3) the *orbits* for housing the eyes; (4) the *otic* or *auditory capsules* for lodging the ears; (5) the *upper jaw,* which is fused with the ventral margin of the skull; (6) the *lower jaw,* which is movably articulated to the upper jaw; and (7) the *hyoid apparatus,* supporting the floor of the mouth cavity.

The **cranium** forms the narrow central part of the skull. It is tubular in shape and pierced by a large hole at the posterior end. This is the **foramen magnum** (*foramen*=bore) through which the brain joins the spinal cord. The foramen is bordered on each side by a cartilage bone called **ex-occipital,** which bears a convex *occipital condyle.* The two occipital condyles fit into depressions on the first vertebra. The roof of the cranium is formed by a pair of thin flat bones called **frontoparietals** which extend from the ex-occipitals behind to the nasal capsules in front. The floor of the cranium is supported by a dagger-shaped **parasphenoid,** the blade of which points forwards. At the anterior end of the cranium is a ring-like cartilage bone, the **sphenethmoid,** which is completely covered over by the frontoparietals. Besides the foramen magnum, the brain case is perforated by several small openings for the exit of the cranial nerves.

The **nasal capsules** are attached to the anterior end of the cranium and the sphenethmoid extends into them from behind. Each half of the nose is covered over by a triangular membrane bone called **nasal,** and its floor is supported by another membrane bone, the **vomer.**

The **orbits** are represented by two large cavities, one on each side of the cranium. They are meant for lodging the eyes. The

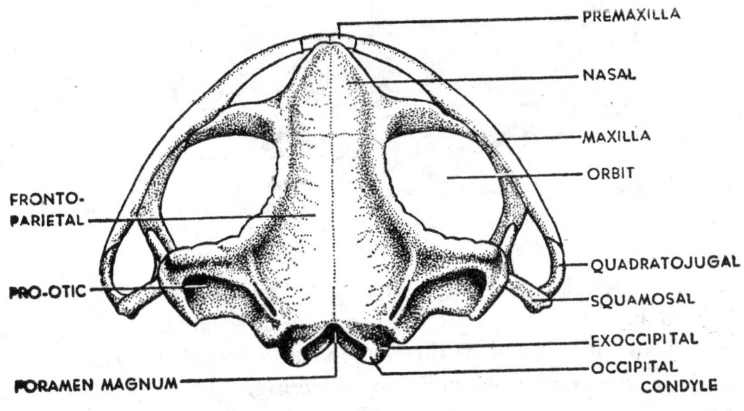

Fig. 8. Skull of toad (dorsal view).

cartilaginous *optic capsules* are not fused with the cranium and therefore they are not found in a prepared skeleton.

The **auditory** or **otic capsules** are situated posteriorly, one on each side of the cranium, just in front of the ex-occipitals. Each capsule is a cartilaginous box which is covered over by a single cartilage bone called **pro-otic.** The floor of the capsule is supported by a lateral expansion of the parasphenoid. On the outer side of the pro-otic is a thin hammer-shaped membrane bone, the **squamosal,** which connects the otic capsule with the posterior end of the upper jaw.

The **upper jaw** forms the outer margin of the skull. It consists of two halves which are united to the cranium, both in front and behind. Each half of the jaw is composed of an outer series of *three* membrane bones, the **premaxilla,** the **maxilla** and the **quadratojugal.** The premaxillae are small bones at the anterior part which unite with one another in the middle line and are fused to the nasal capsules. Behind the premaxilla is a long slender maxilla which is joined to the quadratojugal. At the posterior end of the quadratojugal is a small cartilaginous piece, the quadrate, which

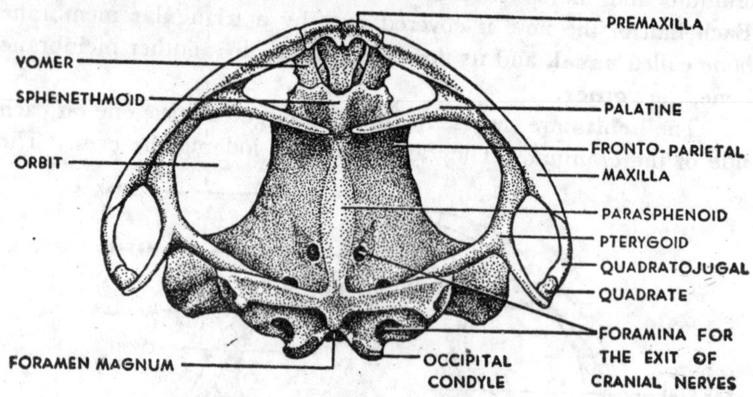

Fig. 9. Skull of toad (ventral view).

bears a concave facet for articulating with the lower jaw. Extending between the cranium and the outer series of three bones just described, there is an inner series of three supporting bones. These are the **palatine,** the **pterygoid** and the **quadrate.** They serve to brace up the upper jaw with the cranium. The palatine is a rod-like bone in front of the orbit which connects the maxilla with the spheneth-moid. The pterygoid is a Y-shaped bone, the foot of which is attach-ed to the junction between the palatine and the maxilla. The two arms of the pterygoid are directed backwards; the inner arm sup-ports the floor of the otic capsule and the outer arm projects towards and fuses with the quadrate. Thus the pterygoid and the quadrate form an efficient girder for the articulation of the lower jaw.

The **lower jaw** consists of two halves which are united an-teriorly, in the middle line, by a short ligament. Each half is de-rived from a *Meckel's cartilage* and is composed of three bones, the **mento-meckelian,** the **angulo-splenial** and the **dentary.** The mento-meckelian is a small cartilage bone at the extreme anterior end. Behind the mento-meckelian is a long slender angulo-splenial which encloses the inner margin of the Meckel's cartilage. The pos-terior end of the angulo-splenial articulates with the quadrate of the upper jaw. The dentary is a membrane bone which covers the outer surface of the angulo-splenial but does not extend up to the posterior end. Both jaws are toothless.

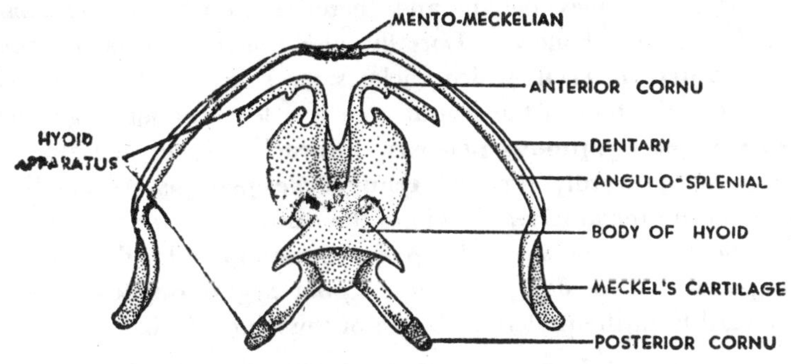

Fig. 10. Lower jaw and hyoid apparatus of toad.

The **hyoid apparatus** is a cartilaginous plate lying on the floor of the mouth cavity. It forms the surface for the attachment of the tongue. The main bulk of the apparatus is known as the **body of the hyoid.** A pair of horns called **anterior cornua** extend from the anterior part of the body to the auditory capsules. Similarly, a pair of **posterior cornua** project backwards from the body of the hyoid and enclose the laryngotracheal chamber or voice box. The anterior cornua are much longer than the posterior cornua.

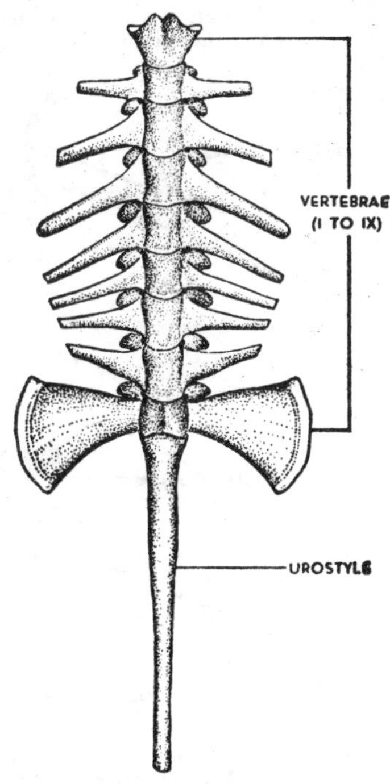

Fig. 11. Vertebral column (ventral view).

Vertebral column. The vertebral column is situated dorsally and extends from the base of the skull to the posterior end of the trunk. It is composed of a series of *ten* separate bones, the first nine of which are the ring-like *vertebrae* and the last is a long rod-like *urostyle* (*oura*=tail; *style*=rod).

A **typical vertebra,** taken from the middle of the series, consists of a solid cylindrical part, the body or **centrum,** which is concave in front and convex behind and therefore said to be *procoelous* (*pro*=front; *coel*=hollow). Dorsally, the centrum bears a bony ring, the **neural arch,** which encloses the spinal cord in a *neural canal.* On the roof of the neural arch is a low median ridge, the **neural spine** or **spinous process.** On each side, the neural arch carries an outwardly directed **transverse process.** Extending forward from the anterior margin of the neural arch are two spoon-like articulating processes called **pre-zygapophyses;** and similarly, two posterior articular processes or **post-zygapophyses** project backward from the posterior margin of the neural arch.

When the vertebrae lie in their natural position, the convex posterior surface of one centrum fits into the concave anterior surface of the following centrum, and the post-zygapophyses of the vertebra in front overlap the pre-zygapophyses of the vertebra behind. The vertebrae cannot separate, because they are held together by ligaments. A flexible vertebral column is produced in this manner. It encloses the spinal cord in a continuous neural canal formed by the ring-like neural arches of the successive vertebrae. There are *intervertebral foramina* between consecutive neural arches, through which spinal nerves come out.

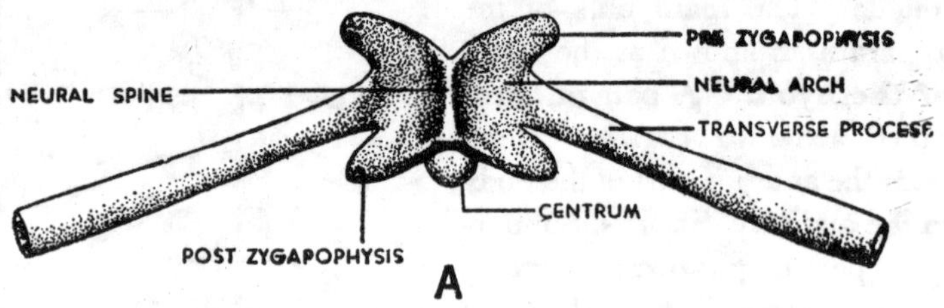

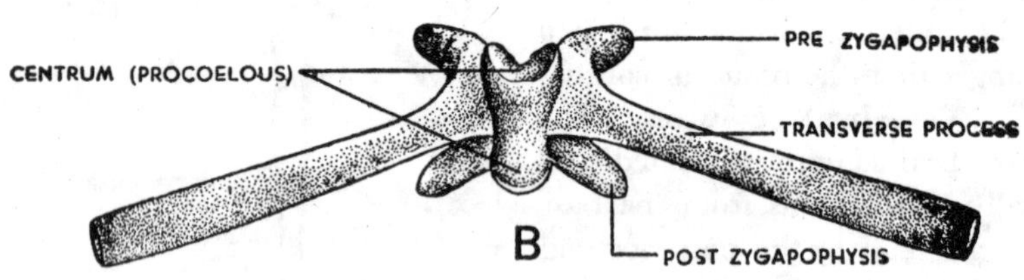

Fig. 12. A typical vertebra of toad
(A=dorsal view; B=ventral view).

The **first vertebra** or **atlas** articulates with the skull. It has neither transverse processes nor pre-zygapophyses. Instead, it carries on its anterior face two concave facets for articulating with the occipital condyles of the cranium. Its centrum is greatly reduced and flattened, and its neural arch bears a pair of post-zygapophyses for articulating with the pre-zygapophyses of the second vertebra.

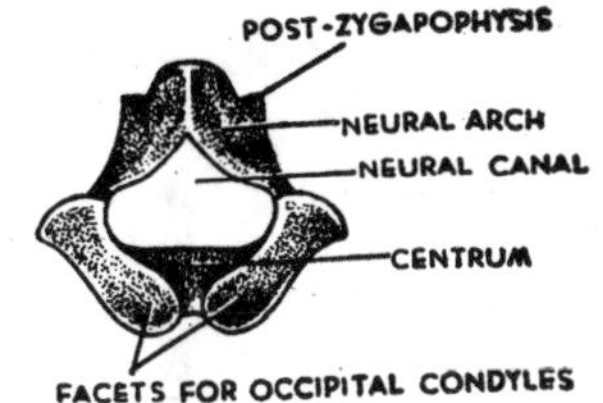

Fig. 13. First vertebra of toad.

The following seven vertebrae resemble the typical vertebra that has already been described.

The **ninth** or **sacral vertebra** is the largest of the series. Its centrum is procoelous; but the posterior end of the centrum bears two rounded knobs, the *condylar facets*, which fit into corresponding concavities on the urostyle. Its transverse processes are stout and fan-shaped; sloping outwards, they articulate with the pelvic girdle. Since the ninth vertebra is the last, it has no post-zygapophyses.

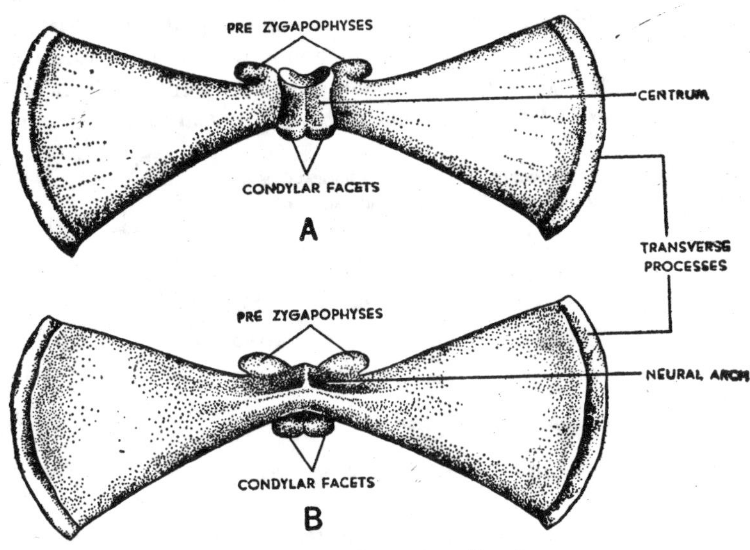

Fig. 14. Ninth vertebra of toad (A=ventral view; B=dorsal view).

The **urostyle** represents a number of fused vertebrae. It is a long slender rod which extends from the ninth or sacral vertebra to the posterior end of the trunk. It bears a median neural crest dorsally, and a pair of socket-like concavities at its anterior end for articulating with the centrum of the ninth vertebra.

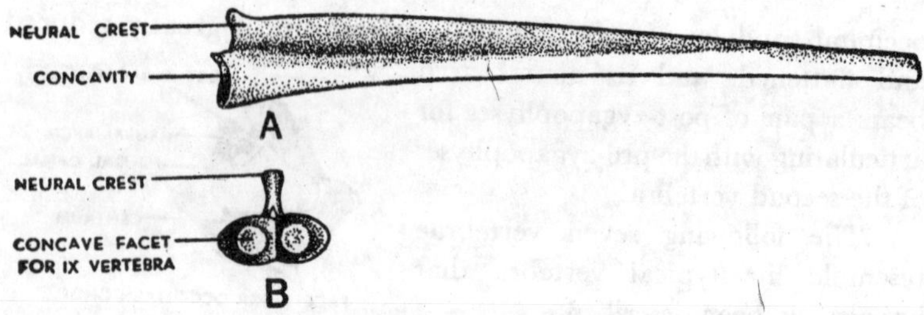

Fig. 15. Urostyle of toad
(A=side view; B=anterior end on view).

Pectoral girdle. The pectoral girdle is a bony framework encircling the anterior part of the trunk. It consists of two symmetrical halves which are united with one another in the mid-ventral line, but separated dorsally. Each half of the pectoral girdle is composed of the following structures. Lying dorsally and above the vertebral column is a broad, partly cartilaginous plate called **supra-scapula.** Attached to the supra-scapula is a stout **scapula** or shoulder blade, which forms the outer side of the girdle. Connected with the lower end of the scapula and lying ventral to the vertebral

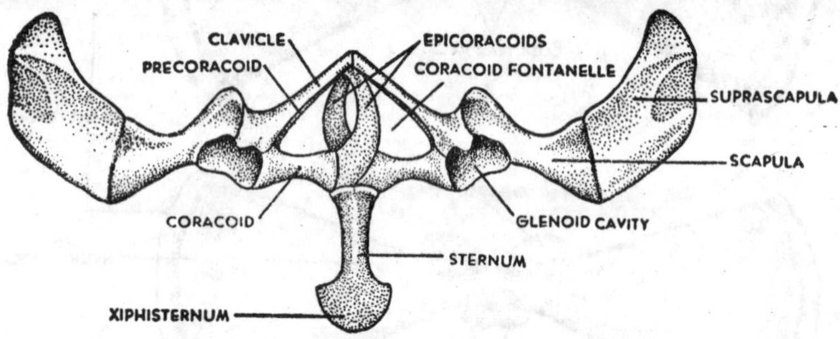

Fig. 16. Pectoral girdle of toad (ventral view).

column are two rod-like bones, a **clavicle** in front and a **coracoid** behind. The clavicle is a thin membrane bone which encloses the **precoracoid cartilage.** The precoracoid being ensheathed is not easily visible. The inner ends of the clavicle and the coracoid are joined by a curved piece of cartilage, called **epicoracoid.** The right epicoracoid is slightly overlapped by the left one. The clavicle, the coracoid and the epicoracoid are separated from one another by a wide gap which is known as the **coracoid fontanelle.**

There is a cup-shaped **glenoid cavity** at the junction of the scapula, the clavicle and the coracoid. The head of the arm bone fits like a ball into the socket of the glenoid cavity.

The two halves of the pectoral girdle are united by the epicoracoids in the mid-ventral line, and the breast-bone or **sternum** projects backward from the united posterior ends of the two epicoracoids. The **xiphi-sternum** is a flattened cartilaginous plate forming the terminal part of the sternum.

It is to be noted that no part of the pectoral girdle is directly articulated to the vertebral column. The girdle is kept in normal position by strong muscles, and if these are cut the whole framework may be slipped along the long axis of the body. The pectoral girdle performs the following functions: (1) besides supporting the forelimbs which can move freely in the glenoid cavities, the girdle encloses and therefore protects important organs, such as

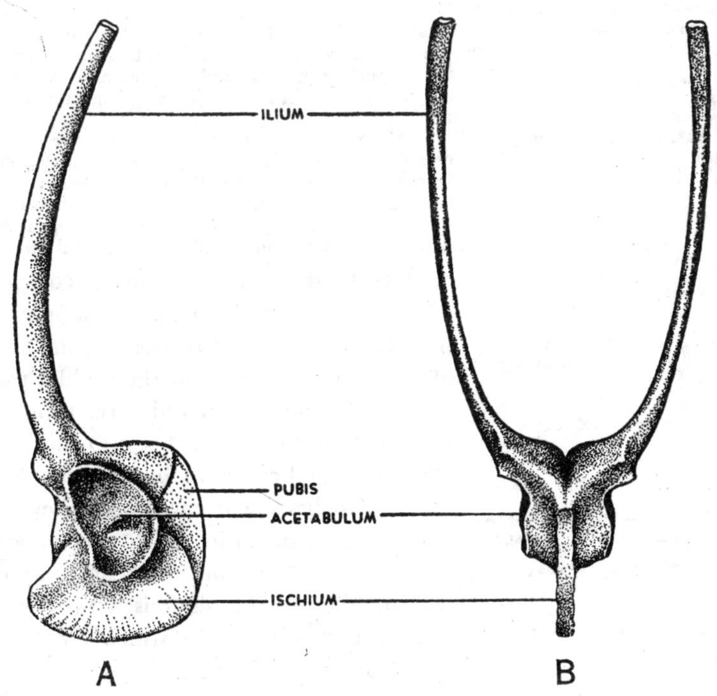

Fig. 17. Pelvic girdle of toad (A=side view; B=dorsal view).

heart and lungs; (2) it serves for the attachment of muscles which are concerned with respiration.

Pelvic girdle. The pelvic girdle is a V-shaped structure at the posterior end of the trunk. The apex of the V is directed backwards and the two free ends of the V form a complete hoop by articulating anteriorly with the transverse processes of the ninth vertebra. The posterior end of the pelvic girdle is disc-like, with a cavity called **acetabulum** on either side. The acetabulum supports the hindlimb by articulating with the thigh bone. The disc itself is formed by the fusion of three bones on each side, the **ilium,** the **ischium** and the **pubis.** The two ilia form the anterior and dorsal sectors of the disc; they extend anteriorly as slender rods, one on each side of the urostyle, and articulate with the ninth vertebra. The two ischia are completely fused to form the posterior sector of the disc. The ventral sector of the disc is formed by a triangular cartilage which represents the two pubes.

There is a general resemblance between the pectoral and the pelvic girdles, the scapula corresponding with the ilium, the coracoid with the ischium, and the precoracoid surrounded by the clavicle with the pubis. At the junction of the three elements of the pectoral girdle there is a cup-shaped glenoid cavity for supporting the forelimb. Similarly, there is an acetabulum in the pelvic girdle for supporting the hindlimb.

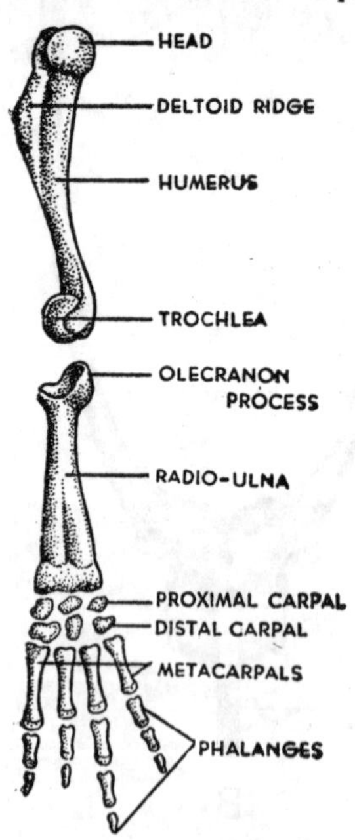

Fig. 18. Bones of the left forelimb of toad.

A very important point of difference is the direct articulation of the pelvic girdle with the vertebral column. The ilia are attached to the fan-shaped transverse processes of the ninth or sacral vertebra. Such direct articulation with the vertebral column is not found in the pectoral girdle.

Forelimbs. Each forelimb is composed of several bones. The arm is supported by a long bone called **humerus,** the ends of which are swollen and the middle part or *shaft* is slightly curved. Its rounded proximal end, the *head*, fits into the cup-shaped glenoid cavity to form the shoulder joint. Its distal end carries a pulley-like *trochlea* for articulating with the forearm bone. Extending from the head to about the middle of the shaft is a prominent crest called *deltoid ridge* which serves for the attachment of muscles.

The forearm is supported by a single bone called **radio-ulna** which is formed by the fusion of two bones, radius and ulna. The proximal end of the radio-ulna is concave and articulates with the trochlea of the humerus to form the elbow-joint. It is drawn out into a backwardly directed projection, called *olecranon process*, which extends beyond the trochlea and lies behind the elbow. The distal end of the radio-ulna is expanded and flattened for joining with the wrist. The wrist or carpus is composed of *six* small pieces of **carpal bones** which are arranged in two rows, a proximal row and a distal row, each consisting of three carpals. The palm of the hand is supported by four slender rod-like bones, the **metacarpals.** There are four fingers or digits, each of which is composed of bones called **phalanges** (singular=*phalanx*). The innermost finger corresponding to our thumb is absent. The first and second fingers have two phalanges in each, but the third and fourth have three in each.

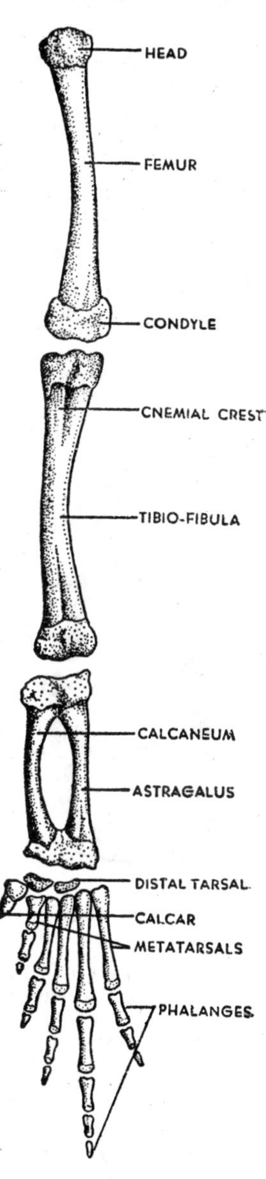

Fig. 19. Bones of the left hindlimb of toad.

Hindlimbs. Each hindlimb is a system of bones. The thigh is supported by a long bone called **femur,** with a slightly curved *shaft* and swollen ends. Its rounded proximal end or *head* articulates with the acetabulum forming a ball and socket hip joint, and its distal end is slightly expanded to form a *condyle* for articulating with the shank bone. The shank or *crus* is supported by a single bone called **tibio-fibula** which is formed by the union of two bones, tibia and fibula. The tibio-fibula is expanded at both the ends; its proximal end articulating with the femur forms the knee joint, and the distal end articulating with the tarsals forms the ankle joint. In the ankle or *tarsus* there are two rows of tarsal bones. The two **proximal tarsals** are the **calcaneum** and the **astragalus.** Both are elongated and united with each other at their ends, but widely separated in the middle. The **distal tarsals** consist of two or three small pieces, which owing to their smallness are easily overlooked. The foot or *metatarsus* is composed of five slender **metatarsals.** There are *five* toes or digits. The first and second toes have two **phalanges** in each, the third has three, the fourth four, and the fifth three. The great toe or *hallux* has a small projection on its inner side. This is known as the *calcar* or *pre-hallux,* and is composed of one or two small nodules.

The two pairs of limbs are built on the same plan. Both are essentially **pentadactyle,** that is 'five-fingered'; but the thumb or *pollex* is absent in the forelimbs. A comparison between the forelimb and the hindlimb reveals that the humerus corresponds to the femur, the radio-ulna to the tibio-fibula, the carpals to the tarsals, the metacarpals to the metatarsals, and the phalanges of the fingers to those of the toes. The bones of the hindlimb are of course longer than the corresponding bones of the forelimb and the proximal tarsals are very much longer in comparison to the proximal carpals.

Joints. In the construction of the skeletal framework bones are joined with one another in various ways. The term *joint* is applied to the place of union between bones. The joints are planned in such a manner that the animal derives the best mechanical advantage out of them. The bones of the cranium need not move at all. They therefore are joined by interlocking **sutures** which permit no movement. The bones of the girdles are also locked by sutures. In the limbs, the bones are to move freely. In a **movable joint** the articulating surfaces of the bones are covered by smooth *articular cartilages*. The joint as a whole is enclosed by a loose bag of connective tissue which is known as the *articular capsule.* The capsule is strengthened by bands of fibrous tissue called *ligaments,* and lined internally by a smooth, delicate membrane, the *synovial membrane.* The joint-cavity, between the two bones, contains a colourless *synovial fluid* which serves for lubrication. There are *two* main types of movable joints: (1) **ball and socket joint** and (2) **hinge joint.** The shoulder and the hip are ball and socket joints which permit movement of bones in every plane. The elbow and the knee are hinge joints permitting movement in one plane only. A hinge joint does not allow side-to-side movement.

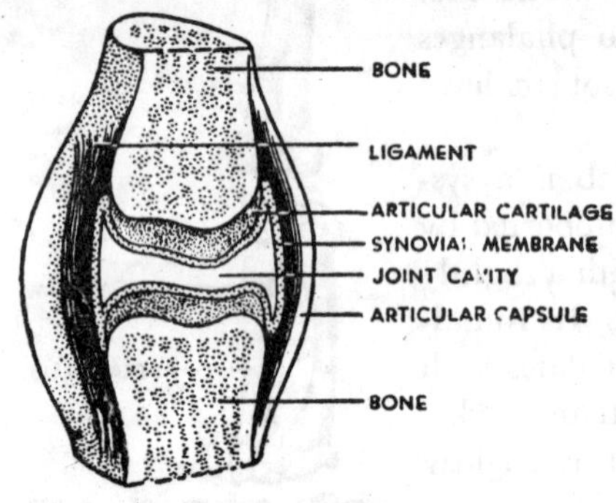

Fig. 20. Diagram of a hinge joint.

Muscular System

It has already been pointed out that the muscular system is composed of muscles which are used for movements. There are two kinds of muscles: (1) the **voluntary muscles,** which can be contracted at will and are attached to the skeleton; and (2) the **involuntary muscles,** which are not controlled by desire and occur as sheets on the wall of organs such as heart, blood vessels, digestive canal, etc. The minute structure of the various kinds of muscles has been described in Chapter XVIII.

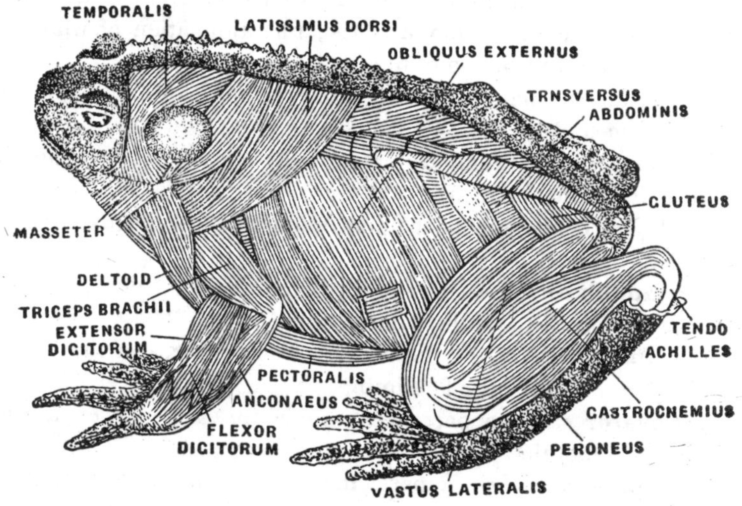

Fig. 21. Skeletal muscles of toad (side view).

The main mass of external musculature of toad is formed by the voluntary muscles. They are attached to the bones and are hence designated as *skeletal muscles.* There are about two hundred skeletal muscles in the body of a toad and the more important ones are shown in the adjoining diagram (Fig. 21). Each is composed of bundles of muscle fibres bound together by connective tissue. Some muscles occur as sheets, such as the *obliquus externus* covering the belly. Others are elongated and spindle-shaped, such as the *deltoid* and the *gastrocnemius.* A few have fibres arranged circularly round an opening; these are the *sphincter muscles.*

A muscle has at least two points of attachment fastened to separate bones. The end of the muscle by which it is attached to a relatively fixed bone is known as its *origin.* The other end, attached to a more movable bone, is its *insertion.* A muscle may or may not be connected to a fibrous *tendon* at its ends. When both the ends are tendinous, the fleshy central part of a muscle is distinguished as the *belly*.

Muscles possess the power of contraction and relaxation. When a muscle contracts, its two ends are apposed and the bone to which it is inserted is brought nearer to the bone from which it originates. It is in this way that movements are effected. Each muscle is supplied by a motor nerve, the stimulation of which is followed by the contraction of the muscle.

According to their mode of action, the muscles may be of the following types:—

(a) An **extensor** straightens or extends a part; *Triceps* muscle straightens the forearm.

(b) A **flexor** bends one part upon another; *Biceps* bends forearm upon the arm.

(c) An **adductor** draws a limb towards the axis of the body; the *latissimus dorsi* adducts the forelimb.

(d) An **abductor** draws a limb away from the axis of the body; the *deltoid* abducts the forelimb.

(e) A **levator** raises a part; the *masseter* raises the lower jaw to close the mouth.

(f) A **depressor** lowers a part; the *mylohyoid* lowers the floor of the buccal cavity.

(g) A **rotator** rotates a part; the small *pyriformis* rotates the hind-limb in the acetabulum.

Muscles are usually arranged in opposing groups, so that when one set is contracting the opposite set is kept in a relaxed state. To be more precise, when biceps, which is a flexor muscle, is contracting, the triceps, which is an extensor muscle, is kept relaxed. This co-ordination is effected through the agency of the nervous system.

Alimentary System

The alimentary system is concerned with nutrition. It consists of: (1) *alimentary canal* or *gut* or *enteron* or *digestive tube*, and (2) the associated *digestive glands*. The alimentary canal serves for the ingestion, digestion and absorption of food, and for the egestion of the indigestible residue. The digestive glands secrete substances which help in the digestion of the food.

Alimentary canal. The alimentary canal is a long muscular tube running through the body from end to end. It begins anteriorly in the *mouth* and ends posteriorly at the *cloacal aperture*. The tube is swollen in some places and at others it is narrowed and coiled, so that one can easily recognise different distinct parts. The various parts of the alimentary canal are *mouth, buccal cavity, pharynx, oesophagus, stomach, small intestine, large intestine, cloaca* and *cloacal aperture.*

The **mouth** is a wide opening at the anterior end of the snout. It extends from corner to corner as far back as the tympanic membrane. Mouth leads into a spacious **buccal cavity** which is bounded by two jaws. The upper jaw is immovable but the lower jaw can be moved up and down, closing and opening the mouth. The jaws are toothless and covered by hard sheaths. The buccal cavity is lined by mucous membrane. On the roof of the buccal cavity, and situated anteriorly, are two small openings called **internal nares** which communicate through the nose with the external nares. Just behind the internal nares are the impressions of the eyeballs. These are increased by pushing the eyes into the orbits. Still further behind, near the angle of the jaws, are the **eustachian apertures,** one on each side. These are large openings leading into the tympanum of the same side. Attached to the floor of the buccal cavity is a fleshy **tongue.** It is a broad mucous-covered flap which is fixed by the anterior margin. The free posterior margin of the tongue may be flicked out of the mouth and quickly retracted in, so as to form an efficient apparatus for catching insects. Behind the tongue is a raised area with a median longitudinal slit. This is the **glottis** which communicates with the lungs. Beyond this, the buccal cavity narrows down to form an indistinct **pharynx** which leads directly into the *gullet* or *oesophagus.* The **oesophagus** is a short wide tube which runs into the body cavity and connects with the stomach. Internally its mucous membrane is thrown into numerous longitudinal folds, so that it may be expanded during the passage of a big bolus of food. The **stomach** is a thick-walled, slightly bent, muscular bag which lies a little to the left of the middle line. Its broader anterior end is connected with the oesophagus. This is the *cardiac end* (*kardia*=heart), because it lies nearest to the heart. The opposite end of the stomach is the *pyloric end* which opens into the small intestine through the *pylorus* (*pylorus*=gate-keeper). The opening is guarded by a circular band of constricted sphincter muscle, the *pyloric valve*, which regulates the departure of food from the stomach. Embedded in the internal mucous membrane of the stomach there are branched and tubular gastric glands secreting **gastric juice** for the digestion of food. Following the stomach is the **small intestine.** The first part of the small intestine is known as the

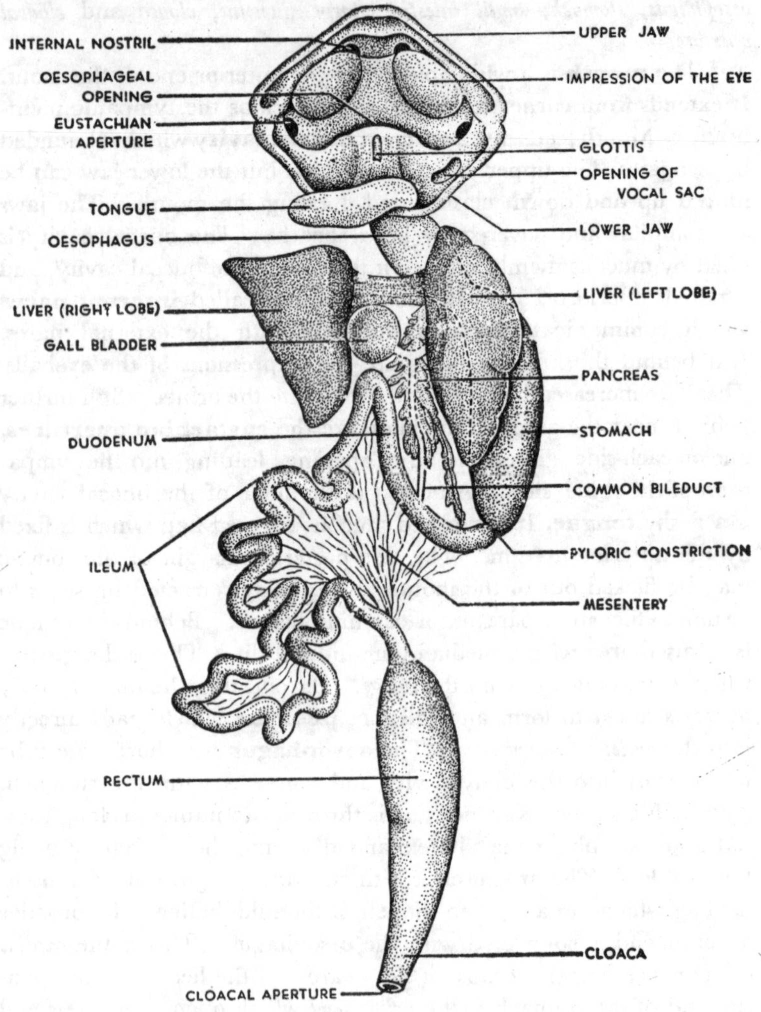

Fig. 22. Alimentary system of toad.

duodenum and the second part as the *ileum*. The **duodenum** is a relatively short tube which runs forward from the pylorus and lies nearly parallel to the body of the stomach to which it is attached by a fold of peritoneum. It receives the ducts from the liver and the pancreas and is continued backwards as the much-coiled ileum. The **ileum** is the longest part of the alimentary canal. The coils of the ileum are held together and fastened to the body wall by a fold of *mesentery.* The mucous membrane of the small intestine contains a large number of intestinal glands which secrete **intestinal juice.** Moreover, the mucous membrane is thrown into folds which hang freely within the lumen of the gut for absorbing the digested food. The

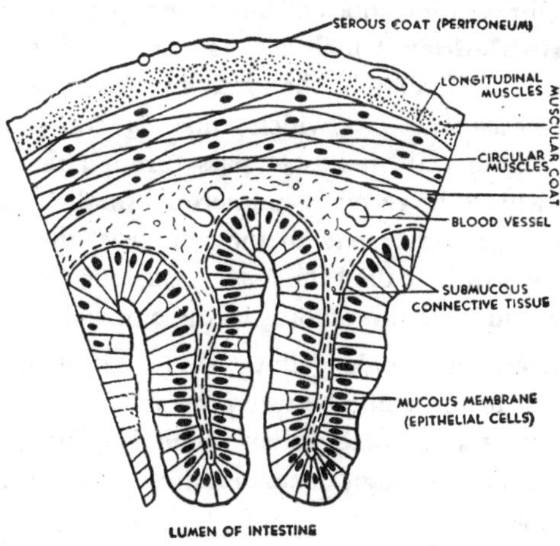

Fig. 23. Part of a cross-section through the ileum (diagrammatic).

small intestine is abruptly enlarged to form the **large intestine** which consists of the *rectum* and the *cloaca.* The **rectum** is a wide flask-shaped tube about 2·5 cm. in length. Posteriorly it is joined to the *cloaca.* The **cloaca** receives the ducts from the kidneys, the urinary bladder and the reproductive organs and finally opens to the exterior through the *vent.* The **vent** or **cloacal aperture** is a small opening at the posterior end of the trunk.

Histologically, the wall of the alimentary canal is composed of five layers: (1) a thin outer layer of peritoneum forming the **serous coat;** (2) an outer layer of longitudinal muscles and (3) an inner layer of circular muscles, the two together forming the **muscular coat;** (4) a connective tissue layer containing blood vessels and forming the **submucous coat;** (5) the innermost **mucous coat** or **mucous membrane** composed of epithelial cells. These layers are best seen in a cross-section of the ileum (Fig. 23). It is to be noted that by the contraction of the muscular coat, the alimentary canal drives the food onwards in peristaltic waves (*peri*=around; *stalsis*=constriction).

Digestive glands. Besides the minute gastric and intestinal glands already mentioned, there are two important digestive glands which are associated intimately with the gut. These are the **liver** and the **pancreas** (*pan*=all; *kreas*=flesh). The **liver** is a large, dark-red organ suspended by a mesentery in the anterior part of the body cavity. It consists of two main lobes, right and left, connected by means of a narrow portion. The left lobe of the liver is incompletely subdivided into two masses. The liver secretes **bile** which comes out by tubes called *hepatic ducts* (*hepar*=liver). Between the right and left lobes of the liver is a dark-green, spherical sac, the **gall-bladder,** which acts as reservoir of the bile. The *hepatic ducts* from the liver join the *cystic duct* from the gall-bladder to form the *common bile duct* which runs backward towards the duodenum. On the way, it has to pass through the substance of the pancreas where it receives numerous small *pancreatic ducts.* Coming out from posterior end of the pancreas, the common bile duct, which may now be called the *hepato-pancreatic duct,* finally opens into the duodenum near the pylorus of the stomach.

The **pancreas** is a pale yellow, elongated structure of irregular shape lying in the angle between the stomach and the duodenum. It secretes **pancreatic juice** which is carried, along with bile, to the duodenum through the common bile duct. The path has already been described.

Events during feeding and digestion. The toad is a carnivorous animal fond of ingesting living insects, caterpillars, earthworms and snails. It is attracted by the movement of the prey. Insects are snapped in by the tongue, and earthworms are crushed by the jaws till they become quiet, and then swallowed as whole.

The body of the prey is composed of: (1) **proteins,** (2) **fats,** (3) **carbohydrates,** (4) **mineral salts,** (5) **vitamins,** and (6) **water.** These are known as the **proximate principles** of the food. Vitamins, mineral salts and water are directly absorbed into the body and need not undergo digestion. Carbohydrates, fats and proteins, however, are very complex organic compounds which cannot be absorbed unless they are broken down and prepared into simpler substances. The process of preparing the food for absorption into the body is known as digestion. The ingested food cannot be considered as a part of the body unless it is digested and absorbed by the cells. Digestion of the food is effected by the digestive juices which contain among other things chemical substances called *enzymes* or *ferments.* The **enzymes** are responsible for the digestion of proteins, fats and carbohydrates. Each enzyme is **specific,** that is, it can act upon only one particular kind of food. An enzyme which splits proteins will not break down carbohydrates or fats. Moreover, the digestive enzymes act by causing **hydrolysis.** This means that a molecule of water is chemically added to a molecule of the substrate upon which the enzyme is acting. Lastly, the enzymes are **catalysts** and hence they are not spent up in the course of the chemical reaction.

Considering these properties, the digestive enzymes are classified into *three* main groups: (1) **proteolytic** enzymes such as *pepsin* of the gastric juice, *trypsin* of the pancreatic juice, and *erepsin* of the intestinal juice; (2) **lipolytic enzyme** such as *lipase* of the pancreatic juice; (3) **amylolytic** or **diastatic enzyme** such as *amylase* of the pancreatic juice. The proteolytic enzyme **pepsin** breaks down proteins into *peptones* which are further broken down to *amino acids* by **trypsin** and **erepsin.** The lipolytic enzyme **lipase** splits fat into *fatty acids* and *glycerine.* The amylolytic enzyme **amylase** acts upon carbohydrates like *starch* and breaks them down to complex sugars like *maltose.* The maltose is then further broken into simple sugars like *glucose* by the action of **maltase** of the intestinal juice.

No digestion takes place in the mouth of the toad. The food is mixed with mucus and propelled into the stomach by the peristaltic contraction of the oesophagus. The food is churned and thoroughly mixed with gastric juice in the stomach. The gastric juice contains hydrochloric acid which helps the pepsin to convert proteins into *peptones.* The food is temporarily stored in the stomach for a short time and then partially digested food passes by peristalsis through the pylorus into the duodenum. Here the *acid chyme* is treated with bile, pancreatic juice and intestinal juice. The bile is alkaline in reaction. It neutralises the acid and helps in the digestion and absorption of fats. There is, however, no digestive enzyme in the bile. The pancreatic trypsin then converts peptones into amino acids. The process is hastened by the erepsin of the intestinal juice. The proteins are thus converted completely into soluble amino acids. The pancreatic lipase digests fat into soluble fatty acids and glycerine. The amylase and maltase together convert starchy matter into glucose. The soluble food is then absorbed into the blood stream through the folds of the mucous membrane of the ileum. The indigestible residue accumulates in the rectum and then egested as *faeces* through the cloacal aperture. The absorbed food is assimilated, that is, changed into the protoplasm of the toad. It is then utilised by the animal for maintaining its growth and repairing its tissues. A part of the assimilated food is spent as a readily available source of energy for carrying out vital activities. The surplus is stored in the body for future use. The glucose is converted into *glycogen* or animal starch and stored as such in the **liver** and **skeletal muscles.** The fatty acids are reconverted into *fat* and stored in the **fat bodies.** The amino acids are not stored, the surplus being converted into *urea* and excreted as such in the urine.

Circulatory or Vascular System

It is the transport system by which materials such as food, oxygen and wastes are carried from one part of the body to other parts. The conveying medium is a liquid called **blood.** This is forced by a central pumping organ, the **heart,** through channels called blood vessels. The large blood vessels carrying blood away from the heart are the **arteries.** These break up into smaller branches as they pass farther away from the heart. The smallest branches penetrate into the tissues of the body and break up into very fine, thin-walled channels called **capillaries** (*capillus*=hair). The capillaries unite with one another to form a close network surrounding the tissue cells. The blood is collected from the other end of the network by another kind of vessels called **veins.** A vein receives other veins as tributaries, and thus increasing in size the veins bring the blood back to the heart. Although the blood circulates through the tissues, yet it is not liberated into them because it is conveyed through a closed system of vessels. Only the liquid part of the blood carrying the material to be transported, passes out through the thin walls of the

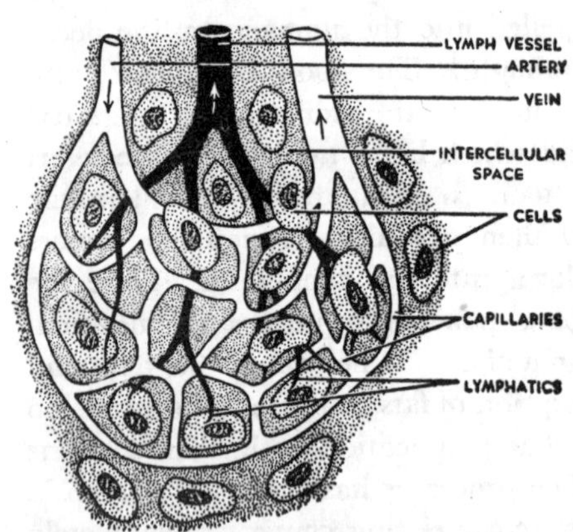

Fig. 24. The capillaries and lymphatics in relation to the body cells (diagrammatic). The arrows indicate the direction of circulation.

capillaries into the intercellular spaces. This fluid is known as the **lymph.** It is collected by a special kind of vessels called **lymphatics** which ultimately return the lymph to the veins. The total volume of the blood is restored in this way (Fig. 25).

Thus the **circulatory system** is actually made up of two separate systems: the *blood vascular system* and the *lymphatic system.* The arteries carrying blood away from the heart, and the veins carrying the same towards the heart make up the blood vascular system. The lymphatics carrying the lymph away from the tissue spaces constitute the lymphatic system.

The arteries and veins differ not only in function, but they also differ in structure. The wall of an **artery** consists of *three* layers: (1) the outermost layer is composed of fibrous tissue and is, therefore, known as the fibrous coat or *tunica adventitia;* (2) the middle or muscular coat consists of involuntary muscle fibres running round the vessel; this is the *tunica media;* (3) the innermost layer forms the *tunica interna;* it is composed of elastic fibres and endothelium which make up the internal lining membrane of the artery. A **vein** has all the three coats; its fibrous coat is thicker and the muscular coat is thinner. Moreover, a vein has no elastic fibre in the tunica interna. Thus a vein has a delicate, thin and non-elastic wall, whereas an artery has a tough, thick and elastic wall. The lumen of a vein is larger in diameter than the lumen of the corresponding artery. There are valves inside larger veins for regulating and directing the blood streams towards the heart.

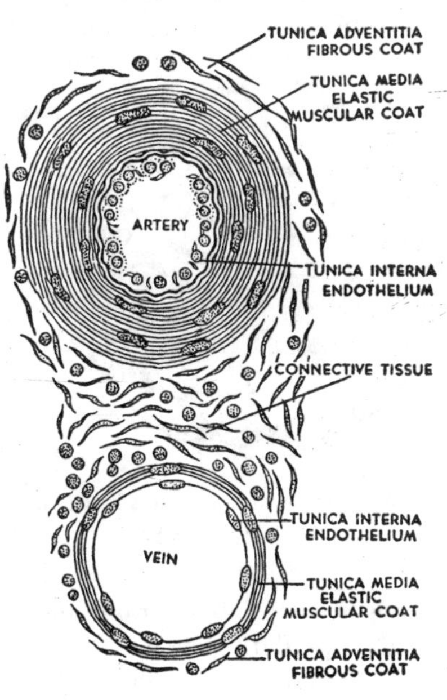

Fig. 25. Cross-section of an artery and a vein.

The **blood-vascular** system has three main components: (1) the *blood*, (2) the *heart*, and (3) the *blood vessels*, *i.e. arteries* and *veins*.

Blood. The blood is a reddish opaque liquid flowing through the blood vessels. It consists of a straw-coloured ground substance called *blood plasma* and solid cells called *blood corpuscles*. The corpuscles are suspended in the plasma.

The **blood plasma** is mostly watery. Many substances are found in solution in the blood plasma. These include mineral substances, food, waste products, hormones, gases, etc. It is the chief transporting medium.

The **corpuscles** are of three kinds: (a) red blood cells or **erythrocytes** (*erythros*=red), (b) white blood cells or **leucocytes** (*leucos*=white), and (c) blood platelets or **thrombocytes.** The **erythrocytes** are oval, biconvex, nucleated cells about 15 to 20 micra in size (one micron=1/1000 mm.). They carry the colouring matter of the blood, an iron-containing protein called **haemoglobin.** The haemoglobin has a remarkable affinity for oxygen and is, therefore, useful in the transport of this gas during respiration. There are about 4 to 5 lakhs of erythrocytes in a cubic millimetre of blood which serve for transporting oxygen. The **leucocytes** are less numerous than the erythrocytes, numbering about 4 to 5 thousands in a cu. mm. of blood. They are small, colourless nucleated cells capable of changing their shape like an amoeba. They may creep out through the thin-walled capillaries to engulf and remove foreign bodies from the tissues. Thus the leucocytes are the scavengers of the body. They destroy bacteria and protect the toad against invading microbes. Moreover, they can ingest fat globules from the small intestine and carry them into the circulating blood. Leucocytes, therefore, are scavengers, soldiers and porters of the body. The **thrombocytes** are small spindle-shaped nucleated cells which are often designated as the *blood platelets*. They break down when the blood is shed, and release an enzyme which helps in the clotting or *coagulation* of blood. Further bleeding is thus prevented when a blood vessels is injured. In the adult toad, the blood corpuscles are mainly manufactured by the *bone-marrow*.

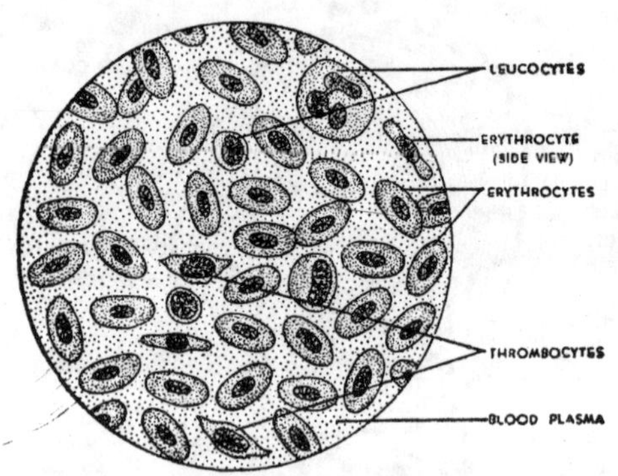

Fig. 26. A drop of toad's blood (highly magnified).

Heart. The heart is the central pumping station for the circulation of blood. It is a hollow, pear-shaped, muscular organ which is situated in the anterior part of the body cavity, in front of the liver. It is completely enclosed in a transparent bag of membrane, the **pericardium.** The broad *base* of the heart is directed forwards, whereas the narrow *apex* points towards the posterior end and lies between the two main lobes of the liver. The heart is mainly composed of *three* chambers: a thick-walled, conical *ventricle* and two thin-walled *auricles*, right and left. There are two other smaller chambers: a thin-walled triangular *sinus venosus* on the dorsal side, opening into the right auricle, and a thick-walled tubular *conus arteriosus* ventrally, connected to the base of the ventricle.

The **sinus venosus** is a thin-walled sac-like chamber which is situated on the dorsal side of the heart. It is more or less triangular in shape with the three *caval veins* opening into its three corners. These veins carry deoxygenated blood into the sinus. It

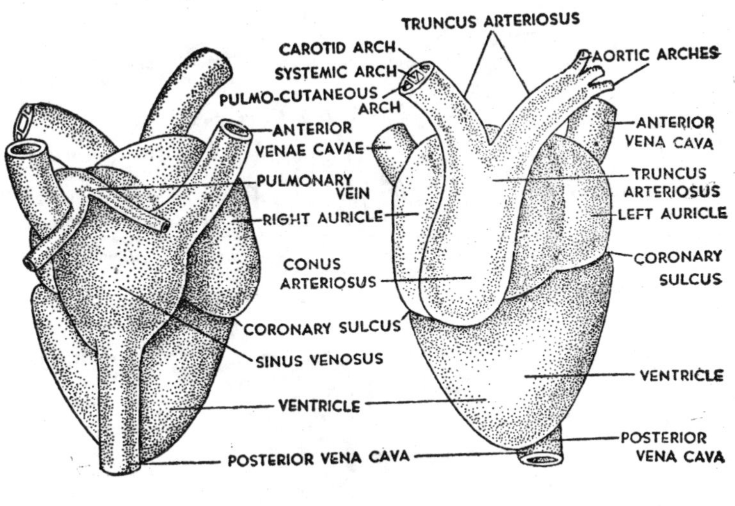

Fig. 27. Heart of toad.

communicates with the right auricle through a slit-like *sinuauricular aperture*, the edges of which are guarded by the *sinuauricular valves.* The valves permit entry of blood from the sinus into the right auricle, but no backflow or regurgitation is allowed.

The two **auricles,** right and left, form the base of the heart. They are separated from the ventricle by a narrow groove, the *coronary sulcus.* The auricles are completely separated from one another by a strong vertical partition, the *interauricular septum.* The right auricle is larger than the left; it receives deoxygenated blood from the sinus venosus through the *sinuauricular aperture.* The left auricle receives oxygenated blood from the lungs through a small opening, the *aperture of the common pulmonary vein.* The two auricles communicate with the ventricle by a common opening, the *auriculo-ventricular aperture,* which is guarded by the *auriculo-ventricular valves.* The membranous cusps of this valve hang down like curtains to allow the passage of blood from the auricles into the ventricle, but not in a reverse direction. The free edges of the cusps are attached to the ventricular wall by means of a number of fine threads called *chordae tendineae.*

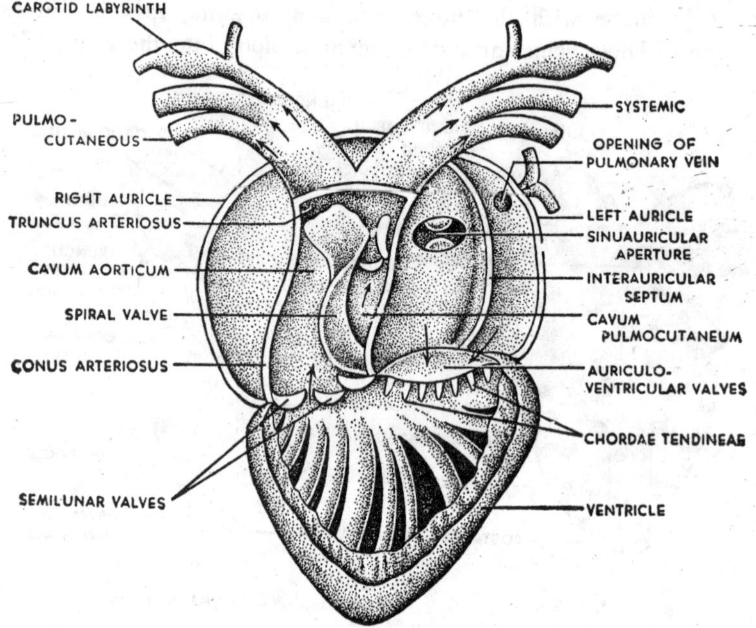

Fig. 28. Internal structure of the heart
(Arrows indicate direction of blood-flow).

The **ventricle** is the thick-walled conical chamber which is situated behind the auricles. Its posterior bluntly pointed portion forms the apex of the heart. The ventricular cavity is greatly reduced by a number of interlacing muscle fibres which arise from its own wall. The ventricle, therefore, is spongy and when it is cut, its interior looks like a honeycomb. The sponginess of the ventricle prevents admixture of the two kinds of blood in the ventricular cavity, the oxygenated kind from the left auricle and the deoxygenated kind from the right auricle.

The **conus arteriosus** is the stout tube which arises ventrally from the base of the ventricle and passes obliquely towards the left. It is continued forwards as the *truncus arteriosus* which is the base of the main artery for carrying the blood away. The truncus however is not a part of the heart; it belongs to the arterial system. The conus is separated from the truncus by a set of pocket-like *semilunar valves.* A similar set of three *semilunar valves* guards the opening between the ventricle and the conus. These valves prevent backflow into the ventricle. Inside the conus is a twisted, longitudinal flap, the *spiral valve.* This divides the cavity of the conus into a right channel, the *cavum aorticum,* and a left channel, the *cavum pulmocutaneum.* The spiral valve, the semilunar valves and the spongy ventricular cavity co-operate with one another in guiding the two kinds of blood. The deoxygenated blood is pumped through the cavum pulmocutaneum and the oxygenated kind through the cavum aorticum. The two kinds of blood enter different arterial arches and are carried to different places.

Arterial system. The arteries and their branches form the *arterial system.* The **truncus arteriosus** is the main artery originating from the conus. It at once divides into right and left branches. Each of these trunks splits into *three* arterial arches: (1) the **carotid** anteriorly, (2) the **systemic** in the middle, and (3) the **pulmocutaneous** posteriorly. Thus, there are three pairs of arterial arches for supplying blood to different parts of the body. The carotids supply the head region, the systemics supply the trunk and limbs, and the two pulmocutaneous supply the lungs and skin.

(1) **Carotid arches and their branches.** The carotid arch, on each side, proceeds outward and forward. It soon bifurcates into an inner branch called *external carotid artery* and an outer branch, the *internal carotid artery.* Immediately near the point of bifurcation is a small swelling called **carotid labyrinth.** Inside the labyrinth, the carotid arch breaks up into numerous minute vessels which reunite at the other end. The carotid labyrinth offers an extra resistance to the passage of blood through the carotid arch. The **external carotid artery** supplies blood to the floor of the buccal cavity, tongue and outer side of the head. The **internal carotid artery** turns backward and comes very close to the systemic arch

of the same side, to which it is tied by a small amount of fibrous tissue, the *carotid ligament.* Finally, it turns forward to enter the skull through a foramen and is distributed to the brain and its coverings.

(2) **Systemic arches and their branches.** The systemic arch sweeps outward. It then curves round the oesophagus to reach the dorsal side, where it joins with its fellow of the opposite side to form a median artery called *dorsal aorta.* Thus, the two

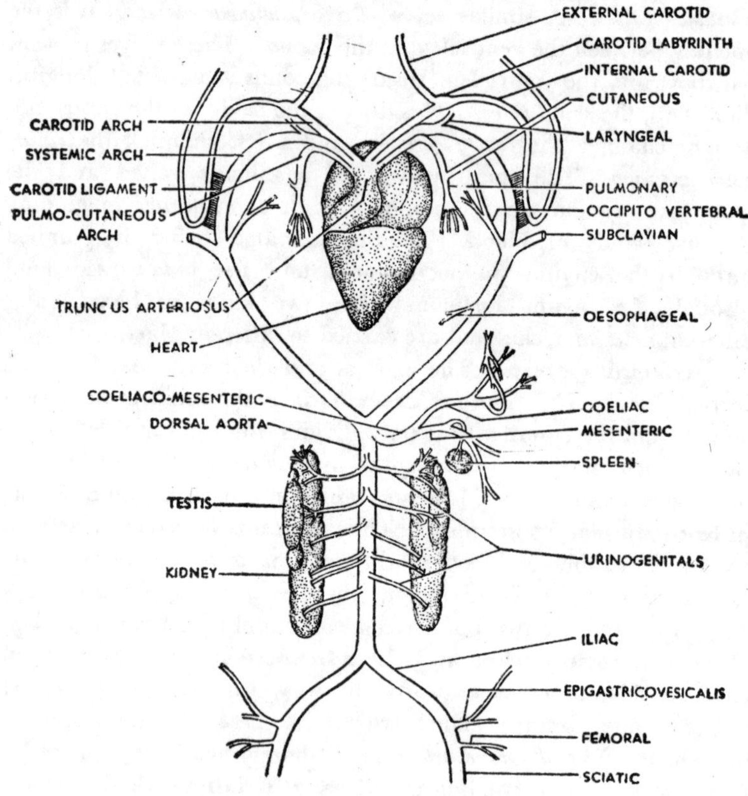

Fig. 29. Main arterial vessels of toad.

systemics form an arterial ring round the oesophagus. Each systemic gives out the following branches: (i) a short **laryngeal artery** to supply the voice-box; (ii) an **occipito-vertebral artery** which breaks up into branches for supplying the pharynx, the back of the head, the vertebral column and the spinal cord; (iii) a stout **subclavian artery** which proceeds outward to supply the shoulder and forelimb of the same side; (iv) the left systemic gives off an additional twig, the **oesophageal artery,** to the oesophagus. *The right systemic has no oesophageal branch.*

The **dorsal aorta** is formed in the mid-dorsal line by the union of the right and left systemic arches. It extends backward, lying in front of the vertebral column, and terminates at the posterior end of the body cavity by dividing into two *iliac arteries.* A stout **coeliaco-mesenteric artery** is given out from the commencement of the dorsal aorta. This breaks up into a *coeliac* branch for supplying the stomach, pancreas, liver and gall-bladder, and a *mesenteric* branch for supplying the intestines, cloaca, spleen and mesentery. In passing through the space between the kidneys, the dorsal aorta gives off four to five pairs of **renal arteries** for supplying the urinogenital organs.

The **iliac arteries** are the terminal branches of the dorsal aorta. Each iliac gives off an *epigastrico-vesical* branch to supply the urinary bladder and the ventral body-wall. After this the iliac artery enters the hindlimb of the same side where it divides into *femoral* and *sciatic* branches.

(3) **Pulmocutaneous arches and their branches.** Two pulmocutaneous arches carry deoxygenated blood to the lungs and skin for aeration. They are the hindermost and the shortest of the arterial arches. Each arch passes outward and then backward. A very slender branch is given out to supply the skin. This is the **cutaneous artery.** The main trunk then runs into and supplies the lung of the same side as the **pulmonary artery.**

Venous system. The veins and their tributaries constitute the *venous system.* The arteries break up into smaller and smaller branches, and these branches merge into a network of thin-walled, hair-like vessels, called *capillaries.* The capillaries merge into small veins, which join to form larger veins.

The venous system of toad may be subdivided into three separate groups: (1) the *pulmonary,* (2) the *systemic,* and (3) the *portal.*

(1) **Pulmonary veins.** Two pulmonary veins carry oxygenated blood from right and left lungs. The right and left pulmonary veins unite to form a **common pulmonary vein,** which opens into the left auricle on the dorsal side.

(2) **Systemic veins.** These are represented by the three large veins or *venae cavae* draining into the corners of the triangular sinus venosus. They carry deoxygenated blood from all parts of the body except lungs. Two of the vanae cavae are situated anteriorly; these are the right and left *precavals.* A single *postcaval* is found posteriorly, opening into the apex of the sinus venosus.

Each **precaval vein** is formed by the union of *three* tributaries: (i) *external jugular*, (ii) *innominate*, and (iii) *subclavian.*

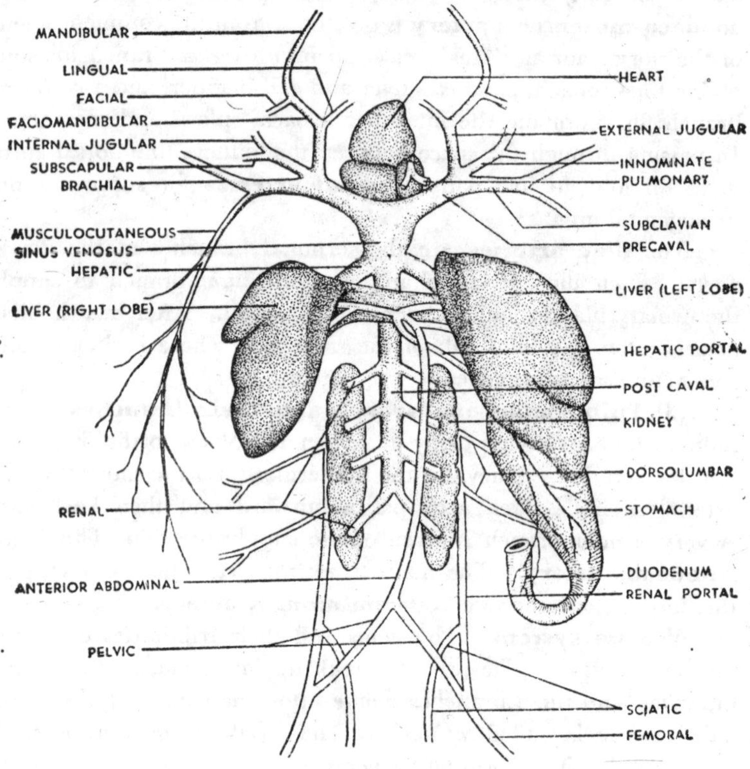

Fig. 30. Venous system of toad.

(i) The **external jugular vein** is formed by the union of two tributaries—a **lingual vein** bringing blood from the tongue, and a **faciomandibular vein** from the jaws and snout.

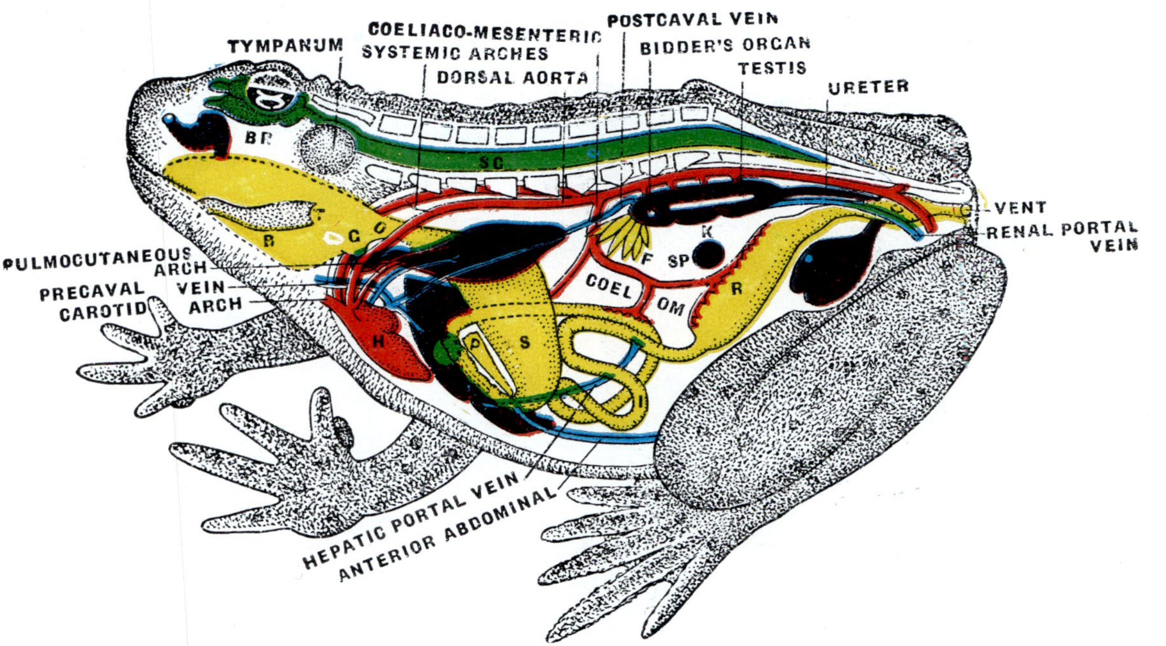

Organ system of a male toad.

B=buccal cavity; BR=brain; C=cloaca; F=fat body; G=glottis; H=heart; I=small intestine; K=kidney; L=liver; LU=lung; N=nasal chamber; O=oesophagus; P=pancreas; R=rectum; S=stomach; SC=spinal cord; SP=spleen; T=tongue; U=urinary bladder.

(ii) The **innominate vein** is also formed by the union of two tributaries—an **internal jugular** from the interior of the skull, and a **subscapular** from the back of the shoulder.

(iii) The **subclavian vein** is similarly formed by the union of two veins—a **brachial** from the forelimb and a **musculocutaneous** from the skin and muscles. In toad, the skin is an accessory respiratory organ. Hence the blood carried by the musculocutaneous vein is comparatively rich in oxygen.

The **postcaval vein** has its roots in the kidneys. It is formed by the union of four to five pairs of **renal veins** which collect blood from the kidneys. The *genital veins* from the reproductive organs drain into some of the renal veins. The postcaval runs forward to enter the liver substance, and receives a pair of hepatic veins one from each lobe of the liver. It finally terminates by joining the posterior end of the sinus venosus.

(3) **Portal veins.** A portal vein begins in capillaries and ends in capillaries before the blood which travels through it is returned to the heart. There are two portal systems in the toad: (A) *hepatic portal system,* and (B) *renal portal system.* In the hepatic portal system, venous blood collected from capillaries at the posterior part and from the gut filters through capillaries in the liver on its way back to the heart. In the renal portal system, venous blood collected from capillaries in the posterior part of the body passes through capillaries in the kidneys on its way back to the heart.

(A) **Hepatic portal system.** In this, the two principal veins are the *hepatic portal vein* and the *anterior abdominal vein.* The **hepatic portal vein** is formed by the union of several small veins returning blood from the stomach, intestines, pancreas, spleen, etc. The main trunk of the hepatic portal comes to the under surface of the liver, where it receives the anterior abdominal vein. Thus enlarged, it enters the liver substance and breaks up into hair-like sinusoids which finally drain into the hepatic veins.

The **anterior abdominal vein** is formed in the following manner. Blood from the hindlimb is returned by two large veins, the *femoral* and the *sciatic.* On entering the body cavity, the femoral vein gives off a median branch called *pelvic vein.* The right and left pelvic veins unite in the mid-ventral line to form the anterior abdominal vein, which runs forward along the middle line under cover of the ventral body-wall. It receives several small tributaries from the ventral body-wall and from the urinary bladder.

At the level of the liver, it turns inward to join the hepatic portal vein. The conjoined trunk then divides into vessels which enter the lobes of the liver and break up into sinusoids. The hepatic veins originate at the other end of these networks and thus the blood is finally drained into the postcaval.

(B) **Renal portal system.** Blood is returned from the hindlimbs by **femoral** and **sciatic veins.** The femoral vein comes from the front of the thigh. On entering the body cavity, it gives off the median *pelvic* branch which joins with its fellow of the opposite side to form the *anterior abdominal vein.* The main trunk of the femoral now receives the sciatic vein from the back of the thigh and proceeds towards the kidney as the **renal portal vein.** At the outer margin of the kidney, the renal portal is joined by two or three **dorsolumbar veins** bringing blood from the body-wall. It now enters the kidney and breaks up into sinusoids which ultimately drain into the renal veins, thus returning the blood into the postcaval.

It is to be noted that the postcaval vein of the toad is the product of the renal portal and hepatic portal system, and the anterior abdominal vein is joining the two portal systems.

Circulation of blood. The heart continues to beat, taking short rest between successive contractions, and drives the blood into the arteries. The heart muscles have an innate tendency to contract and relax with a definite rhythm. Each period of contraction or **systole** is followed spontaneously by a shorter period of relaxation or **diastole.** The contraction starts at the sinus venosus, spreads to the auricles, then to the ventricle, and finally to the conus arteriosus. The heart of a freshly killed toad beats rhythmically for a long time even when it is taken out.

During diastolic period, the sinus venosus receives deoxygenated blood from two precavals and the single postcaval vein. The left auricle, at the same time, is filled with oxygenated blood through the common pulmonary vein.

With the commencement of the systole, the sinus contracts and the deoxygenated blood is pumped into the right auricle, through the sinuauricular aperture. This is followed by the auricular systole. The two auricles contract simultaneously driving their contents into the single ventricle, through the auriculo-ventricular aperture. Regurgitation of deoxygenated blood into the sinus venosus is prevented by the closure of the sinuauricular valves. Thus both kinds of blood enter the ventricle at the same time. As the ventricular cavity is spongy, there is not much mixing. During the very short stay in the ventricle the deoxygenated blood is stored on the right side, the oxygenated blood on the left side, and there is a mixed column of blood in the middle of the ventricular cavity.

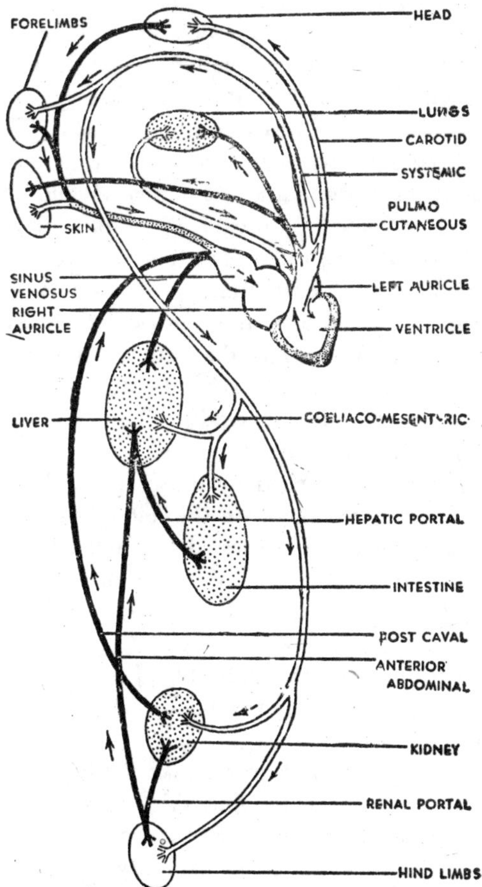

Fig. 31. Course of circulation of blood in toad (diagrammatic).

The ventricle now contracts. Blood cannot flow back into the auricles because the auriculo-ventricular aperture is tightly closed by the auriculo-ventricular valves. Since the conus arises from the right side, the first blood to enter the conus will be the deoxygenated kind in the right side of the ventricle. This is now guided by the spiral valve into the cavum pulmocutaneum and from there, through the pulmocutaneous arches, to the lungs and skin. Moreover, the pulmocutaneous arches are the shortest of the arterial arches. They, therefore, offer least resistance to the entry of deoxygenated blood into them at the very beginning of the ventricular systole. The deoxygenated blood is aerated in the lungs and brought back to the left auricle through the pulmonary veins—thus completing the **'pulmonary circuit'.** The portion of blood, which becomes oxygenated in the skin, is returned to the sinus venosus through the musculocutaneous vein. So this portion of oxygenated blood is mixed up with the deoxygenated blood stream before it is returned back to the heart.

As the force of the ventricular contraction is increased, the mixed blood in the middle of the ventricular cavity is pumped through the cavum aorticum into the systemic arches. This mixed blood cannot enter the carotid arches because of the high resistance offered by the carotid labyrinths. Thus the mixed blood is distributed to the trunk and limbs.

Lastly, the pressure exerted by the carotid labyrinths is overcome and the oxygenated blood in the left side of the ventricle is now forced into the carotid arches, through the cavum aorticum. It is distributed to the head region of the animal. It is to be clearly understood that the spiral valve directs the various courses of the three kinds of blood streams.

During ventricular diastole, the blood cannot flow back from the arterial arches because of the pocket-like semilunar valves at the root of the conus.

The systemic and carotid arches take the blood away to every part of the body, except lungs and skin. Food and oxygen are carried to the tissues in this way. Exchange of materials occurs by osmosis through the thin wall of the capillary networks. The blood is finally brought back to the heart by the three caval veins—thus completing the **'systemic circuit'.** While returning, the blood collects carbon dioxide and nitrogenous waste products from the tissues. Further, the part of the blood stream which is returning from the alimentary canal becomes loaded with absorbed food. The carbon dioxide is mixed up with the general stream of systemic circuit and eliminated by the lungs and skin. Blood rich in nitrogenous waste products is directed through the **renal portal**

system into the kidneys for exertion. Lastly, the blood rich in food materials is diverted to the liver via the **hepatic portal system.** The liver cells are thus provided with a better chance of extracting food and storing the same for future use.

Vandervael and Foxon have experimentally demonstrated that there is no separation of blood streams in the frog's heart and the blood flowing out through the main arterial arches is nearly of the same composition. They injected a radio-opaque substance called **thorotrast** into the sinus venosus of a frog. Subsequently a series of X-ray photographs were taken, which revealed that the opaque substance was uniformly distributed in all the three arterial arches. Similar results were obtained by injecting thorotrast through the common pulmonary vein.

It appears that there is no separation of the oxygenated and deoxygenated blood streams in the heart of frogs and toads. As aeration of blood occurs through the skin and mouth, the blood in the sinus venosus cannot be completely deoxygenated, because the cutaneous veins open into it. Although a separation of the oxygenated and deoxygenated streams would be far better and more efficient, toads and frogs can somehow manage without it.

Lymphatic system. It has already been pointed out that the blood is not directly poured into the tissues. While circulating through capillary networks, the blood plasma exudes into the intercellular spaces, coming into direct contact with the cells. This fluid is known as the **lymph.** It is a colourless liquid containing a few leucocytes, but no erythrocyte. The lymph is collected from the intercellular spaces by small vessels called **lymphatics.** These drain into **lymph sacs.** Some of the lymph vessels are contractile and are known as **lymph-hearts.** The lymph-hearts pump the lymph back into the veins. There are a pair of lymph-hearts near the urostyle, opening into the femoral veins; another pair, below the scapulae, drain into the subscapular veins.

The lymph transports food and oxygen from the blood to the cells of the body. Waste products are carried by the lymph into the blood stream.

Respiratory System

Respiration is the exchange of oxygen and carbon dioxide between the toad and its environment. Oxygen is taken into the body and carbon dioxide is given off. Every cell of the body consumes oxygen. With the help of oxygen, the assimilated food which is stored within the cells is slowly oxidised. The result is the *liberation of heat energy*, and formation of carbon dioxide and water. Thus when glucose is oxidised, one molecule of glucose combines with six molecules of oxygen to form six molecules of carbon dioxide and six molecules of water. This is not all; because one gramme of glucose, when completely oxidised in the body, yields about four calories of heat, which is utilised by the animal for carrying out its other activities. Respiration, therefore, is not merely exchange of oxygen for carbon dioxide, but it also involves transformation of assimilated food into energy. When no reserve food is available, the animal transforms its own protoplasm into energy and thereby loses weight.

It is to be noted that the source of the oxygen is the air and the carbon dioxide is formed in the tissues. The transport of the two gases occurs through the agency of the blood. Acting as a 'middleman' the blood carries oxygen from the air to the tissues, and carbon dioxide from the tissues to the outside air. The exchange of oxygen of the air for the carbon dioxide of the blood takes place in the *respiratory organs* and is known as **external respiration.** The true respiration, resulting in the release of energy, occurs in the *tissues* by the exchange of oxygen of blood for the carbon dioxide of the tissues. This is known as **internal** or **tissue respiration.**

While in blood, the oxygen is actually carried by the haemoglobin of the erythrocytes. Haemoglobin combines with oxygen to form an unstable compound called *oxyhaemoglobin*. In this form, oxygen is carried to all the body cells. As the blood is circulating through the capillary networks of deoxygenated tissue, the unstable oxyhaemoglobin releases the bulk of its oxygen and is reduced to its former state. It is then sent back to the respiratory surface for reloading with fresh oxygen. The oxygen that is set free diffuses out through the capillaries into the lymph, which comes into direct contact with the body cells. Tissue respiration now takes place. Oxygen combines with food materials which are stored in the tissues for this purpose. Carbon dioxide is formed along with release of energy. The carbon dioxide is carried by the lymph and diffuses out into the blood. It is transported chiefly by the blood plasma in the form of *bicarbonates* and finally released and expelled through the capillary networks on the respiratory surface.

Types of respiration. As the toad is an amphibian, it is able to receive oxygen and given out carbon dioxide from *three* kinds of respiratory organs. These are: (1) the **lungs,** (2) the **skin,** and (3) the mucous membrane of the **buccal cavity.** The moist epithelium of these organs are richly supplied with blood capillaries. Atmospheric oxygen dissolves in the moisture of the respiratory surface and diffuses into the blood, through the thin wall of the capillary networks. Carbon dioxide of the blood diffuses out in the opposite direction.

Pulmonary respiration. The organs concerned in the pulmonary respiration are: (1) the *external nares,* (2) the *internal nares,* (3) the *buccal cavity,* (4) the *glottis,* (5) the *laryngo-tracheal chamber,* (6) the *bronchi,* and (7) the *lungs.*

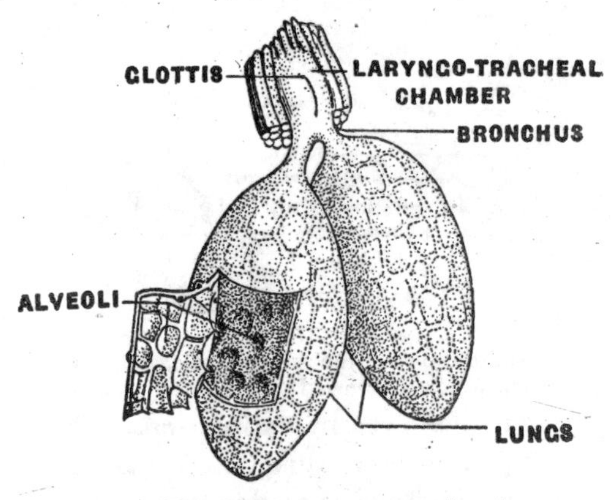

Fig. 32. Respiratory organs of toad.

The **lungs** are thin-walled spongy structures which extend backward into the anterior part of the body cavity. There are two lungs, one on each side of the heart. Each lung is a pink-coloured elastic sac which is connected to the base of the laryngo-tracheal chamber by a short tube called **bronchus.** The **laryngo-tracheal chamber** or voice-box communicates with the floor of the **buccal cavity** by a slit-like **glottis.** The buccal cavity opens into the exterior by the **internal** and **external nares.**

The outer wall of each lung is hollowed out internally to produce innumerable simple chambers which are called **air-sacs** or **alveoli.** Each alveolus receives its own blood supply from a minute branch of the pulmonary artery. The alveolar epithelium, presenting a moist surface and being richly supplied with capillary network, carries out O_2—CO_2 exchange.

Mechanism of respiration. Ventilation of the lungs is effected by two processes, namely, breathing in of air or **inspiration** and breathing out of air or **expiration.** The two processes occur alternately in three stages: (1) *aspiration,* (2) *expiration,* and (3) *inspiration.*

(1) The toad closes its mouth and lowers the floor of the buccal

cavity, thus drawing in air through the external nares. This air is sucked into the buccal cavity through the internal nares. As the glottis remains closed, the air cannot pass into the lungs. This is known as **aspiration.**

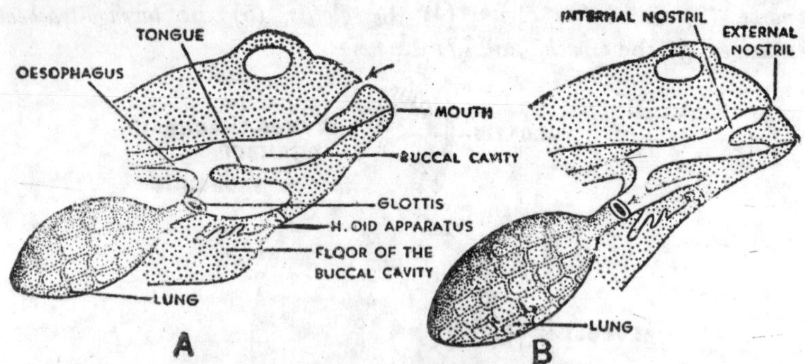

Fig. 33. Diagram to illustrate the mechanism of pulmonary ventilation (A=aspiration which is followed by expiration; B=inspiration).

(2) Aspiration is followed by **expiration.** The trunk muscles forcibly contract upon the elastic lungs. This expels air rich in CO_2 through the glottis, which opens at this stage. The buccal cavity, which is already full of inhaled atmospheric air, becomes very much distended with the exhaled air from the lungs. A mixture of the two kinds of air now rushes out through the open nostrils.

(3) Expiration is quickly followed by **inspiration.** The external nares are tightly closed and the floor of the distended buccal cavity is forcibly raised. The mixed air is thus pumped into the lungs through the glottis.

In a living toad, the floor of the buccal cavity is constantly moved up and down, and the mouth is kept closed. This means that the animal is breathing. The respiratory movements are caused by the protraction and retraction of the hyoid apparatus which lies on the floor of the buccal cavity.

The animal appears to use its lungs only when oxygen is urgently needed. Otherwise it is satisfied with **buccal respiration.** The buccal cavity is always distended with air and its mucous membrane is very richly supplied with blood vessels. Hence, O_2—CO_2 exchange can occur freely in this part.

Since the skin is exposed to air, no movement is necessary to carry on **cutaneous respiration.** This goes on at all times but is specially useful in the winter when the animal hibernates.

The tadpoles breathe through **gills.** These are slender epithelial outgrowths containing capillary networks. As the gills are constantly bathed in water, they can absorb oxygen and give out carbon dioxide.

Voice. The laryngo-tracheal chamber is a short and stout box which is strengthened by cartilages. Stretching across the cavity of the box, there are two elastic bands called **vocal cords.** Croaking is produced by the vibration of the vocal cords when air is expelled through the glottis. Intensity and pitch of the sound depends on the elastic tension of the cords. In a male toad, the inflated **vocal sac** acts as a resonating loudspeaker.

Urinogenital Organs

The urinogenital organs belong to the **excretory system** and the **reproductive system.** It is customary to describe the two systems together, because the products of the reproductive organs pass out through excretory channels. Functionally, however, the two systems are completely separate.

Excretory System

Various substances are manufactured within a toad's body during its routine metabolic activities. Some of these are of no use and may cause harm if they are retained for a long time. Only these end products of metabolism are the waste products. Faeces is not a waste product, because it is not produced by metabolism in the tissues. The principal waste products are urea, uric acid, carbon dioxide, excess of water and mineral salts. These are collected from the tissues by the blood and transported to the excretory organs, which promptly eliminate them from the toad's body. The process by which metabolic waste products are separated and removed from the body is known as **excretion.**

Excretory organs. Of the waste products, CO_2 and H_2O are formed in all the tissues as by-products of internal respiration. These are expelled by the **lungs** and **skin.** Urea and uric acid are nitrogenous waste products. Urea is formed in the **liver** by the decomposition of amino acids. The amino acids are first broken down to ammonia which is then converted into urea by the liver cells. Urea, uric acid, excess of water, and mineral salts are excreted in the form of urine. The organs concerned with the removal of the urine are: (1) the **kidneys,** (2) the **ureters,** (3) the **urinary bladder,** (4) the **cloaca,** and (5) the **vent.**

The **kidneys** are a pair of flattened elongated bodies, reddish-brown in colour, attached to the dorsal wall of the body cavity, one on each side of the vertebral column. They lie behind and are completely covered by the peritoneum, that is, they are *retro-peritoneal*

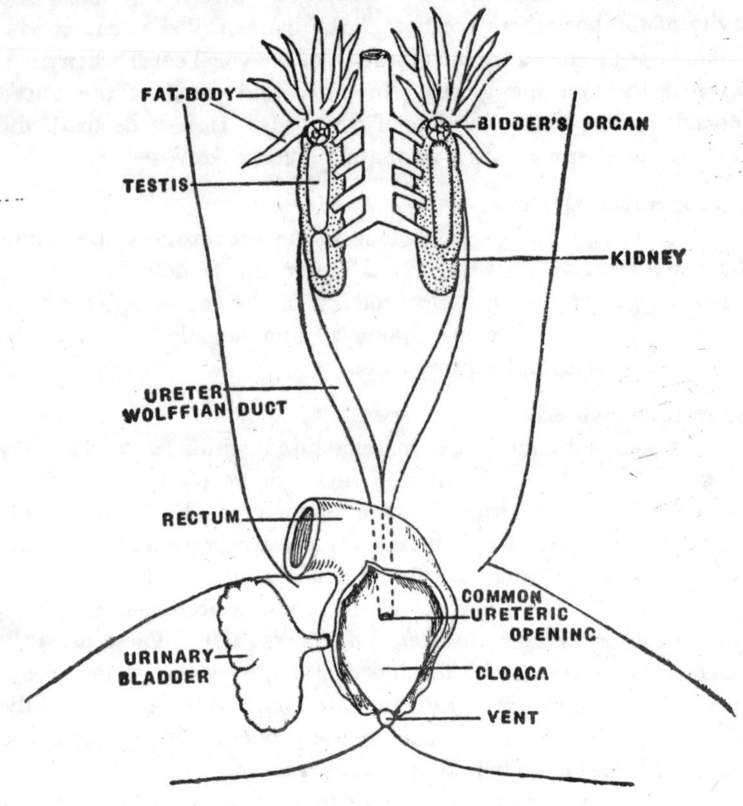

Fig. 34. Urinogenital organs of a male toad.

in position. The outer border of each kidney is somewhat convex, and the edges are indented by notches. The toad's kidney is recognised as the **Wolffian body** or **mesonephros** (*mesos*=middle; *nephros*=kidney), because it is more highly developed than the *pronephros* or the primitive kidney, but not as highly developed as the *metanephros* which is found in higher animals. The **ureters** or **wolffian ducts** are thin-walled whitish tubes found on the outer border of the kidneys. The two ureters, one from each kidney, run backward. They join to form a single median tube at the posterior end of the trunk, and this tube finally opens into the dorsal wall of the **cloaca.** Arising opposite, from the ventral wall of the cloaca, there is a thin-walled bilobed sac called **urinary bladder.** Its narrow opening is guarded by a sphincter muscle. The cloaca communicates with the exterior by the **vent.**

Structure of the Kidney. A kidney is composed of a large number of minute tubes called **uriniferous tubules.** Each tubule begins blindly in a two-walled cup with a narrow mouth. The cup is known as **Bowman's capsule.** A small afferent vessel from the renal artery enters the capsule and breaks up into a bunch of

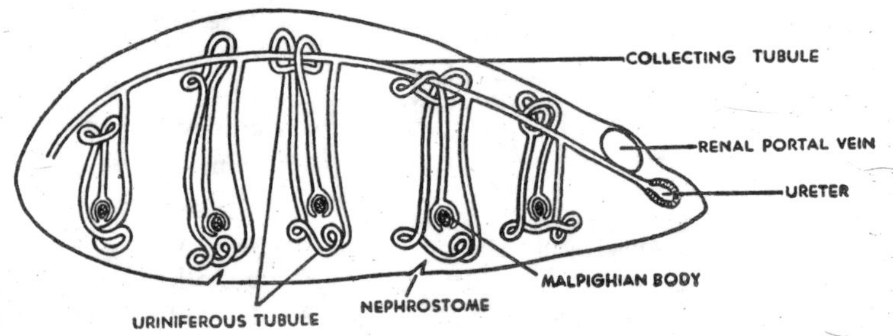

Fig. 35. A diagram illustrating the arrangement of the uriniferous tubules in the kidney.

capillaries which fits into the cup. This is the **glomerulus** (*glomus* =a ball). An efferent vessel emerges at the other end of the glomerulus and drains into the renal vein. A capsule with its glomerulus is recognised as the **Malpighian body.** The other end of Bowman's capsule is connected with the tubule proper, which extends as a much convoluted narrow channel. Ultimately the tubules join with one another to form larger tubes called **collecting tubes** and these in turn open into the ureter. The convoluted part of each tubule is surrounded by capillary networks derived from the renal portal vein. On the ventral surface of the kidney, there are numerous microscopic openings called **nephrostomes** (*nephros*= kidney; *stoma*=opening) which drain into the renal veins.

Mechanism of urine formation. Blood loaded with waste products is brought to the kidney by the renal artery and the renal portal vein. While passing through the glomerulus, nitrogenous waste products along with water and other substances are filtered out of the blood into the capsular space. The filtrate enters the convoluted part of the tubule where useful substances are reabsorbed and sent back into the circulating blood. Oxygen that was brought by the renal artery is used up by the cells of the tubules, so that deoxygenated blood devoid of waste products now flows out of the kidney by the renal veins. Probably the nephrostomes drain away excess of water and other wastes from the body cavity.

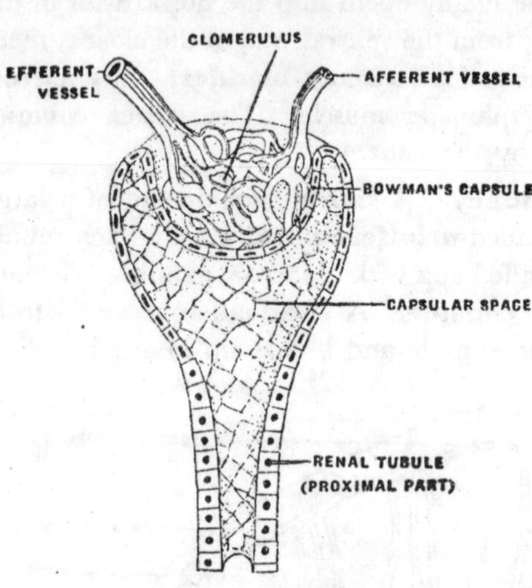

Fig. 36. Enlarged diagram of a Malpighian body.

The fluid which runs into the ureters via the uriniferous tubules is the urine. It drips into the cloaca continuously and gravitates into the urinary bladder for temporary storage. When filled with urine, the bladder contracts and the content is ejected through the vent.

Reproductive System

The sexes are separate. Male and female toads are very much alike in external feature. During the breeding season, however, the female bulges out, being distended by her eggs, whereas the male remains slender. The male, on the other hand, produces cushion-like thumb-pads at the bases of his first two fingers. This thickening is used for grasping the female. Lastly, the male can croak loudly because it has a vocal sac under its throat.

The reproductive system consists of the **gonads** (*gono*=that which generates) which produce the germ cells, and the ducts through which the germ cells are let out. The function of this system is the perpetuation of the race.

Male reproductive organs (Fig. 34). In a male toad the gonads are a pair of yellowish-white, elongated bodies called **testes**

(singular=testis). Each testis is attached to the ventral surface of the corresponding kidney by a fold of peritoneum called *mesorchium.* Sometimes a testis may consist of two or three pieces joined end to end. Each testis is a collection of very fine tubes called **seminiferous tubules** which manufacture the male germ cells or *spermatozoa.* A spermatozoon is a highly specialised cell, with an oval head containing the nucleus, a short middle piece containing the centrosome, and a long wavy tail composed of cytoplasm. The seminiferous tubules of a testis are connected to the collecting tubes of the kidney by several fine ducts called **vasa efferentia** which pass along the mesorchium. The sperms are expelled through the vasa efferentia into the ureter which, therefore, serves as the urinogenital duct. The sperms are finally let out through the cloacal aperture at the time of mating.

Attached to the anterior end of the kidney, and in front of the testis, is a small-rounded body called **Bidder's organ.** It is an undeveloped female gonad and serves no useful function in the normal male.

In front of the gonads and in both the sexes, there are several yellowish outgrowth called **fat bodies.** These are the reservoirs of fatty substances which are utilised in the manufacture of germ cells. They also provide nourishment during hibernation, when no feeding is possible.

Female reproductive organs. In a female toad the gonads are the two **ovaries.** Each ovary is a much-folded sac of irregular shape fixed to the ventral surface of the corresponding kidney by a fold of *mesovarium.* In the breeding season the ovaries are much enlarged and distended with eggs. The eggs or **ova** (singular=ovum) are the female germ cells. Each ovum is a spherical cell with a blackish *animal pole* containing protoplasm, and a whitish *vegetative pole* full of yolk. The egg is surrounded by a thin *vitelline membrane.* The **oviduct** for each ovary is a long much-coiled tube situated laterally. The oviducts are not connected with the ovaries in any way. The anterior end of each oviduct lies far forward near the base of the corresponding lung, and opens directly into the body cavity by a funnel-shaped *ostium.* The middle part of the oviduct is thick-walled, narrow, and convoluted, whereas the posterior part widens out to form a thin-walled *ovisac* or *uterus.* The two uteri join to form a single median tube which opens into the dorsal wall of the cloaca, just anterior to the urinary opening. When mature, the eggs escape into the body cavity by rupturing the ovarian wall. They then find their way into the oviducts through the funnel-shaped ostia. As the eggs are passing down the convoluted middle part of the oviducts, each egg receives a coating of jelly-like albumen.

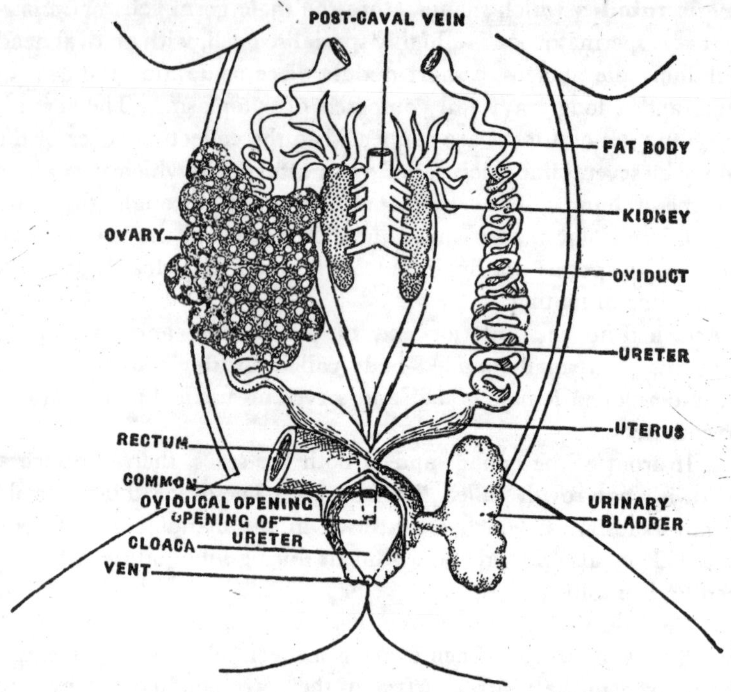

Fig. 37. Urinogenital organs of a female toad (left ovary removed).

The eggs accumulate in the dilated ovisacs or uteri and are voided through the vent during mating.

The process of reproduction will be described along with the life-history.

Nervous System

The nervous system is concerned with the complicated processes by which the toad responds to changes in its environment. It is responsible for the control and co-ordination of the various organs so that the animal can function as a whole.

For the sake of convenience, the nervous system is divided into *three main parts*. These are (1) the **central nervous system** including *brain* and *spinal cord;* (2) the **peripheral nervous system,** composed of *cranial nerves* originating from the brain and *spinal nerves* from the spinal cord; (3) the **autonomic nervous system** consisting of a number of *sympathetic ganglia*, two chains of which are found, one on each side of the vertebral column.

The Central Nervous System

This is composed of a hollow median tube, the anterior enlarged part of which is known as the **brain** and the posterior narrow part as the **spinal cord.** The cavity of the brain is continuous with that of the spinal cord and is filled up with a fluid called *cerebro-spinal fluid.* The solid parts of the central nervous system are composed of *neurones* or nerve cells, together with their processes, the *nerve fibres.* The part of the neurone in which the nucleus lies is the cell-body. The extended processes of the neurones which run together within a connective tissue sheath form a *nerve.* Aggregated masses of nerve cells with their contained nuclei has a greyish appearance and are, therefore, recognised as the *grey matter.* Bundles of nerve fibres, on the other hand, appear to be white and are, therefore, called *white matter.* The grey matter lies on the outer surface of the brain and covers up the white matter. In the spinal cord, however, the condition is reversed; the white matter is on the outer side and the grey matter is on the inner side. The brain and the spinal cord are enclosed in two protective coverings or *meninges:* (1) an outer and thicker *dura mater,* and (2) an inner and thinner *pia mater.*

Brain. The brain is housed within the cranium of skull. It is composed of three primary regions: the *forebrain,* the *midbrain,* and the *hindbrain.* Subsequent changes in forebrain and hindbrain result in a further subdivision of these two parts, so that ultimately, the brain consists of five portions which may be tabulated antero-posteriorly in the following manner:—

I. **Forebrain** or **Procencephalon** (*Pros* = fore; *kephalon* = brain)	(1) Telencephalon (*Tel* = end) (2) Diencephalon (*Di* = between) or Thalamencephalon
II. **Midbrain**	(3) Mesencephalon (*Mes* = middle)

III. Hindbrain or **Rhombencephalon** (*Rhomb* = like spinning top)	(4) Metencephalon (*Meten* = after) (5) Myelencephalon (*Myelon* = near spinal cord)

I. The **forebrain** is composed of **telencephalon** and **diencephalon.** At the anterior end of the telencephalon there is a pair of small **olfactory lobes,** which are concerned with the sense of smell. The rest of the telencephalon consists of two elongated lobes, the **cerebral hemispheres.** The floor of each hemisphere is thickened to form a mass called *corpus striatum.* The thin roof and sides of the hemispheres are smooth and constitute the *pallium.* The cerebral hemispheres are the seat of intelligence, consciousness, and voluntary motion.

The posterior part of the forebrain is the diencephalon or thalamencephalon. It is a depressed region behind the cerebral

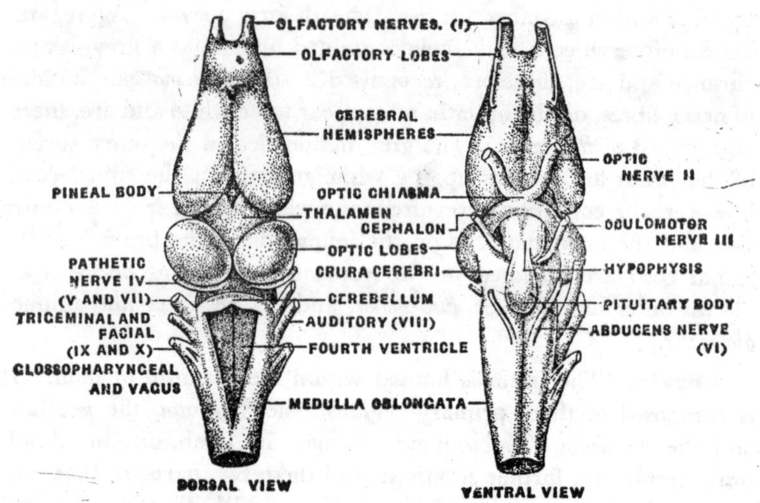

Fig. 38. Brain of toad

hemispheres which gives out dorsally a thin projection called **epiphysis.** To this is attached a small **pineal body** of unknown function. In front of the epiphysis is a highly vascular membrane, the *anterior choroid plexus.* On the ventral surface of the diencephalon is an X-shaped structure, the **optic chiasma,** formed by the crossing of the two optic nerves. A little behind the chiasma is a projection called **hypophysis** or **infundibulum,** and to this is attached a rounded glandular structure, the **pituitary body.** Each side of the diencephalon is thickened to form the **optic thalamus.**

II. The **midbrain** or **mesencephalon** is the widest part of the brain. A pair of hollow oval bodies grow out from its dorsal surface. These are the **optic lobes** or **corpora bigemina.** Ventral to the optic lobes, there are two longitudinal bands or peduncles, the **crura cerebri,** which connect the forebrain with the hindbrain.

III. The **hindbrain** is composed of **metencephalon** and **myelencephalon.** The metencephalon is represented dorsally by a thin transverse shelf called **cerebellum,** which lies behind the optic lobes. It is the centre for the co-ordination of movements. The myelencephalon connects the hindbrain with the spinal cord. Its floor and sides are greatly thickened to form the **medulla oblongata.** Dorsally, on the roof of the myelencephalon, is a thin membrane richly supplied with blood vessels. This is known as the *posterior choroid plexus.* The medulla contains the centres for the control of important vital processes such as regulation of heart beat, respiration, and metabolism. It has been proved by experiment that if all the parts of the brain be destroyed and only the medulla is retained the toad can hop, swim, and breathe. Removal of medulla, however, is at once followed by death.

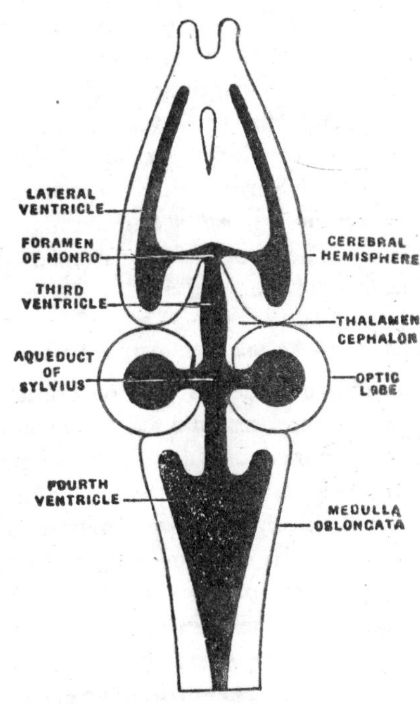

Fig. 39. Ventricles of the brain (diagrammatic).

Cavities of the brain. As stated previously the brain is not a solid mass. It has cavities within, which are known as the **ventricles.** The hollow of the cerebral hemispheres constitute the first and the second ventricles which are also recognised as the **lateral ventri-**

cles. The cavity of the diencephalon is the **third ventricle.** Each lateral ventricle communicates with the third ventricle by a small opening called **foramen of Monro.** The **fourth ventricle** is the shallow cavity of the medulla oblongata. Its thin roof is composed of the highly vascular posterior choroid plexus. The third ventricle communicates with the fourth ventricle by means of a narrow passage called **iter** or **aqueduct of Sylvius** which extends throughout the length of the midbrain. Posteriorly, the fourth ventricle communicates with the central canal of the spinal cord.

Spinal cord. It is a hollow tube encased within the neural canal of the vertebral column. Extending backwards from the medulla oblongata, the spinal cord ends within the urostyle as a slender filament, the *filum terminale.* Two median grooves run along the whole length of the cord. The one running in the mid-ventral line is the *ventral fissure* and the other in the mid-dorsal line is the *dorsal fissure.* The cavity of the cord, the *central canal* or *neurocoel,* is continuous with the fourth ventricle of the brain. The spinal cord is covered with two protective meninges, outer *dura mater* and inner *pia mater,* which are continuous with those of the brain. Surrounding the central canal is a zone of nerve cells forming the *grey matter* and arranged like a butterfly with extended wings. The grey matter is enclosed in a cylinder of nerve fibres forming the *white matter.*

The Peripheral Nervous System

This includes the paired cranial and spinal nerves originating from the cerebrospinal axis. A **nerve** is a white thread-like structure composed of a bundle of microscopic *nerve fibres.* The nerve fibres originate from the cell bodies of the neurones which are situated within the grey matter of the central nervous system. Sometimes the cell bodies may aggregate to form a small swelling in the course of a nerve trunk. Such a swelling is known as a **ganglion.** The **nerve fibres,** are of two kinds: (1) *afferent* or *sensory fibres,* and (2) *efferent* or *motor fibres.* The sensory fibres convey messages from **receptor organs,** such as eyes, ears, nose, tongue, and skin, to the central nervous system. Motor fibres are those which carry impulses from the central nervous system to the **effector organs,** such as glands and voluntary muscles. A **sensory nerve** is composed entirely of sensory fibres, and similarly a **motor nerve** is composed of motor fibres. A nerve may contain both sensory and motor fibres and in that case it is known as a **mixed nerve.**

Cranial nerves. Ten pairs of cranial nerves originate from the brain. Some of the cranial nerves are entirely *sensory*, some are *motor* and some are *mixed*. The cranial nerves are numbered in sequence to their origin. They are known by their respective names as well as by number. The following is a detailed account of the *origin*, *course*, *distribution*, and *function* of the cranial nerves.

I. Olfactory nerve. The olfactory nerves originate from the olfactory lobes of the telencephalon. Extending forward into the nasal capsule, each nerve is distributed to special sensory cells on the mucous membrane of the nose of the same side. It is a purely *sensory nerve* concerned with the sense of smell.

II. Optic nerve. Arising from the optic lobes, the two optic nerves run across the ventral surface of the diencephalon in front of the pituitary body. At this point they form the X-shaped **optic chiasma** in which the two nerves cross each other. Thus the left optic nerve crosses over to the right side and *vice versa*. Emerging from the chiasma, the optic nerve enters the orbit and is distributed to the retina of the opposite eye. It is a purely *sensory nerve* concerned with vision.

III. Oculomotor nerve. The oculomotor nerves arise from the ventral surface of the mesencephalon. Proceeding forward, it enters the orbit and supplies four of the **extrinsic muscles** which move the eye, *namely*, superior rectus, inferior rectus, internal rectus and inferior oblique. As the name implies oculomotor is a purely *motor nerve*.

IV. Trochlear or **Pathetic nerve.** It is a slender *motor nerve* which arises from the dorsal surface of the mesencephalon just behind the optic lobe. It follows the course of the third nerve and innervates another extrinsic eye muscle, called superior oblique.

V. Trigeminal nerve. The trigeminal is a stout *mixed nerve* arising from the lateral side of the medulla oblongata. It bears a **Gasserian** or **pro-otic ganglion** near its origin and comes out of the skull through a small foramen in front of the auditory capsule. Outside the skull the nerve is characteristically divided into *three* main branches: the *opthalmic*, the *maxillary*, and the *mandibular nerves*.

The **opthalmic nerve** is the smallest of the three branches. It proceeds forward through the orbit and supplies *sensory fibres* to the skin round the anterior part of the head, the upper eyelid and the mucous membrane of the nose.

The **maxillary nerve** is the main branch to the upper jaw, supplying *sensory fibres* to the upper lip, lower eyelid and mucous membrane of the upper jaw.

The **mandibular nerve** curves downward and backward to the corner of the mouth. It now proceeds along the outer side of the lower jaw supplying *sensory* fibres to the skin and *motor* fibres to the muscles on the floor of the buccal cavity. Thus the mandibular branch is a *mixed* nerve.

VI. Abducens nerve. The abducens is a slender *motor nerve* arising from the ventral surface of the medulla oblongata. It runs through the orbit and supplies the external rectus muscle of the eye.

VII. Facial nerve. The facial is a *mixed nerve* arising from the medulla oblongata just behind the V nerve. It runs into the Gasserian ganglion and leaves the skull through the same foramen along with the trigeminal. Outside the skull the facial nerve divides into two branches: the *palatine*, and the *hyomandibular*.

The **palatine nerve** proceeds forward beneath the mucous membrane of the roof of the buccal cavity and is distributed mainly to the roof of the mouth. It is joined anteriorly by a small branch from the maxillary division of the V nerve.

The **hyomandibular nerve** turns outward and then proceeds forward along the floor of the buccal cavity. It receives an anastomosing branch from the glossopharyngeal nerve and then breaks up into branches which supply the tympanum, the mandibular joint, and the mucous membrane and muscles on the floor of the buccal cavity.

VIII. Auditory nerve. It is a short and stout nerve which arises from the lateral side of the medulla oblongata posterior to the facial nerve. It runs outward to supply the internal ear. The auditory is a purely *sensory nerve* concerned with hearing and balancing.

IX. Glossopharyngeal nerve. This is a *mixed nerve* originating from the medulla posterior to the auditory nerve. It immediately runs into the vagus ganglion at the base of the X nerve. Separating from the ganglion, the glossopharyngeal gives off an anastomosing twig which joins the hyomandibular branch of the facial. The main trunk of the nerve curves round the pharynx and then turns forward, pursuing a tortuous wavy course along the floor of the buccal cavity. It supplies sensory fibres to the floor of the buccal cavity and the tongue. Motor fibres innervate muscles.

X. Vagus or Pneumogastric nerve. The vagus is a large *mixed nerve* arising from the medulla oblongata by a number of roots. It bears a **vagus ganglion** near its origin. Coming out of the cranium through a small foramen in the ex-occipital bone, the vagus nerve gives off a few twigs to the pharyngeal wall. It then turns back and divides into *four* main branches: (i) **laryngeal** to the voice-box, (ii) **cardiac** to the heart, (iii) **pulmonary** to the lung of the same side, and (iv) **gastric** to the alimentary canal. Because of its long course, the vagus is recognised as the *wandering nerve*. The X nerve is perhaps the most important cranial nerve for it controls heart beat, respiratory movements and peristalsis of the alimentary canal.

Spinal nerves. There are ten pairs of spinal nerves. Each spinal nerve is connected with the spinal cord by two roots: a dorsal *sensory root*, and a ventral *motor root*. The dorsal root bears a *ganglion* near its origin. The two roots soon unite to form a common trunk, which comes out through the intervertebral foramen. Soft chalky masses, called *calcareous bodies*, surround the openings by which the spinal nerves come out. Coming out, each spinal nerve divides into three branches: (i) a small *dorsal branch* supplying the skin and muscles of the back; (ii) a large *ventral branch* forming the main trunk of the nerve and supplying the skin and muscles on the ventral surface of the body; in some spinal nerves, the ventral branch runs into the limb; (iii) a very short *ramus communicans* which at once joins the nearest sympathetic ganglion. All the spinal nerves are *mixed nerves*.

The first spinal nerve is the **hypoglossal nerve.** It leaves the vertebral column behind the atlas and accompanies the glosso-pharyngeal nerve to supply the muscles of the tongue.

The second and third spinal nerves join to form a network, called **brachial plexus.** From this arises the *brachial nerve* for supplying the forelimb of the side.

The fourth, fifth and sixth spinal nerves supply the skin and muscles of the trunk region.

The seventh, eighth, ninth and tenth spinal nerves join with one another to form the **sciatic plexus,** from which arises the large *sciatic nerve* for supplying the hindlimb. The tenth spinal nerve is very small and is often absent at one side.

The Autonomic Nervous System

The autonomic (*auto*=self; *nemo*=distributing) nervous system consists of two minute cords, the *sympathetic trunks*, one

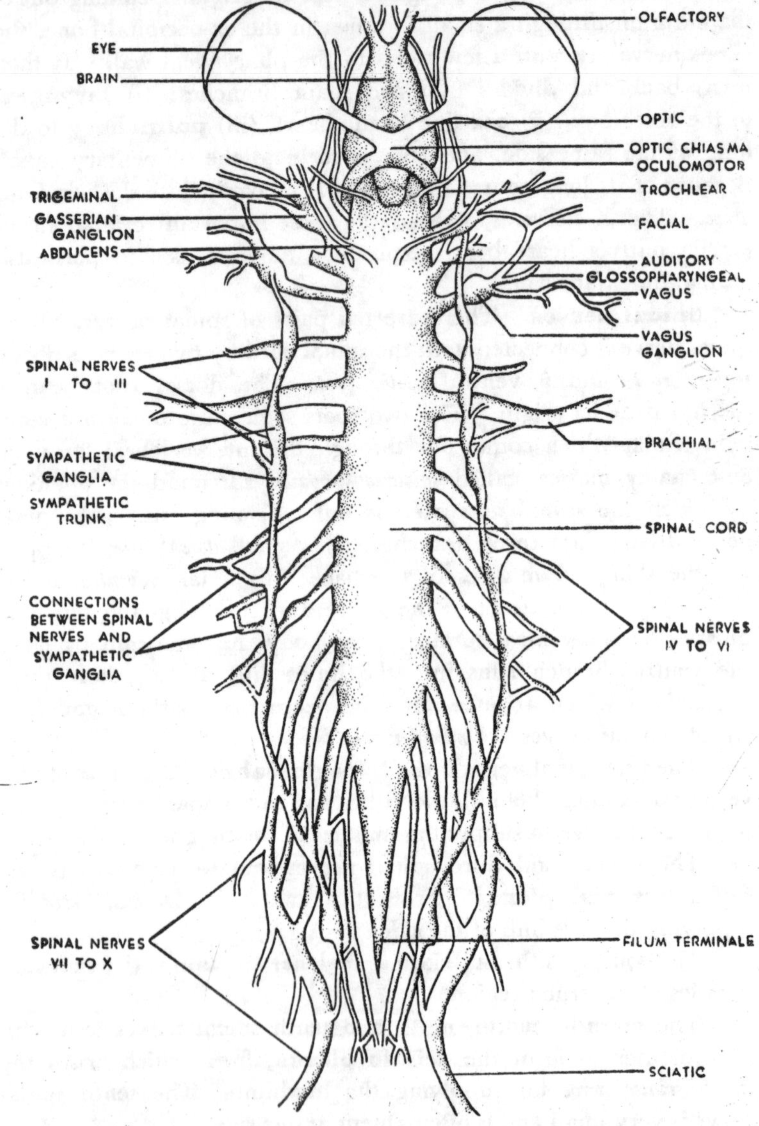

Fig. 40. Nervous system of toad (ventral view) (diagrammatic).

on each side of the dorsal aorta. The sympathetic trunks begin posteriorly near the point of bifurcation of the dorsal aorta. They diverge anteriorly to accompany the systemic arches. Further forward, each splits to encircle the subclavian artery of its own side. It now enters the cranium, supplies a communicating twig to the vagus ganglion, and ends by joining the Gasserian ganglion of the V nerve. Each sympathetic trunk bears ten small ganglia in its course and each sympathetic ganglion is connected to the corresponding spinal nerve by a short *ramus communicans*.

The sympathetic trunks send out slender branches for the innervation of involuntary muscles on the wall of heart, blood vessels, digestive tube, etc. These branches often unite to form ganglionated nerve plexuses. Thus there is a **cardiac plexus** supplying the heart, bearing Bidder's ganglion in the inter-auricular septum and Remak's ganglion on the sinus venosus. Similarly, there is a **solar plexus** surrounding the coeliacomesenteric artery, which innervates the stomach and the intestines.

The autonomic nervous system is mainly concerned with the regulation of *involuntary actions* such as beating of the heart, peristaltic movements of the alimentary canal, and the activities of the viscera in general. It is not completely independent of the central nervous system, because most of the sympathetic ganglia are connected with peripheral nerves. It is autonomic in the sense that its actions are not disturbed by the central nervous system.

Receptor Organs

Associated with the nervous system are certain organs of special senses called receptor organs. Each *receptor* is affected by a particular kind of *stimulus* which is transmitted by sensory nerves to the central nervous system and produces its own peculiar kind of *sensation*. The receptor organs themselves do not perceive anything but merely serve as a gateway to the central nervous system. The central nervous system is the seat of all sensation.

There are receptors for (1) touch, (2) pain, (3) heat, (4) cold, (5) smell, (6) taste, (7) sight, (8) hearing, and (9) balancing (equilibration). Our knowledge of the receptor organs of the toad is based upon a comprehensive knowledge of these structures in man.

The receptors for **touch, pain, heat,** and **cold** are scattered microscopic structures situated beneath the epidermis of the skin. They, therefore, are referred to as the *cutaneous receptors*. Some cutaneous receptors consist merely of free endings of sensory nerves in the deeper layers of the epidermis (Fig. 41, A). In others the nerve endings may be ensheathed in special connective tissue capsules of various shapes (Fig. 41, C).

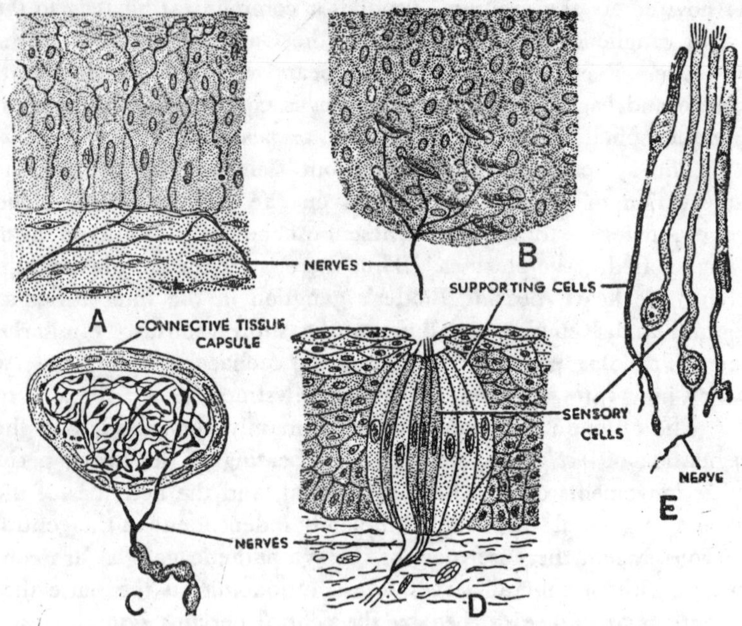

Fig. 41. Receptor organs.
(A–C=cutaneous receptors; D=taste bud; E=olfactory receptors).

Receptors for **smell** are located in the nasal passages which open into exterior by the external nares and communicate with the buccal cavity by the internal nares. Each nasal passage is lined by mucous membrane containing peculiar sensory cells called **olfactory cells** and slender **supporting cells** (Fig. 41, E). The olfactory cells are the actual receptors of smell. The free ends of these are brush-like in appearance, each cell bearing several delicate hairs at its tip. The other end of the olfactory cell is connected with a fibre of the olfactory nerve. The brush-like free ends of the olfactory cells are stimulated by chemical substances, and the impulse is conveyed to the brain via the olfactory nerve, but substances must be in solution before they can be detected.

Receptors for **taste** are microscopic structures distributed upon the mucous membrane of the mouth and tongue. These are called *taste buds.* A **taste bud** (Fig. 41, D) is composed of two kinds of cells: sensory cells or *taste cells* and *supporting cells.* Each taste cell bears a minute bristle at its tip, and its other end is connected with a sensory fibre of the glossopharyngeal nerve. Taste buds are particularly numerous on the tongue, in association with small elevations on the surface epithelium, the *papillae.*

The taste cells and the olfactory cells are recognised as **chemoreceptors** because they are stimulated by chemical substances in solution. Over and above the minute and scattered receptors just described there are two pairs of compact structures, the **eyes** for *vision,* and the **ears** for *hearing* and *balancing.*

Eye. The eye is the organ of vision. Each eyeball is lodged in its own orbit. It can be turned and rotated within limits by the action of *six* **extrinsic muscles** which are attached on its outer surface. These are: (1) *superior rectus,* (2) *inferior rectus,* (3) *external rectus,* (4) *internal rectus,* (5) *superior oblique,* and (6) *inferior oblique* muscles. The eye can be thrust out by a muscle called *levator bulbi* and withdrawn into the socket by another muscle called *retractor bulbi.*

The eye is a hollow spherical box fitted with a rounded lens in front and a light-sensitive screen behind. It is light-proof and works like a photographic camera. The eyeball is composed of *three* coats, which surround the transparent parts for transmitting light upon the sensory screen.

The outermost coat is thick and tough. It consists of two portions: (1) a transparent anterior portion, the **cornea,** and (2) an opaque posterior portion, the **sclera,** the anterior part of which is commonly referred to as the 'white' of the eye. The cornea is practically circular in outline; it permits entry of light into the eyeball. A thin membrane, the **conjunctiva,** which is continuous with the inner surface of the eyelids, covers the outer surface of the cornea. The sclera is composed of fibrous tissue and cartilage; it protects the soft parts and gives shape to the eyeball.

The middle coat is the **choroid.** It is a highly vascular black membrane forming an internal lining for the sclera. Anteriorly, the choroid turns sharply inward from the junction of the sclera with the cornea. This part of the choroid is modified to form a thin, circular, pigmented disc, the **iris.** There is an opening in the centre of the iris which is spoken of as the **pupil** of the eye. The iris contains radial and circular muscle fibres which help the pupil to dilate or contract. Thus, a regulated amount of light is permitted to enter the eye through the pupil. Barring this, the eyeball is perfectly light-proof.

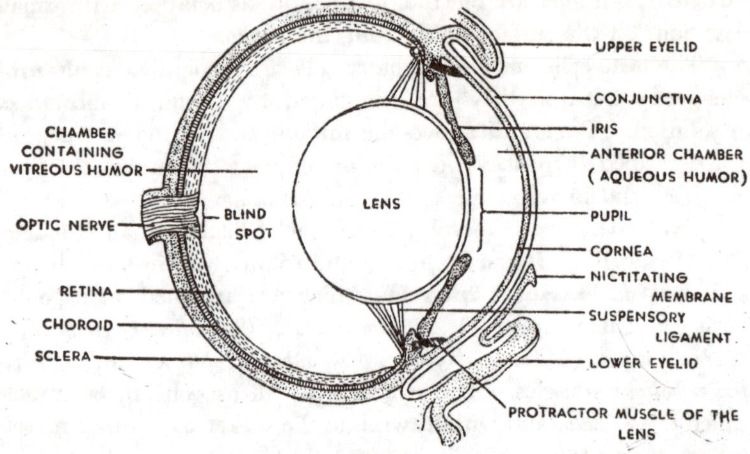

Fig. 42. Sectional view of the toad's eye.

The innermost coat is the **retina.** It lies against the choroid and forms the light-sensitive screen of the eye. The retina contains peculiar cells, called *rods* and *cones,* which are the receptors of the visual impulse. The rods are sensitive to dim light, and the cones to bright light and colours. The rods and cones are connected to the fibres of the optic nerve.

The crystalline **lens** is nearly rounded in shape. It lies just behind the pupil, and is held in position by the **suspensory ligament.** The cavity of the eye between the lens and the cornea is known as the **anterior chamber.** It is filled with a clear, watery fluid, the *aqueous humor.* The space between the black of the lens and the retina contains a transparent jelly-like substance, the *vitreous humor.* The toad cannot change the shape of its lens. Two small muscles, one dorsal and the other ventral, join the suspensory ligament of the lens with the inner surface of the cornea. These are the *protractor lentis* muscles; when contracted, they draw the lens closer to the cornea.

Vision. Light rays, entering the eye at the cornea, pass through the pupil, and being refracted by the crystalline lens, converge to a small part of the retina. A reduced and inverted image is caught on the retina in the same way as in a photographic camera. Focussing of the image, that is *accommodation*, is effected by drawing the lens forward or backward through the protractor lentis muscles. The retina contains rods and cones which are the actual *photoreceptors*. The reduced and inverted image caught by the rods and cones is transmitted to the brain via the optic nerves. During transmission. the inverted image is mysteriously transformed into a corrected one.

Toad's eyes are placed laterally and each eye covers a different visual field. Vision, therefore, is **monocular.** An animal which can focuss both eyes on the same object is said to enjoy **binocular vision.** Birds of prey, such as owls, have binocular vision which enables them to judge distance accurately when swooping down upon an object. Man too has binocular vision. Animals with laterally displaced eyes have monocular vision; they enjoy a wider visual field but are unable to judge distance.

Ear. The ear serves as the organ of hearing and balancing. The toad's ear is composed of three parts: the *external ear* represented by the *tympanum* or eardrum, the *middle ear* or tympanic cavity, and the *internal ear* including the membranous labyrinth.

The ear is represented **externally** by the **tympanum,** one on each side of the head. This is a tightly stretched membrane covering the middle ear.

The **middle ear** is the funnel-shaped cavity beneath the tympanum. It is connected with the buccal cavity by the **eustachian tube** which acts as a safety valve by equalising atmospheric pressure on the two sides of the eardrum. A small rod-shaped bone, the **columella,** extends across the middle ear, from the centre of the eardrum. It is joined to a membranous partition separating the middle ear from the internal ear.

The **internal ear,** including the **membranous, labyrinth,** is enclosed within the auditory capsule of the skull. The labyrinth floats in a fluid called **perilymph** which fills the cavity of the auditory capsule. The labyrinth itself is filled with another fluid called **endolymph.** The auditory capsule is sealed off from the middle ear by a membranous partition which is connected to the eardrum by the columella. The **membranous labyrinth** consists mainly of two chambers, one placed above the other. The upper chamber is known as the **utriculus,** and the lower chamber as the **sacculus.** The two chambers are connected by a narrow canal. The utriculus gives out three narrow tubes, the **semicircular canals,** all opening into the utriculus at both ends. Each canal bears an enlargement or **ampulla** at only one end. One of the canal is horizontal in position, but the other two are vertical and at right angles to one another. Thus the semicircular canals lie in three different planes. The sacculus is produced into a short **lagena.**

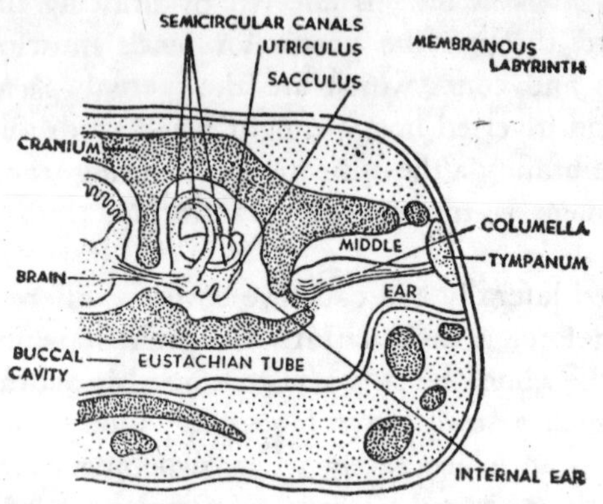

Fig. 43. Diagram of the ear.

There are patches of *sensory receptors* in the inner wall of the membranous labyrinth. Each patch consists of *supporting cells,* and *sensory cells* with long hair-like bristles extending into the endolymph. The receptors are found within the ampullae of the semicircular canals, and within the utriculus, sacculus as well as lagena. Each receptor cell is connected with a fibre from the auditory nerve.

Hearing. Sound waves impinge directly upon the eardrum. Vibrations of the eardrum are conveyed to the internal ear by the columella. These vibrations are transmitted to the endolymph and stimulate the sensory cells. The receptors receive the impulses by their bristles and transmit the same to the brain via the auditory nerves. The impulses are translated in the brain as sound. Sound impressions are recorded by the sensory receptors within the sacculus and lagena.

Balancing. The semicircular canals are responsible for the maintenance of equilibrium. There are calcareous particles within the canals which impinge upon the bristles of the sensory receptors as soon as the animal loses its balance. Moreover, being arranged in three different planes and at right angles to one another, the canals can detect changes in the direction of the centre of gravity, when the toad is moving.

Endocrine System

The endocrine (*endos*=within; *krine*=separate) system includes a number of **ductless glands** which secrete chemical substances called **hormones** (*hormao*=excite) directly into the blood stream. They are thus distinguished from the *exocrine glands*, such as liver, which pour out their secretions externally through ducts.

The hormones are carried by the circulating blood to every part of the body. They control the function of distant organs and regulate the activities of the animal as a whole. Moreover, any single hormone does not usually function alone; it either stimulates or depresses the production of some other hormone. Thus, the endocrine glands act as a team which is responsible for the **chemical co-ordination** of the vital processes. It is for this reason that the endocrine glands, though small in size and scattered in distant parts of the body, are recognised as an organ system. The endocrine system shares with the nervous system the important task of co-ordinating the parts of the body. The nervous system exerts its control through nerves which are directly connected with all the body structures. The endocrine system does the same job by a more simple method; its means of control is purely chemical, the hormones being carried to distant parts of the body by the blood stream.

The various glands included under this system are: (1) thyroid, (2) parathyroid, (3) thymus, (4) pituitary, (5) adrenal, (6) pancreas, (7) gonads, and (8) spleen. The pancreas plays a dual part by giving off both exocrine and endocrine secretions. Similarly, the gonads manufacture germ cells and at the same time release sex hormones. The spleen, although it is a ductless gland, has no endocrine function.

Thyroid glands. These are a pair of small oval structures located on the floor of the buccal cavity, one on each side of the hyoid apparatus. The glands are deeply concealed and often overlooked. They produce a hormone, called **thyroxine,** which regulates general metabolism in the adult and metamorphosis in the tadpole. When small tadpoles are fed with thyroid gland substance, they stop growing, undergo an early change, and are quickly transformed into tiny toads. Ecdysis, or periodic shedding of the horny layer of the skin, is also regulated, at least partly, by the activity of the thyroid gland.

Parathyroid glands. Parathyroids (*para*=near) are yellowish, ovoid bodies lying near the thyroid, upon the external jugular veins. They regulate the amount of calcium salts in the blood stream.

Thymus glands. These are small yellowish structures situated on either side of the head just behind the tympanum. Functioning in young tadpoles, the thymus regulates growth and retards premature formation of germ cells. The glands are lost by atrophy in adult toads. Nowadays, the thymus is not regarded as an endocrine organ.

Pituitary body. This is a glandular structure attached to the hypophysis and located on the floor of the diencephalon. The gland has two lobes: an *anterior lobe* and a *posterior lobe.*

The posterior lobe of the pituitary body secretes a hormone called **pituitrin** which regulates the *intake of water by the skin.* Considering the fact that a toad hardly drinks water by the mouth, this is perhaps the only means for the intake of water. Moreover, pituitrin enables the toad to change the colour of its skin by shifting the position of the chromatophores.

The anterior lobe of the pituitary secretes a number of hormones. In tadpoles and young toads, it secretes a **growth-stimulating hormone** and a **thyroid-stimulating hormone.** It has been proved conclusively that the anterior pituitary, in collaboration with the thyroid, regulates the process of metamorphosis by which a tadpole is changed into a toad. In adult toads, the anterior lobe secretes: (1) a **gonad-stimulating hormone** which hastens the liberation of germ cells, (2) a metabolic hormone which stimulates metabolism in general, and (3) several other hormones for regulating the activity of other endocrine glands. It is for this reason that the anterior pituitary is regarded as the captain of the endocrine team.

Adrenal glands. These are narrow orange-coloured bands on the ventral surface of the kidneys. The outer portion or **cortex** of the gland produces a hormone called **cortin** which is essential for life. Removal of both adrenals is followed by the death of the animal in the course of a week.

The inner portion or **medulla** of the adrenal secretes another hormone called **adrenaline** which regulates the activity of the involuntary muscles by stimulating the autonomic nervous system. It controls blood pressure and stimulates metabolism in general.

Pancreas. Scattered within the substance of the pancreas are special groups of cells called **islets of Langerhans.** The cells of the islets are quite different from the ordinary pancreatic cells which manufacture digestive enzymes. Moreover, they are not connected with the pancreatic duct in any way. The secretion of the islets, the **insulin,** is poured directly into the blood stream.

Insulin regulates carbohydrate metabolism and favours storage of glycogen in the liver.

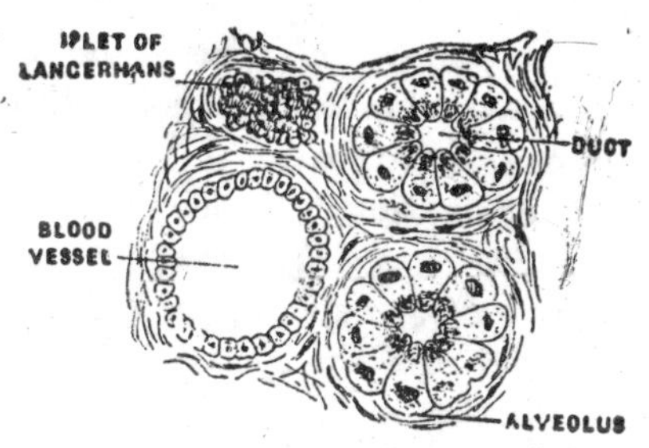

Fig. 44. Cross-section of pancreas showing an islet of Langerhans.

Gonads. The testes and the ovaries, in addition to manufacturing germ cells, produce *sex-hormones* which regulate the development of the *secondary sexual characters.* The male sex-hormone is known as the **testosterone.** Swelling of the thumb pad, inflation of the vocal sac, and the clasping habit of the male toad are regulated by the male sex-hormone.

The endocrine secretion of the ovaries is known as the **oestradiol.** It controls the behaviour of the female toads at the time of breeding. In the male toads, a peculiar structure called **Bidder's organ** is located at the anterior end of each testis. It has been proved, that if the testes are removed, Bidder's organs will soon develop into functional ovaries. A complete reversal of sex may thus be induced, the transformed Bidder's organs helping the process by secreting female sex-hormone.

Spleen. The spleen is the only ductless gland which does not secrete any hormone. It is a small dark-red structure of spherical shape which is situated in the mesentery near the junction of the ileum and the rectum. The spleen serves as a storehouse for blood and manufactures one kind of leucocytes. Moreover, old and worn-out erythrocytes are destroyed in the spleen.

Life-history

Breeding. Toads breed during the rainy season. Males come by the side of a suitable pond or ditch and begin to croak loudly by inflating their vocal sac. Invited by the song, the females come to the water in a large number. This is followed by *mating* or **sexual reproduction.** A male clasps a female, his grip on her being tightened by the thumb pads which are particularly well-formed during the breeding season. This state may continue for two or three days. The female now lays her *eggs* or **ova** in shallow water and the male pours his *milt* or **spermatozoa** over the eggs as they are extruded.

The eggs look like small black spots placed upon white strings of jelly, thus forming a convoluted ribbon which floats freely on the surface of the water. This is the toad **spawn.** The spermatozoa swim actively by lashing their tail and a swarm of them encircles every egg on the spawn.

Fertilization. Each egg is *fertilized* by a sperm, the process taking place *externally*, that is outside the body, in water. After the entry of one sperm, the outer membrane of the egg becomes impervious to other sperms. The sperm which has succeeded to enter the egg loses its tail; its head is now transformed into the **male pronucleus.** In the mean time, the nucleus of the egg changes into the **female pronucleus.** In **fertilization** the male and female pro-nuclei fuse to form a single nucleus. The fertilized egg is known as the **oosperm** or **zygote.** It is a single cell which is destined to develop into a toad.

Under abnormal circumstances, unfertilized eggs may develop by *parthenogenesis* and behave like fertilized eggs.

Embryonic development. Zygotes are left in water without and parental care. Each zygote undergoes a period of rest and then begins to develop by repeated cell division. The first **cleavage** results in two cells or **blastomeres.** A large number of blastomeres are rapidly formed, all enclosed within the outer membrane of the egg. The blastomeres arrange themselves into groups in a complicated way and eventually produce a young **embryo** with three **germinal layers:** (1) an outer **ectoderm,** (2) an inner **endoderm,** and (3) a **mesoderm** in between the two. The formation and fate of the germinal layers will be fully considered in Chapter XIX.

The embryonic development goes on for about two weeks. At the end of this period a small **embryo** is seen to wriggle within the egg. *An embryo may be defined as the developing young of an animal inside the egg.* During the course of its development the embryo draws its nourishment from the yolk, which is stored in the egg for this purpose. Eventually the embryo hatches out by rupturing the egg membrane.

Larval stage. The freshly hatched young tadpole bears no resemblance to its parent. It has an ovoid head, a short trunk, and a slender compressed tail with a small vent near its root. There is no limb, but there is an adhesive **sucker** on the ventral side, by which it attached to some submerged object. As there is no mouth it cannot ingest any food and depends on the residue of the yolk. Three pairs of branched processes, called **external gills,** develop on the sides of the head. These are highly vascular folds of skin which serve as the *first set of respiratory organs.*

After resting for a few days a **mouth** is formed near the sucker and a pair of **horny jaws** encircle the mouth. The tail grows longer and develops a dorsal and a ventral fold. It is thus converted into a **fin.** Moreover, V-shaped bands of muscles appear

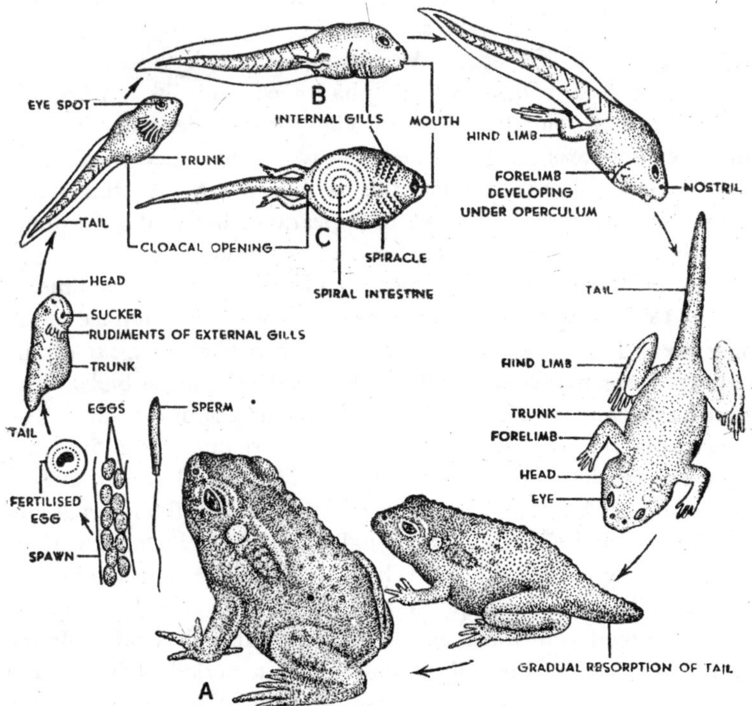

Fig. 45. Life-history of toad from eggs to adult. A=adult; B=older tadpole (side view); C=central view of dissected B showing internal gills and spirally coiled intestine.

on both sides of the tail which are now used for active swimming. The free-swimming young scrapes algae with its jaws and ingests small bits of water-weeds through its mouth. Its alimentary canal grows enormously long, and to accommodate the same within the shor trunk, the intestine is coiled spirally like the spring of a watch.

Evidently the newly-hatched young, though self-supporting does not resemble the toad either in form or in habits. It is known as the **tadpole larva.** *A larva is defined as the self-supporting immature young of an animal which bears no resemblance to the adult.*

At a later stage, the pharynx becomes perforated at the sides by **gill-slits. Internal gills** are now formed in between the gill-slits. The internal gills are vascular outgrowths from the pharyngeal wall. A fold of skin called **operculum** appears and covers up the gills and gill-slits. The operculum or *gill-cover* fuses with the trunk ventrally and on the right side, leaving a small opening, called **spiracle,** on the left side. Water enters through the mouth and while passing out through the pharyngeal gill-slits washes the internal gills and finally escapes through the spiracle. The internal gills are the *second set of respiratory organs* of the tadpole and with their formation the external gills wither away. At this stage, the tadpole is not only fish-like in appearance but it also resembles a fish in the manner of locomotion and respiration.

When the internal gills have been functioning for a time, a pair of **lungs** appear as outgrowths from the ventral surface of the pharynx. The lungs are the *third set of respiratory organs* of the tadpole. Meanwhile, the limbs have been developing as buds. The **hindlimbs** appear, one on either side of the root of the tail. The **forelimbs** remain under cover of the operculum for a time and then emerge by bursting through it. The tadpole now comes to live in shallow water. It can still breathe through the internal gills and comes to the surface occasionally for taking in a gulp of air into its newly-formed lungs. When the lungs are fully formed the internal gills rapidly disappear and the tadpole now looks like a miniature toad with limbs and everything but with a tail. As the limbs grow in size the animal enters into a period of fasting and the tail is gradually absorbed into its body, the material of the tail being used as the source of nutrition. The young toad now leaves its watery home and hops on the land. Finally the mouth widens and the horny jaws are replaced by true bony jaws. The animal changes its food and becomes a carnivore; consequently its gut becomes short and more or less straight. The development is completed in about twelve weeks.

Metamorphosis. The young tadpole larva which hatches out from the toad's egg hardly resembles its parents. It is more like a fish than like a toad and lives a free-swimming independent life.

Though immature, it can take care of itself and procure its own food. The term metamorphosis is applied to the series of changes which a larva has to undergo for altering its structure and mode of life before it comes to resemble its parents. During metamorphosis of the tadpole, fish-like characters are given up and toad-like characters are taken on. The larva is thoroughly changed in its mode of respiration, locomotion, and nutrition. The metamorphosis occurs *rapidly* towards the end of the tadpole's aquatic life. There is, however no abrupt transition from one stage to another.

Metamorphosis is hastened if tadpoles are fed artificially with the substance of the thyroid gland. In the American bull-frog the normal larval period is about two years; experimental feeding with thyroid reduces this long larval stage to a very short period. On the other hand, removal of thyroid from a tadpole completely stops metamorphosis although the larva may grow and attain a huge size. Anterior lobe of the pituitary body plays an important part in the metamorphosis by stimulating the thyroid gland.

General notes on toad. The young toad is a *terrestrial* creature. It feeds voraciously and continues to grow. Since toads destroy insects, they are *beneficial to man.* The animal is habitually shy but when attacked by a snake it puffs up and assumes an aggressive posture. This curious habit is a means for its *protection*, but it seldom succeeds to escape in this way. The animal requires a damp environment to keep its skin moist for the purpose of cutaneous respiration. It always keeps close to a collection of water because of its peculiar breeding habit. The toad is a cold-blooded or **poikilothermal** animal; its body temperature changes with that of the environment. In this respect, toads differ from birds and mammals which are warm-blooded or **homoiothermal** animals having a constant body temperature irrespective of the environment. The cold-blooded condition has a profound influence over its activities and behaviours. Summer warmth increases the animal's metabolism and its feeds, grows and reproduces its kind. The cold winter days, on the other hand, are spent in **hibernation.** During this period, metabolism drops down to the minimum level; the animal is sleepy and torpid. The heart-beat slows down and its lungs being inactive, respiration occurs mainly through the skin.

Young toads attain **sexual maturity** in two years. They are now fit to reproduce their own kind. A female lays thousands of eggs during a single spell of mating and the milt poured by the male contains even more spermatozoa. Had all these eggs been fertilised and the resulting tadpoles developed into toads the whole world would have been full of only these creatures. Some of the eggs fail to develop and many are eaten by enemies. Larvae are devoured by fishes and tortoises. The adults are taken by snakes and herons. The few that survive are sufficient for perpetuating the race.

Though fertilisation is the rule yet eggs may develop into tadpoles without being fertilised by a male sperm. This peculiar mode of sexual reproduction is known as **parthenogenesis.** Tadpoles resulting from parthenogenetic eggs have been observed to grow and reach the adult stage.

There is a controversy about the longevity of the toad. It has been said to live for more than thirty years but there is no definite proof to support this statement. With the progress of years the metabolic activity gradually becomes slow. The body of the animal wastes steadily due to failure of co-ordinated action on the part of the organ systems. At last comes death which ends all.

CHAPTER III

PROTOZOA : THE LOWLIEST OF THE LOWLY

We have learned the details of the morphology, physiology, habitat and life-history of a higher animal. As we are already acquainted with the principal phyla of the animal kingdom, it is better, now, to start from the lowest group—the **protozoa** (*protos*=first; *zoon*=animal).

The protozoa are the simplest animals and are usually microscopic in size. A protozoan body is regarded as a single cell but functionally it is an entire individual, performing all the vital activities which characterise a living animal. It is for this reason that they are regarded as **acellular** animals rather than as **unicellular** ones. Moreover, there may be more than one nucleus in the protozoan body, but none of them is in sole charge of a particular part of the animal. In the cellular animals (**metazoa**), each nucleus is in absolute charge of a particular part of the body, known as the cell. In the acellular protozoa, the body is not subdivided into cells, and the portions of protoplasm set apart for carrying out special functions are regarded as **organelles.** At present about 30,000 species of protozoa are known to us. They occur throughout the earth, in fresh and salt water, in damp places, and in decaying organic matter. Many are **free-living** but a few occur in **colonies.** Some protozoa live as *unwanted guests* or **parasites** within the body of other animals; they often cause serious diseases, such as malaria, kala-azar, sleeping sickness, dysentery, etc., in their *host.*

Classification of Protozoa

The phylum protozoa has been divided into four **classes.** The classification is based principally on their mode of locomotion.

Class I. Rhizopoda (*Rhiza*=root; *podus*=foot). Protozoa having peculiar *temporary* organelles for locomotion, called **pseudopodia** or false feet. Examples—*Amoeba proteus; Entamoeba histolytica* (causing dysentery).

Class II. Mastigophora (*Mastix*=whip; *phoros*=bearer) or Flagellata (*flagellum*=whip). Protozoa that move by the lashing of whip-like organelles called **flagella.** Example—*Euglena viridis;*

Trypanosoma gambiensi (causing African sleeping-sickness); *Leishmania donovani* (causing kala-azar).

Class III. SPOROZOA (*Spora*=seed; *Zoon*=animal). Parasitic protozoa possessing no locomotor organelles and reproducing by means of spores. Example—*Monocystis gregarina; Plasmodium vivax* (causing malaria).

Class IV. CILIOPHORA (*Cilium*=eyelash). Protozoa that move by hair-like **cilia.** Example—*Paramoecium caudatum* (the slipper-animalcule); *Vorticella campanula* (the bell-animalcule).

CLASS RHIZOPODA : AMOEBA PROTEUS

Habitat

Amoeba proteus lives on the bottom of fresh water ponds, streams and ditches, gliding on the algae-covered mud or crawling on the surface of green submerged plants.

Morphology

The name of this animal is derived from two Greek words—*Amoeba* meaning change, the *proteus* meaning a mythological sea-god who had the power of constantly changing his shape.

To the naked eye the Amoeba is just visible as a minute speck of chalk particle. A large specimen, when fully expanded, measures about 0·01 inch (0·25 mm.) in length. Under the microscope, the Amoeba appears to be an irregular mass of active jelly which is continually changing its shape by thrusting out and withdrawing blunt finger-like projections called **pseudopodia** (*pseudo*=false; *podus*=foot).

The protoplasm of the body consists of the following parts: (*i*) A very thin and elastic limiting membrane called **plasmalemma** which encloses the entire surface of the animal. (*ii*) A narrow outer zone of clear transparent **ectoplasm** which is totally free from granules and which lies just beneath the plasmalemma. The plasmalemma may be regarded as the outermost layer of the ectoplasm. (*iii*) Surrounded by the ectoplasm is the main mass of protoplasm called **endoplasm.** The endoplasm is granular and translucent. It consists of an outer stiff *plasmagel*, and an inner fluid *plasmasol* exhibiting streaming movements. The protoplasm varies in consistency according to the state of activity of the animal. Plasmagel may change into plasmasol and even ectoplasm may change into endoplasm. (*iv*) Lying within the endoplasm there is a disc-like **nucleus** which is not easily seen in a living Amoeba. In a killed and stained specimen the nucleus is distinct. It is surrounded by a thin *nuclear membrane* and contains masses of *chromatin* granules which are suspended in the nuclear sap. (*v*) A spherical vesicle, the

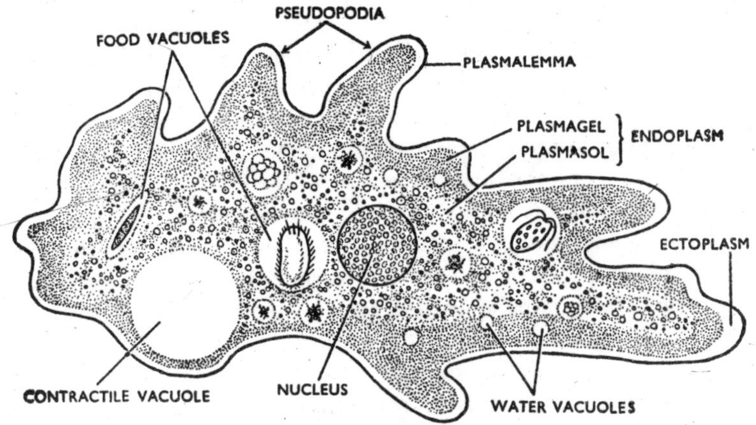

Fig. 46. *Amoeba proteus*.

contractile vacuole, is found within the endoplasm. It is filled with water and bubbles of gas. On being carefully observed for a time the contractile vacuole seems to expand slowly and then moving to the surface it suddenly bursts, discharging the contents into the surrounding water. A new contractile vacuole then re-appears in its original position. (*vi*) Lying scattered in the endoplasm there are several **food vacuoles.** These are of various sizes and contain morsels of food undergoing intracellular digestion. (*vii*) Spherical **water vacuoles** are also seen scattered throughout the endoplasm and appearing as transparent drops of water. They never change their shape like the contractile vacuole. (*viii*) **Cytoplasmic inclusions** such as mineral crystals, oil globules and granules of various sizes lie scattered in the endoplasm. (*ix*) Blunt finger-like projections or **pseudopodia** are protruded out from the surface of the body and drawn into it when not in use. Pseudopodia, therefore, are mere temporary structures.

The function of the various structures, just described, may briefly be stated as follows: (*i*) The plasmalemma retains the protoplasm within the cell body. It is permeable to water, oxygen and carbon dioxide. (*ii*) The ectoplasm *protects* the more soft inner parts. (*iii*) The endoplasm takes important part in movement by forming the pseudopodia; it is the structure which includes the nucleus and the vacuoles. (*iv*) The nucleus regulates all the vital activities of the animalcule. It is particularly concerned with reproduction. (*v*) The contractile vacuole is the organelle for respiration, excretion and fluid-regulation. (*vi*) The food vacuoles serve as temporary stomachs where food is digested. (*vii*) Water vacuoles along with the contractile vacuoles are responsible for fluid regulation. (*viii*) The various inclusions are either reserve food materials for future use or excretory matter in the process of elimination. (*ix*) The pseudopodia are used for locomotion and for capture of their prey. They are very sensitive and help the animal to avoid injurious objects.

It will not be out of place to mention here the relative importance of the nucleus and the cytoplasm. It is possible by micro-dissection to take out the nucleus from the cytoplasm. It has been observed in such experiments that both the parts perish. If on the other hand, an Amoeba is cut into two pieces, one-half containing only cytoplasm and the other containing the nucleus as well as a portion of the cytoplasm, then both the halves survive for a time. The part composed of only cytoplasm is soon exhausted and dies; but the other part actively feeds, grows and leads a healthy life. This proves that although the nucleus governs the various vital activities of the cell body yet the nucleus and cytoplasm must function together.

Vital Activities

All the vital activities that characterise a living animal may be observed in Amoeba on a miniature scale. It moves about in search of food which is captured, ingested and digested. It absorbs and assimilates the digested food and egests undigested particles. It secretes useful substances and excretes waste products. It breathes, repairs itself, grows, reproduces its own kind and responds adequately to changes in its environment. An Amoeba is immortal and has no natural death.

Locomotion. The Amoeba moves from place to place by forming temporary finger-like projections called pseudopodia. A pseudopodium may form at any place upon the surface of the body. It appears as a blunt projection of ectoplasm. The sheet of plasmagel at the root of the ectoplasmic protrusion slowly thins out and is converted into plasmasol. The central granular part of the endoplasm, that is plasmasol, than flows into the ectoplasmic outgrowth by streaming movement through the gap in the plasmagel sheet. Next to this, the stiff plasmagel at the opposite end of the animal is gradually converted into fluid plasmasol which rushes forward into the pseudopodium. While returning back along the sides of the finger-like tube, the plasmasol is again converted into plasmagel. Many pseudopodia appear simultaneously but usually one of them is extended and the others are retracted. This sort of peculiar motion by thrusting out pseudopodia and then flowing into one of them is known as **amoeboid movement.** Effective locomotion occurs only when the animal is in contact with the substratum and not when it is floating on the surface of the water. The tip of the pseudopodium may adhere to the substratum by secreting a sticky juice and the entire animal glides forward into the pseudopodium. The speed of locomotion usually ranges from 0·5 to 5 micra per second. In favourable condition the speed may be as high as 1 inch (25 mm.) per hour. It varies with the temperature, increasing slowly up to a temperature of 30°C. Movement totally ceases at 33°C.

Nutrition. The **food** of Amoeba consists of small algae and other actively moving protozoa. The animal seems to have a choice in the selection of its food. It is always attracted by the movement of its prey. Its dishes include Desmids, Oscillatoria, Paramoecium, Colpidium, and a small flagellate called *Chilomonas.* Food may be **ingested** anywhere on the surface of the body but it is usually taken in at the anterior end. The Amoeba sends out pseudopodia which engulf the prey by forming a *food-cup.* When the prey is very swift, a large food-cup is formed and several victims are thus enclosed in a single cup without being touched. Gradually the food is completely surrounded on all sides along with a drop of water. The result is a *food vacuole* whose walls are formed by the plasmamembrane and which contains the food suspended in water. The entire process of food-taking requires only a minute or two if the food is a favourite one. Chilomonas is ingested more rapidly than any of the others and the next choice is Colpidium. Feeding occurs only when the animal is attached to a solid object and the rate of feeding is best at 25°C. The amoeba cannot feed when it is floating. When the temperature of the surrounding water is raised to 38°C or above, food-taking totally stops. The **food vacuole** serves as a temporary stomach secreting digestive juice. The digestive juice contains free hydrochloric acid and appropriate enzymes. Later on, the reaction of the digestive juice becomes alkaline. Chilomonas and Colpidium remain alive in a food vacuole for about 5 to 15 minutes

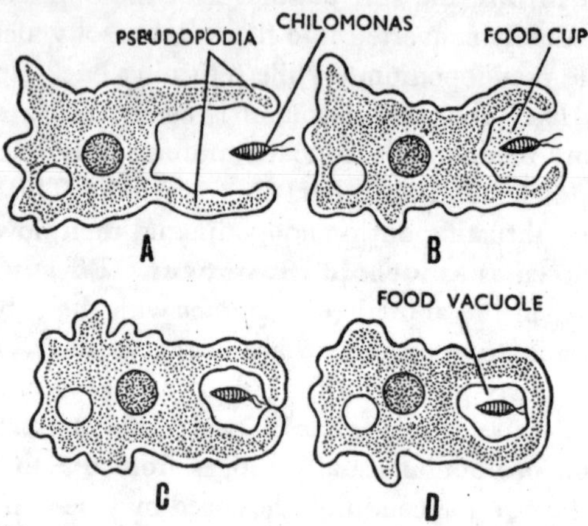

Fig. 47. Amoeba running after and engulfing a Chilomonas.

and are usually **digested** in the course of a day. The body of the prey swells up and gradually diminishes in size as the various enzymes break down the insoluble proteins, fats and carbohydrates to soluble forms. These are now **absorbed** into the cytoplasm and **assimilated.** The assimilated food is spent in providing energy for locomotion and the surplus is stored for future use. The food vacuole gradually decreases in size with the progress of digestion, and at the end only indigestible residue is left behind. This is now **egested** out of the body. Indigestible particles are usually heavier than the protoplasm of Amoeba. As such they lag behind as the animal moves forward and finally pass out from any point on the surface of the body near the posterior end. Thus in Amoeba, digestion takes place *intracellularly*, that is within the cell. As solid food is ingested, the mode of nutrition is *holozoic.*

Excretion. The Amoeba forms nitrogenous waste products such as ammonia and urea, and other mineral substances in the course of its metabolism. A portion of these useless substances along with the excess of water in the body gradually accumulate in the contractile vacuole. The latter dilates and ultimately bursts like a bubble, liberating its contents into the surrounding water. A new contractile vacuole soon appears in its original position. Major part of the waste products are, however, eliminated through the permeable plasmalemma from the surface of the body.

Respiration. The Amoeba requires oxygen for the physiological burning of stored food, so that it may derive energy for its locomotion. The oxygen necessary for this purpose is procured from the dissolved air in the surrounding water. The oxygen diffuses into the cytoplasm through the plasmalemma. The resulting CO_2 collects as bubbles in the contractile vacuole and is expelled from the body when the latter bursts into the exterior. A part of the CO_2, however, diffuses out through the plasmalemma. Thus, in Amoeba, the mode of respiration is **aerobic.**

Behaviour. Any change in the environment acts as a *stimulus* and produces a response or *reaction* in Amoeba. As the animal has no nervous system these reactions are entirely due to the inherent irritability of its protoplasm.

The Amoeba reacts *negatively*, that is, it moves away from harmful stimuli like injurious chemicals, pricking with a needle, strong beam or light, increased temperature above 30°C, and lowered temperature nearing the freezing point. Beneficial stimuli, on the other hand, produce *positive* reactions and the animal moves towards the source of such stimuli.

Response to light (*phototaxis*) is negative when the rays are very strong; but the animal may react positively to very weak light. The Amoeba moves away and repeatedly changes its course when strong light is flashed on it from different directions.

Response to contact (*thigmotaxis*) is variable. When touched or pricked by a needle, a creeping Amoeba reacts negatively by withdrawing its pseudopodia, and then moving away in a different direction. A floating Amoeba, on the other hand, adheres to a solid object when it is lightly touched by the same, and the reaction, therefore, is positive. If its forward progress is hampered by an obstacle, it is able to avoid the same by a process of repeated 'trial and error'.

Response to changes in the temperature of the medium (*thermotaxis*) has already been mentioned in connection with amoeboid movement. The speed of locomotion increases steadily with a rise in temperature up to 30°C, but at 33°C locomotion abruptly ceases. Similarly, there is no locomotion when the temperature is brought down to the freezing point.

Response to chemical changes in the medium (chemotaxis) depends upon the nature of the stimulus. The animal reacts negatively and moves away from substances such as acetic acid, common salt, and cane sugar. There is, however, a positive response to weak chemicals diffusing from the surface of suitable food substances. These are quickly engulfed by the pseudopodia, and contact with the food is sometimes unnecessary. It is in this way that the Amoeba can discriminate a Chilomonas from a Colpidium and select its own food.

Sharp tapping of the coverslip produces rapid vibration of the medium, and to this the animal responds by withdrawing its pseudopodia (negative reaction). Weaker vibrations, such as those produced by a running ciliate, elicit formation of pseudopodia on the stimulated side (positive reaction). This is followed by the formation of a wide food-cup which is suitable for encircling a moving prey.

When Amoeba is subjected to weak electric current, the animal moves towards the cathode, and all pseudopodia are withdrawn from the side of the anode. Stronger currents produce cessation of movement, and ultimate disintegration of the body.

Thus the Amoeba *behaves* exactly like other higher animals and reacts to states of pleasure, pain and hunger.

Reproduction. Amoeba reproduces *asexually* by two different methods: (1) binary fission, and (2) multiple fission.

Binary fission is the usual mode of reproduction by which a full-grown *parent* individual divides into two *daughter* Amoebae. During binary fission, the Amoeba becomes spherical and is covered with small pseudopodia (Fig. 48 A). Its nucleus swells up to divide by a special process of mitosis. The nucleus seems to elongate then becomes dumb-bell-shaped (Fig. 48 B, C.). Finally it divides into two daughter nuclei. Meanwhile the body of the animal narrows down in the middle. Lastly, this constriction cuts through completely and the two halves separate, each containing a nucleus of its own (Fig. 48 D).

Usually binary fission is completed within half an hour at 24°C. If the temperature, is raised to 30°C the entire process is finished in 21 minutes. Fission is greatly retarded by low temperature, and the time required is slightly more than 4 hours at 4°C.

Each amoebula feeds voraciously and grows to normal size within three days, when it again becomes ready for binary fission. As the animal reproduces by fission, there is no death from old age and the Amoeba, therefore, is potentially immortal. But the amoebae, like all other living organisms, are liable to death by accident or other means.

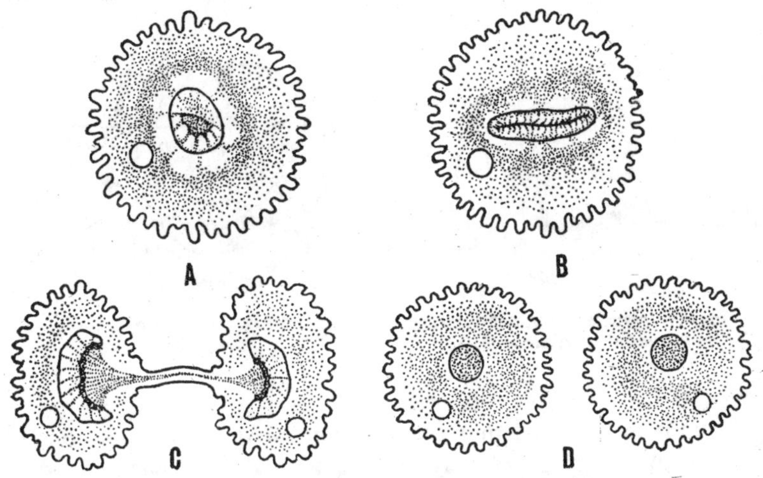

Fig. 48. Amoeba undergoing binary fission.

When water and food become scarce or when unfavourable environment threatens with death, *Amoeba* reproduces by **multiple fission** or **sporulation.** This, therefore, is a means for tiding over a critical period in its life-history, and not a normal method of reproduction. **Sporulation** is preceded by **encystment.** The Amoeba retracts all pseudopodia and becomes spherical. The plasmalemma secretes a *three-layered* **cyst** within which the spherical body is safely housed. The cyst is very light and may be carried by wind to a favourable environment. In any case, the tough and impervious wall of the cyst affords shelter and protection till the return of normal condition. In the mean time, the nucleus of the encysted individual undergoes rapid division to produce a large number of daughter nuclei. Later on, the cytoplasm divides, and a fragment of cytoplasm surrounds each daughter nucleus. A large number of **pseudopodiospores** are formed in this way. With the return of favourable conditions, the cyst wall breaks and the pseudopodiospores come out by extending pseudopodia. Each pseudopodiospore is thus converted into a small amoebula. In this manner, a large number of amoebulae are formed out of a single encysted parent individual. They feed actively and are soon converted into adult amoebae which reproduce by binary fission.

Encystment followed by sporulation have been described in other free-living Amoebae, but this process of reproduction has not yet been confirmed in *A. proteus*.

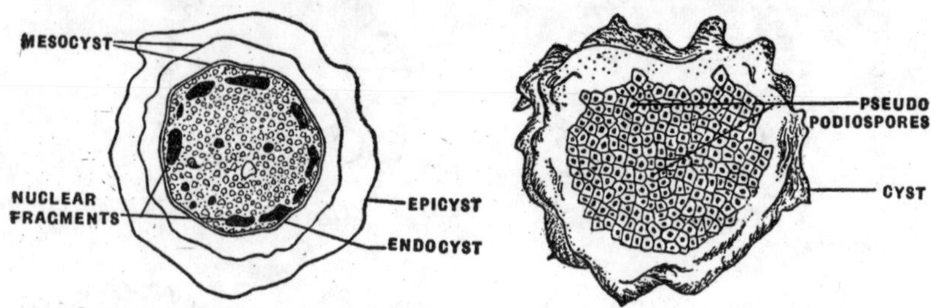

Fig. 49. Nuclear fragmentation in an encysted amoeba.

Fig. 50. Formation of pseudopodiospores.

It is to be noted that reproduction in Amoeba is not effected by the union of germ cells or *gametes*, and only one individual can reproduce its own kind. Amoeba, therefore, is said to reproduce asexually.

Animals closely resembling *Amoeba proteus* live as parasites in the alimentary canal of man and other higher animals. They are included in the genus *Entamoeba.* They feed on the nutritive materials in the gut of their host and occasionally invade and destroy tissues. Thus, *E. histolytica* causes amoebic dysentery and liver abscess in man.

ENTAMOEBA

Habitat

Entamoeba histolytica is a parasite and lives in the mucous and submucous layers of the large intestine of man. It may occur in the liver and lungs. Rarely it invades brain, spleen etc. causing ulcers. The disease caused by the parasite is known as "Amoebic dysentry."

Morphology

It occurs in two forms—Trophic and Cystic.

Trophic. The trophic forms vary in size from 15-40 micra average being 25 micra. The cell body is divisible into two distinct portions—**ectoplasm** and **endoplasm.** The ectoplasm is clear and translucent while the endoplasm is granular. The endoplasm often contains ingested red blood corpuscles. The pseudopodia may be long, finger like or rounded in shape. In freshly passed stool the parasite shows a single pseudopodium (Fig. 50A).

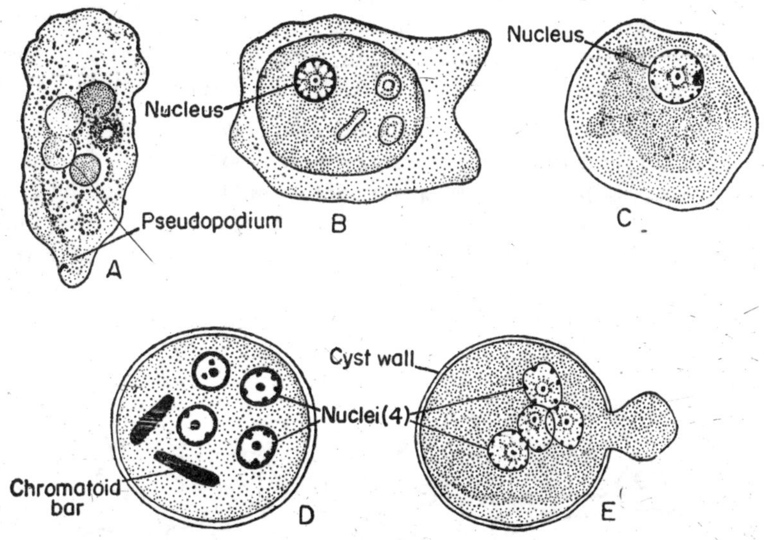

Fig. 50A. *Entamoeba histolylica*—Trophozoite—living (A) and Stained (B), Precystic stage, (C), Cyst (D), Emergence (E).

The nucleus is indistinct in living condition but in stained preparation it shows a central **Karyosome** or **Endosome** and a uniform ring of peripheral chromatid granules.The nuclear membrane is very delicate. The nucleus is 4-6 micra in diameter.

Cyst. Cysts vary in diameter from 10-20 micra. The cysts are spherical. The cyst wall is double and the cytoplasm usually bears four nuclei. The cytoplasm is clear and often contains black rod-like **chromatoid bar** or bodies. Presence of these bodies is the characteristic of *E. histolytica* cyst. They occur singly or in multiples of two.

Nutrition

The parasite gets its nutrition from the red blood cells and tissue cells of their host. They produce a toxic substance which dissolve the mucous and submucous layers of the large intestine. From the resultant absess it derives its nutrition. This is evident from the fact that the stool of persons with amoebic dysentry contain pure blood and mucus.

Life-cycle

Entamoeba histolytica multiplies by **binary fission** in the trophic stage. They have the capacity to encyst. Prior to encystment the parasite rounds up and eliminates food vacuoles. A cyst wall develops and the nucleus divides first into 2 and then into 4.

The quadrinucleate cyst is the infective stage. It comes out with the feces of the host. Through contaminated food or drink the infective cysts pass into the lower portion of the small intestine of the new host. There the cyst wall dissolves by the action of the intestinal enzymes and a 4 nucleate amoeba emerges. Division of both the cytoplasm and the nuclei of the amoeba results 8 small amoebulae. They become active and transform into trophic form (Fig. 50B).

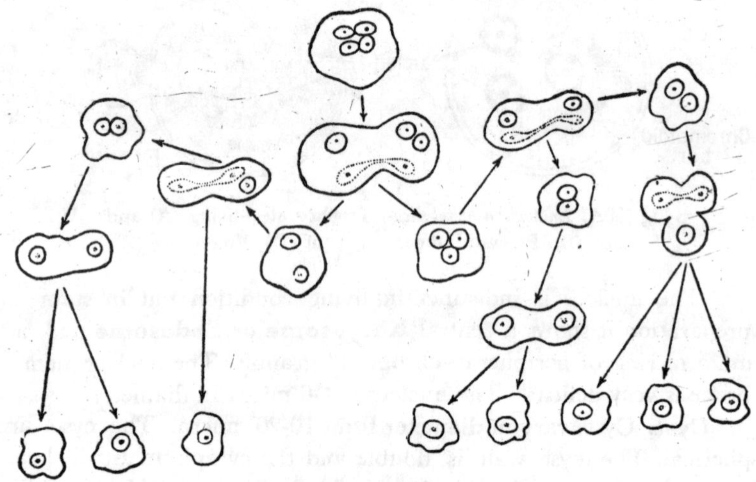

Fig. 50B. Transformation of a quadrinucleate cyst into eight uninucleate trophozoite

CLASS MASTIGOPHORA : TRYPANOSOMA

Habitat

Trypanosoma gambiense is a haemoflagellate parasite. It resides in the *blood plasma* of man. Often they are found in the cerebro-spinal fluid, spleen and lymph nodes. In its vector host Tse-tse fly, *Glossina palpalis* it is found in the intestine. The disease caused by Trypanosoma in man is known as Trypanosomiasis or Sleeping sickness. This fatal disease is most prevalent in West and Central Africa.

Morphology

The Trypanosomes show polymorphism presenting different morphological forms under different conditions: In the blood

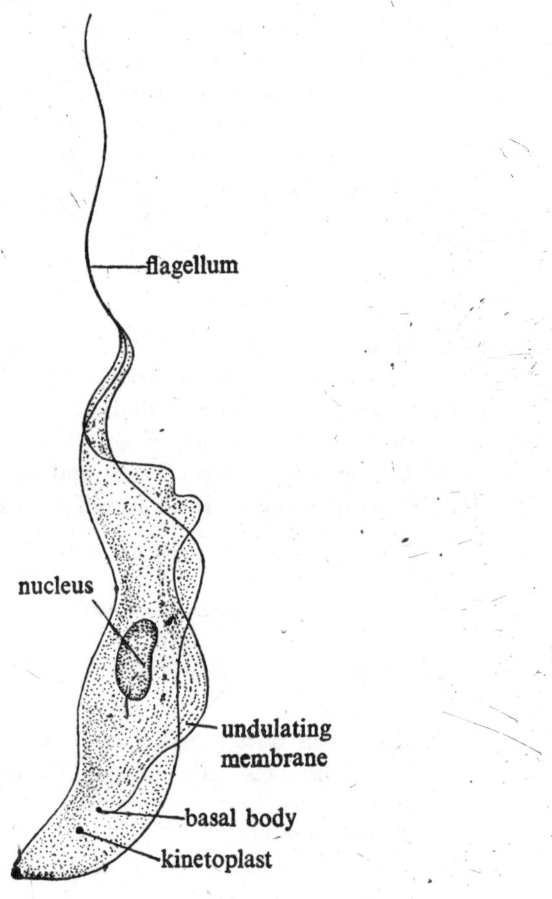

Fig. 50C. Morphology of Trypanosoma.

plasma of man is found the Trypanosome form which looks like a thin, flattened and leaf-like body tapering at both ends. The pointed end is considered as anterior end and the blunt end is considered as posterior end. The length of the body is 15-32 micra and the breadth in the middle is about 2·5 micra. The body is covered by a thin, elastic and firm **pellicle.** The pellicular part to which the flagellum is attached draws out as **undulating membrane** during flagellar movement (Fig. 50C). A whip-like long flagellum arise from a basal body (also called blepharoplast or centriole) located at the posterior end. The flagellum skirts the whole body and projects out of the anterior end. Just posterior to the basal body there lies a small, spherical or disc-shaped **Kinetoplast.** The **nucleus** lies almost in the centre of the cell body as a large, oval, vesicular body. The nucleolus is large. Some vacuoles and volutine granules are present in the cytoplasm

Nutrition

Nutrition is saprozoic. It feeds on the blood and tissue fluids of the host. It is believed that feeding occurs by pinocytosis.

Reproduction

Trypanosoma divides by **longitudinal binary fission.** The basal body takes the lead during division. It is followed by nuclear division. The cytoplasm divides longitudinally. The old flagellum is retained by one of the daughter cells and a fresh one is formed for the other individual. There is no record of sexual reproduction in Trypanosoma.

Life-cycle

The life-cycle of Trypanosoma needs two hosts—man and Tset-tse fly for completion. Man is the primary host and the fly is intermediate host or vector. (Fig. 50D.)

Life-cycle in man

When the vector bites a man it inoculates few parasites along with its salivary secretions. The entry of the parasites in the blood stream of man results Trypanosoma fever. Fever and headache reccur at irregular intervals.

After their entry into the blood stream the parasites attain the characteristic **long slender forms** and undergo multiplication by longitudinal binary fission.

After multiplying a number of times the parasites assume the characteristic **short stumpy form** and lose the free part of the flagellum during the process. The stumpy forms do not feed and if not sucked up by the fly, they die.

Some of the parasites shorten their body length and flagellum and become the characteristic **intermediate form.** The intermediate forms go to lymph gland or spleen and spend their **period of incubation** there. The period lasts for about three months and multiplication occurs during this period. Now they go to the cerebrospinal fluid.

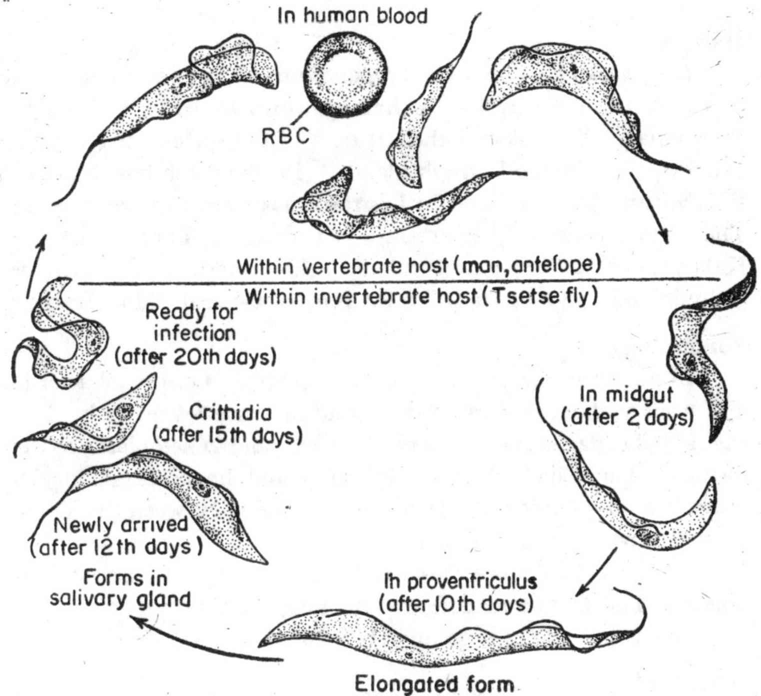

Fig. 50D. Life-cycle of Trypanosoma.
Note the change of shape in invertebrate host.

Life-cycle in fly

When the fly bites an infected man some **stumpy forms** are sucked up. The stumpy forms reach the midgut of the fly and undergo rapid multiplication. The resultant parasites assume the shape of **slender forms.** These forms reach the salivary glands of the fly *via* the proventriculus. They get attached to the wall of the gland and undergo a phase of multiplication resulting **crithidial stage.** Crithidial forms are characterised by shorter flagellum and undulating membrane. The crithidia form undergo multiplication and produce the **metacyclic trypanosomes** which infect man along with the bite of the fly.

The whole life-cycle (both in man and fly) takes 20-30 days for completion.

LEISHMANIA

Habitat

Leishmania donovani like Trypanosoma is a haeomoflagellate parasite. In man it resides in **Leishmanial form** in lymphoid—macrophage (Reticuloendothelial) cells of the spleen, liver, bone-marrow, intestine and lymph glands. In its vector host Sand fly, **Phlebotomous** it is found in **Leptomonad form** in the intestine. The disease caused by leishmania is known as Leishmaniasis or Kala-azar. The disease is prevalent in the Eastern hemisphere specially in India, Southern U.S.S.R., Burma, Central China, Iraq etc.

Morphology

In man the intracellular parasite occurs as Leishmanial forms. These are oval or spherical in shape and measure 2-4 micra by 1·5-2 micra. The flagallum is absent. Nucleus and kenetoplast are well formed. The kinetoplast is elongated and lie perpendicular or tangential to the nucleus. An axoneme ranging between the kinetoplast and periplast may remain present..

In the intermediate host the parasite occurs as Leptomonad forms. The leptomonad forms are long, slender measuring 7-20 micron. It has a long flagellum (Fig. 50E).

Life-cycle

In man the intracellular parasites undergo repeated binary fission resulting over crowdedness inside the host cell. The host cell ultimately bursts and parasites are released.

The fly ingests Leishmanias along with the blood of man. The parasites after reaching the midgut of the fly increase in size, develop flagellum and becomes metamorphosed into leptomonad form. The leptomonads multiply by longitudinal binary fission and reach the proventriculus of the fly. Repeated multiplication inside the proventriculus causes complete obstruction of the organ. In this condition when the fly tries to ingest fresh blood, the meal can go no further than the oesophagus. This causes a regurgitation of the sucked blood during bite. Along with the regurgitated blood leptomonad forms are introduced in the blood stream of man.

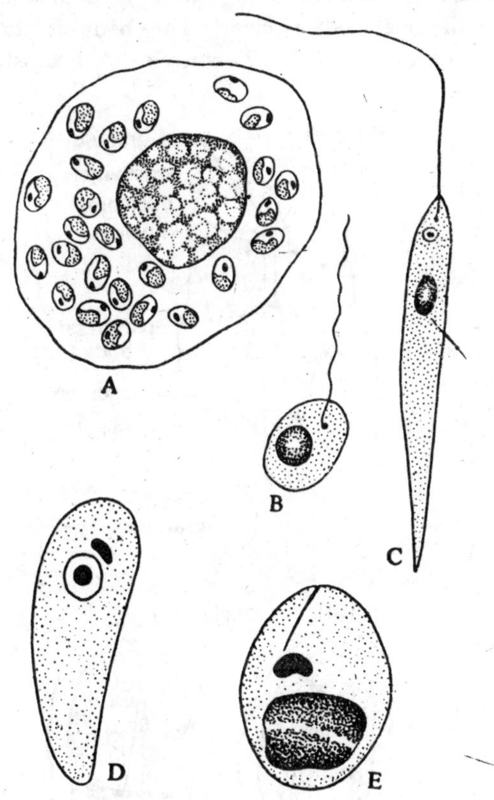

Fig. 50E. *Leishmania donovani* in WBC (A), Flagellate forms (B, C), Two different forms of *L. tropica* (D, E).

GIARDIA

Habitat

Giardia lamblia (also known as *G. intestinalis*) lives as a flagellate parasite in the small intestine of man. It causes a disease called *giardiasis* which causes digestive disturbances. Heavy infection interfere with normal absorption as the flagellates adhere to the mucosa.

Morphology

Giardia presents both trophic and cystic forms. **Trophozoite.** The trophozoites measure 9-20 micra by 6-20 micra. The protoplasm of the oval body is clear. The body is flattened dorsoventrally. The dorsal surface is convex. A bean-shaped sucking

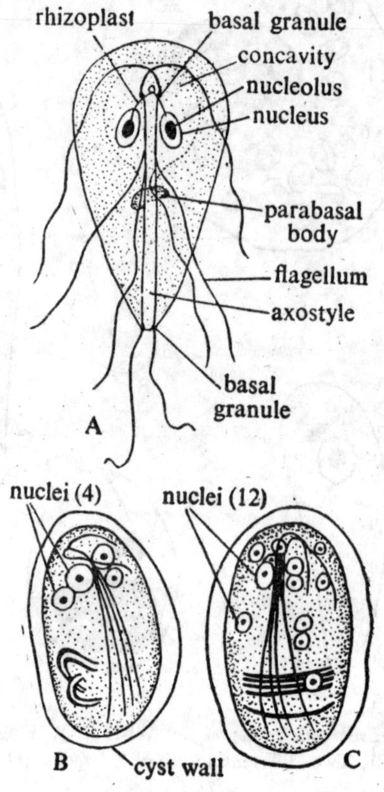

Fig. 50F. *Giardia lamblia* —Dorsal view of Trophozoite (A), Cyst with 4 nuclei (B), with 12 nuclei (C).

disc is present on the ventral surface of the broad or anterior end. Two flagella emerge from the posterior end while three other pairs of flagella extend from the lateral and anterolateral surfaces. There are two nuclei with prominent nucleoli. The axostyles are paired.

Two parabasal bodies, sometimes fused, lie close to the axostyle and in the posterior third of the body. **Cyst.** The cysts measure 8-14 micron by 6-10 micron. The cysts contain 4-16 nuclei, axostyles, parabasal bodies and fibril (Fig. 50F).

Reproduction

The trophozoites transform to cysts and reproduction within the cyst occurs by binary fission.

Transmission.

Cysts come out with stool. Infection occurs through contaminated food or drink.

CLASS SPOROZOA : MONOCYSTIS

Habitat

Monocystis is a **parasite** and belongs to the class sporozoa. It lives *within* the seminal vesicles of the common earthworm, which, therefore, is its **host.** As it lives *within* the body of the host, it is an **endoparasite.** Monocystis causes injury by destroying the sperms which are formed and nourished in the seminal vesicles of its host. But the amount of injury seems to be very slight, since the earthworm thrives quite well in spite of heavy infection. Almost every earthworm dissected in the class-room is found to be infected with Monocystis.

Morphology of the Adult Individual

In the full-grown stage the parasite is known as the **trophozoite.** Its body is elongated and more or less fusiform in shape. It is composed of a *nucleus* lying within a mass of *cytoplasm*, th whole thing surrounded by a thick *cuticle*.

The **nucleus** is spherical in shape. It is surrounded by a delicate nuclear membrane and is filled with nuclear sap. The chromatin is aggregated into a solid mass called endosome which is either attached to a side of the nuclear membrane or suspended in the nuclear sap.

The **cytoplasm** is divided into a clear stiff outer portion or **ectoplasm,** and a granular soft inner portion or **endoplasm.** In the deeper layers of the ectoplasm there are a number of fine contractile threads called **myonemes.** The endoplasm includes reserve food materials, chiefly glycogen, paraglycogen and droplets of fat. Vacuoles of any kind are totally absent.

The **cuticle** surrounding the cell body is thick, porous and permeable.

The animal is now in its *feeding* or **trophic** stage and hence the name **trophozoite.**

Nutrition

The adult trophozoite is suspended in the nutrient fluid meant for the nourishment of the earthworm's spermatozoa. It is surrounded by *soluble food* which is absorbed into the endoplasm by slow diffusion through its porous cuticle. The young trophozoite lives within the sperm mother cell of the earthworm. It secretes digestive enzymes from its outer surface which dissolve the protoplasm of the sperm mother cell. The digested product is then absorbed by diffusion through the cuticle. The parasite requires no food vacuole, because it either gets digested food or digests food outside its own body.

As the parasite absorbs soluble food through the outer surface of its body, the mode of nutrition is *saprozoic*.

Respiration

The parasite gets a poor supply of oxygen within the seminal vesicles of its host. Respiration, therefore, is mainly **anaerobic.** Release of energy occurs by breaking down of glycogen and paraglycogen molecules into lactic acid with the help of enzymes. Carbon dioxide, formed as by-product, passes out by diffusion through the porous cuticle.

Excretion

Being slow and inactive, the animal produces very little nitrogenous waste products. This is excreted in solution through its cuticle. This explains the absence of contractile vacuole in Monocystis.

Locomotion

Movement is sluggish because the animal has no necessity to run after its food. It is capable of slow wriggling movements by bending the cell body to one side or the other. This is due to rhythmical contractions of the myonemes in its ectoplasm. It can also glide along by producing series of wave-like contractions of the myonemes, which consist of both longitudinal and circular fibrils. Expansion and contraction of the cell body is effected by using the two sets alternately.

Reproduction and Life-history

Monocystis reproduces *sexually* and the process of reproduction is complicated. Two mature trophozoites associate with each other for this purpose. They are called **gametocytes** or **gamonts.** They grow shorter and secrete round them a two-layered cyst called **gametocyst.** The individuals, thus enclosed, are identical and they never fuse with one another. Within the gametocyst, each gametocyte undergoes *multiple fission.* Its nucleus divides rapidly to produce a number of daughter nuclei which migrate towards the periphery. Each bit of the nucleus is surrounded by a fragment of cytoplasm, and each of these nucleated fragment is a **gamete.** The gametes adhere to a central lump of undivided cytoplasm called **residual body.** Usually a large number of gametes are produced by each gametocyte and these are nourished by the residual body which is full of reserve food. Fully formed gametes are finally set free within the gametocyst. They are more or less rounded or oval bodies which now fuse in pairs. The gametes produced by the two gametocytes are exactly similar and are, therefore, **isogametes.** But during fusion, the two in each couple are derived from different parents. The fusion of the gametes is complete, involving both nucleus and cytoplasm. The product is a larger cell called **zygote** or sporoblast which clothes itself with a resistant boat-shaped covering called **sporocyst.** The sporocyst is also known as the *pseudonavicella*, because it appears to resemble a unicellular plant called

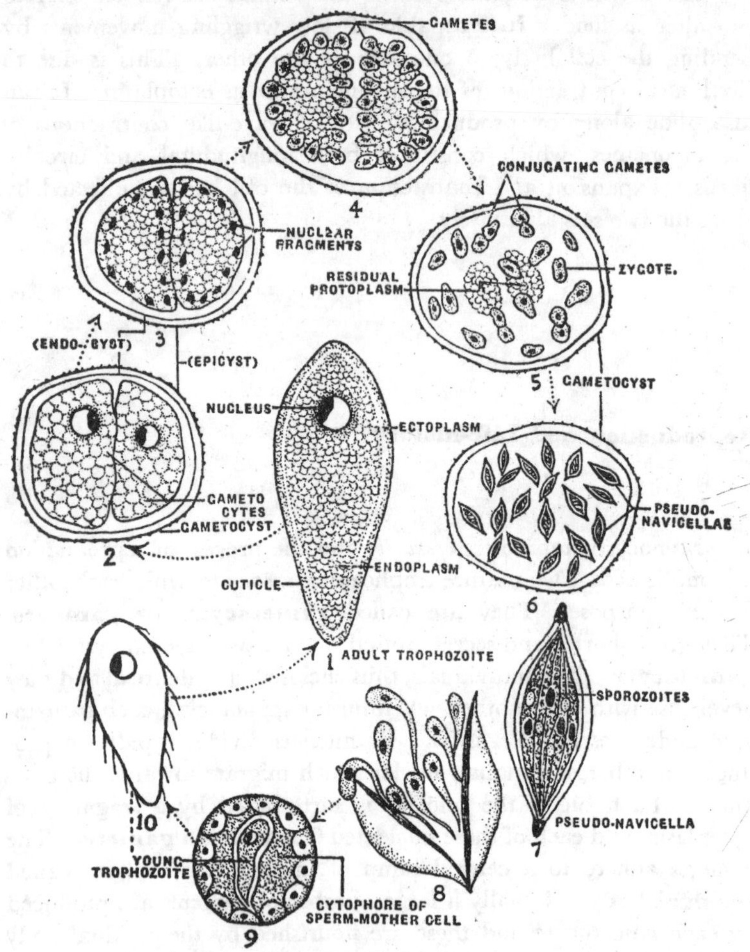

Fig. 51. Life-history of *Monocystis*.

1—Adult trophozoite. 2—Two gametocytes within a gametocyst. 3-4—Formation of gametes. 5—Fusion of gametes to form zygotes. 6—Pseudonavicellae within gametocyst. 7—Formation of sporozoites. 8—Bursting of pseudonavicella to liberate eight sporozoites. 9—A young trophozoite as intracellular parasite within a sperm-mother cell of its host. 10—A young trophozoite with tails of spermatozoa attached to its outer surface.

Navicella. Enclosed within the pseudonavicella the zygote undergoes three successive divisions, producing eight sickle-shaped falciform youngs called **sporozoites.** At this stage the original gametocyst is loaded with a large number of pseudonavicellae, each containing eight sporozoites.

Further development cannot take place until the sporozoites are transferred to a new host. The actual method of transmission is not known. The earthworm may die and disintegrate, liberating the sporocysts in the soil; or the worm may be devoured and digested by a bird, and in this case the sporocysts, being resistant, escape digestion and pass with the bird's faeces into the soil. By chance, a sporocyst may be eaten along with soil by another earthworm. In the alimentary canal of the new host the sporocyst dissolves, liberating eight falciform sporozoites. The sporozoites bore through the wall of the gut and ultimately find their way into the seminal vesicles. Each sporozoite now enters a sperm-mother cell and becomes an *intracellular parasite.* It grows rapidly by digesting and absorbing the cytoplasm of the sperm-mother cell and is converted into a young trophozoite. The young trophozoite digests the heads of the spermatozoa which the sperm-mother cell is producing. Consequently the slender wavy tails of the sperms remain attached to its outer surface and the young animal appears to be covered by cilia. In course of time it is converted into a full-grown adult trophozoite and the cycle, just described, is repeated.

There are *two* important features in the life-history of Monocystis which require special consideration.

Firstly, the remarkable simplicity of its structure and the easy-going life of the adult trophozoite; and *secondly,* the complex nature of its mode of reproduction. Both are associated with the parasitic mode of life.

Compared with a *free-living Amoeba,* the trophozoite of *Monocystis* appears to be a degenerated individual. The Amoeba, possesses several organelles such as food vacuoles for digestion, contractile vacuole for excretion and pseudopodia for active locomotion. None of these are found in Monocystis; because, in the seminal vesicle of its host, ready-made food is abundant and competition is absent. Consequently structures to carry on normal activities are unnecessary.

This comfortable life has at least one drawback. The parasite must be transferred from host to host for perpetuation of its race. This involves extra risks, for the animal has to pass through the outside world which is full of danger. Further, actual transmission depends upon mere chance, because a sporocyst may never be eaten by a new host. This explains the complexity of its reproductive method. The parasite, reproducing *sexually,* releases a large number of resistant sporocyst. Most of the sporocysts may perish but the few that survive will perpetuate the race.

Life-cycle

There are two generations in the life-cycle of Monocystis. The zygote or sporoblast represents the **diploid generation** because it possesses a double set of chromosomes represented by number as '2n'. The zygote divides thrice to produce eight sporozoites. As one of these divisions is *reduction division*, each sporozoite contains a reduced or haploid number of chromosomes represented by number as 'n'. Sporozoites are converted into trophozoites and the latter into gametocytes. The gametocytes produce gametes by multiple fission. All these forms represent the **haploid generation.** With the fusion of gametes, that is, sexual reproduction, the *diploid* or double number of chromosomes (represented by '2n') is restored. **Thus, in Monocystis the zygote alone is *diploid*** and all the other stages of the **life-cycle are *haploid* as in higher plants.**

PLASMODIUM VIVAX

Plasmodium vivax causes benign tertian malaria, so called, because the paroxysm of fever occurs every other day and the disease is rarely fatal. *Plasmodium* is an endoparasite which lives within the red blood corpuscles of man to whom it is transmitted by the bite of an infected female mosquito (*Anopheles sp.*). Thus the parasite has *two hosts*—man and *Anopheles*—and there is an alternation of the hosts in its life-cycle. Laveran discovered *P. vivax* in 1890 in the blood of man but its connection with the insect-host and its mode of transmission was workedout in Calcutta by Ross in 1899. *Plasmodium* is the most important sporozoon which infects man. It causes wide-spread havoc in tropical and subtropical regions, especially West Bengal, Bangla Desh, Behar and Assam. There are three species of malarial parasite—*P. vivax*, *P. malariae* and *P. falciparum.* Their life-cycles closely resemble one another but the symptoms which they produce vary. Thus *P. vivax* is responsible for benign tertian malaria, the meaning of which has already been explained. *P. malariae* causes quartan malaria, so called, because the outbreak of fever sets in every fourth day. *P. falciparum* produces malignant malaria, in which the fever is irregular, and the disease proves fatal unless the victim is promptly treated by antimalarial drugs. In untreated cases the parasites choke up the capillaries of the brain.

Life-history

There are three main phases or cycles in the life-history of *Plasmodium:* (1) an *exo-erythrocytic phase* of growth and asexual multiplication within the liver cells of man lasting for ten to fourteen days; (2) a phase of growth and asexual reproduction by multiple fission (*schizogony*) in the red blood cells of man which is completed in about forty-eight hours; (3) a sexual phase which starts in man and is continued within the stomach of the mosquito and is followed by asexual reproduction by multiple fission (*sporogony*)—the cycle being completed in about two weeks.

Life-cycle in Man. The parasite is transmitted to man by the female *Anopheles.* The male insect does not bite and therefore plays no part in transmission. Infection starts by the injection into the blood stream of the victim of a few spindle-shaped individuals along with the saliva of the infected female mosquito. The parasite, known at this stage as sporozoites, circulates in the blood stream for about half an hour.

At the end of this period they disappear from the blood and enter the liver cells. Here each sporozoite grows rapidly and then divides by multiple fission to produce about one thousand small individuals called *cryptozoites.* The **cryptozoites** are released into the liver sinusoids by rupture of the liver cell. This phase of rapid growth and reproduction in the human liver has been discovered only recently and is known as the **pre-erythrocytic** or **exo-erythrocytic phase.** It is completed in about ten to fourteen days, during which the man usually does not suffer from fever. It is, therefore, known as the **incubation period** of the disease. The exo-erythrocytic phase may continue and the liver may thus be converted into a reservoir of parasites which can prolong the disease in a *latent form* for indefinite period.

From the liver sinusoids the parasites pass either into the general circulation or re-infect fresh liver cells. Those which pass in the general blood stream attack and penetrate into red blood corpuscles where each is converted into a **trophozoite.** The young trophozoite is ring-like in shape. It grows at the expense of the erythrocyte, absorbing fluid nutrition by the surface of its body, and ingesting haemoglobin particles by throwing out pseudopodia like an Amoeba. The ingested haemoglobin is converted into a brown pigment called melanin which is deposited in the cytoplasm of the trophozoite. In about 48 hours the trophozoite is full-grown. It is now called a **schizont,** which soon reproduces asexually by **schizogony** and divides into about twenty-four **merozoites.** The infected red blood corpuscle now bursts, liberating the merozoites. Each escaping merozoite attacks a fresh red cell of the host and is again converted into a young trophozoite. Soon after it has settled down in the red cell, a vacuole appears in its centre, pushing the small nucleus to one side. The parasite now looks like a **signet ring.** This ring-form is a resting stage of the parasite, because in its active state it is amoeboid and constantly changes shape by throwing out pseudopodia. Most of the merozoites become schizonts, which repeat the sexual cycle; a few, however, develops into

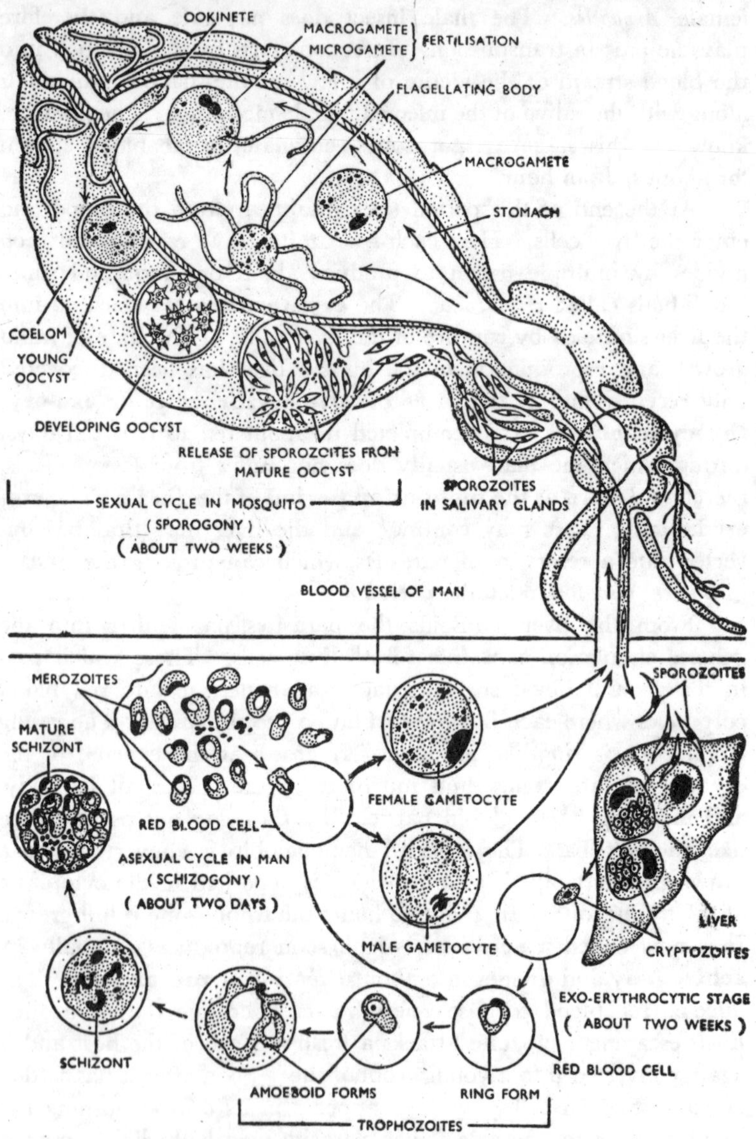

Fig. 52. Life-history of *Plasmodium vivax*.

male and female **gametocytes** which remain latent till an *Anopheles* sucks them up along with the blood of the host. Unless this occurs within a reasonable time the gametocytes degenerate and perish. Formation of gametocytes marks the onset of the **sexual phase** in the life-history of the parasite. A merozoite, which is destined to produce a gametocyte, rounds up and grows slowly within a red blood corpuscle. It is converted into a compact spherical body without a vacuole and is loaded with pigment granules. In *P. falciparum* the gametocytes appear to be crescentic in outline.

The male gametocyte is smaller than the female gametocyte.

Life-cycle in Mosquito. When the female *Anopheles* sucks blood from a malaria-victim, the **gametocytes** continue further development in her stomach while the **schizonts** are destroyed and digested. The male gametocyte is converted into what is called a **flagellating body.** Its nucleus breaks up into a number of chromatin granules each of which collects a long whip-like mass of cytoplasm and projects out from the surface. The flagella-like structures are the actual **microgametes,** and they soon break away from the parent body and swim about actively in search of a macrogamete. The female gametocyte matures into a **macrogamete** within the stomach of the mosquito. A macrogamete is non-motile and spherical in shape. One of the microgametes now fuses with a mature macrogamete—the process is known as *fertilisation* and the product of fertilisation is a **zygote.** This is a rounded body containing two nuclei, the male and female, which mix up intimately. The **zygote** puts out a pseudopodium and is converted into the **ookinete** which starts penetrating the gut-wall of its host. Ultimately the ookinete protrudes into the body cavity of the mosquito and encysts. The resulting **oocyst** gradually expands and the zygote reproduces asexually by **sporogony.** Its nucleus breaks up into a number of fragments by multiple fission of the cytoplasm becomes vacuolated. Eventually a large number of spindle-shaped **sporozoites** are formed within the expanding oocyst which now bursts and liberates its contents into the coelom. The sporozoites find their way into the salivary glands and the mosquito is now ready to infect other human hosts. The cycle in the mosquito is completed in about two weeks. The sporozoites are injected into the wound along with saliva when the mosquito bites man. They are carried, through the blood stream to the liver where they start the pre-erythrocytic phase, and the entire life-cycle is repeated in the new host.

CLASS CILIOPHORA : PARAMOECIUM

Habitat

Paramoecium, known popularly as the slipper-animalcule, is a very small creature about 0'3 mm. in length and just visible to the naked eye. The animal is abundantly found in stagnant water containing decaying vegetable matter, feeding voraciously on the bacteria that swarm in such a polluted environment. Microscopic examination is necessary to reveal details of its structure.

Structure

Paramoecia are elongated and slipper-shaped. The anterior end of the animal is bluntly rounded but the posterior end is pointed. Thus the anterior end resembles the heel and the posterior end the toe of a slipper. On the ventral surface, near the anterior end, is a depression leading into the **oral groove** or **peristome** which runs obliquely and becomes continuous through a small mouth, the **cytostome,** with a narrow tubular gullet or **cytopharynx.** The gullet ends blindly inside the endoplasm. The whole surface of the animal is covered with minute, hair-like, protoplasmic projections called **cilia** which arise from granular structures called basal bodies.

The animal is composed of soft jelly-like protoplasm which is divisible into an outer, clear **ectoplasm** and an inner granular **endoplasm.** The ectoplasm is covered by a distinct pellicle. The **pellicle** is divided into a number of roughly hexagonal pits, from the centre of each of which arises a cilium. Embedded in the ectoplasm are a large number of spindle-shaped bodies called **trichocysts** (*thrix*=hair). When irritated the trichocysts are discharged through pores which are situated on the opposing walls of the hexagonal areas already referred to. When discharged they form an effective jelly-like barrier against an attacking enemy. Moreover, the jelly-like substance holds the paramoecium to the substratum when the animal is feeding, otherwise it rotates like a rifle bullet and fails to collect its food. They are also used to entangle minute bacteria and other bodies upon which the slipper animalcule feeds. Embedded in the endoplasm, near the centre of the animal, there is a large bean-shaped **macronucleus.** A very much smaller **micronucleus** is closely applied to and lodged in a depression on the concave surface of the macronucleus. The macronucleus governs all the vital activities of the animalcule and the micronucleus plays a very important part during its reproduction. In *P. aurelia*, there are two micronuclei and one macronucleus.

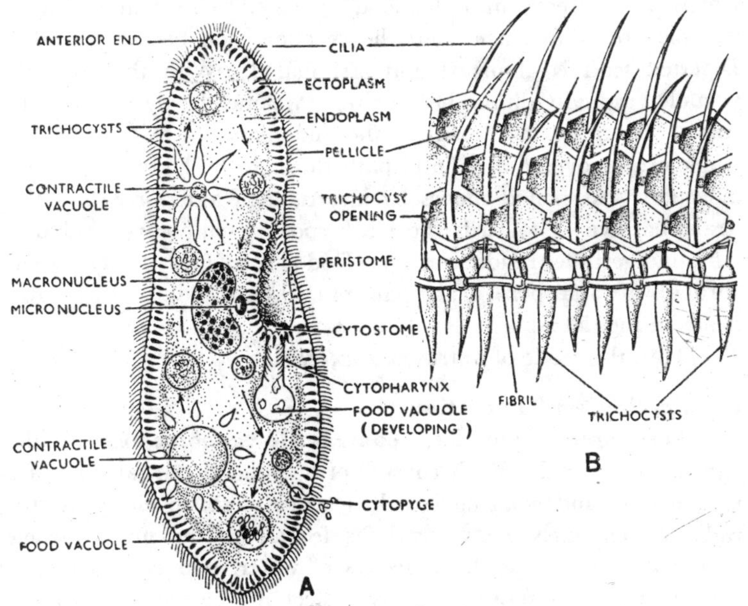

Fig. 53. A=*Paramoecium caudatum;* B=Structure of the pellicle and associated organelles.

Lying within the endoplasm there are numerous spherical **food vacuoles** for digesting food and *two* **contractile vacuoles** for eliminating waste products. Granular bodies are scattered throughout the endoplasm. Some of these are protoplasmic granules; others are granules of reserve food materials and still others are excretory or secretory products. There is a weak spot posterior to the mouth, the *temporary* **anal pore** or **cytopyge,** for egesting undigested residue.

Nutrition

The food of Paramoecium consists of bacteria, yeasts, algae, and small protozoa. The cilia of the oral groove are more or less fused to form an undulating membrane which sweeps the food particles down through the cytostome into the cytopharynx. At the posterior end of the cytopharynx the particles are surrounded by a watery vacuole. This, when fully loaded, constricts off, and begins to circulate in the endoplasm as a food vacuole by the rotary streaming movement called **cyclosis.** The food is digested by enzymes which are secreted into the food vacuole. The contents of the vacuole are at first acid, but the reaction soon becomes alkaline. Digested food is absorbed and assimilated. With this, the food vacuole gradually diminishes in size. Meanwhile, by following the rotary streaming movement of the endoplasm (cyclosis), the food vacuole has covered a definite path through the body of the paramoecium. At first it moves backward, then anteriorly, and then again backwards towards the weak spot or temporary anal pore. The indigestible residue in the food vacuole is finally egested through the anal pore. The path of cyclosis is indicated by arrowheads in Fig. 53 A.

Thus the mode of nutrition is *holozoic.*

Respiration and Excretion

There are two prominent **contractile vacuoles,** one at either end of the animal. Each consists of a rounded central space surrounded by and connected with a series of six to eleven narrow radiating channels which serve as feeders to the main vacuole. Nitrogenous waste products, excess of water, and carbon dioxide formed during respiration slowly collect in the vacuoles through the feeding channels. The vacuoles increase in size and at the limit of dilatation they contract and burst, discharging their contents to the exterior. They are again slowly formed at their original site and the whole process is repeated, hence the name pulsating vacuoles. These vacuoles are, therefore, excretory and respiratory in function. Oxygen required for respiration is absorbed through the pellicle from the dissolved air in the surrounding water.

Locomotion

Locomotion is rapid. It is effected by cilia which can beat either forwards or backwards. The animal, therefore, can swim in both directions. In normal circumstances, however, it runs forward. As the cilia beat obliquely backwards, the animal rotates on its long axis during its forward progression. The cilia in the oral groove are longer and more or less fused. They beat obliquely against the axis of the groove and hence the anterior end of the animalcule is swerved to one side or the other alternately. As a result of the combination of forward motion, rotation, and lateral swerving, the animal is compelled to follow a spiral path. It can never swim in a straight line. The ciliary motion, when viewed under the microscope, presents an undulating appearance comparable to the waves passing over a paddy field during a strong breeze.

Behaviour

Paramoecium, when it comes in contact with a solid object, recedes back and changing its course runs again till it can avoid the obstacle in its forward path. This is effected by a series of *trial and error* experiments. It is always kept informed about its environment by taking a sample of water in its oral groove and ejecting out the same. Any change of condition is thereby constantly and carefully noted. The animal avoids injurious environment and by repeated efforts brings itself to a suitable position. Even food is rejected when there is no hunger. When hungry it roams about aimlessly in search of food. When violently stimulated by chemicals or an injurious enemy, the Paramoecium responds by throwing out trichocysts, which forms an effective barrier round the animal and protects it.

Reproduction

Paramoecium usually reproduces *asexually* by dividing into two daughter animalcules in a plane at right angles to the long axis of its body. This simple process is known as **binary fission** and it always occurs in a transverse plane. The division of the body is

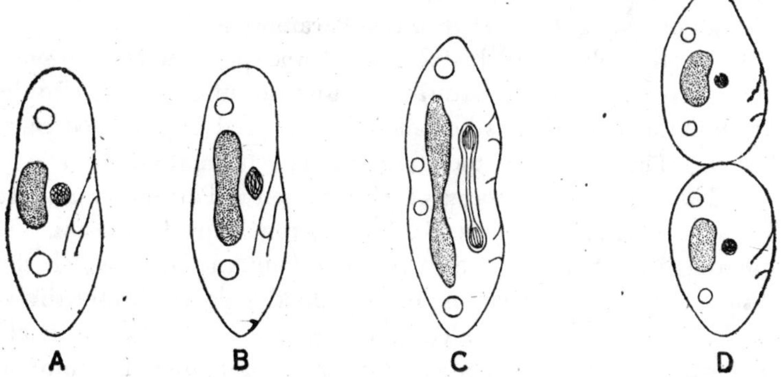

Fig. 54. Paramoecium undergoing binary fission.

preceded by division of both the nuclei. The micronucleus divides mitotically into two exactly equal halves, but the macronucleus divides directly by amitosis. The old gullet divides into two, so that each daughter is an exact but smaller replica of the original parent. The entire process is completed, in about two hours. If they are kept at a temperature of 15°C. to 17°C. the daughter Paramoecia are ready to divide again at the end of twenty-four hours. Raising the temperature to 20°C. results in two divisions during the course of a single day. When well-fed, Paramoecium may multiply asexually by *fission* for several months or even years.

Occasionally binary fission is interrupted by **conjugation.** This is a *temporary union* between two individuals which come together by opposing their ventral surfaces. A protoplasmic bridge is formed between them. The macronuclei of the two conjugants disappear and are absorbed into the cytoplasm. The micronucleus of each conjugant divides twice to produce four nuclei—of which three degenerate. The remaining nucleus divides unequally into a larger *stationary pronucleus* and a smaller *migratory pronucleus.* There is mutual exchange of the migratory pronucleus between the two individuals and the migratory pronucleus of one fuses with the stationary pronucleus of the other. The conjugants now separate and the single nucleus of each *ex-conjugant* divides thrice to produce eight nuclei, three of which are absorbed into the cytoplasm. Of the remaining five, four becomes macronuclei and undergo no further change whilst the other divides twice to produce four micronuclei. In the mean time the ex-conjugant divides twice to produce four individuals, each with a *macronucleus* and a *micronucleus.* In all, therefore, eight new individuals are produced as a result of conjugation between two Paramoecia.

In all probability this occasional and peculiar *sexual reproduction* by conjugation is a natural provision to infuse new energy into an individual which has become old and tired by repeated binary fission. The process is, therefore, one of rejuvenation.

There is another process of rejuvenation in Paramoecium. It is known as **endomixis.** In this the macronucleus disappears. The micronucleus divides twice to produce four nuclei, of which three disintegrate. The remaining nucleus divides along with the division of the body—so that two Paramoecia, each with a single nucleus is formed. The single nucleus in each daughter Paramoecium divides twice to produce four nuclei. Two of these are converted into macronuclei and the other two into micronuclei. The daughter Paramoecium now divides into two, each having one macronucleus and one micronucleus. Endomixis, then, produces four rejuvenated individuals from a single tired Paramoecium.

If condition of life becomes very hard and there is chance of impending death due to scarcity of food and water, the animal secretes a tough coat round it called *cyst*. In this encysted stage it tides over the difficult period in its life-history. With return of favourable condition, the cyst wall breaks and the animal comes out to begin a new life. **Encystment,** however, is very rare in Paramoecium.

CHAPTER IV

CNIDARIA : THE SIMPLEST METAZOA

The phylum **Cnidaria** (*coel*=hollow; *enteron*=gut) includes the simplest of the *cellular* animals or **metazoa.** The coelenterate body is a two-layered sac, opening to the exterior by a single aperture called **mouth.** The mouth is surrounded by a number of **tentacles** and leads into the *gastrovascular cavity* or **coelenteron.** The body wall consists of *two cellular* layers, outer **ectoderm** and inner **endoderm.** Between these two cellular layers, there is a *non-cellular* jelly-like plate called **mesogloea.** The cnidarians , therefore, are **diploblastic** (*diplos*=double; *blastos*=layer). Peculiar *stinging cells* or **nematocysts** are typically present in their body wall. Most of the coelenterates are radially symmetrical, that is, their body can be divided into a number of similar parts around a median axis passing through the mouth. The colonial cnidarions are usually polymorphic (*poly*=many; *morphe*=form). The same individual may bear different forms, some of which are tube-like feeding stage or *polyps* and others are saucer-shaped reproductive stage or *medusae.* Alternation of generations or **metagenesis** is frequently observed. The polyps, reproducing *asexually*, bring forth the medusae, and the medusa, reproducing *sexually*, brings forth the polyp.

Classification

The phylum includes nearly 9,500 known species which are divided into three main classes.

Class I. **Hydrozoa** (*Hydra*=a sea-serpent). This includes simple forms like the common *Hydra* and complex colonial form such as *Obelia*, exhibiting *polymorphism* as well as *metagenesis*.

Class II. **Scyphozoa** (*Skyphos*=cup). Predominantly *medusoid* forms such as the jelly-fishes which swim about in the sea. *Aurelia* and *Chrysaora* furnish good examples.

Class III. **Anthozoa** (*Anthos*=flower). Predominantly polypoid forms with no medusoid stage and fixed to the bottom of the sea. The beautiful, highly coloured sea-anemones and the corals are included in this class.

We shall discuss only two types of coelenterates—a simple fresh water form, the Hydra and a complex marine form, the Obelia.

HYDRA

Habitat

Hydra is one of the simplest of the metazoa. It is a solitary *polyp* of microscopic size which lives in clean fresh water ponds, attached to stones or water weeds. When the surface water becomes

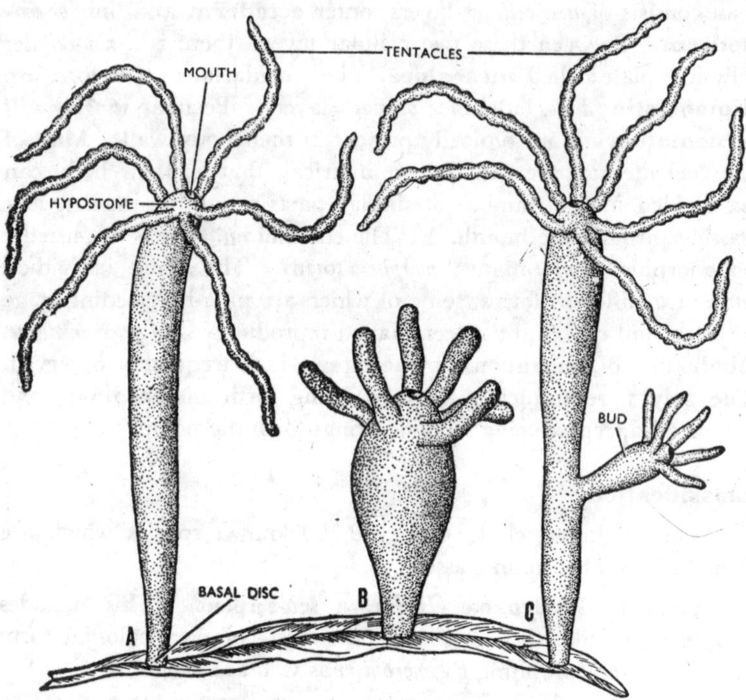

Fig. 55. Hydra (A=fully extended; B=partly contracted; C=a hydra with bud).

foul and hot, the animal retreats to the comparatively clean and cool bottom, where it lives in a peculiar depressed state. Depression is associated with lowered metabolism, shortening of the body and gradual disintegration. Recovery may occur if circumstances become favourable.

There are several species of Hydra. The grayish-brown *Hydra vulgaris* (formerly called *H. grisea*) is the common Indian species. *Pelmatohydra oligactis* (formerly *Hydra fusca*) is another common species. Its tentacles are longer than those of *H. vulgaris*. The green-coloured *Chlorohydra viridissima* (formerly *H. viridis*) is not commonly found in India.

External Feature

In appearance, Hydra resembles a small needle-like cylinder, about 10 mm. to 30 mm. in length, with a varying number of fine threads radiating from its open end. The hollow tubular body is closed at one end by the **basal disc** which is used for attachment and locomotion, and is, therefore, known as the **foot.** The free end of the tube bears a small opening called **mouth,** upon a raised conical **hypostome,** the base of which is surrounded by a circlet of about half a dozen hollow threads, the **tentacles.** The mouth serves for the ingestion and egestion of food. It leads through a narrow passage directly into the hollow of the tubular body which is continuous with the slender cavities of the tentacles. The enclosed cavity is known as the *enteron* or more precisely **coelenteron** and serves as the digestive tract. The animal is elastic and flexible. It may extend fully as a slender tube and bend in any direction. When contracted, Hydra assumes a spherical shape with the tentacles appearing as knobs at the free end. In well-fed and fairly grown specimens the solitary polyp may bear lateral **buds** (Fig. 55 C). In the breeding season, *gonads* may be found as projections from the surface of the body. When present, the male gonads or **testes** are found at the oral end, whereas the single female gonad or **ovary** occurs near the basal disc. Hydra is **radially symmetrical,** that is, the parts of its body are arranged in a circle around a central median axis passing through the mouth.

Histology of the Body wall

The body wall is composed of two layers of cells, an outer **ectoderm** and an inner **endoderm.** Between the two *cellular* layers, there is a thin *non-cellular* **mesogloea** of jelly-like consistency. The ectoderm forms the thin outer *epidermis* which is chiefly concerned with protection and sensory function. The endoderm, which is two or three times as thick as the ectoderm, forms the inner *gastrodermis* lining the coelenteron. It is mainly concerned with nutrition. The jelly-like non-cellular mesogloea serves as an elastic-supporting framework and connects the two cellular layers. Being composed of the same layers, the tentacles are structurally identical with the body wall.

The cellular layers of the body wall are composed of several types of cells for carrying out different functions. Various types of cells in the ectoderm and the endoderm are: (1) epithelio-muscular cells, (2) gland cells, (3) interstitial cells, (4) cnidoblasts or stinging cells with nematocysts, (5) sensory cells, (6) nerve cells, and (7) nutritive-muscular cells.

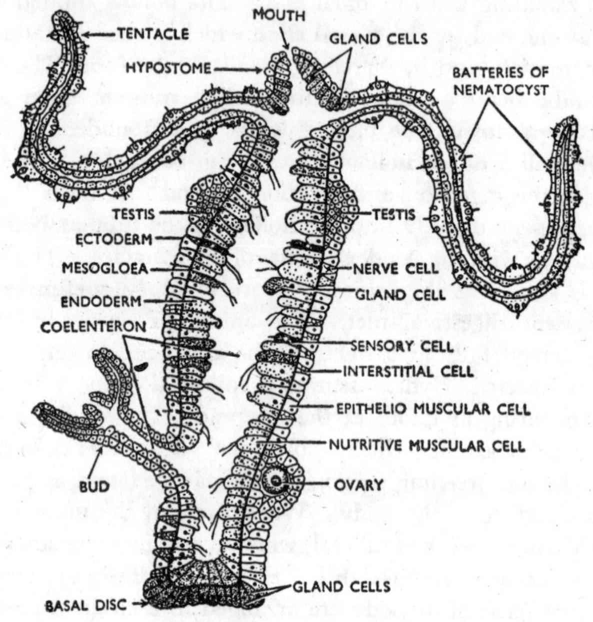

Fig. 56. A diagrammatic longitudinal section of Hydra.

Ectoderm or Epidermis. The ectoderm is chiefly composed of large **epithelio-muscular cells.** These are roughly conical in shape with their broad cuboidal surface directed outwards and fused to form a layer of protective *cuticle* over the outer surface of the body. The narrow ends of the cone touch the mesoglea and are produced into two or more contractile *muscular processes* which run along the length of the animal. These processes act as longitudinal muscles, the contraction of which shorten the length of the body and tentacles of Hydra. Each cell has a prominent nucleus and several small vacuoles. The protective cuticle is absent from the basal disc and the epithelio-muscular cells are replaced here by narrow and tall **gland cells** which secrete a sticky material for fixing the animal firmly to its substratum. Sometimes, the gland cells secrete bubbles of gas which help the animal to float.

In between the narrow ends of the epithelio-muscular cells there are intercellular spaces filled with groups of small, rounded **interstitial cells.** These cells are undifferentiated embryonic structures which may be modified to produce other kinds of cells, such as cnidoblasts, germ cells, etc.

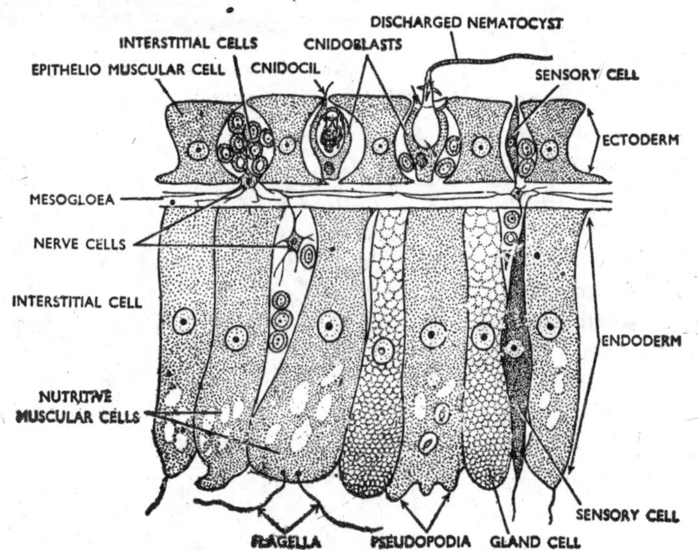

Fig. 57. Part of a cross-section through the body wall of Hydra (magnified).

Embedded in between the epithelio-muscular cells there are **cnidoblasts** (*cnidos*=nettle) or stinging cells containing *nematocysts*. These are particularly numerous on the tentacles and at the free end of the body, but none are found on the basal disc. A fully formed cnidoblast is a pear-shaped cell from the outer end of which a short trigger-like **cnidocil** projects out. Within the cell is a fluid-filled sac called **nematocyst,** the narrow outer end of which is produced into a long hollow thread. This thread is inverted and lies coiled up inside the sac. Surrounding the nematocyst there is a layer of special contractile protoplasm containing the nucleus of the cnidoblast. When the cnidocil is stimulated, the contractile layer of protoplasm squeezes the sac, and by compressing the liquid causes the long thread to be everted out and vigorously expelled. It has been demonstrated experimentally that direct stimulation of the cnidocil by contact with a glass rod does not provoke the discharge of the thread. Chemicals, such as, acetic acid, or substances released in the water by small animals on which Hydra feeds, effectively stimulate the cnidocil and the eversion of the thread is caused by increased pressure of fluid within the sac. There are *three* main types of nematocysts: (1) the large **penetrant** type with a long straight thread bearing three barbs at the base and rows of small spines or nettles; (2) the small **volvent** type, containing a short thick thread which is coiled

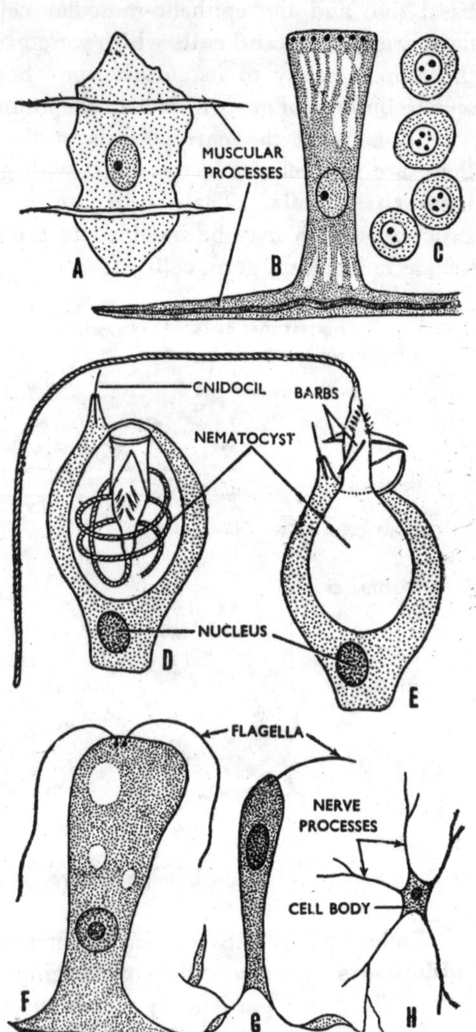

Fig. 58. Types of cells in the body wall of Hydra (A=epithelio-muscular cell in surface view; B=same, side view; C=interstitial cells; D=Cnidoblast; E=same with nematocyst discharged; F=nutritive-muscular cell of endoderm; G=sensory cell; H=nerve cell).

spirally into two or more loops; (3) the small **glutinant** type having a straight but somewhat sticky thread devoid of barbs or nettles.

In the tentacles the stinging cells occur in groups to form small swellings on the outer body wall. Such a group or battery of nematocysts usually consists of one or two of the large penetrant type in the centre surrounded by several of the other two kinds.

The three kinds of nematocysts carry on different functions. The penetrant type is used for offence and defence as well as for capturing and paralysing the prey. When discharged, it penetrates into the tissue of the victim. The fluid within the sac contains a poison called *hypnotoxin* which has a paralysing effect when injected into the body of the victim through the hollow tube. The prey is thus held the easier, while the Hydra swallows it.

The volvent or spiral type of nematocyst coils round the victim and the sticky thread of the glutinant type adheres to its body. These are therefore used for immobilising and capturing food. Glutinant nematocysts which produce sticky secretion are also used in locomotion.

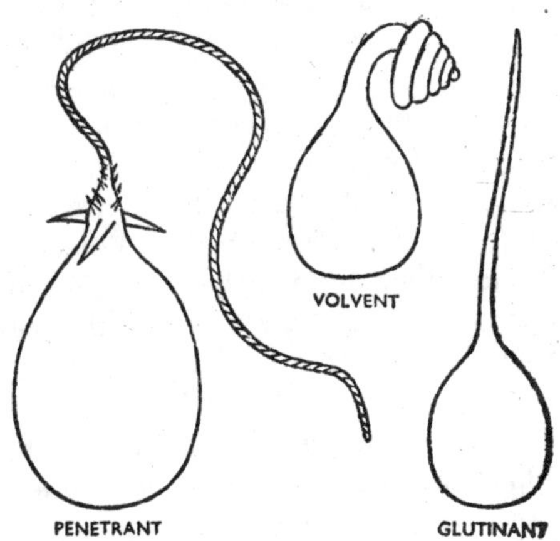

Fig. 59. Three types of nematocysts.

Nematocysts that are once shot out, cannot be withdrawn into their cnidoblasts and are lost. They are replaced by new cnidoblasts formed by modification of interstitial cells in the epidermis of the body wall. From here, they migrate to their final position upon the epidermis of the tentacles.

Sensory cells are scattered throughout the superficial part of the epidermis and are abundantly present upon the tentacles, hypostome and basal disc. A sensory cell has an elongated slender cell body with delicate tips, one end of which is connected to a nerve cell.

Nerve cells are spider-like in appearance due to the presence of fine processes which originate from the cell body. The processes are joined directly to the sensory cells and to the muscular processes of the epithelio-muscular cells. Nerve cells are scattered in the deeper part of the epidermis adjacent to the mesogloea. The processes of neighbouring nerve cells join freely to form a network in this part of the body, thus forming a very simple apparatus for effecting nervous co-ordination. The sensory cells receive stimuli from outside, the nerve cells conduct the impulse and the epithelio-muscular cells react to the latter.

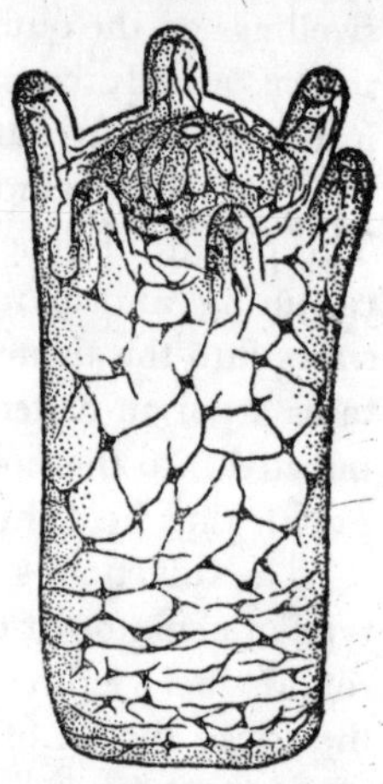

Fig. 60. Hydra-showing nerve-net.

Endoderm or Gastrodermis. The endoderm is chiefly composed of the **nutritive-muscular cells.** They are tall and columnar, with their broad ends towards the coelenteron and narrow ends touching the mesogloea. Their narrow ends are produced into contractile muscular processes which extend transversely and, therefore, encircle the body. When contracted, they act as a band of circular muscle to reduce the diameter and extend the length of the polyp. The broad ends of the columnar cells project into the coelenteron. In some, the broad ends are **amoeboid** and engulf particles of food by pseudopodia, digesting them like an *Amoeba*. Others bear whip-like **flagella** which are used for cutting food into bits. In the green species, *Chlorohydra viridissima*, the nutritive muscular cells harbour a kind of rounded unicellular algae called *Zoochlorella*. The association is mutually beneficial to both. Hydra affords protection to the *Zoochlorella* and supplies it with CO_2 and nitrogenous waste products, which are utilised by the plant as raw materials for manufacturing food. The plant, on the other hand, supplies the animal with carbohydrate food manufactured by photosynthesis. This kind of 'living together', where the association is mutually beneficial, is known as **symbiosis** (Fig. 61). Scattered in between the nutritive-muscular cells there are narrow **gland cells** which secrete digestive juice into the coelenteron. The endodermal gland cells are larger in comparison to those on the epidermis of the basal disc and are mostly present in the hypostome near about the mouth. They are coarsely granular and devoid of vacuoles.

Spider-like *nerve cells*, narrow *sensory cells*, and a few *interstitial cells* are also found scattered throughout the gastrodermis, in between the tall nutritive-muscular cells.

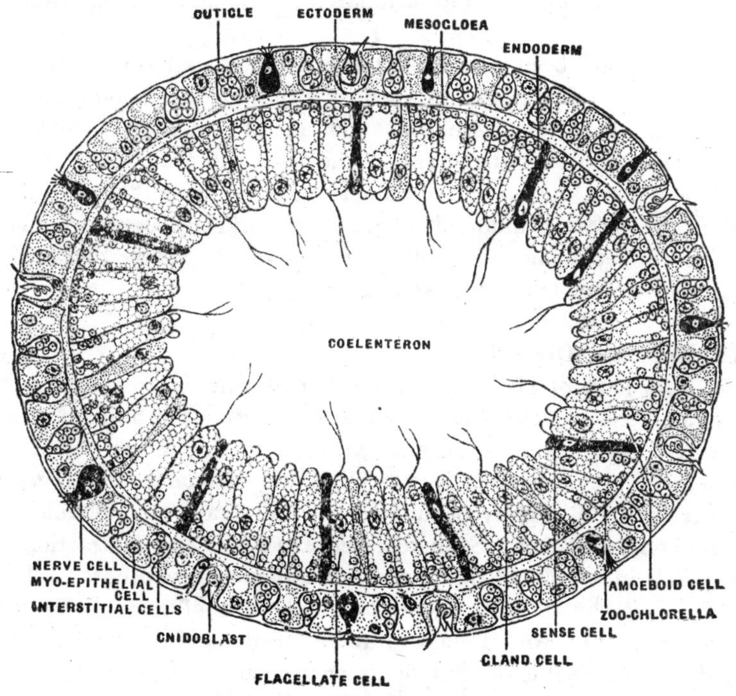

Fig. 61. Transverse section of a green hydra, showing *Zoochlorella* within the endodermal cells.

In Hydra, the cellular differentiation is associated with **physiological division of labour.** Each kind of cell is destined to perform a definite function in co-operation with other cells for welfare of the individual as a whole. Thus, the ectodermal cells, together form a *protective* **epidermis,** whereas the endodermal cells comprise the *nutritive* **gastrodermis.** The nerve cells and the sensory cells produce a delicate **nerve-net** which enable the Hydra to *respond to stimuli.* The stinging cells or **cnidoblasts** are useful for *securing food* and in *defending the animal* from its enemies. The **germ cells,** when formed, are meant for *reproduction.* **Interstitial cells** are the *embryonic structures* which can change into any other kind of cells. They produce buds, repair tissues, and manufacture germ cells. The **epithelio-muscular cells** are not only *protective* but their contractile processes are used for *shortening the body*. The contractile processes of the **nutritive-muscular cells** *lengthen the body;* the **flagellate cells** *cut food* into small bits, and the **amoeboid cells** engulf and *digest food* intracellularly. The gland cells secrete *mucous* and *digestive enzymes*.

It has now been established that subhypostomal zone of Hydra serves as growth region. It is believed that in this region cells are produced and from here cells flow in two directions—towards tentacles and towards basal disc. The older cells are thrown out of the body from the tip of the tentacle and from near the basal disc. Thus the Hydra is unique in the animal world in having a hidden system of replacing older cells which causes no visible change in external morphology of the animal.

Feeding and Digestion

The food of Hydra consists of small aquatic creatures such as water-fleas like *Daphnia*, and larvae of insects. The animal, when hungry, expands its body and tentacles to the fullest extent in search of prey. If food is not found, the animal contracts to a spherical knob, slightly bends its body and again expands fully in a different direction. Such repeated expansion and contraction of the body in search of food is known as **hunger movement.** When a Daphnia, by accident, comes near a battery of cnidoblasts, it is captured. The volvent and glutinant nematocysts adhere to the body of the victim preventing its escape, and the penetrant type paralyses it by injecting hypnotoxin into its tissues. Other tentacles bend down upon the paralysed Daphnia and carry it to the mouth. The latter expands to receive the food which is now pushed in. The food enters the coelenteron where it is cut into bits by the beating of the whip-like flagella of the **flagellate cells.** The **gland cells** now pour *alkaline* digestive juice containing *proteolytic enzymes.* Proteins are thus broken down into amino acids, which pass by diffusion into the endodermal cells and thence transferred to the ectodermal cells. Small bits of food which escape digestion in the coelenteron are engulfed by the **amoeboid endodermal cells** and digested within food vacuoles in typical amoeboid fashion. The contents of the food vacuoles are at first acid, then alkaline. Fat-splitting, protein-splitting, and sometimes carbohydrate-splitting enzymes are secreted into the vacuoles. Food is thus completely digested and the excess is stored within the endodermal cells chiefly as fat.

Digestion of food in Hydra is partly *extracellular* occurring in the digestive tract, and partly *intracellular* occurring within the amoeboid endodermal cells. Extracellular digestion is purely proteolytic. Fats, and sometimes carbohydrates, are digested intracellularly. Proteins are quickly digested in the coelenteron within a few hours but intracellular digestion requires a longer time, usually a few days.

Indigestible matter, such as the shell of a water-flea, is egested through the mouth.

Locomotion

Although attached to the substratum by the basal disc and usually standing erect, the Hydra has several methods of locomotion. Usually it bends down and attaches the tentacles to the substratum with the help of the glutinant nematocysts. It now releases and removes the basal disc to a new position and again stands up by disengaging its tentacles. This type of locomotion is known as *walking*. It may bend its body and perform *looping movement* like that of a leech (Fig. 62 A). When in a hurry, the animal runs by performing a series of *somersaults* (Fig. 62 B).

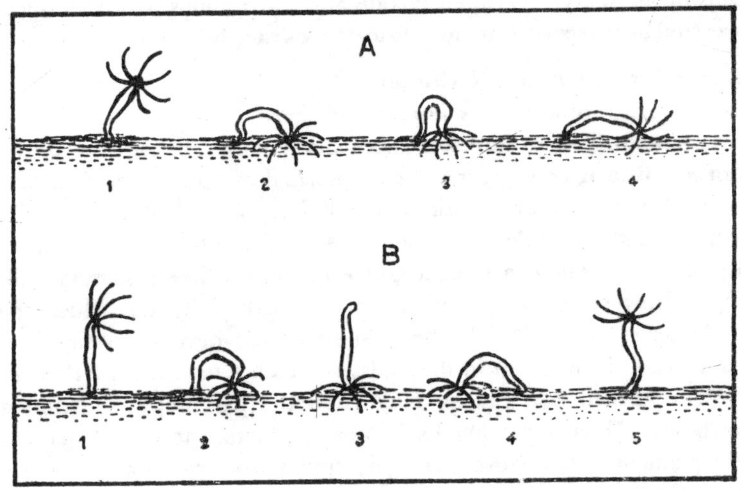

Fig. 62. Diagram representing methods of locomotion in Hydra (A=Looping; B=Somersaulting).

Sometimes it walks *inverted* by using the tentacles as legs. Again it may *glide* considerably by dragging the basal disc--the cells of which have special power of throwing out pseudopodia like an Amoeba.

Occasionally Hydra secretes a bubble of gas under its basal disc and *floats* about by the hydrostatic action of the gas bubble. A Hydra may *climb* by attaching its long tentacles to a distant object. Then releasing the basal disc and contracting the tentacles the body is drawn up, and it resumes its normal position on the new substratum.

It is to be noted that: (1) *Fixation* is effected by the mucus-secreting gland cells in the epidermis of the basal disc and the glutinant nematocysts with sticky threads on the tentacles. (2) *Shortening* the length of the body is due to the contraction of the muscular processes of the ectodermal epithelio-muscular cells. (3) Increasing the length of the body is produced by contracting the muscular processes of the endodermal nutritive-muscular cells.

Respiration and Excretion

Respiration and excretion are presumably carried out through the surface of the ectoderm and endoderm. The necessary oxygen passes by diffusion from the surrounding water directly into all the cells of the body. Carbon dioxide and nitrogenous waste products are similarly passed out by cell-to-cell **osmosis.**

Reproduction and Life-history

Hydra reproduces by *asexual* and *sexual* methods.

Asexual reproduction takes place either by *budding* or by *fission*. **Budding** appears to be the normal method of reproduction in Hydra and occurs at all times of the year. A bud is usually formed in the middle of the body by rapid multiplication of interstitial cells. These are soon converted into other necessary cell-types, thus producing a bulging of the body wall into which the coelenteron extends. The bud elongates and develops a circlet of tentacles at its free end in the midst of which a mouth is perforated. A fully formed bud breaks away from the parent by constricting its base. Then attaching itself to a new substratum it starts an independent life. More than one bud may form simultaneously upon the same parent.

Occasionally a Hydra will reproduce asexually by **fission.** The animal splits, either *transversely* or *longitudinally*, into two halves. Each half grows the wanting parts, through the activity of the interstitial cells.

Sexual reproduction occurs ordinarily in autumn. Most of the Indian species of Hydra are hermaphrodite or **monoecious,** and bear both male and female gonads upon the same individual. **Dioecious** species, in which sexes are separate, commonly occur in the United States. The gonads are temporary structures, developing on the sides of the body during the breeding season.

There is only one **ovary** in each individual at the proximal end near the middle of the body. The same individual bears several **testes** at its distal end. Both kinds of gonads are formed by rapid proliferation of interstitial cells which are later modified into the germ cells.

The male gonad or **testis** is a blunt conical swelling of the ectoderm. It is composed of a heap of rapidly multiplying interstitial cells covered over by a protective capsule of musculo-epithelial cells. The interstitial cells thus enclosed are converted into **spermatocytes,** each of which then divides twice to produce four **spermatozoa.** The sperm has a conical head, a short neck and a wavy tail. The covering capsule splits when the testis matures. Spermatozoa, thus liberated, swim about in the water where they remain active for a day or two.

Development of the female gonad or **ovary** is similar to that of a testis at the initial stage. The interstitial cells multiply and push out the musculo-epithelial cells to form a slight bulge near the proximal end of the animal. One of the interstitial cells, the *future ovum*, becomes amoeboid. It is the **oocyte.** It increases in size rapidly by ingesting the neighbouring interstitial cells with its pseudopodia. After finishing off the other interstitial cells, the oocyte becomes spherical. Its cytoplasm is now loaded with dark yolk granules. Two maturation divisions quickly follow, resulting in three small **polar bodies,** and the ripe **ovum.** The ovum is the female germ cell and the polar bodies are without any function. The capsule of the ovary ruptures and shrinks away, thus exposing the mature ovum, which, however, remains attached to the ovarian wall. The ovum secretes a gelatinous substance by which a swarm of sperms are attracted to it. One of them fertilises the ovum. If not fertilised within a short time it perishes. During fertilisation the nucleus of the sperm fuses with the nucleus of the ovum. The product of fertilisation is known as the oosperm or zygote which is destined to give rise to a young Hydra.

Normally, testes of an individual mature before the ovary.

Therefore the sperms of one have to find the ovum of some other, who is older than itself. Self-fertilisation is prevented in this manner. It must, however, be admitted that inbreeding occurs occasionally.

Life-history. The fertilised egg or **zygote** soon begins to divide. The **cleavage** is total and equal. Eventually a hollow ball, bordered by a single layer of cells, is formed. This is the **blastula** stage and the cavity of the blastula is the *blastocoel*. New cells are cut off from the inner ends of the older cells. The new cells migrate into and fill up the cavity of the blastula which in this manner, is converted into two layers of cell. This is the **gastrula** or strictly speaking *stereogastrula* stage. Its outer cell layer is destined to produce the *ectoderm* and the solid inner layer the *endoderm*. A jelly-like *mesogloea* is soon formed separating the two cellular layers, and in course of this, the embryo secretes a horny capsule

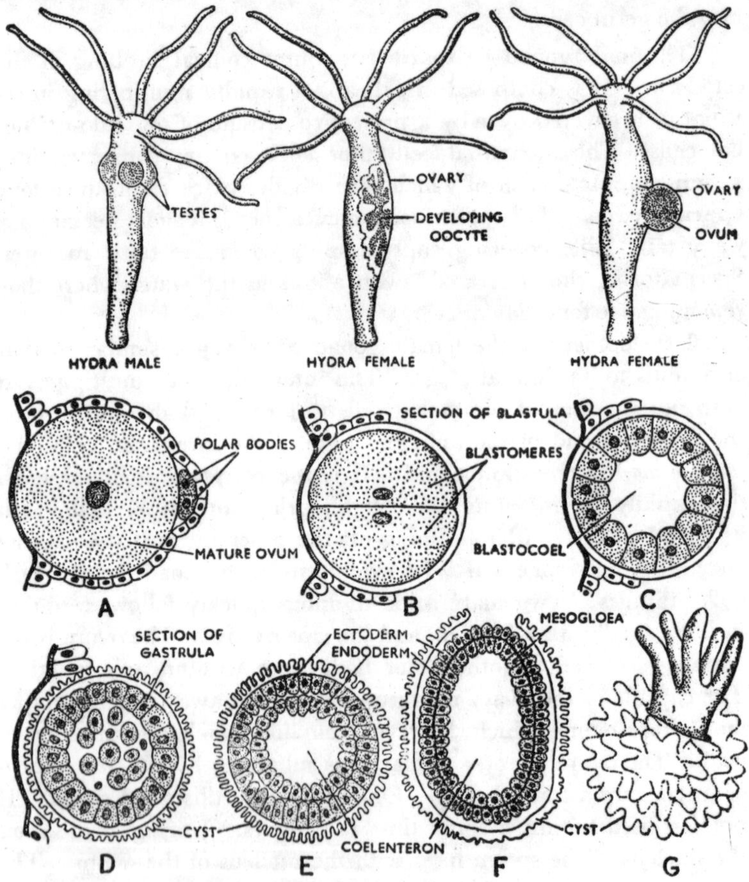

Fig. 63. Stages in the development of Hydra (diagrammatic). A=Fertilised ovum, B=Cleavage, C=Blastula in section, D & E=Gastrula in section, F=Fully formed young in section, G=Young Hydra hatching out.

or **cyst** with spiny outer surface. Thus enclosed it drops away from the body of the parent to the bottom of the pond.

In about two months, the capsule bursts and a young Hydra hatches out, with short tentacles and a mouth hidden amongst them. The coelenteron is formed by a split in the middle of the solid endodermal layer. The young Hydra now settles down on a water weed and begins to feed and grow. There is no larval stage.

It is evident that the sexual phase in the life-history of Hydra, with production of a resistant cyst, is primarily necessary for survival. It is a means to tide over difficulty, and possibly has an invigorating effect on individuals which are tired by repeated budding.

Hydra in relation to higher metazoa

It has been said that Hydra is the simplest of the metazoa. This is proved by the following facts:

(*a*) Hydra has developed cellular differentiation and physiological division of labour. But still it is chiefly a *tissue-animal*, having no organ-systems like those of a leech or a frog.

(*b*) It is *diploblastic*, having only *two cellular layers*—an outer ectoderm and an inner endoderm with an intervening *non-cellular* mesogloae; higher metazoa are all *triploblastic*, having *three cellular layers*.

(*c*) Hydra possesses a number of undifferentiated interstitial cells which can be changed into any other kind of cells in the body. In higher metazoa, the body cells are highly specialised and such undifferentiated cells are usually absent.

(*d*) The sac-like body has a *single gastrovascular cavity* or coelenteron which serves as a body cavity as well as a digestive tract in higher metazoa, the digestive tract is surrounded by a separate body cavity or coelom.

(*e*) Digestive tract communicates with the exterior by only one opening—the mouth; in higher metazoa, there are two openings—a mouth in front and an anus behind.

(*f*) Digestion is both *extracellular*, as in higher metazoa, and intracellular, as in some protozoa.

(*g*) Hydra *reproduces* asexually by budding or by fission; higher metazoa reproduce by the sexual method.

(*h*) Respiration and excretion occur by cell-to-cell osmosis from the body-surface; higher metazoa have definite respiratory and excretory organs.

(*i*) Nervous co-ordination is effected by a *simple-nerve-net* formed by a few scattered nerve cells and sensory cells; higher metazoa possess well-developed nervous system.

OBELIA

Habitat

Obelia is a marine colony of hydroid organisms. It is a moss-like animal found on submerged rocks or sea weeds, in shallow water off the sea coast.

Structure

It may be regarded as a Hydra which has undergone repeated budding but the buds have not separated from the parent. The supposed budding must be visualised to have occurred in regular order, producing polyps upon a common axis called **hydrocaulus.** The entire colony is whitish-brown in colour. It is fastened to the substratum by a root-like base, called **hydrorhiza.** The polyps are of two kinds: *hydranths* and *blastostyles*. The **hydranths** are more numerous and constitute the feeding zooids. Each of them resembles a Hydra in appearance, having about twenty solid tentacles. A hydranth, is encased and protected in a transparent vase-like *hydrotheca*. It captures minute prey by using its tentacles which contain batteries of nematocysts. The coelenteron of the hydranth communicates with the cavity of the hydrocaulus and thus the absorbed food is distributed to the entire colony.

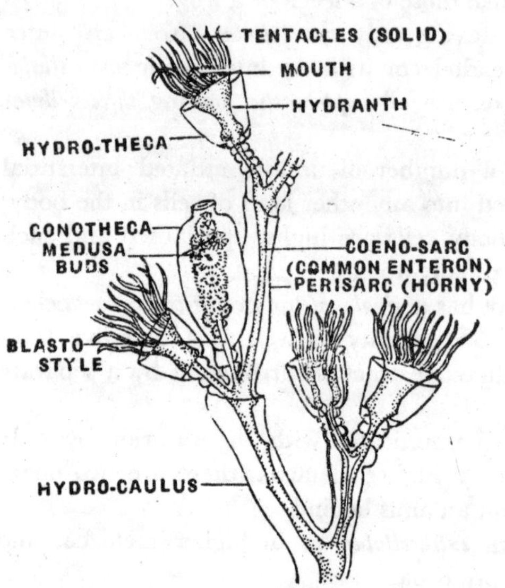

Fig. 64. Part of an Obelia colony.

Blastostyles are the reproductive zooids which appear only when the colony has reached a mature size. Each is covered by a transparent *gonotheca*, thus forming a cylindrical *gonangium* which has no mouth or tentacles. A gonangium occurs in the axil between the hydrocaulus and the axis of a hydranth. On the axis of the blastostyle are formed lateral buds which ultimately develop into the *medusae.*

The hydrocaulus is invested in a horny, transparent case or *perisarc* which is continuous with the hydrotheca and gonotheca. The soft parts of the hydrocaulus is known as the *coenosarc.* The minute structure of the body wall conforms to that of the Hydra, and the various cell-types found in the latter are also present in Obelia in a slightly modified form.

The medusae are budded off from the axis of the blastostyle. They escape into the sea through an opening in the gonotheca. A **medusa** is popularly known as the jelly-fish. Each medusa is a minute umbrella-shaped body bearing numerous tentacles arranged round its margin or *velum.* Its convex surface is known as the *ex-umbrella* and the concave surface as the *sub-umbrella.* The short handle-like structure hanging down from the centre of the sub-umbrella is the *manubrium.* The tip of the *manubrium* is perforated by the *mouth* which leads through a *gullet* into the coelenteron of the medusa. Four *radial canals* radiate from the middle of the coelenteron to join a circular *ring canal,* situated round the margin of the umbrella.

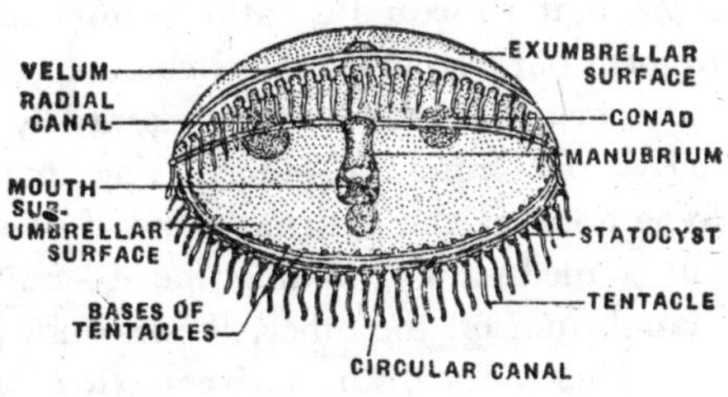

Fig. 65. The medusa of Obelia.

There are two nerve rings around the velum of the medusa connected with a nervous network within the body. A gelatinous mesogloea fills the space between the ectoderm covering the outer surface and the endoderm lining the coelenteron. Gonads develop on the sub-umbrella, whence ova and sperms are released into the water. The sexes are separate, so that medusa may either produce sperms or ova, but not both. There are four **gonads,** one upon each radial canal.

Locomotion is effected by contraction of the sub-umbrellar surface, so that water is driven out, and the organism is propelled forwards with its ex-umbrellar surface in front, or float and feed in the sea.

When mature the medusa reproduces sexually. **Fertilisation** takes place in water. The fertilised ovum, by repeated cleavage, produces a hollow blastula, the cavity of which becomes obliterated by migration of cells from the bordering wall to form a solid mass of endoderm. The outer layer of the blastula may now be considered as the ectoderm. Later on a split appears in the middle of the endoderm which represents the coelenteron. The latter communicates by a mouth with the exterior. Instead of secreting a horny case, as in Hydra, the ectoderm becomes ciliated and a **planula larva** is formed, which freely swims about in water. Subsequently, the planula settles down, fixes to a substratum, develops tentacles at the free end, and is thus converted into a polyp. The polyp produces buds in a regular and orderly manner and the result is a moss-like colony of Obelia bearing the two kinds of hydroid zooids: hydranths and blastostyles.

The medusa is a means for dispersal. It would appear, at first sight, that a medusa is entirely different from a hydranth, not only in form but also in minute structure. This idea is erroneous. The basic structure of the two is essentially the same and we can easily derive one from the other. The manubrium corresponds to the hypostome of the hydranth and if in imagination, we lightly pull out a medusa by holding the ex-umbrella in one hand and the manubrium on the other, the medusa is converted into a polyp.

There is a distinct **alternation of generations** in the life-history of Obelia. The fixed colony represents the *diploid* or **asexual generation,** reproducing by budding. The medusae derived from the polyp are independent units, living apart, and feeding themselves. They represent the *haploid* or **sexual generation,** reproducing by gametes. Product of fertilisation is the free swimming planula larva, which subsequently settles down to form the fixed asexual colony. This phenomenon is known as **metagenesis.**

An Obelia colony exhibits **polymorphism.** It includes three types of individuals: (1) hydranths or feeding polyps; (2) blastostyles or reproductive polyps, reproducing asexually by lateral budding; (3) medusae, reproducing sexually by germ cells and serving as means for dispersal.

CHAPTER V

PLATYHELMINTHES : THE ACOELOMATE METAZOA

THE LIVER FLUKE

(Fasciola hepatica)

Habitat

The liver fluke is a flat worm belonging to the phylum **platyhelminthes.** It is a parasite living within the bile ducts of the sheep, or of other cattle, and rarely of man. It may invade other organs and causes a disease known as 'liver rot'. Heavy infection often ends in the death of the sheep. The parasite requires two hosts to complete its life cycle. One stage of its life-history is spent within the tissues of a snail (*Limnaea truncatula*). Consequently liver fluke thrives in marshy meadows where the appropriate snails are abundant.

External feature

The liver fluke is a flat leaf-shaped worm about 2·5 cms long and 1·5 cms broad, with a blunt triangular projection, the **head lobe,** at its anterior end. A cup-shaped **anterior sucker** is situated at the apex of the head lobe. The **mouth** is an oval opening in the middle of the anterior sucker. A short distance behind the mouth there is an imperforate **posterior sucker** on the ventral surface of the body. The small genital aperture or **gonopore** lies ventrally between the two suckers. The single **excretory aperture** is situated at the narrow and bluntly pointed posterior end of the body.

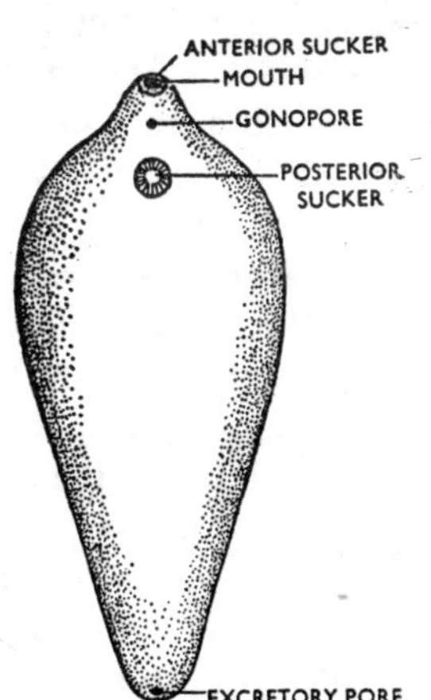

Fig. 66. The liver fluke (ventral view).

Body wall

The animal is covered with a cuticle in which are embedded small backwardly directed spinules. Beneath the cuticle there are successive layers of circular, longitudinal and diagonal muscle fibres. The epidermal cells lie among the longitudinal muscle fibres. There is no Coelom. The space between the body wall and the gut is filled up by a tissue called parenchyma which is composed of a meshwork of protoplasm with scattered nuclei and a few oval cells. Muscle fibres are found to pass across the parenchyma from the dorsal to the ventral side.

Alimentary System

The **mouth** leads into an oval muscular **pharynx** with a short **oesophagus** which at once divides into left and right branches. These are the **intestines,** which extend backwards along the two sides of the middle line to the posterior end of the body giving off a number of Caeca on either side which in turn may be much-branched. The intestines end blindly and there is no anus. The liver fluke feeds mainly on the blood of its host, which is sucked from the wall of the bile duct with the help of its sucker. It can also absorb predigested sugars such as glucose *saprozoically* through its cuticle.

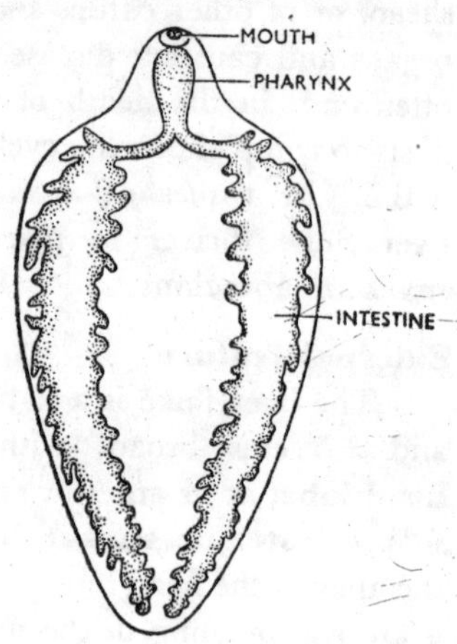

Fig. 67. Alimentary system of liver fluke.

Respiratory System

There is no respiratory organs in the liver fluke. The bile of its host is saturated with oxygen which is absorbed by the general body surface. CO_2 is released by exosmosis through the cuticle.

There is no evidence to believe that the liver fluke is anaerobic because a reduced oxygen tension leads to the death of the parasite.

Excretory System

The excretory system consists of a number of much-branched capillary tubules ending in **flame cells.** These are the **protonephridia** which ultimately drain into a single median tube, the excretory vessel, opening into the excretory aperture at the posterior end of the body. The flame cells extract fluid waste products and drive them out via the excretory channels.

Nervous System

There is a ganglionated nerve ring surrounding the oesophagus at the anterior end of the body. From these arise a number of slender nerves which innervate the different parts of the body, thereby maintaining nervous co-ordination. Two large lateral nerves, one on either side, run backwards to the posterior end giving off slender branches to the parts nearby.

Reproductive System

The liver fluke is *hermaphrodite,* which means that the male and female reproductive organs are found simultaneously in the same individual.

The **male reproductive organ** consists of: Two much branched tube-like **testes,** one lying behind the other, in the middle part of the body. A **vas deferens** arises from each testis, and the vasa deferentia run side by side towards the anterior end. Close to the posterior sucker, the vasa deferentia join with one another to produce a large pear-shaped **seminal vesicle.** A twisted narrow tube, the **ejaculatory duct,** connects the seminal vesicle with a hollow muscular structure, the **penis** or **Cirrus,** which opens at the gonopore.

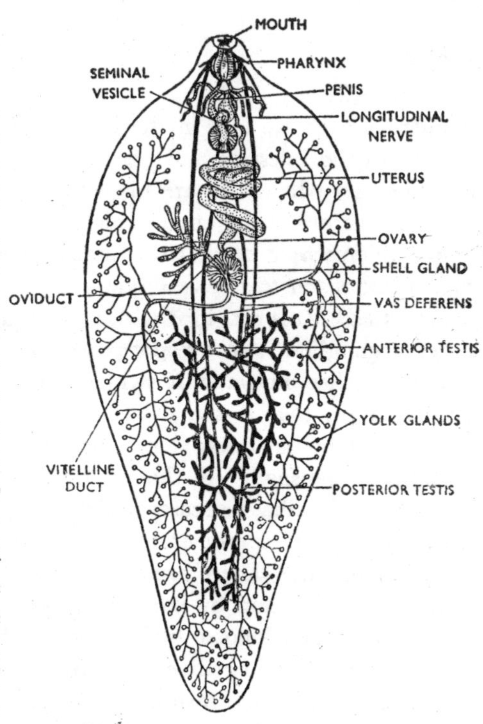

Fig. 68. Reproductive system of liver fluke.

Normally the Cirrus is enclosed in a *Cirrus-sac* but it can be everted and thrust out through the genital pore.

The **female reproductive organ** consists of: A single much-branched **ovary** lying on the right side in front of the testes. A narrow **oviduct** comes out from the ovary and passes towards the middle line to join the median **vitelline duct** from the **yolk glands** which lie laterally along the sides of the body. The junction of the oviduct and the vitelline duct is surrounded by a sac-like **ootype,** and the rounded **shell-gland** which secretes the egg-shell. A wide twisted tube, the **uterus,** runs forwards from the ootype to open at the gonopore. Arising from the ootype, there is a narrow channel, **Laurer's canal,** which is connected to a small aperture on the dorsal surface of the fluke near the anterior end. The animal is usually self-fertilised but cross-fertilisation may occur through Laurer's canal.

Life-history

Liver fluke's life-history is an extremely interesting process. Numerous eggs are laid in the bile ducts of the sheep. They are carried to the intestine and passed outside with the faeces. They develop only when they are placed in damp marshy places and if the temperature of the environment is above 50°F. If placed suitably a microscopic larva, called **miracidium,** hatches out of each egg in about ten to twelve days. The miracidium is multicellular and ciliated. It is provided with a pointed knob at its anterior end, a nerve ganglion, two eyespots, two flame cells and a small gut. If it is hatched in water it swims actively in search of the intermediate host, that is the snail *Limnaea truncatula.* It dies within about eight hours unless it can bore its way into the pulmonary chamber of the snail. In the soft tissues of the new host the miracidium loses its cilia, increases in size and is converted into a sac-like **sporocyst.** The sporocyst may divide transversely and produces other sporocysts. There are germ cells within the sporocyst each of which can develop by parthenogenesis, that is without fertilisation, into another kind of larva called **redia.** A sporocyst usually produces four to eight rediae. A **redia** is a hollow elongated structure with a mouth and a short gut, a pair of blunt projections near the posterior end and a minute birth pore. Within a short time the rediae burst out of the sporocyst and migrate into the liver of the snail. A redia has germ cells within its cavity which now develop parthenogenetically to produce other rediae. The daughter rediae escape from the parent through the birth pore. Several generations of rediae are produced in this way during the summer months. Eventually, the redia produces the fourth type of larva, called

cercaria. It has a flat heart-shaped body, with a slender tail which is twice as long, with two suckers and a forked gut. The cercariae emerge from the redia by the birth pore, work their way out of the snail and swim through water by the lashing of the tail. Very soon it settles upon the blade of a grass, near the water surface,

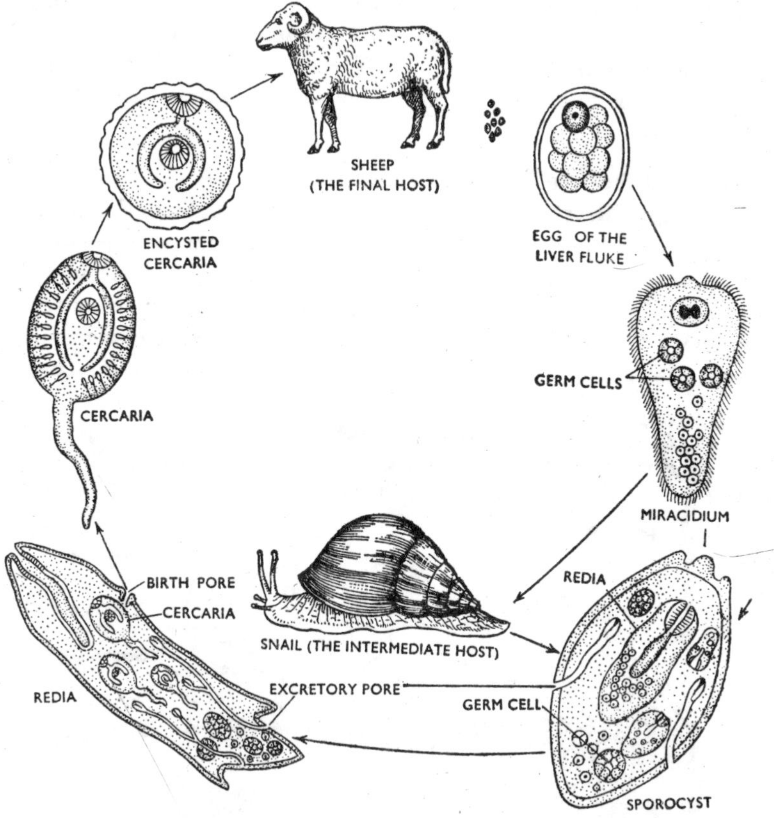

Fig. 69. Life-history of liver fluke.

loses its tail, secretes a cyst around itself and becomes a **meta-cercaria.** The encysted metacercariae remain alive for several months in damp places, if not subjected to very high temperature or drying. When infected vegetation is eaten by sheep or other suitable final host, the cyst wall dissolves and the larvae burrow through the intestinal wall into the body cavity. The bore their way into the liver in a day or two and are converted gradually into the adult fluke.

TAPEWORM

(*Taenia solium*)

Habitat

The name tapeworm is due to its elongated tape-like appearance. It may attain a length of several metres. Sexual forms of Taenia occur as endoparasites in the intestine of man. Asexual forms are encountered in the muscles of pig or in exceptional cases in the muscles of man. Man is its **primary host** and pig is its **secondary host.** Distribution of Taenia is world-wide and is most common in European countries. Nowadays *Taenia solium* occurs rarely in man.

External feature

A fully-developed Taenia may attain a length of 3-5 metres. Its antero-posterior ends are clearly distinguishable but it is difficult to differentiate the dorsal from the ventral surface. The body is ribbon-like (Fig. 69A) and consists of a distinct head or **scolex** at the anterior region. The tip of the head bears a conical elevation—the **rostellum** which can be retracted or extended. The rostellum bears

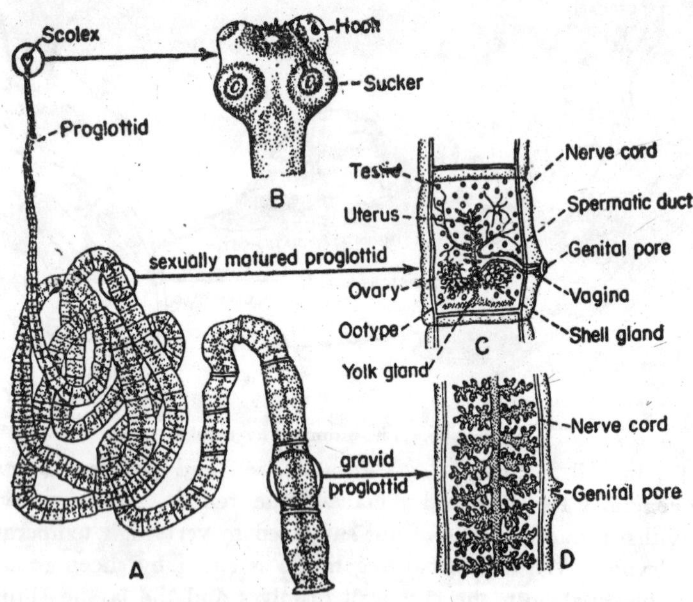

Fig. 69A. External morphology of Taenia. (A), Enlarged views of Scolex (B), mature proglottid (C) and gravid proglottid (D).

28 to 33 **hooks.** The hooks are of two types—larger and smaller and they alternate with each other. Each hook (Fig. 69B) has three

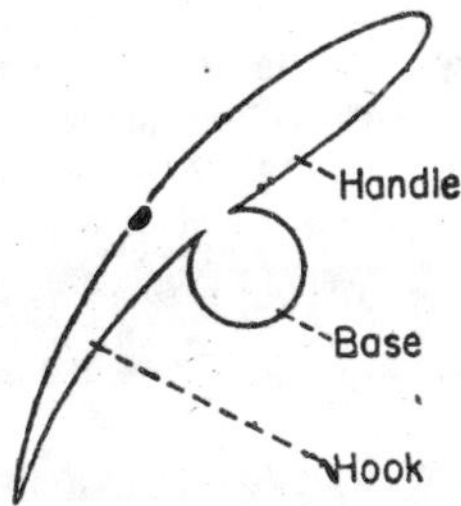

Fig. 69B. Structure of a large hook of Taenia.

parts—a base or guard, a conical blade at the tip and a handle projected from the middle. The hooks are arranged in two rows. When the contractile rostellum is withdrawn the hooks become anteriorly directed and get fixed into the host tissue.. The head bears in the middle four cup-like suckers or **Acetabulum.** Rostellum and suckers act as organs of attachment to the intestine of the host. Behind the head there is a narrow and small tubular region—the **neck** or the zone of proliferation. The rest of the body or tape is called **strobila.** The strobila is segmented in appearance. The chain-like strobila is made up of numerous segments or sexual units called **proglottids.** The proglottids progressively increase in size and mature towards the posterior extremity. The youngest or newly formed proglottid occupies a position just beneath the neck while the oldest one is at the posterior end. The number of proglottids varies from 800-850 in a full-grown worm.

Structure of a Proglottid

A proglottid from the middle region of the strobila offers a rectangular outline. The surface is lined by **cuticle** (recently renamed as epidermis). It is thick and perforated at intervals by fine canals, at the bottom of which either gland cells or nerve endings are situated. It is followed by longitudinal and circular layers of muscles. The circular muscles divide the **parenchyma** cells into an outer **cortex** and inner **medullary** regions (Fig. 69C). Towards each lateral margin is found the **longitudinal nerve** and just median to them lies the longitudinal pair of **excretory vessels.** A transverse excretory canal is situated at a posterior position of the proglottid. The anterolateral borders of the medulla are housed with **testes** and posterior lateral borders are with **ovary.** The male and female genital ducts open to a chamber called **genital atrium**

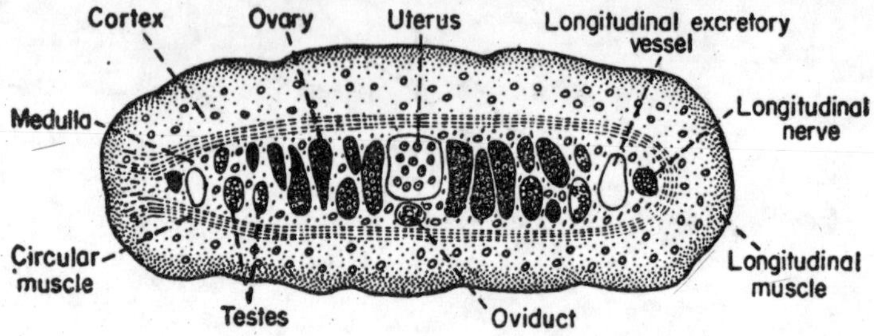

Fig. 69C. Transverse section of Taenia under microscope. Note principle organs are inside the medulla.

which is situated in the middle of one of the lateral margins. The atrium opens to the exterior through an aperture called **genital pore.**

Body wall

The body is covered by a thick **epidermis.** The epidermis is many-layered and perforated. It is impregnated with calcium carbonate. The epidermis remains sunk in the parenchyma. **Longitudinal muscles** run under the epidermis. The parenchyma is divided into an outer cortical zone and inner medullary zone by the **circular muscles.** Nervous, reproductive and excretory organs are situated within the medullary zone.

For a long time it was regarded that the body of tapeworm is covered by a thick cuticle. But recently electron microscopic studies have revealed that the outer layer of the body of all cestodes contains mitochondria and remains continuous with processes emerging from the cytoplasm of some underlying cells. There is a recent trend to replace the term 'cuticle' by epidermis or tegument. The ontogenetic development of the epidermis is not exactly known. The epidermis bears microvilli which increase the absorptive area of the tapeworms where gut is absent. Beneath the epidermis lies the prominent basement membrane.

Alimentary System

It is completely absent. This absence is due to the parasitic life of the tapeworm. It absorbs nutrients from the intestinal contents of the host.

Excretory System

There are two pairs of main **longitudinal trunks.** Each pair runs along the lateral margins of the strobila. The paired state of the longitudinal trunks is well recognised at the anterior part of the strobila and in the posterior part one from each of the pairs becomes lost. The two pairs of the longitudinal trunks are connected with each other in the head region by a ring-like vessel. Similar connections by straight transverse vessels are seen in the posterior regions of each proglottid. In the last and penultimate proglottids the longitudinal trunks open into a pulsatile **caudal vesicle** which opens to the outside by a single median aperture. When the last proglottid is cast off at maturity, the longitudinal trunks open to the exterior separately and independently. The main trunks give off branches from which numerous fine canaliculies each terminating in a **flame cell** arise.

Nervous System

The nervous system consists of a pair of ill-defined **ganglia** situated in the head or scolex. The ganglia are connected to each other by a broad transverse commissure of slender nerves. Each sucker is provided with a pair of nerves arising from the ganglion. **Two longitudinal nerves** of considerable thickness arise, one each from the ganglion and run along the lateral margin up to the last proglottid. Each longitudinal nerve gives a pair of **accessory nerves.**

Reproductive System

Taenia is **hermaphrodite.** Male and female reproductive organs occur in the same individual. Each proglottid behind the first 200 is equipped with a set of reproductive organs. Male reproductive organs develop first in each proglottid and then female organs make their appearance.

Male reproductive organs. There are numerous rounded **testes** distributed along the length and breadth of the proglottid. Each testis is provided with a fine **efferent duct.** The efferent ducts of neighbouring regions join together and form larger ducts. These larger ducts open into the **vas deferens** or main duct of the testes. The vas deferens is convoluted, transverse in position and extends towards the lateral margin (left or right) of the proglottid. The tip of the vas deferens is narrow and it pierces a narrow protrusible process the **cirrus** and opens at its extremity in the **genital atrium** (Fig. 69D). The base of the cirrus is enclosed in a muscular sac—the **cirrus sac.**

Female reproductive organs. There is a pair of bilobed (paired according to some) **ovaries** or **germaria** situated in the posterior region of the proglottid. The two lobes are unequal in size and lie one on each side of the median line. The ovaries are made up of numerous branching tubules which converge to form the

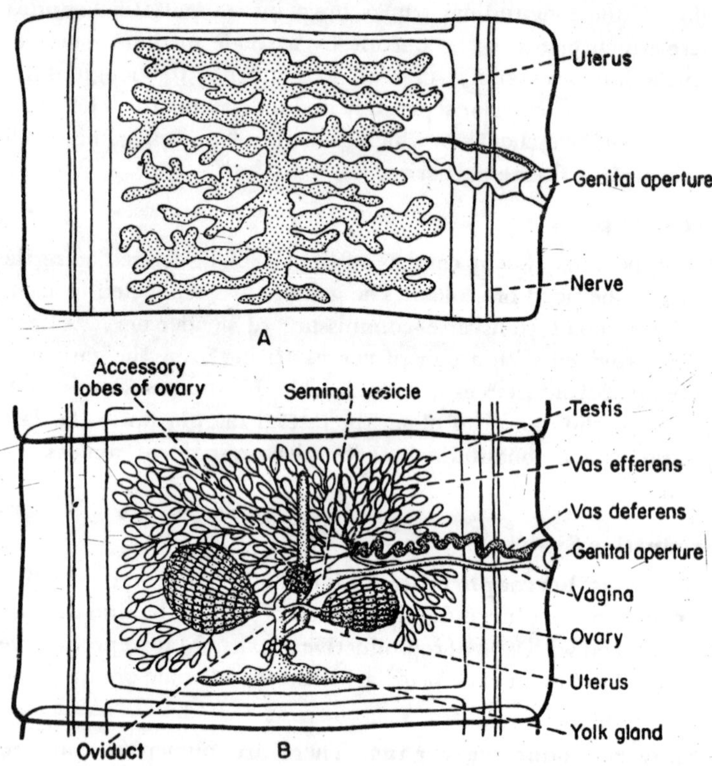

Fig. 69D. Reproductive structure (Male and Female) of Taenia in a mature proglottid (B). Overdevelopment of uterus after fertilization (A).

oviducts. The two oviducts meet and form a common median oviduct. A single **vitelline gland** or yolk gland consisting of few lobules is situated at the posterior border of the proglottid and opens into the median oviduct through yolk duct. Numerous round **shell glands** (also designated as Mehlis' **gland**) are situated round the yolk duct and the shell gland ducts open into the oviduct by the side of the yolk duct. The specialised part of the oviduct where shell gland ducts and yolk duct open is called **ootype.** The ootype passes anteriorly into a median, elongated and blind **uterus.** A fertilizing or **spermatic duct** arises from the ootype. The anterior part of the spermatic duct becomes dilated to form the **receptaculum seminalis.** From the receptaculum seminalis arises the **vagina** which runs forward and lateral and opens into the **atrium.** In a mature or gravid proglottid the uterus becomes distended and branched and occupies the major space inside the proglottid. Consequently other structures become reduced and modified.

Life-history

Taenia practises **self-fertilization,** i.e., eggs are fertilized by sperms from the same proglottid or one proglottid may be inseminated by a proglottid situated anterior to it. This is achieved by the bending of the strobila into folds. The possibility of cross-fertilization is remote since no host will be in a position to house two large tapeworms at a time.

Fertilization occurs inside the ootype and fertilized eggs become surrounded by mass of yolk cells secreted by the yolk gland. The fertilized ovum and yolk cells are subsequently enclosed in a thick egg shell secreted by the shell glands. Finally the eggs pass to the uterus.

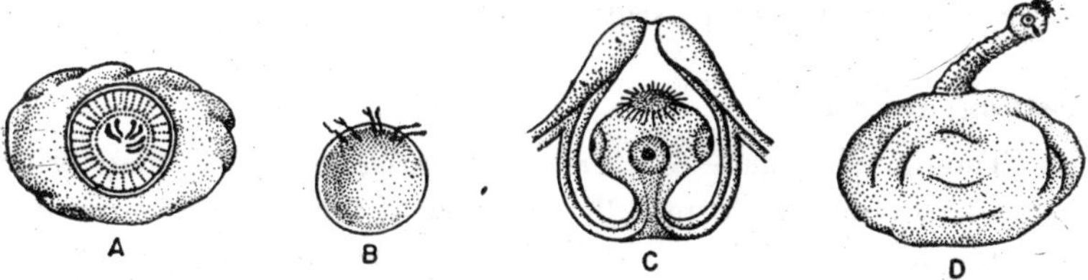

Fig. 69E. Development of Taenia. Hexacanth in the uterus (A-B), cysticercus (C) in Pig, emergence of Taenia (D) in man.

In the uterus the development of the egg starts (Fig. 69E). The first cleavage results a large **megamere** and a small embryonic cell. Both the megamere and the small embryonic cell divide and the descendants of the megamere ultimately form an envelope called **embryophore** round the embryonic cells. From the embryonic cells develop the embryo proper. Later on, six hooks develop in the posterior pole of the embryo and it is now called a **hexacanth embryo** [Fig 69E (A-B)]. The whole structure containing the embryo, embryophore and egg-shell is called **onchosphere.** Terminal proglottid containing onchospheres break off from the strobila (four or five at a time) and pass out of the body of the host along with the faeces. A newly-shed proglottid wriggles for some time but eventually dies and disintegrates. The onchospheres are not affected but further development within them do not occur until they enter the body of the secondary host (pig) through its food. In the stomach of the pig, as the egg-shell and embryophore get digested, the hexacanth embryo is released. The embryo with the help of its hooks bores into the wall of the gut and reaches the blood stream. Then it reaches the voluntary muscles or tongue or any other muscular tissue *via* heart and becomes *encysted.*

Inside the cyst the embryo increases in size. A large cavity filled with watery fluid is formed inside the cyst and the whole structure assumes a bladder-like appearance. At one point in the bladder an invagination occurs and at the bottom of the invagination a small scolex—the **proscolex** is formed [Fig. 69E (C-D)]. The embryo at this stage is called **cysticercus** or bladder-worm. It does not develop further. Pork meat infected with bladder-worms is spotted in appearance and if such imperfectly cooked meat is eaten, the bladder-worms enter the body of man. The invagination is everted and the scolex attaches itself to the wall of the intestine, develops the neck which buds off proglottids.

Pathogenic Effects

Taenia may cause—(i) cerebral cysticercosis, (ii) may bring about reduction and complete occlusion of the lumen of intestine, (iii) anaemia and (iv) gastric disturbances leading to regurgitation of gravid proglottids.

CHAPTER VI

NEMATHELMINTHES : ASCARIS

Habitat

It resides as an **endoparasite** in the small intestine of man. The worm may migrate into other neighbouring channels. The parasites are most frequent in tropical and subtropical countries.

External feature

It is milk-white in colour but presents a reddish-yellow shade when alive. Female worms are larger than the males and are further distinguished by the presence of separate and independent **genital aperture** situated on the ventral surface at about one-third of the body length from the anterior (Fig. 69F). Sexes can also be distinguished externally by the shape of the post-anal portion of the body. The post-anal portion in the male is sharply curved downwards while in the female it is nearly straight. Females are usually 20-25 cm in length and 5 mm in diameter while the males are 15-17 cm in length and 3 mm in diameter.

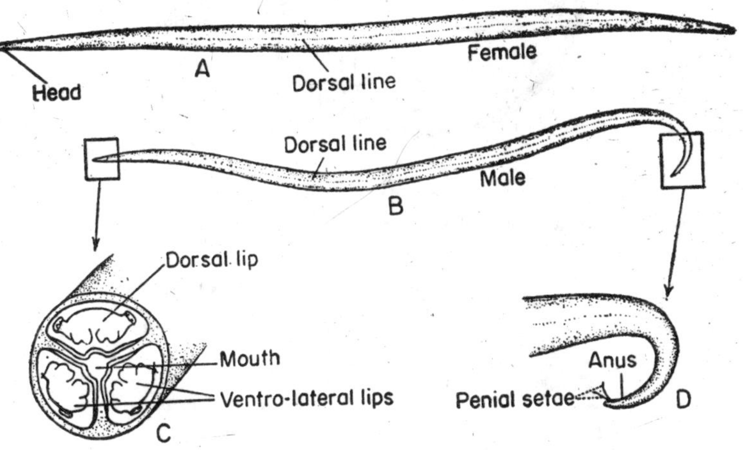

Fig. 69F. External features of *Ascaris*. A. Male. B. Female. C. Enlarged view of head end. D. Enlargzd view of posterior end of male.

The body is cylindrical and tapering at both ends. The **mouth** aperture is anterior and terminal in position and is bounded by **three lips,** one median and dorsal and two ventro-lateral in position. A little up on the ventral side of the posterior end is situated the **anus** which is transverse in position and in males it serves also as the reproductive aperture. In males a pair of needle-like chitinoid bodies project from this aperture which are called **penial setae.** A little down the anterior tip and on the ventral surface is situated an **excretory pore.** The **female genital aperture** is situated on the

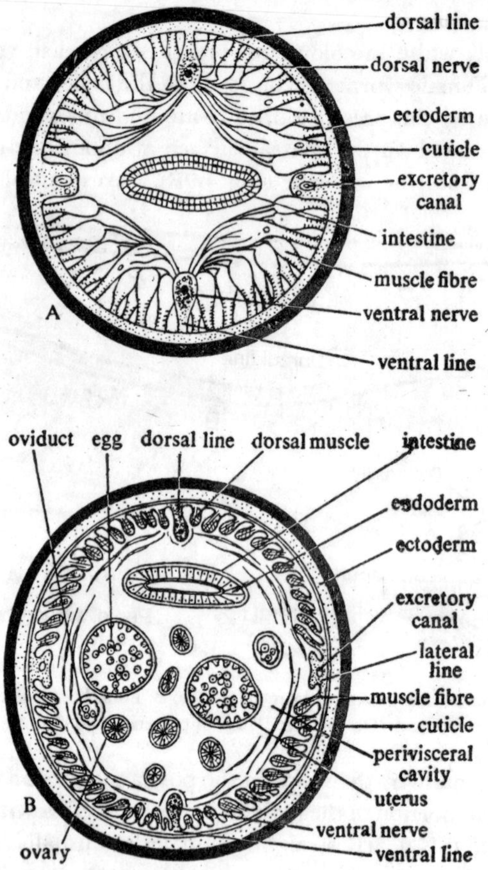

Fig. 69G. Transverse section of *Ascaris* passing through anterior third (A) and through middle region of female (B).

ventral surface at about one-third of the body length from the anterior end. There is no independent **male genital aperture.** The body is marked with four **longitudinal streaks** or **lines** running along the entire length of the body. Of these four streaks, one is **dorsal,** one is **ventral,** and two are **lateral** in position. The dorsal and ventral streaks are narrow and pure white in living condition while the lateral ones are thick and brown in colour.

Body wall

The outer surface of the body is furnished by **cuticle** which is thin, transparent, delicate, and wrinkled transversely (Fig. 69G). The cuticle is secreted by the ectoderm. The **ectoderm** underneath the cuticle forms a syncytial protoplasmic layer. Below the ectoderm is a single layer of **longitudinal muscles.** The individual cells of the longitudinal muscles are very peculiar in appearance (Fig. 69H).

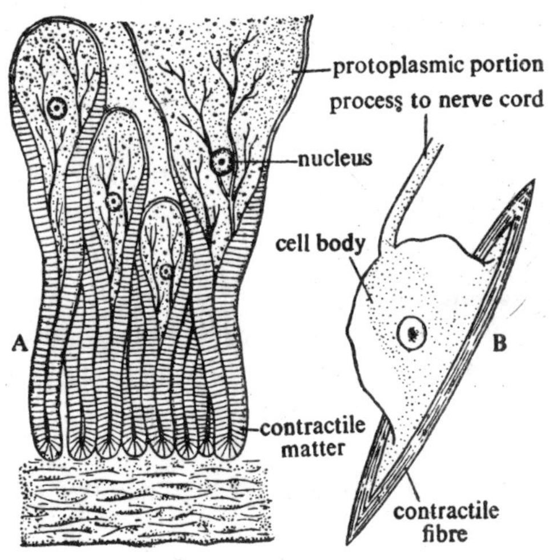

Fig. 69H. Showing the muscle cells in transverse section (A) and an isolated muscle cell (B) of *Ascaris*.

Each cell is spindle-shaped, straightened longitudinally and bears at the middle a bladder-like mass of protoplasm containing the nucleus which faces the side of the body. This peculiar appearance is due to the differentiation of the cell into a contractile and a nuclear part. The longitudinal muscle layer is not continuous and is arranged into four longitudinal bands, two **dorso-lateral** and two **ventro-lateral.** This is due to the fact that the ectoderm projects inwards at the dorsal, ventral and two lateral sides between the muscles to give rise to the streaks on the outer surface of the body.

Body cavity

The body cavity is not a coelom in the true sense as it is not lined by cell layers derived from mesoderm. The body cavity is lined externally by fibrous processes of the longitudinal muscle cells and internally by cuticle encasing the intestine. Fig. A and B depicts the body wall and disposition of major organs as seen in transverse sections.

Alimentary System

The **mouth** is anterior and terminal in position and leads into a straight tube which runs along the entire length of the body (Fig. 69IA). The mouth is guarded by three lips. Behind the mouth is **buccal cavity,** which leads into **pharynx.** The pharynx is dilated and its wall is muscular. The pharynx draws food from the intestinal contents of the host. It leads to the **intestine.** The posterior part of the intestine narrows down to form the **rectum** which opens to the outside through **anus.** The entire alimentary canal is made up of single **epithelial layer** covered internally and externally by **cuticle** derived from ectoderm. Digestive glands of any kind are absent.

Excretory System

The excretory system consists of two **longitudinal excretory canals** one through each lateral line (Fig. 69J). The two canals unite with each other at the anterior end and open to the outside through the single **excretory aperture** situated on the ventral side. Four or six big tufts of cells with ramifying protoplasmic processes remain in close contact with the canals. It is believed that these cells collect, store and pass on the waste matter in dissolved state to the excretory canal. The whole system is neither a ciliated one nor there is any flame cell. It is presumed that the canal is an intracellular tubular extension of a single enormously elongated cell.

Nervous System

The nervous system consists of a **ring of nervous matter** round the pharynx (Fig. 69IB). The ring is swollen at the ventral side and is ganglion-like. The ring gives off six **nerves** to the anterior and six **nerves** to the posterior side. Of the posterior nerves, two are of considerable thickness and run along the dorsal and ventral lines up to the posterior end of the body. The **dorsal** and **ventral longitudinal** nerves are connected with each other by **transverse**

commissures. The tip of the ventral nerve cord swells and forms a **ganglion** just in front of the anus.

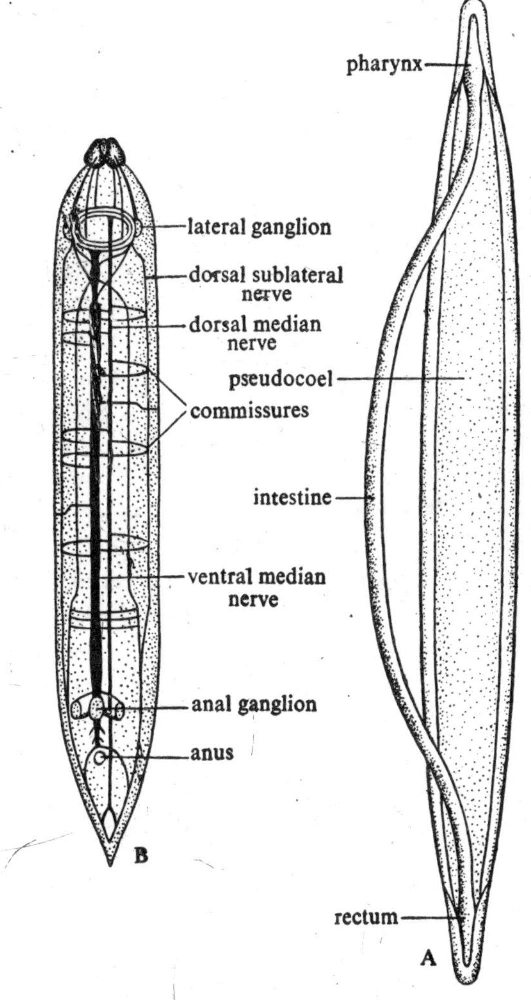

Fig. 69I. A. Alimentary system of *Ascaris*. B. nervous system of *Ascaris*.

The only sense organs are the **sensory papillae** situated as small elevations on the lips.

Reproductive System

The sexes are separate [Fig. 69J(A)] and there exists a considerable degree of sexual dimorphism.

Male reproductive organs. Male reproductive organs consist of a single thread-like, much-coiled structure occupying some portion of the body cavity. This structure is differentiated at the anterior region as the **testis,** the middle region as the **vas deferens** and the posterior region as the **seminal vesicle.** The demarcations between these three parts are very poor. The seminal vesicle continues as the **ejaculatory duct** and opens into the **anus.** The opening is associated with a pair of chitinoid spicules called **penial setae,** each of which is provided with a **muscular penial sac.** The anus, in case of male Ascaris, is used for the elimination of faeces as well as of sperms and in some way it resembles the **cloaca** of toad.

The sperms of Ascaris are very peculiar as they show amoeboid movement inside the gonoduct of the female. A ripe sperm is cone-shaped in appearance having a broad base and an apex [Fig. 69J(C)]. The apex contains the acrosome and the broad base contains nucleus and mitochondria. The sperms remain non-motile in the male gonoduct.

Female reproductive organs. Female reproductive organs consist of a pair of much coiled thread-like **ovaries** which pass into a **uterus.** The two uteri unite and form a muscular **vagina** which opens to the outside by the **female genital aperture** situated on the ventral surface at about one-third of the body length from the anterior end [Fig. 69J(B)].

Life-history

Eggs are produced in huge numbers (20,000 a day) and are fertilized in the upper part of the uterus while the parasites remain in the intestine of the host. After fertilization each egg becomes enclosed in a chitinous egg shell [Fig. 69J(D)].

Shelled eggs without being segmented come out of the body of the host along with the faeces. Formation of embryo becomes complete in about two weeks.

Infection is direct, i.e. intermediate hosts are not involved. Infection results from ingesting fully developed embryos encased in shell along with food or drink.

The embryos are not hatched in the stomach and pass on to the intestine where they hatch within two hours. Hatching is probably due to mechanical injury of the shell and is not comparable to a biological hatching.

A newly-hatched larva burrows the intestine and enters the **lymphatic ducts** or **veinules.** From the lymphatic ducts it is carried

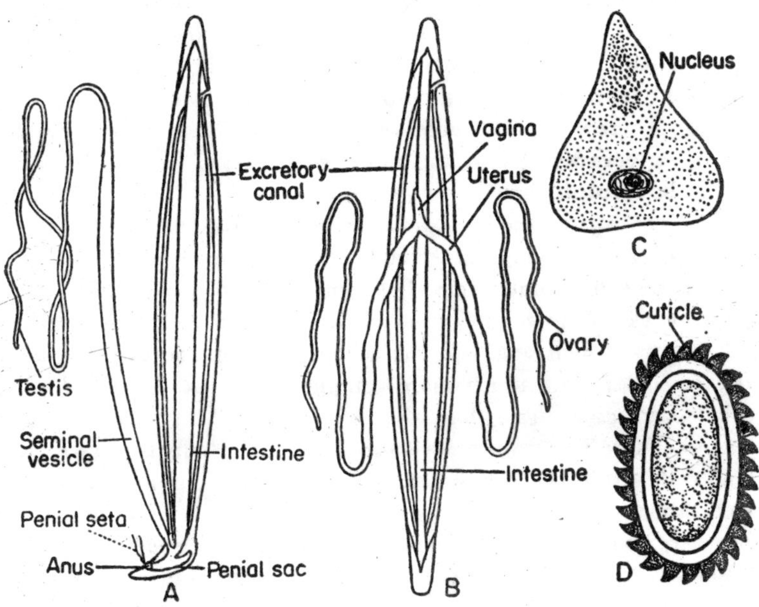

Fig. 69J. A. Showing the exretory system and male reproductives system of *Ascaris*. B. Female reproductive system of *Ascaris*. C. Spermatozoon of *Ascaris*. D. Egg of *Ascaris*.

to the **mesenteries, lymph nodes** and from there to the **portal vein** through circulation. From the portal vein the larva comes to the **liver** through hepatic veins. Here they pass from interlobular to intralobular veins and are carried to the right side of the **heart** from which they are carried to the **lungs.** Development and moulting take place in the alveoli of lungs and then they come down to **intestine** through trachea and oesophagus. The journey takes about ten days and adulthood is reached in about two months time.

CHAPTER VII

ANNELIDA : THE COELOMATE METAZOA

Coelomata Versus Coelenterata

Higher metazoa possess a body cavity or **coelom;** hence the name **coelomata.** All the larger and more specialised animals belong to this division. It includes several *phyla* which can be arranged in a gradually increasing order of complexity. The coelomates have advanced over the coelenterates in possessing *three cellular* body layers. There is a middle layer or **mesoderm** intervening between the outer **ectoderm** and the inner **endoderm;** hence they are said to be **triploblastic.**

In coelomata, the mesoderm splits into two parts—an outer *somatic layer* underlying the ectoderm and an inner *splanchnic layer* surrounding the endoderm. Between the two parts there is a more or less extensive cavity called *coelom.* The coelom is not the only feature which separates the coelomata from the coelenterata. The coelomates are much more complex and have developed not only *tissues* and *organs* but also *organ systems.* In coelenterates, nutritive materials, respiratory gases and waste products can diffuse from cell to cell, because the two cellular layers lie in apposition to each other. In coelomates, however, the ectoderm and the endoderm are widely separated; hence the necessity of a *middle man,* the blood vascular system, to negotiate between the different parts of the complex body. For similar reason, the coelomates have developed separate organ systems to carry on respiration, excretion and reproduction. There is a complex nervous system along with receptor-organs to co-ordinate and rule over the other organ systems. The simple digestive cavity of the coelenterates communicates with the environment by only one opening—the mouth; whereas in coelomates there are two openings—the mouth in front and the anus behind. Gonads in coelenterata originate generally from the ectoderm, but in coelomates the gonads are mesodermal in origin.

We shall discuss in the following pages three *grades* of coelomate organisation: (i) **Annelida,** the *segmented worms;* (ii) **Arthropoda,** the *animals with jointed limbs,* and (iii) **Chordata,** to which man himself belongs.

Phylum Annelida: The Segmented Worms

The **Annelida** (*annelus*=a little ring) are so named because they are segmented, externally as well as internally, into a number of ring-like **somites** or **metameres.** They have often been designated as 'lowly worms'; yet they constitute a well-organised group of animals who have established themselves all over the world. They live in damp soil or in underground burrows, in fresh water or on sea-shore, and a few are parasites. The phylum includes familiar animals such as earthworms, leeches and sand-worms.

The annelids are triploblastic, bilaterally symmetrical and metamerically segmented worms. There is an extensive coelom or body cavity surrounding the alimentary canal. Body wall is provided with well-developed circular and longitudinal muscle fibres. Alimentary canal is complete with two openings, a mouth in front and an anus behind. Excretory organs are present in the form of tubes called **nephridia.** Appendages, when present, are either rod-like **setae** or fleshy **parapodia** both of which are used in locomotion. Development occurs directly or by metamorphosis of a **trochophore larva.**

Classification

There are at present 7500 known species of segmented worms which are divided into the following classes:

Class I—**Archiannelida** (*Archi*=primitive): Primitive marine worms: *Polygordius.*

Class II—**Chaetopoda** (*Chaite*=hair or bristle) : These are worms with setae which are used in locomotion. There is an extensive coelom : the earthworm *Pheretima posthuma;* marine sandworm *Nereis.*

Class III—**Hirudinea** (*Hirudo*=leech) : Flattened worms, with suckers at both ends of the body: *Hirudinaria granulosa, Hirudo medicinalis.*

Morphology and life-history of the earthworm, *Pheretima* is now discussed as an example of the class *Chaetopoda.* This will be followed, in the next chapter by a description of *Hirudinaria*, which is employed in India for medicinal purpose.

THE EARTHWORM

(*Pheretima posthuma*)

Habitat

Earthworms are soft-bodied creatures which live under damp soil within burrows which may extend two (5 cms) or even three feet (7·5 cms) underground. They come out of their holes at night for feeding and return back at daybreak. During rains, when their burrows are flooded, the animals crawl about in damp places, hiding amongst heaps of dried leaves. They bore their way through the soft earth, eating the soil as they go in. The swallowed earth passes through the alimentary canal and deposited as small pellets of mud lying at the opening of the burrow on the surface of the soil. These pellets are popularly known as 'worm-casts'. In doing this the earthworm makes the soil more porous and their *castings* supply a fine top-dressing for the surface. It is for this reason that Charles Darwin called them *the natural tillers of the soil*. The earthworms, therefore, are real friends of the farmers.

External characters

The body of *Pheretima* is nearly circular in cross-section and varies from 7 to 8 inches (18–19 cms) in length. The general colour of the body is brown but the dorsal surface is darker. A dark line extends from end to end in the middorsal line. The anterior end of the body is pointed and the posterior end is blunt. The animal is elongated and divided into a series of ring-like segments or *metameres* or *somites*, which are separated from one another by narrow transverse grooves. Usually there are 120 true somites. A fleshy lobe, the **prostomium,** projects over the mouth in front of the first segment. The skin of segments, fourteenth, fifteenth and sixteenth, is swollen and pale in mature worms to form a saddle-shaped structure called **clitellum** or **cingulum;** it secretes the material for producing cocoons. Every segment of the body, excepting the clitellar, and the first and last, bears numerous chitinous bristles called **setae;** these are implanted on the body wall and are arranged in the form of rings. The setae serve as hold-fasts when the worm is moving over the ground or resting in its burrow. The entire body is covered by a thin transparent cuticle which is secreted by the epidermis. The cuticle is porous and protects the body from injury.

A number of external openings are found in the body to carry on different functions: (1) The **mouth** is a small crescentic aperture situated at the anterior end of the first segment just beneath the prostomium. It is meant for ingestion of food. (2) The oval **anal aperture** for the exit of faeces lies at the posterior end in the last segment. (3) The paired **male gonopores** are situated ventrally on the eighteenth segment, one on each side of the midventral line. These are the openings of the sperm-ducts. (4) The **genital papillae** are two pairs of cup-like depressions on the ventral surface of the seventeenth and nineteenth segments. The two papillae of a side, are placed one in front of and the other behind the corresponding male gonopore. (5) The single **female gonopore** is a median aperture on the ventral side of the fourteenth segment; it serves as an exit for the eggs.

(6) Four pairs of **spermathecal** apertures are situated in the intersegmental grooves between Vth and VIth, VIth and VIIth, VIIth and VIIIth, and VIIIth and IXth somites. Each of these leads into a pouch called **spermatheca** which acts as the receptacle of sperms received from another worm at the time of mating.

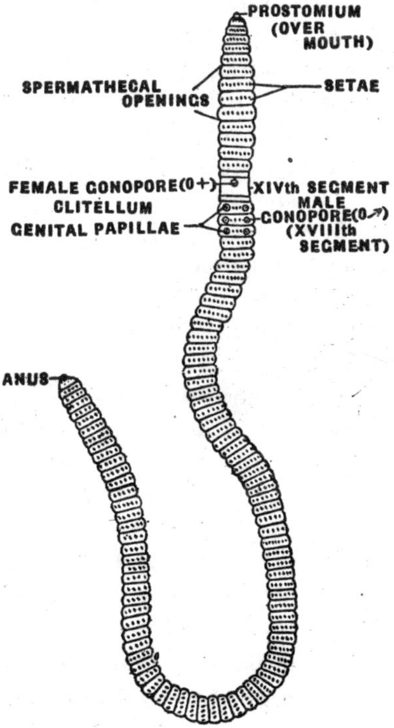

PROSTOMIUM (OVER MOUTH)
SPERMATHECAL OPENINGS
SETAE
FEMALE GONOPORE(♀+)
XIVth SEGMENT
CLITELLUM
MALE GONOPORE(♂) (XVIIIth SEGMENT)
GENITAL PAPILLAE
ANUS

Fig. 70. The earthworm (ventral view).

(7) The body cavity or coelom communicates with the exterior through **dorsal pores.** These occur in the intersegmental grooves along the middorsal line. The first dorsal pore is in the groove between the XIIth and XIIIth segment, and there is one in each subsequent intersegmental groove, excepting the last.

(8) The **nephridiopores** or the apertures of the excretory organs are minute openings found scattered all over the body, excepting the first three segments and the last.

Internal Anatomy. If a worm is cut open from the anterior to the posterior end by an incision through the body wall in the middorsal line, the internal structures may easily be studied. It is to be noted that the body of the earthworm is essentially a double tube. The body wall is the outer tube and the alimentary canal is the inner tube. The two tubes are separated by an extensive space, the body cavity or **coelom.** External segmentation of the body corresponds to internal partition of the coelom into compartments by a number of **septa** which lie beneath the intersegmental grooves. Thus, the body cavity is divided into 100 or more compartments corresponding to the number of external somites. Each compartment is lined internally by epithelium and is filled with a milky fluid which flows from compartment to compartment, because each septum is perforated by a large opening.

The alimentary canal remains suspended in the coelom and is held in position by the septa, which appear to arise from its wall. As the first septum lies between the IVth and Vth segments the coelom anterior to this region is a continuous cavity.

The milky **coelomic fluid** may spurt out through the dorsal pores when the worm is irritated. It contains large granular *phagocytes*, small *yellow cells*, and a few circular nucleated cells of intermediate size. It keeps the skin of the worm moist, and helps in the excretion of waste products. The phagocytes devour bacteria which are injurious to the earthworm, and thus protect it.

Running along the dorsal surface of the alimentary canal is the **dorsal blood vessels,** beneath which is the **supra-intestinal duct.** There is a **ventral blood vessel** ventral to the gut and overlying the **ventral nerve cord.** Beneath the ventral nerve cord is a small **subneural vessel.**

The **reproductive organs** are suspended in the coelom and extend from the VIth to the XIXth segments.

The body wall and the gut wall are best seen in a cross-section passing through the middle of the worm (Fig. 71).

The **body wall** is composed of the following layers:

(*i*) The outermost layer is the thin and transparent cuticle which is a porous non-cellular membrane covering the epidermis. The epidermis secretes and replaces the cuticle from time to time.

(*ii*) The epidermis is composed of a single row of epithelial cells. Some of these are modified into *gland cells* secreting mucus which is poured out through the cuticle and cleanses the body surface. A few epidermal cells are modified as *receptors* and act as sense organs. Others are merely *supporting cells*. These are sacs in the epidermis for implantation of the setae, and minute muscle fibres for moving them.

(*iii*) Beneath the epidermis, there is the musculature of the body which consists of an outer circular layer and an inner longitudinal layer. The former produces a continuous sheath running round the body and the latter runs in parallel bundles along the length of the worm. Pigment cells and blood capillaries are scattered among the muscle fibres.

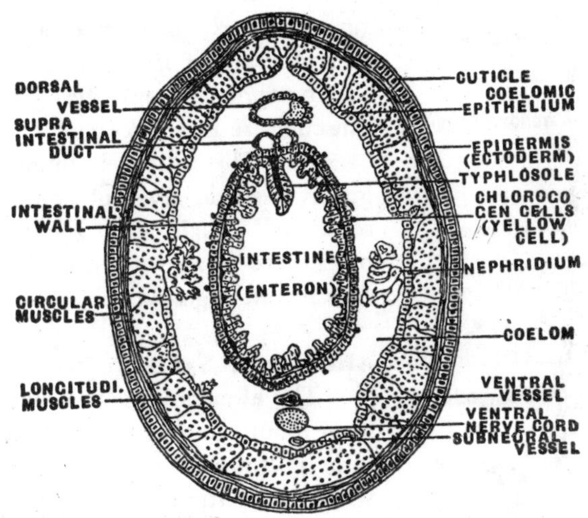

Fig. 71. A cross-section passing through the middle of an earthworm.

(*iv*) A thin layer of coelomic epithelium, consisting of a single row of flat cells, lies just beneath the longitudinal muscles. It forms the innermost lining of the body wall.

The body wall is not only *protective* but it also carries on *respiration* through its moist outer surface. The musculature along with the setae are responsible for *locomotion*. The contraction of the circular muscles elongates, whereas contraction of the longitudinal muscles shortens the length of the body. The two kinds of muscles are brought into play alternately. The setae are used for fixation on the substratum.

The skin contains the receptor organs and therefore serves as a *sensory* membrane.

The **gut wall** is composed of the following layers:

(I) An outer serous coat formed by a layer of tall narrow cells. They are seen to be loaded with minute yellowish granules, specially in the intestinal region of the gut, and are therefore called **chloragogen cells** or **yellow cells.** They collect waste products, and when fully loaded, drop down into the coelom and are discharged through the dorsal pores, thus helping in *excretion.*

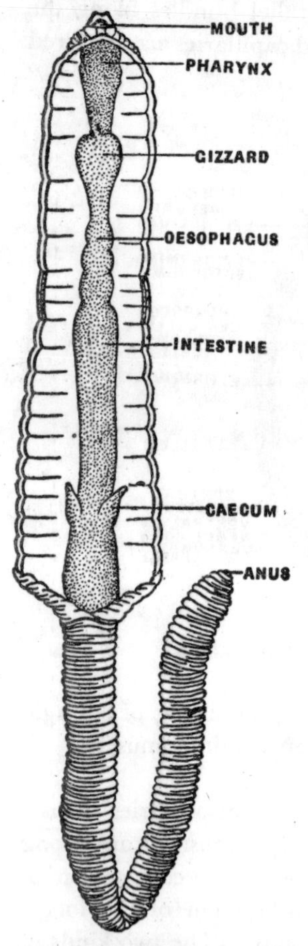

Fig. 72. Alimentary system of earthworm.

(II) Beneath the serous coat are two layers of muscles—the outer one being longitudinal and the inner one circular in disposition. The musculature is poorly developed and consists of non-striped fibres. In the region of the oesophagus and the gizzard the musculature is well-formed.

(III) The internal lining of the gut is a single layer of epithelial cells some of which are glandular and others are absorptive. The glandular cells secrete digestive juice.

Alimentary System

The alimentary canal runs straight from mouth to anus. The **mouth** leads into a thin-walled **buccal chamber** which is continuous with a wide and muscular **pharynx** in the third segment. The pharyngeal wall is thickened dorsally to produce glandular mass called *pharyngeal bulb*. The pharynx is followed by the **oesophagus** which runs back from the fourth to the fourteenth segment. A part of the oesophagus dilates into an oval **gizzard** in the eighth segment. This is a thick-walled muscular organ lined internally by a strong cuticle. The next portion of the alimentary canal is the thin-walled **intestine** which extends from the fourteenth to the ninety-fifth segment. The dorsal wall of the intestine is infolded to form an internal ridge or **typhlosole,** which increases the digestive surface and hangs like a curtain into the lumen of the gut. The intestine bears a pair of small, finger-like, blind pouches or **caeca** in the 26th segment. The rest of the alimentary canal, lying in the last twenty-five segments of the body, is known as the **rectum.** It is devoid of typhlosole and opens by the **anus** at the posterior end of the last segment.

Feeding and digestion. Food of earthworm consists mainly of organic matter in the soil, such as bits of leaf and particles of animal matter. Soil may be taken at any time but the leaves, etc., must be secured at night, when the worm crawls out into the air. The food is sucked into the mouth by the alternate expansion and contraction of the muscular pharynx. It is moistened by saliva which is secreted from glands of the pharyngeal bulb and passed into the oesophagus. The secretion of oesophageal glands neutralises the organic acids of the soil. The food is then ground up in the gizzard and broken into minute fragments by repeated squeezing against the hard internal cuticle. A series of peristaltic waves drive the food backwards along the alimentary tube. Meanwhile, digestion is effected by enzymes. Ultimately, the food is passed into the intestine where most of the digestion and absorption take places. The 'castings' are voided through the anus in the form of small rounded pellets or balls, each distinct from the others.

Circulatory System

The circulatory system of earthworm is a complicated system of closed tubes. The blood circulating in these vessels consists of a liquid plasma and a number of colourless amoeboid cells called corpuscles. The red colour of the blood is due to the pigment haemoglobin which remains dissolved in the plasma. There are no red blood corpuscles. The blood is propelled by a sort of peristaltic contraction of the muscle fibres in the wall of the vessels. Some of the vessels contain valves to prevent leakage and backflow. A number of vessels are large and centrally located. They run along the length of the body, and give off branches which ultimately break up into capillaries. There is no *true* heart; but there are four pairs of dilated pulsatile loops, one pair in each of the 7th, 9th, 12th and 13th segments. These are the so-called **lateral hearts,** because they possess the power of rhythmic contractions.

There are *three* main longitudinal vessels, running parallel to one another along the whole length of the body. (1) There is a large **dorsal blood vessel** in the mid-dorsal line between the gut and the body wall. It extends from the hinder end of the body to the pharynx, where it breaks up into branches which ramify over and supply blood to that organ. The muscular wall of the dorsal vessel contracts and expands rhythmically, driving the blood towards the anterior end. This vessel is provided with valves in its interior which allow blood to pass forwards, but prevent back-flow. (2) The **ventral vessel** runs along the midventral line just beneath the gut. It contains no valves and the blood in it flows backwards. (3) A slender **subneural vessel** runs along the mid-ventral line just beneath the ventral nerve cord. In the 13th segment of the body the subneural vessel bifurcates to form the two **lateral oesophageal vessels,** running along the sides of the oesophagus up to the anterior end of the body (Fig. 73).

It is to be noted that the dorsal vessel is a collecting tube receiving blood through smaller branches. The ventral and the subneural vessels, on the other hand, are distributing channels supplying branches to various parts and organs.

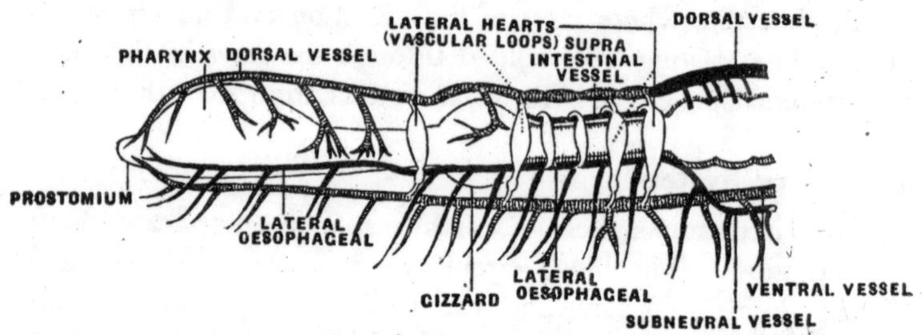

Fig. 73. Blood vessels in the anterior region of an earthworm.

Besides the three main longitudinal trunks, there is a slender **supra-intestinal** vessel lying dorsal to the oesophagus but placed ventral to the dorsal blood vessel. It extends from the 9th to the 13th segment. In the 10th and 11th segments, the supra-intestinal vessel is connected to each of the two lateral oesophageal vessels by slender loops.

The four pairs of **lateral hearts,** one pair in each of the segments 7th, 9th, 12th and 13th, connect the dorsal vessel with the ventral vessel. They are provided with valves so as to direct the blood to flow downwards, that is, from the dorsal to the ventral vessel. The two posterior pairs are joined to the supra-intestinal vessel.

In each segment in the intestinal region, there are two ring-like **transverse vessels** surrounding the intestine, and a **commissural**

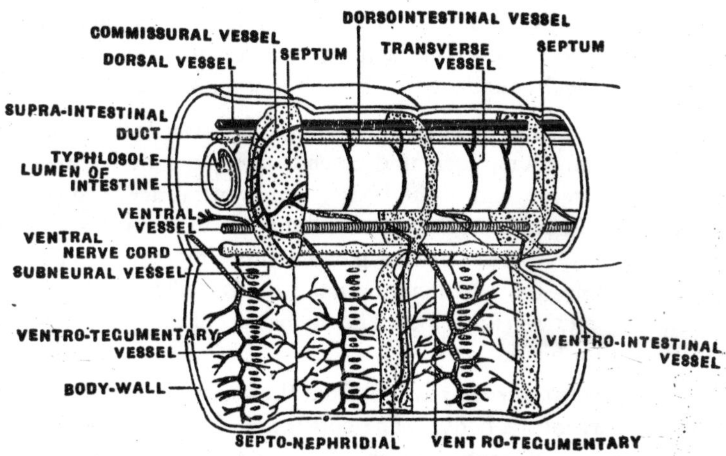

Fig. 74. Blood vessels in the intestinal region of an earthworm.

vessel runs along the posterior face of the septum in front. The dorsal vessel receives, on each side, in each segment, two *dorso-intestinal vessels* from the ring-like transverse vessels. The commissural vessel is connected by a branch with the subneural vessel lying beneath the nerve cord.

In each segment, the ventral vessel gives off a pair of **ventro-tegumentary vessels** to supply the body wall and the nephridia of the segment. A median **ventro-intestinal** branch is supplied to the intestine from the ventral vessel, in each segment.

Respiratory System

The earthworm has no separate respiratory organ. It receives oxygen and gets rid of carbon dioxide through its moist skin. The highly vascular epidermis is a permeable membrane through which O_2—CO_2 exchange is possible. Respiration takes place through the whole surface of the body.

Nervous System

The central nervous system consists of a nerve ring encircling the pharynx and a ventral nerve cord with ganglionic swellings.

The nerve ring is composed of: (1) a pair of **suprapharyngeal ganglia** or *brain* lying on the dorsal surface of the pharynx in the third segment; (2) a **subpharyngeal ganglionic mass** lying ventral to the pharynx in the fourth segment, and (3) two **peripharyngeal connectives,** one on each side of the pharynx, joining the suprapharyngeal ganglia with the subpharyngeal ganglionic mass.

The ventral **nerve cord** runs straight backwards from the subpharyngeal ganglionic mass and extends along the midventral line up to the posterior end of the body. The ventral nerve cord bears a **ganglion** in each segment behind the fourth. Each segmental ganglion really consists of two ganglia fused together. The cord itself is double, the two longitudinal strands being fused together beyond recognition (Fig. 75). The major part of the nervous system lies ventral to the gut.

Peripheral nerves pass out from the various ganglia to supply different parts of the body. The nerves are of two kinds—**afferent** or sensory and **efferent** or motor.

Receptor organs. The sensitiveness of earthworm to touch, smell, light and other stimuli is due to the presence of a number of epidermal receptor organs. These are groups of specialised epidermal cells connected with the nervous system by afferent nerves. They occur more at the anterior and posterior ends than in the middle part of the body.

Excretory System

The excretory matter is carried outside by a number of coiled tubes, called **nephridia,** which are present in every segment of the body excepting the first two.

Excreta are probably stored up as granules within the yellow-coloured **chloragogen cells** situated superficially round the intestine. The granules are discharged into the coelom, from where they are either collected by the nephridia or thrown out by the **dorsal pores.**

The **nephridia** are networks of fine tubules. Some nephridia open into the coelom by funnel-shaped **nephrostomes.** Some open to the exterior through minute openings or **nephridiopores** scattered throughout the skin, and some communicate by ducts with the alimentary canal. The nephridia are of three kinds: (1) integumentary, (2) pharyngeal, and (3) septal.

(I) The **integumentary nephridia** are extensively distributed over the inner surface of the body wall. These are small U-shaped tubes, about 200 of which are present in each segment of the body. They have no nephrostomes but open to the exterior individually, each by its own nephridiopore.

(II) The **pharyngeal nephridia** occur in tufts near the pharynx in the 4th, 5th and the 6th segments. They have no nephrostomes or coelomic funnels. There are no nephridiopores. The terminal ends of the nephridia of a segment join to form a common duct which opens into the lumen of the pharynx.

(III) The **septal nephridia** are attached to all intersegmental septa behind the 15th segment. About 50 of them are found on either surface of each septum, hanging freely into the body cavity. A septal nephridium is the largest and represents the typical form. It consists of a funnel-shaped *nephrostome* opening into the coelom. The funnel is connected by a small tube with the main body of the nephridium. The latter consists of a long spirally twisted loop and a terminal nephridial duct which leads into the septal excretory canal. There is one pair of septal excretory canals on each septum and they communicate with a pair of supra-intestinal excretory ducts which run longitudinally in the middorsal line of the intestine below the dorsal blood vessel. Each supra-intestinal duct receives all the septal excretory canals of the same side and open segmentally by short vertical tubes into the lumen of the intestine.

All the nephridia are bathed in coelomic fluid and very richly supplied by fine blood capillaries. They are composed of glandular cells and partly lined by ciliated epithelium the cilia of which work towards the excretory canals. Waste products are selected out by the nephridial cells from the blood and the body fluid. These are gathered within the nephridial ducts. Thence the waste products are conveyed to the gut and eliminated through the anus along with the faeces. The waste products collected by the integumentary nephridia, however, are discharged directly to the exterior through nephridiopores situated on the outer surface of the body. The septal and the pharyngeal nephridia are said to be *enteronephric* because they open into the gut. The integumentary nephridia belong to the *exonephric* type because they open outside by nephridiopores.

Reproductive System

The earthworm is *monoecious* or hermaphrodite. Male and female reproductive organs occur in the same individual.

The **male reproductive organs** consist of two pairs of **testes** lying beneath the alimentary canal in the 10th and 11th segments. Each testis is a small whitish body which manufactures sperm cells. A ciliated **seminal funnel** lies behind each testis, so that one pair of funnels occur in the 10th and another pair in the 11th segment. Each funnel is continued behind as a slender **vas deferens.** The two vasa deferentia of one side run together along the ventral body wall up to the 18th segment where they join with the **prostatic duct** of the same side.

In the 10th and 11th segments, the testes and the seminal funnels of each side are enclosed in a hollow testis-sac. The cavities of the two testis-sacs in the 10th segment, as well as of the two in the 11th segment, are continuous. There is a pair of horn-like **seminal vesicles** in the 11th and 12th segments which open into the testis-sacs. The testis-sacs of the 10th segment are continuous with the seminal vesicles of the 11th segment, and the testis-sacs of the 11th segment are continuous with the seminal vesicles of the 12th segment. The spermatozoa are matured and nourished within the cavities of the seminal vesicles.

A pair of white elongated bodies called **prostate glands** extend from the 16th to the 21st segment of the body. They are irregular in shape and lie one on each side of the intestine. A *prostatic duct* comes out from each prostate and is at once joined by the two vasa deferentia of the same side to form a horse-shoe-shaped **spermatic duct.** The two spermatic ducts open ventro-laterally on the body wall through the **male gonopores,** situated in the 18th segment.

There are two pairs of **accessory glands** on the ventral body wall, one pair in the 17th and the other in the 19th segment. They lie close to the inner side of the prostate glands and open externally as genital papillae in front of and behind the male gonopores. The exact function of the prostates and the accessory glands are not known.

The **female reproductive organs** consist of a pair of

ovaries, one on each side of the nerve cord, and attached to the posterior surface of the septum between the 12th and 13th segments. Each ovary is a mass of elongated white threads containing the eggs or ova. An **oviduct,** with a funnel-shaped opening, is placed behind each ovary. The short oviducts pass backwards and join with one another in the midventral line of the 14th segment to communicate with the **female gonopore** which is situated there.

There are four pairs of *sperm-sacs* or **spermathecae,** one pair in each of the 6th, 7th, 8th and 9th segments. Each spermatheca is a small white sac which stores spermatozoa received from another worm during copulation. They communicate with the exterior through the spermathecal openings, situated ventro-laterally in the intersegmental grooves of the corresponding segments.

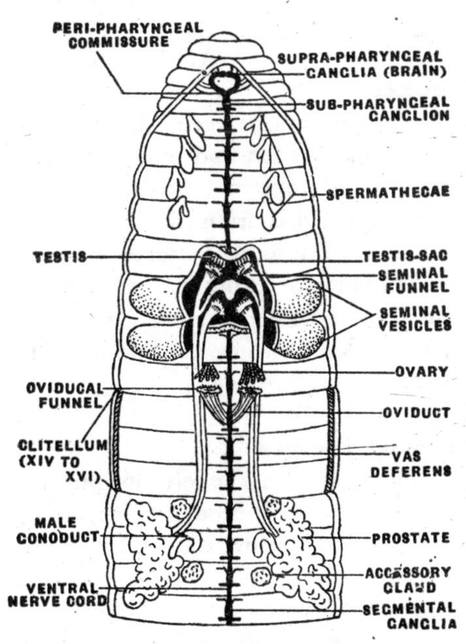

Fig. 75. Dissection of earthworm showing the reproductive organs and the nervous system.

Life-history. Mating takes place during summer within underground burrows. Two worms come together in a head-to-tail position, so that the spermathecal segments of the one are pressed against the clitellar segments of the other. A mutual exchange of spermatozoa takes place, after which the worms separate.

A membranous girdle is formed by the secretion of the clitellar glands surrounding the clitellar segments. This hardens slowly on exposure to air, and the worm endeavours to pull itself out of it. As the girdle is pushed over the female gonopore and the spermathecal openings, it receives ova and sperms. The two ends of the girdle are sealed as soon as it is cast off by the worm and the resulting **cocoon** is deposited in damp soil. Cocoon formation takes place during the summer months. The ova are fertilized inside the cocoon. Development is direct. There is no larval stage. As a rule, one embryo hatches out of each cocoon.

A **cocoon** is a small chitinous structure, more or less spherical in shape and is of light yellow colour. It is swollen in appearance and is filled with albumen which nourishes the developing embryo.

ANNELIDA (CONTINUED) : HIRUDINEA

THE LEECH

(*Hirudinaria granulosa*)

Habitat

The leech, *Hirudinaria,* is a segmented worm inhabiting fresh-water tanks, ponds and shallow weedy lakes. They grow abundantly in swamps and pools near paddy fields, where they can easily suck blood from the bare-footed farmers as well as from the cattle. The animal is an *ectoparasite.*

The blood-sucking habit of the leech had been frequently utilised by the older physicians for the purpose of letting out blood. This practice has now become obsolete and a leech is rarely employed for blood-letting.

External Feature

The animal has a soft elongated body with a sucker at each end and when fully expanded it measures about 2·5 cms in length. It is flattened dorsoventrally but the dorsal surface is a bit more convex than the ventral. When contracted, the animal appears to be elliptical and does not exceed three or four inches in length. The skin is moist and slimy due to constant secretion of mucus, and bears beautiful colour patterns. The dorsal surface is dark-green and the ventral surface is pale yellow in colour. The two flanks are striped with black, yellow and orange bands. The entire body of a leech is divided externally into a large number of rings or *annuli* separated by transverse grooves. There are more or less 102 such rings. These external rings, however, are only superficial structures and do not represent the actual number of true segments or **somites** in the body, the number of which is only 33. A typical segment or **metamere** from the middle of the series includes five rings, of which the first bears the **segmental receptor organs** arranged in a ring round the body. The receptors are the sense organs of the leech and they appear as small black spots on the surface of the skin. Each metamere bears four pairs of segmental receptors on its dorsal surface and three pairs on the ventral surface. The limits of the true segments are determined by the position,

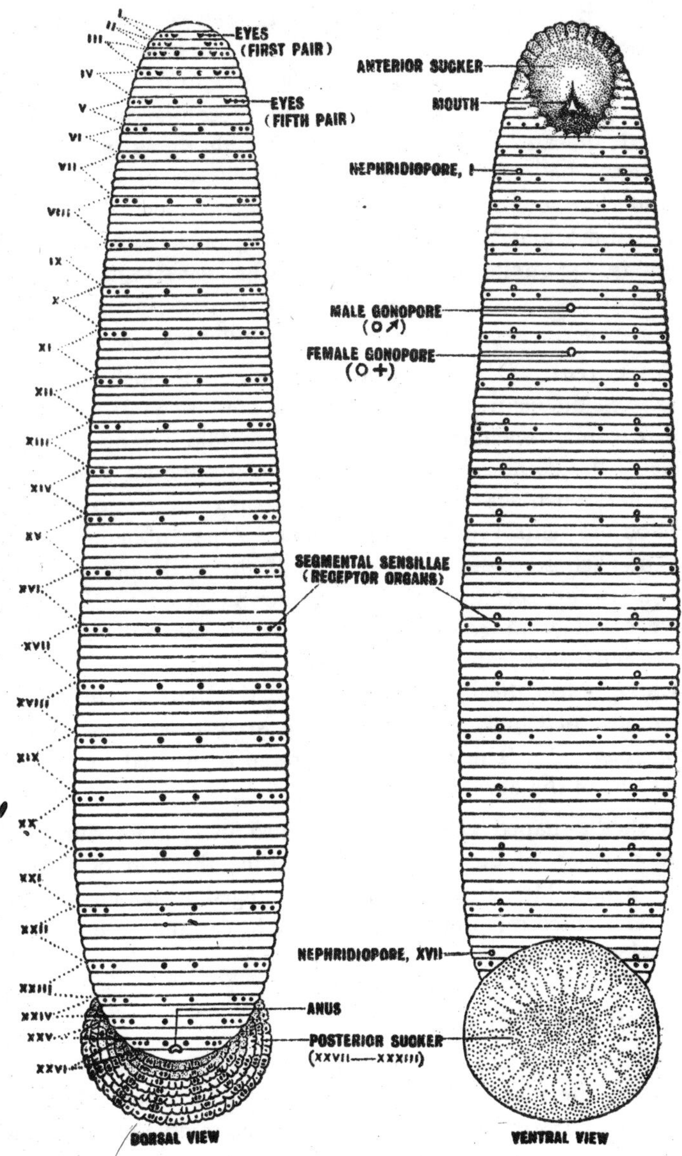

Fig. 76. The leech *Hirudinaria*.

and a number of the segmental receptor organs and not on the basis of the superficial rings. It will be found that at both ends of the body, a segment may be composed of only one or two rings.

There is an adhesive sucker at either end of the leech. The **anterior sucker** is composed of the first five segments. It is oval and cup-shaped, with its hollow towards the ventral surface. There are five pairs of prominent segmental receptors on the dorsal surface of the anterior sucker. These are the **eyes** of the leech. The ventral hollow of the anterior sucker leads into the mouth. The anterior border of the sucker forms an upper lip which is highly sensory and bears the receptors for touch as well as taste. The **posterior sucker** is composed of the last seven segments. It is a muscular disc, circular in outline, and is larger in size than the anterior sucker. It is used as an organ for adhesion and locomotion.

Apertures. (1) The **mouth** is a triradiate opening situated ventrally in the centre of the cup-shaped anterior sucker. (2) The **anus** is a small opening situated in the middle line on the dorsal surface of the 26th segment, in front of the posterior sucker. (3) The urinary openings are the seventeen pairs of minute **nephridiopores** on the ventral surface of the body. A pair of nephridiopores occur in each segment, beginning from the sixth and extending up to the twenty-second. (4) Leech is hermaphrodite. The reproductive openings or *gonopores* are situated in the midventral line, and are separated from one another by five rings. (a) The **male gonopore** lies in the mid-ventral line of the tenth segment and a filamentous *penis* may occasionally be seen protruding through this opening. (b) The **female gonopore** is situated in the mid-ventral line of the eleventh segment.

A saddle-shaped **clitellum** is formed in summer during the breeding season of the leech. At this time the skin of segments 9th, 10th and 11th swells up and is changed in colour and texture. The clitellum produces a protective capsule or **cocoon** for the developing embryo.

Body wall and Body cavity

The body wall of leech consists of a thin cuticle, the epidermis, the dermis, the musculature and the botryoidal tissue. It is best studied in a cross-section passing through the middle of the animal (Fig. 77).

(1) The **cuticle** is a transparent non-cellular membrane which forms a delicate protective covering for the entire body. It is secreted by the underlying epidermis and is cast off periodically to be replaced by a new one.

(2) The **epidermis** is a single layer of columnar cells with their broad ends directed outwards. The spaces between the narrow inner ends of the epidermal cells are filled with a network of capillaries, a small amount of connective tissue and a large number of pigment cells, all of which are derived from the underlying dermis. The epidermis, therefore, forms a vascular membrane through which exchange of gases during respiration may readily take place. Some

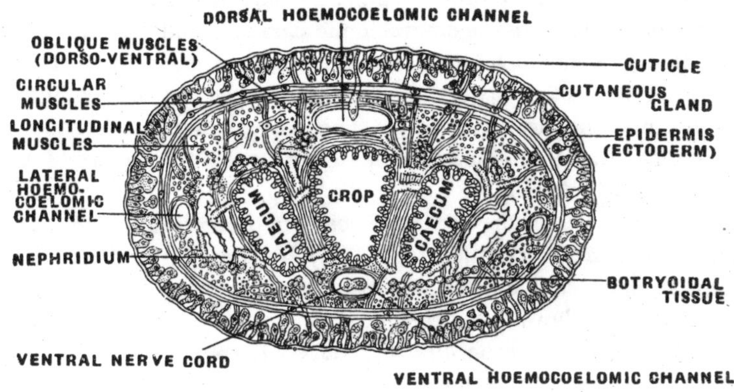

Fig. 77. A cross-section passing through the middle of a leech.

of the epidermal cells are converted into pear-shaped glands which secrete mucus for moistening the skin; others are modified to form the various receptors which act as sense organs.

(3) The **dermis** is a thin basement membrane beneath the epidermis. It consists of fibres, capillary loops and brightly coloured pigment cells, which extend into the epidermis. The cuticle, epidermis and dermis together form the **skin** of the leech.

(4) The **muscles** lie beneath the dermis. This constitutes the thickest part of the body wall. There is an outer layer of circular muscle fibres running round the body, and an inner layer of longitudinal muscle fibres extending along the length of the body. Besides these, there are other bundles running obliquely and from dorsal to ventral surface. The muscles are composed of highly elastic spindle-shaped cells which help the leech to carry on locomotion.

The musculature encloses the body cavity which is filled up by a peculiar kind of spongy connective tissue called *botryoidal tissue*. With the help of this, the layer of longitudinal muscles is firmly adherent to the wall of the underlying alimentary canal.

The **coelom** or body cavity is distinct in a leech embryo. In the adult leech there is no distinct body cavity. The space is obliterated by the botryoidal tissue and reduced to certain longitudinal channels filled with blood. These are known as the hoemocoelomic channels.

The **botryoidal tissue** lies just beneath the longitudinal layer of muscles. It consists of a network of minute canals which are loaded with a dark brown pigment. The free ends of the canals communicate with the hoemocoelomic channels and they are filled with hoemocoelomic fluid. Several canals are enclosed in a thin sheath of connective tissue which adheres to the gut internally and longitudinal muscles externally. The botryoidal tissue is believed to remove waste products but a definite proof in favour of this assumption is wanting.

The gut lies in the centre enclosed by botryoidal tissue. It is represented in the middle of the body by the crop and its caeca. Four hoemocoelomic channels are found running longitudinally between the body wall and the gut wall.

Dorsal to the crop lies the dorsal hoemocoelomic channel, whilst the ventral hoemocoelomic channel, enclosing the nerve cord, is situated ventrally. Lateral to the caeca, there is a lateral hoemocoelomic channel on each side of the body.

Locomotion

There are two principal methods of locomotion—(1) **Swimming,** and (2) **Crawling.**

(1) The leech **swims** gracefully through water by undulating its body in a snake-like fashion. During swimming the dorsoventral muscles are tonically constricted, circular muscles are relaxed, whilst waves of contraction pass along the longitudinal muscles at the two sides of the body alternately.

(2) **Crawling** on a substratum is comparable to that of a looping caterpillar. The suckers play their part alternately as the organs for adhesion. Actual locomotion is effected by changing the length of the body. At first the posterior sucker is attached to the substratum. This produces a wave of contraction to pass forwards

along the circular muscles, whilst the longitudinal muscles are kept in a relaxed state. The result is the extension of the body forwards to its greatest length. The animal now fixes its anterior sucker upon the substratum. This produces contraction of the longitudinal muscles and simultaneous relaxation of the circular muscles. The posterior sucker is released from the substratum and the contracting longitudinal muscles drag the posterior sucker forwards, whereby the body becomes arched and assumes the shape of a loop (Fig. 78—5). The posterior sucker is then fixed to its new position just behind the anterior sucker and the whole process is repeated over and over till the animal reaches its destination.

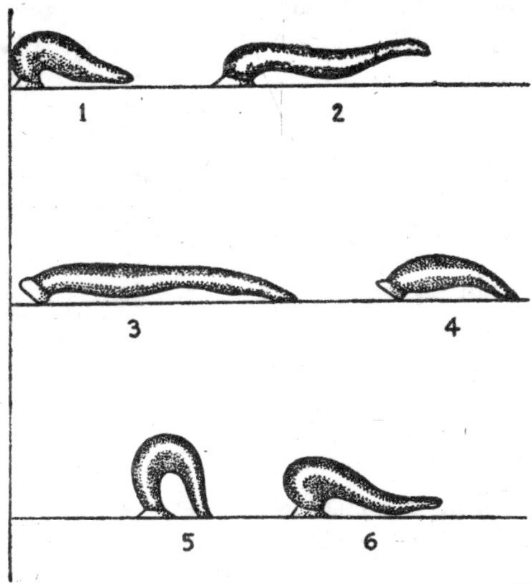

Fig. 78. Successive stages illustrating the crawling movement of a leech.

Alimentary System

The alimentary canal is a straight tube and extends from the mouth in front to the anus behind.

The triradiate **mouth** is guarded by a three-lipped partition called velum. In front of this, the hollow of the anterior sucker forms a *pre-oral chamber.* The mouth leads into the **buccal cavity,** lying just behind the velum. Embedded in the mucous membrane of the buccal cavity there are three *cuticular* **jaws** which can be moved by muscles. One of the jaws is mid-dorsal in position, and the other two are ventro-lateral. The jaw is crescentic in shape and its bears a row of minute teeth on its free semicircular margin. On the sides of the jaw are numerous small button-like **salivary papillae.** The buccal cavity leads into an oval sac called **pharynx** which extends from the fifth to the eighth segment. The pharyngeal wall is supported by well-developed muscles and the interior of the pharynx contains innumerable **salivary glands.** The ducts of the salivary glands open on the salivary papillae. The pharynx leads into the crop through a short and narrow tube, the **oesophagus.**

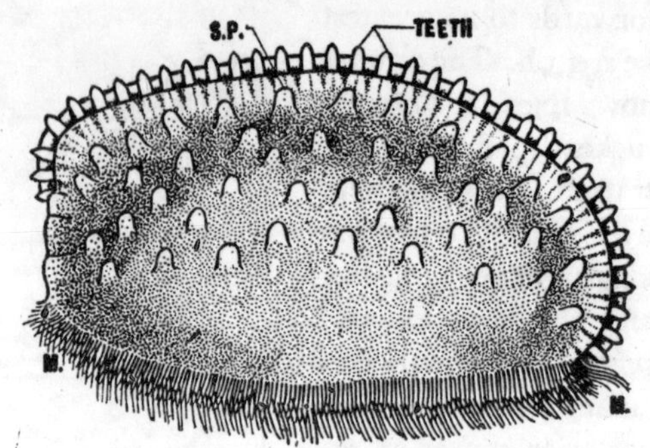

Fig. 79. A jaw of the leech (M—Muscles; S. P.—Salivary papilla).

The **crop** extends from the 9th to the 18th segment and consists of ten thin-walled chambers, each of which is provided with a pair of lateral pockets or **caeca.** The chambers of the crop and the caeca associated with them are serially arranged. The caeca increase in size from before backwards. The first pair are the smallest and the last pair are the largest. All the caeca point backwards and the last pair of caeca may run as far back as the 22nd segment or even beyond—one lying on either side of the intestine. In a recently fed specimen the crop and the caeca are found engorged with blood which has been sucked out of the victim. The crop communicates through a narrow sphinctered opening with a small stomach situated in the 19th segment.

The **stomach** passes imperceptibly into a straight narrow tube, the **intestine** which extends from the 20th to the 22nd segment. The posterior end of the intestine is constricted and becomes continuous with the **rectum** which is a thin-walled tube opening at the **anus,** in the mid-dorsal line of the 26th segment.

Feeding and digestion. The leech is a sanguinivorous animal. It sucks blood from vertebrates like man and cattle. A leech attaches to its victim by the posterior sucker. The cup-shaped anterior sucker is then firmly applied to a suitable part of the skin. The pre-oral chamber flattens out and the jaws are protruded with their serrated margins pressed against the skin. The jaws are moved together by powerful muscles so as to make a triradiate wound, from which blood oozes out freely and passes into the buccal cavity The blood is sucked into the crop by alternate contraction and expansion of the muscular pharynx. The process is similar to pressing and releasing the rubber teat of a dropper for sucking ink to fill a fountain pen. Meanwhile, the blood is mixed with saliva containing **hirudin.** Admixture with hirudin prevents coagulation, so that liquid blood passes easily into the crop and caeca which act as a huge reservoir. The animal sucks three or four times its own weight of blood at a single meal. After this it remains inactive in a hiding place for several months. The blood which is stored in the crop is converted into a jelly-like mass. The erythrocytes swell up and disintegrate and their haemoglobin comes into solution in the blood plasma.

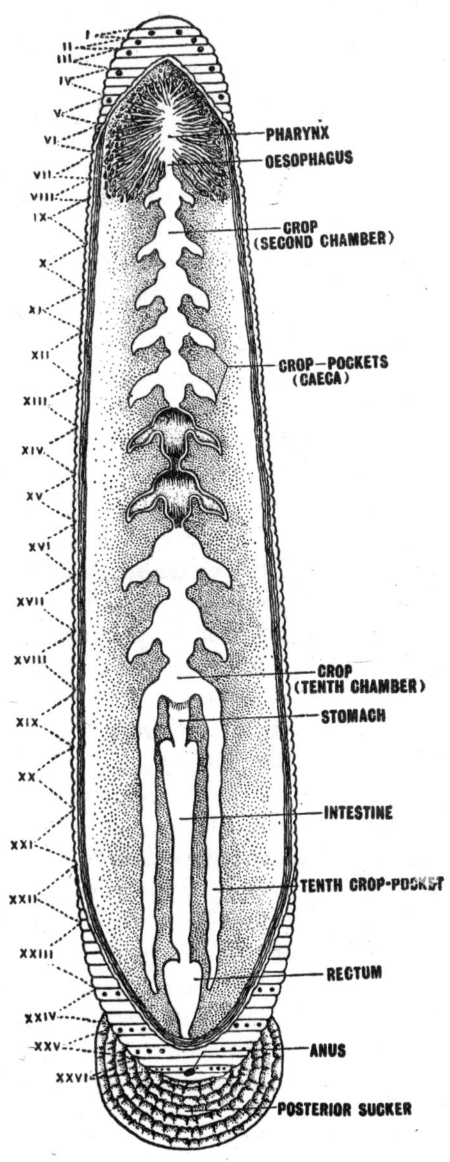

Fig. 80. Alimentary system of the leech.

The animal is very economical as regards the expenditure of its food store. A drop or two of the blood may pass slowly through the sphincter into the stomach, where it is digested and changed into a green mass by the action of proteolytic enzymes. The digested food is absorbed by the folds of the intestine and assimilated. The residue passes into the rectum and voided as a black mass through the anus. A full meal requires about one year to be digested and absorbed.

Respiratory System

The skin forms a large respiratory surface. The cutaneous respiration is facilitated by the secretion of mucus which keeps the skin in a moist state. The skin is profusely supplied with capillaries which penetrate between the epidermal cells. The one-layered epidermis serves as a permeable membrane through which oxygen dissolved in the water is taken in and the carbon dioxide circulating in the hoemocoelomic fluid is given out.

Vascular System

There are no true blood vessels in the leech. As substitutes for blood vessels, there are coelomic channels filled with hoemocoelomic fluid. The **hoemocoelomic** fluid is tinged red with dissolved haemoglobin and contains amoeboid leucocytes. It circulates through *four* hoemocoelomic channels which run along the length of the leech. There is a **dorsal channel** above the alimentary canal, a **ventral chamber** below the alimentary canal, and two **lateral channels,** one on each side of the alimentary canal. They meet one another in the posterior sucker where the ventral channel dilates into a sac to receive the others.

There is no heart for propelling the hoemocoelomic fluid but the lateral channels, having muscular wall and being contractile, perform the function of the heart. Moreover, the presence of valves within the lateral channels prevent back flow. The dorsal and ventral channels are thin-walled and non-contractile. They serve mainly as distributing channels. The lateral channels, on the other hand, serve both as distributing and collecting channels.

The hoemocoelomic fluid flows from behind forwards in the dorsal and lateral channels; but in the ventral channel it flows backwards. The dorsal channel mainly supplies the alimentary canal and the dorsal body wall. The ventral channel supplies the nephridia and the ventral body wall. The lateral channels supply the nephridia and the reproductive organs.

Excretory System

Excretion is mainly effected by a series of glandular bodies called **nephridia.** There are 17 pairs of nephridia—the first pair lying in the sixth segment and the seventeenth pair in the twenty-second segment. The nephridia have no internal openings but they open externally by 17 pairs of **nephridiopores** arranged metamerically on the ventral surface of segments VI to XXII. The nephridiopores are the so-called urinary openings.

A **typical nephridium,** from the middle of the series, is a horse-shoe-shaped mass with two limbs. The main bulk of the nephridium constitutes the **main lobe.** It is situated ventrally on the outer side of the lateral hoemocoelomic channel. Of the two

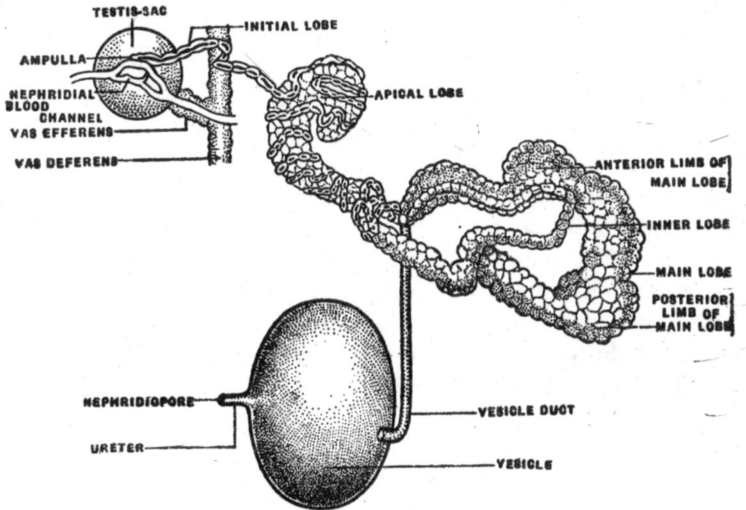

Fig. 81. A typical nephridium of the leech.

limbs of the main lobe, the anterior is longer than the posterior. The *posterior limb* of the main lobe is continued forwards to join a stout **apical lobe.** A narrow **inner lobe** extends between the two limbs of the main lobe fitting into the concavity of the horse-shoe-shaped mass. Twined and coiled round the apical lobe, there is a slender cord-like **initial lobe** which extends from the posterior limb of the main lobe to the nearest testis-sac. The ventral hoemocoelomic channel gives out branches which form a dilated **ampulla** on the wall of the testis-sac. The initial lobe of the nephridium ends blindly over the outer surface of this ampulla. Inside the ampulla, there is a peculiar structure called **ciliated organ,** which corresponds to the nephridial funnel in other annelid worms. But the ciliated organ in leech has lost all connection with the nephridium and takes no part in the extraction of waste products. A tube called **vesicle-duct** arises from the inner end of the *anterior limb*. This tube turns backward to join a thin-walled pyriform bladder called **vesicle.** The bladder communicates with the corresponding **nephridiopore** on the ventral surface of the animal by a short **ureter.**

There is no testis-sac in segments VI to XI. Consequently the initial lobes of the first six pairs of nephridia end blindly in the connective tissue, by the side of the ventral nerve cord. These are the *pretesticular* nephridia. The following eleven pairs of nephridia are connected to the testis-sacs by their initial lobes and are therefore known as the *testicular* nephridia.

A nephridium is a mass of loosely packed glandular cells. There are minute intracellular spaces which join with one another to form fine canaliculi. The vesicular duct arises by the union of these canaliculi at the anterior limb of the main lobe. The nephridia are profusely supplied with branches from ventral and lateral hoemocoelomic channels. The glandular cells extract excess of water and nitrogenous waste products by a process of selective filtration. These pass by the vesicular duct to the bladder and thence eliminated through the nephridiopore.

The **botryoidal tissue** filling up the body cavity is believed to be accessory excretory structures.

Nervous System

The nervous system consists of three parts—the central nervous system, the peripheral nervous system and the sympathetic nervous system.

The **central nervous system** is enclosed within the ventral hoemocoelomic channel and lies ventral to the gut. It consists of a pair of suprapharyngeal ganglia, a pair of peripharyngeal connectives, a subpharyngeal ganglionic mass, and a ganglionated ventral nerve cord with a large terminal ganglionic mass at the posterior end of the body. (a) The **suprapharyngeal ganglia** or *brain* lie close to one another on the dorsal surface of the pharynx in the fifth segment. (b) The **peripharyngeal connectives,** one on each side, form a stout nerve collar surrounding the anterior part of the pharynx in the fifth segment. They connect the *supra-* with the *subpharyngeal ganglionic mass.* (c) The **subpharyngeal ganglion** is situated ventral to the pharynx in the fifth segment. It consists of several ganglia fused together.

(d) The **ventral nerve cord** connects the subpharyngeal ganglionic mass, in the fifth segment, to the terminal ganglionic mass, lying in the middle of the posterior sucker. Throughout its whole course it is situated in the mid-ventral line. The cord bears 21 ganglia, one in each segment, from the sixth to the twenty-sixth. The ventral nerve cord and the ganglia associated with it are really double and consists of two longitudinal chains closely apposed to one another.

(e) The **terminal ganglionic mass,** in the middle of the posterior sucker, is a large oval body formed by the fusion of the last seven pairs of ganglia.

The **peripheral nerves** are paired structures arising from the ganglia. The suprapharyngeal ganglia supply nerves to the first pair of eyes, to the anterior lip and the roof of the buccal chamber. The subpharyngeal ganglionic mass innervates the remaining 4 pairs of eyes and the muscles of the body wall of the first five segments. Each segmental ganglion

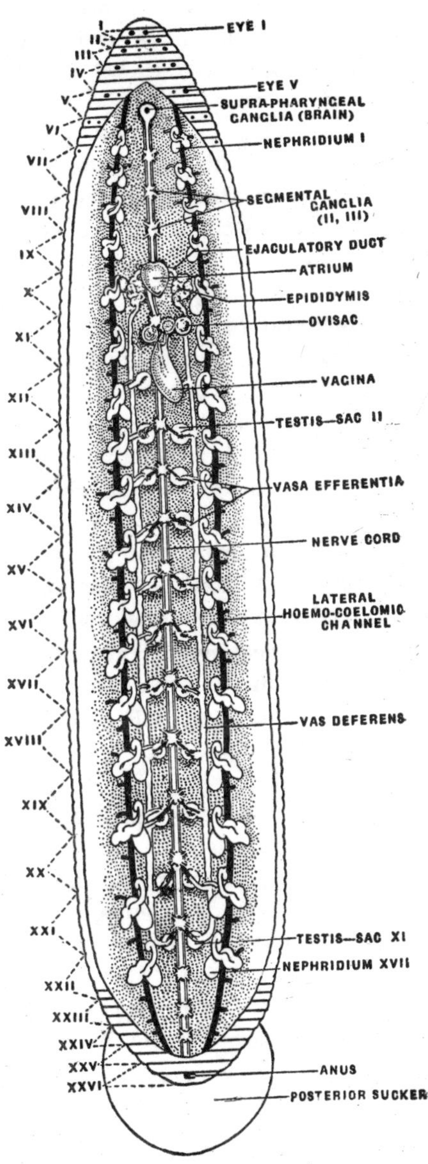

Fig. 82. A dissection of *Hirudinaria* showing excretory, reproductive and nervous systems.

supplies two pairs of lateral branches, which are distributed to the muscles of the body wall and the segmental sense organs. The terminal ganglionic mass sends out branches to the various structures of the posterior sucker. The nerves are of two kinds: (1) **afferent** or sensory which connect the receptor organs to the central nervous system, (2) **efferent** or motor which connect the central nervous system with the effector organs, that is, muscles and glands.

The **sympathetic nervous system** is poorly developed. It consists of an extensive but delicate nerve plexus lying on the dorsal wall of the alimentary canal and supplying the involuntary muscles throughout the body.

Receptor organs. There are 4 kinds of receptors in the leech. These are composed of modified epidermal cells connected with sensory nerve endings.

(*i*) The simplest receptors are the **free nerve endings** found scattered everywhere on the surface of the animal. They occur as tufts of delicate fibres between the epidermal cells, and help the leech to detect changes in the quality of the surrounding water.

(*ii*) The **annular receptors** occur as small papillae arranged in a ring round each annulus of a segment. They are probably concerned with the sense of touch.

(*iii*) The **segmental receptors** are found as raised dots round the first annulus of each segment throughout the body. Most of these are *tactile organs* but there are a few which carry on light-perceiving function.

(*iv*) There are five pairs of **eyes** which occur as black dots on the dorsal surface of the anterior sucker. Each eye is cylindrical in shape, and consists of a number of refractive cells arranged and to end in a longitudinal rows. Each cell is a photo-receptor and contains a crescentic lens surrounded by a thin layer of cytoplasm and a small nucleus. The photo-receptor cells are enclosed in a thick black-pigmented sheath. The top of the eye is covered by a cap of transparent epidermal cells. A nerve enters each eye at its base and distributes branches to the photo-receptor cells. The eyes can distinguish light from darkness. They can also locate the source of light, but most probably they cannot form images of external objects.

Reproductive System

The leech is *monoecious* or *hermaphrodite*, that is, both male and female reproductive organs are found in the same individual.

Male reproductive organs consist of: (1) *testis-sacs*, (2) *vasa efferentia*, (3) *vasa deferentia*, (4) *epididymes*, (5) *ejaculatory ducts*, and (6) *atrium*.

There are eleven pairs of **testis-sacs,** one pair in each of the segments XII to XXII. These are situated ventral to the alimentary canal, between the nerve cord and the nephridia. A testis-sac is whitish in colour and spherical in shape. Being a part of the coelom, its cavity is filled with coelomic fluid. Each testis-sac gives off laterally a short narrow duct called **vas efferens.** All the eleven vasa efferentia of one side open into a long wavy tube known as **vas deferens.** Beginning in the twenty-second segment, the two vasa deferentia run forwards on either side of the nerve cord. At the posterior end of the tenth segment each vas deferens becomes greatly convoluted to form a compact mass called **epididymis.** A narrow but muscular **ejaculatory duct** comes out from the anterior end of each epididymis. The right and left ejaculatory ducts lead inwards to open into a median sac called atrium. The **atrium** is a pear-shaped muscular organ lying anteriorly in the mid-ventral line of the tenth segment. The atrium consists of two parts: a broad anterior part containing the **prostate glands,** and a narrow posterior part or **penis-sac** containing a thin eversible **penis** which is capable of being protruded through the male gonopore on the mid-ventral line of the tenth segment.

Sperm mother cells are budded off from the inner wall of the testis-sacs and develop into spermatozoa. The sperms are carried along the vasa deferentia to be stored within the epididymes. They pass into the base of the atrium where they are glued together to form packets or **spermatophores** by the sticky secretion of the prostate gland. The spermatophores are transferred during copulation into the female gonopore of another leech through the narrow canal of the tubular penis.

The female reproductive organs consist of a pair of *ovisacs*, a pair of *oviducts*, the *common oviduct* and the *vagina*.

There is only one pair of **ovisacs** situated in the eleventh segment. The **ovisacs** are coelomic in origin and each contains a coiled ribbon-shaped **ovary** with dilated ends. The base of each ovisac is continued as a short and narrow **oviduct.** The two oviducts unite in the mid-ventral line of the eleventh segment to form a **common oviduct.** The anterior part of the common oviduct is surrounded by **albumen glands** which open into it.

The posterior part of the common oviduct takes a curved course to open into the vagina. The **vagina** is a pear-shaped muscular bag in the posterior part of the eleventh segment. Its narrow neck leads into the female gonopore which is situated on the mid-ventral line of the eleventh segment.

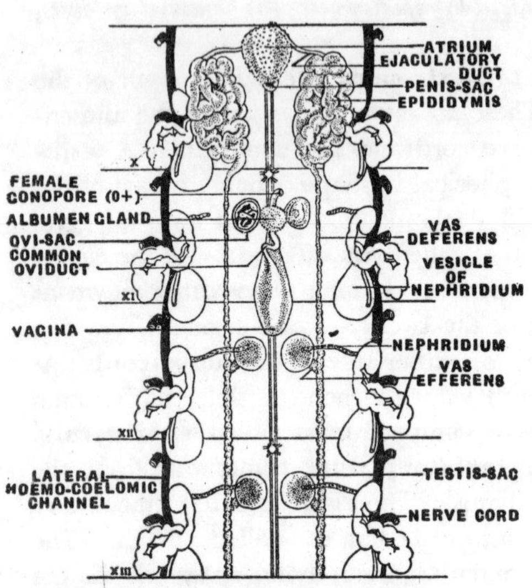

Fig. 83. Reproductive organs of *Hirudinaria* in X to XIII segments.

The ova are budded off from the ovaries. They receive a coating of albumen as they pass through the common oviduct into the vagina where they are stored in readiness for fertilization.

Life-history. The usual breeding season is early spring. Two worms meet and become applied to each other in a head-to-tail position, so that the male gonopore of one lies apposed to the female gonopore of the other. There is a mutual exchange of spermatophores between the two worms and thus **cross fertilization** is ensured. Mating may take place in moist land or under water and the whole process usually lasts for an hour, after which the worms separate. The spermatophores received from the other worm pass into the vagina, where fertilization of the ova takes place.

In the meantime the clitellar segments (9th, 10th and 11th) swell up and secrete a snow-white frothy substance which gradually sets into a jelly-like girdle. The leech now tries to wriggle out and free itself from the soft mass. During the struggle fertilized ova are discharged through the female gonopore, coming to lie in the jelly. The leech ultimately frees itself from the girdle, the two ends of which are quickly plugged. The soft mass, on exposure to air, hardens to form a **cocoon** which contains fertilized ova floating in the albumen secreted by the clitellar glands. The cocoon formation is completed in about six hours, after which they are laid in moist places by the side of a pool.

Each **cocoon** is a barrel-shaped structure about 30 mm. long and half as broad. It is light yellow to amber in colour and is provided with a hard chitinous wall. A cocoon contains about two dozen embryos. Development takes place within the cocoon directly, and young leeches hatch out in about a fortnight. The polar plugs give way to let out the young, which in the meantime have completely used up the albumen. There is no metamorphosis.

CHAPTER VIII

PHYLUM ARTHROPODA : THE JOINTED-LIMBED ANIMALS

The **arthropoda** (*arthron*=joint; *podus*=foot) are the largest group of animals living at the present time. They differ from all other invertebrate animals in possessing a pair of **jointed appendages,** usually in each segment of the body. The arthropod body typically consists of three parts: *head, thorax* and *abdomen;* but in many, the head and thorax unite to form a peculiar *cephalothorax.* The arthropoda are **bilaterally symmetrical** and **metamerically segmented** animals. Each segment of the body is covered with a thick chitinous cuticle which hardens to form a protective armour or **exoskeleton.** Between the segments, the cuticle remains thin and flexible, so that movement is possible. The exoskeleton provides a supporting framework for the soft parts and serves for the attachment of muscles. It is cast off periodically and is then reformed, so that the animal may grow in the meantime. The body cavity is a greatly reduced **hoemocoele** (*haima*=blood, *coel*=cavity) filled with blood. The blood vessels open directly into spaces within the body cavity. Alimentary canal is complete with different types of *mouth parts* in different species, adapted for piercing or biting or sucking or chewing. Respiration is effected either by **gills** or by **tracheal tubes** or by **lung-books. Heart,** when present, lies dorsal to the gut. Excretory system consists of either **green glands** or **Malpighian tubules.** Nervous system is situated ventral to the alimentary canal. Sense organs include **compound eyes.** Sexes are separate. Life-history is often complicated by a larval stage.

The phylum includes diverse forms such as prawns, crabs, centipedes, millipedes, bees, butterflies, scorpions, spiders and kingcrabs. They live wherever life is possible and are adapted to all sorts of habitats.

Classification

There are, at present, 7,50,000 known species of arthropoda. They far exceed the total species of all the other phyla in the animal kingdom. The phylum is divided into the following classes:

Class I. **Crustacea** (*crusta*=a hard shell). Mostly aquatic forms such as prawns, shrimps, crabs, waterfleas and barnacles. A few, such as *Sacculina*, are parasites. The wood-louse, *Oniscus*, is terrestrial.

Class II. **Onychophora** (*onychos*=claw; *phoros*=bearing). Primitive worm-like forms, the only living genus being *Peripatus* and its allies.

Class III. **Myriapoda** (*myrios*=ten thousand; *podus*=foot). Forms with many legs, such as the centipedes and the millipedes.

Class IV. **Insecta** (*insectum*=cut into) or **hexapoda** (*hexa*= six; *podus*=foot). Forms such as butterflies, moths, ants, bees, mosquitoes and cockroaches, with three pairs of legs for walking and usually two pairs of wings for flight.

Class V. **Arachnida** (*arachne*=spider). Spiders, scorpions and kingcrabs.

THE PRAWN

(*Palaemon carcinus*)

Habitat

The fresh-water prawn *Palaemon* is commonly found in rivers, streams and ponds all over India, Bangladesh and Pakistan. It differs somewhat from the marine prawn *Penaeus* which belongs to a separate genus. Both are not commonly available in Calcutta market and relished as food.

Palaemon is habitually a shy animal, hiding by the day and coming to the surface of water at night.

External Feature

A full-grown prawn usually measures about 6 to 8 inches (15 to 20 cms) in length. When freshly caught it is pale blue in colour. The body is elongated and cylindrical, but slightly compressed from side to side. It is composed of two parts: (1) an anterior, rigid, unjointed **cephalothorax** and (2) a posterior, movable, jointed **abdomen** which can be flexed forwards under the cephalothorax.

The **abdomen** is divided into six distinct segments, each bearing a pair of appendages on its ventral surface. At the posterior end of the abdomen, there is a terminal conical piece called **telson** which is devoid of appendages.

The **cephalothorax** is formed by the fusion of the head and thorax. It bears thirteen pairs of appendages on its ventral surface.

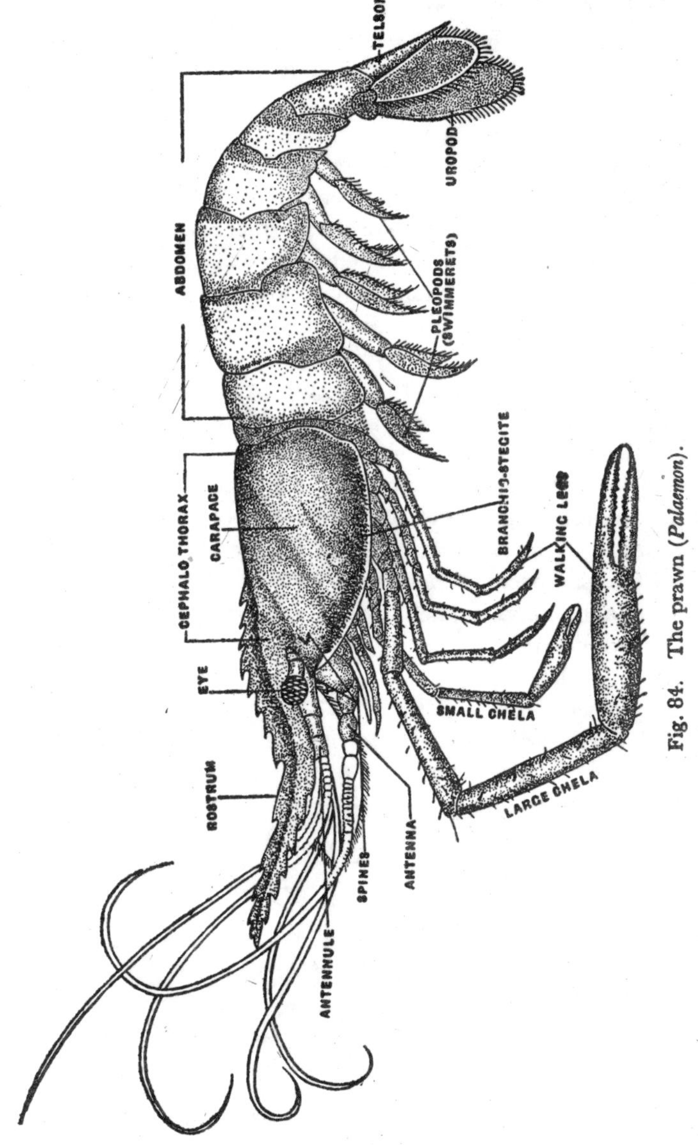

Fig. 84. The prawn (*Palaemon*).

Taking into account that each abdominal segment typically bears a pair of appendages it can easily be inferred that the cephalothorax is composed of at least thirteen appendage-bearing segments. It has been proved conclusively that the first five cephalothoracic segments belong to the head and the remaining eight to the thorax.

Exoskeleton. The soft parts of the prawn are protected by an armour of thick cuticle which is hardened by the deposition of lime salts ($CaCO_3$). The resulting structures constitute the **exoskeleton** of the prawn.

The cephalothorax is covered dorsally by a huge-shield-shaped **carapace** and ventrally by a number of hard calcified sheets called **sternal plates.** The carapace or dorsal shield is produced forward as a long median outgrowth called **rostrum** which bears teeth-like serrations on its upper and lower margins. At the base of the rostrum, on each side of the carapace, there is a distinct *orbital* notch for accommodating the eye. The paired **eyes** are mounted on movable eye-stalks and are lodged in the notch when turned outwards. There are two small **spines** below and behind each eye-stalk; of these two, the anterior spine is longer and is known as the *antennal spine* and the shorter posterior one is the *hepatic spine.* The carapace hangs down freely on either side of the cephalothorax, thus forming a convex outer covering for the narrow-gill-chamber. Each of the gill-covers is known as a **branchiostegite.**

Each abdominal segment is covered by a ring-like calcified piece called **sclerite.** The six successive sclerites are connected with one another by thin uncalcified sheets called **arthrodial membranes** which permit movement of the sclerites.

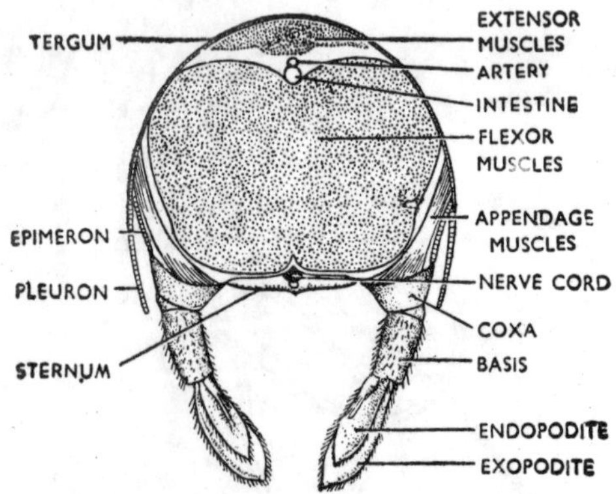

Fig. 85. Cross-section through the abdomen of prawn (diagrammatic).

A **sclerite** is composed of the following parts. The thick arch-like dorsal part of the sclerite is the **tergum,** and the thin plate-like ventral part, placed transversely beneath the abdomen, is the **sternum.** Each side flap of the tergum hangs freely down and is known as the **pleuron.** The pleuron is connected to the appendage of the same side by a small chitinous plate called **epimeron.**

The tergum and pleura of each sclerite slightly overlap those of the succeeding sclerite, thus producing an *imbricate* arrangement. The sclerites are articulated with one another by five hinge-joints which are distinctly visible on each side of the body as orange-coloured spots. The hinge-joints permit flexion and extension of the abdomen in a vertical plane.

Typically there is one pair of appendages in each segment of the body. The prawn carries, in all, nineteen pairs of appendages of which the first five are cephalic, the following eight are thoracic, and the remaining six are abdominal. Each of the first two segments bears a pair of prehensile antennae or feelers. The **first antenna** or **antennule** lies behind the eye. The **second antenna** or **antenna** is

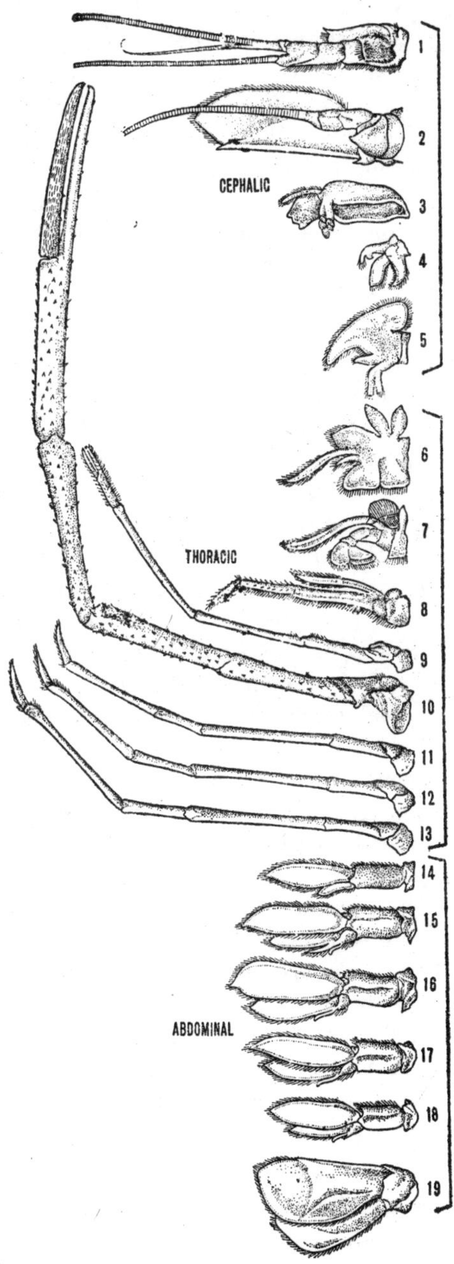

Fig. 86. Appendages of the prawn (taken out from the right side and arranged serially).

just posterior to the antennule. The antennules and the antennae are pre-oral in position, that is, they lie in front of the mouth. Following the antennae, there is a toothed appendage called **mandible** on each side of the mouth; these are used for cutting and crushing food. Behind the mandibles, there are two pairs of leaf-like jaws, the **first maxilla** or **maxillula** followed by the **second maxilla** or **maxilla.** Next to the jaws, there are three pairs of **maxillipedes** and five pairs of **walking legs.** The first and second maxillipedes are flat and foliaceous, while the third maxillipedes and the walking legs are elongated and pediform. The jaws and the maxillipedes are employed for pushing food into the mouth, whereas the walking legs are used for crawling on the river bed. Of the five pairs of walking legs the first two pairs bear pincers at their free ends and are, therefore, called **chelate legs;** these along with the rostrum and the spines are regarded as organs of offence and defence.

The six pairs of abdominal appendages are known as **pleopods** or **swimmerets;** they help the prawn in swimming.

Apertures. There are two median and three paired apertures, opening on the surface of the body.

The **mouth** is a slit-like opening on the ventral surface of the cephalothorax between the third and the fourth segment; it serves for the ingestion of food. The **anus** is a longitudinal slit placed upon the ventral surface of the telson near its base; it serves for the. egestion of faeces. The **gonopores** or reproductive openings differ in position in the two sexes of the prawn. The **male gonopores** are paired apertures, one at the base of a fifth walking leg. The **female gonopores,** however, are found at the bases of the third walking legs. As the prawn is unisexual, the male and female gonopores are never found in the same animal. The paired **renal apertures** or urinary openings are situated internally at the bases of the antennae (second antennae). Besides these, there are two minute openings, one on each antennule (first antenna), by which the statocysts communicate with the exterior. The statocysts are the organs by which a prawn maintains its equilibrium.

Appendages

The appendages of the prawn differ widely for performing different functions. The modification takes place either by variation of shape or by suppression of one or more parts. But all the appendages are built on the same general plan. An appendage is said to be *biramous* because it bears two branches or rami.

A typical **biramous appendage** conists of: (1) a two jointed stalk of **protopodite,** the two segments of which are the **coxa** and the **basis;** the coxa lies proximal to the basis; (2) attached to the basis, there are two branches or *rami*, an outer **exopodite** and an inner **endopodite.** The proximal end of the coxa is lodged into a depression on the ventral surface of the animal, thus forming a ball and socket joint (Fig. 89 A).

There are, as already indicated, nineteen pairs of appendages. According to their position on the body, the appendages are divided into *cephalic, thoracic* and *abdominal.* Of the thirteen pairs in the cephalothoracic region, the first five are cephalic and the following eight are thoracic. The remaining six pairs on the abdomen are abdominal. Appendages of one side are described below.

Cephalic appendages. There are five pairs of cephalic appendages: the antennules, the antennae, the mandibles, the maxillulae and the maxillae.

The **antennule (first antenna)** is situated in front of the mouth just behind the base of the eye-stalk. Its *protopodite* consists of three segments: *precoxa, coxa* and *basis.* The *procoxa* is very large and carries a spine on its outer side. There is a shallow depression on its dorsal surface, bearing the opening of the statocyst. The *statocyst* is a small cuticular sac which lies within the precoxa; it is the balancing organ of the prawn. The *coxa* is short; it is provided with sensory hairs or *setae.* The *basis* is slightly longer than the coxa and carries a pair of many jointed feelers or *flagella.* The outer feeler is divided into two branches, of which the smaller inner branch contains *olfactory setae* for the purpose of smelling. The two feelers are modified *exo-* and *endopodites* of the protopodite.

The **antenna (second antenna)** is also pre-oral in position. Its peduncle or protopodite consists of two segments: *coxa* and *basis.* The excretory organ or *green gland* is lodged inside the *coxa*, and the urinary opening or *renal aperture* is situated on its inner margin. The *basis* gives off two rami—a leaf-like scale or *squama* and a long many-jointed *flagellum* or feeler. The squama represents the exopodite and help the prawn as a balancer in swimming. The flagellum bears tactile setae and represents the endopodite.

Mandible. It is the third appendage and lies on the outer side of the mouth. Its spoon-shaped proximal part is called the *apophysis*, while the solid distal part is the *head.* The head consists of a stout *molar process* and a thin *incisor process.* The molar process is placed at right angles to the apophysis and carries 5 to 6 yellowish-brown teeth. The incisor process is placed in front of the oral aperture and bears three closely set whitish teeth. A three-jointed *mandibular palp* arises from the outer side of the head; it bears sensory setae and lies against the outer side of the incisor process.

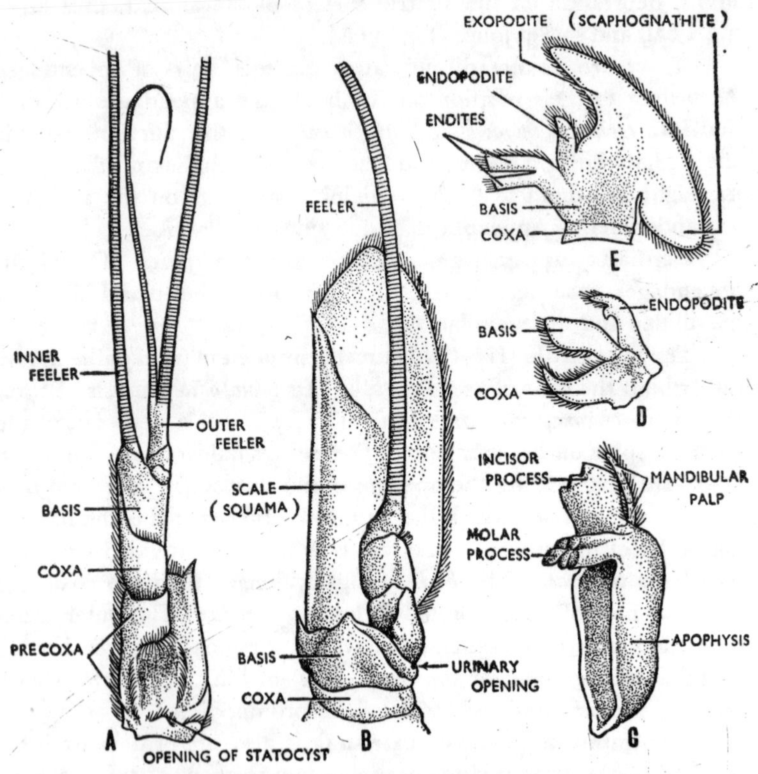

Fig. 87. Cephalic appendages of the right side; A=Antennule, B=Antenna, C=Mandible, D=Maxillula, and E=Maxilla.

The *coxa* of the protopodite is modified to form the apophysis and the head of the mandible. The *basis* of the protopodite corresponds to the proximal segment of the palp and the *endopodite* is represented by the two distal segments of the same. The *exopodite* is absent. The mandible serves as a cutting and crushing apparatus.

Maxillula (first maxilla). It is the smallest of all the appendages and is placed behind the posterior lip of the mouth. The maxillula consists of three foliaceous plates, the margins of which are covered with sensory setae. Of the three plates, the two directed inwards are the jaws or *gnathobases;* these are the coxa and basis of the protopodite. The third, directed outwards represents the *endopodite.* The *exopodite* is absent. The maxillula helps to push food into the mouth.

Maxilla (second maxilla). It is a thin foliaceous appendage placed behind the maxillula. Its peduncle consists of a much reduced *coxa* and a larger bifurcated *basis* which projects inward as the *endites* or jaws. The large fan-shaped *exopodite* is known as the *scaphognathite;* it lies in front of the gill-chamber. The small *endopodite* is placed between the exopodite and the endites. The maxilla has a two-fold function. The endites are used to push food into the oral cavity. The scaphognathite helps in respiration by maintaining a constant flow of water through the gill-chamber.

Thoracic appendages. These consist of three pairs of *maxillipedes* or foot-jaws and five pairs of *walking legs.*

The **first maxillipede** has a *protopodite* consisting of two segments, *coxa* and *basis.* These are flattened leaf-like structures carrying stiff setae on their inner margins. They push food into the mouth. Attached to the outer side of the coxa, there is a bilobed structure called *epipodite* which helps in respiration. The *endopodite* is short but the *exopodite* is long and provided with a thin expansion at its base.

The **second maxillipede** is less foliaceous than the first. Its *protopodite* consists, as usual, of *coxa* and *basis.* The *coxa* is short and covered with setae on its inner margin; it carries a small *epipodite* and a *gill* on its outer margin. The *basis* is immovably articulated to the endopodite. The *exopodite* is long, slender, unjointed and whip-like. The *endopodite* consists of five segments, namely, *ischium, merus, carpus, propodus* and *dactylus.* The propodus and the dactylus are bent and turned backwards forming a knife-like cutting plate.

The **third maxillipede** is pediform or leg-like in appearance. Its protopodite consists of *coxa* and *basis.* The coxa carries a thin epipodite on its outer side. There is a slender unjointed *exopodite* attached to the outer side of the basis and bearing stiff setae. The *endopodite* is three-jointed—*ischium* being fused with the *merus,* and *propodus* with the *dactylus*, while the *carpus* remains free, forming the middle segment of the endopodite.

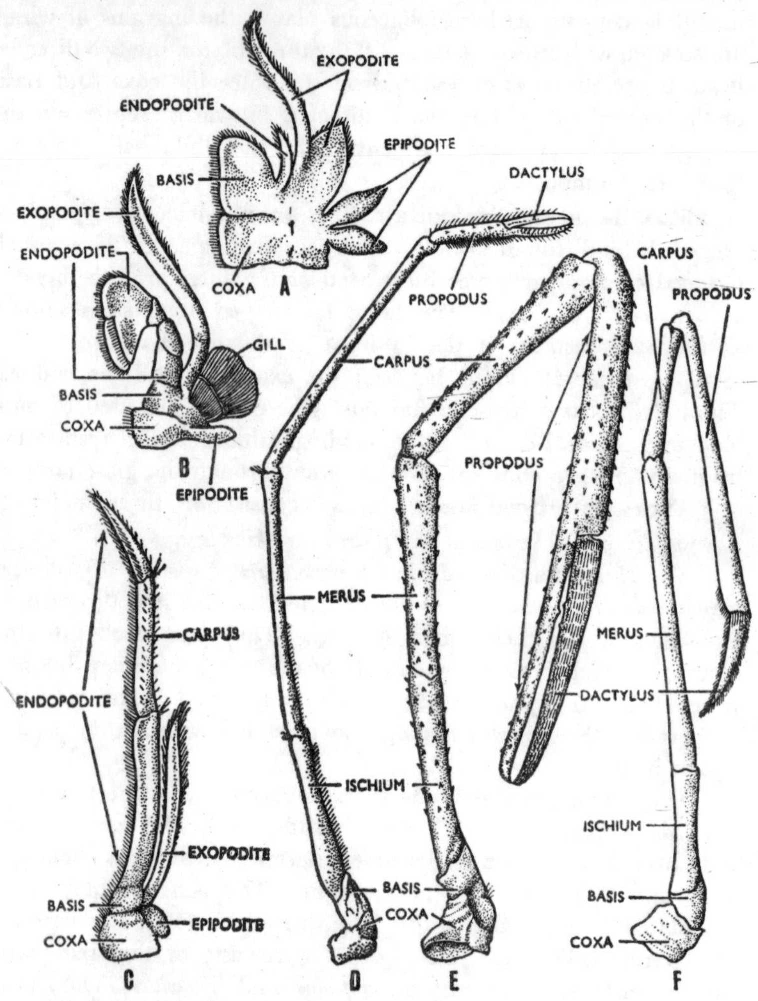

Fig. 88. Thoracic Appendages of the right side; A=first maxillipede, B=second maxillipede, C=third maxillipede, D=first walking leg (small chela), E=second walking leg (large chela), F=a non-chelate walking leg. Third, fourth and fifth walking legs are similar to F.

Walking legs. The prawn is a decapod (*deca*=ten) bearing ten walking legs. Each walking leg consists of seven cylindrical segments joined end to end. The first two segments represent the *coxa* and *basis* of the protopodite and the remainder represent the five segments of the *endopodite*. The *expodite* is absent. The five segments of the endopodite are: *ischium*, *merus*, *carpus*, *propodus* and *dactylus*, the last being clawed. The first and second walking legs carry pincers at their proximal ends; hence they are known as the *chelate legs*. The last three legs are *non-chelate*. The second walking leg is very much larger and bears spines and setae. It is, therefore, known as the **large chela,** whereas the first walking leg is known as the **small chela.** The walking legs help the prawn to crawl upon the river bed. The chelate legs are organs for offence and defence and help in holding the food before being taken. *Gonopores* are situated at the bases of the *third walking legs* in *female*; but in *male*, they lie at the bases of the *fifth walking legs*.

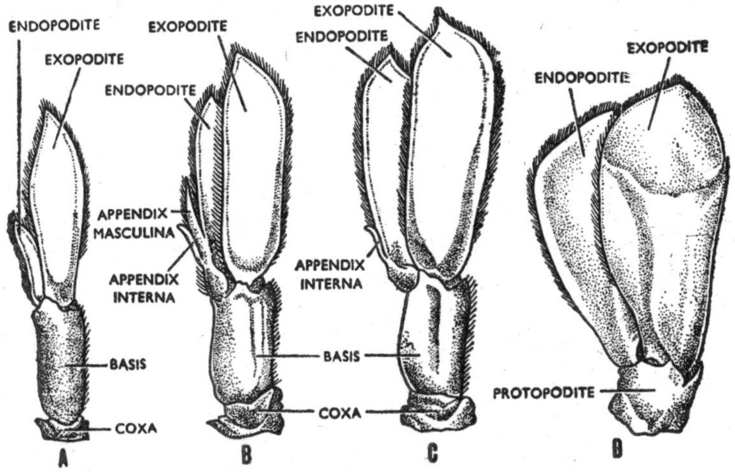

Fig. 89. Abdominal appendages of a male prawn (right side, ventral view); A=first pleopod, B=second pleopod, C=third pleopod, D=uropod. Fourth and fifth pleopods are similar to C.

Abdominal appendages. There are six pairs of abdominal appendages called **pleopods** or **swimmerets** which help the prawn in swimming. The last pair are specially called **uropods** because they form with the telson a powerful tail-fin which is used for leaping backward.

Each **pleopod** is a typical biramous type of appendage. Its protopodite consists of a short proximal *coxa* and an elongated distal *basis*. The basis carries a pair of leaf-like rami, the outer *exopodite* and the inner *endopodite*. Both the rami are covered with tactile setae. The endopodite is smaller than the exopodite. Each of the second, third, fourth and fifth pleopod bears on the inner side of its endopodite a hook-like rod called *appendix interna*. The appendices internae of the two sides interlock with one another during the breeding season, thereby forming a efficient basket for carrying eggs. The **second pleopod,** in male, bears another additional rod-like process, called *appendix masculina* (Fig. 89 B).

The **uropod** has a *protopodite* of only one segment. The *rami* are oval and fan-like; *exopodite* is divided by a fine suture and bears a spine near its base, but the *endopodite* is undivided. Both the rami are covered with setae along their margins.

Movement

Prawn is capable of three kinds of locomotion: **walking, swimming** and **darting.**

In **walking,** the body is held straight, the abdomen is extended, the five pairs of walking legs remain in contact with the substratum and moved in an orderly manner. The pincers are held in front and the feelers of the antennules and the antennae are in continuous motion making a survey of the environment.

In **swimming,** the animal progresses through the water by rapid paddle-like movement of its swimmerets.

In **darting,** the prawn flexes its abdomen under the cephalothorax. It then violently extends the abdomen putting pressure on the water by its fully expanded oar-like tail-fin. This is effected so quickly that the entire animal jerks backward, covering a considerable distance. The movement is repeated frequently whenever the animal apprehends danger. It is, therefore, known as the **escape movement.**

Alimentary System

The alimentary system is composed of the alimentary canal and the digestive gland.

The alimentary canal extends from mouth to anus, and consists of three distinct parts: (1) foregut or *stomodoeum,* (2) midgut or *mesenteron,* and (3) hindgut or *proctodoeum.* The foregut and the hindgut are lined internally by a thick cuticle which is continuous with the outer covering of the body; the midgut, however, is lined by a soft layer of endoderm. The **foregut** includes *mouth, buccal cavity, oesophagus,* and *stomach;* the **midgut** is the long and narrow *intestine;* the **hindgut** consists of the terminal bulb or *rectum* and the *anus.*

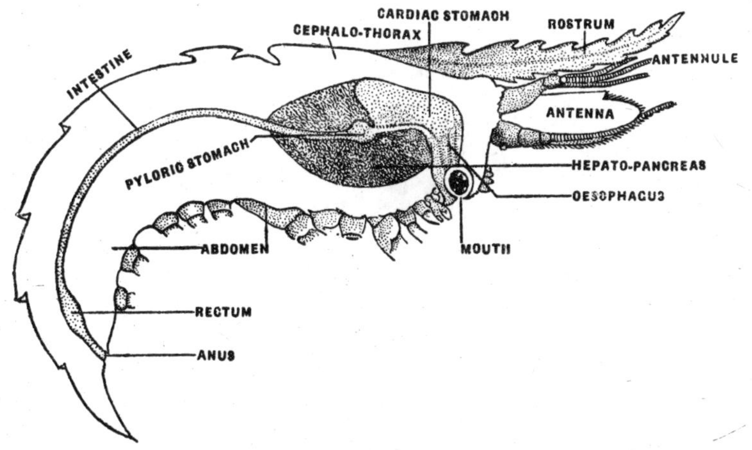

Fig. 90. Alimentary system of prawn.

The **mouth** is a large opening on the ventral side of the cephalothorax between the third and the fourth segment. It is bounded anteriorly by a shield-shaped *labrum,* posteriorly by the *labium,* and on each side by the incisor process of a *mandible.* The **buccal cavity** is a short and narrow chamber which connects the mouth with the oesophagus. The **oesophagus** is a wide rectangular tube running vertically upwards to join the stomach. The **stomach** occupies the greater part of the cephalothorax; it consists of two parts: (1) a large bag-like **cardiac stomach,** and (2) a small narrow **pyloric stomach** placed behind and beneath the former. Two parts of the stomach are separated from one another by a number of valves. The inner wall of the **cardiac stomach** is raised into ridges and strengthened by hard cuticular plates covered with bristles; these help in grinding and guiding food towards proper channel. Thus, there is a *circular plate,* covering the oesophageal opening in the anterior part of the cardiac stomach and immediately behind this, but on the roof, there is a *lanceolate plate.* On the floor of the cardiac stomach and lying in the midventral line, there is a triangular *hastate plate* resembling the head of a spear. The broad base of the hastate plate extends posteriorly, forming the anterior margin of the cardio-pyloric opening. A curved *combed plate* is found on each side of the hastate plate; it carries delicate bristles on its inner margin, thereby resembling a comb. A narrow tunnel called *lateral groove* lies on either side of the hastate

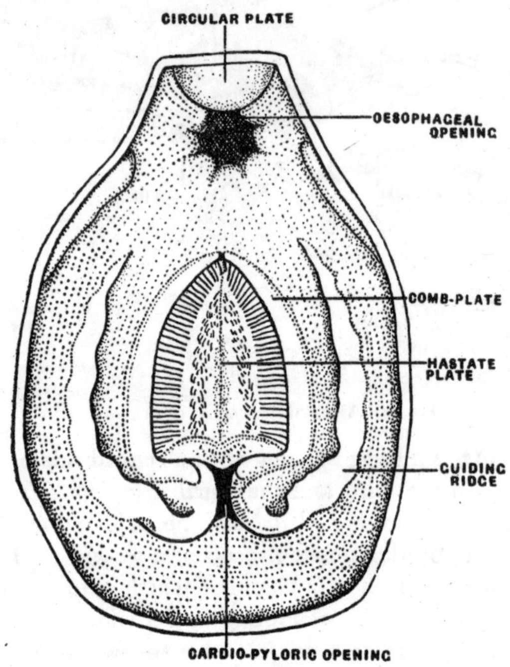

Fig. 91. Diagrammatic view of the floor of the cardiac stomach.

plate; it is a drain pipe covered by the thickly set bristles of the combed plate and leading into the cardio-pyloric opening. Thus a *filter bed* is formed on each side of the hastate plate, permitting the passage of finely divided food particles into the underlying drains. On the outer side of each combed plate, there is a raised longitudinal fold called *guiding ridge*. They are so named because they serve to direct the contents of the cardiac stomach towards the cardio-pyloric opening. The **pyloric stomach** is very small in comparison to the huge cardiac stomach. It receives food either in solution or in a fine state of division through the cardio-pyloric opening. Digestive juice is poured into the pyloric stomach by two hepatopancreatic ducts, one on each side. The pyloric stomach leads into the midgut or **intestine** which is the longest part of the alimentary canal. It is usually found as a narrow black tube running towards the posterior end as far back as the sixth abdominal segment. It lies in the middorsal line, just beneath a thin layer of extensor muscles. The supra-intestinal artery accompanies the intestine throughout its course and lies dorsal to it. The midgut is continued into the short hindgut, the first part of which is dilated to form an intestinal bulb or **rectum.** The rectum narrows down to the **anus**—a slit-like opening on the ventral surface, near the base of the telson.

Hepatopancreas is the only *digestive gland* of the prawn. It is a large orange-coloured body with soft and fragile tissues, occupying a considerable portion of the cephalothorax. The pyloric part of the stomach and the commencement of the midgut are completely surrounded by the hepatopancreas. The gland has two lobes, from each of which arises a hepatopancreatic duct. These open separately into the pyloric stomach and carry digestive juice secreted by the gland. The hepatopancreatic juice contains enzymes for digesting proteins, fats and carbohydrates. Digested food is partly absorbed by the gland through its ducts and stored for future use. The hepatopancreas, therefore, not only secretes digestive enzymes but also absorbs digested food and serves as an important storage-organ.

Mechanism of feeding and digestion. Prawn feeds upon mosses, algae, insects and decaying organic matter. It feeds actively at dusk and just before daybreak. Food is seized by the chelate legs and pushed into the buccal cavity by the maxillipedes, maxillulae and maxillae. The food is cut into small bits and crushed by the mandibles. A series of wavy peristalsis push the food along the oesophagus towards the cardiac stomach, the walls of which contract and expand alternately. The food is thereby sucked into the stomach.

The food is thoroughly churned and to a certain extend crushed by the contraction and expansion of the sac-like cardiac stomach against the cuticular plates in its interior. Small particles of food are strained through the bristles of the combed plate into the lateral grooves, which act as drain pipes and guide the same into the pyloric stomach. The food is digested by the action of the digestive juice from the hepatopancreas. Absorption of the digested food takes place in the intestine and residue is egested through the anus. A part of the digested food is directly absorbed by the hepatopancreas and stored there for future use.

Respiratory System

Prawn is adapted for utilising oxygen dissolved in water for the purpose of breathing. The blood of this animal contains a respiratory pigment called **haemocyanin,** which acts in the same way as haemoglobin of other animals.

The **respiratory organs** of the prawn consist of: (1) the *lining membrane of the branchiostegite,* (2) three pairs of *epipodites,* and (3) eight pairs of *gills.* These are situated in the gill-chambers, one on each side of the cephalothorax. The **gill-chamber** is bounded externally by the branchiostegite and internally by the cephalothoracic wall; it is closed dorsally but communicates with the exterior along its anterior, posterior and ventral borders. Respiratory organs of one side are now described.

(1) The highly vascular and thin **lining membrane of the branchiostegite** forms a huge respiratory surface through which O_2 is readily taken in and CO_2 is given out.

(2) The **epipodites** are small leaf-like membranes attached to the bases of the maxillipedes. They are thin and vascular and therefore take part actively in carrying out respiration.

(3) The **gills** are situated in relation to the thoracic appendages. Out of the eight gills, seven are directly exposed by cutting the branchiostegite away, but the third remains hidden under cover of the second gill. The gills are more or less crescent-shaped in outline. They increase in size antero-posteriorly, so that the first gill is the smallest and the last one is the largest. Each gill has a long narrow base supporting two rows of flat and rectangular gill-plates which are arranged like the leaves of a book. With reference to its position and point of attachment, a gill may be either *podobranch,* or *arthrobranch,* or *pleurobranch.* The first gill is attached to the coxa of the second maxillipede; hence it is a podobranch. The next two are attached to the arthrodial membrane articulating the third maxillipede with the body; hence, these two are arthrobranchs. The remaining five gills are pleurobranchs, because they

are attached to the outer side of the cephalothorax above the articulation of the walking legs.

Each gill receives deoxygenated blood through an afferent branchial channel, and aerated blood leaves the gill by an efferent branchial channel. The afferent and efferent channels are joined by narrow connectives which are distributed on the gill-plates.

Mechanism of respiration. Constant supply of oxygenated water to each gill-chamber is maintained by the paddle-like movement of the **scaphognathite** of the maxilla. Water rushes in

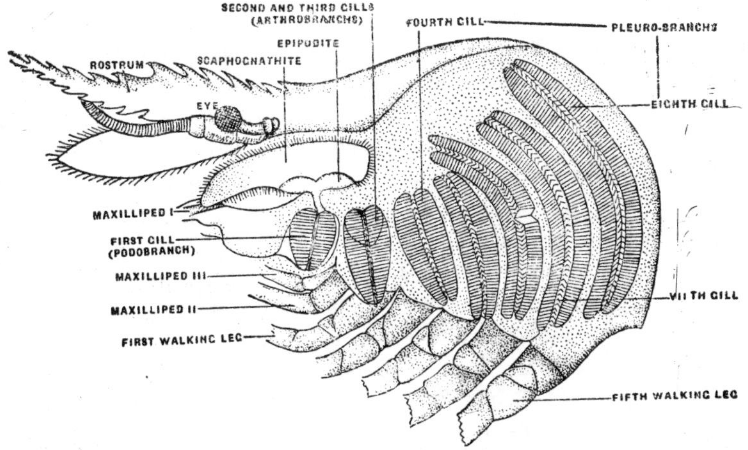

Fig. 92. Respiratory organs in the left gill-chamber of prawn.

along the posterior and ventral borders of the branchiostegite, and is expelled from the anterior end of the gill-chamber by the baling action of the scaphognathite and by the vibratory motion of the exopodites of the maxillipedes.

The gills, the epipodites, and the lining membrane of the branchiostegite are richly supplied with blood. All of them are constantly bathed in fresh-water containing dissolved oxygen. The gaseous exchange between the blood and the water takes place by diffusion—oxygen being taken in and carbon dioxide given out.

Vascular System

As in all arthropoda, the vascular system is open and the blood vessels freely communicate with spaces surrounding the viscera. These spaces together form the body cavity which is full of blood and is, therefore, not a true coelom but a **hoemocoele.** The true coelom, however, is restricted to the cavities of the gonads.

The vascular system of prawn consists of : (*i*) the *heart* surrounded by the *pericardium*, (*ii*) the *arteries* or true blood vessels with definite muscular wall, and (*iii*) the irregular *sinuses* or *blood lacunae* into which the arteries open directly.

Heart. The heart is roughly triangular in shape. It is placed in the middorsal line at the posterior part of the cephalothorax. The pointed apex of the heart is directed forwards and the broad base is directed backwards. It lies beneath the carapace, dorsal to the gonads, and is held in position by a muscular cord, the *cardio-pyloric* strand, which binds it down to the pyloric stomach. The heart is enclosed by a special *hoemocoele*, the *pericardial sinus*, the thin wall of which forms the **pericardium.** The muscular wall of the heart is pierced by five pairs of slit-like openings called **ostia** (*ostium*)=door). Each ostium is a one-way passage permitting blood to flow from the pericardial sinus into the heart. The lips of the ostia act as valves and prevent leakage in a reverse direction. Of the five pairs of ostia, one pair is found on the dorsal surface, another pair on the ventral surface, a third pair on the posterior border, and the remaining two pairs pierce the lateral borders of the heart.

The cavity of the heart is traversed by a large number of interlacing muscle fibres; it is not a continuous cavity and appears to be spongy in structure.

Arteries. Receiving blood from the pericardial sinus, the heart pumps out the same into the arteries which are vessels with strong muscular wall.

One median and two paired arteries originate from the apex of the heart. These include (a) a *median ophthalmic artery*, (2) paired *antennary arteries*, (3) paired *hepatopancreatic arteries*.

The slender **median ophthalmic artery** runs forward up to the roof of the oesophagus, where it ends by anastomosing with the antennary arteries; it lies beneath the carapace and along the mid-dorsal line. The paired **antennary arteries** arise from the apex of the heart, one on either side of the median ophthalmic. While passing forward, each antennary artery supplies: (1) a *pericardial branch* to the pericardial sinus, (2) a *gastric branch* to the stomach, and (3) a *mandibular branch* to the mandibular muscle. Giving off these three branches, the antennary artery extends up to the base of the antenna of its own side and breaks up into a dorsal and a ventral division. The dorsal division turns inward, supplies an *optic artery* to the eye, and finally ends by anastomosing with its fellow of the opposite side and with the median ophthalmic. An arterial circle is thus formed anteriorly which gives off a pair of *rostral arteries* to the rostrum. The ventral division of the antennary artery breaks up into two branches, one supplying the antennule (first antenna) and the other supplying the antenna (second

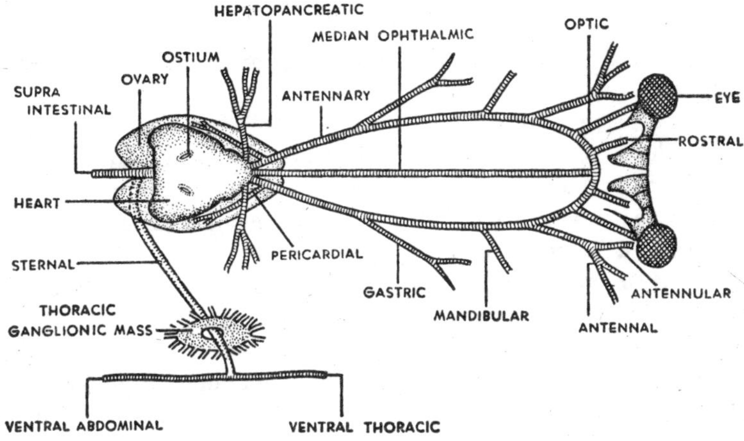

Fig. 93. Heart and main arterial vessels of prawn.

antenna). The paired **hepatopancreatic arteries** originate from the ventral surface of the apex of the heart. Each lies on the outer side of the corresponding antennary artery and plunges directly downwards to supply the hepatopancreas.

A single median artery arises from the postero-ventral surface of the heart. It soon divides into two branches: (1) a slender *supra-intestinal artery*, and (2) the stout *sternal artery*. The **supra-intestinal artery** runs towards the posterior end, lying dorsal to the midgut; it extends as far back as the rectum and supplies branches to the intestine and dorsal muscles. The **sternal artery** is the stoutest of all the blood vessels; it plunges obliquely downwards and passing through an aperture in the thoracic ganglionic mass of the ventral nerve-cord it comes to lie on the ventral side. The artery now divides into two branches: (1) a large *ventral thoracic artery* which extends forwards up to the mouth, supplying blood to the ventral part of the cephalothorax, and (2) a *ventral abdominal* *artery* which runs backward up to the anus, supplying blood to hindgut and the ventral region of the abdomen.

The arteries ramify into all the parts of the body, forming small hair-like vessels which open directly into the hoemocoelic body cavity. These hair-like vessels with open mouth are not true capillaries, because true capillaries are always closed vessels occurring in the form of a network.

Blood-sinuses and blood channels. Blood flows through irregular spaces or *lacunae* and is finally received into a pair of *ventral sinuses.* Each **ventral sinus** is an irregular space beneath the hepatopancreas and extends lengthwise for a short distance into the abdomen. The two sinuses communicate with one another by several connecting channels.

From each ventral sinus blood is supplied to the gills of the same side by six **afferent branchial channels.** Of the six, the last five carry blood to the pleurobranchs, whereas the first supply blood to the two arthrobranchs and the single podobranch.

Aerated blood is brought back from the gills by six pairs of **efferent branchial channels** which drain into the **dorsal** or **pericardial sinus.** From the pericardial sinus the blood flows through the ostia into the heart.

Blood. The blood is a clear fluid containing amoeboid *leucocytes.* Coloured erythrocytes are total absent. The respiratory pigment is a compound of copper with a protein substance, and is known as **haemocyanin.** It remains in a dissolved state in the

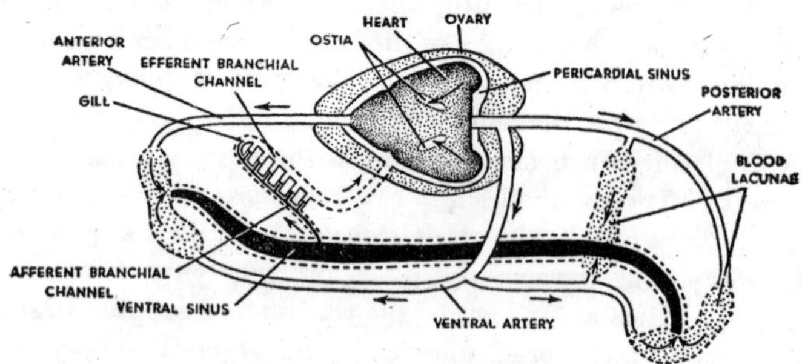

Fig. 94. Diagram illustrating the course of blood-circulation in prawn.

blood plasma. When oxygenated, the blood appears to be pale blue; deoxygenated blood is colourless.

Course of circulation. Deoxygenated blood from blood-lacunae in different parts of the body is collected into two ventral sinuses. It flows to the gills through six pairs of afferent branchial channels. Blood is aerated in the gills and brought back into the pericardial sinus by six pairs of efferent branchial channels. From there, the blood enters the heart through five pairs of ostia. The heart, by systolic contraction, pumps the aerated blood into the arteries. Leakage of blood into the pericardial sinus is prevented by the tightly closed lips of the ostia. Arteries ramify into distant parts of supplying the tissues with O_2. There is no capillary network; the hair-like branches of the arteries open directly into the blood lacunae. By this time, the blood has become deoxygenated. It slowly collects in the ventral sinuses and then sent back to the gills for aeration.

Excretory System

The excretory organs of the prawn consist of a pair of *green glands*, one inside the coxa of each antenna (second antenna). Because of their position, the green glands are also known as the **antennary glands.**

A **green gland** consists of three parts: (1) an *end-sac*, (2) a *labyrinth*, and (3) a *bladder*. The three parts are tightly compressed to form a small opaque body about the size of a pea-seed. The bean-shaped **end sac,** containing a blood lacuna, is placed in the middle, between the labyrinth and the bladder. The **labyrinth** is the gland proper. It is situated on the outer side of the end-sac and is composed of a mass of excretory tubules opening into the bladder. The **bladder** is a thin-walled sac lying on the inner side; it communicates with the urinary opening by a short **ureter.** The urinary opening or **renal aperture** is situated on a raised papilla at the base of the antenna (second antenna).

The green glands are joined to a large median **renal sac,** which lies on the dorsal surface of the cardiac stomach just beneath the carapace. The renal sac is thin-walled and transparent. It is connected to the bladder of each green gland by a narrow tube called **lateral duct.**

Ammoniacal compounds along with excess of water are separated from the blood in the end-sac, and collected in the form of urine by the excretory tubules of the labyrinth. The tubules carry the urine into the bladder. The urine is voided through the renal apertures, the renal sac serving as a temporary reservoir.

The integument of prawn actively separates nitrogenous waste products which are stored in the hard outer shell of its body and eliminated at the time of moulting. The skin, therefore, is an accessory excretory organ.

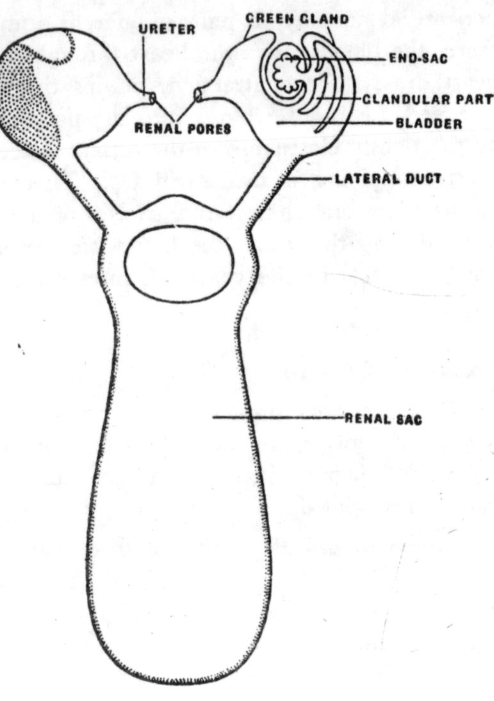

Fig. 95. Excretory organs of prawn.

Nervous System

The nervous system of the prawn is built on the same plan as that of the leech. It is composed of three parts: (1) central nervous system, (2) peripheral nerves, and (3) sympathetic nervous system.

The **central nervous system** consists of (*i*) a pair of *supra-oesophageal ganglia* forming the *brain* and connected by (*ii*) a pair of *circumoesophageal commissures* with (*iii*) a ventral *thoracic ganglionic mass*, and (*iv*) a median ganglionated *ventral nerve-cord* extending backwards from the thoracic ganglionic mass up to the base of the telson (Fig. 96).

The **brain** is a bilobed mass slightly cut in the middle to form the two **supra-oesophageal ganglia.** It lies dorsally beneath the base of the rostrum, at the junction of the oesophagus with the cardiac stomach. Each supra-oesophageal ganglion supplies nerves to innervate the eye, the antennule, and the antenna of its own side. (1) The **optic nerve** enters the eye-stalk from the outer side of the ganglion and supplies the retinal region of the eye. (2) The **antennular nerve** arises anteriorly; it enters the base of the antennule and breaks up into branches to supply the different

parts of the appendage, including the statocyst which is lodged inside the pre-coxa. (3) The stout **antennary nerve** originates posteriorly; it runs obliquely backward and the curves forward to enter the base of the antenna. It sends a branch to the green gland and then breaks up to innervate the different parts of the appendage.

The **circum-oesophageal commissures** are stout cords which connect the brain with the thoracic ganglionic mass. Arising from the posterior parts of the brain, the two cords run backwards and slightly downwards to embrace the oesophagus. They meet posteriorly in the midventral line to join the thoracic ganglionic mass. There is a slender nerve loop connecting the two commissures just before they meet behind the oesophagus. Because of its position, this

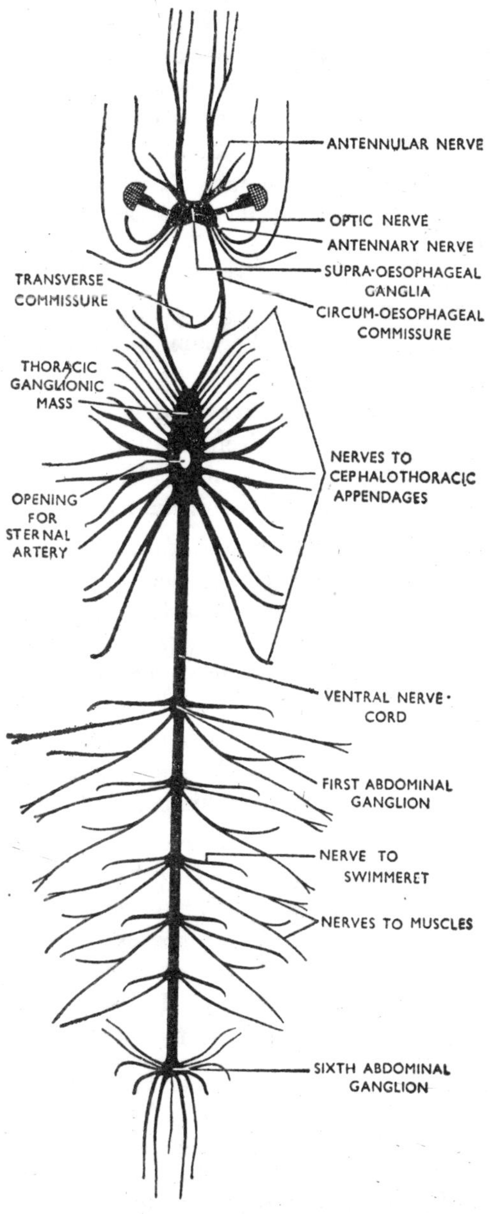

Fig. 96. Nervous system of prawn.

transverse commissure is known as the *postoesophageal loop.* A small nerve originates from the posterior part of each commissure to supply the mandible of its own side.

The **thoracic ganglionic mass** is a large oval body lying at the commencement of the ventral nerve-cord, in the midventral line of the cephalothorax. It is composed of a number of ganglia which are fused together and cannot be distinguished as separate structures. As many as eleven pairs of nerves arise from the ganglionic mass and supply the cephalothoracic appendages excepting the antennules and the antennae. The sternal artery passes through a small opening in the posterior part of the ganglionic mass.

The **ventral nerve-cord** lies beneath the mass of abdominal muscles. Beginning from the posterior end of the thoracic ganglionic mass, it extends along the mid-ventral line and terminates posteriorly near the base of the telson.

Corresponding to the number of abdominal segments, there are six ganglionic enlargements in the course of the abdominal part of the nerve-cord. Each such ganglion is really composed of two fused ganglia. In fact the venrtal nerve-cord, although it appears to be outwardly single, is the product of fusion of two cords placed side by side. Each abdominal ganglion supplies branches to innervate the muscles and the appendages of its own segment. The sixth abdominal ganglion is the largest of the lot and supplies, in addition, the hindgut and the muscles of the telson.

The **peripheral nerves** originate from the ganglia of the central nervous system. They have been already described. Each nerve is composed of a bundle of nerve fibres. The nerve fibres are of two kinds: *afferent* or **sensory** and *efferent* or **motor.** The afferent fibres are distributed to the receptor organs such as eyes, etc. The efferent fibres, on the other hand, innervate the voluntary muscles and the glands. Most of the nerves are **mixed** containing both kinds of fibres.

The **sympathetic nervous system** consists of a few small ganglia connected by slender nerves. These are situated chiefly on the dorsal surface of the cardiac stomach and supply branches to innervate the involuntary muscles of the viscera.

A fine *sympathetic nerve* arises from the posterior part of the brain and runs backwards along the roof of the cardiac stomach. It bears two small ganglia called *visceral ganglia,* one behind the other. The anterior visceral ganglion

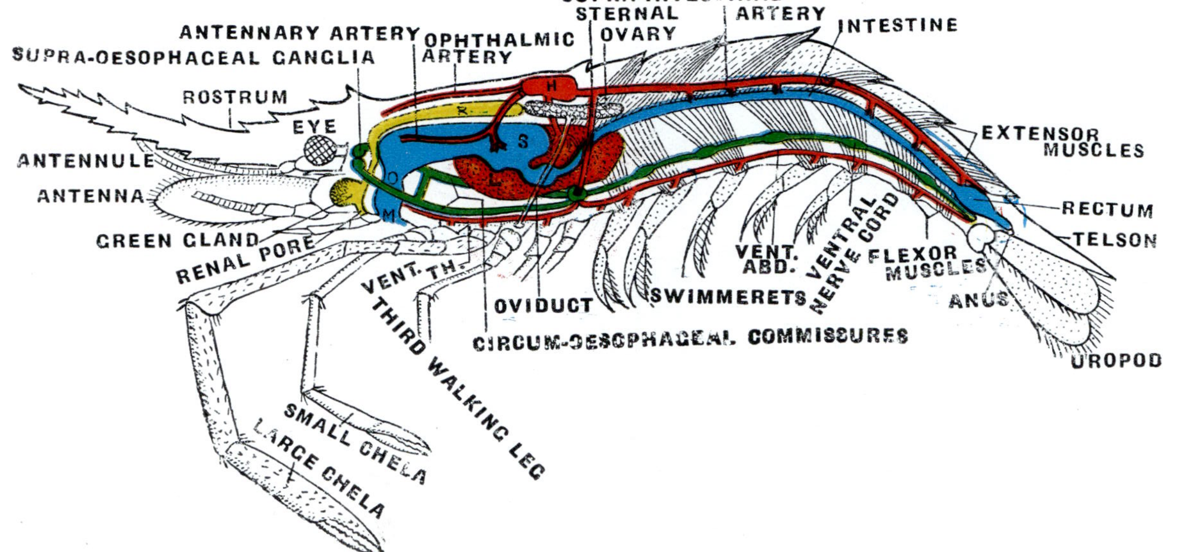

Internal structure of a female prawn.

H=heart; L=hepatopancreas; M=mouth; O=oesophagus; R=renal sac; S=stomach; Vent. Th.=ventral thoracic artery; Vent. Abd.=ventral abdominal artery.

OUT. BIO.

is connected by a slender *visceral loop* with a pair of small *commissural ganglia*, lying along the course of the circumoesophageal commissures. The posterior ganglion is free and gives out paired nerves.

It is to be noted that the brain and the ventral nerve-cord of the prawn are solid structures without cavities. They, therefore, differ entirely from the hollow brain and the dorsal, tubular spinal cord of the toad.

Receptor Organs

The sense organs consist of: (1) paired eyes, (2) paired statocysts, (3) tactile organs, and (4) olfactory setae.

The **eyes** are hemispherical in shape and situated at the anterior end of the cephalothorax, one on either side of the carapace, near the base of the rostrum. Each eye is mounted at the end of a two-jointed **eye-stalk** which can be moved forwards and backwards, as well as upwards and downwards. The eye is **compound,** because it is composed of a large number of a simple visual elements called **ommatidia** or **ocelli.** Such compound eyes are almost universally found in all arthropod animals. The outer surface of the eye is convex and roughly divided into a large number of square facets placed in juxtaposition like the squares on a graph paper. Each facet corresponds to a single ommatidium or ocellus (Fig. 97A). The ommatidia are arranged regularly along the radii of the hemispherical eye. (Fig. 97B).

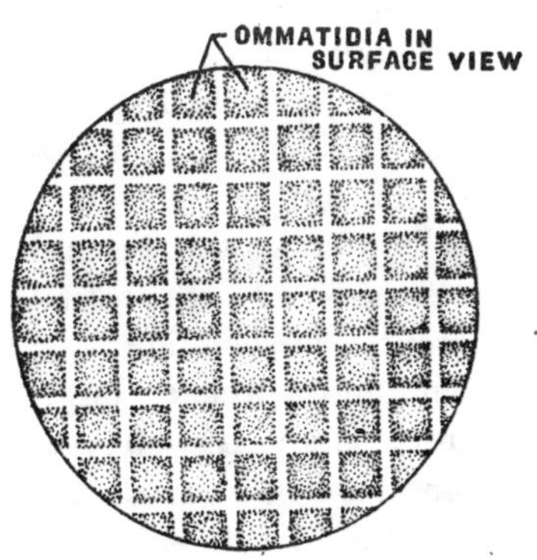

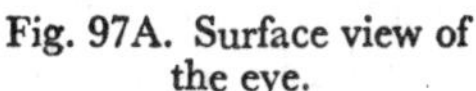
Fig. 97A. Surface view of the eye.

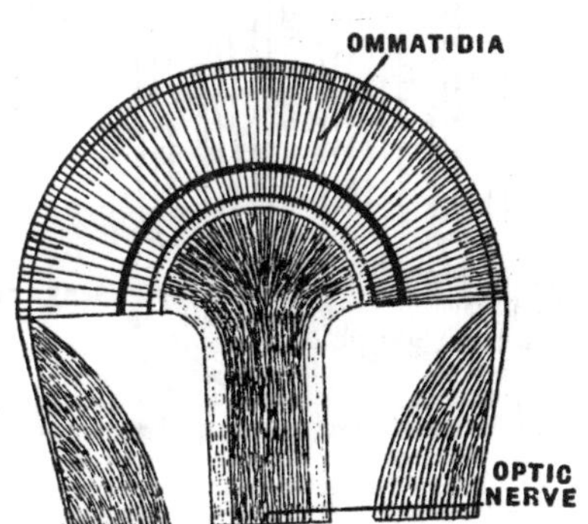

Fig. 97B. Diagrammatic longitudinal section of the eye of prawn.

Each **ommatidium** is a complete apparatus for the reception of light rays and is composed of a number of cells placed end to end along a central axis. Immediately beneath the cuticular **corneal facet,** there is a pair of flat **corneagen cells** which can secrete a new cornea as soon as the old one is lost or damaged in any way. Beneath the corneagen cells lies a transparent body called **crystalline cone** or **lens.** The crystalline cone is surrounded by four tall cells, the **cone cells** or **vitrellae,** which secrete and nourish the lens. The proximal part of the ommatidium is composed of a spindle-shaped, transversely striated body called **rhabdome,** which lies beneath the lens and is completely surrounded by seven long cells, the **retinular cells.** The latter secrete and nourish the rhabdome.

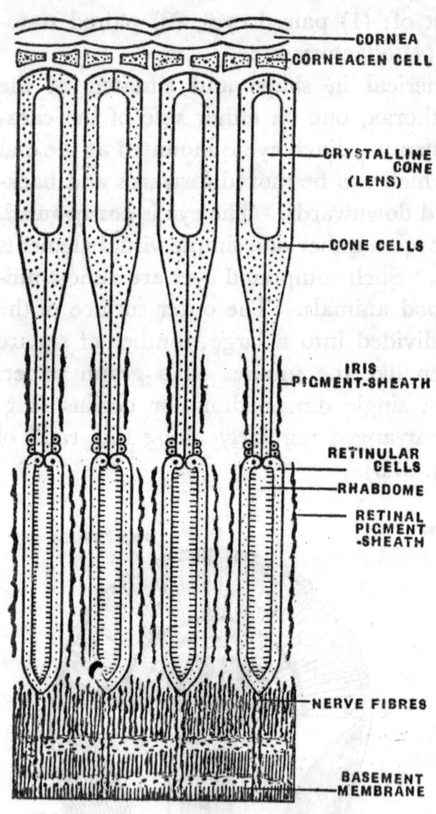

Fig. 98. Diagram showing structure of four ommatidia.

The optic nerve, on entering the eye-stalk, breaks up into branches which supply the retinular cells and the rhabdome. Each ommatidium is optically isolated from the neighbouring ommatidia by dark pigment sheaths composed of chromatophores. These are arranged in two groups: (*i*) the proximal group surrounding the rhabdome and retinular cells is known as the **retinal sheath;** (*ii*) the distal group surrounding the lens and vitrellae constitutes the **iris-sheath.** The pigments sheaths are capable of extension and retraction according to the intensity of the available light. In dim light and during the night the pigment sheaths are retracted, but in bright light and during the day the sheaths are fully extended.

The rhabdome and the retinular cells, together with the retinal sheath constitute the *receptor* or **retinal part** of the eye. The cornea, corneagen cells, crystalline cone, vitrellae and the iris sheath are concerned with focussing of light on the retinal part. They, therefore, form the **dioptrical region** of the eye. The light received by the corneal facet is focussed on the retinal part and the latter transmits the same to the brain.

It is to be noted that each ommatidium is capable of producing an image; when the ommatidia are optically isolated from one another by extension of the pigment sheaths, a composite image is formed, being made up of a large number of separate bits, each of which is the contribution of a single ommatidium. This is known as the *mosaic vision* and the **apposition image** thus produced is formed during the daytime in the presence of bright light. In dim light and during the night, the pigment sheaths are retracted and all the ommatidia work together to form a single complete image. The image in this case is indistinct and blurred. It is called a **superposition image.**

The **statocysts** or **otocysts** are paired organs for *maintaining balance*. Each statocyst is a nearly spherical cuticular sac about 1·5 mm. in diameter, and is lodged within the precoxa of an antennule. It is attached to the lower surface of the concave roof of the precoxa, and communicates with the exterior by a small opening. Its essential part consists of an oval ring of delicate cuticular setae which are situated in the centre of the sac. Each seta consists of a swollen base and a filamentous shaft, which is bent at the middle of its length. The free ends of the setae point towards the centre of the sac and are covered with fine bristles. Each seta receives at its base a fine branch from the statocystic nerve. (Fig. 100A).

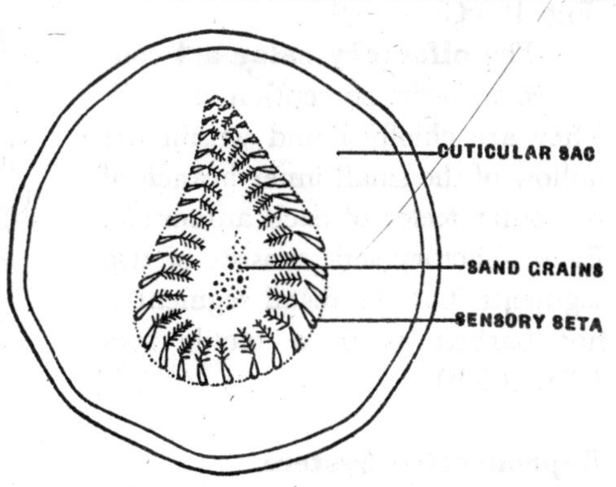

Fig. 99. Otocyst of prawn.

There are numerous sand grains within the cavity of the statocyst which are introduced by the animal at the time of moulting. With alteration of position, the sand grains come in contact with some of the sensitive setae and the stimulus of contact produces an impulse which is transmitted to the brain. The animal, coming to know of the changed circumstances, quickly re-orients itself and regains its normal balance. That these organs are concerned with equilibration is evident from the fact that if they are removed the animal appears to lose all sense of balance and fails to perceive the direction of the force of gravity.

If freshlv moulted prawns are kept in an aquarium in which sand grains are replaced by iron filings, then lacking anything else, the animals put fine iron dust within their statocysts. Such animals when subjected to the influence of a powerful electro-magnet, always swim with their ventral surface turned towards the instrument. This conclusively proves that the statocysts are the organs for perceiving the direction of the force of gravity.

The **tactile organs** are concerned with the perception of touch and contact. The many-jointed elongated feelers of the antennules and antennae are solely used for this purpose. Besides these, the borders of the flattened appendages, such as the rami of the pleopods, are covered with *tactile setae*. *Each* **tactile seta** is a cuticular rod the shaft of which consists of two segments: (1) a swollen base, and (2) a plumose distal part tapering at its free end and bearing two rows of barbs (Fig. 100 C).

The **olfactory setae** are concerned with the perception of smell. They are chiefly found within the hollow of the small inner branch of the outer feeler of each antennule. Each olfactory seta consists of two segments, but the distal segment is not barbed as in a tactile seta (Fig. 100 B).

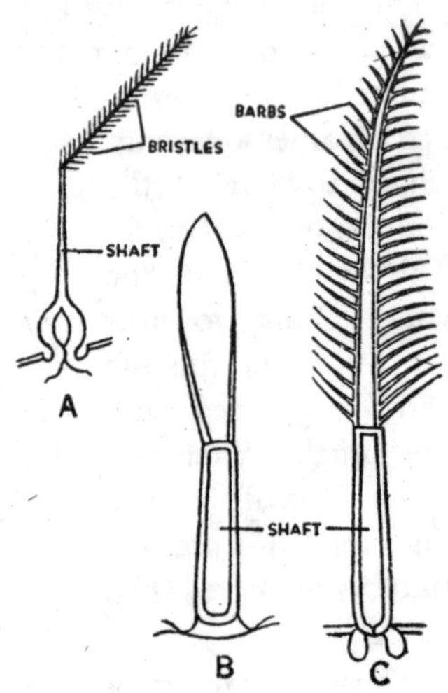

Fig. 100. A—Statocystic seta; B—Olfactory seta; C—Tactile.

Reproductive System

Prawns are *dioecious*, that is, the sexes are separate. The males are distinguished from the females by the following external characters:

(1) the males are larger in size than the females of the same age;

(2) bases of the walking legs are more closely approximated in the males than in females;

(3) second walking legs are stouter and covered with greater number of spines in the male than in females;

(4) second pleopods, in the male, carry an additional stylet called *appendix masculina* which is absent in the female.

(5) the paired *gonopores* are situated at the bases of the fifth walking legs in the male, whereas in a female the gonopores are situated at the bases of the third walking legs.

The **male reproductive organs** consist of: (1) a pair of *testes*, (2) a pair of *vasa deferentia*, (3) a pair of *seminal vesicles*, opening to the exterior by (4) the *male gonopores*.

The **testes** are soft elongated structures lying above the hepatopancreas but beneath the heart. Anteriorly, they extend as far as the renal sac and posteriorly as far as the first abdominal segment. The two testes are united at their anterior ends to form a common lobe, while posteriorly they remain free. The cardio-pyloric strand, which binds the heart to the pyloric stomach, passes through the gap between the posterior parts of the two testes. A testis is composed of a large number of lobules held together by connective tissue. Spermatozoa are budded off from the inner wall of the hollow lobules. A mature spermatozoon looks like an open umbrella.

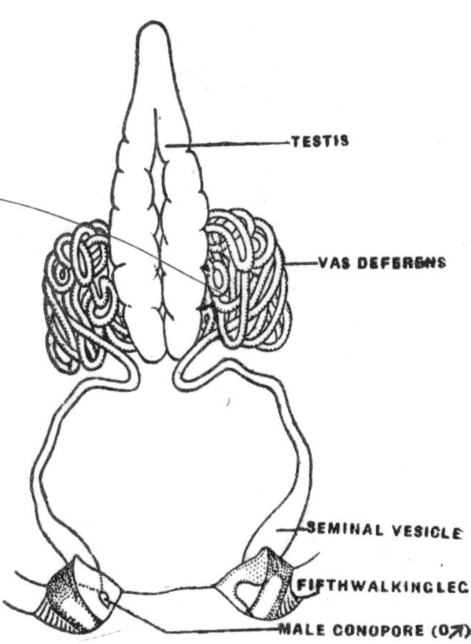

Fig. 101. Reproductive organs of a male prawn.

A **vas deferens** arises from the outer side of each testis. It is a closely coiled tube at its commencement, which soon straightens up and extends backwards and downwards to the base of the fifth walking leg of the same side where it dilates to form a club-shaped **seminal vesicle.** The spermatozoa are stored within the seminal vesicle which finally opens into the male gonopore of the side.

The **male gonopores** are paired; each is a small opening covered by a tongue-shaped flap of integument and is situated at the base of the fifth walking leg.

The spermatozoa are carried by the vasa deferentia into the seminal vesicles where they are stored in the form of compact pockets called *spermatophores* which are released through the gonopores.

The **female reproductive organs** consist of (1) a pair of *ovaries* and (2) a pair of *oviducts* opening to the exterior by (3) the *female gonopores.*

The **ovaries** are situated on the dorsal surface of the hepatopancreas and just beneath the pericardial sinus of the heart. The shape and size of the ovaries vary according to the age of the individual and the season of the year.

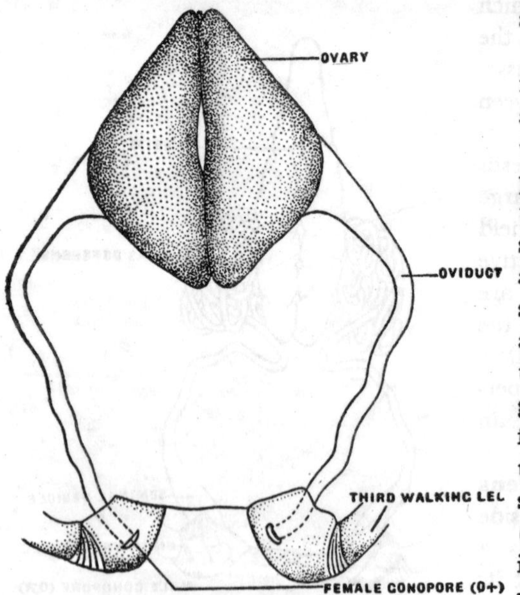

Fig. 102. Reproductive organs of a female prawn.

In a fully grown female they extend anteriorly up to the renal sac and posteriorly up to the first abdominal segment. The ovaries are brown sickle-shaped bodies which are united at both the ends, leaving a gap in the middle for the passage of the cardio-pyloric strand. The **ova** (eggs) are formed in the ovaries. A mature ovum is a rounded cell with a large nucleus, and its cytoplasm is full of yolk granules.

During the breeding season, the ovaries increase in size to hold the mature eggs.

An **oviduct** arises from the outer side of each ovary. It is large and funnel-shaped at its commencement, and then gradually becomes narrow and tubular. The oviducts are shorter in length but broader than the vasa deferentia; they plunge vertically downwards to the female gonopores.

The **female gonopores** are paired; each is a small opening situated on the inner side of the base of the third walking leg.

Breeding and Life-history

Prawns breed during the rainy season. Females carry fertilized eggs on the under surface of her abdomen, in the egg-carrying basket formed by the interlocking of the appendices internae of the second, third, fourth, and fifth pleopods.

The fertilized eggs develop directly into young individuals, which unlike most other crustacea, resemble small edition of their parents. They cling to the setae on the swimmerets of their mother for some time, and then undergo moulting. Immediately following the moulting, the young grow in size. Ultimately they separate and live independently.

In the marine prawn, *Penaeus*, and in most other crustacea, the development is indirect. There is a **zoea larva** which undergoes metamorphosis and ultimately develops into an young individual resembling its own species.

CHAPTER IX

ARTHROPODA (continued) : INSECTA

Insecta : General Consideration

Insecta are arthropods with a segmented body divisible into three distinct parts, namely, head, thorax and abdomen. Most of them are terrestrial but a few are aquatic and others are parasites. The soft parts are enclosed in a thick chitinous exoskeleton. The appendages are jointed and arranged metamerically—each segment typically carrying a pair of appendages. The head bears only one pair of antennae. Mouth parts are built differently in different species in adaptation to the habit of biting, piercing, sucking or licking food. Biting mouth parts are found in locust, beetle, cockroach, dragon-fly, termite and the praying mantis. Piercing and sucking mouth parts are found in bedbug, louse, mosquito, rat-flea and the tsetse-fly. Sucking and licking mouth parts are present in house-fly, butterfly, moth and the honey-bee. The thorax bears three pairs of walking legs and usually two pairs of wings; sometimes there is only one pair of wings as in the mosquito, or rarely none as in the common silver fish.

Most insects have simple as well as compound eyes. Respiration occurs through a system of branched tubes called trachea, and excretion through Malpighian tubules. Nervous system is like that of an annelid worm.

Nearly one million species of insects are already known and others are waiting to be named and described. It is assumed that the number of insects species is six times more than that of all other animals combined together. They are widely distributed all over the world. This surpassing success of insects are due to their prolific reproduction, hard skeleton, small size, variable form and association, instinctive behaviour, power of flight and migration and lastly for the mechanism of respiration.

THE COCKROACH

(*Periplaneta americana*)

Habitat

The cockroach is a common household pest found in warm places such as godowns, kitchens and storerooms which offer protection as well as food. It hides in dark corners of the cellar during the daytime and comes out at night. There are several species of domestic cockroaches of which *Periplaneta americana* is the commonest and the largest. In this species, both sexes are winged and the wings are longer than the length of the body. In *P. orientalis* (*Blatta orientalis*), the female has rudimentary wings and the male has short wings not reaching up to the end of its body. The young of both species are wingless.

External Morphology

The cockroach is a bilaterally symmetrical and metamerically segmented animal. Its body is covered by a brown chitinous cuticle which forms the rigid *exoskeleton.* The flat and elongated body enables the animal to creep into narrow crevices. The body is divisible into three distinct parts: *head, thorax* and *abdomen.* The **head** is the shortest part of the body and is composed of six segments fused together with no trace of external segmentation. It is connected with the thorax by a narrow neck and carries important sense organs as well as mouth parts for feeding. The **thorax** is composed of three segments which bear the organs for locomotion, namely, three pairs of walking legs and two pairs of wings. The **abdomen** lies behind the thorax and is composed of ten segments of

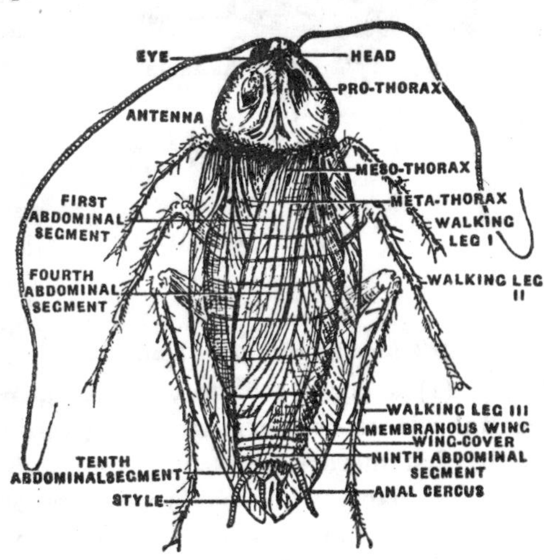

Fig. 103. The male cockroach.

which the first seven are distinctly visible. It carries only one pair of modified appendages called **anal cerci.**

Exoskeleton. In annelid worms the epidermis of the skin is covered by a thin lining cuticle. In insects the cuticle is greatly thickened and forms a dead protective covering on the outer side of the body. Such a structure is known as **exoskeleton.** It is largely composed of two substances, *chitin* and *cuticulin*, both of which are nitrogenous compounds insoluble in water, acids and alkalies. The exoskeleton supports and protects internal organs and provides attachment to muscles.

In cockroach, the cuticle is thickened in every segment of the body, forming hard plates called **sclerites** which are jointed to one another. In between the sclerites, narrow strips of cuticle remain thin and flexible so that the thickened plates on either side may be bent and moved.

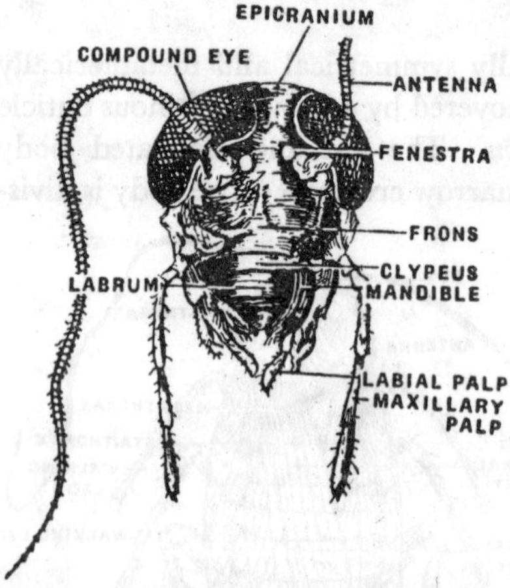

Fig. 104. Anterior view of the head.

The exoskeleton of the head is known as the **head capsule.** It consists of: (1) a pair of **epicranial plates** joined by a median suture and covering the dorsal and posterior parts of the head; (2) a **clypeus** in front, to which is hinged; (3) the upper lip or **labrum;** (4) a **gena** on each side, forming the cheeks.

The exoskeleton of each segment of the thorax and the abdomen consist of two parts: (1) a dorsal sclerite called **tergum** or **notum,** and (2) a ventral sclerite or **sternum.** The **pronotum,** that is the tergum of the first thoracic segment is the largest of the lot; it is triangular in shape, with the broad base directed backwards. The thoracic sterna are smaller in comparison to the terga. The terga and sterna of the abdomen are transversely placed arched plates; the terga project laterally beyond the sterna. The posterior margins of the sclerites of one segment slightly overlap the anterior margins of those of the next. The tergum of the tenth abdominal segment is notched posteriorly.

Head and its appendages. The head is (Fig. 104) triangular in shape, with the broad end turned upwards and the narrow end downwards. It is held at right angles to the thorax. On each side of the head is a large bean-shaped eye. The eye is said to be *compound* because it is composed of a large number of visual units called *ommatidia* which are marked on the surface by numerous *hexagonal* facets. There is a whitish patch on each side of the head just internal to the base of the antenna. This is known as the fenestra.

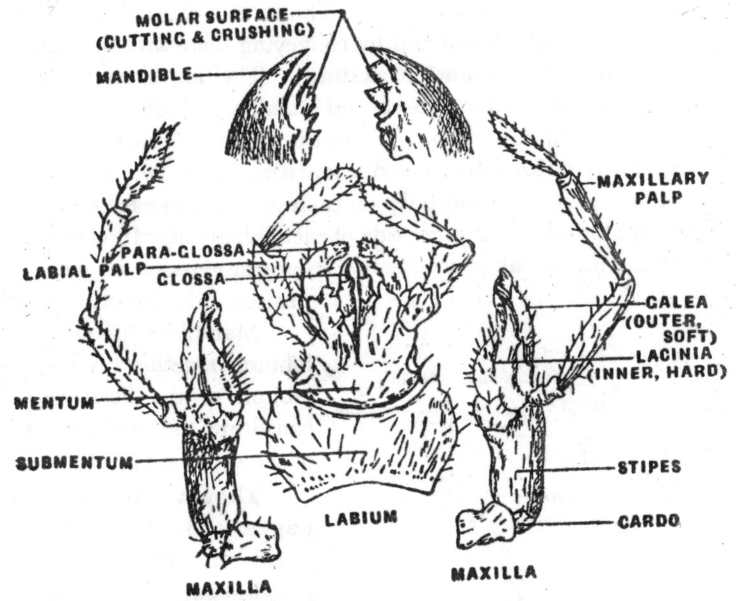

Fig. 105. Mouth parts of cockroach.

The head is enclosed in the head capsule and bears four pairs of jointed appendages: (1) antennae, (2) mandibles, (3) maxillae, and (4) labium.

(1) **Antennae.** These are a pair of long many-jointed thread-like feelers attached to the head just in front of the eyes. The cockroach can move its antennae in every direction and uses them for feeling the way as well as for smelling the food (Fig. 103).

(2) **Mandibles.** The mandibles are strongly built toothed structures placed on either side of the mouth, immediately behind the labrum. They are moved from side to side and their

opposing inner edges work against each other. The mandibles are employed for cutting and crushing food.

(3) **Maxillae (First Maxillae).** The two maxillae lie behind the mandibles. Each maxilla consists of two basal segments, a proximal **cardo** and a distal **stipes,** bent at an angle to one another. A five-jointed **maxillary palp** arises from the outer side of the stipes and is used as a feeler. On the inner side of the palp there is a bipartite piece composed of an outer **galea** and an inner **lacinia.** The galea is hooded and helmet-like, while the lacinia terminates in a claw and bears stout bristles. The lacinia assist the mandibles by holding morsels of food and by conveying them into the mouth.

(4) **Labium (Second Maxillae).** The unpaired labium is really composed of the two second maxillae which are partially fused to form the lower lip. It consists of a basal part of two segments, called **submentum** and **mentum.** Distally, the mentum bears a four-jointed **labial palp** on each side. A **paraglossa** and a **glossa** are found on the inner side of each labial palp. Internally, the labium carries a small process called **hypopharynx** or tongue which projects into the oral cavity.

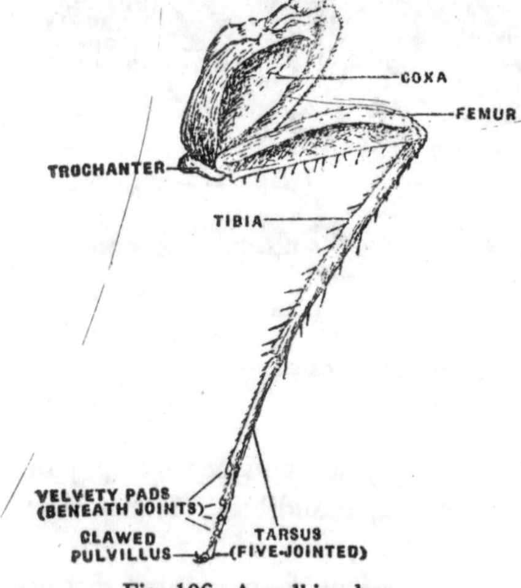

Fig. 106. A walking leg.

Mandibles, maxillae and labium together with the labium of the head capsule constitute the **mouth parts** of the cockroach.

Thorax and its appendages. The thorax is composed of three segments: **prothorax, mesothorax** and **metathorax.** The narrow neck, joining the head with the prothorax is concealed by the large pronotum. Each segment of the thorax carries a pair of **walking legs.** There are thus three pairs of walking legs. A leg is composed of the following five parts: (1) a stout and flat **coxa,** (2) a small **trochanter,** (3) an elongated **femur,** (4) a slender **tibia,** and (5) a five-jointed **tarsus** ending in a pair of curved claws. The terminal segment of the tarsus bears a pad called *pulvillus* on its ventral surface.

The prothorax has no **wings** The mesothorax and the **meta-thorax** have each a pair of wings in the adult cockroach. The wings are expressions of the cuticle supported and strengthened by vein-like *nervures*. The anterior wings are oval, horny structures not useful in flight; they form a protective cover for the hind-wings when the animal is resting. The anterior wings are, therefore, known as the wing-covers or **elytra.** The hind-wings are delicate and membranous; they are larger than the anterior pair and are folded like fans when not in use. The young cockroach has no wings.

Abdomen and its appendages. The abdomen is flattened dorsoventrally and clearly indicates the segmented nature of the animal. There are ten complete segments but the posterior segments are modified and telescoped under the preceding ones. The abdomen is broader in the female than in the male. In the female cockroach, the seventh sternum is enlarged to form a boat-shaped structure which is divided into two halves. In the male cockroach, the ninth segment carries a pair of short unjointed **styles** (Fig. 103) which are absent in the female. The tergum of the tenth segment is divided into two lobes by a notch at its posterior end. Below the tergum of the tenth segment there are two triangular processes called **podical plates.** A pair of jointed feeler-like structures project backward from the under surface of the tenth segment in both the sexes. These are the **anal cerci** (Fig. 103); they represent the only true appendages of the abdomen.

External apertures. The following openings can be detected on the external surface of the cockroach.

(1) **Mouth** is located anteriorly beneath the labrum and between the mandibles; the mouth parts are adapted for biting.

(2) **Anus** is situated posteriorly below the tergum of the tenth abdominal segment between the two podical plates.

(3) **Genital opening** differs in the two sexes. In the male cockroach, it lies between the tergum of the tenth and the sternum of the ninth abdominal segment, and is surrounded by chitinous rods called **gonapophyses** which are concerned with copulation. In the female cockroach, it lies on the ventral surface in the sternum of the eighth segment.

(4) **Respiratory opening** or **stigmata** are ten pairs of minute apertures leading into the respiratory tubes or trachea. Of the ten, only two pairs are thoracic and the remaining are abdominal. The stigmata are situated in the soft cuticle between the terga and

sterna on both sides of the body. The first thoracic stigmata lie laterally in the mesothorax and the second similarly in the metathorax. The abdominal stigmata are found in the first eight abdominal segments.

Body cavity

The body cavity is filled with a loose tissue called **fat body.** It is a restricted space occupied mainly by the digestive and the reproductive organs. Blood vessels open directly into the body cavity bathing the various organs with blood. The body cavity is therefore a **hoemocoele** and not a true coelom. Moreover it is not lined by a coelomic epithelium as in earthworms.

Alimentary System

The digestive system of the cockroach includes the alimentary canal and the associated digestive glands.

The alimentary canal is longer than the body of the animal and is, therefore, coiled. It extends from the mouth in front to the anus behind and consists of three parts: (1) foregut, (2) midgut, and (3) hindgut. The foregut and the hindgut are lined internally by a thick cuticle which is derived from the ectoderm of the skin. The midgut, on the other hand, is lined by a soft glandular epithelium derived from the endoderm.

The **foregut** or **stomodaeum** consists of the mouth, buccal cavity, oesophagus, crop and proventriculus or gizzard. The **mouth lies** between the mandibles and the maxilla, with the labrum forming an upper lip, and the labium a lower lip. It leads into a **buccal cavity,** on the floor of which is a small tongue-shaped projection called **hypopharynx.** The buccal cavity opens into a narrow **oesophagus,** which passes through the thorax into the abdomen, where it expands to form a long and thin-walled **crop.** The crop is an organ for storage of food; oil and sugar may be stored here for months before they are digested and absorbed. The crop opens into a small thick-walled conical **chamber,** the **proventriculus** or **gizzard.** This is a muscular organ containing in its cavity a set of six chitinous teeth for breaking up solid particles of food; behind the teeth is a ring-like cushion carrying hairs which act as a sieve and hold back larger particles from passing further ahead. The gizzard marks the termination of the stomodaeum and leads into the midgut.

The **midgut** or **mesenteron** is a narrow tube lined by columnar epithelium. It receives at its anterior end seven to eight hollow tubular structures called **hepatic caeca,** which secrete digestive juice into the mesenteron. It is followed by the hindgut. At the junction of the midgut and the hindgut, there are numerous fine tubes called Malpighian tubules for excretion of waste products.

The **hindgut** or **proctodaeum** is differentiated into a short

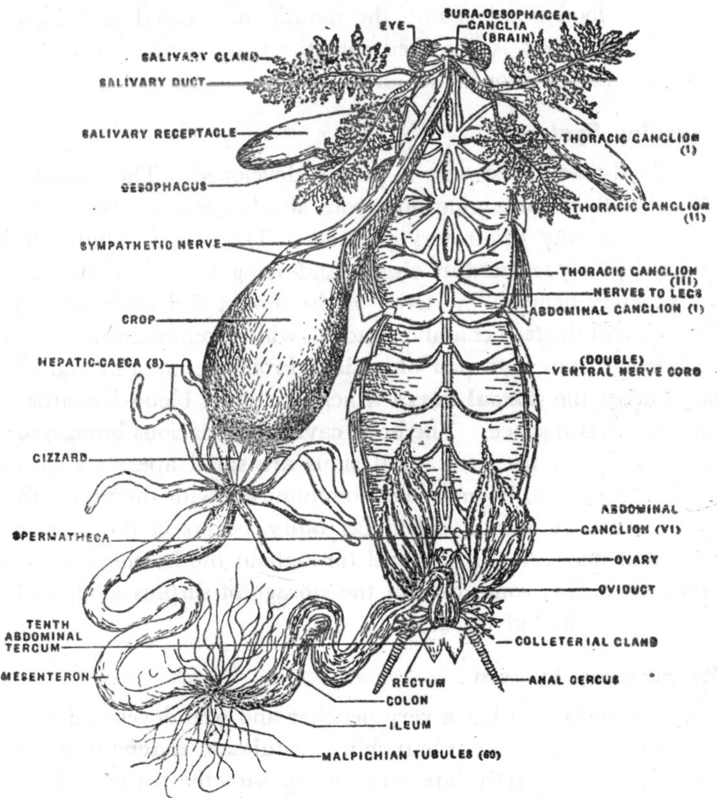

Fig. 107. Dissection of a female cockroach showing alimentary system, nervous system and reproductive system.

and narrow anterior part called **ileum,** followed by a wide and somewhat coiled **colon** and a short dilated **rectum** opening into the anus. The **anus** is situated posteriorly beneath the tergum of the tenth abdominal segment between the two podical plates.

The **digestive glands** lie in the thorax and consist of a bilobed

salivary gland with a reservoir or **receptacle** on either side of the crop. The ducts of the receptacles join to form a common tube, which receives the common ducts from the salivary glands and finally opens into the buccal cavity on the elevated hypopharynx. The secretion of the glands is known as the saliva which contains amylolytic and lipolytic enzymes for digesting starch and fat. Proteins are digested by the secretion of the hepatic caeca.

Solid food is taken into the mouth, masticated and digested. The digested food is absorbed into the body, while the residue is voided as faeces through the anus.

Vascular System

Blood-vascular system is poorly developed. The **blood** is a colourless fluid containing numerous *amoeboid* **leucocytes.** It does not contain any respiratory pigment. There is a tubular **heart** enclosed in a **pericardial sinus,** and lying in the middorsal line immediately beneath the terga of the thorax and abdomen. It is composed of thirteen chambers each of which communicates with the pericardial sinus by a pair of **ostia.** From the anterior end of the heart arises the **dorsal aorta** which drives the blood forwards into the perivisceral space. The body cavity is a spacious **hoemocoele.** In the floor of the pericardial sinus are small apertures through which perivisceral hoemocoeles communicate with the pericardium.

A white substance, called **fat body,** occurs in the pericardial sinus and may remain scattered throughout the hoemocoele. It is believed to play some part in the storage of surplus food, such as albumen, fat and glycogen.

Respiratory System

The cockroach has a very peculiar and well-developed respiratory system, consisting of branching air-tubes or **tracheae** by which air is brought directly into relationship with the tissues. The air-tubes open into the lateral sides of the body by a series of respiratory pores called **spiracles** or **stigmata.** There are ten pairs of spiracles—the first pair in the mesothorax, the second pair in the metathorax and the remainder in the dorsolateral corners of the first eight abdominal segments. Each spiracular opening is bounded by an annular sclerite and leads into a cavity which is connected to the main **tracheal trunk.** There are two large trachea trunks, one on each side, which run lengthwise and are connected to one another by transverse branches. These main tubes divide into dorsal and ventral trunks, the branches from which anastomose freely and penetrate into all parts of the body.

The air-tubes are thickened spirally on the inner side, and thus kept open. The ultimate branches ramify as fine **tracheoles** with no thickening. They actually run inside the tissues and organs. Oxygen is thus conveyed to all parts directly and carbon dioxide is similarly removed. This explains the poor development of the blood vascular system, as there is very little necessity of blood for the transport of respiratory gases.

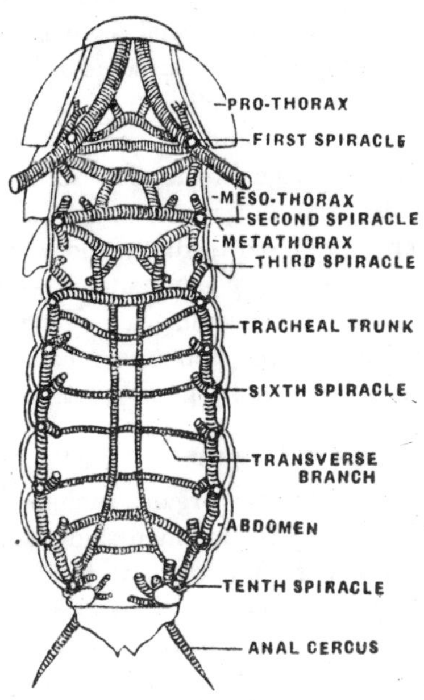

Fig. 108. Tracheal system of the cockroach.

Expiration is effected by contraction of the abdominal muscles resulting in the compression of the elastic tracheal tubes. When the abdominal muscles relax, the tubes regain their normal size and air flows into them through the spiracular openings. This is **inspiration.** The inlet and outlet of gases through spiracles may be controlled by a peculiar method—the animal closing all its breathing pores for a considerable time.

Excretory System

The main excretory organs of the cockroach are the **Malpighian tubules** which float in the hoemocoelic fluid. The number of tubules vary from sixty to seventy. Each tubule is a hollow outgrowth from the proctodaeum and is lined by glandular epithelium which secretes *nitrogenous* waste products. The excretory materials are principally *urates* and *uric acid*. These pass into the proctodaeum and thence out of the body, with the faeces.

Nervous System

The nervous system of cockroach (Fig. 107) resembles that of the prawn. There is in the head a pair of **supra-oesophageal**

ganglia or *brain* which supply nerves to the eyes and the antennae. They are joined by two **circum-oesophageal connectives** with a **suboesophageal ganglion,** from the posterior end of which a double **ventral nerve-cord** runs backward. This nerve-cord is situated, as in prawn, in the midventral line, and becomes enlarged to form a ganglion in each of the three thoracic segments and in the first six abdominal segments. The sixth abdominal ganglion is the largest and represents several ganglia fused to form a mass. From each ganglion, nerves arise and pass out to innervate the muscles, skin and appendages of the segment in which it lies. The suboesophageal ganglion supplies the mouth parts. It is to be noted that the ganglia and the nerve-cord are really double in origin. They are fused medially and therefore appear as single.

There is a ganglionated **sympathetic nerve** connected with the circum-oesophageal connectives to supply the involuntary muscles of the alimentary canal. It lies on the dorsal surface of the crop (Fig. 107).

Receptor Organs

The sense organs include: (1) a pair of compound eyes, (2) a pair of antennae said to be olfactory, tactile and gustatory in function, and (3) scattered sensillae in the palps, walking legs and anal cerci.

The **compound eyes** are not borne on eye stalks like those of the prawn. Each consists, essentially, of a large number of **ommatidia.** The surface of the eye is covered over by a transparent cuticle, the **cornea,** and subdivided into about 20,000 hexagonal facets—each of which represents an ommatidium. The structure of an ommatidium is similar to that already described for the prawn, excepting that the pigment sheaths are not retractile. Consequently, the cockroach utilises only the mosaic image. Though each ommatidium may form an image of the same object, these will be translated during transmission through the optic nerves, so that only one impression is registered.

The receptor cells or **sensillae** are isolated or collected patches of modified epidermal cells situated in various parts of the body. There are olfactory, gustatory and tactile receptors found on the antennae, palps, walking legs and anal cerci. Each sensilla is connected to a sensory nerve. The eye may be regarded as a collection of complex sensillae specially modified for vision.

The sense of smell is well-developed in the cockroach and is located mainly in the antennae. The gustatory receptors concerned with taste are found in the maxillary and labial palps. Tactile receptors for touch are located particularly on the antennae but are also distributed generally all over the body in a scattered state. There is no definite ear in the cockroach; but certain hairs in the anal cerci are set vibrating by sound waves. These may possibly act as very simple auditory organs.

Reproductive System

The cockroach is dioecious, that is the sexes are separate. The male can be distinguished by its **anal styles** which are not present in the female.

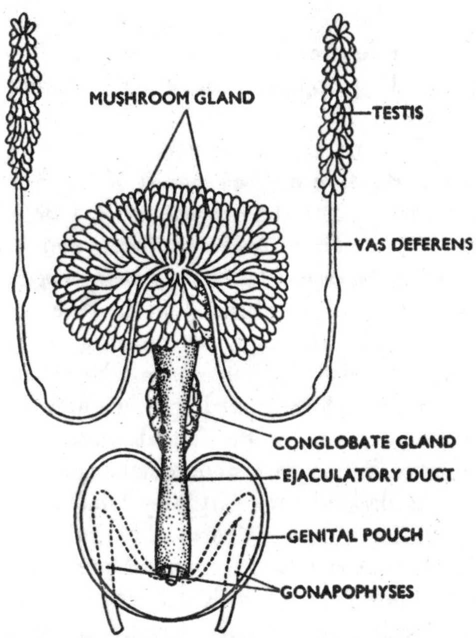

Fig. 109. Reproductive organs of a male cockroach.

Male reproductive organs consist of: (1) a pair of *testes*, (2) a pair of *vasa deferentia*, (3) a pair of *seminal vesicles* forming the *mushroom gland*, (4) an *ejaculatory duct*, (5) a *conglobate gland*, and (6) the *genital pouch*, containing the *male gonopore*.

The **testes** appear like two small bunches of grapes, one on each side, beneath the terga of the fifth and sixth abdominal segments. The testes are active only in the young cockroach; after producing a stock of spermatozoa they become inactive and consequently they are small in the adult. A **vas deferens** arises from each testis and leads into the seminal vesicle of the same side. A number of blind tubular pockets are given out from the surface of both the vesicles, so that the whole structure appears like a mushroom and is, therefore, known as the **mushroom gland.** Spermatozoa formed during the early life of the animal are stored in the pockets of the latter. The seminal vesicles open into a median muscular tube called **ejaculatory duct** which communicates with the genital pouch by the **male gonopore.** The **genital pouch** is a wide space between the sternum of the ninth and the tergum of the tenth abdominal segment. Lying beneath the ejaculatory duct is the **conglobate gland** opening separately into the genital pouch near the male gonopore. There are a number of curved chitinous hooks within the genital pouch. These are known as the **gonapophyses** which assist in copulation.

Female reproductive organs consist of: (1) a pair of *ovaries*, (2) a pair of *oviducts*, (3) a median *vagina*, (4) a pair of *spermathecae* of unequal size, (5) a pair of *colleterial glands*, and (6) the *genital pouch*, containing the *gonapophyses* and the *female gonopore*.

Each **ovary** consists of eight beaded tubules, the ovarioles, which are distended posteriorly with eggs, but tapering at their free ends. The ovarioles of each ovary join posteriorly to form an **oviduct.** The two oviducts unite to form a median **vagina,** which communicates with the genital pouch by the **female gonopore** (Fig. 107).

The sternum of the seventh abdominal segment is large and boat-shaped. It is divided into two lateral halves which can be opened and closed like a pair of doors. The eighth and ninth abdominal sterna are tucked into the segments previous to them. A **genital pouch** is thus formed which is bounded anteriorly by the eighth sternum, dorsally by the ninth sternum and ventrally by the seventh sternum. Thus the female gonopore pierces the eighth sternum to open into the genital pouch. A pair of **spermathecae** open by a median aperture into the genital pouch on the ninth sternum; of the two, only one is sac-like and the other is filamentous; they receive sperms from the male during mating. The **colleteral glands** are branched tubes opening near the female gonopore; they secrete the material for the formation of the egg-case.

There are small chitinous rods, the gonapophyses, near the opening of the genital pouch, which serve to deposit the egg-cases.

Life-history

Spermatozoa are introduced into the genital pouch of the female at the time of pairing and stored in her spermathecae. In the meantime, mature eggs pass into the genital pouch along the oviducts. The eggs is rich in yolk which is concentrated in the centre, so as to push the cytoplasm towards the periphery; such an egg is said to be **centrolecithal.** Sixteen eggs pass out one after another into the genital pouch, one from each ovariole. All are fertilized by thes perms from the spermathecae. The fertilized eggs are finally enclosed in a purse-shaped egg-case or **ootheca** which is formed by the secretion of the colleterial glands. The eggs are arranged in two rows within the ootheca and the ootheca is carried by the female in her genital pouch for several days, till she deposits it in a warm and dark corner. The young hatches by rupturing the egg-case after completion of the embryonic development. The freshly hatched cockroach is known as the **nymph;** it is small and white with black eyes and soft skin, but without wings and gonads. As the nymph grows, its exoskeleton, being non-living, and therefore unable to grow, becomes too small to hold it. The old exoskeleton splits down the back and the enlarged nymph works its way out. It is now a pale soft-skinned creature known as an **instar;** its cuticle soon hardens and then it goes about feeding and growing. In due course, the instar undergoes another moulting and there are altogether five to seven such moults before the adult stage is reached. Wings appear at the last moult. The adult never grows in size after the last moulting. The life-cycle is completed in six to eight months and the striking change that is observed in the young nymph as it grows into the adult consists of the growth of the wings. The metamorphosis in this case is only partial and therefore known as **heterometabolic.** In insects like silverfish, the young closely resembles the parent and the change into the adult form merely consists in the development of the reproductive organs; such insects, having no metamorphosis, are said to be **ametabolic.**

In mosquito, the young larva is completely different from its parent; the larve transforms into a quiet resting stage called pupa, and the latter changes into the adult; the metamorphosis in this case is complete, *i.e.* **holometabolic.**

ECONOMIC IMPORTANCE OF INSECTS

Some insects are useful to man, while many are harmful and the injury is to such an extent that it has threatened the existence of man on this globe. For both the reasons, insects are of great economic importance and its study constitutes the subject matter of entomology, which assures—"future comfort, happiness, prosperity and even the existence of mankind".

Insects which are useful

1. *Help in pollination*—many insects like bees, butterflies, ants, etc., visit flower for collecting nectar and pollens. While doing it they move from one flower to the other. By this act they assist in the dispersal of pollen grains from male to female flower and thus help in cross-fertilization.

2. *Help to get products of utility*—many insects like bees, silk-moth and lac insects are responsible for producing various products which are of great use to man. The bees produce honey and wax, the silk moths reel silks and the lac insects supply lac. All these products are of much utility to mankind.

3. *Help as scavengers*—many insects like flies eat debrises and human wastes. Through this habit they help a lot to the mankind by keeping the earth's surface clean.

4. *Help as food*—many insects like large caterpillars, grass-hoppers and locusts are regarded as edible by some groups of human beings.

5. *Help in decoration*—blisters and ornaments are made from various beetles and butterflies and are used by many people as decoration of house and body.

Insects which are injurious

1. *Direct enemies*—many insects like mosquitoes, bedbugs, wasps, ants and caterpillars are notorious for their biting, piercing or irritating ability. Some have piercing mouth parts (mosquito), powerful jaws (ants), poisonous stings (wasps) and poisonous hairs (caterpillars). All of them produce considerable irritation and uneasiness to the human body.

2. *Parasites on man*—some insects like lice, maggot, etc., may live as parasite on human body.

3. *Vectors of disease-producing microbes*—In addition to their act as direct enemy many insects transmit different pathogenic microbes, which cause different diseases. The mosquitoes are well-known for carrying parasites which produce malaria, filaria etc. The tsetse-fly transmits a parasite which produces sleeping sickness. The house-fly is notorious for bringing germs of numerous diseases through their external and internal parts and contaminate the human food.

4. *Enemies of domestic animal*—The insects attack the domestic animals of man in all the three ways—direct enemy, parasite and vector.

5. *Enemies of plants*—Locusts, caterpillars and various other insects eat voraciously different plants and thus cause a great damage to human economy.

6. *Enemies of human stores*—Many insects, termites, beetles, etc., cause great damage to different furnitures and other stored products. They pose a great problem to the food which are kept for future use.

In the following pages the Biology of two beneficial and two harmful insects will be discussed.

BENEFICIAL INSECT : HONEY-BEE

Introduction and Economic Importance

These insects belong to the order *Hymenoptera.* The most common species is *Apis mellifica.* These are well-known for their organised social life and great economic importance. The nest of honey-bee is called bee-hive, which is commonly found on tall trees. It is possible to rear honey-bees in artificial hives and the cultivation of bees for honey and wax is very common throughout the world.

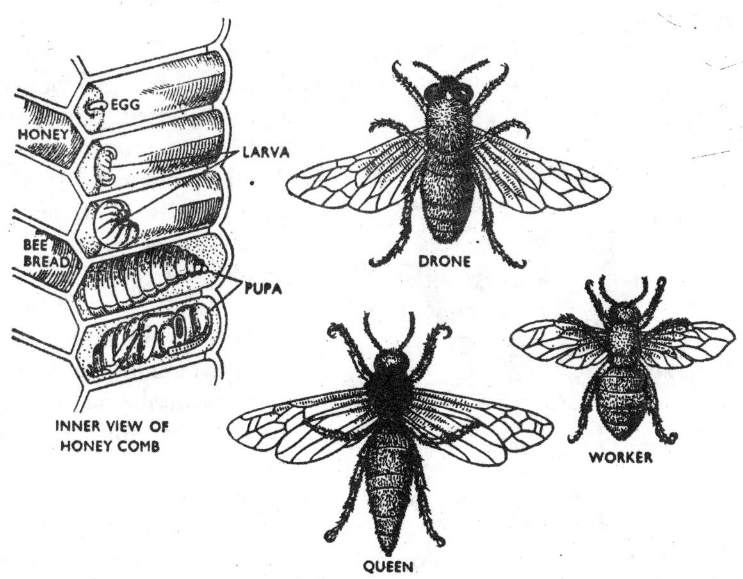

Fig. 110. Three types of bees.

Polymorphism in Honey-bee

Within each hive reside thousands of bees of three different forms—(1) **Workers,** (2) **Drones,** and (3) **Queen** (Fig. 110).

1. **Worker bee** forms the majority within a hive; size is very small; work in number of capacities—cleaner, nurse, builders and architects, technician, soldier and porter.

2. **Drone bee**—slightly larger than the workers; noisy and idle; comes out of the hive only at the time of nuptial flight.

3. **Queen bee**—very large with a tapering abdomen; generally a single matured queen is present in each hive; responsible for the laying of eggs.

For the presence of three different morphological forms in one species, these insects are called polymorphic and the phenomenon is called polymorphism.

External feature

Like a cockroach, the body is divided into three parts—(*a*) **Head,** (*b*) **Thorax,** and (*c*) **Abdomen.** Each part is provided with various structures which are adapted to the peculiar habit of these insects.

(*a*) **Head.** This region contains (1) three ocelli in the middle, (2) a pair of much prominent compound eyes, (3) one pair of long many jointed antennae, and (4) a number of appendages around mouth, which are specially built for the collection of nectar and building of comb.

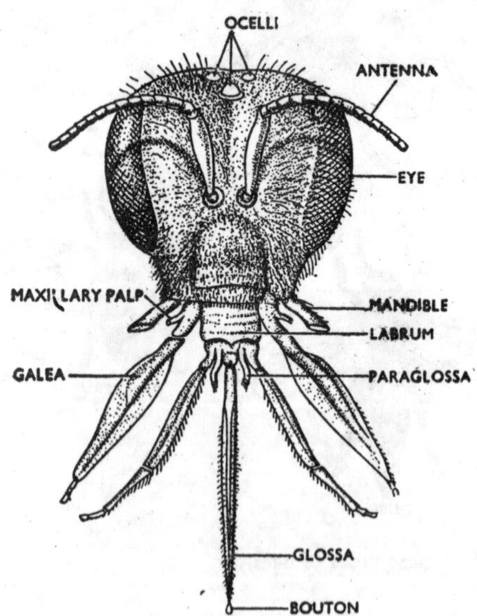

Fig. 111. Mouth parts of a bee.

Mouth parts (Fig. 111)

(*i*) Mandible—one pair; very powerful in worker bees; used during comb building.

(*ii*) First maxilla—one pair; each consists of an *lamina* or *galea* on a basal piece which is made up of stipes and cardo. The *maxillary palps* are rudimentary.

(*iii*) Second maxilla—one pair; forms the lower lip; well-formed in workers; lower part united; from the outer side of each maxilla hangs a long *labial palp* and the inner side is divided into two parts —*glossa* and *paraglossa*. The two glossae, one from each second maxilla are fused together to form *tongue* or *ligula*. The paraglossae remain tightly fixed with ligula at its base. The labial palp, ligula and paraglossae, can be set in such a fashion that it forms an airtight tube for sucking nectar. When not in use, the elongated mouth parts are kept folded on the back. This is unlike butterflies, where the funnel remains coiled during unuse.

(*b*) **Thorax.** Thorax is divided into three segments—(1) Prothorax, (2) mesothorax, and (3) metathorax. Three pairs of *legs* and two pairs of *wings* are the appendages of the thoracic region.

Legs of honey-bee. Legs are built up in the same pattern as that of cockroach, but it has certain modifications which help these insects to their peculiar way of life (Fig. 112).

(*i*) *Legs of prothorax*—Soft hairs called *eye brush* are present on the tibia, which help in cleaning pollens and debris from the eye. On the **first** tarsal segment, number of stiff bristles called *pollen brush* are present for removing pollens from the body hairs. Tibia holds a *spur*, and first tarsal segment contains a *notch*, both of which act as *antenna cleaner*.

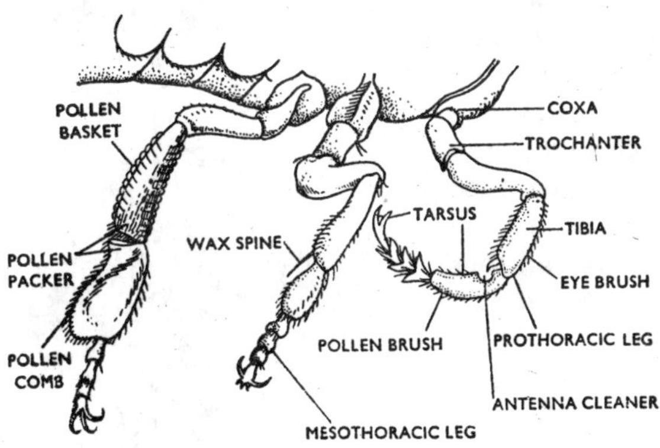

Fig. 112. Legs of honey-bee.

(*ii*) *Legs of mesothorax*—Pollen brushes are also present on the first tarsal segment. From the tibia hangs a *wax spine stick* for removing wax from the wax glands.

(*iii*) *Legs of metathorax*—Inner part of the first tarsal segment contains *pollen combs*. These combs collect pollens from the pollen brush of other legs. Near the junction of tibia and first tarsal segment, lies another structure called pollen packers which clean the pollen combs. On the outer side of the tibia, there is a basket, called *pollen basket*, within which pollens are deposited by the *pollen packers* from the pollen combs.

Wings of honey-bee. The dorsal side of the thorax carries two pairs of wings. Near the base of each wing lies an aperture called stigma. Drones possess the largest wings and queens have relatively smaller wings.

(*c*) **Abdomen.** In workers and queen, the abdomen is divided into six segments but in drones there are seven segments. In drones the abdomen is broad but shorter than the wings, while it is elongated and narrow in the queen. Abdomen is devoid of any appendage but bears a number of apertures called stigmata on it (5 pairs in queen and workers, and 6 pairs in drones). Only in workers the abdomen contains on its ventral side four pairs of wax-producing glands called wax pockets or wax glands. The posterior-most tip of the abdomen in queen and workers can be protruded into a *sting*. With this sting are connected the poison glands. In the queen the sting is also used for the laying of eggs and thus functions as *ovipositor*.

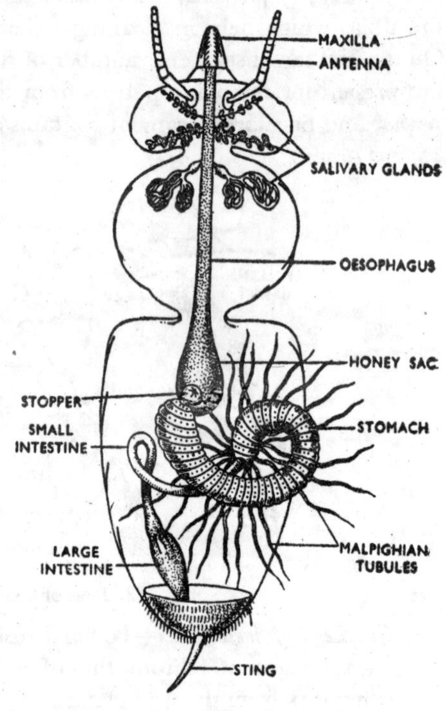

Fig. 113. Alimentary system of a bee.

Internal Structures (Fig. 113)

Alimentary System. Within the body the alimentary canal is divisible into following parts—*oesophagus*, *honey sac*, *chyle stomach*, *small intestine*, *large intestine* and *anus*. The oesophagus is a narrow straight tube passing through the thoracic region. The honey sac is formed by the dilation of the oesophagus within the abdomen. The honey sac leads into the chyle stomach through a passage which is guarded by a complex stopper arrangement. The chyle stomach is a broad, half-bent tube which opens into the coiled small intestine. The opening is separated by a pylorus. Number of chitinous teeth are arranged longitudinally along the inner wall of the small intestine. Large intestine is dilated and short and is guarded by valves of six plates.

In all the three types, there are three pairs of salivary glands which produce saliva. It mixes with the nectar and pollen grains. The nectar is converted by enzymatic action into honey, which either may be regurgitated into the comb foı future use or may be digested. Only the worker bees possess another pair of coiled gland secreting a nitrogenous fluid called 'Royal jelly'.

Respiratory System. Respiratory system consists of *stigmata*, *trachae*, and *air-sacs*. The stigmata are lateral openings, which may be kept closed externally. The trachae are branched tubes, which ramify all around the body. The air-sacs are large vesicles, which are in connection with trachae.

Excretory System. It is represented by innumerable thread-like Malphigian tubules which open into the small intestine.

Circulatory System. The circulatory system is called as open type of circulation, because there is no definite blood vessels with distinct walls. Blood being pumped by tubular hearts (placed along the middorsal region of thorax and abdomen) passes through the body spaces called hoemocoel. The blood which is known as haemolymph, contains a few amoeboid cells with distinct nucleus.

Nervous System. The nervous system consists of nerve ganglia, ventral nerve cord, and peripheral nerves. The ganglia are represented by—(1) Large supra-oesophageal ganglia with broad projection to each compound eye, (2) suboesophageal ganglia, which is formed by the union of three pairs of ganglia, (3) three pairs of ganglia in the thorax, and (4) five pairs (in workers) and four pairs (in drones and queen) of abdominal ganglia. Double ventral nerve cord begins from the suboesophageal ganglion and runs posteriorly along the midventral line. The thoracic and abdominal ganglia are placed along the ventral nerve cord. Different peripheral nerves are given out from these ganglia to different parts of the body like, eyes, antennae, mouth parts, legs, wings, etc.

Following sense organs are present in the honey-bee—(1) Simple eyes—for detecting the intensity of light. (2) Compound eyes—for vision. (3) Antennae—for measuring distance and for smell.

Reproductive System. Sexes are separate, the male bees are called Drones and female bee is called Queen. The worker bees are sterile females.

In Drones—*Testis* one pair; each testis is connected by a narrow tube called *vas deferens*. The outer end of it dilates into seminal vesicle, which in turn opens into the *ejaculatory duct*. The copulatory apparatus is connected with the terminal end of ejaculatory duct. A pair of prominent mucous glands open near its union. When matured the testes reduce and the spermatozoa crowd near the end of the ejaculatory duct. There the sperms are transferred in packets called spermatophores. It may be noted here that drones develop from unfertilized eggs, therefore they have haploid number of chromosomes. Thus, here during spermatogenesis, no meiosis occurs.

In Queen—Large and prominent paired *ovaries*. Each with numerous tubes containing eggs at different stages of maturity. Each ovary opens into an *oviduct*. The two oviducts unite to constitute a common oviduct, which opens through a copulatory pouch. The inner end of the oviduct is connected with a round vesicle called *spermatheca*.

Process of reproduction—When matured the queen takes a nuptial flight with a number of drones. The drone which is able to copulate the queen, looses its copulatory apparatus and dies. The sperm cells remain stored in the spermatheca of the queen. At the end of nuptial flight when the queen returns to the hive it starts laying eggs. The fertilization occurs only at this stage and is believed to be either under the 'voluntary' control of the queen or depends upon the size of the cell where the egg is laid.

Life-history. The queen lays fertilized as well as some unfertilized eggs. The fertilized eggs develop either into queen or worker but unfertilized eggs develop parthenogenetically into drones. Within three days of laying, the eggs develop into larvae. These larvae are fed differentially. The larvae which will become drones are fed with a mixture of honey and pollen grains, called bee-bread. The worker larvae are fed only with honey. The larvae which are destined to be queen are fed with a special food called 'Royal jelly'. This jelly is prepared within special glands of the workers, which open into their pharynx. It is said that workers decide the number of queens to be developed. Accordingly, the workers remove certain larvae before the third day of development in a special large chamber called queen's chamber. The feeding of larvae continues for five days, which results into tremendous growth. Then the larvae spins a cocoon around its body and enters into a pupa stage. Rapid transformation of body takes place within the cocoons and ultimately the adult bee cuts its way out of the cell. The worker bee takes twenty-one days to develop, drones require twenty-four days and the queen leaves the cocoon on the fifteenth day. If several queens are produced, only one survives while others are eliminated.

Swarming of bees—With the emergence of new queen, the old queen vacates the hive. She, being accompanied by a large number of workers fly to a suitable place for building a new hive. This phenomenon is called *swarming*.

Duties of worker bees—The painstaking observations of Rosch and others have revealed the changing duties of worker bees. The workers start working immediately after coming out of the pupal case, which may be described as below.

1*st*–3*rd day*—clean the cell for reuse.

4*th*–9*th day*—act as nurses, feeds the larvae with their own secretion or honey produced by others.

10*th*–16*th day*—act as builder and architects from the wax produced by their own wax glands they make new cells in comb.

17*th*–19*th day*—receive nectars from the bees which are acting as porters and convert it into honey. By the fanning of wings they can also either cool or warm up the hive. Remove all debris from the comb.

21*st*–25*th day*—work as a soldier to protect the hive from foreign intruders including bees from other hives.

25*th day onwards*—work as foreagers. Begin to collect nectar and pollen from flowers. In the European countries, it is detected that life span of a worker bee as adult lasts only for six weeks.

Language of honey bee—It has now been unequivocally demonstrated that honey-bees are capable of passing information, regarding the location of food; by a peculiar dancing they not only indicate the direction but also give an approximate measurement of the distance. For their intricate and well-co-ordinated activities, the honey-bees have been long well-known to the students of animal behaviour.

BENEFICIAL INSECT : SILK MOTH

Introduction and Economic Importance

The silk moths are related to butterflies and belong to the order *Lepidoptera*. From the very ancient times, these insects are well-known for the ability to produce silk. The rearing of silkworm for silk originally started in China, but now it is cultivated throughout the world. Generally, these moths lay their eggs on a particular tree, e.g. mulberry, and the caterpillars devour the leaves of that tree. But the most common silk moth, *Bombyx mori*, which is reared for silk, is absolutely a domesticated form and is very rarely found in wild state. The adults are very short-lived and lay eggs on filter papers. The larva eats mulberry leaves and produce cocoon. Due to domestication, various parts of the body are degenerated. Another moth *Antherareamylita* is available in India, producing silk of little inferior quality. The application of biology which deals with the production of silk is called sericulture.

General Organisation of Adult Moth

The moths are nocturnal and possesses a slender body which in silk moth is 2·5 cm. The body is divided into three regions —*head*, *thorax* and *abdomen*. The *head* contains a pair of compound eyes, a pair of bipectinate antennae. In the butterfly, several appendages around the mouth form a proboscis which is coiled and funnel-shaped. But in silk moth the proboscis is absent and the first maxillae and labial palps are reduced. No mandible for cutting or piercing is present. The *thorax* contains three pairs of *walking legs* and two pairs of *wings*. The wings are fairly large and cream-coloured with brown lines on it. When at rest, it folds the wings over the body. The *abdomen* is well marked, segmented and contains no appendage. In females, it is slightly broader than males.

Life-history (Fig. 114)

The mating takes place in shade and occurs immediately after the coming out of adult from cocoon. After mating the male dies and female dies after the laying of eggs. In the wild state the female moths lay eggs on the leaves and branches of mulberry trees, but in domestication the females lay eggs on filter papers. The eggs, popularly known as seeds, are minute and remain attached by means of a gelatinous covering produced by the female. The eggs are white in the beginning but in course of development becomes brown.

Within a short period, a small, pale yellow-coloured larva wriggles out of each egg. The larva, called a caterpillar, is a great eater and devours green mulberry leaves voraciously by cutting with sharp jaws. It grows rapidly and shades its external skeleton four times. This process is called moulting. The caterpillar grows up to 7·5 cms. and possesses a segmented body with three pairs of thoracic legs and five pairs of stumpy prolegs on the ventral side. Ten pairs of spiracles are present for respiration. At the end of fourth moulting, a pair of silk glands develop on the lateral ride of the body cavity.

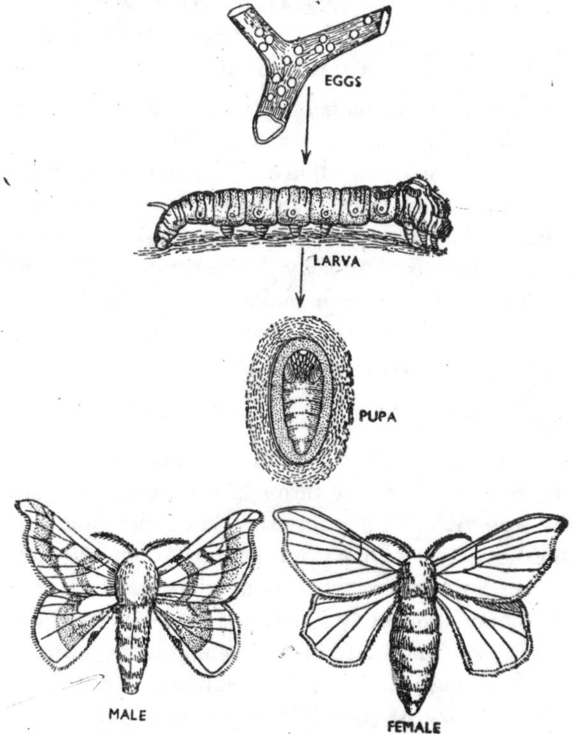

Fig. 114. Life-history of silk moth.

The caterpillar finally undergoes fasting and becomes fixed at one corner. At this stage it secretes silk from its silk glands and by the help of its mouth spins the silk around its body. The silk is sticky in the beginning but immediately after coming in contact with air stiffens and becomes fibrous. Thus the body gets enclosed within a silken case called cocoon and is known as pupa. Within the cocoon the pupa transforms rapidly, at first it becomes *chrysalis* and finally changes into a fully formed moth. When fully developed, the moth softens the one end of the cocoon and comes out of it.

Silk is obtained from the pupal cases which contain the pupa. First of all the pupa is killed by boiling and then the silk is unreeled from the case. Generally fibres are twisted together to form a thread. From a single pupal case nearly thousand yards of silken thread may be obtained.

BENEFICIAL INSECT : LAC INSECT

Introduction and Economic Importances

The lac insects Tachardia belong to the order Hemiptera of the class insecta. Lac insects and its economic importance is known to mankind for a long time. The Atharba Veda contains an account and description of female lac insects. In Ain-E-Akbari specific account of the economic importance of lac insect is given. Two species under the genus Tachardia namely *T. lacca* and *T. chinensis* are very common. *T. lacca* is the common Indian variety. India tops the lac producing countries and Thailand comes next to India. Lac insects live on certain specific trees like Kusum, Khair and Ber.

General organisation

Nymphs developing directly from the fertilized eggs within the body of the female come out through an aperture called **anal tubercular aperture.** They come out in batches and the coming out from the older crust is known as *swarming*. About 150-400 nymphs come out from a single female. The nymphs are boat shaped and reddish in colour. The body is divided into three regions—*head*, *thorax* and *abdomen*. The body of the nymph is short and measures 0·65–0·70 mm in length and 0·25–0·30 mm in breadth. Each nymph possesses three pairs of **thoracic legs,** one pair of **antennae** and a pair of **caudal setae.** Some of the nymphs become winged or wingless males and others become wingless females. The mouth parts being piercing and sucking type the **maxillae** and **mandibles** become specialized.

The males are short lived compared to the females.

Life-history

The nymphs after coming out of the old crust roam about in search of soft branches of the tree in which they live. The nymphs settle in a suitable spot. It sucks the cell sap. It liberates a type of exudate. With repeated moulting the nymph gradually lose most of its body structures. The thrown out skin together with the exudate form a crust around it. Each crust contains a pair of **branchial pores** for respiration and a big **anal tubercular opening.** In the male, the tubercular opening is provided with an **operculum.**

After three **moultings** the males come out by a removing the operculum and copulate with the female. Most of the males die

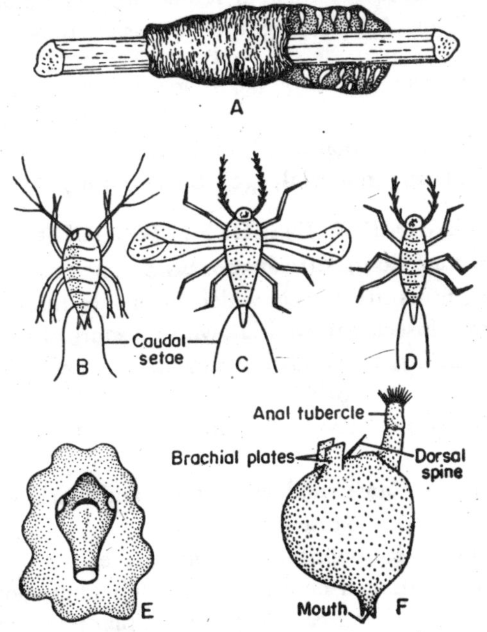

Fig. 114A. Lac Insect—Incrustation around a twig (A), Nymph (B), Winged male (C), Wingless male (D), adult female in incrustation (E), adult female (F).

soon after copulation. The female insect do not come out of its crust. The body of the female after being fertilized by the male assumes the shape of a hollow ball as most of the appendages are lost. It starts laying eggs which remain buried under the body. The development is direct. From fertilized eggs emerge the nymphs which come out through the anal tubercle. The females die after the coming out of the nymphs.

The Nature of Lac

Lac is a resinous exudate secreted from glands situated beneath the skin of lac insects. On chemical analysis it has been found that lac contains resin, carbohydrate, protein, water soluble salts, sand particles, volatile oils and pigments.

Lac Extraction

The twig containing mature incrustation is cut and washed thoroughly. The secreted materials from the twig is then scraped. The materials thus collected constitute *granular lac.* The granules are taken in suitable pot and the pot is heated by open charcoal fire. Dye is used at this stage to colour it suitably. The molten mass is stretched into sheet. After drying, these sheets are broken into pieces.

Use of Lac

The commercial value of lac is great. It is used for the following:

(a) Varnishes and polishes, (b) Lithographic ink, (c) Sealing wax, (d) Insulating material, (e) Shoe polish, toys and ornaments etc.

HARMFUL INSECTS : MOSQUITO

Introduction and Economic Importance

The mosquitoes are insects which belong to the order *Diptera.* These nocturnal insects are well-known as vectors of various disease —producing parasites. Following genera of mosquitoes are well-known throughout the world.

A. *Anopheles* sp.—Carrier of malarial parasite which causes malaria.

B. *Culex* sp.—Causes malaria in birds and filariasis in man.

C. *Stegomyia* sp.—Causes yellow fever.

D. *Aedes* sp.—Responsible for dengue fever.

It may be mentioned here that only the female mosquitoes are of blood-sucking type, whereas males feed on plant saps and are innocent of carrying any disease.

General Organisation of Mosquito

External feature. The body of these small size insects are divided into three parts—**Head, thorax** and **abdomen.** In between the head and thorax, is a short neck. Head contains (*i*) a pair of prominent bean-shaped compound eyes; (*ii*) a triangular lobe clypeus, bearing a pair of elongated antennae (bushy in males) and (*iii*) number of appendages around the mouth.

Appendages around the mouth. In the female mosquito the mouth parts are planned for piercing and sucking, but in the male it is only for sucking. The appendages around the mouth may be broadly grouped into two parts—(1) *Palp*, and (2) *Proboscis.*

1. *Palp*—Paired processes, one on each side of the proboscis. In male, the palps are much longer than the female.

2. *Proboscis*—Well-developed in female; consists of six elongated pieces united together at the base by a sheath. It consists of following structure.

(*a*) *Mandible*—one pair, needle-like with smooth sharpened tip.

(*b*) 1*st* *Maxilla*—one pair, needle-like with serrated sharpened tip.

(*c*) 2*nd Maxilla*—one pair, fused together to form *labium;* the tip of the labium is bifurcated into two white lobes called labellae, which possesses numerous sensory hairs at the free end.

(*d*) *Labium and Epipharynx*—These are unpaired processes, united together by chitinous hypopharynx. The free end is very sharp and the ventral side is provided with a groove. The salivary ducts run along this groove and opens at the tip. The labium, epipharynx and hypopharynx together form a tube for sucking the fluid.

In males the mandibles are absent and the hypopharynx is united with the labium.

Thorax. It is divided into three segments of which one is very large and other two are very small. It carries three pairs of walking legs and one pair of wings. The legs are built up in the same plan of an insect. Only difference is that tarsus is long and coxa is short. The wings are meant for flight and in the case of *Anopheles* these are spotted. The second pair of wings are reduced to two stumpy processes called *halters* which act as balances and sensory structure. Breathing aperture or stigmata is present on the lateral side.

Abdomen. Long and narrow abdomen is divided into nine segments and is covered by imbricate scales. No appendages, excepting a pair of knob-like ovipositor in females, are seen in the abdomen. Anal opening is on the eighth segment and the ninth segment carries the reproductive opening.

Internal Structures. Of the different internal structures (as the process of nutrition is very peculiar in mosquito), only the alimentary system will be discussed.

Alimentary system. It consists of alimentary canal and associated glands. The alimentary canal is built up in such a fashion that it helps to have a heavy meal. It contains following structures:

1. *Mouth*—associated with several appendages which have already been described. It leads to pharynx.

2. *Pharynx*—acts as a pumping organ to suck the fluid.

3. *Oesophagus*—narrow straight tube with three reservoirs at two posterior end. Oesophagus opens into the stomach and the opening is guarded by an oesophageal valve to check the flow of blood to the stomach.

4. *Stomach or midgut*—a straight tube continuous with the hindgut. At the posterior-most end thread-like Malphigian tubules open.

5. *Intestine or hindgut*—near the opening of Malphigian tubules originte the hindgut. It runs straight up to the anus.

In the thoracic region and under the alimentary canal, there are two sets of salivary glands. Each set consists of three lobes. The secretion of the glands, saliva, contains anti-coagulating enzymes which prevent the coagulation of blood in the mouth parts. At the time of feeding, the female mosquito sets the forked tip of the labium on the skin to support the entrance of the six piercing stylets. The saw-like tip of the first maxillae cuts the skin and all the six stylets penetrate within it. The blood is sucked by the action of pharyngeal wall through a food-tube which is formed by labium-epipharynx and hypopharynx. Immediately after piercing, the mosquito drains saliva which contains enzymes for the prevention of coagulation. Sporozoites of malarial parasites also enter into the blood stream at this stage. The meal of blood which is drawn is stored within the oesophageal diverticula.

Life-history (Fig. 115)

Fertilization is internal. After mating the female mosquito comes near water to lay eggs. It prefers stagnant water which is of low depth. Eggs of anopheles remain scattered and are provided with lateral air floats. The female culex mosquito arrange her eggs by hind legs and glue them together to form a raft.

From the eggs develop short larvae which can dive into the water. The body of larva is divided into three parts—*head*, *thorax*, and *abdomen*. Head contains a pair of lateral compound eyes, a pair of antennae and mouth with a pair of maxillae, a pair of mandible and a bunch of hair-like feeding brushes. The larvae eat voraciously different micro-organisms which are driven towards the mouth by the active movement of feeding brushes. Thorax is the broadest part and contains only lateral tufts of long hairs. The narrow abdomen is provided with bundles of hairs on lateral side, long pointed siphon, former leaf-like tracheal gills and a bundle of elongated rudder bristles. The siphon, which contains the air tubes, is for aerial respiration and contains in its free end two spiracles with five terminal flaps. The air tubes of siphon are in connection with the internal tracheal network. The tracheal gills are for aquatic respiration. A secretion from a special gland, called pre-spiracular gland, present around spiracle helps the larvae to alight on the surface of the water. The rudder bristles help in guiding movement under water. The larva of anopheles mosquito floats parallel to the surface of the water and possesses a very short siphon, whereas the culex larva remains under water with its head downwards and possesses a long pointed siphon.

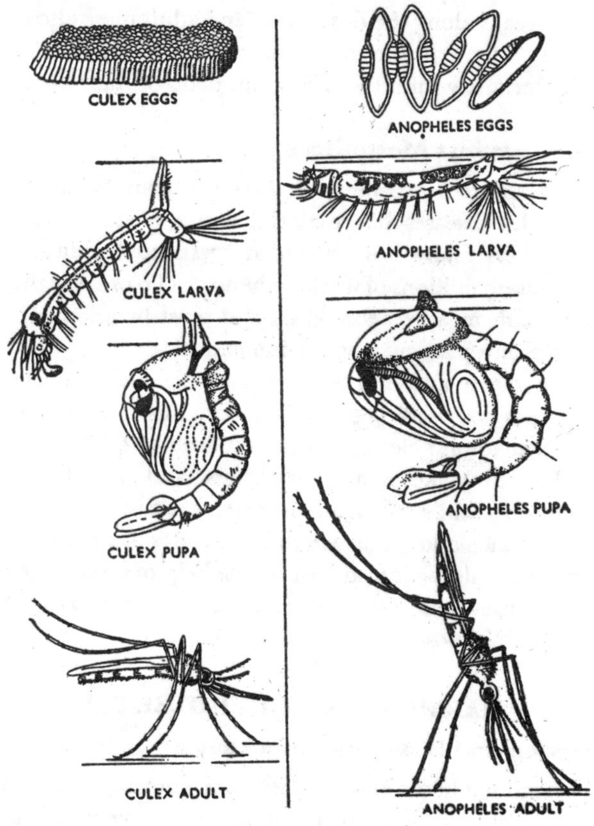

Fig. 115. Life-histories of mosquitoes.

Gradually, the larva changes into pupa. The pupa of mosquito is very mobile but do not eat anything. The anterior part of the body is broad and massive and is known as cephalothorax. The posterior part is represented by a narrow abdomen having nine segments. A pair of short respiratory trumpets are placed on the dorsal side of the thorax and are guarded by numerous hairs which prevent the entrance of water into it. These trumpets act as respiratory siphon of the larvae. On the eighth abdominal segment there is a pair of large fins which work as organs of swimming.

Within the body of the pupa rapid transformation takes place. Most of the pupal structures are used up as nutritive material of the cells which are forming adult structures. When the adult is fully formed, the pupa comes near the surface of water. The pupal skin breaks along the dorsal side and adult mosquito comes out. Immediately after exit, the mosquito rests for a while on the floating shell and after a few minutes drying up, takes off into the air.

Prevention against Mosquitoes

The mosquitoes are menace to the human being. By transmitting deadly diseases like malaria, filaria, yellow fever, dengue, it not only kills many human and domesticated animal lives, but also incapacitate millions of working hands. To prevent the national damage suitable measures should be and must be adopted. Following measures are particularly recommended.

1. To clean and remove the swamps and stagnant water, so that there is no breeding ground of mosquitoes.
2. To maintain cleanliness in and around the house.
3. To spray kerosene, detergents and insecticides in all ditches, crevices and small stagnant watery areas.
4. In ponds and pools where insecticides are poisonous, biological control should be introduced by the help of larvae-eating fishes.
5. To use mosquito-nets and curtains for preventing the entrance of adult mosquitoes.

HARMFUL INSECT : HOUSE-FLY

Introduction and Economic Importance

The common house-fly is scientifically known as *Musca domestica*. The body is grey with yellowish tinge on the ventral side. Four elongated lines are seen on the dorsal side of the thorax and a single streak is present on the middorsal line of the abdomen. These flies are notorious for carrying the germs of various diseases. This they do by carrying microbes through the external and internal parts of their body from garbage, etc., to the foods and drinks of the human beings.

External features

In the characteristic fashion of insect the body is divided into three parts—head, thorax, and abdomen.

Head. Semicircular in outline with two very large compound eyes. Three simple eyes or ocelli are placed on the dorsal side. Head contains following appendages.

1. *Antennae*—one pair, short and flexible. The distal segment is the largest and contains a branch called *artista.*

2. *Mouth parts*—arranged in a fashion to facilitate its mechanism of feeding. This type of mouth parts is called *spongin type,* which includes two structures (*a*) *Proboscis,* and (*b*) *Food channel.* Mandibles are absent and maxillae are represented only by a pair of maxillary pulps.

(*a*) *Proboscis.* It is formed by the labium and is fleshy and retractile. The proboscis may be separated into three parts—(*i*) basiproboscis (basal part), (*ii*) mediproboscis or haustellum (middle portion), and (*iii*) distioproboscis (distal part). The middle part is strongly built and movable. The distal part is extended as a pair of membranous, spongy *labellae.* The labellae are traversed by numerous channels opening externally through the margin.

(*b*) *Food channel.* It is formed by the union of pseudotrachae and consists of three parts—labium, epipharynx, and the hypopharynx. Two styles persist, but are incapable of piercing.

Thorax. Divisions are indistinct. Mesothorax is larger than pro- and metathorax. Three pairs of *walking legs* and one pair of *wing* are the appendages in the thoracic region.

Walking leg of House-fly. Consists of usual divisions of insect leg. A pair of curved *claws* are present in the terminal end of each tarsus. A pair of glandular pads, pulvillae with numerous hollow hairs are present between the claws. A sticky juice is secreted from these hollow hairs which help them in walking.

Wing of House-fly. Presence of one pair of wing is the feature of this order. The second pair of wing is modified with a stick-like structure called halters, which act as sensory and balancing organ. The wings are broad, transparent, hairy and triangular. During flight the wings are vibrated rapidly and a sound is produced by them.

Abdomen. Oval, hairy and segmented (8 in male and 9 in female). First and second segments are much reduced. In the female flies seventh to tenth segments serve as retractile *ovipositors.*

Mechanism of feeding. Lives only on liquid foods. Solid foods like sugars are moistened by its saliva. At the time of feeding the fly extends its proboscis and sets the labella on the food. The pseudotrachae on the labella get filled with food by capillary action and finally it is drawn through food channel by the sucking action of the muscular pharynx

Life-history (Fig. 116)

Reproduction takes place by copulation, which lasts only for a few minutes. Male grasps the female on her dorsal side with the help of its pro- and mesothoracic legs. The female fly inserts its ovipositor within the genital apparatus of the male and the male structures aid in the entrance.

After fertilization, the female fly begins to lay eggs. Eggs are generally laid on organic manures, debris and decaying fruits and vegetables. A single fly lay a large number of eggs at one time. The eggs are white and elongated, carry a curved thickening on its dorsal side. The hatching takes place within 8–24 hours and the process is temperature dependent.

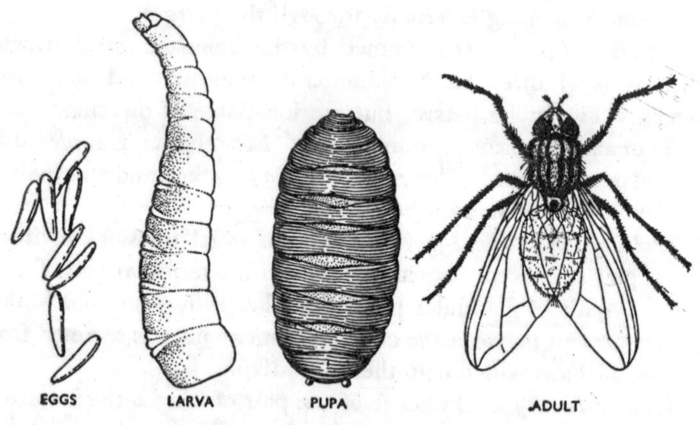

Fig. 116. Life-history of house-fly.

Larva is called *maggot.* It is white and elongated with anterior end narrow and posterior end broad. Head, legs and wings are absent. The body is divided into one anterior segment and twelve trunk segments. The anterior or cephalic segment is rudimentary and contains a pair of *oral lobe*. The oral lobes are provided with sensory 'oral tubercles' (2 in each lobe) and a number of branched food channels. Mouth is present in between the two oral lobes. A hood-like *mandibular sclerite* is present on the anterior side of the mouth and helps only in locomotion.

From sixth to twelfth segment, there are pad-like structures on the ventral side, called *seminiferous pads*. These help in locomotion. A pair of spiracle is present on the middorsal position of the twelfth segment, which opens into the trachae. The larvae feeds on liquid food and grows rapidly. It sheds its skin twice (the process is called moulting) and gradually develops a pair of fan-shaped spiracles.

After certain period, the full-grown maggot contracts and passes into the next stage called pupa. A hard case is formed around the body from which only the spiracles remain projected outwards. Within the pupal skin, rapid transformation takes place. Old organs excepting nervous system are replaced by new, which are formed by special groups of cells called *imaginal discs*. Young fly comes out of the pupal case after knocking the anterior end like a lid.

A newly grown fly differs much from a full-grown fly in having colourless body, rudimentary wings, short legs. But within half of an hour swelling and elongation of different parts take place and it transforms into an adult one.

Preventive measures

As the microbes carried by these flies endanger human lives, great care must be taken to prevent the contamination of foods and drinks by these flies. Following measures are usually recommended.

1. To observe cleanliness in and around house.

2. To keep all the foods and drinks covered, so that flies cannot sit on it.

3. To use different insecticides for killing the maggots and the adults.

CHAPTER X

ARTHROPODA (continued) : MYRIAPODA

SCOLOPENDRA

Scolopendra generally lives under stones crevices, broken wall and feed on small insects and other animals. The body is flattened dorso-ventrally and is divided into numerous segments. Each segment, excepting the last one, bears a pair of legs which are distinctly set apart. Due to the presence of numerous pairs of walking legs, these animals are commonly called *centipedes* (Fig. 117).

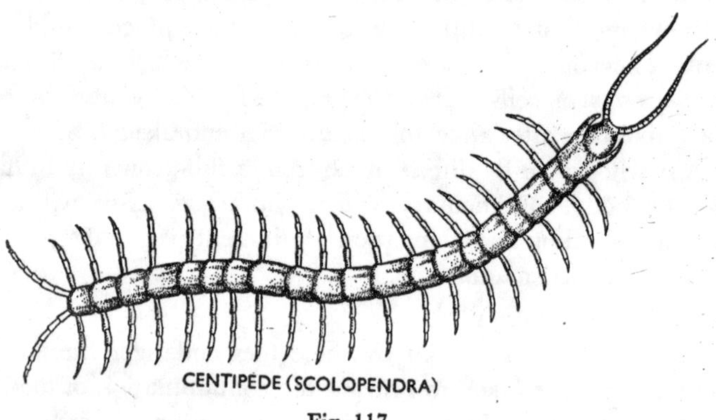
CENTIPEDE (SCOLOPENDRA)

Fig. 117

The body is divisible into two parts—head and abdomen. A soft cuticle serves as external covering of the body. A pair of antennae is present on the anterior end of the head. Each antenna is made up of fourteen segments. Number of simple eyes are present on the head region. Mouth is present in the anterior end of the head and is provided into two-jointed mandibles. All the segments of the abdomen are alike. Each leg is made up of seven parts and the terminal part is called claw. In the first pair of legs, lies the opening of poison gland. Last pair of legs are the largest. The terminal end of the food canal, *anus*, and the openings of the reproductive ducts, genital opening are present in the last segment.

CHAPTER XI

ARTHROPODA (continued) : ARACHNIDA

SPIDER

A great number of spiders of different varieties are seen in our country. They have diverse habitats. Some of them spin net, while others do not. One common spider *Nephilia* sp. (Fig. 118) may be studied as a typical example of spider. Like most other spiders, here also *sexual dimorphism* is present—females have much larger body than males.

The body is distinctly divided into two parts—(1) *Cephalothorax*, and (2) *Abdomen*. These two parts are connected by a slender *waist*. The entire body is covered by hard *cuticle*, which is shedded periodically. The surface of the cuticle is generally hairy and coloured. In the body of spider six pairs of appendages:

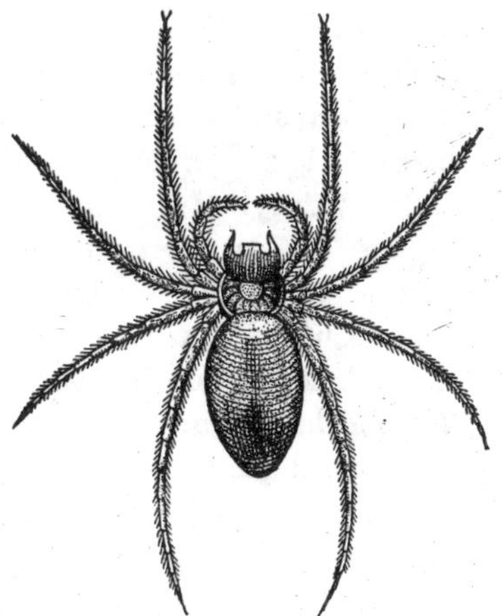

Fig. 118. Spider (*Nephilia*).

(1) *Chelicerae or falces*—one pair, two jointed and placed one on each side of the mouth. The duct of poison gland open through it.

(2) *Pedipalps*—one pair, leg-like, six jointed, help in mastication and in males also serve as reservoir of spermatozoa and copulatory organ.

(3) *Walking legs*—four pairs, each one is seven jointed and clawed. The first pair of walking legs also act as feeler.

Mouth is placed in the anterior and ventral side of the cephalothorax. On the dorsal side of the head region there are four pairs of simple *eyes*. *Spinning glands* are placed on the ventral side of the abdomen, near the base of the legs. At the end of the abdomen, there are *spinnerets* for wearing the net. The spinnerets are modified abdominal appendages.

CHAPTER XII

MOLLUSCA : SOFT-BODIED ANIMALS

The molluscs (*mollis*=soft) are soft-bodied animals with a hard external or internal shell. The phylum includes forms like chitons, snails, slugs, clams, oysters, sepia, loligo, octopus, etc. Generally *aquatic*, some *terrestrial* and *amphibious* forms are also present. A muscular *foot* is present as the thickening of the ventral body wall. The foot is meant for locomotion, but is modified in different forms, i.e., squids, the foot is modified as arms. The body is *bilaterally symmetrical* and this symmetry is modified in snails. Excepting tentacles in some no other appendages are present. The internal organisation is well-developed. The body cavity is represented by a small *coelom* and *hoemocoel*. The digestive system includes an alimentary canal consisting of mouth, coiled digestive tube, anus and a liver as digestive gland. The circulatory system is well-developed and contains heart, blood vessels and hoemocoelomic spaces. The excretory functions are carried by specialised organ called kidney. In aquatic molluscs the gills are the respiratory organs, but in terrestrial forms there are lung-like structures for respiration. The nervous system includes well-developed ganglia, commissures and connectives. Certain sensory organs are also present. The phylum includes both monoecious and dioecious forms. The fertilization may be external or internal. Two larval stages, Trocophore and Veliger, are seen in the life-history of molluscs.

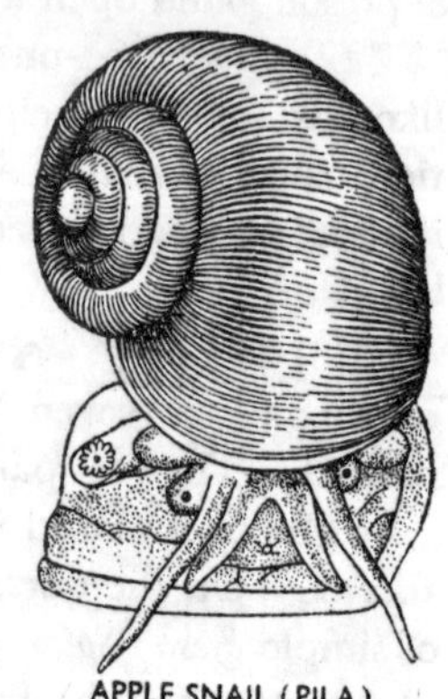
APPLE SNAIL (PILA)

Fig. 119

Two Common Snails

Two snails—one aquatic and one terrestrial, are very common in our country. The aquatic one is popularly called Apple snail (Fig. 119) is scientifically known as *Pila globosa* and the terrestrial one, garden snail is called *Achatina fulica*. In both, the soft parts are covered by a hard, right-handed spiral shell. In *Achatina* there are eight coils and 5–6 coils are present in *Pila*. In the body of the *Pila*, the opening of the shell is guarded by a flat plate called *operculum*, which remains attached with the foot. The operculum is absent in *Achatina*. In *Achatina* the outer part of the shell may possess stripes of various colours. In both these forms the body consists of a distinct head, which is more prominent in *Achatina*. The head contains in both, 2 pairs of tentacles and an opening called mouth. On the right side of the body in *Achatina*, there is an aperture called pulmonary opening which is in communication with the respiratory organs. Food of both these snails are leaves and other parts of plants.

CHAPTER XIII

CHORDATA

The **Chordata** (*chorde*=cord) includes well-known animals such as lampreys, fishes, amphibians, reptiles, birds and mammals. They also include **protochordates** such as balanoglossus, ascidian and amphioxus.

All chordates possess a notochord at least for some time in their life-history. It is the notochord that has given the name chordata to this group. The **notochord** is an elastic supporting rod covered by one or two sheaths of connective tissue and is composed of large vacuolated cells. It is situated in the middorsal line and typically extends along the whole length of the animal, being placed under the nerve cord but above the alimentary canal. It persists throughout life in the lower forms, such as amphioxus and lamprey. In the adults of higher chordates, the notochord is replaced, either wholly or partially, by a skeletal structure known as the **vertebral column.** Such chordates are the **vertebrates.** We are already familiar with the structure of one vertebrate animal, the toad, which possesses a distinct vertebral column. But back-bone is not all. There are other essential features by which the vertebrates like toad, may be distinguished from the **invertebrates** such as earthworm, leech and prawn. These are now presented in a tabular form.

Vertebrates	Invertebrates
1. Vertebrates possess, at least in early life, a stiff-supporting rod, the **notochord,** which is placed in the middorsal line and extends from end to end: it lies beneath the nerve cord but above the gut.	1. **Notochord** is never present.
2. Pharynx is perforated on either side by a series of **pharyngeal clefts** or **gill-slits.** This is an efficient apparatus for carrying out respiration and persists throughout life in lower forms such as the fishes. Gill-slits appear in the embryo of reptiles, birds and mammals, although they never breathe by gills.	2. **Pharyngeal clefts** are never formed, although some of the aquatic invertebrates may have gills for the purpose of breathing.

Vertebrates	**Invertebrates**
3. The **central nervous system** is a hollow tube placed in the middorsal line above the notochord. Its cavity is known as the **neurocoel.**	3. The **central nervous system** is a solid cord placed in the midventral line. There is no **neurocoel.**
4. A living **endoskeleton,** composed of skull and vertebral column, is always found. This framework grows in size along with the growth of the animal.	4. Living **endoskeleton,** that is skull and vertebral column, is never found. There may be an armour of exoskeleton composed of non-living matter which never grows.
5. The **heart** is located ventrally and beneath the gut.	5. **Heart,** when present, is located dorsally and above the gut.
6. Blood circulates through a closed system of vessels—arteries, capillaries and veins.	6. Blood flowing through the arteries may pass freely into sinuses or open spaces within the body cavity.
7. **Hepatic-portal** system, conveying the blood from the alimentary canal into the liver, is always present.	7. Such **hepatic-portal system** is never met with.
8. There are **red blood corpuscles** containing the respiratory pigment haemoglobin in their cytoplasm.	8. **Red blood corpuscles** are not usually found. Haemoglobin, which is of rare occurrence, is found in solution in the blood-plasma.
9. A true **tail,** consisting of a post-anal part of the body-axis, is usually found. This may persist throughout life or it may be an embryonic structure which is lost in the adult stage.	9. True post-anal **tail** is not found. Anus is terminal and opens at the posterior end of the body.
10. **Appendages** or **limbs** are never more than two pairs.	10. **Appendages,** when present, are more than two pairs.
11. **Skin** consists of two distinct layers, an outer epidermis overlying a thick dermis.	11. **Skin** consists of only one layer.
12. **Eyes** originate as outgrowths from the brain. Compound eyes are never found.	12. **Eyes** develop from the skin and not from the brain. Compound eyes are often present.

Classification

Subphylum I. **PROTOCHORDATA** (*protos*=first). Borderline chordates with no true brain or skull.

Class I. HEMICHORDA (*hemi*=half) or ADELOCHORDA (*adelos*=concealed; *chorde*=cord). Marine, worm-like chordates with pharyngeal gill-slits and a stiff rod, suggesting notochord, at the anterior region of the body. *Balanoglossus*, *Dolichoglossus*, etc.

Class II. UROCHORDA (*oura*=tail; *chorde*=cord) or TUNICATA (*tunica*=envelop). Marine sac-like animals in which the adults are sedentary (except in one group) and lost most of the chordate characteristics, but the free-swimming larvae possess (*i*) a notochord in its tail (hence urochorda), (*ii*) pharyngeal clefts, and (*iii*) a dorsally placed hollow nerve tube. During metamorphosis the tail is lost and the notochord along with the dorsal nerve tube is thus sacrificed. *Salpa*, *Ascidia* or seasquirt, *Botryllus*, etc.

Class III. CEPHALOCHORDA (*Kephale*=head; *chorde*=cord) or ACRANIA. Marine, fish-like creatures with a notochord extending from the tip of the tail to the tip of the snout (hence cephalochorda). *Amphioxus* (*Branchiostoma*) or the lancelets.

Subphylum. II. **VERTEBRATA** (*Vertebra*=a joint) or **CRANIATA** (*Kranion*=skull). Higher chordates with a definite brain enclosed in the skull (hence craniata). Notochord is replaced in the higher forms by a metameric vertebral column. There is a living endoskeleton composed of bones and cartilages, a true coelom, and a hollow tubular central nervous system placed in the middorsal line. Lampreys, fishes, amphibians, reptiles, birds and mammals.

There are two branches of the vertebrates: (*i*) the round-mouthed **Agnatha** in which teeth and jaws are absent, and (*ii*) the **Gnathostomata** in which the mouth is surrounded by upper and lower jaws. The cyclostomata are the living agnatha and the rest are gnathostomata.

Class I. CYCLOSTOMATA (*Kyklos*=circle; *Stoma*=mouth). Primitive marine vertebrates with round, suctorial mouth and rasping tongue bearing horny epidermal teeth. Notochord persists throughout life. Endoskeleton is poorly developed and cartilaginous. Paired fins and girdles are absent. Skin is slimy and without any scales. The modern lampreys (*Petromyzon*) and the hags (*Myxine*) are the two living representatives.

Class II. CHONDRICHTHYES (*Chondros*=Cartilage; *icthys*=fish). This class includes sharks, rays, etc. The most striking character of fishes of this class is its entirely *cartilaginous skeleton*. No truc bone is present. An operculum is absent (except in Holocephali—a group of chondrichthyes). There are five gill-slits, on each side, which open directly to the exterior. The body is covered with placoid scales, which are provided with minute spines at the base. The cloaca receives the rectum, excretory and reproductive ducts. Several species are viviparous. Examples:—*Scoliodon* (Indian shark), *Narcine* (electric ray), saw fishes, etc.

Class III. OSTEICHTHYES or ACTINOPTERGII. This class includes most of the common fishes. **Bony skeleton** is present throughout the class. An operculum on each side covers the filamentous gills. The fins are supported by fin-rays.

Examples:—Bhetki, Ruhu, Sole, Lata, etc.

Class IV. CHOANICHTHYES. Dipnoi, one of the existing groups of this class, possess distinct lungs and are of great evolutionary significance in giving rise to land vertebrates.

Examples:—*Epiceratodus* or *Ceratodus* (found in Australia). *Lepidosiren* (found in South America), and *Protopterus* (found in Africa).

The *Latimeria chalumnae*, which is the only living member of a nearly extinct side branch, the Crossopterygii, of the class, Choanichthyes, was found recently near South Africa.

These three existing fish groups, namely Chondrichthyes, Actinopterygii and Choanichthyes along with an extinct group, the Aphetohyoidea, are placed under the **super-class Pisces** (*pisces*= fishes). The entire super-class represents the largest stock of vertebrates, who have dominated the water of the earth. The following are the characteristics of the fishes in general:

(*i*) 'Cold-blooded' vertebrates breathing by means of gills (except in Dipnoi, where lung respiration is also found).

(*ii*) They have a streamlined body in accordance with the aquatic mode of life.

(*iii*) Skin is usually covered by scales.

(*iv*) Pectoral and pelvic fins are usually paired, and there are unpaired median fins. All the fins are supported by fin-rays.

(*v*) There is no tympanic membrane.

(*vi*) The heart consists of: sinus venosus, one auricle and one ventricle. It contains only deoxygenated blood (except in Dipnoi, where the auricle shows indication of partition and also receives oxygenated blood).

Class V. AMPHIBIA (*amphi*=both; *bios*=life). The skin of amphibia is moist due to secretion of mucus and serves as a huge respiratory surface. Scales are usually lacking. As the name implies, they lead a double life. The larva is always aquatic and breathes by gills, whereas the adults are lung-breathing terrestrial creatures with two pairs of limbs. Amphibia includes: (*i*) the tailed **caudata** (*cauda*=tail) or **urodela** (*oura*=tail; *delos*=visible) such as *Necturus* (Mud-puppy) and *Salamandra;* (*ii*) the tailless **Anura** or **Salientia** (*salio*=leap) such as frogs and toads, and (*iii*) the limbless **Apoda** (*A*=no; *podus*=foot) such as the Coecilian *Icthyophis.* In some of the tailed amphibians gills persist throughout life (*Proteus; Necturus*). Fins, when present, are never supported by fin-rays.

Class VI. REPTILIA. The reptiles are craniates with complete adaptation to life on dry land. They have dry skin with epidermal scales. Lungs are the sole respiratory organs even in aquatic reptiles such as tortoises and crocodiles.

The development is direct and a larval stage is absent. The modern reptiles consist of the following four orders : (*i*) **Chelonia** (*chelone*=tortoise) including tortoises and turtles, (*ii*) **Rhynchocephalia** including the Newzealand lizard *Sphenodon;* (*ii*) **Crocodilia** including gavials, crocodiles and alligators; (*iv*) **Squamata** including the snakes, lizards and chameleons.

Class VII. AVES (*avis*=bird). Birds are provided with a covering of feathers and the forelimbs are modified to form the wings. The jaws are toothless in all modern birds. The living birds are divided into two groups : (*i*) **Ratitae** including the flightless running forms, such as Ostrich, Emu and Kiwi; (*ii*) **Carinatae** or the flying birds, such as pigeons, parrots, crows.

Class VIII. MAMMALIA (*mamma*=breast). Mammals are the highest of the vertebrates and include man. The skin is covered by a coat of hair though this may be reduced to a few bristles as in the whales. The young are nourished by milk, a secretion from the breast-gland of their mother. The mammals comprise three subdivisions: (*i*) the egg-laying **Prototheria** (*protos*=first; *therion*=mammal) including the duck-billed mole, *Ornithorhynchus* and the spiny ant-eater, *Echidna;* (*ii*) the pouched **Metatheria** (*meta*=beyond; *therion*=mammal) including the marsupials such as Kangaroos; (*iii*) **Eutheria** (*Eu*= true; *therion*=mammal) including the true placental mammals whose embryo remain attached by means of a structure called placenta, to the womb of their mother. Respiration and nutrition of the embryo is carried on through the placenta. Examples:—Moles, bats, sloths, armadillos, pangolins, beavers, rats, cats, seals, walruses, horse, elephants, rhinoceros, sea-cows, whales, dolphins, lemurs, monkeys, apes and man.

CHAPTER XIV

CHORDATA (continued) : PISCES

BHETKI FISH

(*Lates calcarifer*)

Habitat

The Bhetki is a bony fish found in saline estuaries. It occasionally visits fresh water streams. The animal is one of the relished food fish and is abundantly available during the winter season.

External Morphology

The spindle-shaped body is compressed from side to side and perfectly streamlined, so as to facilitate swift movement through water. The outer surface of the body is covered with a soft mucus-producing epidermis. Beneath this are dermal scales which are arranged in rows in an imbricate fashion like the tiles on a roof. The free posterior margin of each scale is provided with sharp spine-like outgrowths or ctenes. This type of spine-bearing scale is known as the **ctenoid** scale. A second type, known as the **cycloid** scale, is found in other common fishes such as Rohu (*Labeo rohita*) and Catla (*Catla catla*).

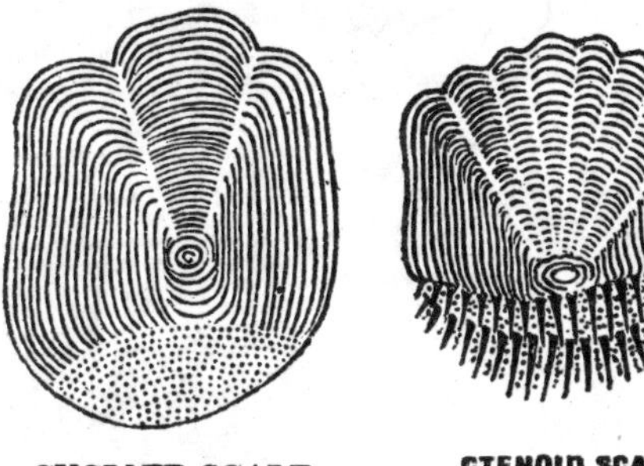

Fig. 120. Two types of scales.

A cycloid scale has no spinous outgrowths and its free posterior margin is smooth. The scales constitute the **exoskeleton** for protecting the body from external violence. The dermis, in which the scales are embedded, is composed of connective tissue, pigment cells and sensory receptors. It is thicker than the epidermis and the two together form the skin of the fish.

The body of the fish is divisible into three regions; (*i*) *head*, (*ii*) *trunk*, and (*iii*) *tail*. The **head** extends from the tip of the *snout* to the posterior margin of the *operculum* (gill-cover), the **trunk** from this point to the anal opening, and the remainder constitute the post-anal **tail** (Fig. 121). There is an **eye** on each side of the head.

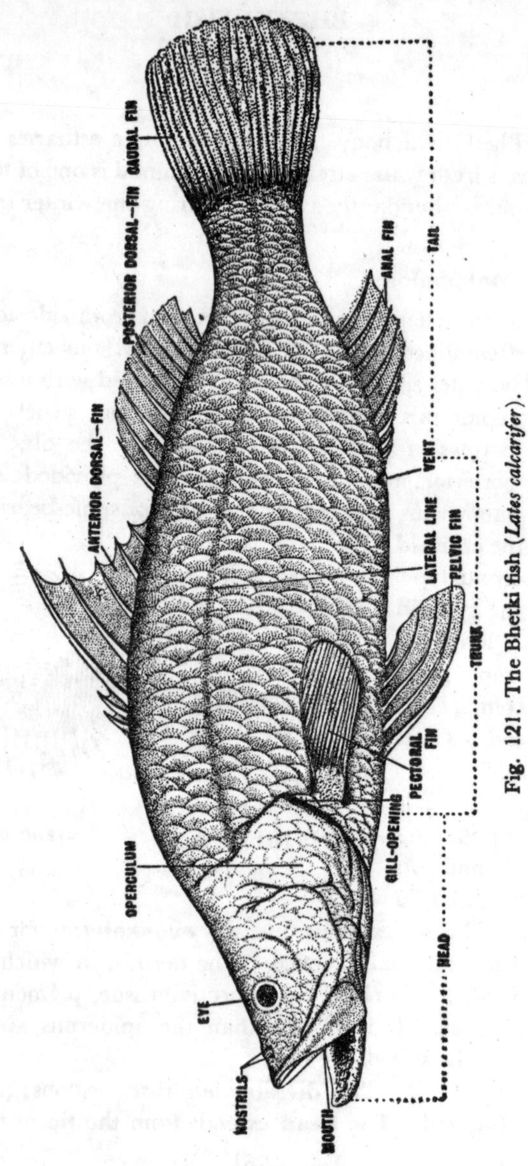

Fig. 121. The Bhetki fish (*Lates calcarifer*).

The eyes are without eyelids and covered by a protective transparent membrane. In front of each eye, there is a pair of **nostrils** which do not open into the buccal cavity and hence serve no purpose in breathing. The large **mouth** is terminal; it is surrounded by well-developed jaws which bear fine teeth. There is a bony **operculum** or *gill-cover* on each side of the head. Attached to the margin of the operculum, there is a flap-like *branchiostegal membrane* supported by bony rods. The operculum and the branchiostegal membrane are movable structures which form the outer boundary of the *gill-chamber*, one on each side of the body. Each chamber contains four comb-like *gills*. Behind the operculum is an oblique slit, the *opening of the gill-chamber*, which can be tightly closed by the branchiostegal membrane. This opening serves for the exit of the respiratory water.

There is a dark **lateral line** on either side of the body, extending backwards through the trunk and tail regions. This is composed of a row of minute pores which communicate with deep-seated sensory cells, thus forming the **lateral line sense organs** by which the fish can understand vibrations of low frequency in its watery environment. There is a depression, called **vent,** on the ventral border of the fish at the junction of the trunk and the tail. Into this open the **anus** in front, **urinary opening** behind, and the two **gonopores** in between them at the two sides. The **tail** is supported by powerful muscles and constitutes an efficient oar for propelling the fish forwards.

The fish bears certain **appendages** known as the fins. The **fins** are membranous extensions of the skin and are supported by **fin-rays.** The fins are either (*i*) median *unpaired* fins, or (*ii*) lateral *paired* fins.

The **median fins** are : (*a*) anterior and posterior **dorsal fins** placed on the dorsal surface of the fish and united at their bases; the anterior dorsal fin is supported by seven spinous rays of which the third one is the largest, and the posterior dorsal fin is supported by twelve rays of which the first one is spinous and the rest are soft; (*b*) the **caudal fin,** at the end of the tail, is a fan-shaped expansion supported by 10 to 20 bony rays; (*c*) **anal** or **ventral fin,** placed on the ventral border of the post-anal tail, is supported by about a dozen fin-rays of which the first three are spinous; of the three spinous rays the third one is the largest.

The **paired fins** are : (*a*) **pectoral fins,** one on each side of the body, and placed behind the opercula; each pectoral fin is supported by 14 bony rays; (*b*) **pelvic fins** are placed on the ventral side of the body below and behind the pectoral fins; each bears a spinous rod in front followed by four bony rays.

The fins serve chiefly for steering, balancing and checking of speed, whereas the muscular tail is the effective organ of locomotion.

Skeletal System

The *dermal scales* and the *fin-rays* constitute the **exoskeleton.** The **endoskeleton** consists of the *skull*, *vertebral column*, *ribs*, *pectoral girdle*, and the small plate-like *pterygiophores* supporting the fin-rays.

(A) The **skull** of Bhetki is a complex structure and is composed of the *cranium* or brain-case housing the brain, the paired *sense capsules* (olfactory, optic and otic), the *jaws*, the *hyoid*, and the *branchial arches* supporting the tongue and gills. The skull is tightly fixed to the vertebral column so that the fish cannot turn its head from side to side. The roof of the cranium is formed posteriorly by the **parietals** and anteriorly by the **frontals.** The posterior part of the cranium is composed of four bones: (*i*) a **basi-occipital** with a *single occipital condyle*, (*ii*) a **supra-occipital** forming the occipital crest and (*iii*) two **ex-occipitals,** one on each side of the **foramen magnum** through which the brain becomes continuous with the spinal cord. Ventrally, and in front of the basi-occipital there is (*i*) a small Y-shaped **basi-sphenoid,** (*ii*) a large **ali-sphenoid** on each side of the basi-sphenoid and (*iii*) a pair of **orbito-sphenoids.** The **orbits** housing the eyes, are surrounded by a ring of scale-like bones the anterior of which is called the **lacrimal.** The **nasal capsule** includes the ethmoid which lies anterior to the frontals. Its roof is formed by the **nasals** which partially overlap the ethmoid, and its floor is formed by the **vomers** which lie in front of the parasphenoid. The **otic capsule,** containing the ear, is composed of five bones : **epi-otic, opisthotic, pro-otic, pterotic** and **sphenotic.**

The **upper jaw** is composed of an outer series of membrane bones and an inner series of cartilage bones. The true jaw, on each side, is formed by the *outer series* of three bones—(*i*) **premaxilla** and (*ii*) **maxilla,** bearing teeth; and (*iii*) a scale-like **jugal** attached to the posterior end of the maxilla. The *inner series* is the **palato-quadrate bar,** which consists of (*i*) a tooth-bearing **palatine** articulated to the ethmoid; (*ii*) a **pterygoid** in the middle, represented by *ecto-*, *meso-*, and *meta-pterygoid* bones; (*iii*) a **quadrate** posteriorly which articulates with the lower jaw.

The **visceral arches** consist of seven pairs of cartilaginous bars which support the jaws, tongue and gills. All of them are more or less ossified in later life.

The **mandibular arch** (first arch) is composed of two parts on each side : (*i*) the dorsally placed *palatoquadrate* which has been already described, and (*ii*) the ventrally placed *Meckel's cartilage* which forms the lower jaw.

Each **lower jaw** is composed of **Meckel's cartilage,** which is ensheathed by three membrane bones: (*i*) an anterior tooth bearing **dentary,** (*ii*) a posterior **articular** which articulates with the **quadrate** of the upper jaw and, (*iii*) a small **angular** at the posterior end of the articular.

The **hyoid arch** (second arch) is in part a support for the tongue and the gill-cover. Each hyoid consists of a **hyomandibular** in front and a **symplectic** behind. Ventral to these two, there are several cartilaginous rods—**epihyal, interhyal, ceratohyal** and **basihyal,** which support the tongue as well as the floor of the buccopharynx. Behind the symplectic and connected with the hyoid, there are certain bones supporting the gill-cover and branchiostegal membrane. These are: (*i*) **opercular,** (*ii*) **pre-opercular,** (*iii*) **sub-opercular** and (*iv*) **interopercular.** There are seven **branchiostegal rays** supporting the branchiostegal membrane on each side.

The third, fourth, fifth and sixth arches constitute the

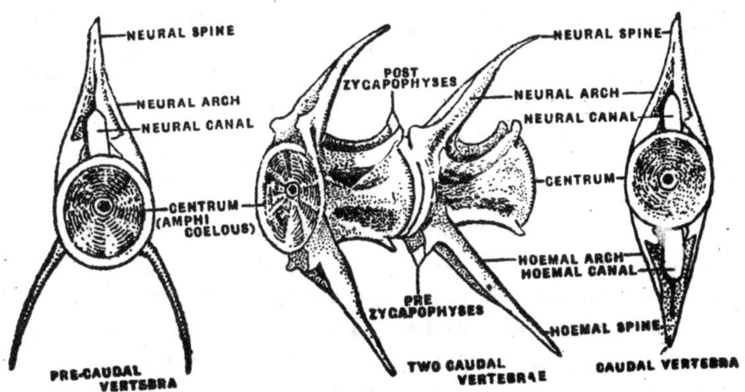

Fig. 122. Vertebrae of Bhetki.

branchial arches which support the gills on their outer curvatures, and spiny gill rakers on their inner curvatures.

The small seventh arch does not bear any gill but only pharyngeal teeth are present on its inner curvature.

(B) The **vertebral column** of Bhetki is composed of numerous *amphicoelous* vertebrae, and is divided into two regions, precaudal and caudal. A typical **precaudal** vertebra from the trunk region consists of : (*i*) a spool-like **centrum,** concave at both ends (*amphicoelous*) ; (*ii*) a dorsally placed **neural arch** to lodge the spinal cord ; (*iii*) a slender **neural spine** in the middorsal part of the neural arch and (*iv*) paired **transverse processes** or *parapophyses* projecting out from the sides of the centrum. Articulated to the distal ends of the transverse processes, there are ventral *ribs* curving downwards to enclose the viscera. In addition to these, there are paired **zygapophyses** or *articular processes* by which successive vertebrae articulate with one another.

In a **caudal vertebra** from the tail region, the *centrum* bears a ventrally placed **haemal arch** crowned by a slender **haemal spine** in addition to the dorsally placed *neural arch* and *neural spine*. The haemal arch protects the caudal vessels. Ventral ribs are not found in caudal vertebrae.

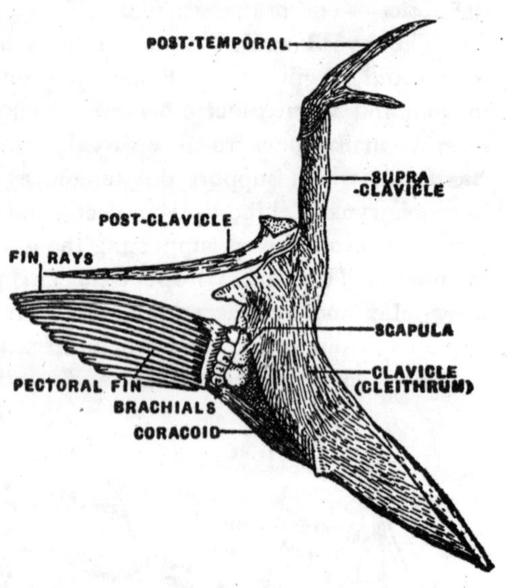

Fig. 123. Pectoral girdle and pectoral fin.

(C) The **pectoral girdle** consists of several bones which support the *pectoral fins* on either side. Each half of the girdle is composed of a dorsal **scapula** and a ventral **coracoid** which are articulated to a large **clavicle** or **cleithrum.** The clavicle slopes downwards to meet its fellow of the opposite side in the midventral line. From the dorsal end of the clavicle, a slender **post-clavicle** extends backwards and a stout **supra-clavicle** extends upwards. A forked

post-temporal connects the supra-clavicle with the back of the skull. Attached to the posterior surface of the scapula and coracoid, there are four or five small **brachial ossicles** or *pterygiophores* which support the *dermotrichia* or fin-rays of the *pectoral fin.*

The **pelvic girdle** is represented on each side by a thin plate—the **basipterygium,** which bears the dermotrichia of the *pelvic fin.*

Muscular System

The muscles of the fin, gill region and head are small. The trunk and tail, however, bear powerful muscles arranged in the form of V-shaped segments called **myotomes.** Successive myotomes are separated by delicate partitions of connective tissue. These muscles are utilised for swimming.

Coelom

Bhetki has a large body cavity to accommodate the visceral organs. This is divided into two parts—an anterior **pericardial cavity** and a posterior **perivisceral cavity.** The small pericardial cavity encloses the heart and is lined by the *pericardial membrane.* The large perivisceral cavity contains the digestive organs, swim-bladder and the urinogenital organs. It is lined by the *peritoneum* which encloses most of the organs between its two folds, thereby fixing them in their proper place within the body cavity (see Plate).

Alimentary System

The alimentary system consists of : (1) a digestive canal and (2) the associated glands. The alimentary canal is complete with a *mouth* in front and an *anus* behind. The **mouth** is a wide opening bounded by upper and lower jaws which bear sharp conical teeth. The mouth leads into the **buccal cavity,** which in turn is continued backwards into the **pharynx.** The tongue is small and non-muscular; it is firmly attached to the floor of the buccal cavity and supported by the basi-hyals. Numerous *mucous glands* pour their secretion into the mouth but salivary glands are absent. There are no internal nares on the roof of the buccal cavity. The pharynx is supported on each side by bony gill-arches. There are five gill-slits on either side in between the gill-arches by which the pharynx communicates with the gill-chambers. Inner margin of each gill-arch is provided with horny **gill-rakers** which act as strainers and prevent escape of food into the gill-chambers. The pharynx leads through a short **oesophagus** into the stomach. There is a strong

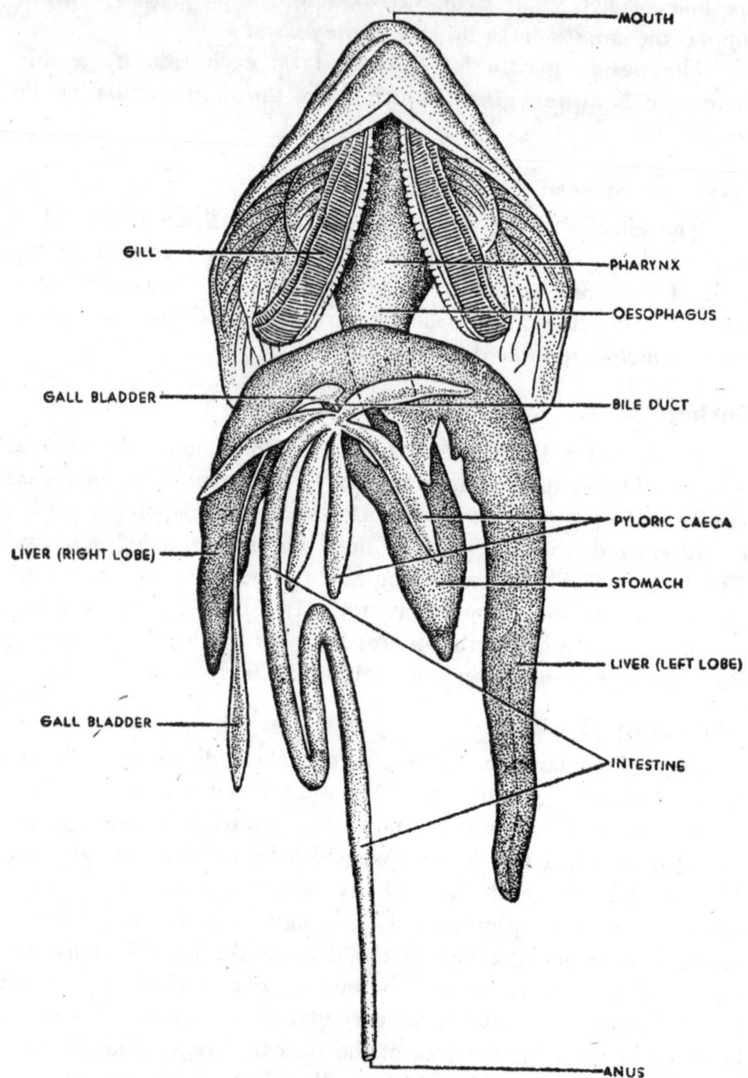

Fig. 124. Alimentary system of Bhetki.

oesophageal sphincter which prevents entry of water into the stomach.

The **stomach** of the fish is a recurved U-shaped sac. The

portion nearer to the oesophagus is the *cardiac stomach* and the opposite portion is the *pyloric stomach.* The cardiac stomach is directed backwards and ends blindly, whereas the pyloric stomach is directed forwards and is continued into the intestine. There is a *pyloric constriction* separating the pyloric stomach from the intestine.

The **intestine** is long and narrow. Five blind tubular pouches called **pyloric caeca** originate from the intestine close to its junction with the pyloric stomach. These caeca are either secretory structures producing digestive juice or they may help in the absorption of the digested products. The **intestine** extends backward and then bends forward forming a loop. Finally, the tube again curves backward to open at the **anus.** The anus is situated in the anterior part of the **vent** in front of the anal fin. A short posterior part of the intestine is known as the **rectum** but this cannot be sharply demarcated from the intestine. The mesentery supporting the gut is darkly pigmented and loaded with fat.

Digestive glands. The large **liver** lies in the anterior part of the body cavity. It has two lateral lobes connected by a flat median portion. The elongated lateral lobes extend backwards into the posterior part of the body cavity. Bile formed within the liver passes into the gall-bladder and thence by the bile-duct into the first part of the intestine. The **gall-bladder** consists of two thin-walled sacs connected by a narrow tube. One of the sacs lies attached to the posterior extremity of the right lobe of liver and the other one is placed on the median portion. There is no separate existence of **pancreas.** Pancreatic cells are present in a diffused state amongst the tissues of the liver. Thus the pancreatic secretion is mixed with bile and is conveyed into the intestine by way of the bile-duct.

Digestion. Its principal food is the small fry of fishes. The wide mouth, and the jaws with sharp teeth act as a trap for capturing the prey, and not for chewing food. The swallowing is helped by the mucous secretions, which do not contain any enzyme. The swallowed food is conducted through the short oesophagus into the stomach. The stomach and the first part of the intestine are the digestive regions. By the movement of the wall of the stomach the food substances are *triturated* and then partly converted into simpler forms by the action of the gastric secretions. In this condition the food gradually passes into the first part of the intestine through the pyloric constriction. Now, it is digested by the different juices secreted from the liver, pancreatic cells, wall of the intestine and probably from the pyloric caeca.

As a result of the enzymatic action the protein food substances are converted into amino acids, fat into fatty acids and glycerol, and carbohydrate into simple sugars such as glucose. Now, the completely digested food is ready for absorption. The absorption takes place through the numerous intestinal folds into the blood, and the residue is egested through the anus via rectum.

Swim-bladder. There is a large swim-bladder or *air-bladder* in the dorsal portion of the body cavity under the vertebral column. This is a thin-walled elongated sac filled with gases, such as oxygen, nitrogen and carbon dioxide. Originally the swim-bladder arises as an outgrowth from the anterior part of the alimentary canal, and its connection with the latter is often maintained by means of a narrow tube, the **pneumatic duct;** but a pneumatic duct is not found in Bhetki. A pneumatic duct is found in the common carp, Rohu. The air-bladder consists of two chambers of which the anterior one is smaller than the posterior. The inner wall of the bladder is vascularised by bunches of capillaries derived from branches of the mesenteric artery. These capillaries are recognised as the *rete mirabele.* The epithelial covering of the *rete mirabele* can isolate molecules of oxygen, nitrogen and carbon dioxide from the circulating blood and secrete the gases into the anterior chamber. The *rete mirabele* of the posterior chamber can reabsorb the gases. The volume of the gas in the bladder may thus be controlled by alternate secretion and reabsorption. A full air-bladder means decreased specific gravity and consequent floating of the fish. Reabsorption of the gases increases the specific gravity and the fish sinks to a greater depth. The fish makes this adjustment slowly when it moves up and down in a vertical plain. Thus the air-bladder serves as the hydrostatic organ of the fish.

Vascular System

The **blood** of the fish is pale and scanty in comparison with that of the toad. The blood plasma contains nucleated, elliptical **erythrocytes** and amoeboid **leucocytes.**

The **heart** is enclosed within a thin *pericardium* and is situated in the anterior part of the body cavity ventral to the oesophagus. It consists of a thin-walled **sinus venosus,** and **auricle** and a muscular **ventricle.** The **sinus venosus** is situated dorsally. It receives deoxygenated blood through right and left *caval veins* or **ductus cuvieri.** The **auricle** is situated ventral to the sinus venosus. Ventral to the auricle and partly surrounded by its indented outer margin, lies the thick-walled conical **ventricle.** The deoxygenated blood passes from the sinus to the auricle through a **sinu-auricular opening** and thence to the ventricle through the **auriculo-ventricular opening.** These openings are guarded by valves to prevent back-flow. Thus all the chambers of the heart contain deoxygenated blood.

A stout median artery, the **ventral aorta,** originates from the base of the conical ventricle. The base of the aorta is dilated to form a whitish, non-contractile **bulbus aorta.** The ventral aorta runs forward in the midventral line along the floor of the pharynx. It supplies paired **afferent branchial arteries** to the gills, thereby distributing the deoxygenated blood to the capillaries in the gill-filaments. The ventral aorta forks anteriorly at its terminal end to form the first pair of afferent branchial arteries which run into the first pair of gills. The second pair of afferent branchial arteries originate from the aorta a little behind the first. They enter the ventral ends of the second pair of gills. Proximal to the second afferents, two stout branches arise from the ventral aorta, one on each side. Each of these vessels divides to form the third and fourth afferent branchial arteries of the side which enter the corresponding gills. Evidently the ventral aorta and its branches are concerned with sending deoxygenated blood to the respiratory organs for the purpose of oxygenation; hence they constitute the **afferent branchial system** (Fig. 125).

The oxygenated blood is not returned into the heart but is conveyed directly from the gills to different parts of the body. The set of arteries carrying blood away from the dorsal ends of the gills constitutes the **efferent branchial system** (Fig. 126). Corresponding to the four pairs of afferent arteries, there are four pairs of efferent arteries. The *first* and *second* **efferent branchials** unite with one another to produce an **anterior epibranchial artery** on each side. The anterior epibranchials join in the middorsal line to form a median artery, the **dorsal aorta.** The *third* and *fourth* **efferent branchials** unite with one another to form a **posterior epibranchial artery** on each side. These drain into the commencement of the dorsal aorta. A **cephalic artery** is given out anteriorly from each first efferent branchial. This soon divides into an **internal carotid** for supplying the brain, and an **external carotid** to the tongue, hyoid arch, and orbital region. The internal carotids run side by side anteriorly. Ultimately, they enter the

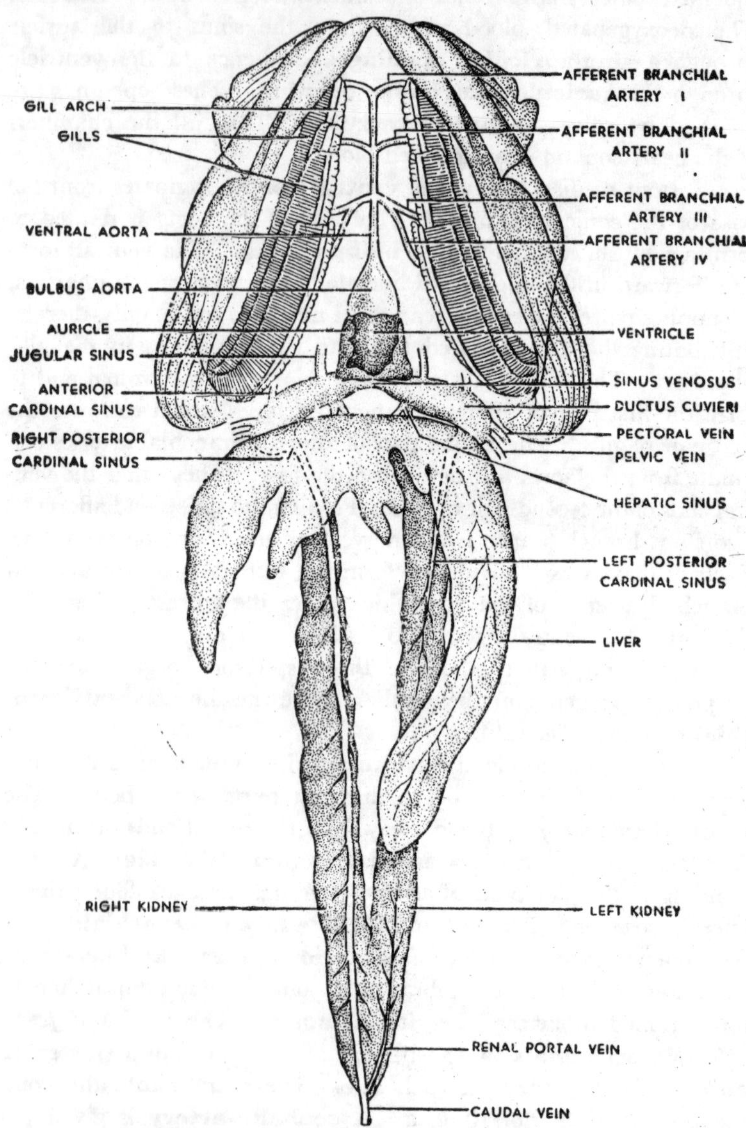

Fig. 125. Afferent branchial vessels and principal veins of Bhetki.

cranium to supply the brain. All the vessels described above are situated on the roof and lateral walls of the pharynx.

The **dorsal aorta** lies ventral to the vertebral column and pursues a backward course along the middorsal line. It extends through

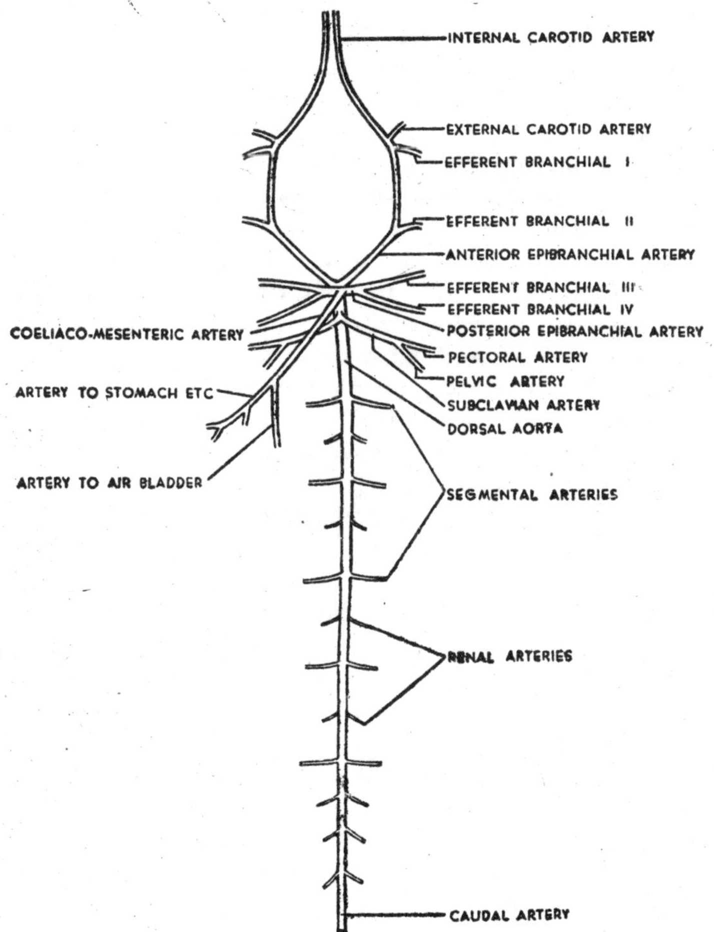

Fig. 126. Efferent branchial vessels and principal arteries of Bhetki.

the trunk to the tail, where it is known as the **caudal artery.** A stout **coeliacomesenteric** artery arises from the dorsal aorta near the region where the posterior epibranchials open into it. Th coeliacomesenteric breaks up into branches to supply the swim-bladder, stomach, intestine, spleen, liver, etc. Distal to the coeliaco-mesenteric, arises a pair of **subclavian** arteries which supply the pectoral and pelvic fins. During its backward course through the trunk, dorsal aorta gives off paired **renal arteries** to the kidneys, **genital arteries** to the gonads and **segmental arteries** to the body muscles. The continuation of the dorsal aorta enters the tail through the haemal canal and is known as the caudal artery. It extends as far as the root of the caudal fin and supplies the muscles of the tail.

Veins of Bhetki (Fig. 125) are of two kinds: (*i*) **systemic veins** which carry deoxygenated blood directly into the Cuvierian ducts, and (*ii*) **portal veins** which break down into capillaries is some intermediate organ (either liver or kidney).

The principal systemic veins are the Cuvierian ducts which open into the sinus venosus. Each **ductus Cuvieri** is mainly formed by the union of: (*i*) an **anterior cardinal sinus,** (*ii*) a **jugular sinus,** and (*iii*) a **posterior cardinal sinus.**

Anterior cardinal and jugular sinuses of each side bring blood from the dorsal and ventral regions of the anterior part of the body, whereas the posterior cardinals bring blood from the posterior part. **Pectoral** and **pelvic veins** from the paired fins and **hepatics** from the liver are slender veins draining into the ductus Cuvieri of their own side.

Blood from the tail is returned by a **caudal vein** which on entering the trunk, divides into two parts. The right division enters the right kidney. Passing through the kidney substance, it emerges at the anterior end as the right posterior cardinal sinus and joins the right Cuvierian duct. The left division of the caudal vein enters the left kidney and breaks up into capillaries, thus forming a one-sided **renal-portal system.** The capillaries of the renal-portal vein reunite to produce the left posterior cardinal sinus at the anterior end of the left kidney which drains into the left Cuvierian duct. Both the posterior cardinal sinuses receive renal veins from the kidneys, genital veins from the gonads and segmental veins from the body muscles.

Blood from the alimentary canal and other associated structures is returned by small veins which unite with one another to produce a large **hepatic-portal vein.** This enters the posterior surface of the liver, and breaks up into capillaries which reunite to produce the hepatic veins.

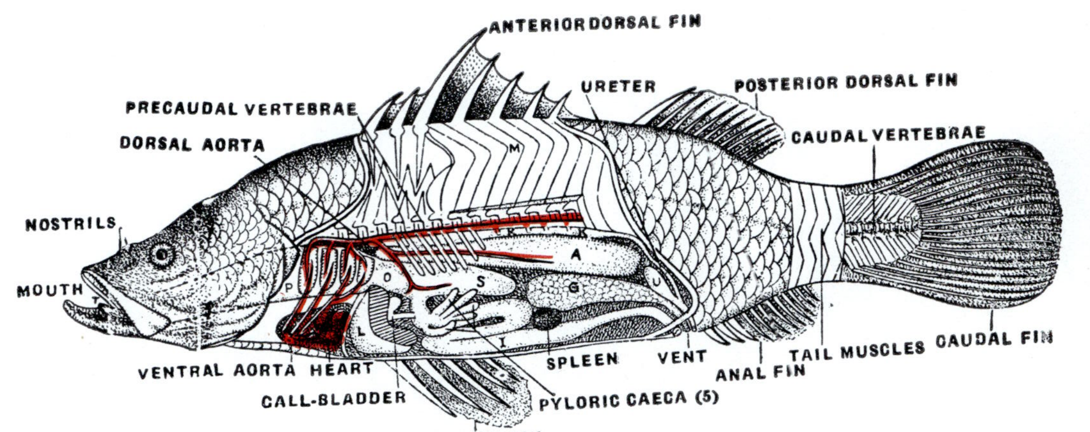

Internal structure of the Bhetki fish : operculum and pectoral fin cut away. A=air-bladder; G=gonad; I=intestine; K=kidney; L=liver; M=myotomes; O=oesophagus; P=pharynx; S=stomach; T=tongue; U=urinary bladder.

OUT. BIO.

Circulatory mechanism. The blood in the fish travels only in

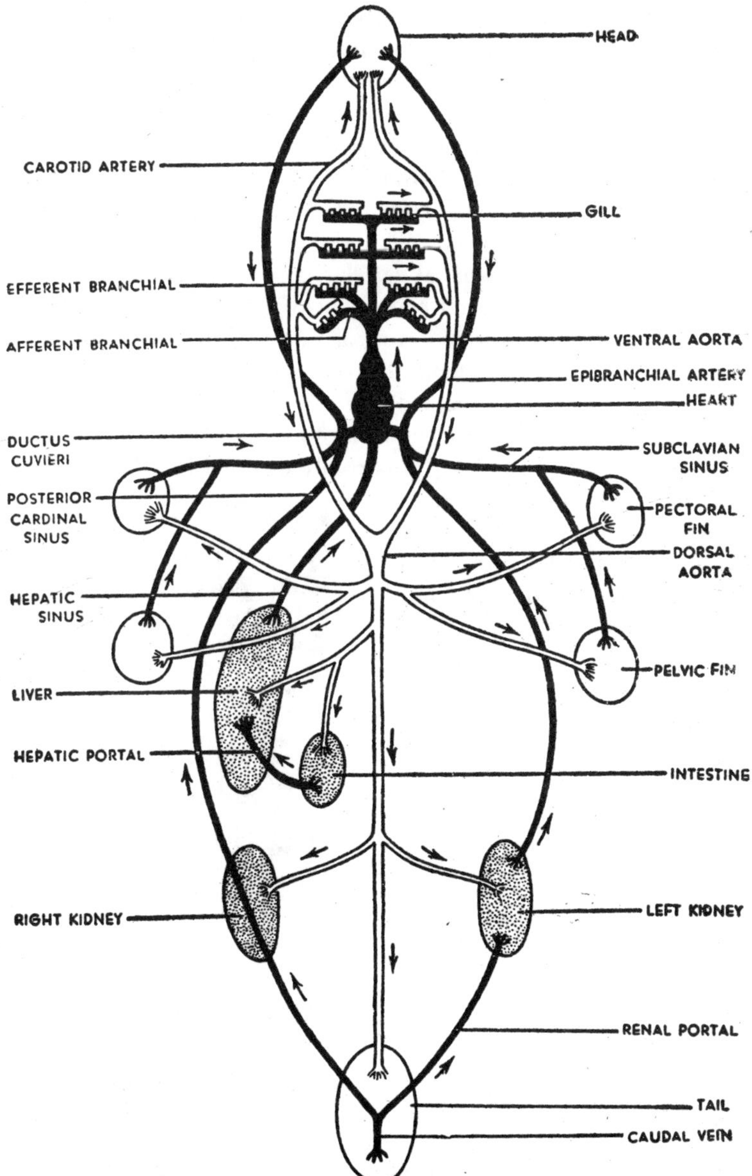

Fig. 127. Course of circulation of blood in Bhetki fish (diagrammatic).

one direction. The circulation, therefore, occurs in a *single circuit.* The heart maintains the circulation by rhythmic contraction (systole) and expansion (diastole). During a diastolic phase, deoxygenated blood comes back to the sinus venosus along the Cuvierian ducts. It is then pumped through the single auricle and ventricle by the rhythmic contraction of the heart. The sinu-auricular and the auriculo-ventricular valves prevent leakage and direct the blood stream to flow forwards. Ultimately the deoxygenated blood is, pumped into the gills through efferent branchial arteries. The latter break up into capillaries in the gills and the blood is aerated while circulating through the branchial capillaries. This is *branchial circulation.* The oxygenated blood comes out of the gills via the efferent branchial arteries and circulates through the body of the fish supplying oxygen and nourishment. In course of circulation through the tissues the blood is deoxygenated. It is now returned by the veins to the sinus venosus. While returning to the heart, the blood from the posterior part of the body is partially diverted into the left kidney, thereby forming a one-sided renal-portal system. Similarly, the blood from the alimentary tract is led into the liver forming a hepatic-portal system. Thus the heart of the fish has to pump only deoxygenated blood. The oxygenated blood never enters the heart.

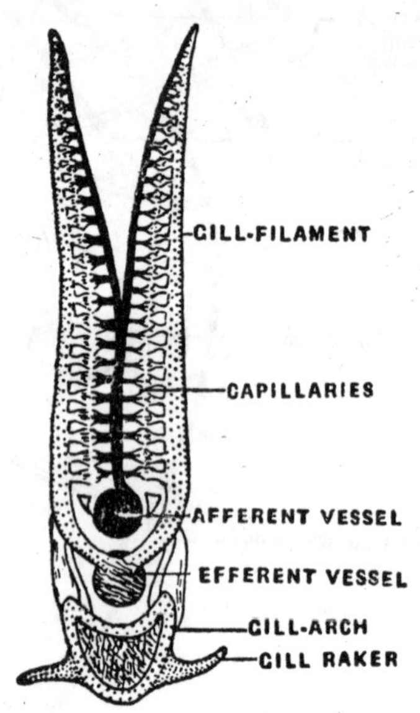

Fig. 128. Diagrammatic sectional view of a gill.

Respiratory System

The fish breathes through four pairs of **gills.** The gills are situated in two **gill-chambers,** one on either side of the pharynx. The gill-chamber is covered externally by a bony plate, the **operculum.** A *branchiostegal membrane* is attached to the posterior margin of the operculum.

The pharyngeal wall is perforated on each side by five **gill-slits,** which are separated from one another by four thickened strips. These are known as *interbranchial* *septa* or gill-arches. A **gill-arch** bears two rows of fine comb-like **gill-filaments** on its convex outer border, and teeth-like **gill-rakers** on the concave inner border. The gill-rakers are pointed towards the pharynx and they prevent escape of food into the gill-chambers. Such a typical gill, bearing two rows of filaments hanging from the convex outer border of the gill-arch, is known as a *holobranch*. In Bhetki all the four gills are holobranches. In other fishes a gill-arch may bear only one row of gill-filaments; such a gill is known as a *demibranch*. Deoxygenated blood is carried to the gill by the afferent branchial artery, which lies superficially on the outer side of the gill-arch. This blood is aerated while flowing through the capillaries in the gill-filaments—O_2 being taken in and CO_2 expelled through the delicate epithelium by diffusion. The aerated blood is conveyed through the efferent branchial artery which lies deeply on the inner side of the gill-arch. The fish requires a constant supply of oxygen charged water to flow through its gill-chambers, and cannot live long when it is taken out of water.

Mechanism of respiration. During **inspiration,** the opercula bulge out slightly but the branchiostegal membranes remain tightly apposed to the body wall, thereby sealing off the outer openings of the gill-chambers. The gill-arches are pushed out by the

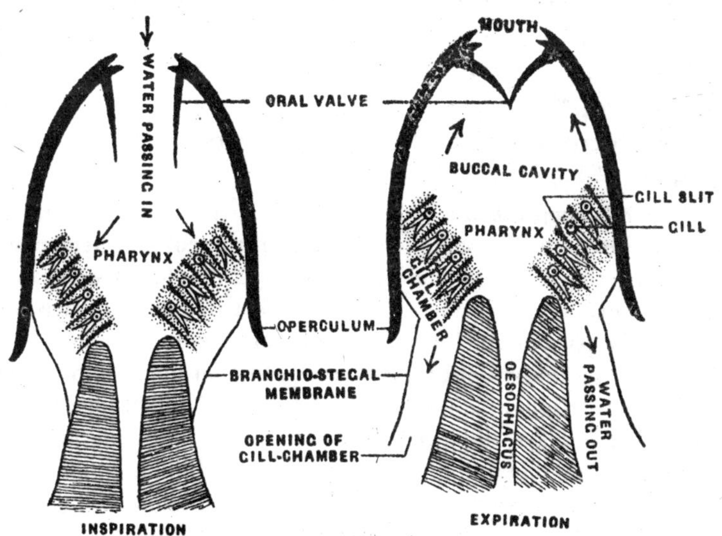

Fig. 129. Diagram explaining mechanism of respiration.

dilatation of the pharynx and buccal cavity. As a result of this increased space, water rushes in through the mouth and fill up the gill-chambers through the pharyngeal gill-slits. The gills are thus washed in fresh-water current, and O_2-CO_2 exchange takes place. In the mean time the buccal and pharyngeal cavities contract and press upon the contained water. The oral valves are pushed out and the mouth-opening is sealed, thereby preventing escape of water through the mouth. The opercula are now lifted and the branchiostegal membranes give way. The water in the gill-chambers is thus forced out through the openings under the opercula. This is **expiration.** Expansion and contraction of the buccal and pharyngeal wall occur alternately in quick succession due to the retraction and protraction of hyoid arch. In a fish, inspiration is always associated with ingestion of food into the mouth; expiration, on the other hand, is associated with swallowing of the food. The food cannot escape into the gill-chambers due to combined action of the gill-rakers. There is a strong *oesophageal sphincter* which prevents entry of water into the stomach.

Nervous System

The nervous system is built on the same general plan as in the toad and consists of *central*, *peripheral* and *sympathetic nervous systems*.

The **central nervous system** is composed of the *brain* and the *spinal cord*. The **brain** of Bhetki consists of: (*i*) *forebrain*, (*ii*) *mid-brain*, and (*iii*) *hindbrain*.

The **forebrain** is divided into an anterior *telencephalon* and a posterior *diencephalon* (*thalamen-cephalon*).

The **telencephalon** is large and undivided. This enlargement is due to development of its ventral region, the **corpus striatum.** Anteriorly, the telencephalon bears two well-developed **olfactory lobes.** This can be correlated with the fact that the fish depends more on the sense of smell than on sight. Posterior to the olfactory lobes are the **cerebral hemispheres** with thin non-nervous roof. Thus the fish has no *pallium* and therefore the large telencephalon of the fish is merely a *smell-brain* and not a centre for higher acquisitions such as intelligence, etc., as in higher vertebrates.

The **diencephalon** is poorly developed. The minute **pineal body** attached to a slender **epiphysis** projects forward from the roof of the diencephalon. Its floor projects downwards as the

infundibulum (*hypophysis*) to which is attached the **pituitary body** and **saccus vasculosus.** The saccus vasculosus is a highly vascular structure attached to the tip of the infundibulum of the fish. It secretes cerebrospinal fluid. There is a **lobus inferior** on each side of the pituitary body which extends backward and meets in the midventral line. The *saccus vasculosus* and the *lobi inferiores* are not found in toad and guinea-pig. The fish has no *optic chiasma* in front of the pituitary body.

The **midbrain** or **mesencephalon** is composed of two **optic lobes** dorsally, and the **crura cerebri** ventrally. This is the largest part of the brain. The two optic lobes from the *corpora bigemina*.

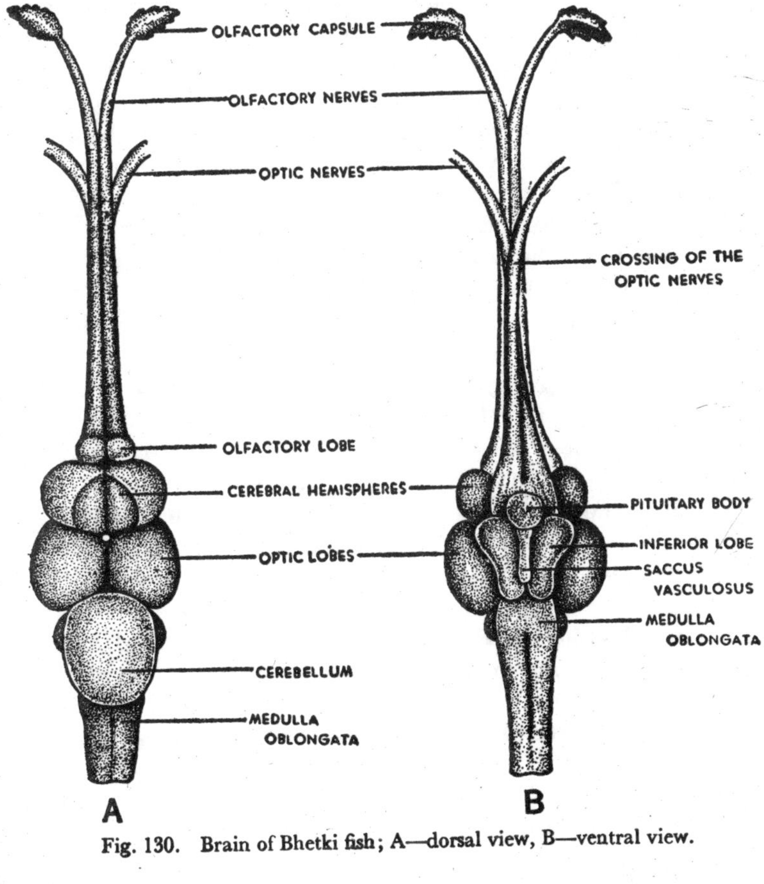

Fig. 130. Brain of Bhetki fish; A—dorsal view, B—ventral view.

The **hindbrain** is composed anteriorly of **cerebellum** (**metencephalon**) and posteriorly of **medulla oblongata** (**myelencephalon**). The cerebellum is located dorsally and covers the major part of the medulla; it is large and somewhat oblong in shape. The enlargement of the cerebellum is explained by the great muscular activity exercised by the fish.

The **medulla oblongata** is continued backwards as the **spinal cord** which extends throughout the entire length of the neural canal formed within the vertebral column.

The **peripheral** nervous system consists of the **cranial** and **spinal nerves.** They originate in the same manner and from the same places as in toad.

There are ten pairs of **cranial nerves.** (I) The **olfactory** nerves are long. They are distributed to the olfactory capsules. (II) The **optic** nerves have no chiasma. They cross one another much anteriorly, the right nerve going to the left eye and the left nerve to the right eye. (III) **Oculomotor,** (IV) **trochlear,** and (V) **abducens** innervate the *extrinsic muscles* of the eyeball. (VI) The **trigeminal** has, as usual, three divisions of which the *ophthalmic* divides into the two sensory branches: (*i*) *superficial ophthalmic* innervating the top of the head and snout; (*ii*) *deep ophthalmic* innervating the skin covering the eyeball. (VII) The **facial** has a *cutaneous* branch in addition to its usual branches. (VIII) The **auditory** is distributed to the membranous labyrinth of the internal ear. (IX) The **glossopharyngeal** sends out a branch to the first gill-slit of its own side. (X) The **vagus** divides into two main portions: (*i*) *Lateralis*, and (*ii*) *Visceralis*. The lateralis runs along the lateral line of the side innervating the lateral line sense organs. The visceralis sends out three branchial branches which supply the second, third and fourth gill-slits of the side. The cardiac and the gastric branches for supplying the heart and the alimentary canal respectively arise from the branch to the fourth gill-slit.

Paired **spinal nerves** emerge between the neural arches of successive vertebrae to supply the body segments of the trunk and tail.

The **sympathetic nerve trunks** are devoid of ganglia.

Receptor Organs

These consist of sense organs such as eyes, ears, olfactory organs, and receptor cells for touch and taste.

Eye. The eyes are located on the two sides of the head. They are devoid of *eyelids* and *eyelashes*. The transparent **cornea** is flat, and the rounded **lens** lies in apposition to the cornea. The **anterior chamber,** containing *aqueous humor*, is greatly reduced. Each eye is composed of three coats: (1) the cartilaginous **sclera** on the outer side, (2) the vascular **choroid** in the middle, and (3) the sensitive **retina** internally. There is the usual **iris** with a central hole, the **pupil.** The choroid layer is pigmented and contains vascular **choroid glands.** A peculiar silvery layer called **argentea** is placed between the sclera and the choroid. There is a **falciform** process from the choroid extending into the **posterior chamber** which is filled by the jelly-like *vitreous humor.* A *retractor lentis muscle* is attached to the falciform process posteriorly and to the back of the lens anteriorly. There is no *ciliary muscle.*

The eye is normally focussed for near vision. Accommodation for distant vision is not due to change of shape of the lens

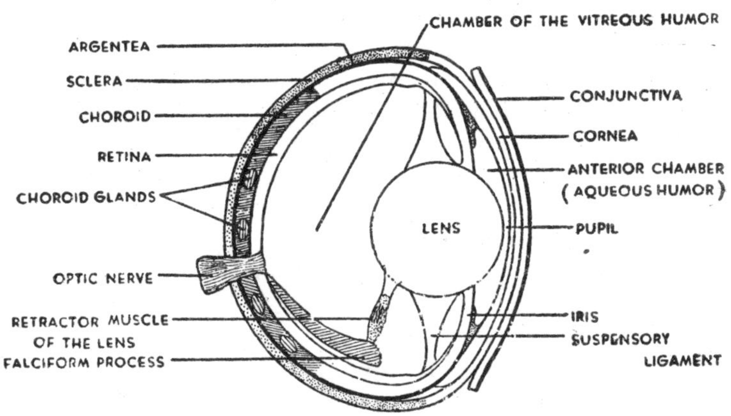

Fig. 131. Eye of a teleostean fish (diagrammatic).

but due to change of position of the lens. Contraction of the retractor lentis muscle pulls the lens closer to the retina, so that the image is brought to the proper focus. But how far this accommodation is perfected is not known.

The vision is **monocular.** While the right eye visualises one object, the left eye can visualise another object. The fish cannot close its eye even during sleep.

Ear. Ear of fishes is comparable with the internal ear of higher vertebrates. It is more concerned in maintaining equilibrium than in hearing. The middle ear and the external ear are not developed in fishes.

The ear sac or *membranous labyrinth* consists of three **semicircular canals** and three other chambers: **utriculus, sacculus,** and **lagena.** In each chamber there is an *ear-stone* or **otolith.**

Experimentally it has been found that some fishes can respond to sounds of different wave-length. The seat of hearing is located in the sacculus and lagena. How these structures carry out their function is not known. The ear of the fish receives all its sound vibrations through the body surface, because there is no eardrum. The lateral line sense organs help in perceiving low-frequency vibrations of the watery medium.

Olfactory organs. These are represented by two nasal sacs which do not communicate with the buccal cavity. Each nasal sac has two openings to the exterior. Olfactory cells of the nasal sacs are chemoreceptors for the sense of smell.

Touch. Tactile cells are usually found on the lips around the mouth and on the surface of the body.

Taste. Taste cells are present in the mucous membrane of the mouth and over the whole body surface. These can discriminate food of different taste.

Endocrine System

Of the ductless glands, the following require special mention.

Thyroid. It is a ductless gland producing thyroxin, a compound rich in iodine. In fish, it is a small reddish mass found on the floor of the buccal cavity along the ventral aorta. The thyroid gland is the metabolic regulator of the body.

Thymus. It is a paired mass found, one on each side, near the last branchial arch.

Spleen. It is the largest ductless gland found attached to the intestine. It is principally concerned with the destruction of worn-out blood corpuscles and to serve as a storehouse of blood.

Excretory System

The greatly elongated **kidneys** lie ventral to the vertebral column but dorsal to the air-bladder. They extend through the coelom, from the gill-arches in front, to a point just anterior to the vent. The two kidneys are partially fused posteriorly but their anterior parts are separate. Each kidney is composed of a large number of **uriniferous tubules** held together by connective tissue. The ventral surface of the kidney bears numerous minute apertures, the **nephrostomes,** which communicate internally with the uriniferous tubules. A duct or **ureter** arises from the posterior part of each kidney. The two ureters unite behind to form a single tube communicating with a thin-walled sac—the **urinary bladder.** The bladder discharges the urine through a narrow canal which communicates with the **urinary opening** at the posterior part of the **vent,** behind the anus. The urine is formed by **selective filtration** at the **glomeruli** of the uriniferous tubules.

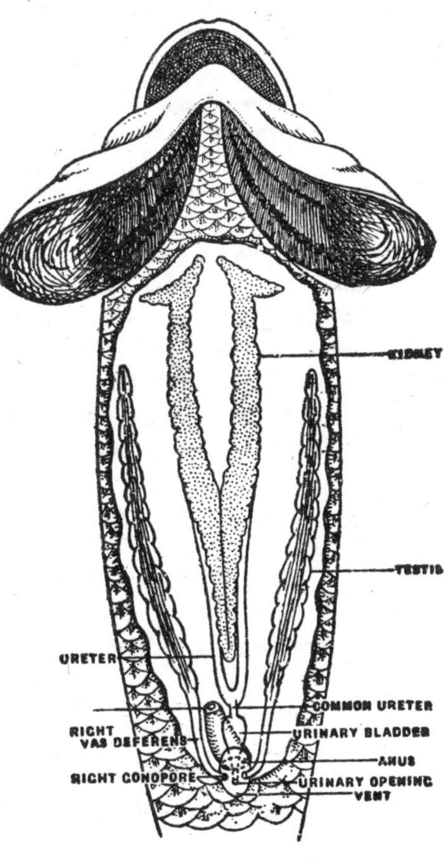

Fig. 132. Urinogenital organs of a male Bhetki.

Reproductive Ssystem

Sexes are separate. **Gonads** are of enormous size in sexually matured individuals, particularly during the breeding season. It is difficult to distinguish testes and ovaries in the young fish.

In a *male* Bhetki, the gonads consist of a pair of smooth elongated **testes** situated in the posterior part of the body cavity. Each testis communicates by a duct, the **vas deferens,** with the reproductive opening of its own side. There are two **reproductive openings** which lie laterally at the two sides of the vent.

The **ovaries** found in *female* Bhetki are much bigger than the testes of a male of the same age and contain numerous small eggs or *ova*. Matured eggs are carried by the **oviducts** and discharged outside through the reproductive openings. There are **two** reproductive openings lying laterally at the two sides of the vent.

Life-history: Spawning takes place in winter. Seminal fluid or *milt* containing spermatozoa are shed by the male fish in water. Similarly, the female lays her eggs. Fertilization of the ova by sperms is a matter of chance and takes place in water. A large quantity of the *spawn* is destroyed—being devoured by aquatic animals or by other natural factors and only a few of the fertilised eggs develop into the tiny *fry*. The young fry hatches out of the fertilized egg in about 20 hours. It has a small *yolk-sac,* containing reserve food, attached to its ventral surface. In the course of a few days the yolk is exhausted and the sac disappears. The fry has, by this time, enlarged in size. It feeds on algae, insect larvae and small crustaceans. It becomes predatory and devours fry younger than itself. The fish attains sexual maturity in about a couple of years.

LATA FISH

Account of Some Systems of Lata

It is often difficult to procure Bhetki fish in large numbers for dissection purpose particularly outside big cities and during off-season. The *Lata fish* (**Ophicephalus punctatus**) it easily available all over India and Bangladesh throughout the year. Anatomical pecularities of the Lata fish are therefore represented diagrametically (Fig. 133) for the guidance of the students in their practical work along with a few important systems. Its generic name has now been changed to *Channa*.

Habit

The Lata or Murrel or Goroi is a fresh water fish and is found in ponds, ditches, jheel etc. They are very common in West Bengal, and Bangla Desh. Being a **carnivorous** fish it takes smaller fish and aquatic animals as food. Because of the presence of **accessory respiratory organs** it can live outside water for a considerable period.

External Features

The body is spindle shaped and laterally compressed. A full grown fish is 15-18 cm in length. The colour of the dorsal surface is blackish grey while the ventral surface in yellowish grey. Above and below the **lateral line sense organ** are found eight alternate patches. The body is covered with scales. The scales are of two types—**Ctenoid** and **Cycloid**. The body is divided into three regions—**Head, Trunk** and **Tail** (Fig. 133.). The dorsal side of the head is flat and the scales here are large and plate like. The **dorsal fin** is broad and undivided, the anal fin is extended. The tail fin looks like that of Bhetki. The lateral line is interrupted at frequent intervals. The other structural aspects resemble that of Bhetki.

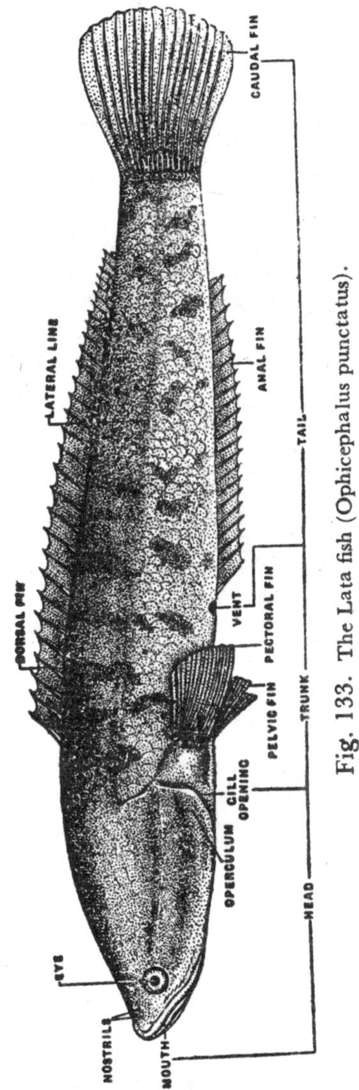

Fig. 133. The Lata fish (Ophicephalus punctatus).

Alimentary System

It consists of the following parts—**Mouth.** The mouth aperture is wide and terminal. It is bound by upper and lower jaws. The jaws are lined with sharp and conical teeth. Mouth leads to the **Buccal cavity.** The floor of the buccal cavity is provided with a tongue which is devoid of muslces. It leads to the **Pharnyx.** The

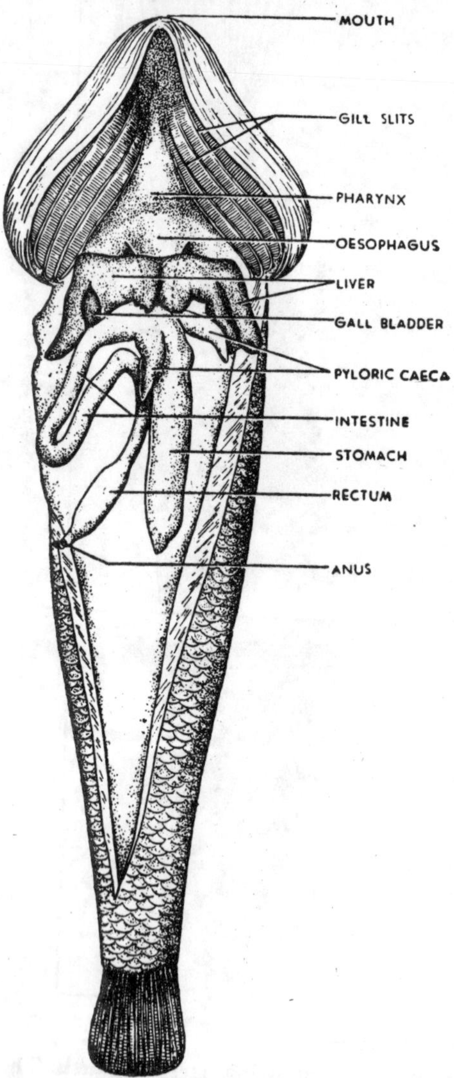

Fig. 134.. Alimentary system of Lata fish.

lateral walls of the pharynx bear 4 gill-slits in line as rows on either side. The pharynx leads to the **stomach** *via* a short **oesophagus**. A sphincter is present between oesophagus and stomach. The **cardiac** and **pyloric** part of the stomach is identifiable. (Fig. 134.).

The pyloric part continues as **intestine.** The intestine is a coiled tube. At the line of demarkation between the pylorus and intestine there is present a pair of finger like projections called **Pyloric Caeca.** The posterior portion of the intestine is inflated and forms the **Rectum.** The rectum opens to the outside through the **Anus** located at the posterior ventral end of the trunk.

Liver is the largest digestive gland. The reddish brown liver is located beneath the heart. It is made up of two distinct lobes—left and right. Liver secrets bile. The bile remains stored in a **gall bladder** located in a notch on the right lobe of the liver. A **Bile duct** arises from the gall bladder and opens into the intestine through an aperture close to the pyloric caecum.

Respiratory System

Lata being aquatic takes oxygen that remains dissolved in water. The respiratory organs are **Gills.** There are two gill chamber one on either side of the pharynx. Each chamber bears 4 gills arranged along the anteroposterior axis of the body. Each chamber is covered by **operculum.** The mechanism of respiration resembles that of Bhetki.

Lata can remain alive for some time on land and by using aerial oxygen with the help of its **Accessory respisatory organ**. The accessory respiratory organ is a sac-like structure derived by the out pushing of the pharynx. On the dorsal side of each gill chamber is located one such organ. The internal lining of the organ being highly vascular acts as respiratory surface.

Heart of Lata

The circulatory system is similar to that of *Lates* already discussed. But a detailed study reveals minor variations which are discussed below:

The heart is composed of a thin-walled **sinus venosus,** an oval muscular **ventricle** and an **auricle.** The conus arteriosus is absent and this region, like that of Bhetki, is marked by the presence of a pair of valves guarding the entrance of the bulbus aorta. The bulbus aorta is non-contractile and represents the dilated base of the ventral aorta. The bulbus aorta and the ventral aorta are covered by a tough membrane.

Afferent branchial system. The deoxygenated blood from the ventricular cavity is conveyed through a stout median **ventral aorta** to the gills for aeration by four pairs of **afferent branchial arteries.** All the four pairs of the afferent branchial arteries have

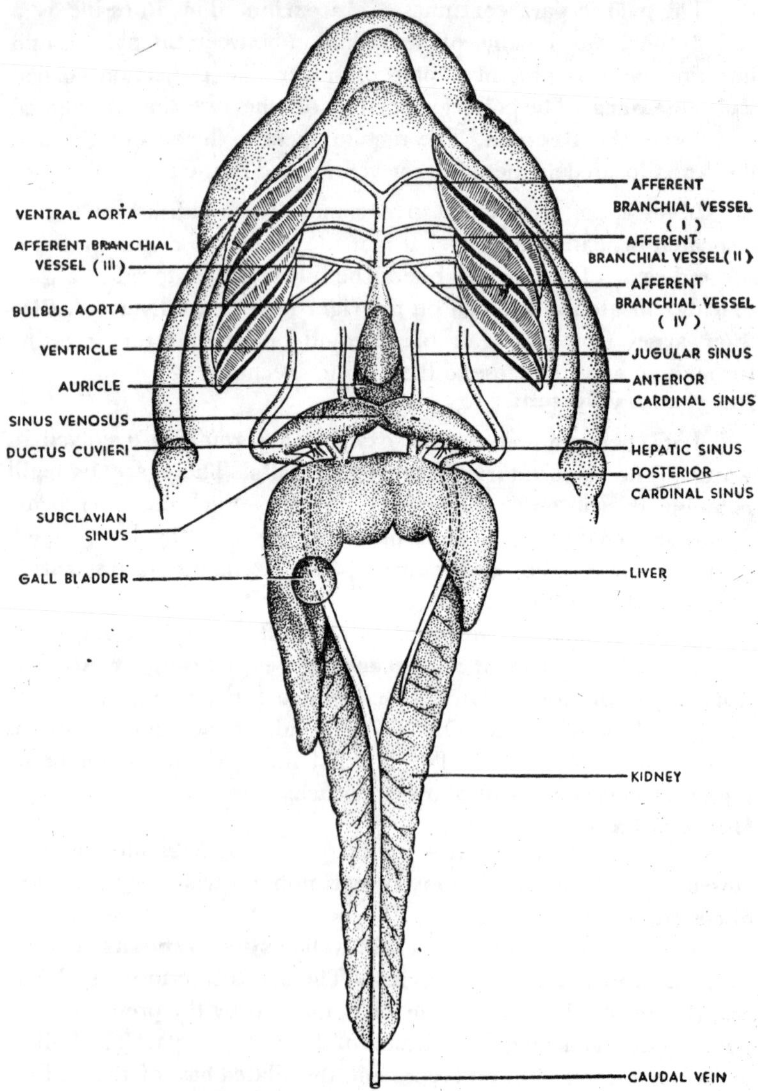

Fig. 135. Afferent branchial system and principle veins of Lata fish.

independent origin from the ventral aorta. The first and second pairs of the afferent branchial arteries are simple straight arteries leading to the corresponding pairs of gills. While the third and fourth afferent branchial arteries of each side cross each other forming a loop after their origin from the ventral aorta (**Fig. 135.**).

Efferent branchial system. The efferent branchial arterial system is peculiar. The first and second pairs of efferent branchial arteries are lacking (**Fig. 136.**). The third efferent branchial artery

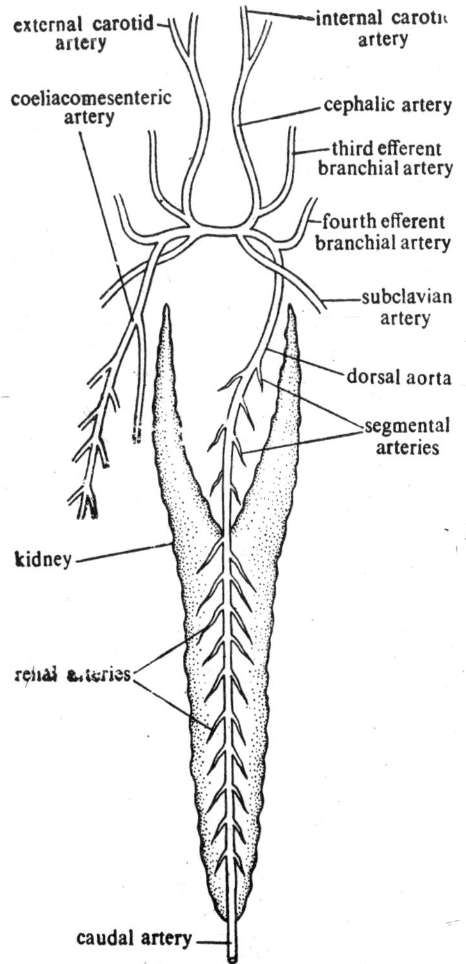

Fig. 136. Efferent branchial system of Lata fish.

on each side receives oxygenated blood from the third gill of that side and gives off a **cephalic artery.** This artery runs forward and bifurcates into an **external carotid artery** and an **inernal carotid artery.** A transverse artery acts as a connecting bridge between the third and fourth efferent branchial arteries of the two sides. The left fourth efferent branchial artery gives origin to the **dorsal aorta.** The **coeliacomesenteric** artery originates from the point of union of the transverse artery and the right fourth efferent branchial artery. Two **subclavian arteries** arise independently on the two sides and each divides into two arteries—the **pectoral artery** supplying the pectoral fin and a **pelvic artery** to the pelvic fin. The dorsal aorta occupies a middorsal position and runs posteriorly up to the tail as **caudal artery.** Along its course it gives off paired **segmental arteries, gonadal arteries** and **renal arteries** (Fig. 136.).

The absence of first and second pairs of efferent branchial arteries in Lata is difficult to interpret. The oxygenated blood from the first two pairs of the gills is possibly drained into the suprabranchial cavity directly through some sort of capillary networks. These capillary networks together with that of the supra-branchial cavity unite to form an artery which joins with the cephalic arteries.

SOME OTHER FISHES

Throughout the world large number of fishes are seen. They are of different varieties and live in oceans, rivers, lakes, tanks and pools. Broadly the fishes are divided into two groups: (1) Fishes with cartilaginous skeleton (chondricthyes), and (2) Fishes with bonny skeleton (osticthyes). The chondricthyes group are represented by sharks, skates and rays, while Bhetki, Rohu, Magur, etc., are the examples of osticthyes. The structure of a fish has been discussed in detail through Bhetki. In this chapter some general account about certain other fishes has been given

SHARK (Fig. 137): A CARTILAGINOUS FISH

Quite a different types of sharks are available in the oceans of our country, of which the most common is *Scoliodon* sp. This shark may attain a size of 60 to 90 cms. or 2 to 3 feet. The dorsal side of the body is dark-grey and the ventral side is white. Both ends of the body are pointed. The entire body is divisible into three regions—*head*, *trunk*, and *tail*. The head is slightly flattened and drawn anteriorly into a pointed snout. Mouth is present on the ventral side of the head and is semilunar in appearance. One nostril and one eye is present on each side of the head. The eyes are prominent and provided with functionless eyelid. Near the junction of head and

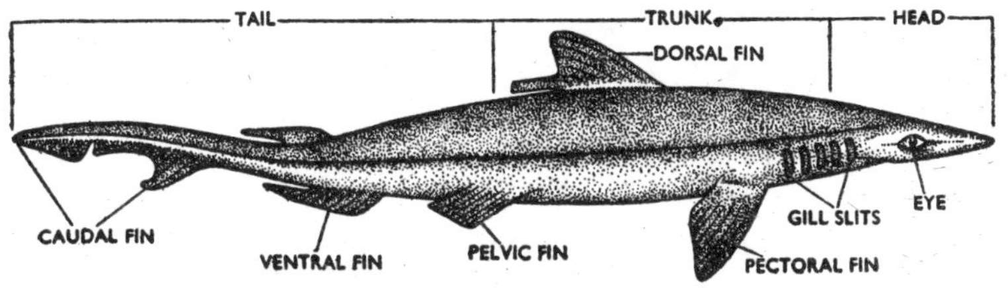

Fig. 137. Shark (*Scoliodon*).

trunk, on each side of the head lie five small transverse openings called gill-slits. The trunk is indistinctly separated from the head. The entire body is covered by a special type of small and pointed scales. These scales are called placoid scales. Like the cycloid and ctenoid scales, these scales do not cover the anterior part of the next row. Lateral line sense organ runs along each lateral side of the body from head to tail. The vent or outlet of the food canal is present on the ventral side of the body. Another pair of small opening, abdominal pore is present slightly posterior to the vent. In the male scoliodon only, a pair of elongated claspers are present in this region. The body contains a number of fins with supporting fin-rays. Some fins are paired, while others are unpaired. Following fins are present in scoliodon.

A. *Paired fins*

1. *Pectoral fin*—one on each lateral side, triangular, present near the junction of head and trunk.

2. *Pelvic fin*—one on each side of the vent. In males the claspers are present close to the pelvic fin.

B. *Unpaired fins*

1. *Dorsal fin.* Divided into two parts—anterior dorsal fin and posterior dorsal fin. Both these parts of the dorsal fin are triangular and the anterior is larger than posterior.

2. *Ventral fin.* Present on the ventral side and posterior to the pelvic fins.

3. *Caudal fin.* Divided unequally into two parts—dorsal lobe and ventral lobe. The dorsal lobe is larger than ventral lobe and contains the last part of the vertebral column. This type of tail is called Heterocircal tail.

ROHU (Fig. 138) : A BONNY FISH

This is the most important edible fish of our country. It is scientifically known as *Labeo rohita.* It grows well in ponds and pools to attain a weight of 20–25 kg., but never breeds in enclosed water. A reddish tinge is visible throughout the body of these fishes.

The body is divided into three parts: (1) head, (2) trunk, and (3) tail. The head is more conspicuous than Bhetki. It is pointed

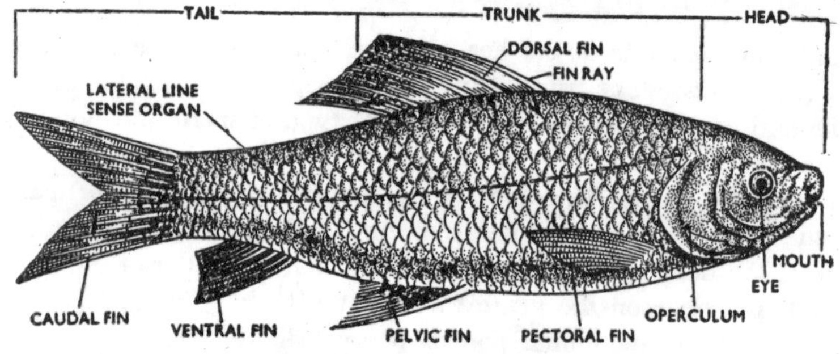

Fig. 138. Rohu (*Labeo rohita*).

anteriorly and is terminated by mouth. Mouth is bounded by jaws which are devoid of teeth. A pair of barbs are present near the mouth. The eyes are large and slightly reddish in colour. The gill openings are broad and covered by strongly built operculum. The body is covered by cycloid scales. Along each lateral side runs the lateral line sense organ, which begins near the opercular opening and ends to the tail region. Following paired and unpaired fins with supporting fin-rays are present.

A. *Paired fins*

1. *Pectoral fin.* One on each ventrolateral side, near the opercular opening. Each fin contains 19 fin-rays.

2. *Pelvic fin.* Present on the ventral side, consists of 9 fin-rays.

B. *Unpaired fins*

1. *Dorsal fin.* Present on the middorsal region of the body. Provided with 13 fin-rays.

2. *Anal fin.* Present on the midventral region of the post-anal part. Supported by 4–6 fin-rays.

3. *Caudal fin.* The caudal fin is divided equally into two parts. Each part is pointed at its free end. Several fin-rays are associated with the caudal fin.

Air-breathing fishes

Fishes are inhabitants of water. To get oxygen which is dissolved in water, the fishes have specialised respiratory organ called **gills.** The gills are inactive on land, which causes the death of the fish when it is out of water. Though aquatic respiration by gills is common in fishes, there are certain members among the fishes which can also breathe air. They have specialised organs for the purpose, in addition to gills. Those fishes are called air-breathing fishes and common examples are Singhi, Magur, Koi, etc.

Singhi (Fig. 139). It is scientifically known as *Heteropneustes fossilis*. It is a common air-breathing fish which in addition to gills

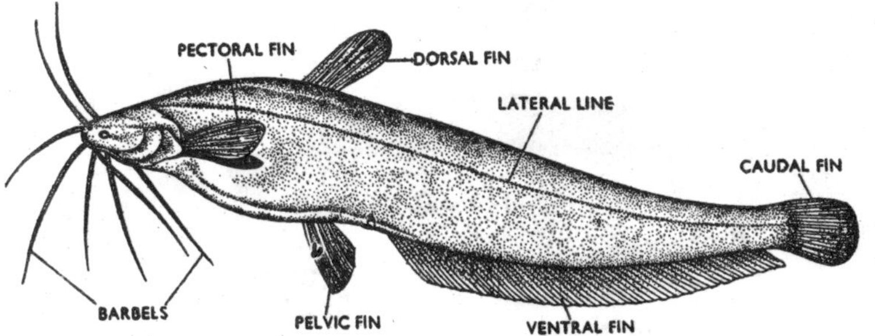

Fig. 139. Singhi (*Heteropneustes fossilis*).

possesses an air tube within its body. While living under water these fishes frequently visit the surface to breathe air. If these fishes are prevented to come near the surface, they die. It shows that mechanism of aquatic respiration alone cannot meet the oxygen requirement in these fishes.

The grey-coloured body is slender and devoid of scales. It has three distinct regions: (1) head, (2) trunk, and (3) tail. The tail is the largest part of the body. The head is small, flattened and its anterior end is marked by an opening called mouth. Around the mouth are present 4 pairs of barbels, which are sensory in function. A pair of nostrils and a pair of eyes are present on the head region. Following fins with supporting fin-rays are present.

A. *Paired fins*

1. *Pectoral fin.* One on each side of the trunk near the opercular opening. Each consists of 9 fin-rays of which the first one is large, sharp and pointed. It acts as defensive organ.

2. *Pelvic fin.* Placed on the ventral side of the trunk and close to the pectoral fin. It is small and has 6 fin-rays.

B. *Unpaired fins*

1. *Dorsal fin.* Very small, placed on the middorsal region of the body.

2. *Anal fin.* This is the largest fin with numerous fin-rays. Present along the entire ventral side of the tail region.

3. *Caudal fin.* Round, small and is present on the posteriormost end of the tail.

Magur (Fig. 140). This is another air-breathing fish. Here the accessory respiratory **organ** is branched and placed near the

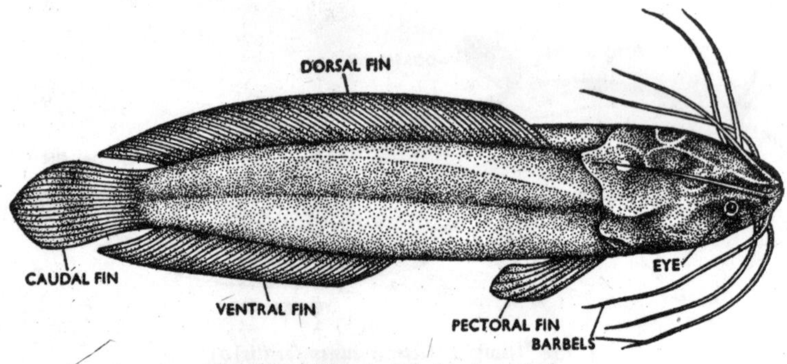

Fig. 140. Magur (*Clarius batrachus*).

gills. This fish is scientifically known as *Clarius batrachus.* The head of this fish is larger than *singhi* and mouth is present on the ventral side. Of the eight barbels, two are much stouter and longer, while the others are similar. Following paired and unpaired fins are present, which differ from the *Singhi*.

A. *Paired fins*

1. *Pectoral fin.* Present near the opercular opening, first fin-ray is modified into a sharp defensive structure.

2. *Pelvic fin.* Present on the ventral side and close to the pectoral fin. Supported by soft fin-rays.

B. *Unpaired fins*

1. *Dorsal fin.* Very large, extends along the middorsal line of entire trunk and tail region. Supported by innumerable soft fin-rays.

2. *Anal fin.* Present along the midventral line of the tail region. Numerous soft fin-rays are present.

3. *Caudal fin.* Semicircular and present on the terminal end of the tail region.

Koi (Fig. 141). It is scientifically known as *Anabus testudineus*. It possesses fan-shaped accessory respiratory organ near the gills. These are small fishes (15 to 20 cms. in length) with flattened body; the dorsal side is slightly curved. The head is prominent, triangular

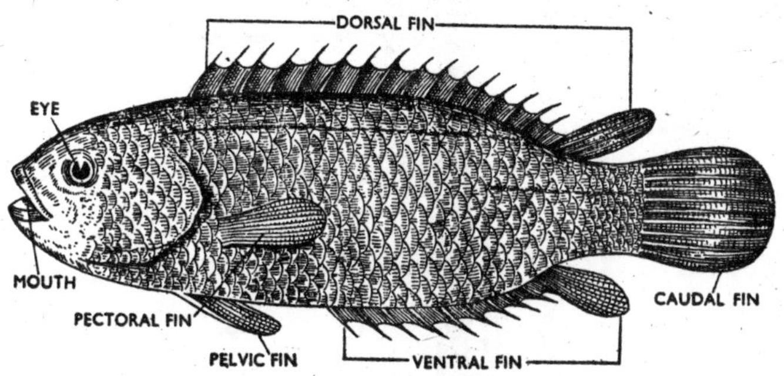

Fig. 141. Koi (*Anabus testudineus*).

and terminated by a broad mouth. Eyes are paired and quite prominent. The body is covered by rows of ctenoid scales and along each side of the body runs the lateral line sense organ. The line is breached near the tail region. Following paired and unpaired fins are present, where the supporting fin-rays are all sharp and pointed.

A. *Paired fins*

1. *Pectoral fin.* Placed near the opercular opening. The free end of the fin is broad and semicircular, while the base is narrow.

2. *Pelvic fin.* Placed midventrally and close to the pectoral fin.

B. *Unpaired fins*

1. *Dorsal fin.* Consists of two parts—anterior elongated part supported by stiff fin-rays and posterior round part supported by soft fin-rays.

2. *Ventral fin.* Like the dorsal fin, consists of two parts—anterior elongated part which is supported by stiff fin-rays, and posterior round part supported by soft fin-rays.

3. *Caudal fin.* Round, short and supported by numerous fin-rays.

CHAPTER XV

CHORDATA (continued) : REPTILIA

Calotes Versicolor

This is the most common garden lizard (Fig. 142) of our country. They belong to the class *Reptilia* and order *Lacertilia*. Following reptilian features are present.

1. Body completely adapted for land life.
2. Body covered by horny epidermal scales.
3. Poikilothermal or cold-blooded animals.

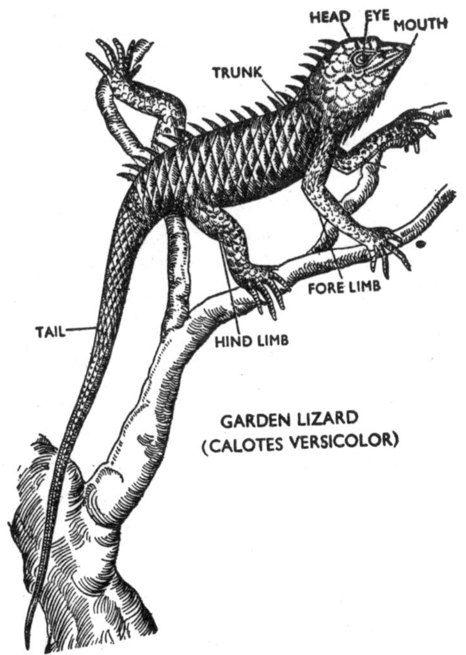

Fig. 142. Garden lizard.

4. Presence of lungs as respiratory organ.
5. Presence of single occipital condyle in the skull.
6. Presence of embryonic membrane.

The calotes generally attains a length of one foot. The dorsal side of the body is dark-grey and the ventral side is pale yellow in colour. The neck region of male calotes turns red during mating season. The body is divisible into three parts—*head, trunk* and *tail.* A short neck is visible between head and trunk. The head is small and pointed anteriorly. The mouth opening is broad and present as a transverse slit on the anterior region of the head. It is bounded by one upper and one lower jaw just above the mouth opening a pair of small nostril is present. A prominent eye is present on each lateral side of the head. Each eye is provided with three eyelids —(1) thick upper eyelid, (2) thin lower eyelid, and (3) a transparent third eyelid or nictitating membrane which runs across the eye. Posterior to each eye, lies the auditory opening which is guarded by a smooth, round skin called tympanic membrane. Entire body is covered by dry scales. On the dorsal side the scales are raised as sharp spines.

The trunk region possesses two pairs of appendages—forelimb and hindlimb. Both these limbs are built up for crawling and climbing. Forelimb is divisible into—(1) Forearm, (2) Arm, (3) Wrist, (4) Palm, and (5) Digits. The hindlimb possesses following divisions—(1) Thigh, (2) Shank, (3) Ankle, (4) Palm, and (5) Digits. All the digits in forelimbs and hindlimbs are provided with claws.

A transverse opening, cloaca is present on the ventral side of the trunk region near the commencement of tail. The tail is long, gradually tapering and covered by large scales. These animals have the wonderful ability to break off the tail at the time of danger. Lost tail grows within a short time by a process called regeneration. Small insects are taken as food by these animals.

SNAKES

Snakes

Snakes are a very specialised group of Reptiles under order **Ophidia.** Nearly 3,000 species of snakes are now found in the tropical and sub-tropical parts of the world. Out of these about 300 are **Poisonous** and others are **Non-Poisonous.** WHO has estimated that thirty to forty thousand people die annually of snake bite. Snakes are not found in New Zealand and Ireland.

Distinctive features

Snakes are slender and elongated reptiles without limbs, eyelids and external ears. The whole body is covered with epidermal **scales** and **shields.** The scales are small and usually overlap each other but the shields are large and join each other by their margins. Shields are prominent on the head. The **occipital condyle** is single and tripartite. The lower jaw is attached to the skull with the help of **quadrate** which articulates in such a movable fashion that the jaw can move laterally and downwards. The two halves of the lower jaw are loosely articulated. All these are special adaptation for its peculiar feeding habit. The **teeth** are conical, recurved backward and fixed on jaw bones. In poisonous snakes two maxillary teath (one on either side) become enlarged and pointed to form the **Fangs.** There are two types of fangs—open type and closed type according to the nature of the groove. In poisonous snakes the **labial glands** have become modified into **poison glands** and these do not help them in digestion. The tongue is bifid at the apex. Left lung is reduced. Urinary bladder is absent. Most snakes are terrestrial. Few are arboreal and aquatic.

Accounts of some poisonous snakes

Cobra

There are about 10 species of cobras of which two are found in India. They are *Ophiophagus hanah* and *Naja bungarus*. Cobra is a very poisonous snake. It is usually found in forests. Usual length is 2-3 metres. The colour is black or buff. The hood is well developed. It is supported by long ribs located just behind the head. The hood bears marks like spectacles. In some cobras there are two spectacles marks while in others there is a single spectacles mark. Under the neck there are found 2 or 3 dark belly plates. The fang is open type.

Krait

Kraits (*Naja bungarus*) is smaller in size compared to the cobras and measures 1-1·25 metre. The colour of the body is glistening black with linear arches. They are found in gardens, on the roof of bunglows. The poison fangs are small. The shields in the belly are broad. The tail is round at the tip. Central rows of scales down the back are large and hexagonal. The plates under the tail are entire.

Vipers

These are rather large and stout snakes. The triangular head is broad, flat and covered with small scales. The neck is narrow.

Eyes are with vertically elliptical pupil. The plates in the belly are broad and extend across (Fig. 142A).

Fig. 142A. Some poisonous snakes—Viper, (A), Cobra (B), King cobra (C), Krait (D).

Effects of poison

Cobra. The effects are observed within half an hour after bite. The venom is neurotoxic type. Symptoms are giddiness, high pulse rate, salivation, partial paralysis of tongue and larynx, vomiting and contraction of pupil.

Viper. The effects are observed within quarter of an hour. The venom affects the circulatory system. Symptoms are swelling of the affected region, discolouration, acute burning pain, pupil dilation, profuse vomiting and watery discharge from the rectum.

POISON APPARATUS

The poison apparatus consists of a pair of **poison glands** and a pair of **fangs.** The glands are situated one on either side of the upper jaw. These are probably modified superior **labial glands.** Each gland is sac like with a narrow duct emerging from its anterior end. The duct passes forward along the side of the upper jaw and opens either at the base of the fang or at the base of the tunnel on the fang. The glands are held in position by ligaments. An anterior ligament attaches the anterior end of the gland to the maxilla. A posterior ligament runs between the gland and quadrate. Many fan-shaped ligaments are found between the side walls of the gland and squamoso quadrate junction.

The fangs are pointed and enlarged maxillary teeth which regenerate when lost. There are two types of fangs. In open type

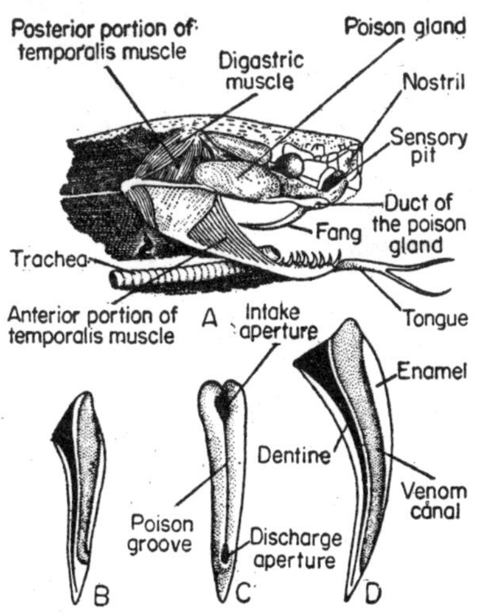

Fig. 142B. Poison apparatus of Snakes (A), Types of fang (B, C, D).

as in cobras the poison groove on the fang is open and in the closed type as in vipers the poison groove forms a tunnel having two openings, (Fig. 142B) one at the base (intake aperture) and the other near the apex (discharge aperture).

The poison apparatus is associated with specialised muscles like **digastric, sphenopterygoid, anterior** and **posterior temporalis** etc. These muscles help in squeezing the gland and opening and closing of the mouth.

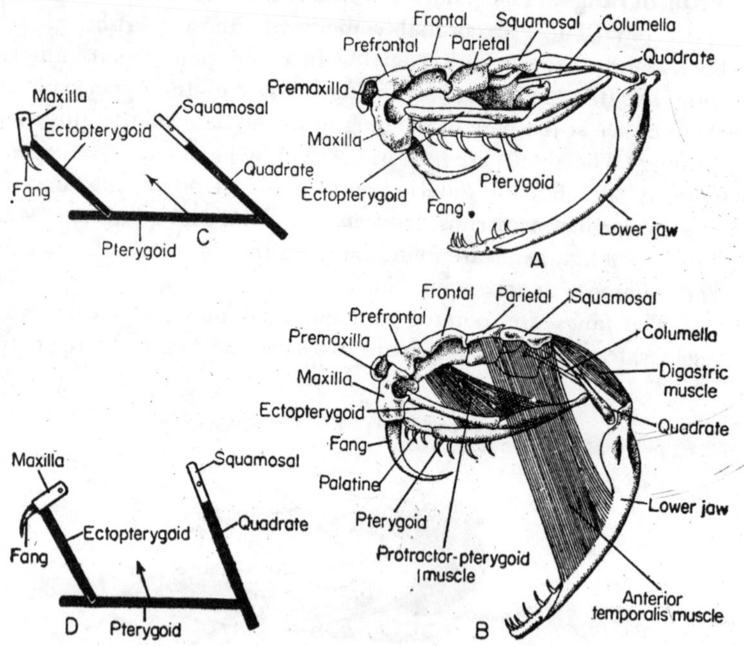

Fig. 142C. Mechanism of Bite of a poisonous snake.

CHAPTER XVI

CHORDATA (continued) : AVES

Columba Livia

Columba livia is the scientific name of common pigeon (Fig. 143). It belongs to the class Aves which includes all the birds of the world. Most birds can fly in air. This peculiar habit has rendered certain changes in them, which are regarded as their characteristics. Some of these features are mentioned below, which are well seen in pigeon.

1. Body streamlined.
2. Forelimbs are modified as wings which help in flying.
3. Heavier parts of the body are centrally placed.
4. Body covered by feathers which help in buoyancy of the birds.
5. Bones are hollow and filled up with air.

Fig. 143. Common pigeon.

6. Well-developed air-sacs are associated with respiratory organs.
7. Tail is short and works both as rudder and balancer.
8. Homoiothermal or warm-blooded.

There are innumerable varieties of domesticated pigeon. In all of them the body is divisible into three parts—*head*, *neck* and *trunk*. The head is small and movable due to the neck. Mouth is placed at the anteriormost end and is provided with two beaks. Each beak is hard and pointed. The beaks are used for picking food and also for defence.

Slightly above the mouth there is a pair of nasal opening. On each side of the head lies a large, prominent eye. Each eye is provided with three movable eyelids—upper, lower, and nictitating membrane. The auditory openings are present posterior to the eye. The forelimb containing three digits is modified into wing. Each wing is provided with 23 long feathers called *remiges*. The hindlimbs are capable of supporting the entire weight of the body and are used during perching and sitting. The digits of the hindlimbs are provided with sharp claws, which are used for defence. Number of scales are present on the hindlimb. Tail is short and provided with 12 long feathers called *retrices*. Near the base of the tail, lies an oil producing gland called *preen gland*, the secretion of which keeps the feather clean and prevents from becoming wet. On the ventral side of the caudal region lies the anal aperture.

CHAPTER XVII

CHORDATA (continued) : MAMMALIA

THE GUINEA-PIG

(*Cavia porcellus*)

Habitat

The guinea-pig, *Cavia porcellus*, is a well-known household pet. The name seems to be a misnomer, because the animal is neither a pig nor does it come from Guinea. It is a herbivorous rodent belonging to the same order as the common rat and the rabbit. It is timid and inoffensive in the domesticated state. In the wild state, guinea-pig is gregarious and lives in underground burrows or in bushes hiding inside thick vegetation. It feeds chiefly at dusk and makes a hasty retreat on slightest provocation.

The guinea-pig is a mammal because: (1) its body is covered with a *coat of hair* which serves to keep it warm; (2) it is warm-blooded or *homoio-thermal* and its body temperature remains constant in spite of changes in the temperature of its environment; (3) it is *viviparous*, *i.e.* its, young is born alive and fed by the mother with milk produced in her mammae.

External Feature

The body of the guinea-pig is divisible into *head*, *neck*, *trunk* and *limbs*. There is no tail, but a rudimentary tail-bone is found at the posterior end of the trunk. All the parts of the body, excepting the tip of the nose and the foot pads, are covered with **hair** which may be diversely coloured, ranging from white to brown or even black.

The **head** is elongated and produced anteriorly into a narrow *snout* bearing a pair of **external nostrils** at its tip. The **mouth** is a small opening situated beneath the nostrils. It is bordered by soft lips of which the upper one is cleft in the middle, producing the so-called hare-lip; two chisel-shaped front teeth are thus exposed. A number of stiff and sensitive whiskers or *vibrissae* are found on the snout by the sides of the external nostrils. The large **eyes,** one on each side of the head, are protected by movable upper and lower eyelids which bear a few eyelashes. There is a third eyelid in the form of a whitish **nictitating membrane** which can be flicked across the eye from the inner corner of the orbit, thereby cleaning the outer surface of the eyeball. The most prominent

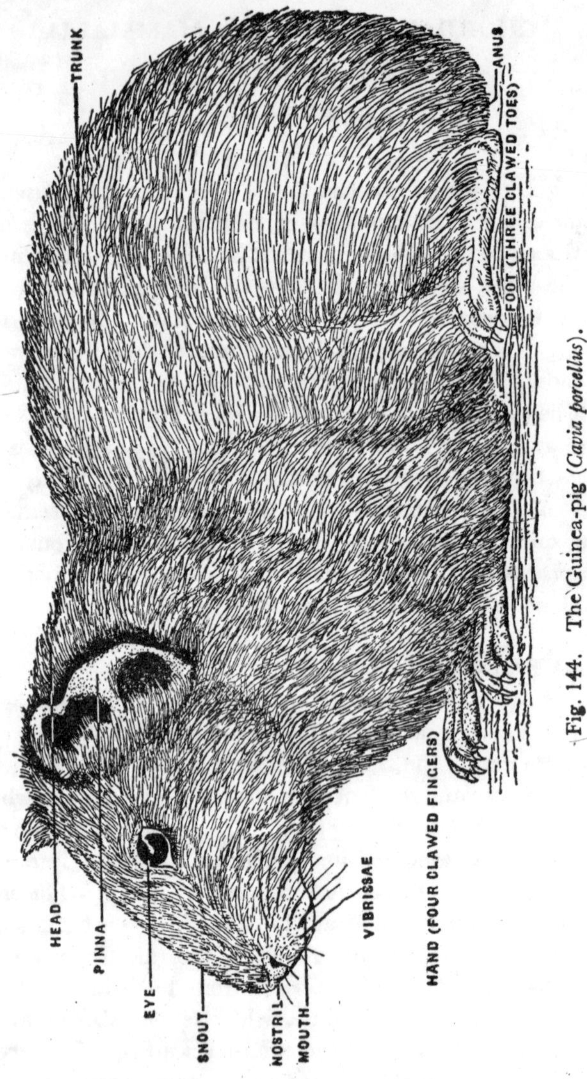

Fig. 144. The Guinea-pig (*Cavia porcellus*).

feature on the head are a pair of large **pinna** or external ears; each pinna is freely movable and can be turned towards the source of the sound. At the base of the pinna is an aperture called

external auditory meatus which leads through a short tube to the *tympanum* or eardrum.

The **neck** is short and flexible. It connects the head with the trunk and permits free movement of the head.

The **trunk** is composed of an anterior part, the **thorax,** protected by ribs and sternum, and a posterior part, the **abdomen,** having no bony support. A pair of *teats* or nipples are seen on the ventral surface of the abdomen near the posterior end of the trunk; they bear the external openings of the *mammary glands.* The teats are small and the mammary glands are undeveloped in male guinea-pigs. The **anus** lies beneath the rudimentary tail-bone at the posterior end of the trunk. There is no cloaca; **urinary** and **genital** apertures lie ventral to the anus. The space between the anal and the urinogenital apertures is known as the **perineum.** A small **perineal gland** opens in the perineum on either side of the anus and the secretion of these glands gives the guinea-pig its characteristic smell. In the male, the urinogenital aperture lies at the tip of a muscular organ called **penis** which is ensheathed by a fold of skin, the *prepuce.* A pouch of skin, the **scrotum,** is found at the root of the penis in the adult male; this pouch encloses the testes. In the female, the genital aperture is the slit-like **vulva** which is bounded laterally by folds of skin called *labia.* A small elevation not larger than the head of a pin, the **clitoris,** is found in the anterior part of the vulva and the urinary opening is situated close to this. The clitoris corresponds to the penis of the male.

The **limbs** are of the *pentadactyle* type and built on the same plan as in the toad; but the digits end in *claws* and are not webbed. The limbs are paired and the forelimbs are shorter than the hind-limbs.

Each **forelimb** consists of arm, forearm, wrist and hand; the hand bears four-clawed digits, the one corresponding to our thumb being absent.

Each **hindlimb** consists of thigh, shank, ankle and foot; the foot bears only three clawed toes, the toes corresponding to our first and second being absent.

The animal walks on all fours. The short forelimbs help in burrowing underground. The long hindlimbs are used for leaping. The planter surfaces of the feet and ankles rest on the ground during locomotion.

The Skin

The mammalian skin has two principal parts, an outer *epidermis* and an inner *dermis*. The epidermis is derived from the ectoderm and the dermis from the mesoderm.

The **epidermis** of the skin consists of a superficial horny layer, the **stratum corneum,** and a deeper germinative layer, the **stratum germinativum.** The former is composed of dead scale-like cells and the latter is formed by living polyhedral cells. The cells of the germinative layer divide actively to regenerate the superficial layers of dead horny cells. The protoplasm of the scale-like superficial cells is gradually replaced by a hard protein called keratin due to

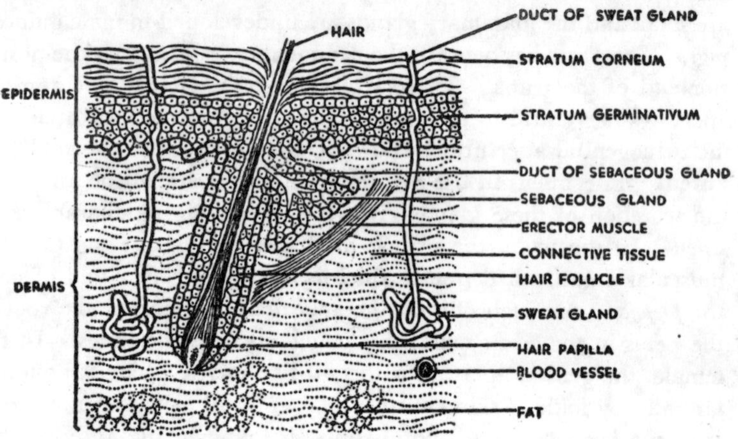

Fig. 145. Histological structure of mammalian skin.

increased pressure of new cells that are constantly forming under them. In parts exposed to pressure and friction, such as palm of the hand and sole of the foot, the horny layer is thickened and hardened to form the so-called friction *tori* (singular—*torus*). The *claws* at the ends of the digits are much hardened and specially modified pieces of the horny layer.

The **dermis** is a layer of fibrous connective tissue lying under the epidermis; it contains blood vessels and nerves.

Embedded in the dermis, but formed by downward prolongation of the epidermis, are small pits called **hair follicles.** At the bottom of each pit, there is a **hair papilla** from which arises a **hair.** Each hair follicle is provided with a bunch of smooth muscle fibres whose contraction erect the hair. Opening into the hair follicle by short duct is a group of flask-shaped structures called **sebaceous gland;** it liberates an oily substance which acts as a natural hair oil and keeps the hair in good condition. There is another set of epidermal glands called **sweat glands.** A sweat gland is a coiled tube-like structure opening by a long narrow duct on the surface of the skin. The sweat glands secrete watery sweat and thereby regulate the body temperature; they also excrete waste products.

In addition to the above, there are other modified glands in the skin. The **perineal glands,** opening at the surface of the perineum on either side of the anus, are specialised sebaceous glands. The **mammary glands** occur in both sexes on the ventral surface of the abdomen near the posterior end of the trunk. Each is a much branched tube, secreting **milk** and opening on the surface of a small prominence called nipple or teat. The glands are well-developed in female guinea-pigs, particularly during the period of lactation, and milk is secreted for feeding the young. In the male, the glands are small and do not function.

Body Cavity

The skin is loosely attached to the underlying muscles by a thick layer of subcutaneous connective tissue. On cutting through the skin and muscular body wall, the body cavity is exposed. It is divided by a transverse partition called *diaphragm* into an anterior thoracic and a posterior abdominal cavity. The **diaphragm** is a dome-shaped muscle which is convex towards the thorax and concave towards the abdomen; its central part is thin and transparent, and radiating from this part are muscle fibres which are attached to the vertebral column, ribs and sternum. The dome-shaped diaphragm is flattened by the contraction of its muscle fibres, so as to increase the size of the thoracic cavity. The **thoracic cavity** is bounded posteriorly by the back-bone, laterally by the ribs, and ventrally by the dagger-shaped sternum. It is lined internally by a membrane called **pleura.** The two lungs, one on each side, are enclosed in the pleural cavities. Each pleural cavity is a double-walled bag. Between the two pleural bags is a small space called **mediastinum** which contains the heart enclosed in the **pericardium.** The thoracic part of the oesophagus and trachea also lie in the mediastinal cavity.

The **abdominal cavity** is larger than the thoracic cavity, and is lined internally by the **peritoneum.** It contains the stomach, intestines, spleen, liver, pancreas, kidneys, urinary bladder and ovaries. The peritneum is folded to form **mesenteries** for supporting the various abdominal organs.

Thus in mammals the body cavity consists of four coelomic chambers: the peritoneal, the pericardial and the two pleural cavities.

Skeletal System

The skeleton of guinea-pig is almost entirely osseous with strips of cartilage persisting at the ends of some bones. Cartilage is also present in a few places such as in the external ears and nose.

The skeleton is divided into *axial* and *appendicular* portions. The **axial skeleton** is composed of the skull, vertebral column, ribs, and sternum. The **appendicular skeleton** consists of the limbs and limb girdles.

Skull. The skull consists of three main parts: (1) the *cranium*, which surrounds the brain; (2) the sense capsules—*olfactory*, *optic*, and *otic*—which enclose the organs of special sense; and (3) the *visceral skeleton* formed by the jaws, hyoid apparatus and part of the larynx. Cranium, sensory capsules and the upper jaw are firmly united.

The **cranium** or brain-box is composed of a number of bones rigidly articulated with one another by well-marked *sutures*. At the posterior part of the cranium, there are two **occipital condyles** for articulating with the vertebral column, one on either side of the **foramen magnum.** Surrounding this foramen there are four bones—a **supraoccipital** above, a **basioccipital** below, and two **exoccipitals,** one on each side. Each exoccipital is produced ventrally into a pointed and curved **paroccipital process** which serves for the attachment of muscles. The roof of the cranium consists, from behind forwards, of the single **supraoccipital,** paired **parietals** and paired **frontals.**

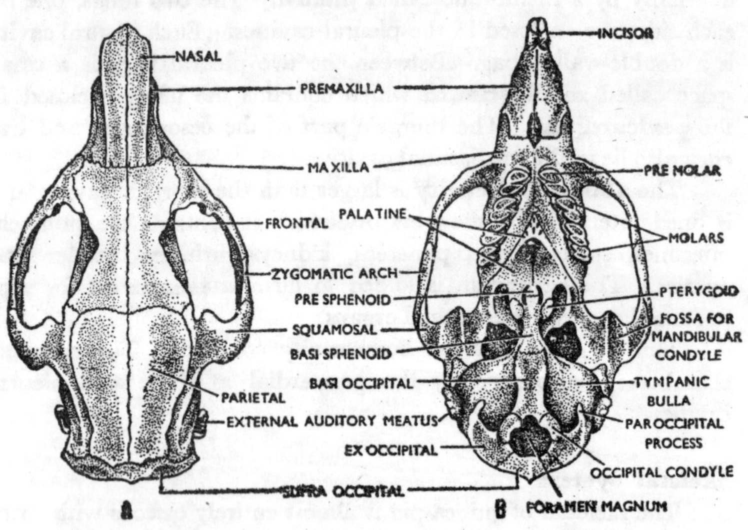

Fig. 146. Skull of Guinea-pig; A—Dorsal view, B—Ventral view.

The floor is composed of the **basioccipital, basisphenoid,** and **presphenoid.** Each lateral wall of the cranium is made up of **squamosal, alisphenoid** and **orbitosphenoid.** Arterial wall of the cranium is formed by a **cribriform plate** pierced by a number of holes. This plate separates the cranium from the olfactory capsule.

The **upper jaw,** on each side, is composed chiefly of a **premaxilla,** a **maxilla,** and a **jugal** which runs backwards to meet a process of the squamosal, thus forming the **zygomatic arch.** The premaxilla bears a cutting or **incisor tooth,** whereas the maxilla bears four grinding teeth, of which the anterior one is the **premolar** and the remaining three are **molars.** A long space, the **diastema,** separates the incisor from the premolar. The anterior part of the floor of the skull is known as the **hard palate.** This is formed by inward expansions of the premaxillae, maxillae, and by two **palatines.** Just behind each palatine is a small **pterygoid.** The quadrate is absent.

The **lower jaw** or **mandible** consists of two halves united in front by a weak symphysis. Each half is composed of a single bone, the **dentary,** which bears teeth similar to those on the upper jaw, and there is a diastema between the incisor and the premolar. Each dentary has an ascending *ramus* which forms the **condyle** for articulating with the undersurface of the squamosal at the posterior end of the zygomatic arch. In front of the condyle is a small curved process called **coronoid process.** The posterior end of each

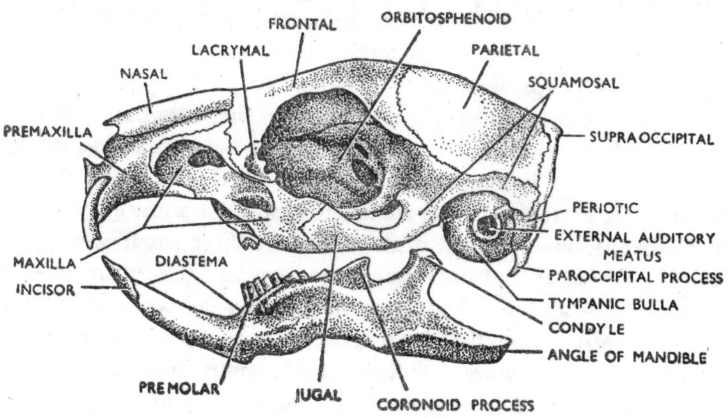

Fig. 147. Skull and lower jaw of Guinea-pig (side views).

dentary is produced behind into an **angular process** which forms the angle of the mandible.

The **olfactory capsule,** housing the nose, is roofed over by two **nasals,** whilst its floor and side walls are formed by the inward expansions of the **premaxillae** anteriorly, **maxillae** posteriorly, and the **vomer** in the midventral line. Posteriorly, the vomer articulates with the presphenoid. There is a porous **cribriform plate** at the posterior part of the olfactory capsule which shuts out the nose from the cranium. The olfactory nerves enter the nose through the holes on this plate. A vertical plate of cartilage, the **mesethmoid,** separates the two nasal chambers, and a few insignificant spongy bones called **turbinals** are found inside each nose to support the nasal epithelium.

The **optic capsule,** housing the eye, is represented on each side by a large **orbit.** At the anterior part of each orbit is a small **lacrimal bone** interposed between the frontal above and the maxilla below; this bone is perforated by a minute lacrimal foramen. The orbit is bounded dorsally by the frontal, ventrally by the zygomatic arch, and posteriorly by the squamosal. The inner wall of the orbit is the interorbital septum. It is formed by the orbitosphenoids above, and the presphenoid below.

The **otic** or **auditory capsule,** housing the internal ear, is placed between the squamosal and the exoccipital on either side of the cranium. Each capsule is principally composed of a *periotic bone*, a *tympanic bone*, and the *auditory ossicles*. The **periotic bone** is formed by the fusion of three small bones; it is irregular in shape and encloses the membranous labyrinth. The **tympanic bone** is firmly applied to the outer surface of the periotic; it is flask-shaped—the opening of the flask is the external auditory meatus, and the swollen base of the flask is the **tympanic bulla** which encloses the tympanic cavity along with the auditory ossicles. The pinna is attached to the rim of the external auditory meatus. The **auditory ossicles** constitute a chain of three small bones, the **malleus,** the **incus** and the **stapes;** they connect the eardrum with the internal ear and are themselves enclosed by the tympanic bulla.

The **hyoid** is a small bone lying at the root of the tongue on the floor of the buccal cavity. It consists of a stout median part, the *body* and two pairs of slender horns or *cornua*. The anterior cornua connect the hyoid with the periotics; the posterior cornua are similarly connected with the larynx.

Vertebral column. The vertebral column is composed of about thirty-seven *vertebrae*, which are separated from one another by pads of cartilage, called *intervertebral discs*. It is divisible into five regions: (1) the *cervical* regions consisting of *seven* vertebrae,

(2) the *thoracic* region of *twelve* vertebrae, (3) the *lumbar* region of *seven* vertebrae, (4) the *sacral* region of *four* vertebrae, and (5) the *caudal* region of about *seven* vertebrae. In each of these five regions the vertebrae possess certain common distinguishing features.

A **typical vertebra,** from the middle of the series, consists of the following parts: (1) a vertebral body or **centrum** with a flat surface at the anterior and posterior ends; the centrum is therefore *acoelous;* (2) a **neural arch,** placed above the centrum, enclosing the *neural canal* and thereby protecting the spinal cord; the part of the neural arch which joins with the centrum is known as the *pedicle*, and the remaining part forming the roof of the neural canal is the *lamina;* the margins of the pedicle are notched in such a way that openings, called *intervertebral foramina*, are formed between successive vertebrae serving as outlets for the spinal nerves; (3) a spinous process or *neural spine*, arises dorsally from the summit of the neural arch; (4) two **transverse processes,** one on each side, arising from the junction of the centrum with the neural arch; they are directed outwards and downwards; (5) a pair of upward-facing anterior articulating processes or **prezygapophyses,** borne upon the inner side of a pair of large forwardly directed processes, called **metapophyses;** (6) a pair of posterior articulating processes or **postzygapophyses** facing downwards and outwards from the posterior end of the neural arch; at the base of each postzygapophysis is a backwardly directed small process, called **anapophysis** (Fig. 148—F, G).

The **cervical region** is composed of *seven* cervical vertebrae. All the cervical vertebrae, excepting the first and the second, are characterised by the shortness of the centrum, the smallness of the neural spine, and the presence of a bifurcated transverse process on each side pierced through its base by a small hole, the vertebroarterial canal, for the passage of the vertebral artery. The *first vertebra* or **atlas** (Fig. 148—A, B) is a bony ring with no distinct centrum; it has no zygapophyses and its neural spine is very small; its transverse processes are flattened and perforated by the vertebroarterial canal. Anteriorly, the atlas bears a pair of concave facets for articulating with the occipital condyles. Posteriorly, there are similar but smaller facets for articulating with the second vertebra. The large ring-like neural canal of the atlas is divided horizontally by a fibrous *ligament* into an upper and a lower compartment. The upper compartment is occupied by the spinal cord, and the lower by an anteriorly directed process of the second vertebra, the *odontoid process.* The *second cervical vertebra* or the **axis** (Fig. 148—C) has a broad centrum which is produced anteriorly into a pig-like **odontoid process;** this is really the centrum of the atlas removed from it

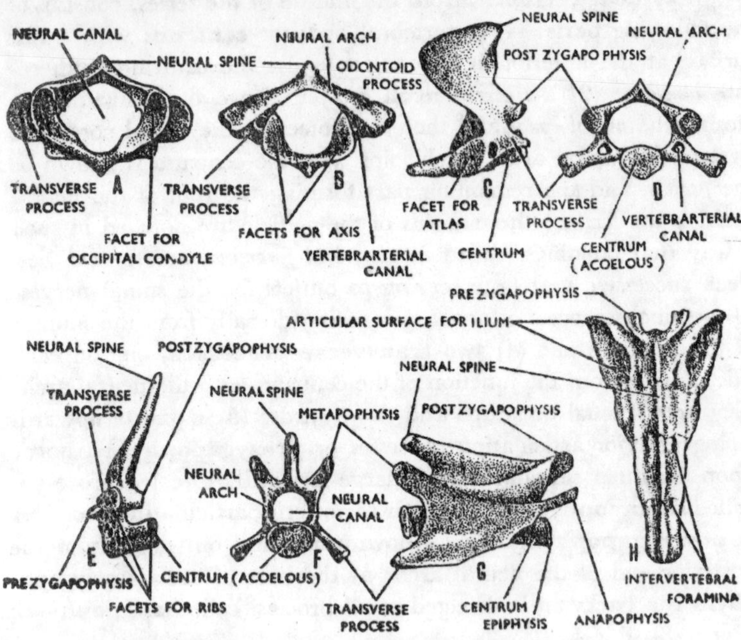

Fig. 148. Vertebrae of Guinea-pig; A=Atlas (anterior end), B=Atlas (posterior end), C=Axis (side view), D=typical cervical vertebra, E=thoracic vertebra, F=lumbar vertebra (anterior end), G=lumbar vertebra (side view), H=Sacrum (dorsal view).

and fused with the axis. The large neural spine is elongated anteroposteriorly and compressed from side to side. There are no prezygapophyses, but a pair of postzygapophyses are present for articulating with the prezygapophyses of the third vertebra. On each side of the odontoid process is an articulating facet for the atlas. Transverse processes of the axis are small and perforated.

The **thoracic region** is composed of *twelve* thoracic vertebrae, each of which typically bears a pair of *ribs.* They are divided into two series: (1) an anterior series composed of the first nine vertebrae; and (2) a posterior series including the last three. An *anterior* **thoracic vertebra** (Fig. 148—E) has a short thick centrum which bears at its sides concave facets for articulating with the *capitulum* of a rib; its neural spine is tall, slender and backwardly directed. The transverse processes are stout and horizontal; each bears a facet ventrally for articulating with the *tuberculum* of a rib. A *posterior* **thoracic vertebra** has smaller transverse processes which do not bear any facet for articulating with the ribs. Its centrum, however, carries well-developed facets for the *capitulum* of the ribs. The neural spine is flattened from side to side and is inclined forwards. Prezygapophyses are supported on forwardly directed *metapophyses*.

The **lumbar region** is composed of *seven* lumbar vertebrae (Fig. 148—F, G). They are the largest in the body, increasing in size from before backwards. The structure of a lumbar vertebra conforms to the typical vertebra which has been already described. In addition to the zygapophyses a lumbar vertebra bears metapophyses anteriorly and anapophyses posteriorly. The neural spine is flat and points forwards. The transverse processes are large and expanded. The centrum of the first and the second lumbar vertebra bears a median ventral ridge, the **hypapophysis,** which serves for the attachment of the diaphragm muscle. In others, the hypapophysis is absent.

The **sacral region** is composed of *four* sacral vertebrae immovably fused together to form a single bone, the **sacrum,** (Fig. 148—H), which is wedged in between the two halves of the pelvic girdle. The first sacral vertebra is the largest; its neural spine is upright; its transverse processes are expanded into wing-like structures which are attached to the ilia of the pelvic girdle. Metapophyses are small and anapophyses are absent. The zygapophyses are greatly reduced. The remaining sacral vertebrae decrease in size from before backwards.

The **caudal region** is composed of about *seven* caudal vertebrae. They decrease in size from before backwards, gradually losing all their processes and finally the neural arch as well. The last one or two are merely rod-like centra.

The **ribs** and **sternum.** The ribs, with the sternum and the vertebral column, form a basket for enclosing the heart and lungs. There are *twelve* pairs of ribs, which are articulated to the vertebral column in such a manner that they are capable of movement. A typical **rib** such as the sixth is a curved rod divisible into two portions: (1) a *vertebral portion* which is bony and articulates with the vertebral column, and (2) a shorter *sternal portion* which is cartilaginous and therefore known as the **costal cartilage;** the costal cartilage joins the vertebral portion to the sternum. The first seven are complete or *true ribs*, because they extend from the vertebral column to the sternum. The following five are *false ribs*, because they do not extend up to the sternum. The costal cartilages of the eighth and the ninth ribs are attached in front to the costal cartilage of the seventh rib and are thus *indirectly* connected to the sternum. The last three ribs remain free at their ventral ends, and are, therefore, called *floating ribs*. There are thus seven true ribs and five false ribs, of which the last three are floating. The first nine ribs have each *two articulations* to the vertebral column: (1) the rounded proximal end or heat or **capitulum** of the rib articulates with concave facets borne on the centra of adjacent vertebrae; (2) the **tuberculum,** which is situated a little distal to the capitulum, articulates with a facet on the undersurface of the transverse process of the following vertebra. The last three ribs do not articulate with the transverse processes of vertebrae and they have no tubercles.

The **sternum** is an elongated narrow rod, lying in the mid-ventral wall of the thorax. It is composed of a series of segments or **sternebrae** with which articulate the costal cartilages of the true ribs. The first sternebra or **manubrium** is the largest and articulates with the greatly reduced clavicle. The last sternebra or **xiphisternum** extends beyond the attachment of the ribs and expands into a large flat cartilaginous plate, the *xiphoid cartilage.*

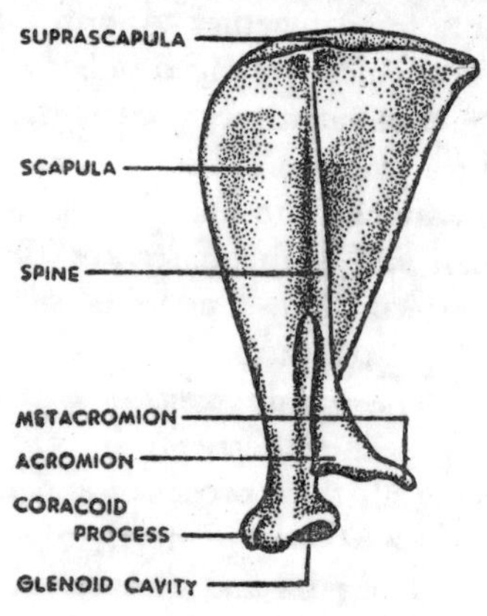

Fig. 149. Left scapula of Guinea-pig.

Pectoral girdle. The pectoral girdle of the guinea-pig consists of two bones on each side: (1) a large flat *scapula,* and (2) a small rod-like *clavicle.* The **scapula** is triangular in scape, with its base turned upwards and the apex pointing forwards and downwards; its flat portion overlies the ribs to which it is attached by muscles. The apex of the scapula is expanded and bears the *glenoid cavity* into which fits the head of the humerus. Overhanging the glenoid cavity is a hook-like *coracoid process.*

Posteriorly, the broad base of the scapula bears a narrow strip of cartilage, the *suprascapula*. On the outer surface of the scapula is a prominent bony keel known as the *spine*. The spine ends in a pointed *acromion* near the apex of the scapula and a long *metacromion* projects downwards from the acromion. The **clavicle** is a short slender bone which is imperfectly developed. It is merely embedded in muscles and has no articulations. Extending from the sternum to the acromion is a strong band of fibrous tissue; the clavicle occupies the middle third of this band. In some mammals such as man the clavicle is a well-developed bone connecting the scapula with the manubrium of the sternum.

Compared with the toad, the pectoral girdle of the guinea-pig is relatively small; it is completely ossified, except for the greatly reduced cartilaginous suprascapula. Two halves of the girdle are not connected by any hard structures; the precoracoid and coracoid are fused with one another to form a slender coracoid process which is permanently fixed to the scapula. The clavicle is also reduced and imperfect. The sternum of the guinea-pig is formed by the fused ventral ends of the ribs and does not correspond to the sternum of the toad.

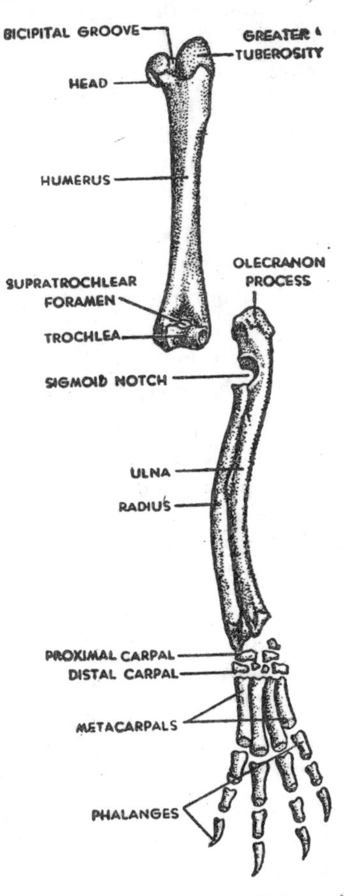

Fig. 150. Left forelimb of Guinea-pig.

Forelimb. The *arm* is supported by a stout **humerus.** The humerus has a rounded *head* at its proximal end, articulating with the glenoid cavity of the scapula. Close to the head is a *greater tuberosity* on the outer side, and a *lesser tuberosity* on the inner side; between the two tuberosities is the *bicipital groove* for lodging the tendon of the biceps muscle. Anteriorly, the shaft of the humerus bears a *deltoid ridge*. Distally, the humerus has a pulley-like surface, the *trochlea*, for articulation with radius and ulna of the forearm. Above the trochlea are two deep depressions, one in front and the other behind. The front one is the *coronoid fossa*, into which fits the radius and the back one is the *olecranon fossa*, into which fits the olecranon process of the ulna. The two fossae are joined by a small hole through the bone, the *supratrochlear foramen*.

The forearm is composed of two separate bones, the **radius** and the **ulna,** which are immovably articulated to each other. The radius is the shorter of the two bones and lies towards the inner side of the forearm. As the two bones are fixed in a prone position with the thumb pointing inwards, the guinea-pig cannot rotate its palm like a man. The ulna is longer than the radius; it bears a *sigmoid notch* at the proximal end for fitting into the trochlea of the humerus. The proximal end of ulna is continued backwards as the *olecranon process* which forms the point of the elbow and acts as a level for the arm. The ulna and the radius form a hinge-joint with the trochlea of the humerus. Distally, both the bones bear articular surfaces for the carpals.

The wrist or *carpus* is composed of *seven* **carpal bones** which are arranged in two rows. The proximal row includes the **radiale,** the **intermedium** and the **ulnare.** The distal row consists of four small bones. In addition to these, a very small sesamoid bone, the **pisiform,** is found on the under side of the wrist.

The palm includes *four* **matacarpal** bones. There are *four fingers*, each of which is composed of *three* **phalanges.** The terminal phalanx ends in a *claw*.

Pelvic girdle. The pelvic girdle of the guinea-pig consists of two halves, the *innominate* bones, which are united with one another in the midventral line by a strip of cartilage. Each **innominate** bone is composed of three parts—the *ilium*, the *ischium*, and the *pubis*, which are completely fused into one. It bears on its outer side a cup-shaped cavity, the *acetabulum*, in which fits the head of the thigh bone of the same side. The acetabulum marks the point of union of ilium, ischium, and pubis. The **ilium** is a blade-like bone which lies anterior to the acetabulum; it is joined to the sacrum by a rough articular surface on its inner side. The **ischium** forms the posterior third of the acetabulum, and is continued backwards along the line of the ilium; it is then continued downwards to meet the pubis of the same side. The **pubis,** the smallest of the three bones, lies ventral to the acetabulum; it joins the ilium anteriorly and the ischium posteriorly. A large *obturator*

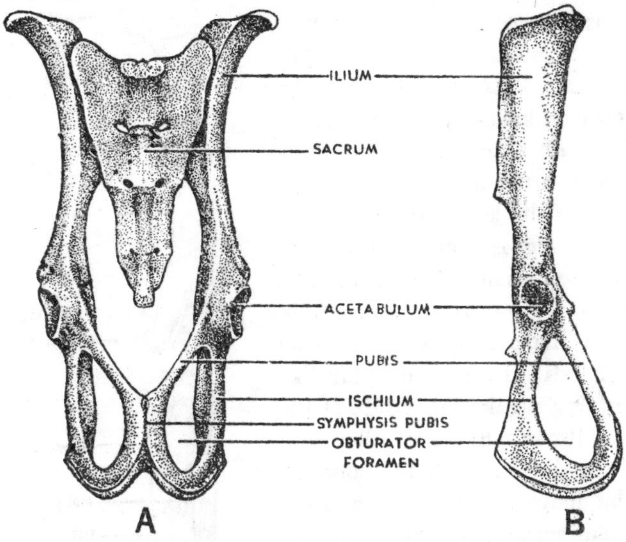

Fig. 151. Pelvic girdle of Guinea-pig; A=ventral view, with sacrum; B=side view of a single innominate bone.

foramen is enclosed between the pubis and the ischium. The pubes of the two sides meet in the midventral line to form the *symphysis pubis.*

Compared with that of the toad, the pelvic girdle of the guinea-pig is more massive. The ilium is relatively short and forms nearly two-third of the acetabulum; the remaining part of the acetabulum is formed by the ischium, and the pubis has practically no share in this. In toad, all the three bones take part in the formation of the acetabulum. In toad, the innominate bones are fused together more or less completely at their posterior ends; only the two ilia remain free. In guinea-pig, the two innominates are united only by the pubic symphysis; ilia as well as ischia remain free.

Hindlimb. The thigh is supported by a long bone, the **femur,** which bears medially a rounded *head* for articulation with the acetabulum, thereby forming a ball and socket hip-joint. Close to the head are three rough elevations: (1) a *greater trochanter* externally, (2) a *lesser trochanter* internally, and (3) a *third trochanter* externally just below the greater trochanter. Distally, the femur has two rounded *condyles* for articulation with the tibia. The condyles are separated by an *intercondylar groove* which is continued upwards for a short distance for accommodating the **patella** or knee-cap. The patella is a small sesamoid bone which develops in the tendon of a muscle in front of the knee-joint.

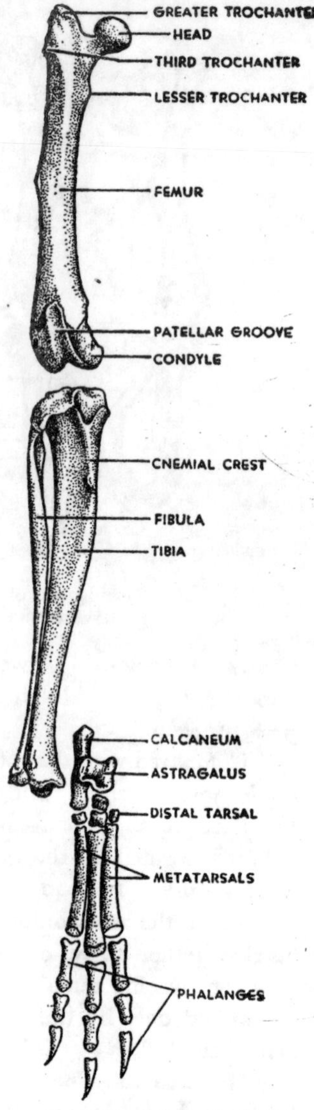

Fig. 152. Right hindlimb of Guinea-pig.

The leg is composed of two bones, the *tibia* and the *fibula*. The **tibia** is a stout bone bearing, in front, a sharp ridge called *cnemial crest*. The **fibula** is a thin splint-like bone lying on the outer side of the tibia. Proximally, the tibia and fibula articulate with the condyles of the femur; distally, they form articular surfaces for the tarsals.

The ankle or *tarsus* includes six bones arranged in *two* rows: a proximal row of two bones called **astragalus** and **calcaneum,** a centrally placed bone called **navicular,** and a distal row of three small tarsals.

The foot is composed of three **metatarsals.** There are three toes, each consisting of three **phalanges.** The terminal phalanx is clawed.

Alimentary System

The alimentary system of guinea-pig consists of the *alimentary canal* and the *glands* associated with it.

Alimentary canal is complete with the mouth in front and the anus behind. **Mouth** is a transverse slit bounded by soft and movable lips, of which the upper is slightly cleft in the middle to expose the incisor teeth. The mouth leads into the **buccal cavity,** the roof of which is the *palate.* The palate is divisible into two parts, an anterior *hard palate* supported by bones, and a posterior *soft palate* without bony support. The palate separates the buccal cavity from the nasal cavities which open by the internal nostrils into the pharynx. The **tongue** is a mobile muscular organ which is attached to the floor of the buccal cavity. Its free anterior end is just visible in the gap caused by the cleft upper lip. The surface of the tongue is covered by numerous white spots, the *papillae*, some of which bear *taste buds.*

The guinea-pig has **teeth** for cutting and chewing food. Teeth are implanted in sockets on the margins of both the jaws and hence the dentition is **thecodont.** It may be recalled that in the fish the teeth are not implanted into sockets and are, therefore, *pleurodont.* Moreover, in guinea-pig, all the teeth are not alike; that is the dentition is **heterodont.** In the fish the dentition is *homodont* (same shape). Lastly, the guinea-pig produces two generations of teeth of which the first set, known as the *milk teeth* or *deciduous teeth,* is *temporary* and is replaced later by the second set of *permanent teeth.* As there are only two sets of teeth, the dentition is **diphyodont.** Thus, the dentition is *thecodont*, *heterodont*, and *diphyodont.* This is characteristic of mammals.

In most mammals, there are four kinds of teeth: (1) the **incisor** or cutting tooth with chisel-like cutting edge and situated in front; (2) the **canine** or biting tooth with pointed edge, situated behind the incisor; (3) **premolar;** and (4) **molar,** with flat edges for grinding and situated behind the canine. The premolars and molars are practically identical—the only difference being that the molars are not present in the temporary set. The number and arrangement of the various kinds of teeth differ according to the feeding habit, and is expressed in the form of a **dental formula.** The types of teeth are denoted by their initial letters. Thus, the letters *i, c, pm,* and *m* stand for incisors, canines, premolars, and molars respectively. Each letter is followed by a fraction, the numerator of which stands for the number of that kind on one side of the upper jaw, and the denominator indicates the number of the same kind on same side of the lower jaw. Thus the dental formula of man is $i\frac{2}{2}$, $c\frac{1}{1}$, $pm\frac{2}{2}$, $m\frac{3}{3}$; this indicates that there are two incisors, one canine, two premolars, and three molars, on each side of each jaw, and there are in all thirty-two teeth.

In guinea-pig, the dental formula is $i\frac{1}{1}$, $c\frac{0}{0}$, $pm\frac{1}{1}$, $m\frac{3}{3}$, indicating that there are in all twenty teeth. There are one incisor, one premolar, and three molars in each side on each jaw, and the canine is absent. The incisors are long and curved and they grow indefinitely as their edges wear away. There is a long gap or **diastema** between the incisor and the premolar on both sides of the two jaws. In carnivorous mammals the diastema is occupied by the canine teeth (Fig. 147).

A tooth consists of two parts: the *crown* projecting above the gum, and the *root* lodged in a socket of the jaw. The bulk of the tooth is composed of *dentine* which is enveloped by a coat of hard *enamel* at the crown, and by a substance called *cement* in the region of the root. The central cavity of the tooth is occupied by a jelly-like pulp, blood vessels, and nerves.

The buccal cavity leads into the **pharynx,** which consists of a dorsal part or *nasopharynx* and a ventral part or *buccopharynx*. The nasopharynx is continuous in front with the nose through the internal nostrils and bears on each side a small *Eustachian aperture* leading into the middle ear. The posterior margin of the soft palate extends into the naso-pharynx to form the *velum*. On either side of the velum is a mass of lymphoid tissue called *tonsil*. The buccopharynx is continuous in front with the buccal cavity and bears on its floor a slit-like *glottis*, situated behind the tongue. The glottis opens into the voice box or *larynx* and is guarded by a flap of cartilage, the *epiglottis*, which shuts the glottis during the act of swallowing.

Posteriorly the pharynx is continued into the oesophagus, and the oesophageal opening lies dorsal to the glottis.

The **oesophagus** is a long cylindrical tube which runs backward, along the neck, just ventral to the vertebral column. The *trachea* or windpipe lies in front of the oesophagus. Passing through the thoracic cavity, it pierces the diaphragm muscle and finally opens into the cardiac portion of the stomach.

The **stomach** is a large sac which lies transversely on the left side of the abdominal cavity. Its concave anterior border is known as the *lesser curvature*, whereas the convex posterior border is the *greater curvature*. As usual, the stomach is divided into two parts: a *cardiac* portion lying towards the left side of the oesophageal opening, and a narrow *pyloric* portion situated on the right side. The pyloric stomach is continued behind into the **small intestine,** the first part of which is the *duodenum*. There is a pyloric constriction between the pylorus and the duodenum containing a circular sphincter, the *pyloric valve*. The mucous membrane lining the interior of the stomach bears innumerable gastric glands which produce gastric juice.

The **duodenum** extends backwards and is then continued

forwards to form a U-shaped loop. The remaining part of the small intestine is the much-coiled **ileum,** the loops of which are held

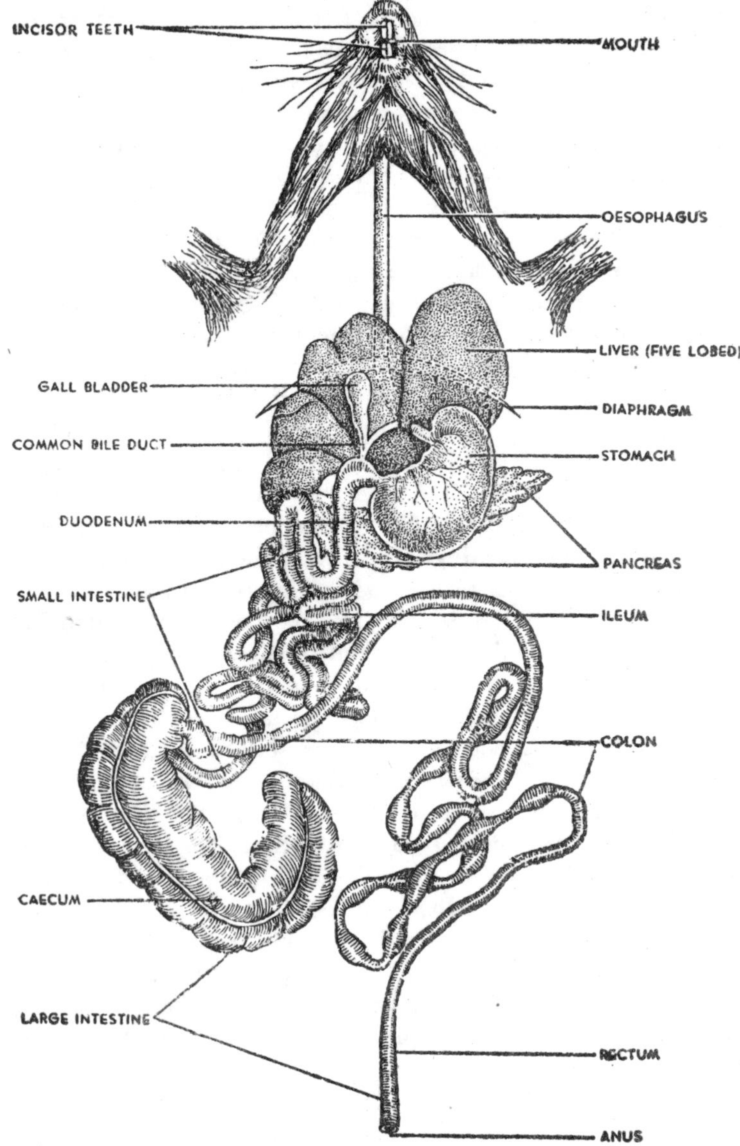

Fig. 153. Alimentary system of Guinea-pig.

together by mesentery. The mucous membrane of the small intestine is very much folded to form finger-like projections called *villi.* Each *villus* is richly supplied with blood capillaries and lymphatics. There are *Brunner's glands* opening in between the bases of the villi for secreting intestinal juice (succus entericus) (Fig. 180).

The **ileum** is directly continuous with the **colon** which is the proximal part of the **large intestine;** but there is, at the junction of the two, a huge, sacculated, blind, diverticulum known as the **caecum.** The aperture between the ileum and the colon is guarded by the *ileocolic valve.* Commencing on the right side, the colon forms several coils before it is continued into the **rectum,** which lies on the left side of the abdomen. The rectum communicates with the exterior by the **anus** which is guarded by a sphincter.

Glands associated with the alimentary canal are: (1) the salivary glands, (2) the liver, and (3) the pancreas.

There are *four* pairs of **salivary glands,** opening into the buccal cavity by separate ducts and secreting saliva for moistening food. The **parotids** are the largest of the salivary glands and are situated in the cheek near the angle of the mandible. The **submaxillary** glands are compact white masses opening on the floor of the buccal cavity. The **sublinguals** lie below the tongue. The **infraorbitals** are small, pale yellow bodies lying beneath the eyes.

The **liver** is the largest gland in the body. It is subdivided into five lobes. The liver lies behind the dome-shaped diaphragm to which it is attached by a band of peritoneum, called *falciform ligament.* Its convex anterior surface fits into the concavity of the diaphragm and the concave posterior surface lies close to the stomach. The **gall-bladder** is pyriform in shape; it is attached to the posterior surface of the liver between the third and the fourth lobes. The *cystic duct* from the gall-bladder joins the hepatic ducts from the liver to form the *common bile duct* which opens into the duodenum behind the pyloric sphincter.

The **pancreas** is an elongated mass of irregular shape situated in the mesentery between the two limbs of the duodenum; its narrow tail extends under the dorsal surface of the stomach and ends near the spleen. The pancreatic duct opens distally into the interior of the duodenum.

Feeding and **digestion.** The young are fed with their mother's milk. There is a milk-curdling enzyme, called **rennin,** in the gastric juice. Rennin coagulates milk-protein before it is digested by proteolytic enzymes. The adult is herbivorous. The vegetable food is thoroughly chewed by the teeth and mixed with saliva. The latter contains an amylolytic ferment, known as **ptyalin,** which splits starch into malt sugar. The masticated food passes into the stomach *via* the oesophagus. In the stomach, the food is acted upon and partially digested by the enzymes of the gastric juice in the presence of free HCl. The pyloric sphincter relaxes after the gastric digestion is over. The acid chyme now enters the duodenum and is immediately neutralised by the Na_2CO_3 which is present there. In the duodenum, the partially digested food is simultaneously treated with pancreatic juice, bile and succus entericus. Bile is an antiseptic and helps in the digestion and absorption of fats. The pancreatic and intestinal juices contain several kinds of enzymes. Proteins are acted upon by proteolytic enzymes such as **trypsin** and **erepsin,** and converted into amino acids. Fats are split into fatty acids and glycerol by the **lipase;** and carbohydrates, such as starch, are converted into monosaccharides such as glucose and fructose by **amylase.** The digested food is absorbed by the villi of the ileum. The residue passes into the caecum and colon where it is solidified by the absorption of water. The solid faecal matter is extruded through the anus as pellets.

Vascular System

This is built on the same plan as that of the toad, and consists of heart, arteries, veins, blood and lymphatics.

(A) Heart. The heart of guinea-pig is ovoid in shape. It lies completely enclosed by a membranous pericardium, in a space called mediastinum, between the two pleural cavities. The broad anterior end of the heart is known as the base and the narrow posterior end is the apex. The apical region is slightly projected towards the left side. The heart is four-chambered and consists of two auricles, right and left, and two ventricles, right and left.

The **auricles** are placed anterior to the ventricles. They form the broad base of the heart. Each auricle communicates with the ventricle of its own side, but the left and right auricles are completely shut off from each other by a strong partition, the *inter-auricular septum*. Compared with the ventricles, the auricles are relatively thin-walled.

The **right auricle** receives deoxygenated blood from the two venae cavae which open separately into its lumen. There is no sinus venosus, because it is completely fused with the right auricle. The **left auricle** receives oxygenated blood from the pulmonary veins. As the two auricles are separated from each other by the inter-auricular septum, there is no chance of admixture of the two kinds of blood. There is a small thin area, the *fossa ovalis* in

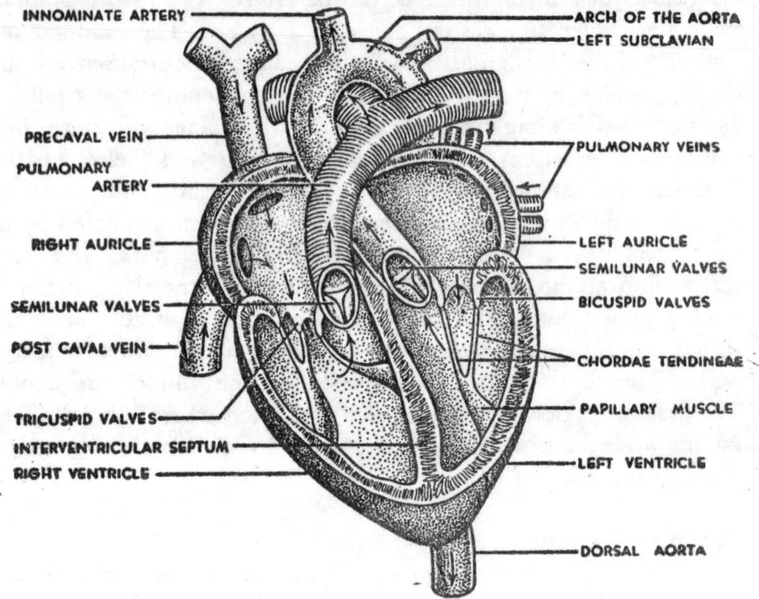

Fig. 154. Heart of Guinea-pig (diagrammatic); arrows indicate the course of circulation.

the centre of the interauricular septum. It marks the position of an embryonic aperture between the two auricles, the *foramen ovale*, which, however, is completely closed before the animal is born.

The **ventricles** are very thick-walled. Strong muscular ridges, known as *columnae carneae*, project from the inner wall into the cavity of both the ventricles. The left ventricle is lightly larger than the right, and is more muscular. The *inter-ventricular septum* is strongly built and shuts off the cavities of the two ventricles from one another. The **right ventricle** receives deoxygenated blood from the right auricle through the right auriculo-ventricular aperture. Similarly, the **left ventricle** receives oxygenated blood from the left auricle through the left auriculo-ventricular aperture. The *auriculo-ventricular apertures* are guarded by flap-like valves, the edges of which are attached to *papillary muscles* by means of cords, known as *chordae tendineae*. The right auriculo-ventricular valve has three cusps and hence it is known as the *tricuspid valve*. The left auriculo-ventricular valve has only two cusps; it is, therefore, known as the *bicuspid* or *mitral valve*.

The main **arterial arches** arise from the bases of the ventricles—the **systemic** or **aortic** from the left, and the **pulmonary** from the right. The entrance from the ventricles into both the arterial arches is guarded by a set of three pocket-like *semilunar valves*. The pulmonary artery carries deoxygenated blood from the right ventricle to the lungs for aeration. It is firmly attached to the under surface of the aortic arch by means of a band of fibrous cord known as the *ductus arteriosus*. The aorta distributes oxygenated blood from the left ventricle to different parts of the body.

(B) **Arteries.** The main arteries of the guinea-pig are the *aorta* and the *pulmonary* artery together with their branches.

The **aorta** arises from the left ventricle. It curves to the left and then turns sharply backward to run behind the heart, where it lies ventral to the vertebral column. From this point onwards it is designated as the **dorsal aorta,** whereas the curved anterior portion is known as the **arch of aorta.** The *dorsal aorta* runs through the thorax in the middorsal line. Passing into the abdomen through the diaphragm, it finally ends by dividing into two common iliac arteries at the posterior part of the abdominal cavity. During its long course the aorta gives out several branches to supply oxygenated blood to the entire body. A pair of very slender **coronary** arteries arise from the root of the arch to run into the tissues of the heart itself. A stout **innominate** artery arises from the arch proper and at once divides into the **right subclavian,** and the **right** and **left common carotid** arteries. A little beyond the innominate and slightly on its left side, arises the **left subclavian** artery as an independent branch from the arch of aorta. Each **common carotid** artery runs forward through the neck along the outer side of the windpipe, and near the level of the voice box divides into two branches—an **external carotid** supplying the outer side of the head, and an **internal carotid** supplying the brain and its meninges. Each **subclavian** artery divides into three branches: (1) a **vertebral** artery supplying to the vertebral column, (2) an **internal mammary** to the ventral body wall including the mamma, and (3) a **brachial** to the forelimb of the same side.

The thoracic part of the dorsal aorta supplies five to six pairs of **intercostal** arteries to the chest wall and the intercostal muscles between the ribs.

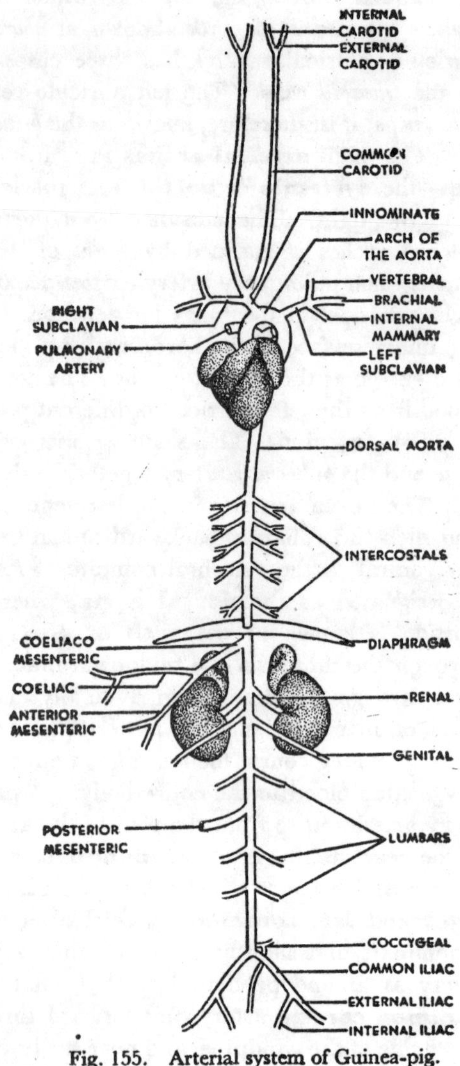

Fig. 155. Arterial system of Guinea-pig.

The abdominal part of the dorsal aorta gives out a stout coeliaco-mesenteric artery from its commencement. It breaks up into two branches: (i) a **coeliac** artery supplying to the liver, stomach and spleen, and (ii) an **anterior mesenteric** artery to the small intestine, pancreas and the mesenteric folds. A pair of **renal** arteries originate behind the coeliaco-mesenteric for supplying to the kidneys. This is followed by a pair of slender **genital** arteries to the gonads and three to four pairs of **lumbar** arteries to the dorsal body wall.

A median **posterior mesenteric** artery arises in between the roots of the first and second pair of lumbar arteries. It is distributed mainly to the large intestine and its mesentery. A small **coccygeal** artery is given out from the distal end of the dorsal aorta which supplies the sacral and coccygeal parts of the vertebral column.

The abdominal aorta terminates by dividing into a pair of **common iliac** arteries near the posterior end of the trunk. Each common iliac divides into an external and an internal iliac arteries. The **external iliac** is continued into the hindlimb of the same side as the **femoral** artery, whilst the **internal iliac** supplies to the ventral body wall, urinary bladder and the uterus.

The **pulmonary** artery arises from the base of the right ventricle. It divides into two branches for distributing deoxygenated blood to the two lungs.

(C) Veins (Fig. 156). As in toad, the veins are of three kinds: (i) pulmonary veins, (ii) systemic veins, and (iii) portal veins.

(i) Two **pulmonary veins** convey oxygenated blood from each lung back into the left auricle.

(ii) The **systemic veins** include the two **venae cavae** which return deoxygenated blood to the right auricle. The **anterior vena cava** or **precaval** vein is formed by the union of the right and left **innominate** veins behind the first segment of the sternum. Each **innominate** vein is formed by the union of five tributaries: a **subclavian** from the forelimb, an **internal jugular** from the brain, an **external jugular** from the outer side of the head and face, an **anterior intercostal** from the first and second intercostal spaces and an **internal mammary** from the ventral body wall. The **azygos** vein, from the posterior thoracic wall, drains directly into the precaval or rarely into the right innominate vein. The **coronary** vein from the heart drains into the precaval vein just before the latter opens into the right auricle.

The **posterior vena cava** or **postcaval** vein is formed by the union of a pair of **common iliac** veins at the posterior end of the trunk. It runs forward along the ventral aspect of the vertebral column by the right side of the dorsal aorta. It enters the liver and passing through the diaphragm muscle ends in the right auricle of the heart. An **external iliac** vein from the hindlimb and an **internal iliac** vein from the pelvic viscera meet each other to produce the **common iliac** vein of the side. The posterior vena cava in its long course receives two or three pairs of **lumbar** veins from the dorsal body wall, a pair of **genital** vein (spermatic in males and ovarian in females) from the gonads, a pair of **renal** veins from the kidneys, a pair of **hepatic** veins from the liver, and **phrenic** veins from the diaphragm. The left spermatic vein of the male guinea-pig usually drains into the left renal vein.

(iii) **Portal veins.** There is no *renal portal system* but well-developed **hepatic portal system** is present. Blood is conveyed from various parts of the alimentary canal by a large **portal** vein (Fig. 156) which is formed by the union of gastric, splenic, intestinal,

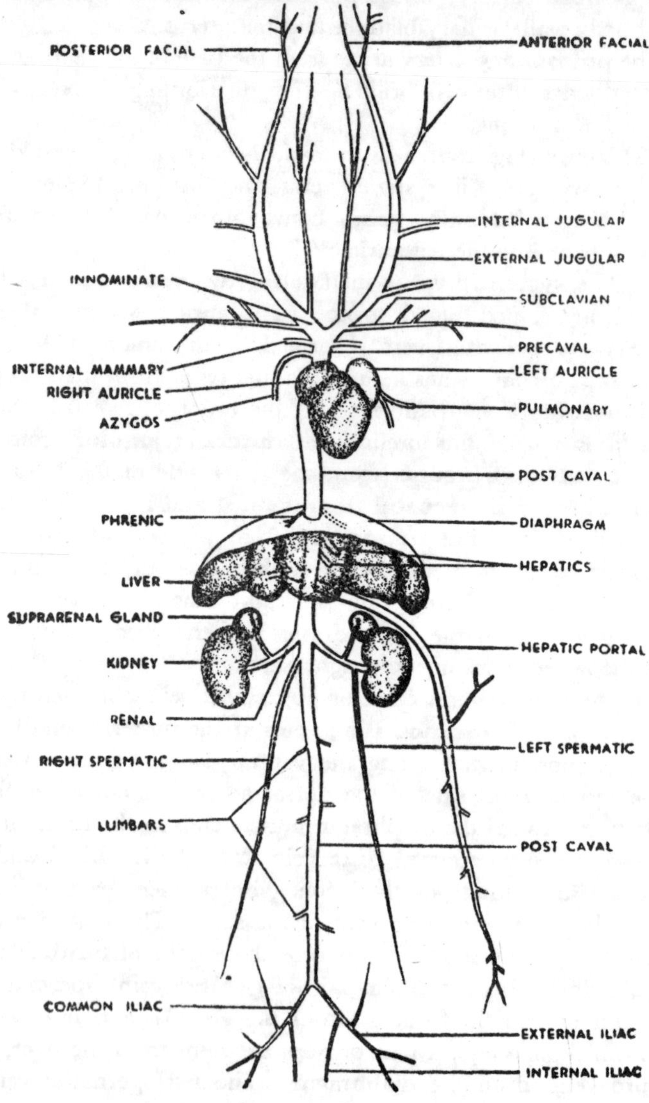

Fig. 156. Principal veins of Guinea-pig.

pancreatic and mesenteric veins. The portal vein on entering the liver, breaks up into sinusoids. A part of the food is thus extracted out of the blood and stored in the liver for future use. Finally, this blood is diverted through the hepatic veins into the postcaval.

(D) Blood. The blood consists, as usual, of **blood plasma** and the **blood corpuscles.** The **erythrocytes** are circular, biconcave, non-nucleated cells charged with haemoglobin. They are better known as the red blood corpuscles. The **leucocytes** or white blood corpuscles are nucleated, amoeboid cells devoid of haemoglobin. **Blood platelets** are small non-nucleated cytoplasmic bits occurring in groups and derived from broken cells in the bone-marrow. They correspond in function to the *thrombocytes* of the toad's blood.

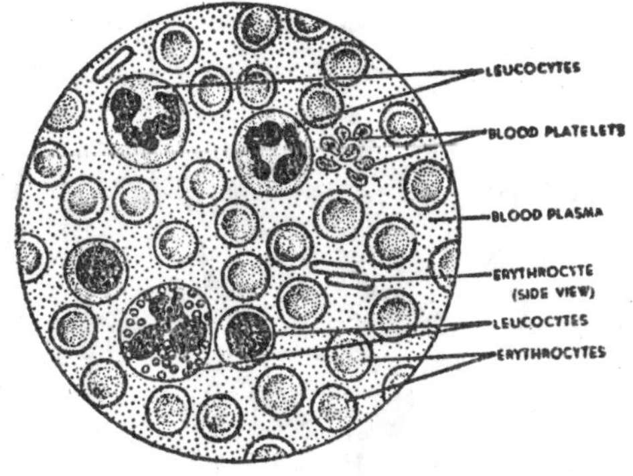

Fig. 157. A drop of Guinea-pig's blood (highly magnified).

(E) Lymphatics. Associated with the main venous vessels of every organ and limb, there are slender **lymphatic vessels** which ultimately drain into two large trunks situated within the body cavity. There is a stout **thoracic duct** on the left side of the dorsal aorta, communicating with the left subclavian vein at its junction with the external jugular vein. Similarly there is a **right lymphatic duct** on the right side, opening into the right subclavian vein.

Course of circulation. The blood is kept circulating through the lungs and the body by the alternate contraction (systole) and expansion (diastole) of the heart. The systole starts simultaneously in the two auricles, thereby driving the blood into the corresponding ventricles. When the ventricles are filled, the blood attempts to flow back, but leakage is prevented by the closure of the **tricuspid** and **mitral valves.** This is followed by a sharp contraction of the two ventricles expelling the blood through the aorta and the pulmonary artery, because the auriculo-ventricular valves (bicuspid and tricuspid) are tightly closed by the contraction of the *chordae tendineae.* The vessels expand with the inflow of blood, but with the commencement of the ventricular diastole their elasticity tends to drive the blood back into the ventricles, a process which is prevented by the rapid closure of the pocket-like **semilunar valves.**

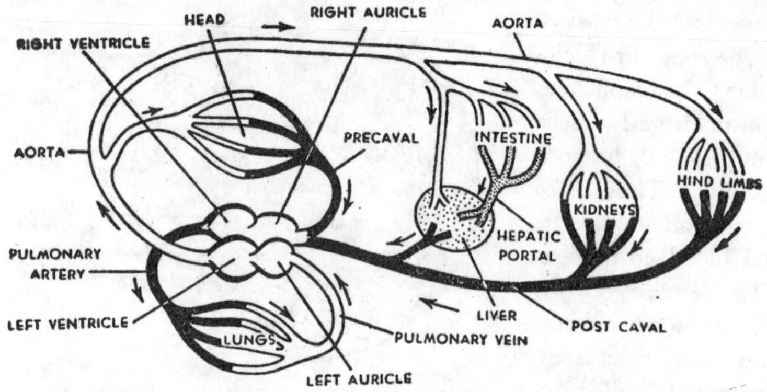

Fig. 158. Course of circulation of blood in the Guinea-pig (arrows indicate the direction of flow).

Deoxygenated blood is brought back to the right auricle by the vanae cavae. This blood is then pumped into the right ventricle and thence through the pulmonary artery into the lungs. The blood is aerated in the lungs and brought back to the left auricle by pulmonary veins, thus completing the **pulmonary circuit.** From the left auricle, the aerated blood is pumped into the left ventricle and then through the aorta it is pumped into every part of the body. It is deoxygenated after circulating through the tissue capillaries and finally brought back to the right auricle by the venae cavae, thus completing the **systemic circuit.** As there is no communication between the two sides of the heart, there is no possibility of admixture between the two kinds of the blood.

Thus, in guinea-pig, any portion of the blood has to pass through the heart twice before it returns to a particular part of the body Here we find an example of complete **double circulation**—a **pulmonary circuit** through the lungs and a systemic **circuit** through all other parts of the body, excepting the lungs. The systemic circuit includes a **hepatic-portal system,** in which blood loaded with food is diverted through the liver *via* the portal vein.

The latter breaks up into capillaries in the liver substance, so as to enable the liver cells to extract food for storage and other purposes. These capillaries ultimately drain into the hepatic veins which join the postcaval. There is no renal-portal system.

Respiratory System

The respiratory system of guinea-pig consists of the lungs and the respiratory passages leading from them into the exterior. Its parts are best described under the following headings: (1) the *accessory respiratory tract*, (2) the *true respiratory tract*, and (3) the *paired lungs*.

The **accessory respiratory tract** includes the external nares, the nasal cavities, the internal nares and the nasopharynx. During breathing air rushes inwards through the *external nares* into the *nasal cavities* which are separated from one another by the *nasal septum*. The nasal cavities are separated from the buccal cavity by the *palate*. The indrawn air passes into the *nasopharynx* through the *internal nares*. The nasopharynx opens into the *buccopharynx* behind the soft palate and the air crosses the buccopharynx to enter into the true respiratory tract through the *glottis*. The glottis is guarded by a hood-like cartilaginous flap, the *epiglottis*.

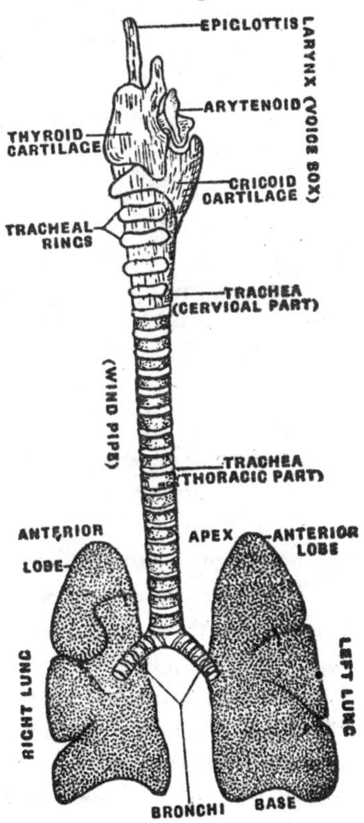

Fig. 159. Respiratory organs of Guinea-pig.

The **true respiratory tract** includes the larynx, the trachea and the bronchi. The **larynx** or *voice-box* is composed of *four* cartilaginous plates. Its ventral and lateral walls are strengthened by the saddle-shaped *thyroid cartilage*, and the dorsal wall by a pair of small *arytenoid cartilages*. A ring-like *cricoid cartilage* supports the posterior part of the larynx which is continued into the *windpipe* or **trachea.**

Stretched across the cavity of the larynx there are two elastic bands of membrane. These are known as the *vocal cords*. The vocal cords lie side by side, being separated from one another by a narrow chink, the *rima glottidis*. When air is driven out through the rima, the vocal cords vibrate and these vibrations produce sound waves. The intensity and pitch of the sound depend mostly upon the degree of elastic tension of the cords and the size of the rima glottidis. The **trachea** is a long tube running along the median line of the neck and lying ventral to the oesophagus. It is stiffened by cartilaginous rings, each of which is incomplete dorsally. Extending into the thorax, the trachea divides into right and left **bronchi.** The *bronchus* enters the lung of its own side, within which, it divides again in a tree-like fashion.

The **lungs** (Fig. 159) of guinea-pig are enclosed within the thorax, one on either side of the heart. They are not simple hollow sacs like those of the toad. The mammalian lungs are highly elastic, spongy, structures with three main lobes—anterior, middle and posterior. The right lung is slightly larger than the left, and has an additional lobe called *azygos lobe*. Each lung is surrounded by a *pleural cavity* which is composed of two thin membranes, an outer *parietal pleura* lining the thoracic wall, and an inner *visceral pleura* invetsing the outer surface of the lung. The two layers of the pleura are continuous with each other near the root of the lung through which the bronchus enters. On entering the lung, the bronchus divides repeatedly. The very small terminal branches of the bronchus are called *bronchioles*. Each terminal bronchiole divides into a number of finer tubes called *alveolar ducts*, which give out sac-like *alveoli* from their walls. The neighbouring alveoli lie pressed against one another, so that they are separated by very thin partitions. The pulmonary artery forms capillary networks which are distributed on these partitions. While the blood is circulating through the pulmonary capillaries, gaseous exchange occurs between the blood and the alveolar air.

Mechanism of respiration. The mechanism of breathing is entirely different from that of the toad, and the buccal cavity takes no active part in the process. The elastic lungs invested by the airtight pleural cavities are located within the thorax. This is a cage surrounded by the vertebral column, ribs and sternum. The thoracic cage is closed posteriorly by the dome-shaped *diaphragm*, a strong muscular partition with its convexity turned forwards. During respiration, the size of the thoracic cage is increased by the raising of the ribs and the flattening of the diaphragm. The ribs are raised by a set of *intercostal muscles* (inter= between; *costa*=rib), thereby increasing the transverse diameter of the chest. The antero-posterior diameter of the thorax is increased by the flattening of the diaphragm. When the thoracic cage increases in size, the pleural cavities are enlarged. Thus, the pressure within the pleural cavities is diminished and the lungs expand. Consequently, air rushes into the lungs through the respiratory tracts. This process is known as **inspiration.** After this, the thoracic cage reverts to its normal size by the relaxation of diaphragm and intercostal muscles. Intrapleural pressure is thus increased and air is forced out of the lungs. This is **expiration.**

Nervous System

The nervous system consists, as in toad, of a cerebrospinal axis, the peripheral nerves and the autonomic nervous system. It is the brain which has undergone marked specialisation, so as to exhibit very considerable advance over that of the fish and that of the toad. Nevertheless, the general relationship of the parts are the same.

(A) Brain. This may be subdivided into the same five parts as in toad. The **telecephalon** consists of a pair of greatly enlarged **cerebral hemispheres,** in front of which there is a pair of relatively small club-shaped **olfactory lobes.** The **cerebral hemispheres** (Fig. 160) are so enormous that they completely overlap the diencephalon and the mesencephalon. The two hemispheres are separated from each other by a deep groove, the **median fissure.** Laterally, there is another groove called **Sylvian fissure** which divides each hemisphere into two parts. The larger anterior part is faintly subdivided into *frontal* and *parietal lobes*, whereas the posterior part into *temporal* and *occipital lobes*.

The superficial part of each hemisphere is composed of nerve cells and is known as the *cerebral cortex* or *pallium*. The great development of pallium chiefly distinguishes mammals from fishes and amphibians. The ventral and lateral walls of each hemisphere is thickened to form the **corpus striatum.** The two hemispheres are connected by a transverse band of nerve fibres, the **corpus callosum,** which is situated on the floor of the median fissure and can be seen from above by pressing the hemispheres apart.

The **diencephalon** is completely hidden by the cerebral hemispheres. Its thick side walls form the **optic thalami** (singular—*thalamus*). **Epiphysis,** ending in the **pineal body,** arises from **its** dorsal surface, whereas the **optic chiasma** and the **hypophysis are** found on its ventral surface. The hypophysis (*infundibulum*) bears

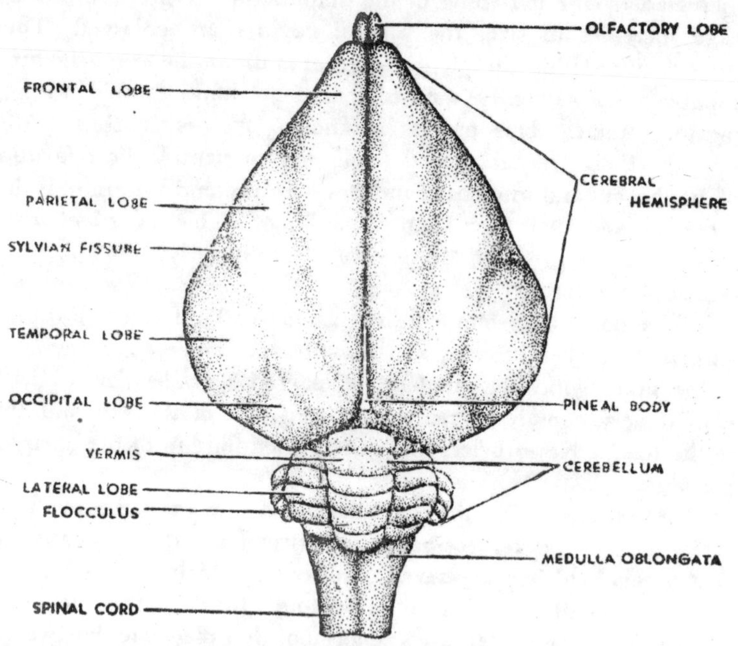

Fig. 160. Brain of Guinea-pig (dorsal view).

the **pituitary body.** Posterior to the pituitary, there is a pair of small rounded masses, the **mammillary bodies.** The roof of the diencephalon is non-nervous and includes a vascular fold, called *anterior choroid plexus*. The floor of the diencephalon, including the optic chiasma, pituitary body and mammillary bodies, is known as the **hypothalamus.**

The roof of the **mesencephalon** is thickened and expanded to form four rounded bodies, the **optic lobes** or **corpora quadrigemina;** of the four, the anterior pair form the *superior colliculi* and the posterior pair, the *inferior colliculi*. They are covered by the cerebral hemispheres. Ventral to the optic lobes are the thickened *cerebral peduncles* or **crura cerebri** which connect forebrain with hindbrain.

The **metencephalon** is formed by the much enlarged **cerebellum** which is subdivided into a median part called **vermis** and two **lateral lobes,** each of which terminates in a small **flocculus.** It is marked dorsally by numerous transverse folds. Ventrally the

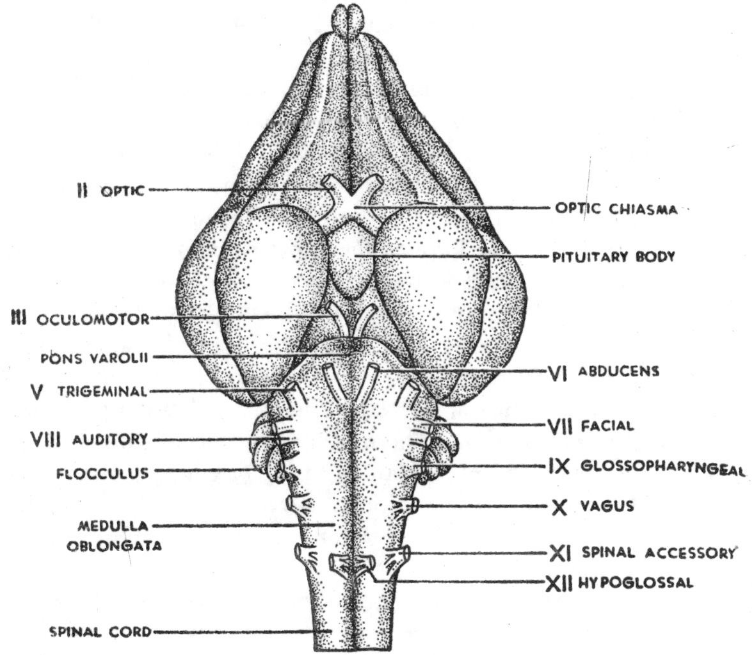

Fig. 161. Brain of Guinea-pig (ventral view).

two halves of the cerebellum are connected by a broad band, the **pons varolii.**

The **myelencephalon** or **medulla oblongata,** gradually tapers behind to join the spinal cord. The roof of the medulla is non-nervous and includes the *posterior choroid plexus.*

The **ventricles** of the brain are named as in other vertebrates. The cavities of the cerebral hemispheres are known as the *lateral ventricles.* They communicate with the cavity of the diencephalon, that is *third ventricle,* and with one another by a transverse slit, the **foramen of Monro.** The narrow third ventricle passes backwards to the equally narrow *iter*, which connects with the cavity of the medulla oblongata or *fourth ventricle.* The wide fourth ventricle communicates behind with the central canal of spinal cord. All the cavities are filled with *cerebrospinal fluid.*

The brain is enclosed in membranes, which are collectively known as the **meninges.** The tough outermost covering is the **dura mater.** The highly vascular innermost covering is the **pia mater.** Between the pia mater and the dura mater, there is a spongy layer known as **arachnoid.**

(B) The **spinal cord** is fundamentally similar to that of the toad. It passes backward without any visible line of demarcation from the posterior end of the medulla oblongata. It is situated within the neural canal and protected by meninges which are continuous with those of the brain. There is a deep longitudinal groove running along its midventral line, forming the **ventral fissure.** The cavity of the cord, the **central canal** or *neurocoel,* is continuous with the fourth ventricle of the brain and filled with cerebrospinal fluid. The spinal cord is slightly swollen in the branchial and lumbar regions and ends posteriorly in a narrow cone, the **conus terminalis** from which arise a tuft of nerves, forming the **filum terminale.**

(C) Cranial nerves. There are twelve pairs of cranial nerves, of which the first ten are similar in name, origin and distribution to those of the toad. The additional nerves are: (XI) the **spinal accessory** and (XII) the **hypoglossal.** Both are motor nerves, arising from the posterior part of the medulla.

XI. The **spinal accessory** is a short nerve which passes almost vertically downwards to innervate certain muscles of the neck region. A ramus internus from spinal accessory joins the vagus of the same side, to be distributed to the muscles of the pharynx and larynx.

XII. The **hypoglossal** (Fig. 162) is a stout nerve which arises by a number of roots. Near the angle of the jaw, the nerve crosses the internal carotid artery and splits into branches for supplying the muscles of the tongue.

The distribution of the **vagus nerve** is slightly different in the neck region of the guinea-pig (Fig. 162). The vagus comes out of the skull and expands to form a large **vagus ganglion.** Just beyond this, it gives off a **depressor branch** to the heart, and an **anterior laryngeal** branch to the larynx. The main trunk of the vagus now runs ventral to the aortic arch on the left side and the subclavian artery on the right side. It then gives off a **recurrent laryngeal branch,**

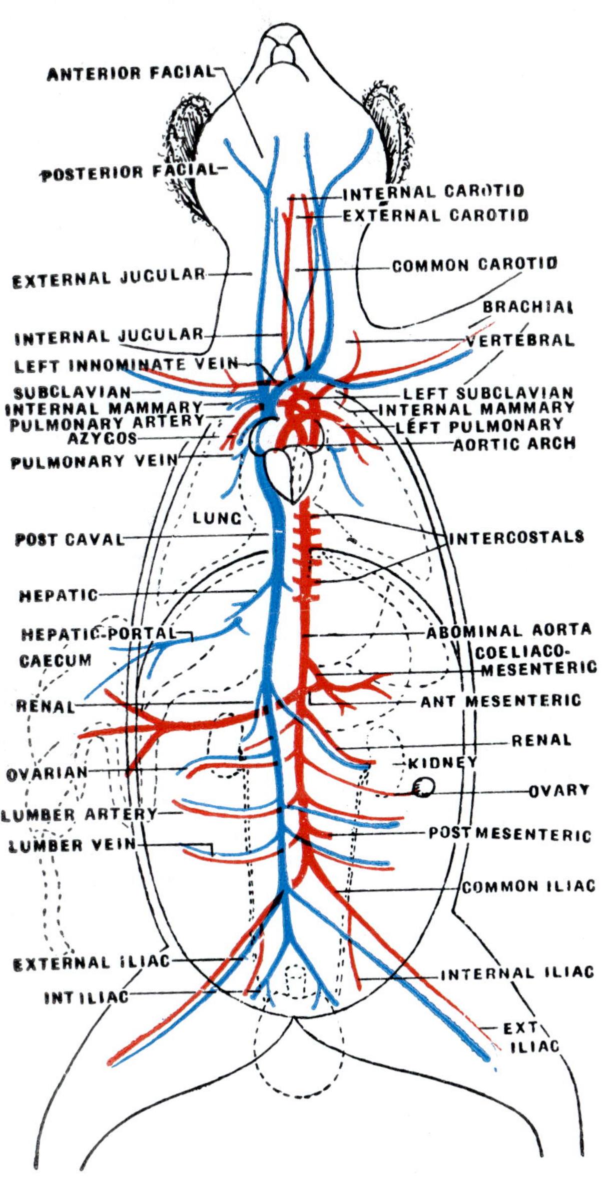

Vascular system of guinea-pig.
Arteries = red ; Veins = blue.

OUT. BIO.

which curves round the aorta on the left side, and round the subclavian artery on the right side. Both recurrent laryngeals ultimately turn forwards to enter the larynx. Supplying a branch

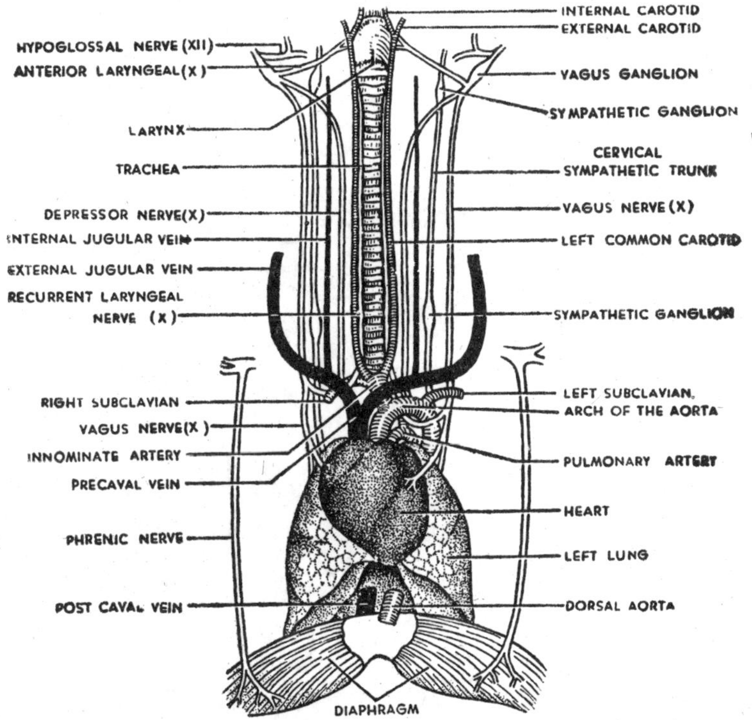

Fig. 162. Nerves and blood vessels in the neck region of Guinea-pig.

to the lung of its own side, each vagus now swings backwards into the abdominal cavity and is distributed to the visceral organs.

(D) Spinal nerves. These are built on the same plan as those of the toad, but are much more numerous. There are about 32 pairs of spinal nerves, each of which arises by two roots. The dorsal root is ganglionated. The last four cervical nerves and the first thoracic nerve unite to form the brachial plexus for supplying the forelimb. Similarly the last two lumbar nerves and the sacral nerves form the sciatic plexus for supplying the hindlimb. The fourth and fifth cervical nerves unite on each side to form a **phrenic nerve** for innervating the diaphragm muscle (Fig. 162).

(E) Sympathetic or **autonomic nervous system.** Running on each side of the aorta, and lying on the ventral surface of the vertebral column there is a ganglionated sympathetic nerve trunk. It receives, at each ganglion, a small **ramus communicans** from the spinal nerve of that segment. It traverses through the thorax and in the root of the neck swells up to bear a **posterior cervical ganglion.** Further forward each sympathetic trunk bears an **anterior cervical ganglion,** which lies just dorsal to the vagus ganglion of the side (Fig. 162). It then accompanies the internal carotid artery and enters the skull to end in the **Gasserian ganglion** of the V nerve. The main trunk on each side is connected with the trigeminal glossopharyngeal vagus and hypoglossal nerves.

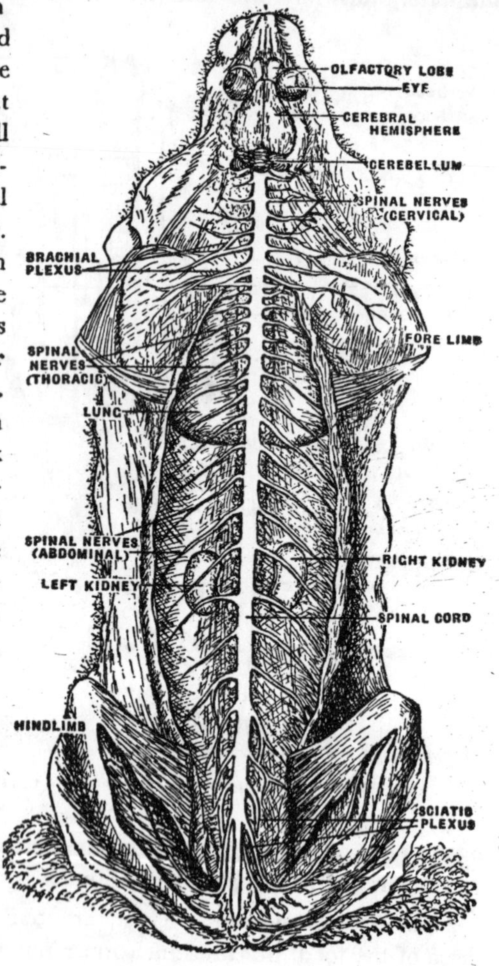

Fig. 163. Nervous system (Dorsal view).

The sympathetic nervous system innervates all the viscera and regulates the activity of the involuntary muscles. It sends communicating branches to the sympathetic plexus on the heart as well as on other viscera.

Receptor Organs

The receptors for *touch, pain* and *temperature* occur in the superficial layers of the skin, either as free nerve endings or as encapsulated corpuscles (Fig. 41 A-C).

Chemoreceptors for *taste* are found on the papillae of the tongue as groups of cells forming the *taste buds* (Fig. 41 D).

Receptors for *smell* are distributed as *olfactory cells* in the mucous membrane of the nose.

Eye. The eye is the organ of *vision*. It is essentially camera type and is built on the same plan as that of the toad.

The eyeball is more or less rounded and is composed of three coats: (1) *sclerotic* or *sclera*, (2) *choroid* and (3) *retina*.

The **sclera** is composed of dense fibrous tissue and forms a protective covering for the eyeball. On the exposed outer surface of the eye, the sclera is transparent and is known as the **cornea.** The cornea covers the *pupil* and the *iris* which lie jump behind. The externally visible part of the eyeball is covered by a delicate transparent membrane, the **conjunctiva,** which is continuous with the internal lining of the eyelids. A **lacrimal gland** is found on the upper part of each eyeball, which secretes *tear* for lubricating and washing the surface of the eye. The excess of tear drains into the nose through a small foramen on the lacrimal bone.

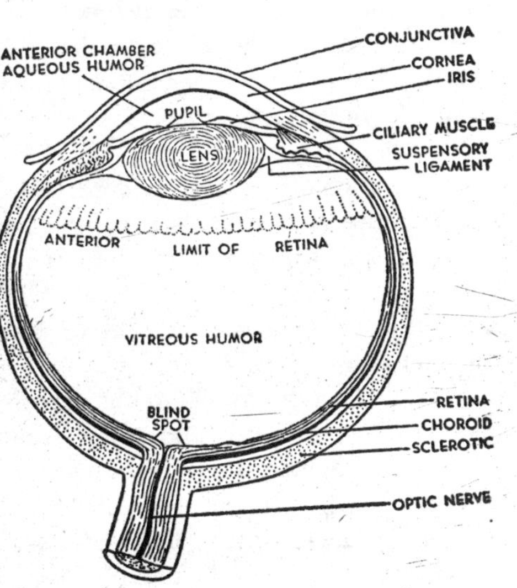

Fig. 164. Eye—The organ of vision.

The **chorbid** is a darkly pigmented membrane which lies internal to the sclera. It makes the eyeball light-proof. In front of the eye and a little behind the cornea, the choroid forms a circular disc, the **iris,** which is perforated by the **pupil.** The pupillary aperture may decrease and increase like the *iris diaphragm* of a microscope, thereby regulating the amount of light entering the eye. The action is effected by radial and circular muscle fibres which are present in the iris. The biconvex **lens** lies behind the pupil. It is held in position by a circular fold, the **suspensory ligament.** *Accommodation* or focussing is brought about by *ciliary muscles* which can alter the curvature of the lens.

The **retina** is the innermost coat which lines the interior of the eyeball. It extends anteriorly up to the ciliary muscles with which it forms the *ciliary body.* It is the light-sensitive screen of the eye and contains photoreceptors, the *rods* and *cones.* There are three kinds of **cone cells** for perceiving red, green and violet colours. The **rods** perceive varying intensity of light by producing a peculiar substance called *visual purple.* The rod cells and cone cells are connected to the fibres of the optic nerve, which spread out as a thin layer on the inner surface of the retina. The point through which the nerve enters the eye is known as the **'blind spot',** because it is devoid of rods and cones.

The eyeball encloses a more or less spherical cavity which is divided into two chambers by the lens and iris. The chamber between the cornea and the lens is filled with a thin watery fluid called *aqueous humor*, and that between the lens and the retina by a thick jelly-like substance called *vitreous humor.* Both the humors are transparent.

Light is focussed upon the retina in the same manner as in toad. When the rods and cones are stimulated, impulses are set up in them which are transmitted along the optic nerves to the brain. Vision is monocular. *Accommodation* or focussing is effected by altering the curvature of the lens and not by changing its position.

Ear. The ear is the organ of *hearing* and *balancing.* It includes the *external* ear, the *middle* ear and the *internal* ear.

The **external ear** is composed of the *pinna* and the *external auditory meatus.* The eardrum or tympanic membrane is stretched at the far end of the auditory meatus. The pinna is movable; it serves to collect sound waves and transmits the same down to the eardrum. The **middle ear** lies behind the tympanic membrane It is enclosed within the tympanic cavity of the tympanic bulla The *Eustachian tube* connects the middle ear to the pharynx, thus equalising the pressure on the two sides of the eardrum. The middle ear includes the three auditory ossicles, namely the *malleus,* the *incus* and the *stapes.* Of these the malleus is connected to the tympanic membrane and the incus. The incus is articulated to the stapes, and the latter is attached to an opening in the wall of the internal ear. This opening is known as the *fenestra ovalis.* **The auditory ossicles** serve to conduct the sound waves from the **tympanic membrane** to the internal ear.

The **internal ear** is enclosed within the tympanic bone. Its essential part is the *membranous labyrinth* which is surrounded by the corresponding *bony labyrinth.* The cavity of the bony labyrinth is filled by a fluid called *perilymph*, whereas the membranous labyrinth itself is filled with *endolymph.* The membranous labyrinth of guinea-pig is essentially built on the same plan as that of the toad. It consists of: the *utriculus*, the *semicircular canals*, the *sacculus* and the *cochlea.* The three semicircular canals, *anterior, posterior* and *horizontal*, are perpendicular to one another. Each semicircular canal opens into the utriculus by both ends, but bears an *ampulla* at only one end. The utriculus and the sacculus are separated by a narrow constriction. The sacculus is connected with a long, spirally coiled

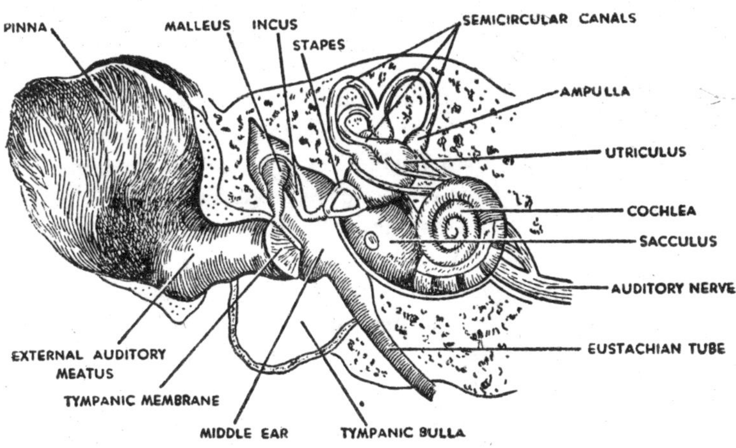

Fig. 165. Mammalian ear (diagrammatic).

cochlea which contains the *organ of Corti.* The latter is composed of receptor cell specially modified for perceiving sound waves. The auditory nerve breaks up into branches which are distributed to the organ of Corti and to other receptor cells within the sacculus, utriculus and ampullae of the semicircular canals.

Changed position of the body in any of the three planes causes flow of the endolymph in one or more *semicircular canals.* The receptors within the *ampulla* perceive these movements and the impulse is transmitted to the brain *via* the auditory nerve. The animal at once assumes correct posture. Receptor cells of the *utriculus* and *sacculus* probably perceive the direction of the gravitational force.

Sound waves are caught by the *pinna* and transmitted to the *tympanic membrane*. The vibration of the latter is transmitted to the internal ear through the *auditory ossicles* and cause corresponding vibrations of the *perilymph*. These are perceived by the *organ of Corti* and transmitted to the brain where they are translated as sounds.

The ear of the guinea-pig differs from the amphibian ear in three important particulars. First, there is an well-developed pinna for collecting sound waves. Secondly, the middle ear is traversed by three ossicles, the malleus, the incus and the stapes; in toad, there is a single columella which serves the same purpose. Thirdly, the sacculus gives out a spirally coiled cochlea containing organ of Corti.

Endocrine System

The principal ductless glands are the *thyroid*, *parathyroid*, *thymus*, *pituitary*, *adrenal* and *spleen*. The *pancreas* and the *gonads* produce endocrine secretions in addition to their exocrine functions.

The **thyroid** is a soft bilobed structure situated at the anterior end of the trachea, behind the thyroid cartilage of the larynx. It consists of a collection of vesicles enclosed in connective tissue. The vesicles are lined by a single layer of cubical epithelium. The thyroid vesicles secrete the hormone **thyroxine** which has now been successfully isolated and synthesised. It is an iodine-containing derivative of amino acid. The thyroxine stimulates growth and metabolism. If the gland is removed from a young animal, the development is defective and the individual becomes a dwarf.

The **parathyroids** are two pairs of globular bodies closely associated with the thyroid, from which they are difficult to separate. It is believed that the parathyroid hormone regulates bone formation. It has a controlling influence over the metabolism of calcium and phosphorus both of which are important constituents of bone.

The **thymus** is a fairly large soft pinkish mass, lying within the thorax encircling the base of the aorta. It is relatively larger in the young and may then partly cover the heart. It is composed mainly of lymphoid tissue and its secretion retards the onset of premature sexual maturity.

The **pituitary body** is attached to the floor of the diencephalon and lodged in a pit at the base of the skull. Its function has been fully discussed when considering this organ in the toad.

The **adrenals** are conspicuously yellow-coloured bodies composed of two distinct zones—the cortex and the medulla. Each adrenal lies in front of a kidney. The cortical hormone is essential for maintaining life, whereas the medullary hormone, adrenaline, controls the activities of the sympathetic nervous system.

The **islet of Langerhans** are microscopic patches of insulin-secreting tissues embedded in the pancreas.

The **gonads,** in addition to producing germ cells, act as ductless glands. The functions of these are the same as the corresponding structures in the toad.

The **spleen,** though not secreting any hormone, is a ductless gland. It is a dark-red, elongated body attached to the dorsal surface of the stomach by a fold of mesentery. Spleen is mainly composed of lymphoid tissue which forms a soft mass, the *splenic pulp*. The spleen is a reservoir for the blood and manufactures blood corpuscles. Worn-out corpuscles are destroyed in the spleen.

Urinogenital System

As in toad, the *excretory organs* are closely related to the *reproductive organs*, and the two sets of organs are described together as the urinogenital system. In the male guinea-pig, urine and germ cells are voided through a common passage, the **urethra.** But unlike toad, this urethra is never associated with the gut to form a cloaca, and opens to the exterior separately.

Excretory System

The excretory organs consist of a pair of *kidneys*, a pair of *ureters*, the *urinary bladder* and the *urethra*.

The **kidneys** are situated in the anterior part of the abdominal cavity, attached to the dorsal body wall, and behind the peritoneum. Each kidney is bean-shaped, having on its inner border a concavity called *hilus*. Projecting into the hilus is a small protuberance, the *pyramid*. The renal artery enters and the renal vein leaves the kidney at the hilus. The cavity of the hilus is continued backwards as the ureter.

Secretion of a kidney reveals that the kidney substance is distinguishable into two parts, an outer dark-red *cortex*, and an inner striated *medulla*. The kidney is essentially composed of *uriniferous tubules* and *collecting tubules*. The collecting tubules are straight and they open into the hilus on the summit of the pyramid. They run radially through the medulla to the cortex where each receives a number of uriniferous tubules. Each uriniferous tubule beignsin the cortex as a cup-shaped **Malpighian body** resembling that of the toad. It then runs a tortuous course to join a collecting tubule in the medulla.

The urine is formed in the same manner as in the toad. It passes through the uriniferous and collecting tubules into the cavity of the hilus which is continuous with the proximal end of the ureter.

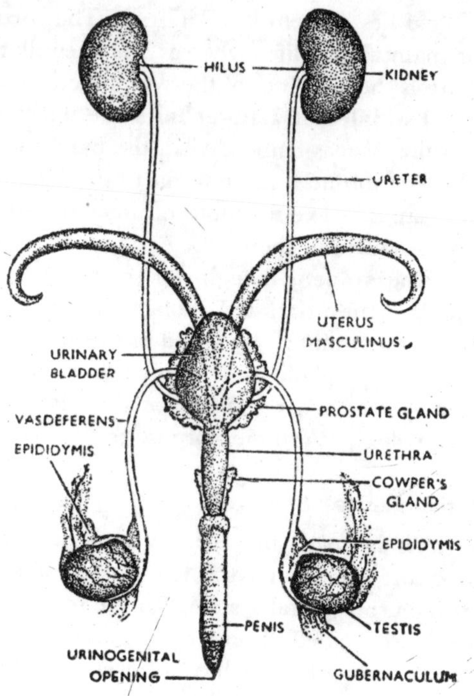

Fig. 166. Urinogenital organs of a male Guinea-pig.

The **ureter** runs backward as a white tube to open dorsally into the neck of the urinary bladder. The **urinary bladder** lies ventrally in the posterior part of the abdominal cavity. It is a pear-shaped muscular sac in which the urine is held until its evacuation is convenient. The broad anterior end or *fundus* of the urinary bladder gradually tapers back into the narrow *neck* which is continued into the *urethra*. The **urethra** is a narrow canal which, in the male guinea-pig, receives the genital ducts dorsally. It then traverses through the *penis* opening to the exterior at the tip of the copulatory organ. In the female guinea-pig, the urethra is short and opens into the *vulva* very close to the *clitoris*. The female urethra is solely urinary in function, whereas the male urethra is a channel for the voiding of urine as well as germ cells.

Reproductive System

The reproductive system of guinea-pig includes the organs of reproduction, namely, the *testes* in the male and the *ovaries* in the female, and their *ducts*.

Male reproductive organs. There is a pair of ovoid **testes** which in the young *guinea-pig* lie close to the kidneys in the anterior part of the abdominal cavity. As the animal approaches mature age, the testes pass backwards and ultimately come to lie within a pouch-like fold of skin at the root of the penis. This pouch is the *scrotal sac*. It communicates with the abdominal cavity by narrow passages, the *inguinal canals*, through which blood vessels and nerves from the abdomen pass into the testis. Each testis is suspended from the scrotal sac by a short cord called *gubernaculum*.

A **vas deferens** arises from each testis. The first part of the vas deferens is much-coiled and is bound up together to form the *epididymis*. The epididymis encircles the inner side of the testis The *spermatozoa* which develop in the testis are released through the convoluted tubules of the epididymis into the vas deferens. From the posterior end of the epididymis the vas deferens leads away as a straight tube. It runs through the inguinal canal where it lies enclosed along with the spermatic vessels and nerves to form the *spermatic cord*. Entering the abdomen, the vas deferens curves round the ureter of its own side and runs to the dorsal surface of the neck of the bladder. Here the two vasa deferentia lie side by side. Finally, they open into the base of a thin-walled, bifurcated diverticulum, the *uterus masculinus*, which is a hollow, median sac arising from the commencement of the *urethra*. The uterus masculinus corresponds to the vagina of the female guinea-pig.

The terminal part of the **urethra,** traverses through the *penis* openings outside at its tip. The penis is the male copulatory organ. It is enclosed in a loose fold of skin, the *prepuce* and projects ventrally in front of the anus. It consists of two *corpora cavernosa* and the single *corpus spongiosum*. All the three are composed of highly vascular erectile tissue. The urethra runs through the corpus spongiosum.

There are a few accessory glands the secretion of which nourishes the male germ cells. Surrounding the neck of the bladder and the base of the uterus masculinus, there is a compact **prostate gland** which opens into the urethra. Behind the prostate are a pair of small **Cowper's glands** which open into the urethra as it enters the penis.

Two **perineal glands** are found near the root of the penis, one on each side. They produce a secretion which gives the animal its characteristic smell.

Female reproductive organs. The **ovaries** are small ovoid bodies, one lying behind each kidney. They are attached to the dorsal body wall by folds of peritoneum and retain their abdominal position throughout life. The surface of the ovary is covered with raised vesicles of different size, the *Graafian follicles*, each of which encloses an egg or *ovum*. Near each ovary lies the funnel-shaped opening of an **oviduct.** Mature eggs burst out from the follicles and are caught by the adjacent oviducal funnel. The first part of the oviduct is narrow and twisted. It is known as the *Fallopian tube*. The second part of the oviduct is wide and muscular. This is the *uterus*, where early development of the *embryo* occurs. The two uteri unite to form a median tube which is continued backwards behind the neck of the urinary bladder as the *vagina*. This is the third part of the oviduct. The uteri and the fallopian tubes are attached to the body wall and held in place by a fold of peritoneum known as the *broad ligament*. The vagina communicates with the exterior through a slit-like *vulva* which is bounded laterally by raised folds, called *labia*. Just internal to the vulval outlet there is a very short *vestibule* containing the urinary opening in its antero-median part and the vaginal opening at the postero-median part. There is a

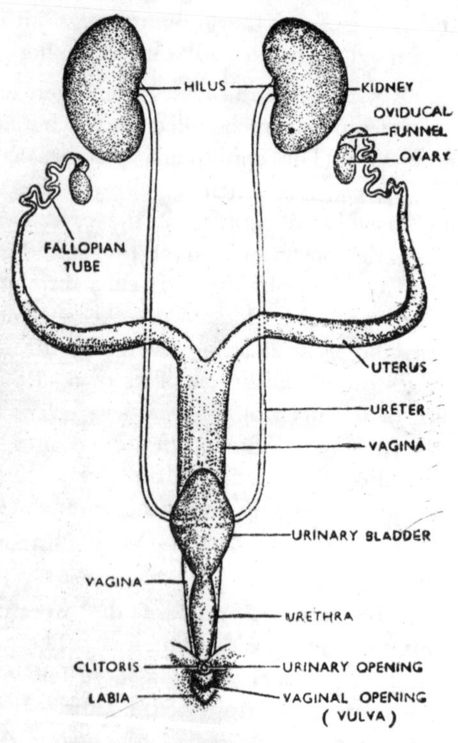

Fig. 167. Urinogenital organs of a female Guinea-pig.

small rod-like *clitoris* closely associated with the urinary opening The clitoris corresponds to the penis of the male guinea-pig.

Breeding and Life-history. There is a periodic ripening of the eggs during the breeding season. When ripe, the eggs are discharged by the bursting of the Graafian follicles. These are caught by the oviducal funnels and led into the Fallopian tubes. This process is known as *ovulation.* It occurs spontaneously after *copulation.*

During copulation sperms are injected into the body of the female by the male. The sperms swim up the vaginal canal into the uteri and thence into the Fallopian tubes where they meet the ova. *Fertilization* now takes place—a sperm uniting with an ovum. The fertilized ova descend into the uteri, getting attached to the uterine wall by a peculiar tissue called *placenta.* In the mean time, a blood clot develops inside each Graafian follicle from which the ovum has been extruded. Each blood clot is transformed into an endocrine gland, the *corpus luteum,* secreting a hormone, known as the *progestin,* which prevents further ovulation and brings about certain somatic changes associated with conception. But if the ova are not fertilized the corpus luteum dries up and is replaced by scar tissue. Development proceeds within the uteri, each embryo becoming attached by a stalk-like umbilical cord to the placenta. The *placenta* is a highly vascular mass of tissue derived partly from the embryo and partly from the mother. It is the organ by means of which nutriment and oxygen are conveyed from the mother to the developing embryo, and CO_2 and nitrogenous waste products are removed. Development is completed in the course of about ten weeks—this interval being known as the period of *gestation.* It is followed by *parturition* during which the young is expelled to the exterior through the vulva by the rhythmic contraction of the wall of the uteri and the vagina. This type of reproduction, wherein the young are retained within the body of the mother, is known as **viviparous** reproduction. Compared to this, the toad and the fish are **oviparous,** because they lay eggs which subsequently develop into the young, outside the body of the mother. Usually two or three young are born in a litter. They are nourished by a milky secretion produced in the mammary glands of their mother for a few days, after which they commence nibbling at green vegetables.

CHAPTER XVIII

CYTOLOGY AND HISTOLOGY

According to the *cell theory* animals are composed of cells and of cell products. The cell is the structural and functional unit of the animal body and the branch of Zoology which deals with cells is known as **cytology.** In unicellular animals the cell and the organism are one, whereas in multicellular animals the cells are integrated for proper functioning and arranged in the form of tissues and the tissues are aggregated into tissue-systems or organs. The study of tissues and tissue systems is known as **histology.** The protoplasm and the cell theory have already been dealt with. The present chapter includes the consideration of animal cell, methods of cell division and the various tissues and tissue systems of animals.

The protoplasm of an animal cell is normally differentiated into two main parts, the **cytoplasm** and the **nucleus.** The cytoplasm is covered by a delicate **cell membrane** which is a living part of the cell. The membrane is 7·5-9 nm thick and is composed of a double layer of lipid with a layer of protein on either side

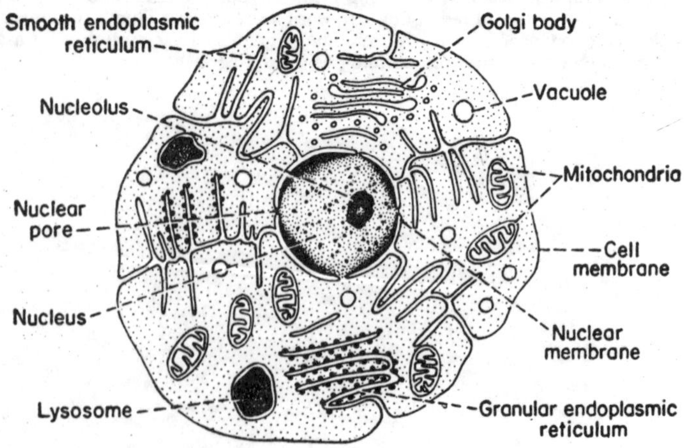

Fig. 168. Diagram of a typical animal cell under electron microscope

of the lipid bilayer. The membrane is perforated by small pores 0·8 nm in diameter through which water and water soluble molecules may pass. In addition to its limiting function, the cell membrane serves to regulate the entrance into and exit of substances out of the cell. It is said to be semi-permeable, because it allows certain dissolved substances and liquids to pass through it by **diffusion** and **osmosis,** but serves as barriers to others.

The **nucleus** is the headquarter of all cellular activities. It is usually spherical or oval in shape and lies in the centre of the cell, enclosed in a distinct **nuclear membrane.** The nuclear membrane is about 40 nm thick and is made up of Lipoprotein arranged in the same fashion as that of the cell membrane. The nuclear membrane bears large pores (50-100 nm in diameter). Some of the pores remain attached to the endoplasmic reticulum while other are free and through them the nucleus maintains a direct communication with the surrounding cytoplasm. The nucleus is composed of a rather specialised kind of protoplasm known as **nucleoplasm,** which consists of a fluid **nuclear sap** and solid threads of varying shape—the **chromosomes.** In most cases, the nucleoplasm of the living cell is perfectly transparent and the only visible structures are one or two rounded bodies called **nucleoli** (singular—*nucleolus*). This is explained by the fact that the chromosomes in the nucleus of a resting cell often remain invisible or only partially visible but they can easily be seen in a dividing nucleus (Fig. 168). In any case, the chromosomes always retain their identity and can be demonstrated in the living cell by microdissection.

The chromosomes are mainly composed of deoxyribonucleic acid (DNA) and a protein and are chiefly concerned with heredity.

The protoplasm of the cell, excluding the nucleoplasm, is known as the **cytoplasm.** It constitutes the cell body and is often divided into outer, clear **ectoplasm** and inner, granular **endoplasm.** Small fluid-containing spaces called **vacuoles** may be detected in the cytoplasm. A distinct **central body** or **centrosome,** including one or two minute *centrioles*, is often seen close to the nucleus. The centriole is usually surrounded by a zone of clear cytoplasm Fig. 168.

The central body plays an important part during the division of the parent cell into two daughter cells. Embedded in the cytoplasm are rod-like or granular bodies called **mitochondria,** (Fig. 169A) occurring either in groups or in a scattered state. They are of diverse size and shape. Average length and breadth of mitochondria is about 2·5 μm and ·5 μm respectively. It consists of two **mitochondrial membranes** surrounding a central fluid **matrix.** Structurally the membranes are similar to that of cell membrane. The two membranes remain separate. The inner membrane forms a number of folds towards the central matrix. These folds are

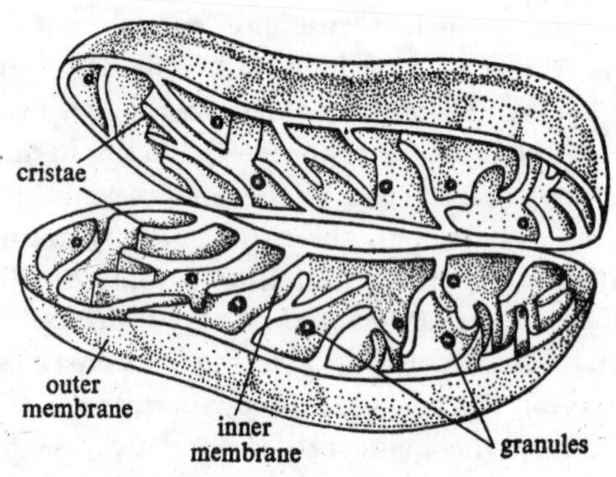

Fig. 169A. Diagrammatic representation of a longitudinally splitted Mitochondrion.

known as **cristae.** The outer mitochondrial membrane contains ribosomes, some DNA and RNA. The central matrix is rich in protein and Ca^{++} binding granules. These bodies are responsible for cellular oxidation and energy liberation. For this reason they are called the 'power-house' of the cell. A **Golgi body**, in the form of a network, often is present surrounding the central body (Fig. 169B). It consists of a system of flat smooth surfaced and single membrane vesicles and cisternae packed close together in the form of a stack. The single membrane forming the wall of the vesicles

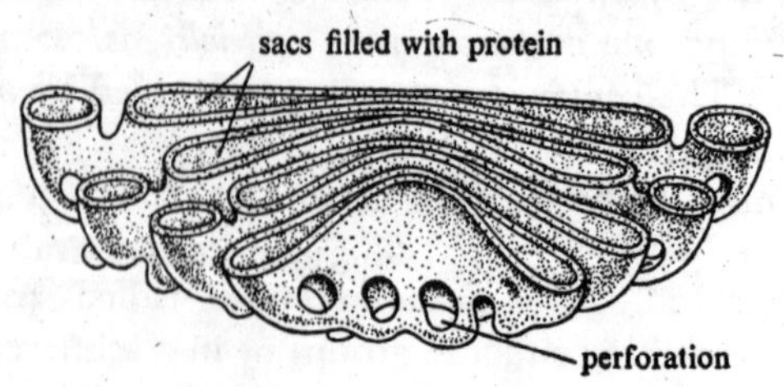

Fig. 169B. Golgi body in cross section.

and cisternae is 6-8 nm in thickness and structurally similar to that of the cell membrane. They are most abundant in secretary cells. The Golgi body is concerned with the secretion, synthesis and storage of certain cell substances.

Between the cell and nuclear membranes there exists an extensive network of minute tubes, sacs and cisterns. This network is known as **Endoplasmic Reticulum** (Fig. 169C). The wall of

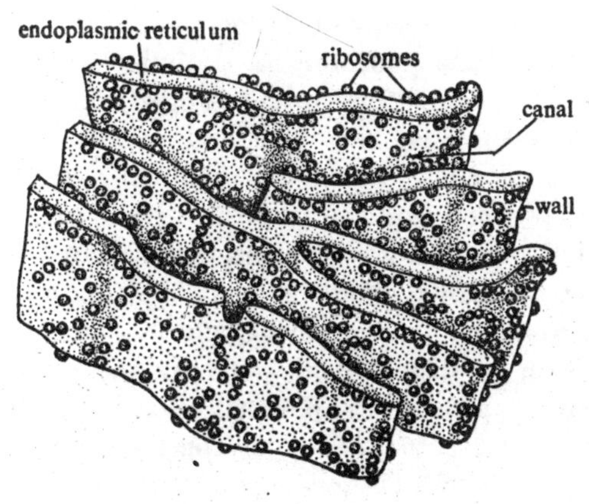

Fig. 169C. Arrangement of Endoplasmic reticulum and ribosomes in cytoplasm.

these tubes is about 6-7 nm thick. The endoplasmic reticulum serves functions like protein synthesis, and transporting secretory materials to the Golgi body.

Many small granular aggregates are found in the cytoplasm. These are called **Ribosomes.** They remain free in the cytosol or remain attached to the wall of the endoplasmic reticulum. The overall diameter of a single ribosome is about 20-25 nm and it is made up of two unequal subunits. Both free and endoplasmic reticular bound ribosomes participate in protein synthesis.

A good number of mitochondria like membrane enclosed organella called **Lysosomes** are found in the cells. These are particularly abundant in phagocytic cells like WBC. A newly formed lysosome is a spherical vesicle having a diameter of 20-80 nm. The wall of the vesicle is 7-8 nm thick and is made up of lipoprotein. The lysosomes contain hydrolytic enzymes. In case of damage or death of a cell the lysosomes undergo dissolution releasing the enzymes. As a result autolysis occurs.

The various cytoplasmic structures just mentioned are detected only by special cytological technique and differential staining. Otherwise they are not visible.

Apart from the living inclusions there are non-living **cytoplasmic inclusions** consisting of granules scattered throughout the cytoplasm. These include substances such as yolk granules, glycogen, globules of fat, secretions and excretions.

Such then is a brief sketch of an animal cell. We are already aware of what a cell can do. It feeds, grows, breathes and reproduces its kind. We can grow cells in the laboratory apart from the parent-body by special methods of tissue-culture. We can alter cells considerably by treating them with X-rays, or kill them with heat and cold, or dissect their parts by the recently discovered microdissecting apparatus. In recent years considerable amount of knowledge has accumulated regarding the structure and function of the cell. It is now well-understood that cells are the units of the living body and all the working of a living system are the manifestations of the cellular function. It is the activity of the cell which determines different complicated life processes—development, growth, maintenance, heredity, diseases and death. Still there are many more unsolved riddles in Biology, the answers of which are expected only to come from the future studies of cell structure and cell function.

CELL DIVISION

A cell never appears spontaneously but is always formed from some pre-existing living cell. It gradually increases in bulk till it reaches a certain fixed size, and then, in response to some mysterious stimulus, divides into two similar daughter cells. There is a fixed relationship between the size of the nucleus and the volume of the cytoplasm in every kind of cell. This is known as the **kern-plasma ratio.** Hertwig, who has introduced this term, claims that whenever there is a disturbance of the K–P ratio by relative increase of the nuclear bulk the cell responds by dividing into two daughter cells. Division of the cytoplasm is always preceded by the division of the nucleus. The nucleus may divide in *three* ways : (1) **amitosis,** (2) **mitosis,** and (3) **meiosis.**

Amitosis

Direct nuclear division or amitosis is a simple process. The nucleus constricts in the middle into a dumb-bell or hour-glass form, and the two swollen ends finally separate to produce two nuclei of approximately equal size. Division of the cytoplasm soon takes place by constriction, so that two new cells are formed. It must be noted that, in amitosis, the sharing of nucleus and cytoplasm between the two daughter cells is only approximate. Amitosis is found in lower forms of animals such as the protozoa. It is also found in metazoa, probably occurring only in very old and abnormal cells.

Amitosis may occur without the division of the cytoplasm, resulting in a multi-nucleate structure called *syncitium*.

Mitosis or Karyokinesis

This is the usual form of nuclear division and occurs universally amongst the somatic cells (*soma*=body) of higher animals. Mitosis involves a complex series of changes by means of which the nucleoplasm, particularly the chromosomes, undergo accurate division into two sets, which are distributed equally between the two daughter cells. For the purpose of description, **mitosis** or **karyokinesis** is divided into four stages: (i) **prophase,** (ii) **metaphase,** (iii) **anaphase,** and (iv) **telophase.** Telophase is usually followed by **cytokinesis** or division of the cytoplasm. It must, however, be borne in mind that mitosis, once started, runs continuously until it is finished and the phases are merely arbitrary divisions, with no clear line of demarcation between them.

Prophase. The first sign at the onset of mitosis is the appearance of thin coiled threads within the nucleus. These are the **chromosomes.** They soon begin to uncoil, and with this become shorter and thicker. This shortening and thickening is due to *dehydration.* In a resting nucleus the chromosomes are in a state of *hydration* and that is the reason for their indistinctness. The number of chromosomes differs in different species, but it is always constant for a particular species. Thus in the British frog,

Rana temporaria, there are 12 pairs, and in man there are 23 pairs. The chromosomes are essentially paired structures and the two

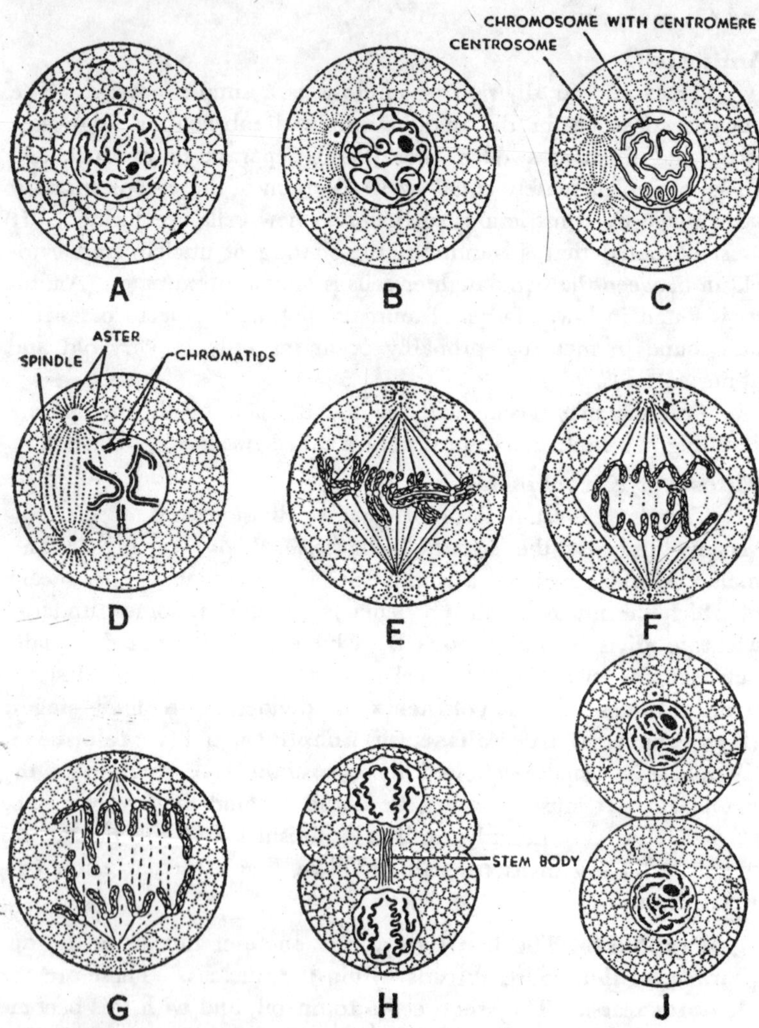

Fig. 170. Diagram of mitosis in the somatic cell of an animal; A=resting stage of interphase, B–D=prophase, E=metaphase, F–G=anaphase, H=telophase, J=cytokinesis.

identical members of a pair are known as **homologous chromosomes.** Though usually elongated and thread-like, the chromosomes may be rounded or U-shaped or they may have other forms.

Even at the early stage of prophase each individual chromosome is distinctly divided lengthwise, the two halves being termed **chromatids.** In the meantime the centriole has split and the two portions start to move away from each other until they take up position at opposite ends of the nucleus. As the centrioles move apart, they seem to drag out little delicate threads which connect them. A spindle-shaped body is formed in this manner. This is the **achromatic spindle,** two two ends of which are marked by star-like system of radiating fibrils. These are called **asters** (Fig. 170 D). The achromatic spindle is a jelly-like mass which can be removed as a whole on the point of a dissecting needle. Meanwhile the chromosomes have undergone further shortening. The nucleolus disintegrates and the nuclear membrane tends to disappear. The chromosomes are now distributed haphazardly on the spindle. These developments mark the end of prophase.

Metaphase. The delicate nuclear membrane completely dissolves. The longitudinal splitting of each individual chromosome into two exactly equal chromatids is complete. The chromatids are, however, joined with one another at a particular point known as the **centromere** or **spindle-attachment.** The chromosomes now start to arrange themselves around the equator of the spindle and each is attached to a spindle-fibre in such a way that both of its chromatids are fixed to the spindle by spinle-attachments.

Anaphase. The two chromatids constituting each chromosome appear to repel each other, so that they now become separated. Each chromatid is pulled out into a loop which tends to move away from its counterpart. This process is assisted by longitudinal extension and transverse constriction of the spindle, forming the so-called **stem body** (Fig. 170 H). In this manner, two constituent chromatids of each chromosome start movement and finally reach opposite poles of the spindle. This ensures equal distribution of chromosomes at the two poles.

Telophase. The collection of chromatids, at each pole, now becomes enclosed in a reformed nuclear membrane. The nucleolus reappears. The chromatids are slowly rehydrated and thus they are gradually transformed into a new set of invisible chromosomes. This is followed by **cytokinesis,** that is division of the cell body. When the nuclear reconstruction is in progress, the cytoplasm divides into two by the formation of a slowly deepening constriction along a plane equidistant between the two centrioles, so that two new cells, each with its own nucleus and central body, are finally formed.

Significance of mitosis. The chromosomes are the means by which hereditary factors are transmitted from parent to offspring. If the daughter cells are to have the same characters as the parent cell, these determining factors must pass equally to both of them. The hereditary or genetical factors have been located in the chromosomes with the help of the modern electron microscope. They are designated as the **genes.** The genes are arranged in a linear series along the body of each chromosome. It is evident that when a particular chromosome splits longitudinally into two daughter chromatids the genes are reduplicated and shared equally by the two. It follows, therefore, that the nuclear changes associated with mitosis are effective means by which *equal, qualitative and quantitative inheritance of the genes* between the daughter cells is ensured.

Meiosis

This is a special type of nuclear division which occurs during the formation of germ cells, that is, spermatozoa and ova. It has been found out that the nucleus of a mature germ cell contains only half the usual number of chromosomes that are found in a somatic cell of the species. To be more precise, the number of individual chromosomes, in the somatic (*soma*=body) cell of a particular species, are expressed as $2n$ and represents **diploid** number; the germ cells of that particular species, on the other hand, contain only n chromosomes, which represents the **haploid** number. It has been repeatedly pointed out that higher animals reproduce sexually and that the essential feature of fertilization is a complete fusion of the two nuclei—one derived from the male and the other from the female. If the male and female germ cells contribute a full diploid number, there will be doubling of the chromosomes at the time of fertilization and the somatic cells of the new-born individual will present double the usual number of chromosomes found in their parent. This is absurd, because the chromosome number is fixed for a particular species. Obviously, therefore, the diploid number is reduced to the haploid state during the formation of the germ cells and the diploid condition is restored by fertilization.

Meiosis involves two divisions, in the course of which the chromosomes are totally reduced to half. For the sake of convenience, each division is arbitrarily subdivided into a number of phases.

In the **first meiotic division** the **prophase** covers a number of changes. It is very much different from the prophase of ordinary somatic mitosis and is itself subdivided into four stages.

(i) At the onset the full diploid number of chromosomes appear in the same way as in ordinary somatic mitosis but they remain single and do not show any sign of division into chromatids. This constitutes the **leptotene stage.**

(ii) Homologous chromosomes, that is chromosomes of the same size and shape, now come to lie side by side in contact with one another. Thus a number of double threads are formed, each component of which is a chromosome, and not a chromatid. The connection between each pair of homologous chromosomes is so intimate that only the haploid number is visible but apparently each chromosome is really double or *bivalent*. This is the **zygotene stage.**

(iii) The chromosomes become shorter and thicker due to condensation of chromatin and dehydration. The relation between the homologous pairs becomes still more intimate and they appear to coil round one another. This is known as the **pachytene stage.**

(iv) This is followed by division of each individual chromosome into two chromatids. Consequently each component of the bivalent splits into two, producing a quadrupling of threads which now tend to separate. This is the **diplotene stage.** During this period of separation which is by no means complete, cross-connection or *chiasma* might develop between the four chromatids of a bivalent chromosome so that *crossing over* or interchange of genes might occur.

This is followed by **metaphase.** A nuclear spindle now appears, as in mitosis, and the threads tend to arrange themselves at its equator. The nuclear membrane disappears. The chromosomes develop spindle-attachments by which they are fixed to the nuclear spindle; but the chromatids of each individual chromosome still remain attached to one another. Metaphase comes to an end with the formation of the *equatorial plate.*

During **anaphase,** the two homologous chromosomes each consisting of its two chromatids are drawn apart due to separation of the spindle-attachments. They travel along the spindle, each towards its corresponding pole. Thus one pair of united chromatids from each quadruplet moves towards one pole of the spindle and the other pair to the opposite pole.

The **telophase** now occurs. Cytoplasmic constriction occurs as in somatic mitosis, resulting in the division of the cell body into two parts, each with only half the number of chromosomes. The nuclear membrane may now appear and the two daughter cells enter a brief period of **interphase** (resting period). In some cases, however, the nuclear membrane never reappears and the daughter cells immediately proceed with the second meiotic division.

In the **second meiotic division,** the prophase is very short. A nuclear spindle reappears and the chromosomes, each consisting of a pair of chromatids, again become visible. The chromatids are attached to the spindle by the spindle-attachments. The metaphase, anaphase, and telophase which follow in quick succession are exactly like those of the ordinary mitosis. It is to be noted that in this case one chromatid of each chromosome travels to one pole and the other chromatid to the opposite pole.

Meiosis, therefore, is characterised by two divisions. In the first division only half of the chromosomes, that is one from each homologous pair, pass into each daughter nucleus. In the second division each chromosome splits into two chromatids such as would occur in an ordinary mitosis. Four cells are thus produced, each cell having haploid or reduced number of chromosomes.

ANIMAL TISSUES

A multicellular animal is composed of different kinds of cells. Each kind is destined to perform a particular function. They may serve for protection or for support or for movement or for conduction of impulses. A collection of similar cells along with their products and performing the same function is known as a **tissue.**

Histology is the study of the structure and mode of arrangement of tissues. The four kinds of animal tissues have already been discussed in an elementary way (page 12). These are : (1) epithelial tissues, (2) connective tissues, (3) muscular tissues, and (4) nervous tissues.

Epithelial tissues

Epithelial tissues cover the surface of the body, outside as well as inside; hence the cells are arranged in the form of a membrane.

The epithelium is said to be **simple** when the cells composing them are arranged in only one layer. It is said to be **stratified** when the cells are several layers in thickness as in the epidermis of the skin of toad (Fig. 6). In an epithelium, the cells are tightly packed and they stick together by their borders. There is no intercellular space and very little ground substance, but the cells are often supported on a structureless basement membrane. The individual cells may be flattened and scale-like, or cube-like or cylindrical, thus forming squamous, or cubical, or columnar epithelium respectively.

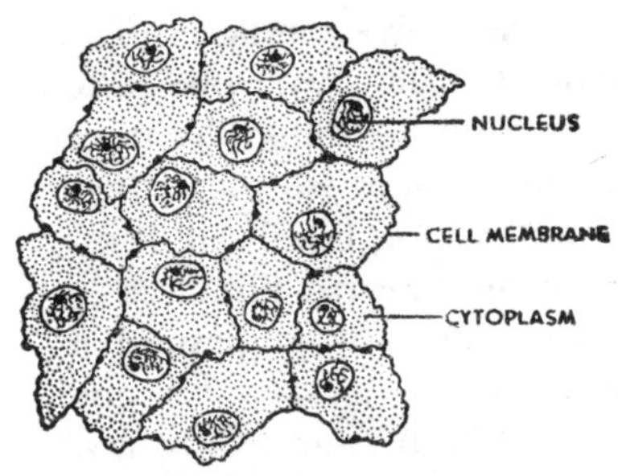

Fig. 171. Squamous epithelium.

Squamous epithelium is composed of thin flat cells, arranged like the tiles on a roof; such cells form the *peritoneum* lining the coelom of a toad and the *endothelium* which lines the inner surface of a blood vessel. (Fig. 171).

Cubical epithelium consists, as the name implies, of cube-like cells which appear to be polygonal in surface view. It is found lining the alveoli of the salivary glands and the vesicles of the thyroid; it forms the liver in part, and the uriniferous tubules of the kidney. (Fig. 3).

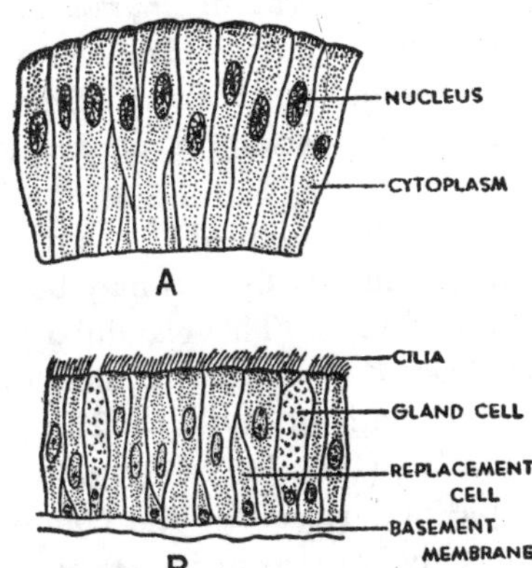

Fig. 172. A=Columnar epithelium, B=Ciliated epithelium.

Columnar epithelium (Fig. 172A) consists of tall pillar-like cells lining the interior of the alimentary canal from the stomach to the rectum. The nucleus is elongated and the free surface of the cells, that is the surface away from the basement membrane, has a striated border for facilitating absorption. The intestinal and the gastric glands are lined by columnar cells containing mitochondria which actively secrete digestive enzymes.

Ciliated epithelium (Fig. 172B) is composed of columnar or cubical cells bearing numerous protoplasmic threads or cilia on their free surface. The hair-like cilia can move back and forth rhythmically, producing a very peculiar *ciliary motion* which has been compared to the waves produced in a paddy-field by the passage of a current of breeze. At the base of each cilium there is a dark basal granule. In between the ciliated cells are found *gland cells*, secreting mucus, and *replacement cells* which are transformed into ciliated cells, if necessity arises. Ciliated epithelium is found in the mucous membrane of the respiratory tract where movement of the cilia serves to remove foreign particles.

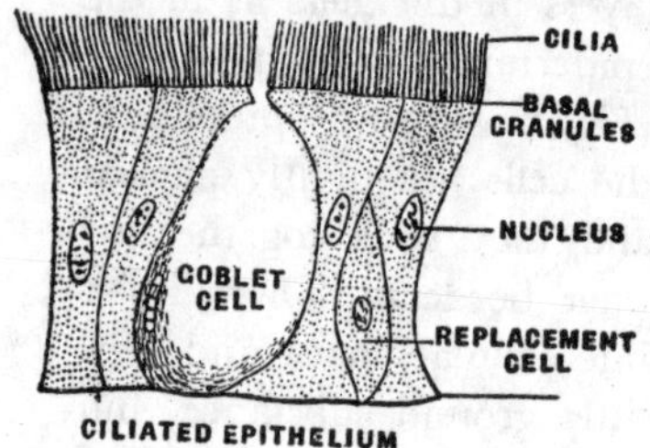

Fig. 173. Ciliated epithelium with goblet cell.

Epithelial tissues may be *protective*, *glandular* or *sensory* in function.

Protective epithelium saves animals from external injury. It is *simple*, that is one-layered, as in the epidermis of lower forms such as leech or earthworm. It is many-layered or *stratified* in the epidermis of higher forms such as toad or guinea-pig, where the basal layer, by repeated mitosis, produces successive layers of cells. The superficial cells flatten out and become *cornified* as they are pushed towards the outer surface.

Glandular epithelium is specially adapted for secreting useful substances or excreting waste products. Glands may be unicellular, *e.g.* **goblet cells,** or multicellular. The glandular cells are either columnar or cubical. They do not form a membrane but are aggregated into masses with or without cavities. The cavities of a gland are known as the *alveoli*. (Introduction, Fig. 7).

Sensory epithelium is simple and resembles the columnar type. Each cell usually bears a sensitive process at its free surface, and is connected by a sensory nerve to the central nervous system. The olfactory epithelium lining the nasal passages is a sensory epithelium (Fig. 41 D, E). Scattered sensory cells, occurring singly or in groups, are abundantly found on the outer epidermis of multicellular animals.

Connective Tissues

The connective tissues serve to support the body, and bind other tissues and organs together. In contrast to the epithelium, in which cells are abundant and the intercellular ground substance is practically wanting, the connective tissues possess only few cells but considerable amount of intercellular matrix. Depending upon the nature of the ground substance, three kinds of connective tissues can be recognised : (1) connective tissue proper, (2) skeletal tissues, and (3) vascular tissues.

(1) **Connective tissue proper** includes the *areolar* and *fibrous tissues* both of which are mainly concerned with binding parts of the animal body. The **areolar tissue** forms a continuous layer beneath the skin and is employed as an effective packing material to fill up gaps between organs. It is composed of a thin jelly-like ground substance in which are scattered a few cells. Fine fibres form an interlacing network throughout the matrix, and the cells are contained within the meshes of the fibres. The cells lying in the ground substance are principally of two kinds—the **fibroblasts** which secrete the fibres, and the *amoeboid* **matrix cells** which probably secrete the jelly. The fibres, too, are of two types : slender **yellow fibres,** branching freely and joining up to form the elastic network, and the wavy **white fibres arranged** in bundles which yield gelatin when boiled with water (Fig. 174).

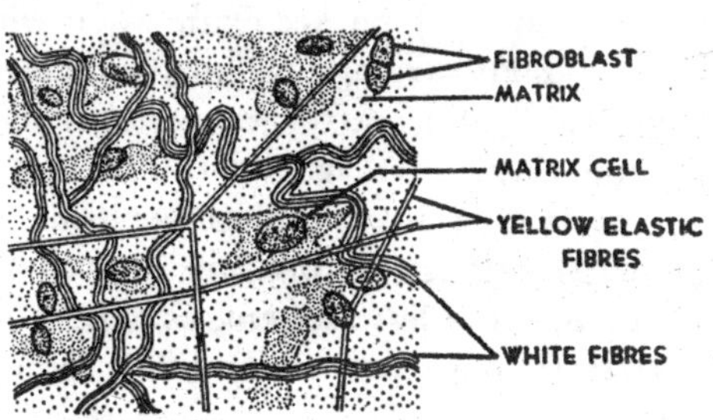

Fig. 174. Connective tissue proper.

Fibrous tissue forms ligaments and tendons which help in connecting the framework of the body. It also forms sheets for enclosing muscles. As its name signifies, it is composed mainly of fibres.

Adipose tissue is a kind of areolar tissue containing fat cells.

(2) **Skeletal tissues** include *cartilage* and *bone* which form the vertebrate endoskeleton.

Cartilage is popularly known as the *gristle*. It is a rigid and elastic connective tissue which can resist compression and absorb mechanical stress. Cartilage consists of a ground substance or matrix which is typically composed of an organic matter called *chondrin*. The *cartilage cells* secreting the matrix are embedded in spaces known as the *lacunae* (singular —*lacuna*). The cells usually occur in groups of two or four, and they have a tendency for multiplying. The entire cartilage is surrounded by a tough coat of fibrous areolar tissue called *perichondrium*.

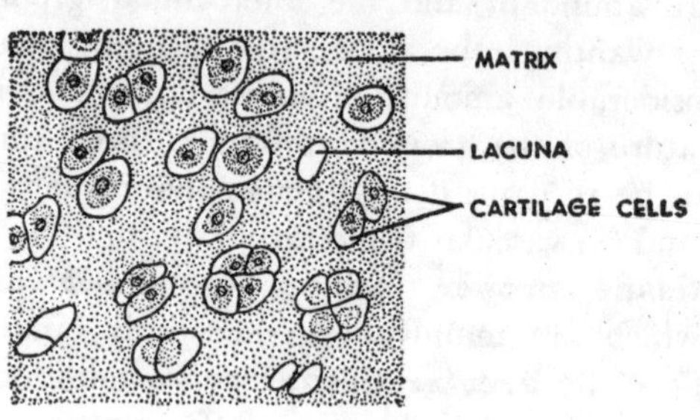

Fig. 175. Cartilage.

There are several kinds of cartilage, the commonest and simplest being **hyaline cartilage.** In this the matrix is clear and homogeneous. In **elastic cartilage** the matrix is permeated with innumerable elastic fibres, and in **white fibro-cartilage** the matrix is densely packed with wavy white fibres running in bundles. Costal cartilage in mammals and the cartilage covering joints are hyaline cartilages, whereas the pinna of the ear is composed of elastic cartilage.

Bone is a hard connective tissue of great strength, offering greater resistance to pressure and to bending stress than the cartilage; but it is less resilient, that is, it does not spring back when released from a stress. Like other connective tissues, bone is composed of cells and of intercellular matrix.

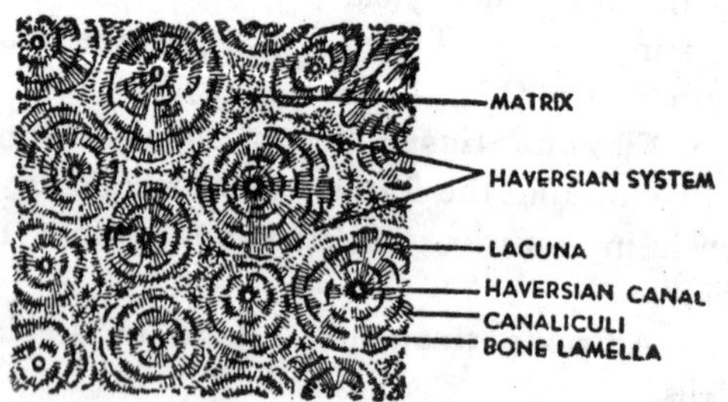

Fig. 176. Bone.

The ground substance or matrix is the hard material of the bone; it always contains calcium phosphate. The *bone cells* are arranged in regular pattern inside *lacunae*. There is a large central cavity containing bone-marrow. This is surrounded by *bone lamellae* which are arranged in concentric layers. The marrow cavity communicates with smaller channels called *Haversian canals*, around which are arranged further lamellae and lacunae, thus forming elongated cylinders called Haversian systems. The lacunae are connected with one another by narrow *canaliculi*. The bone cells or *osteoblasts* are spider-shaped. They send out fine protoplasmic processes through the canaliculi and these interlace with others from adjacent lacunae. The entire bone is surrounded by a fibrous sheath known as the *periosteum*, which carries blood vessels to the interior of the marrow cavity.

(3) **Vascular tissue** includes *blood*, *lymph* and the structures in which these fluids are formed, stored and destroyed. But in the following paragraph only blood will be described. **Blood** differs from all other tissues in consisting of a liquid matrix or *blood plasma*, in which *blood corpuscles* are suspended. It must be noted that the *plasma* is not a product of the corpuscles. It is a straw-coloured liquid consisting of an aqueous solution of salts and organic substances. Floating in the plasma are the blood corpuscles. (See Figs. 26 and 157.)

The red corpuscles or *erythrocytes* contain a red pigment, called *haemoglobin*. In mammals, each red corpuscle is a biconcave, non-nucleated disc about 7.5 μm in diameter. In fishes and amphibians, however, these are oval, biconvex and nucleated.

There are about 5 millions of erythrocytes in a cu. mm. of human blood.

The white corpuscles or *leucocytes* are larger and less numerous. They are devoid of haemoglobin and they have a nucleus which may be of irregular shape. Number of leucocytes varies from 6 to 8 thousand per cu. mm. in human blood. Some of the leucocytes are *phagocytes*, devouring bacteria with their pseudopodia.

The *thrombocytes* or blood platelets are numerous small cells of irregular shape which help in the clotting of blood.

Muscular Tissues

The cells comprising the muscular tissues have become highly specialised in adaptation to their peculiar contractile function. They are concerned with movements of parts and may cause the body to move as a whole. This is effected by contraction of muscles which are composed of *muscle fibres.* There are three kinds of muscle fibres : striped, unstriped, and cardiac.

(1) A **striped** or **striated muscle fibre** (Fig. 177) is elongated and cylindrical in shape. It is invested in a thin membrane known as the *sarcolemma.* Each fibre is a *coenocyte,* that is, a multinucleate mass of protoplasm. The nuclei are elongated and scattered throughout the length of the fibre. A fibre is marked by alternate light and dark bands, which occur transversely. Faint longitudinal striations, passing from end to end, may also be detected. There are the *myofibrils* or *sarcostyles.*

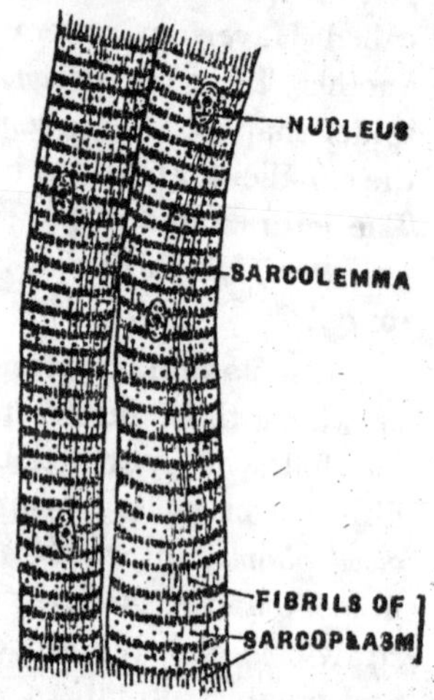

Fig. 177. Striated muscle fibres.

A muscle, in the anatomical sense of the word, is composed of muscle fibres bound together by connective tissue into bundles, called *fasciculi.* Each fasciculus is invested with a muscle sheath, the *perimysium.* The striated muscles are spoken of as **voluntary** muscles because they may be contracted by wish. They are innervated by branches from the central nervous system and constitute the main mass of *skeletal muscles.*

(2) The **cardiac** muscle is a modified form of striated muscle confined to the tissues of the heart. These fibres are short and branched—the branches forming a network or *syncitium.* A cardiac muscle fibre is not a coenocyte, because each is a uninucleate mass of protoplasm.

(3) An **unstriped** or **smooth muscle fibre** (Fig. 178) consists of a spindle-shaped elongated cell. There is no sarcolemma and each fibre contains only one nucleus. Cross striations are absent but the fibre is traversed from end to end by delicate contractile threads, known as *myofibrils.* The unstriped muscles are **involuntary,** since they contract independently of the 'will'. They are innervated by branches from the sympathetic nervous system and restricted to the wall of the visceral organs and blood vessels.

The cardiac muscle, though striated, belongs to the group of involuntary muscles.

Nervous Tissues (Fig. 179)

This is a highly specialised tissue for reception, transmission, and discharge of stimuli. It keeps the animal informed about its environment and conducts impulses from one part of the body to another. Nervous tissue is composed of *nerve cells* and *nerve fibres*.

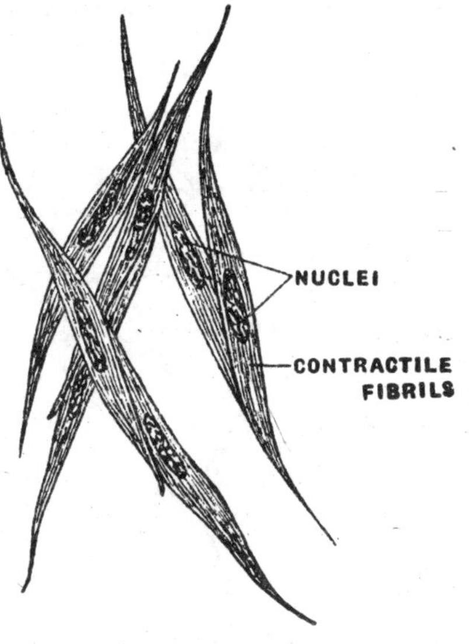

Fig. 178. Unstriped muscle fibres.

A **nerve cell** or **neurone** has a large cell-body with a number of cell-processes. The nucleus is oval and the cytoplasm contains numerous dark-staining *Nissl's granules*. The cell-processes are of two kinds. Of these, the one that is long and unbranched, is known as the *axon;* it carries impulses away from the cell. The much-branched, short processes arising from the neurone are the *dendrites*. They carry stimuli into the cell-body. The axon of the neurone is continued forwards as the central core or *axis cylinder* of a nerve fibre. It either ends in close contact with a muscle fibre or with the dendrites of another neurone. A link is produced in this way for shunting on impulses from cell to cell. This is known as the *synapse*. Scattered amongst the neurones there are undifferentiated supporting cells called *neuroglia*.

A **nerve fibre** consists of a central core or *axis cylinder* formed by the axon of a neurone. Covering the axon is a delicate sheath, known as the *neurilemma*. Just beneath the neurilemma lies a very thin layer of protoplasm containing nuclei at intervals. These are the nuclei of the neurilemma, which are concerned with the formation of the sheath.

Nerve fibres are of two kinds : **medullated** and **non-medullated.** In the former, the axis cylinder is separated from the neurilemma by a layer of fatty substance called *myelin sheath* or *medullary sheath*. The medullated fibres are white-coloured. The medullary sheath is constricted at intervals to form the *node of Ranvier*. In the **non-medullated** fibres the myelin sheath is absent. Consequently they appear to be gray-coloured. **Nerves** are formed by a collection of nerve fibres, bound together by a connective tissue sheath, the *perineurium*.

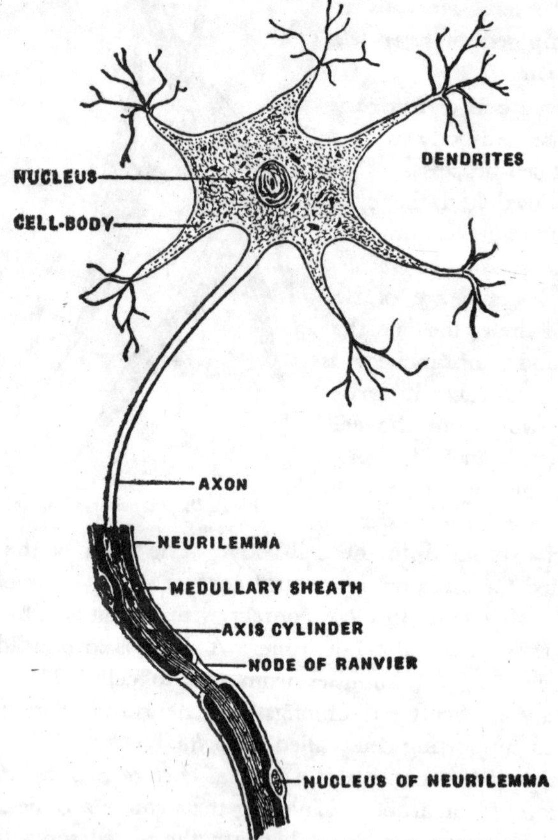

Fig. 179. A neurone with a portion of nerve fibre (diagrammatic).

Non-medullated nerve fibres are restricted chiefly to the sympathetic nervous system, whereas the medullated fibres occur in the various parts of the central nervous system.

TISSUE SYSTEMS OR ORGANS

In higher animals the tissues are associated to form organs. Histology of the organs have been previously discussed while describing them in the animal body. Histological structure of the mammalian small intestine is described here as an example.

The wall of the mammalian small intestine is composed of four layers: (1) The outer layer consists of a row of *epithelial cells* forming the **serous coat.** It is derived from the peritoneum which lines the body cavity. (2) Beneath the serous coat lies the **muscular coat.** It is composed of a mass of *unstriped muscle fibres* arranged in two layers : (*a*) a superficial zone of longitudinal muscles and (*b*) a deeper zone of circular muscles. (3) Just beneath the circular muscles lies the **submucous coat** which consists of *connective tissue* and contains vessels and nerves supplying the intestine. Embedded in the submucosa, there are numerous *Brunner's glands* which open by ducts into the lumen of the gut. In the deeper layers of the submucous coat is a layer of *involuntary muscles* constituting the *muscularis mucosae.* (4) The interior of the gut is lined by a layer of *columnar epithelium* forming, the **mucous membrane** or **mucous coat.** This is raised into finger-like processes or *villi* which project inwards. Brunner's glands open into crypts, in between the villi.

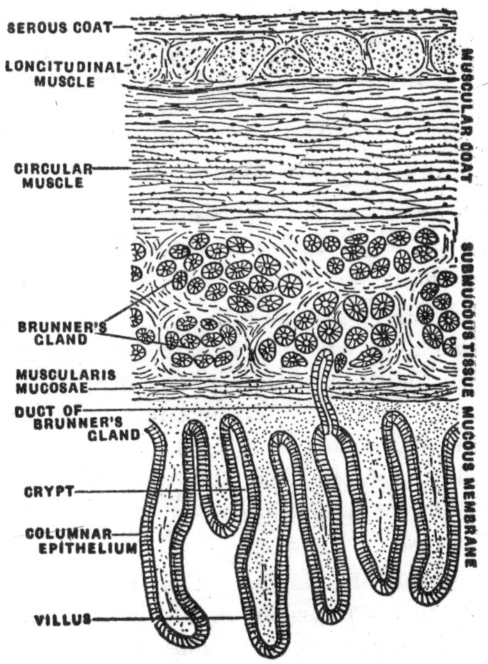

Fig. 180. Part of a transverse section of the duodenum of a mammal (diagrammatic).

CHAPTER XIX

REPRODUCTION AND DEVELOPMENT

(A) REPRODUCTION

A new animal comes into existence by the transformation of some part of a pre-existing animal; this is **reproduction.** The animals reproduce their kind by two fundamental methods : (1) *asexual*, and (2) *sexual*. **Asexual reproduction** involves only *one parent* and no special reproductive structures. **Sexual reproduction** usually involves *two parents* and the union of two germ cells, or of two cells of some kind, or of two nuclei derived from different cells.

Asexual Reproduction

It occurs only in simpler and lower forms of animals. Vertebrates never reproduce asexually. The more important types of asexual reproduction are : ***binary fission***, ***multiple fission***, and ***budding***.

Binary fission : In this an individual divides into two *equal halves* after which each part grows to the original form. When occurring in a protozoon, the division of the cell body is always preceded by the division of the nucleus. Fission may occur in a transverse plane, as in *Amoeba* and *Paramoecium*, or it may be longitudinal, as in *Euglena*. The most important feature of fission is that the parent disappears as an individual, and the two new individuals take its place.

Multiple fission : This is found only in protozoa where the nucleus fragments repeatedly and then the cytoplasm divides, so that a part of it surrounds each of the nuclear fragment. Each bit of nucleated fragment is virtually an asexually produced spore; hence multiple fission is also known as **sporulation.** Multiple fission is usually preceded by encystment as in *Amoeba*. In *Plasmodium*, however, there is no encystment.

Budding : When an animal divides into two *unequal* parts, it is said to reproduce by budding. The larger portion is regarded as the parent and the smaller one as the offspring. A **bud** usually arises as a small protuberance from the body of the parent; it grows larger and develops similar parts. It may remain attached to the parent or it may separate and live independently. Budding is the usual method of reproduction in *Hydra*.

Sexual Reproduction

Higher animals usually reproduce sexually; each new organism originates from the union of two **germ cells.** These are the male and female gametes, and are known respectively as **spermatozoa** and **ova.** The gametes are produced in *gonads*, the ova in *ovaries* and the spermatozoa in *testes*. There is a great interest in the ultimate origin of the gametes, because they serve to transmit hereditary characters from the parent to the offspring. August Weismann (1834–1914) held that the gametes or *germ-plasm* are totally kept apart from the influence of other body cells or *Somatoplasm*, and are transmitted continuously from generation to generation. It has been definitely established that at least there is a continuity of the chromosomes by which the hereditary characters are transmitted from parents to offspring.

The germ cells are derived from the *germinal epithelium* of the gonads. The *primordial* germ cells are known as **spermatogonia** in the male, and as **oogonia** in the female. At the approach of sexual maturity these cells multiply rapidly by mitosis and the nucleus of each cell contains a dual set of chromosomes, which is the **diploid** number and is represented as $2n$. The process by which the spermatogonia become spermatozoa and the **oogonia** become ova is known as **gametogenesis.** It involves two successive divisions—one of which is a reduction division or *meiosis*. As the gametes of the two sexes differ in form, size, and behaviour, the gametogenesis, therefore, occur differently in the two; but the meiotic changes associated with the formation of the gametes are identical.

Spermatogenesis : In male, the spermatogonia are converted into full-grown cells known as the **primary spermatocytes,** each of which has $2n$ chromosomes. Every primary spermatocyte divides twice in quick succession. The first division is meiotic, and produces two **secondary spermatocytes,** each having n **chromosomes** in its nucleus.

During the following mitotic division each secondary spermatocyte divides into two **spermatids.** Thus a primary spermatocyte yields four spermatids, each containing *n* chromosomes, which represents the **haploid number.** It may be pointed out here that from each pair homologous chromosomes in a primary spermatocyte, any one spermatid receives only one. The spermatids undergo a process of metamorphosis and are converted into the **spermatozoa.** The whole process is known as **spermatogenesis.**

Sperm : The **spermatozoon** is small and motile. It is usually thread-like in appearance and consists of a **head,** a **middle piece** and a **tail.** The oval head represents the *nucleus* and the middle-piece encloses the *central-body.* The flagella-like wavy tail is composed of *cytoplasm.*

Oogenesis : In females, the **oogenesis** is slightly different The oogonia are converted into **primary oocytes** with 2*n* chromosomes. Each primary oocyte undergoes two successive maturation divisions of which the first one is meiotic. In the first division it is converted into two unequal cells—a large **secondary oocyte,** and a small **first polar body** with very little cytoplasm. Later division results in the formation of a **second polar body** and the *mature* **ovum.** The polar bodies are useless and never function as gametes. Thus each primary oocyte yields but one mature ovum, containing *n* chromosomes; yet the nuclear changes that produce the ovum and the polar bodies are equivalent to those by which four spermatozoa are derived from a primary spermatocyte.

Ovum : The *mature* **ovum** is a spherical or oval, non-motile cell. It usually contains a variable quantity of *yolk* to nourish the future embryo. The hen's ovum is about 30 mm. in diameter, whereas a toad's ovum is only 2 mm. The human ovum contains very little yolk; hence it is as small as 0.15 mm. in diameter. When the quantity of yolk present in the ovum is very large, it is usually located at one pole of the cell, the *vegetative pole;* such an ovum is said to be **telolecithal.** If, on the other hand, the quantity of yolk is very small, the ovum is said to be **alecithal** or **microlecithal.** The ovum is surrounded by a thin *vitelline membrane* which is derived from the oocyte. There is a prominent *nucleus* embedded in the centre of the *cytoplasm.* A *central body* is absent.

Fertilization : The union of a mature spermatozoon with a mature ovum is known as **fertilization,** and the resulting cell is a **zygote.** In fertilization, the fusion takes place primarily between the two nuclei, each contributing *n* chromosomes. This restores the normal *diploid number* (2*n*) of the species, in the nucleus of the zygote. Meeting of the ovum and spermatozoon always takes place in a fluid medium in order to permit the *active* movements of the motile male gamete. The ovum remains *passive* during the whole period. There is evidence to believe that the outer layer of the ovum secretes a chemical substance called *fertilizin* which attracts spermatozoa by positive **chemotaxis.** On reaching the ovum, the head of the spermatozoon penetrates through the vitelline membrane along with the middle-piece; the tail, however, is left behind. A central body arises from the middle piece of the sperm and soon forms the spindle. The male and female *pro-nuclei* now break up into chromosomes which take their position in the equator of the spindle. During this process the homologous chromosomes lie side by side in pairs. In every such homologous pair, one chromosome is contributed by each *pro-nucleus.* The fertilized ovum or zygote now divides into cells, containing 2*n* chromosomes. The products of division are known as the **blastomeres** which ultimately develop into a young of the same species.

Special types of Sexual Reproduction

In some protozoa, sexual reproduction involves union between two similar cells which are known as *isogametes.* The process is known as **conjugation** and is found in *Monocytis.* In other protozoa, such as *Plasmodium,* the gametes are dissimilar and hence they are *anisogametes.* The smaller and motile *microgamete* fertilises the larger and passive *macrogamete.*

In *Paramoecium,* two individuals *conjugate temporarily* and make an exchange of nuclear material. After this they separate and reproduce by *binary fission.*

Parthenogenesis : Sexual reproduction usually involves two parents and the fusion of two germ cells. An ovum, however, may develop into a young individual without being fertilized by a sperm. This peculiar phenomenon is known as **parthenogenesis** (*parthenos*=virgin; *genesis*=origin).

Parthenogenesis occurs *normally* in ants, bees, wasps and aphids. The female of these animals lay both fertilized and unfertilized eggs. She can control fertilization by releasing or retaining the sperms which are stored in her spermatheca. In bees the males are derived from the unfertilized eggs, and the females from the fertilized eggs. In certain insects, having no males, reproduction is entirely parthenogenetic.

Parthenogenesis can be induced *artificially* in the egg of many animals which ordinarily require fertilization. Thus shaking the eggs vigorously, heating them, or pricking them with a needle, all have started development in certain eggs. Loeb, by pricking thousands of frog-eggs with pointed needles obtained over two hundred tadpoles, and reared nearly one hundred frogs. Recently a rabbit-egg has also been developed parthenogenetically to produce a young rabbit.

Although parthenogenesis involves only one parent, yet by origin the cells giving rise to the new individuals are **ova.** It is for this reason that parthenogenesis is regarded as a kind of sexual reproduction.

Paedogenesis : Sexual reproduction is usually carried on by adult animals. But this is not always the case. There are certain species who have remarkable power of reproducing sexually when they are in the *larval* state. This peculiar sexual reproduction by a larval animal is known *paedogenesis*. The **axolotl larva** of the tiger salamander attains sexual maturity and breeds under certain conditions. Larva of certain flies produce ova which may develop by parthenogenesis. This is an instance of *parthenogenetic paedogenesis*.

(B) DEVELOPMENT

As a result of sexual reproduction a new individual starts life as a single cell. It is the fertilized or activated eggs, or zygote. The zygote by repeated mitotic division produces many cells which in due time differentiate into the tissues and organs of the developing embryo. This process is known as *embryogeny* and the branch of zoology which deals with the development of individuals is known as **embryology.** The following is a generalised account of the early development of an embryo.

Cleavage

Soon after an egg is fertilised, the single-celled zygote begins to divide by mitosis. The first division results in two cells, the two divide into four, and so on. The process is known as **cleavage** and it results in the *segmentation* of the egg into a large number of smaller cells, each of which is a **blastomere.** The solid cell mass, thus produced, is known as the **morula.** If the ovum is *alecithal* or *microlecithal*, the cleavage is total or **holoblastic.** If, on the other hand, the ovum is *telolecithal*, then the cleavage is **meroblastic** and the blastomeres are unequal.

Blastula

As cleavage continues, the blastomeres become arranged in the form of a hollow ball, or **blastula.** The cavity of the blastula is designated as the **blastocoele,** or segmentation cavity.

Gastrula

The blastula now undergoes **gastrulation.** This is a complicated process by which the hollow ball of cells is converted

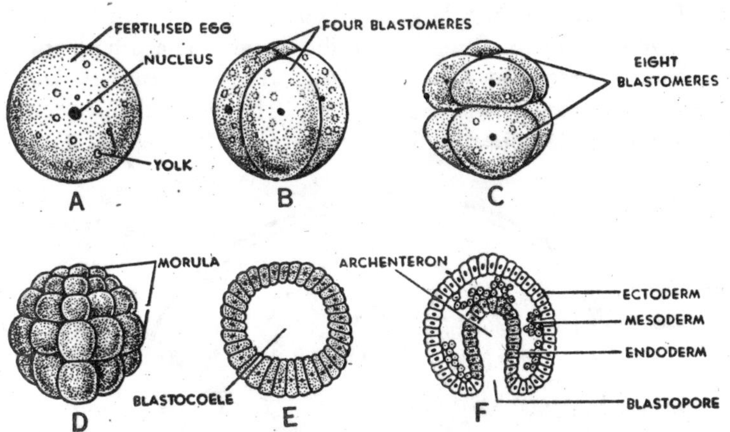

Fig. 181. Stages indicating early development of a microlecithal egg; E—section of blastula, F—section of gastrula.

into a double walled cup, the **gastrula.** The **blastocoele** is gradually obliterated by pushing in of one side of the ball into the other, that is by **invagination.** The cavity of the cup constitutes the primitive gut or **archenteron** which communicates with the exterior by the **blastopore.** When complete, the gastrula consists of (1) an outer layer of cells, the **ectoderm,** (2) an inner layer of cells, the **endoderm,** and between these two layers develops (3) a third cell-layer, the **mesoderm.** These are the three primary **germ layers** which by modification and orientation produce the various tissues and organs. The **ectoderm** forms the outer covering of the body, the nervous system and the sense-organs. The **endoderm** is transformed into the lining membrane of the digestive canal and the other structures associated with it. The **mesoderm** produces the supporting tissues, vascular tissues, muscles, and epithelial lining of the body cavity.

The preceding description of the formation of a new individual is more or less generalised. An outline of the development of a chick-embryo will now furnish a good example to illustrate the principles of embryology.

(C) DEVELOPMENT OF THE CHICK

The Egg

The hen produces a large egg containing an enormous quantity of yolk. This is a single cell—the ovum. The protoplasm of the ovum is restricted to a small area at the *animal pole*, and is known as the *blastoderm* or *germinal disc*. The rest of the ovum consists of yolk and represents the *vegetative pole*. The entire ovum is

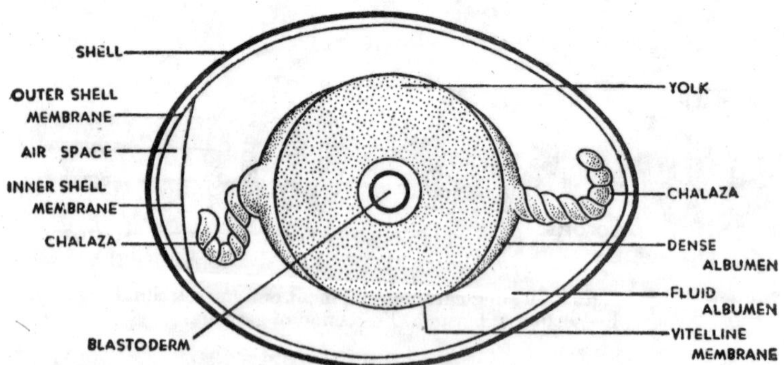

Fig. 182. A diagram of the hen's egg at the time of laying.

surrounded by a thick and porous *vitelline membrane*. Because of the enormous yolk-content, the hen's ovum is an extreme example of *telolecithal* egg.

The mature ovum drops into the body cavity by rupturing the Graafian follicle. It then finds its way into the oviduct where it is fertilised by a sperm.

Fertilization

Fertilization is internal and occurs at the upper part of the oviduct. A sperm penetrates through the vitelline membrane and enters the germinal disc. The nuclei of ovum and sperm then come together and join to form the chromosomes of the *zygote.* The mature germ cells of the domestic fowl is *haploid*, each containing only nine chromosomes. When the male and female pro-nuclei unite, the fertilized ovum or zygote becomes *diploid* and contains eighteen chromosomes.

As the fertilized ovum descends along the oviduct it receives a series of envelops : the *white* or *albumen*, the *shell membranes*, and the *shell*. The **white** is chiefly composed of a mixture of proteins and water. It is deposited in layers, of which the innermost is pulled out at opposite poles to form a pair of spirally twisted cords, the *chalazae* (sing.—*chalaza*). The albumen is covered by two layers of **shell membranes** which separate from one another at the broad end of the egg, thereby enclosing an *air-space* from which oxygen is supplied to the developing chicken. Lastly, there is the hard porous **shell** composed of calcium carbonate which encloses and provides mechanical protection for the soft contents.

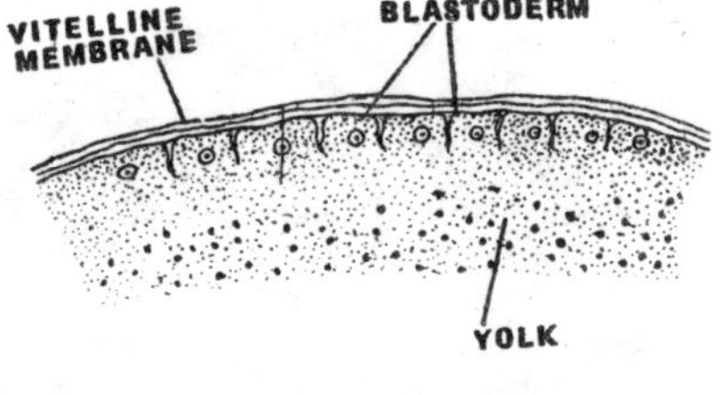

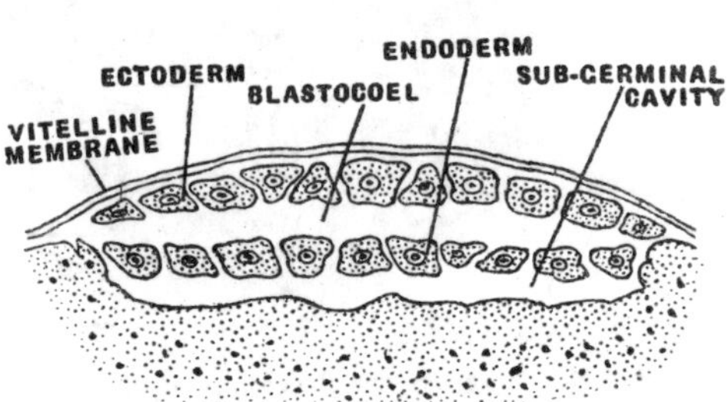

Fig. 183. Sectional view of the blastoderm (above) and formation of blastula (below).

Cleavage

Cleavage starts almost immediately after fertilization, while the egg is still descending along the oviduct; by the time when the egg is laid, it has been in action for nearly twenty-four hours. So great is the quantity of yolk that cleavage is only partial or *meroblastic.* It starts at the animal pole and the *germinal disc* divides into a number of cells or *blastomeres.* The yolk, however, remains undivided. This small group of cells is the **blastoderm.** It is a flat disc whose margin rapidly grows over the surface of the yolk. Beneath the blastoderm is a cavity which separates it from the underlying yolk. This is the **blastocoele** or *subgerminal cavity*. The embryo has now reached the **blastula stage.**

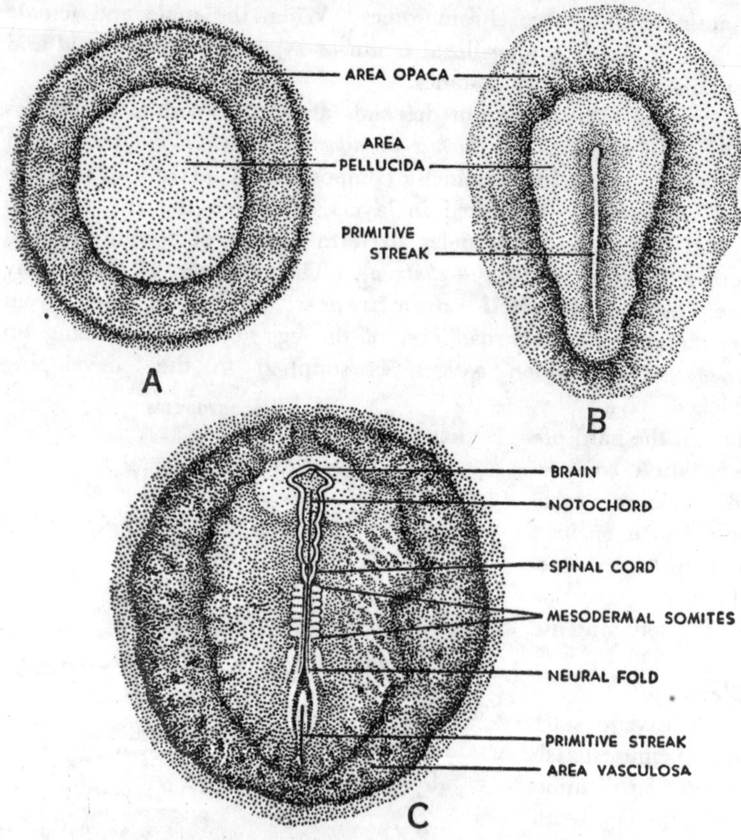

Fig. 184. Stages in the development of chick; A—Surface view of the blastoderm; B—Surface view of area pellucida with a primitive streak (about 18 hours old); C—An embryo with 7 pairs of somites (about 26 hours old).

On surface view, the central region of the blastoderm overlying the subgerminal cavity looks transparent and is, therefore, called **area pellucida;** but the peripheral part of the blastoderm which lies directly on the yolk appears opaque and is, therefore, known as the **area opaca.** The embryo proper is formed from the area pellucida, whereas the area opaca gives rise to accessory structures.

Gastrulation

Gastrulation begins before the laying of the egg. It is a very much modified process. Before gastrulation the area pellucida is a fine membrane, only one-cell thick. From one particular spot, near what will be the posterior end of the embryo, cells migrate into the subgerminal cavity. These cells are joined by others which split and drop down from the undersurface of the area pellucida by *delamination*. In this way a second layer of cells is formed under the area pellucida. The embryo is now *diploblastic* or two-layered and the *gastrulation* is complete. The upper cell-layer of the area pellucida may now be called **ectoderm,** and the lower cell-layer formed by the migration of cells from the area pellucida is the **endoderm.** The two **germ layers** are fused with each other at the margin of the area opaca, which in the mean time is spreading over the yolk and trying to enclose it (Fig. 183). Meanwhile the area pellucida elongates and becomes oval (Fig. 184).

The egg is laid at about this stage. The process of development is now arrested unless the temperature of the egg is kept at about 30°C.; this is usually done by the body-heat of the parents, one or both of which sit upon and *incubate* the egg until the young chicken is hatched. If the temperature falls below 28°C. further development stops resulting in abnormalities even when adequate temperature is restored after a period of interruption. Continued cooling for more than a day is usually fatal.

Formation of Mesoderm and Primitive streak

Early in the first day of incubation, a longitudinal opaque band appears in the posterior part of the area pellucida. This is the **primitive streak.** It is due to migration and accumulation of cells from each side towards the middle line. It extends forwards for a short distance but never up to the anterior margin of the area pellucida. A shallow groove is now formed along the middle of the primitive streak; this is the *primitive groove*. This indicates that the cells forming the primitive streak have a tendency to grow downwards between the ectoderm and the endoderm. These cells now multiply rapidly, and the new cells extend laterally and away from the primitive streak towards the area opaca. A third sheet of cells, the **mesoderm,** is formed between the ectoderm and the endoderm in this way (Fig. 185). It extends through the whole of the area pellucida, excepting the part lying in front of the primitive streak. At the two sides, the mesoderm enters between ectoderm and endoderm of the area opaca. By this time the area pellucida has lost its circular outline and has become pear-shaped; its broad end is the anterior end of the embryo. A solid and stiff rod of vacuolated cells, the **notochord,** is formed beneath the primitive streak. It extends forward between ectoderm and endoderm and is really formed by modification of the endodermal cells. The notochord, when fully formed, serves as the supporting axis of the developing embryo.

Formation of Neural Groove, Somites and Coelom

A wide band of thickened ectodermal cells appears in the middle line dorsal to the whole length of the notochord. This is the *neural plate.* The edges of the neural plate soon rise up as the **neural folds** and the neural plate is thus converted into the **neural groove.** At a much later date the edges of the neural folds fuse in the middorsal line and the neural groove is converted into a hollow *neural tube.* The anterior part of this tube forms the brain and the posterior part forms the spinal cord.

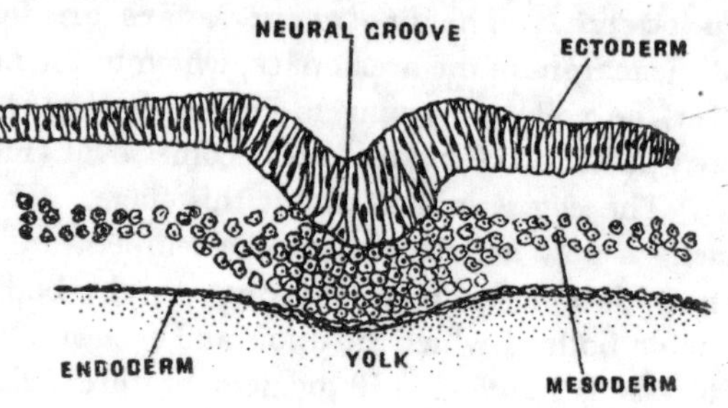

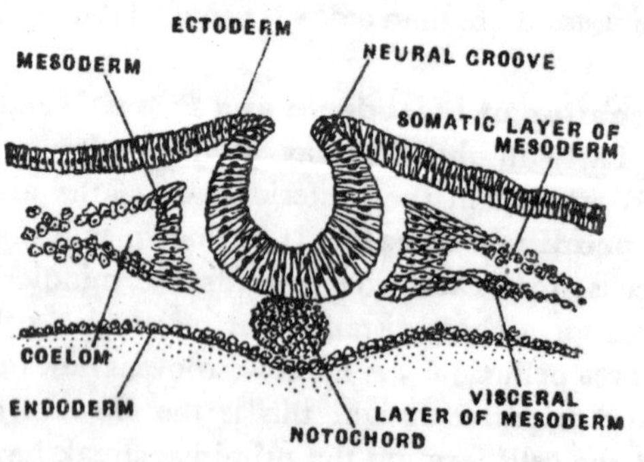

Fig. 185. Transverse sections through developing chicken, showing formation of mesoderm, neural groove, notochord and coelom.

In a transverse section cut through the middle of the embryo (Fig. 185), the notochord is seen medially, and the mesoderm extending laterally from each side of the notochord. After about twenty hours incubation, a transverse cleft appears across the mesoderm on both sides of the notochord. An hour later another similar cleft appears behind the first, and part of the mesoderm between the two clefts appears like a solid cubical block. These blocks are the first pair of **mesodermal somites.** About an hour later a second pair of somites form behind the first and then more and more pairs are cut off. At the end of 26 hours the embryo has developed 7 pairs of somites (Fig. 184 C).

During the course of the next three days about fifty pairs of somites are developed, of which the last ten pairs degenerate at a later date. These somites indicate that the chick-embryo is metamerically segmented.

Lateral extension of each somite now splits into two layers, of which the one nearer to the ectoderm is the *somatic layer* and the other near the endoderm is the *visceral* or *splanchnic layer.* The space between the two is the **coelom** (Fig. 185). The somatic mesoderm and ectoderm together make up the **somatopleur,** whereas the splanchnic mesoderm and endoderm constitute the **splanchnopleur.** The coelom extends laterally into the area opaca, that is into parts beyond the limits of the embryo. This part of coelom is known as the *extra-embryonic coelom.* As the mesoderm grows into the area opaca, its inner part becomes mottled in appearance. This mottling is due to the growth of irregular blood channels and hence the mottled area is called **area vasculosa** (Fig. 184 C).

At the end of the first day the embryo is demarcated from the rest of the area pellucida by the formation of a **head fold.** This is a downward and backward kinking of the area pellucida just in front of the neural plate. A small pocket develops under the head fold. This pocket marks the beginning of the **foregut** and it opens posteriorly into the subgerminal cavity.

The head fold is followed by the development of kinks at the two sides and at the posterior end. Thus the embryo proper is pinched up from the rest of the blastoderm and the latter develops into extra-embryonic structures, such as *yolk-sac, amnion* and *chorion.* In about four days, the embryo is completely folded up from the yolk. But owing to the smallness of space, the embryo turns over, so that it now lies with its left side apposed to the yolk and its right side facing the shell. The body becomes strongly flexed so that the head is placed close to the tail. The head is very large in comparison with the rest of the body and is provided with a pair of bulging eyes (Fig. 187). A mouth develops which is at first square, but soon the margins of the mouth grow out and a beak is formed. Three pairs of gill slits appear on the outer sides of the head; of these, the second ad third are soon covered up, but the first is converted into the Eustachian canal and communicates with the tympanic cavity. Gills, however, are never formed.

Formation of Foetal Membranes

The chick embryo is covered with a series of *embryonic membranes* or *foetal membranes* which protect it from shock and desiccation, and serve in respiration, excretion and nutrition during embryonic life. These are: (1) *yolk-sac*, (2) *amnion*, (3) *chorion*, and (4) *allantois*.

The **yolk-sac** is a thin-walled bag which encloses the yolk of the egg for nourishing the embryo. At the end of the sixth day the *area opaca* has practically spread over the entire mass of yolk and the whole area is soon vacularised and converted into *area vasculosa*. The *extra-embryonic coelom* extends right round the yolk which is thus enclosed in a bag of *extra-embryonic splanchnopleur*. This bag is the *yolk-sac*. It may be recalled that the splanchnopleur consists of the splanchnic layer of mesoderm lined internally by at layer of endoderm (Fig. 186 A). The endodermal cells digest yolk, and the products of digestion are absorbed into the *vitelline vein* which has developed in the mesodermal covering. The yolk-sac is linked with a midgut of the embryo by a narrow stalk (Fig. 186 C). By the time when the embryo hatches, the whole mass of yolk is digested and absorbed.

The **amnion** is a closed sac covering the embryo. Its cavity is filled with watery fluid, and for this reason it has been described as the 'private pond' of the embryo. The amnion protects the embryo from desiccation, friction and shock.

After thirty hours incubation the *extra-embryonic somatopleur*, just in front of the embryo rises up as a crescentic fold. This fold, being somatopleur, is covered with ectoderm and lined internally with somatic mesoderm, and its cavity is a part of the extra-embryonic coelom. It is known as the *head fold of the amnion*. As it grows backwards, it covers the embryo with a double roof of *somatopleur.* During the third day a *tail fold of the amnion* appears. It resembles the head fold in structure but grows forward. The

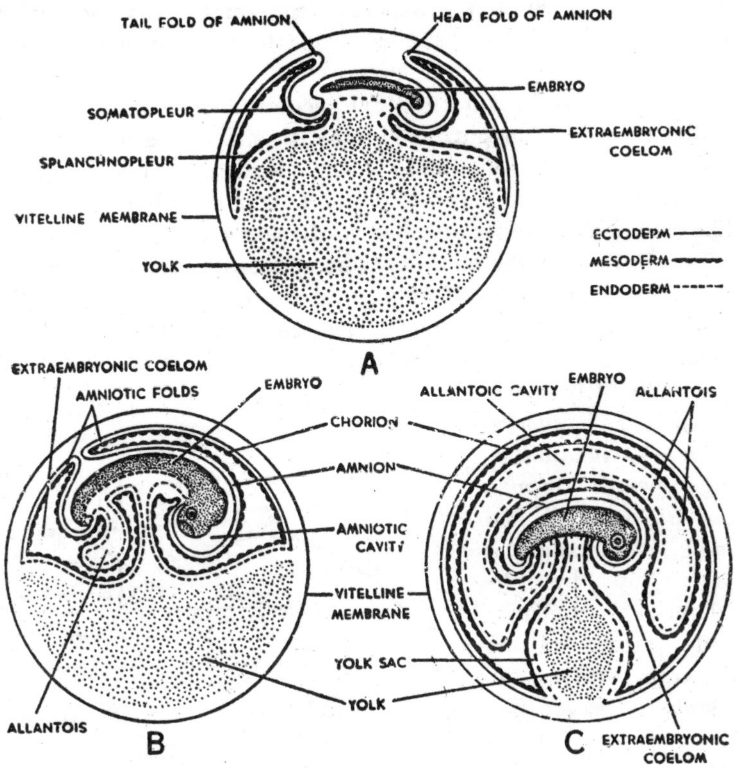

Fig. 186. Diagrammatic longitudinal sections, showing three stages in the formation of foetal membranes in a chick-embryo.

head and tail folds are soon joined by the *lateral folds of the amnion*, and the embryo is now completely covered by a double roof growing over it from all sides. With complete fusion of the folds, the embryo is covered by *two* membranes which are separated from each other by the extra-embryonic coelom.

The *inner* membrane is the **amnion.** It consists of ectoderm on the side towards the embryo, and somatic mesoderm on the side away from the embryo. The space between the amnion and the embryo is the *amniotic cavity*. It is lined by ectoderm and filled with *liquor amnii* or amniotic fluid (Fig. 186 B, C).

The *outer* membrane is the **chorion.** It consists of ectoderm on the side towards the shell and somatic mesoderm on the side towards the amnion. The chorion surrounds the amnion, and the space between the two is a part of the extra-embryonic coelom. Ultimately the chorion is fused with the vitelline membrane of the egg.

The **allantois** is a blind diverticulum which pushes out ventrally from the hindgut, and extends into the extra-embryonic coelom. It is lined internally by the endoderm and is covered externally by the splanchnic mesoderm. Late in the fifth day it appears like a rounded bladder attached to the hindgut by a narrow stalk (Fig. 186 B). By the tenth day, the allantois completely surrounds the embryo and yolk sac, filling up the extra-embryonic coelom (Fig. 186 C). It now comes to lie against the chorion and shell membranes. Blood vessels grow out on its surface and are ultimately developed into the *allantoic vessels*. Oxygen diffusing inwards through the porous shell and shell membranes is taken up by the allantoic capillaries and transported to the embryo. Thus the allantois serves as the respiratory organ of the developing chick. Secondly, the embryonic kidneys pass their excretions into the hindgut, of which the allantois is an outgrowth. Urea, uric acid and other waste products are thus stored in the allantois temporarily to be left behind in the egg-shell

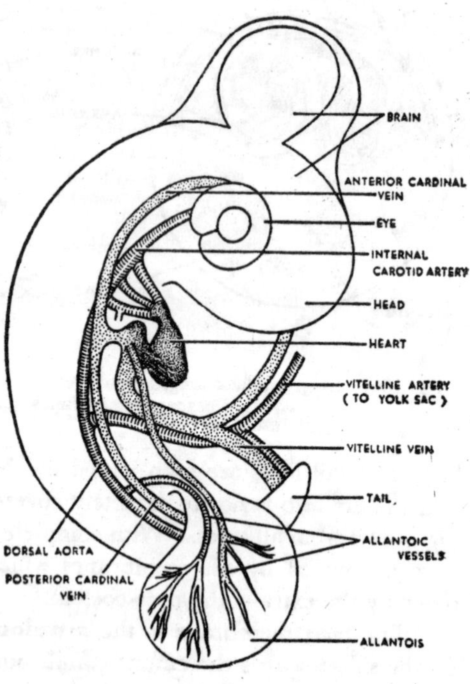

Fig. 187. A five-day old chick embryo drawn specially to show the allantois and main blood vessels.

when the young hatches out. The allantois, therefore, serves as an embryonic urinary bladder.

Hatching

By the twentieth day most of the yolk is digested and the entire yolk-sac is drawn into the body of the embryo. The amniotic fluid is greatly reduced and the large embryo occupies most of the space within the egg, excepting the air-space between the two shell membranes. Feather-papillae have already appeared covering the outer surface of the body. The embryo now pierces the air-space with its beak and takes the first breath of air into its lungs. With this, the allantoic circulation stops and the allantois begins to dry up. At the end of the twentieth day the shell is broken and the young chick struggles out. The embryonic membranes are ruptured and left behind within the egg-shell. The hard shell is broken by means of a horny elevation at the end of the beak.

CHAPTER XX

ORIGIN AND EVOLUTION OF LIVING THINGS

ORIGIN OF LIFE

The establishment of the law of biogenesis led the scientists to believe that living things do not arise *de novo;* they originate from pre-existing living things. This means that the only problem is to find out the origin of the first living thing. Unfortunately it is not a simple task, and the problem still remains unsolved.

The earth we live in came into existence about 2000 million years ago by the condensation of red-hot masses which were torn out from some other heavenly body. It cooled down slowly and in course of time acquired a gaseous atmosphere as well as collections of water. Life could never exist in such a severely hot environment. In all probability, the earth became ready to sustain life upon its surface after 1000 million years, although the centre of the earth was still in a red-hot state. The earliest living things have left no trace since their remains were totally destroyed and cooked by the great heat prevailing at that time.

There is controversy regarding the formation of the first living thing. It is assumed that the earliest living things originated spontaneously out of inorganic matter. This strange phenomenon occurred only once due to the peculiar conditions which prevailed upon the earth at that pre-historic period. The soil was submerged beneath the cooling water and the atmosphere was surcharged with carbon dioxide. Water evaporated from the earth's surface continuously, thus producing a canopy of cloud which covered the sun. Lightening accompanied with torrential rains, subdued light, and frequent volcanic eruptions were the order of the day. These are supposed to be the impelling causes which converted an inorganic nitrogenous compound into a complex molecule of protein which was forthwith transformed into a speck of protoplasm. These pre-historic living things were extremely resistant against the destructive environment and aggregated by rapid multiplication to produce more complex forms. In all probability, they were similar to the filterable viruses. These bodies modified considerably before their transformation into the simplest plants. Animals must have originated later, because they required plant-products for their food. Spontaneous generation of life cannot take place today, because the impelling causes are not present.

ORIGIN OF SPECIES

A globe-trotter in the course of his travel is sure to look at the numerous species of beautiful plants and animals which are marvellously adapted to their respective corners. He may not be a biologist but he is bound to enquire, how these diverse forms have come into existence. Where did they come from? These two questions reacted upon the human mind from the bygone ages. It led to the germination of various theories for solving the mystery—namely, the *origin of species.*

Dead ideas

Each race, each religion, and every nation have some peculiar myth regarding the origin of the world and the origin of the species. These legends are known as *cosmogony.* The latest cosmogony which dominated only about one hundred years ago is the biblical story of creation and this was offered to the scientific world as the *theory of special creation.*

Theory of special creation. According to this view the world was created in six days during 4000 B.C., and each kind of plants and animals originated spontaneously from dust and other non-living matter. It also states that since their creation plants and animals have suffered no change, and that they are incapable of changing, that is they are *immutable.*

The believers in special creation theory got the hardest blow from Louis Pasteur (1822–95) who completely refuted the idea of spontaneous origin of living things by establishing the *law of biogenesis.* His classical experiment is now used throughout the world for preventing the growth of living organisms in preserved food, and the process is known as *Pasteurisation* (page xxv).

Special creationists believed in the *immutability of species* but modern biologists have proved that plants and animals often change by *mutation.* Even in the past wild stocks were changed into useful domesticated forms by cross-breeding experiments.

Lastly, the special creation theory fails to account for the presence of fossil remains of extinct plants and animals which are entirely different from the modern types.

Theory of Catastrophism. Cuvier (1769–1832) examined the fossil remains of ancient plants and animals the like of which are not found at present as living forms. In order to explain their occurrence, Cuvier postulated that there had been in the long history of the world several creations. At the end of certain period, each creation was destroyed by some cataclysmic incident leading to widespread extinction of species. After this, fresh set of plants and animals were created *de novo* on different lines.

Unfortunately there is no evidence of such catastrophic incident. Being himself a believer in the special creation theory, Cuvier failed to grasp the importance of fossils. He realised that living forms alter in adaptation to changes in their environment but tried to explain fossils with his theory of catastrophism, thereby gaining no credence in the modern era.

Present idea

The present is the age of *evolution.* The world evolution means change—constant and progressive. Modern scientists believe that the physical world has originated by *inorganic evolution* and the living species by *organic evolution.* The idea of origin of species by organic evolution is really an old one. It comes from the Greek philosophers five centuries before the Christian era. This germ of evolutionary thought was eclipsed for several centuries and then reappeared in the modern age through the laborious work of several scientists.

Theory of organic evolution. The theory of organic evolution postulates that life began with very simple forms which were simpler than those which are now represented by the lowest organisms. Furthermore, all the existing species of plants and animals are derived by *descent with modifications* from other pre-existing species which were simpler in structure. In other words, modern plants and animals have originated by a gradual *process of evolution* from simple pre-existing forms.

Although the *modus operandi* or the *mechanism of evolution* has raised many controversial opinions yet the *path along which evolutionary changes* have occurred is substantiated by profuse and conclusive evidences.

Path of evolution. The first speck of living substance, which was of complex protein nature, probably originated about one billion years ago in the sea-water. They triumphed over the non-livings for having three important features : (1) maintenance of steady state, (2) power of reproducing their own kind, and (3) ability of transmitting altered characteristics to their progeny.

The first living forms had to depend upon the energy-yielding inorganic compounds. Some of them acquired chlorophyll and thus equipped, they were able to manufacture organic food in the presence of sunlight. The earliest plants originated in the ocean in this way. Simplest of the animals evolved later out of some such plants which lost their chlorophyll by *mutation*. Gradually the earliest plants and animals attained the cellular status. As the ages rolled by, some of the unicellular species fused to form colonies, which in turn organised to produce multicellular species. Constant and progressive changes went on during the following 600 million years. The result was the evolution of the sea-algae, the primitive invertebrates, and the primitive fish-like vertebrates. Up to this stage life was restricted to the sea because the land surface was mostly covered by water.

About 350 million years ago, small marshy spore-producing plants invaded the land. These were the ancestors of the modern bryophytes. They were soon followed by tall spore-producing tree-ferns nearly 30·5 metres high. The primitive seed-bearing plants appeared about 300 million years ago but they gained ascendancy at a later date. The ancient plants spread into extensive forests from pole to pole. They were destroyed by natural calamities, such as extreme cold and severe wind storms, about 200 million years ago. The old forests are now found as coal beds. The primitive seed-bearing plants were replaced by cone-bearing conifers, and the flowering plants. The conifers dominated the earth about 150 million years back and the flowering plants gained ascendancy 50 million years later. The modern flowering plants are only 60 million years old.

Animals which came to live on the land along with the primitive bryophytes were arthropods related to scorpions and insects. The earliest amphibians originated 300 million years ago as descendents of the lobe-finned fishes. They invaded the land and were soon followed by the reptiles. The primitive giant-like reptiles ruled the earth during the following 150 million years and then they became extinct. The mammals and birds evolved out of some small primitive reptilian ancestors about 150 million years back. The mammals gained ascendancy over other animals during the last 60 million years. Man, the highest mammal and the master of the modern world, is only one million years old.

Evidences of Organic Evolution

The fact that evolution has actually occurred is hardly challenged today because of the overwhelming evidences in its favour. The evidences are collected from various sources, and all of them indicate that new species originated by constant and progressive changes from old pre-existing species. The array of evidences are outlined below under the following heads :

Palaeontological evidence. Palaeontology is a branch of Geology which is concerned with the study of fossils. **Fossils** are long-preserved remains of plants and animals which are found buried in the earth. They may be *petrified* skeletons or shells which have been converted into stones. They may be *foot-prints* of animals or *imprints* of other parts such as leaves of a plant. They may be *intact fossils* such as the mastodons preserved in the frozen ice of Siberia. They may be insects preserved in crystal clear amber. They may be petrified *cast* of an individual from the interior of which the individual itself has disappeared by decay. Most of the fossils were formed when a section of earth-containing buried organisms were gradually converted into stone by geological changes. In such a section the oldest layer is evidently the deepest, and the youngest is the topmost. But as the years rolled by, this arrangement may be disturbed by various geological upheavals, such as cracking and erosion of the upper layers, thereby exposing the buried fossils. Embedded in the earth along with fossils there are radioactive substances, such as Uranium, which undergoes decay and is changed into Lead within a definite period of time. Thus by measuring the quantity of these substances in a given layer, the age of the crust can be determined accurately along, with the age of its fossil records. Judged on this basis the first fossils are about 900 million years old, and these were unicellular plants and animals. Calculating in this manner the various crusts of the earth are divided into five *eras*, and each era is further subdivided into a number of *periods*. The fossil treasure of these eras and periods clearly indicate that plants and animals became more complex with the passing of time. In the crust just above the remains of the unicellular organisms, fossils are abundantly represented by diverse forms. This layer is 450 million years old and contains fossils of all the invertebrate groups. Fish-like vertebrates appear in the next higher layer which is 380 million years old. This was followed in successive layers by amphibians (350–300 million years), reptiles (300–150 million years), mammals (180–150 million years), birds (150–120 million years) and man (1 million years).

It is to be noted that members of each group at the beginning of its appearance were different from the plants and animals of today. They were mostly queer forms. Some of them were ancestors of later types, others were intermediate forms, and a few were highly specialised which eventually became extinct without leaving any descendants. Prominent amongst the latter were huge **dinosaurs,** a group of ancient reptiles, some of which measured 24 metres from snout to tail and stood 12 metres high. Some dinosaurs, such as the *Icthyosaurus* were aquatic types; others, such as the *Pterodactylus* took to the air. Where are these monsters which ruled the earth only 150 million years ago? We know them today only by their fossils. Why did they become extinct? Probably the tyrants had to pay the price for becoming overspecialised.

The **intermediate forms** of fossils may be cited as 'missing links' between two different groups because they possessed features common to both. The most famous amongst such transitional forms is the fossilised bird *Archaeopteryx* which was found in Switzerland about 100 years ago. This ancient fossil bird is 150–120 million years old. It possessed wings and feathers which are characteristics of birds, yet it had teeth like the typical reptiles and possessed a lizard-like tail. Evidently *Archaeopteryx* is an unmistakable example of a transitional form linking the birds to the reptiles. Similarly, a group of fossil plants, the *Pteridospermae*, serves to connect the pteridophyta with the spermatophyta. The fossilised Pteridospermae possessed characteristics of both the groups. They therefore link the ferns with the modern seed-bearing plants.

In addition to the history of origin of large groups of animals and plants, palaeontology records the history of origin of certain modern species such as camels, elephants and horses. A comprehensive document of the fossil history of the horse is represented diagrammatically in Fig. 188. The ancestors of the modern horse, *Eohippus*, lived about 60 million years ago. It was only 27 cms. in height, and it had four toes on the fore feet and 3 toes on the hind feet. Every subsequent layer of the earth's crust contains slightly different types of horse fossils. They reveal a steady increase in size, particularly the height of the animal, with a progressive

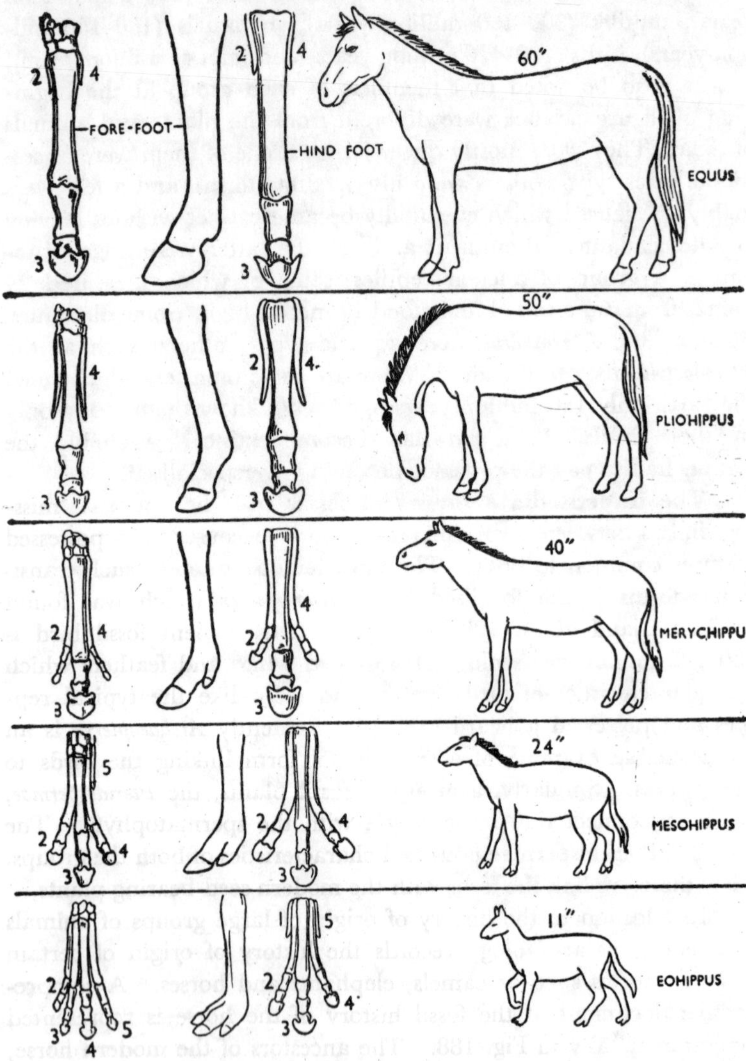

Fig. 188. Diagram of the horse fossils showing the gradual change from the pre-historic *Eohippus* to the modern horse *Equus*. 1, 2, 3, etc., denote the number of toes.

reduction in the number of toes. The *Mesohippus* which was found at a higher layer stood 54 cms. high and had three toes in each foot, the side toes touching the ground. *Merychippus* and *Pliohippus* which were found in successively higher layers had three toes on each foot, side toes no longer touching the ground. Finally came the fossil of the modern horse *Equus* with single-toed foot, the side toes being reduced to mere vestigial flints. *Equus* stood full 150 cms. high. The fossil horses prove that the present-day *Equus* is the child of the pre-historic *Eohippus*. This means that 'descent with changes' has taken place. The documentation of the horse fossils is a direct evidence in favour of the evolution theory.

Since fossils were formed only under certain circumstances and since most of the earlier fossils were destroyed by geological upheavals, the palaeontological record book is more or less incomplete. It is an old book with some of its pages worn out and a few completely torn away, particularly from the first chapters. But in spite of this the palaeontological record book is sufficiently complete to prove: (*i*) that the plants and animals were changing through the past ages; (*ii*) that the earliest were the simplest, and the complex organisms appeared progressively in successive ear; (*iii*) that there had been transitional or intermediate forms connecting two large groups, such as ferns with flowering plants or reptiles with birds; (*iv*) that there were intermediate forms connecting a modern species directly to its pre-historic ancestral form, for example, the intermediate types linking the present-day horse *Equus* to the pre-historic *Eohippus* which lived 60 million years ago.

Geographical evidence. Each corner of the earth has its own contingent of flora and fauna. Different regions are inhabited by different forms, and the climatic condition of a country plays very little part in the present-day distribution of plants and animals. Elephants for instance are found in India and Africa but not in Brazil although Brazil is as much tropical as the other two countries. Furthermore, the African elephants are somewhat different from their Indian brethren. Newzealand and Great Britain are almost alike in climatic condition but the living forms inhabiting the two places are widely different. Again representatives of a group may occur in widely separated places, but not in the regions in between them. For example, Llama and Alpaca, which are related to Camels reside in South America while the Camels are inhabitants of Arabia and Central Asia. It has been proved that the ancestral form of the Camels is really American. It migrated westward to Asia across a land bridge which used to connect America with Asia in the pre-historic time.

From such anomalies in the distribution in space it may be inferred: (*i*) that modern forms originated in the past from a common ancestral type; (*ii*) that the ancient ancestral types, due to their migratory habits, went far away from their original home unless they were prevented to do so by the presence of some obstacle or *barrier* in their path of *dispersal;* (*iii*) that the barriers may be physical or physiological; (*iv*) that having reached a new country they were forced to live in a new environment; (*v*) that they slowly underwent modifications in order to adapt themselves to the new conditions and those which were *completely* successful ultimately changed so much that they were transformed into new species; (*vi*) that those which were *partially* successful changed into weak intermediate forms which ultimately perished and became extinct unless they were protected by the formation of a strong barrier.

It has been observed by travellers that animals residing on islands are similar to but specifically distinct from those of the nearest mainland, and that greater the distance between the two places the greater the difference in their faunas. Thus the Galapagos group of islands, situated at a distance of about 805 kilometres away from the nearest American coast, has a fauna which is typically American but most of the species are restricted only to the islands, and there are some species which are more or less confined to a single island of the archipelago. This fact also suggests 'descent with modification'. Ancestral forms migrated to the islands from the nearest mainland, and having arrived there they were isolated and slowly transformed into new but related species.

Morphological evidence. (1) **Comparative anatomy** of the vertebrates reveals a close similarity in structural planning. The slight modification in detail are the effect of adaptation to different modes of living. For example, if we compare the heart of the fish, toad, lizard and guinea-pig we find that this organ is built on the same general plan in every case. The two-chambered heart of the fish is very simple and admirably adapted to its aquatic enviroment. In the toad, though basically the same, the heart has modified a bit in response to the change of habitat. Thus the two-chambered heart of the fish has been modified into the three-chambered heart in the amphibian, but even this is not sufficient to check admixture of the arterial and venous blood. In the lizard, which is a reptile, the heart is practically three-chambered; there are two auricles,

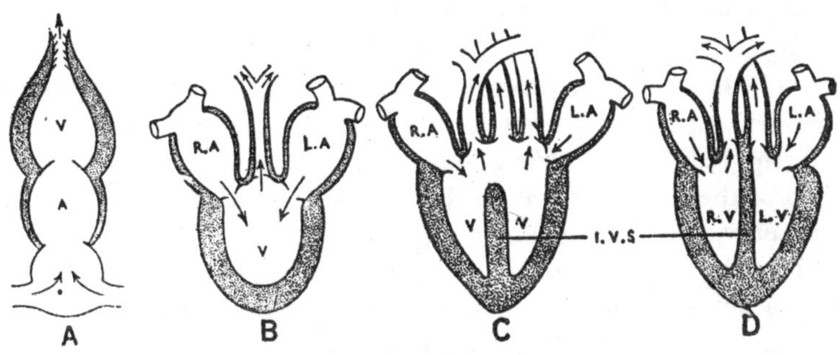

Fig. 189. Diagrammatic representation of the heart in fish (A), frog (B), lizard (C), and guinea-pig (D). In the fish the heart is two-chambered, there is one auricle (A) and one ventricle (V); in the frog the heart is three-chambered, there is a right auricle (R.A.), a left auricle (L.A.), and a ventricle (V); in the lizard the heart is three-chambered but the ventricle is partially divided by an incomplete interventricular septum (I.V.S.); in the guinea-pig the heart is four-chambered, there are two auricles and two ventricles.

but the single ventricle is partially divided by an incomplete partition. In the guinea-pig, which is a mammal, the heart is four-chambered; there are two auricles and two ventricles, and this four-chambered heart is perfectly adapted to separate and direct the two kinds of blood to their respective destination (Fig. 189). The same is true about other organ systems. In all vertebrates the nervous system includes the brain with five comparable subdivisions. The brain, however, exhibits progressive changes from fishes to mammals;

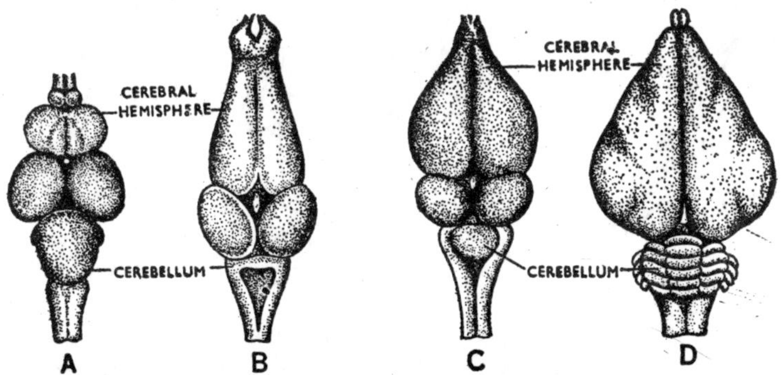

Fig. 190. A—Brain of fish; B—brain of frog; C—brain of reptile; D—brain of guinea-pig.

there is gradual enlargement of the cerebral hemispheres—particularly the pallium which is the seat of intelligence, and also of the cerebellum which is concerned with muscular co-ordination (Fig. 190).

(2) **Homologous structures,** such as the forelimbs of the vertebrates, appear to differ widely from one another. In the horse they are required for locomotion on dry land. In birds they are modified to form the wings for flying. In the whales, the forelimbs take the form of paddle-like flippers for swimming in water. In man the forelimbs are typically pentadactyle and adapted for grasping objects, and for fine manipulations. But an examination of the skeletal architecture of these different types of forelimbs clearly indicates that they are all *pentadactyle* in nature and consist of practically the same bony elements and are all built on the same fundamental plan (Fig. 191).

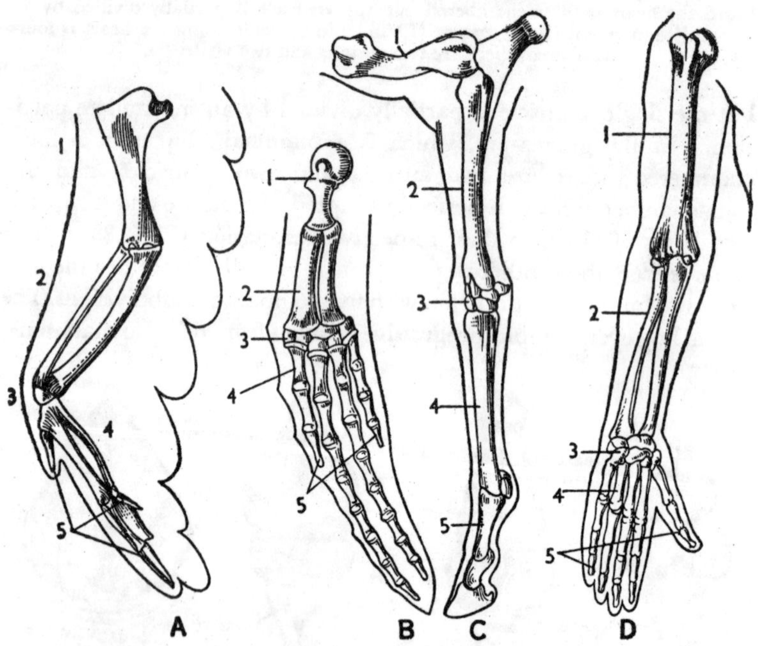

Fig. 191. Diagram to illustrate the homology of the forelimb. A=forelimb of a bird adapted for flying; B=forelimb of a whale adapted for swimming; C=forelimb of a horse adapted for running; D=forelimb of a man adapted for grasping. 1—humerus, 2—radius and ulna, 3—carpals, 4—metacarpals, and 5—fingers with phalanges.

Similarly the green stem of a plant, the tuber of a potato, the thorn of *Duranta* and the tendril of *Passiflora* may appear to differ widely from one another but all of them are *homologous structures* and the difference in appearance is due to difference in function.

(3) **Vestigial organs** are useless structures of reduced size, which are found in many animals. These structures are not functional in a given species but are homologous with functional structures in other closely related species. The muscles concerned with the movement of the pinna in man serve as a good example of vestigial organ. These muscles are useless and small because man does not require to move his ear for catching sound-waves. In related lower forms, such as guinea-pig, these muscles are well-developed useful structures which turn the pinna towards the source of the sound. The useless vermiform appendix in connection with the caecum of man is another example of a vestigial structure which is homologous to the well-formed and useful appendix of the rabbit (Fig. 192).

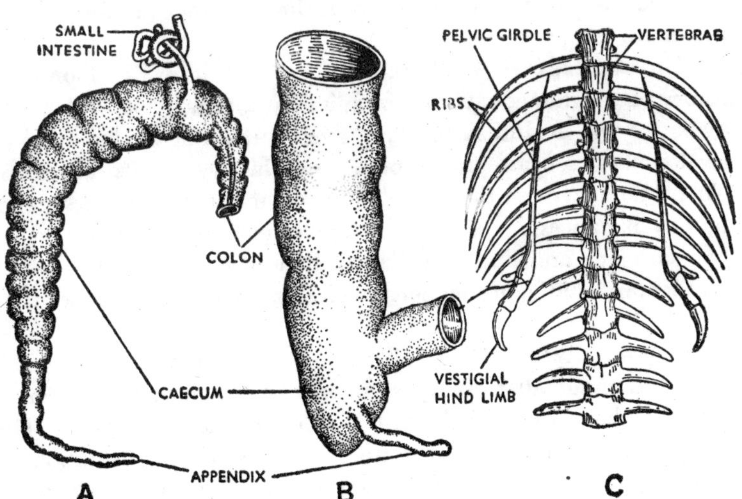

Fig. 192. Diagram to illustrate a few vestigial structures. A=functional appendix of rabbit; B=vestigial appendix of man; C=vestigial hindlimb of python, a kind of snake.

It seems rather difficult to account for the presence of such useless structures in man unless we assume that they represent vestiges of a useful functional organ in some remote ancestral form. The greatly reduced vestigial wings of the running birds such as kiwi and ostrich and the vestigial hindlimbs of the snake python are other examples. The scale leaves on rhizomes and tubers, the staminodes or stamens without anthers, are examples of vestigial organs in plants.

(4) **Intermediate forms** linking closely related groups ought to be present if each individual is to evolve out of the pre-existing lower form. Unfortunately such intermediate forms are invariably weak and often fail to survive. At least one animal, the duck-billed mole (*Ornithorhynckus*) is worthy of mention in this connection. This is found in Australia and Tasmania, along with other ancient forms such as the spiny ant-eater (*Echidna*) and the Kangaroos. It has survived there because competition is very much reduced owing to the early isolation of Australia from the mainland before the origin of the higher mammals. The duck-billed mole is undoubtedly a mammal, since it is hairy and suckles its young. But like the reptiles, it has a large T-shaped interclavicle in its pectoral girdle and it lays heavily yolked eggs. It may, therefore, be regarded as an intermediate form linking the reptiles to the mammals.

Embryological evidence. Study of the developmental history of any individual plant or animal is a convincing evidence in favour of the evolution theory. Every organism during its own development repeats the evolutionary history of its ancestors. We are familiar with the tadpole larva of the toad which resembles the fish-like ancestors of the class amphibia to which toad belongs. Even in chick and human embryos there is a brief stage in which rudimentary and non-functional gill slits appear for a short time (Fig. 193).

The occurrence of a fish-like stage with gill slits in the toad, chick, and mammalian embryos can only be explained on the ground that they have descended from a fish-like ancestral stock. The embryos cannot but *recapitulate* the centuries of slow evolutionary changes through which their ancestors had to pass. They, however, have cut down the process to a few days, quickly adapting themselves, during their embryonic development, to a new mode of life in a new environment.

Thus the adults of two species of animals may resemble each other very little, but their embryos may do so very strongly. For example, early stages of chick and rabbit are almost indistinguishable whereas the adults are easily distinguished from each other.

Similarly the adults of annelid worms and molluscs differ widely, but the trochophore larva is common to both.

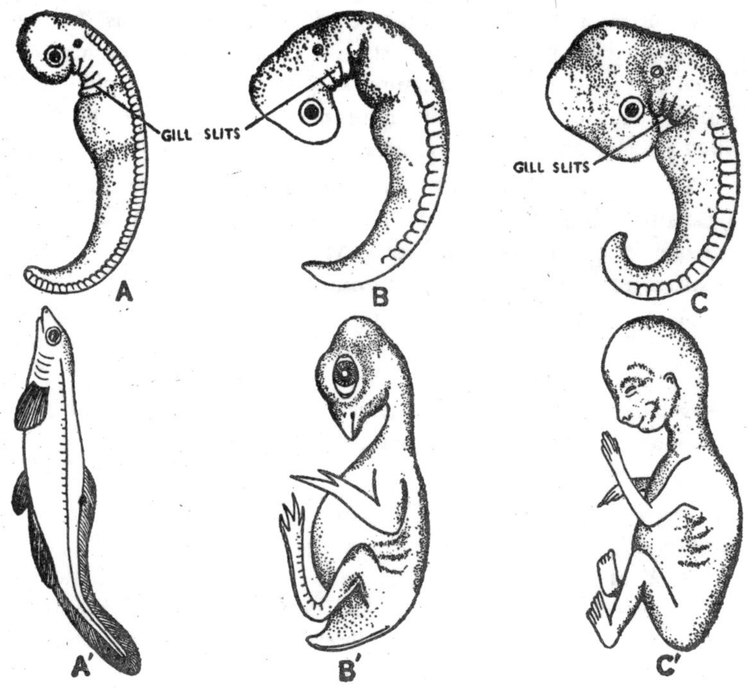

Fig. 193. A and A′—early and later fish embryos; B and B′—early and later chick embryos; C and C′—early and later human embryos.

Sometimes we are able to detect relationship in the embryos, thereby establishing relationship between the adults. *Sacculina*, which is a parasitic crustacea found attached to the abdomen of the crab, is not recognisable as an arthropod animal, but its larva closely resembles other arthropod larvae. Similarly tunicates are recognised as chordates by their tadpole larva.

Embryological studies of the various plant groups support the doctrine of organic evolution. The fern spore, on germination, produces a green *alga-like* structure which quickly develops into a *thalloid* body like the liverworts; finally the prothallus is replaced by the mature fern plant. In *Acacia moniliformis*, the seedling bears foliage leaves like other Acacias; but these are quickly shed off and the petioles expand to produce the leaf-like phyllodes.

Such observations led Haeckel to postulate his **Recapitulation theory** which states that *ontogeny*, that is development history of an individual, recapitulates *phylogeny*, or developmental history of the whole race. Although the recapitulation theory, as such, is not accepted by modern biologists, yet embryological studies have helped to detect vestigial structures such as the transient gill slits of land vertebrates, thereby linking them to the aquatic forms in which gill slits are useful structures.

Taxonomic evidence. The mere fact that we have been able to classify plants and animals into distinct groups is a significant evidence in favour of evolution. Modern classification is based on structural similarity, and this indicates close kinship. Hence the plants and animals belonging to any particular group are related to one another and evolved out of a common ancestral form. The same statement holds good when we consider the inter-relationship between the phyla. It must be noted however that the evidence from classification is really an extension of the morphological evidence.

Physiological evidence. Closely related animals possess the same kind of enzymes and hormones whose physiological actions are identical. Trypsin, for instance, is secreted by the pancreas. It is a proteolytic enzyme which breaks proteins into polypeptides. Trypsin secreted by a monkey is identical in chemical action to that of man, because man and monkey are closely related. The same is true for the hormone insulin which regulates carbohydrate metabolism and prevents diabetes in man.

Chemical studies of blood in closely allied species add further evidence in establishing relationship. It has been proved that haemoglobin crystals prepared from the blood of man and gorilla are identical, but these crystals differ from those derived from a horse or a dog. Thus *comparative physiology* supports the theory of evolution, because closer the relationship between two animals the greater is the resemblance between the chemical processes which occur within them.

Occasionally *comparative biochemistry* helps to establish relationship between greatly divergent groups of animals. For example the starfish, an invertebrate, differs widely from the vertebrates, but the two phyla agree with one another in the chemistry of their muscular contraction. Similarly, the universal occurrence of cytochrome in the protoplasm of all aerobic organisms prove that they are related to one another.

Evidence from artificial selection. Entirely new types of animals are often produced artificially under domestication. By careful selection of types which pleased him most or suited his purpose, man has produced numerous improved types of dogs, pigeons, horses, and cattle, which are structurally different from the wild ancestral stock. The diverse types of domestic pigeons are all derived from the blue rock-pigeon, *Columba livia*, by artificial selection. If it is possible for man to produce changed types of animals within a brief span of time, it is equally possible for the nature to do so within several millions of years.

Evidence from recent progress of evolution. The best evidence for evolution will be to see the process in action. In a few instances, actual progress of evolution has been recorded by men who are even now living. For example, the normal form of the peppered moth has light mottled wings. It was the prevailing type and was found everywhere in the countryside of England less than 100 years ago. Black mutant types would appear occasionally, and these being conspicuous were quickly destroyed by birds. The black melanic types therefore could not get a strong foothold except in industrial towns such as Manchester where they were camouflaged by dust, soot and smoke. With rapid increase of industrial towns, the melanic type has completely replaced the light coloured type. This change-over has occurred recently within the memory of living men.

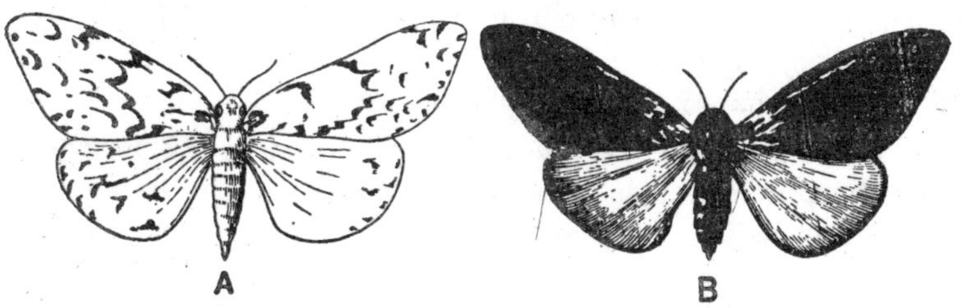

Fig. 194. A—Normal type of moth; B—Melanic type of moth.

Evolution must not be regarded as an event of the past. The process is still in progress, and probably there is no final end. There is no one today who can seriously challenge the fact that evolution has occurred. The occurrence of evolution therefore is not a theory. There are, however, several theories which try to explain the *mechanism of evolution.* How did 'descent with change' actually take place? This question is still debatable.

Mechanism of evolution

History. The concept of evolution started from the time of the early Greek philosophers. Anaximander (611–547 B.C.) first suggested the idea of a change in the living organism. Empedocles (500–430 B.C.) believed in the gradual evolution of different types of organisms. The fact that competition occurred between organisms leading to the *survival of the fittest* was known to him in a crude form, but he had no data to prove this. Aristotle (384–322 B.C.) believed in heredity and recognised relationship between organisms, thereby forecasting a phylogenetic series; but Aristotle did not entertain the idea of the survival of the fittest. The conception of the evolutionary idea was carried through the past ages by Augustine (354–430 A.D.), Thomas Aquinas (1225–1274), Buffon (1707–1788) and Erasmus Darwin (1731–1802). The last named person was the grand-father of the famous naturalist Charles Darwin.

During the past century, modern naturalists have clearly established the concept of evolution, and at least three of them postulated a theory for explaining its *mechanism.* Amongst them may be mentioned the names of J. B. Lamarck (1744–1829), Charles Darwin (1809–1892), A. R. Wallace (1823–1913), Hugo De Vries (1848–1935). The final touch was applied by T. H. Morgan (1866–1945) and his collaborators by clearly establishing the mechanism of inheritance, thereby explaining the *cause* of evolution.

Lamarck's theory. Jean Baptiste Lamarck (1744–1829), a French naturalist, made several valuable contributions to biological science, including the coining of the term 'biology' and using the same in its true sense. He studied comparative anatomy and planned a *tree of life* for explaining the phylogenetic relationship among organisms. He believed in the fundamental unity of living things, and in a progressive development of forms and functions in all organisms. But the most important contribution of Lamarck —his theory of evolution—was framed in 1801 and published in the 'Philosophie Zoologique' in 1809, that is the year in which Charles Darwin was born.

Lamarckism: The essence of the Lamarckian theory or *Lamarckism* may be summarised as follows: (1) Necessity in the organism may give rise to new structures or may lead to the disappearance of certain parts. Lamarck expressed this as the *law of use and disuse.* According to Lamarck an organ which is *used* extensively by the organism would enlarge and become more efficient, while *disuse* or lack of use of a particular organ would lead to its degeneration and ultimate disappearance. For example, the webbed toes of aquatic birds such as swans developed due to constant stretching of the skin at the bases of the toes in some ancestral form which lived on land. The necessity of the web of skin arose when the ancestors migrated into the water in search of food. This led to constant use and stretching, thereby a change was induced and a paddle-like foot evolved. Similarly, the ancestors of the snakes were lizard-like creatures with two pairs of limbs and the modern snakes lost their limbs by constant *disuse* while passing through narrow crevices. Thus by differential use and disuse of various parts, an organism could change a good deal; that is, the organism *acquires* certain *new characteristics.* (2) The second part of Lamarck's theory postulated that *acquired traits* induced by *use* or *disuse* of organs were transmitted to the offspring; this is the *law of inheritance of acquired characters.*

Fig. 195. Jean Baptiste Lamarck (1744–1829).

Lamarckism explains *evolution* of the modern giraffe in the following way. There was a short-necked ancestral stock which used to feed on tree leaves. It stretched its neck further up, to reach higher levels, when the leaves lower down were finished. Due to constant stretching the neck length increased a little and his new trait was inherited by the offspring. The latter in turn kept on stretching their necks and this was continued for many generations. Each successive generation would acquire the gains of the previous generation by inheritance, and would itself add a bit to the neck length. In the course of time, the long-necked modern giraffe evolved out of the short-necked ancestral form.

Criticism of Lamarckism: The first part of Lamarck's theory, that is, the law of use and disuse is acceptable. For example, moderate exercise taken regularly builds big muscles, or a limb put up on splints and not used for a long time undergoes atrophy. But the second part of Lamarckism, that is, inheritance of acquired characters, is not acceptable. It implies that a man who has developed large muscles by lifelong exercise will beget children with big muscles. Lamarckism was chiefly opposed by Weismann (1834–1914) who postulated that germ cells are not affected materially by changes in the body cells.

In spite of the laborious research of neo-Lamarckists such as Guyer, Smith and Cope, Lamarckism is untenable. Acquired characters are *phenotypic variations*. They cannot affect the *genes*. As such they cannot be transmitted to the offspring.

Darwin's theory of evolution. The name of Charles Robert Darwin (1809–1882) is a proverb in the history of science. This illustrious grandson of Erasmus Darwin was born in 1809 and his date of birth coincided with that of Abraham Lincoln. In his early life Darwin, like all other scientists of his time, believed in Lamarckism. As a young man he joined the naval expeditionary ship 'H.M.S. Beagle' and undertook a circumglobal voyage for five long years. He spent his time in collecting numerous specimens of plants and animals from different parts of the world. After returning home Darwin spent 20 years in studying his collections. At this time he was greatly influenced by the publications of Lyell and Malthus. By studying Lyell's 'Principles of Geology' Darwin learnt about the changing

Fig. 196. CHARLES ROBERT DARWIN (1809–1882).

conditions of the earth, and about the fossils which were known at that time. The famous essay on population published by Malthus taught Darwin about overpopulation and consequent competition for food and shelter.

Having completed his study, Darwin was preparing his *theory of natural selection* for explaining the mechanism of organic evolution when he received an essay from a younger scientist, Alfred Russell Wallace (1823–1913), who was working independently on the flora and fauna of Malayan archipelago. To his amazement Darwin found that Wallace's views on the origin of species coincided with his own theory. The natural selection theory was first published as a paper under joint authorship in 1858. Two renowned scientists of that time, Lyell and Hooker, presented the paper at the meeting of the Linnean Society and Darwin was conspicuous by his absence. In the following year, that is in 1859, Darwin published his classical work in the form of a book—"On the origin of species by means of natural selection."

Essence of Darwinism. Darwin's theory is based on intrinsic analysis of *facts* in a scientific spirit by induction and deduction. The following is the essence of Darwinism:

(1) **Prodigality of production:** The plants and animals have a tendency to increase in geometric progression, but the habitable space and the food supply remain constant. Darwin calculated that starting from a pair of elephants, the herd will increase to about 20,000,000 in 1000 years, and elephants are the slowest breeders producing 4 to 6 calves in their life-time. Such enormous prodigality in production results in *struggle for existence*.

(2) **Struggle for existence:** This means a keen competition amongst the living forms for food and shelter. It operates in a three-fold way: (a) *interspecific*, that is, struggle in between different species of organisms, (b) *intraspecific*, that is, struggle between members of the same species, and (c) *environmental*, that is, struggle against the changes of the environment.

(3) **Variation:** Darwin observed that no two living forms were exactly alike. Diversity tends to appear even among members belonging to the same species. Darwin paid particular attention to small, fluctuating and *continuous variations* which appeared randomly. According to him these continuous variations help the organism to win the struggle for existence. Large, *discontinuous variations*, which

appeared suddenly, were considered by Darwin as mere 'sports of nature', and therefore ignored.

(4) **Survival of the fittest:** The organisms possessing *suitable* variations which helped them to win the struggle for existence were better adapted to their environment. They *survived* and propagated their variations to the next generation. The others with *unsuitable* variations *perished.*

(5) **Natural selection:** This is the most important deduction of Darwin. Natural selection is the process by which individuals possessing favourable variations enjoy a competitive advantage over the others. They are better adapted to their environment, and therefore they survive in proportionately *greater* numbers and produce *more* offspring. The rest with disadvantageous variations fail to adapt properly to their environment and therefore eliminated by *natural selection.*

The favourable variations which are the cause of success are handed down to the offspring by inheritance. Thus the number of the favoured individuals increase rapidly, and if *natural selection* operates for a long time, those favourable variations which have attained the *survival value* are intensified successively from generation to generation, until the original ancestral forms are thoroughly changed into a *new species.*

For example, Darwinism explains the evolution of the modern giraffe in the following manner. The original ancestral forms were short-necked, leaf-eating animals. Darwin assumed that as a result of individual variation, some of them had slightly longer or shorter necks in comparison with the population's average neck-length. The longer-necked forms were better adapted to get at foliages situated a bit higher up. Consequently they were better fed than the shorter-necked fellows, and they produced proportionately greater numbers of offspring. As a result of *natural selection* the proportion of the longer-necked population would be doubled in the next generation. This is repeated in successive generations until the entire population would be transformed into individuals with slightly longer necks. Individual variation would occur in the new population and actual neck-lengths would vary more or less on either side of an average. Long necks would again be favoured in a second round of natural selection and then in successive rounds, until the modern giraffe with very long neck evolved out of the short-necked ancestral stock.

Criticism of Darwin's theory: In spite of strong evidences and critical scanning of facts, Darwinism suffers from certain serious drawbacks. A few objections to Darwinism are briefly discussed as follows:

(1) Variations were accepted by Darwin to be the chief tool in the process of evolution of new species, and he believed that small continuous variations of fluctuating type were inherited by the offspring. Unfortunately Darwin had no knowledge about the real cause of variation. At this time the science of genetics was unknown, and the laws of inheritance were unexplored. Most of the fluctuating variations considered by Darwin to be important factors in his theory of natural selection are not *genotypic* and as such they are not inherited.

(2) Darwin, like Lamarck, believed in the inheritance of acquired characters—a fact which is not proved by genetics.

(3) Darwin's natural selection mainly operates in one direction, and often leads to over specialisation and ultimate extinction. The canine teeth of the sabre-toothed tiger and the antlers of the Irish elk increased progressively in size because the characteristics in both the cases were favoured by natural selection. But ultimately, the structures became so large that instead of being helpful they became hindrance in the struggle for existence, and led to the extinction of the species.

(4) Natural selection theory fails to account for the degeneracy which is very often observed in the parasitic forms.

(5) The essence of Darwinian natural selection is the elimination of the unsuitable forms. Hence it is better to name it as the 'theory of natural rejection'.

(6) Darwin actually observed large, *discontinuous variations* or *mutations* to occur in nature. He rejected them as they occurred less frequently. But mutations are genotypic variations and they have now been recognised as important factors in the origin of new species.

In spite of its weakness Darwinism is still accepted as one of the important factors in evolution. Thanks to the untiring efforts of Thomas Henry Huxley (1825–1895), the great champion of natural selection, and others, such as August Weismann, the theory has been firmly established.

De Vries' theory. The *mutation theory* was published in 1901 by the Dutch botanist, Hugo De Vries (1848–1935). His theory is mainly based on his experiments on a plant called evening primrose,

Oenothera lamarckiana. De Vries found that certain strikingly different forms appeared suddenly among a population of normal type of evening primrose. He called them mutants. A mutant is a variant which arises abruptly among normal forms. A mutant always breeds true, that is, it produces offspring like itself. The term *mutation* or *psaltation* is applied to a sudden large change or discontinuous variation in organisms, and this can be inherited. According to the mutation theory, mutations are the real cause for the evolution of a new species. Numerous mutants may be produced in nature. They are then subjected to *natural selection* which determines the types that would *survive.* The mutants which survive in the struggle for existence are responsible for the origin of new species.

Fig. 197. Hugo De Vries (1848–1935).

Criticism: (1) Mutation often produces monsters which have no evolutionary significance.

(2) Mutations occur infrequently and they therefore cannot be regarded as the sole factor in evolution.

(3) Mutation theory accepts *natural selection* as the controlling agent in evolution.

Modern theory of evolution. This is the product of recent researches in cytology, embryology, and genetics. In the opinion of modern scientists, the heritable characters of an individual rest upon particles of nucleoproteins or *genes* in the chromosomes of the gametes. Any variation in the characteristics of an individual, whether *continuous* or *discontinuous,* must come through changes in the genes. Such changes that suit well with the environment are advantageous, and individuals possessing advantageous changes get the better chance of living and multiplying. This will continue for successive generation until a final form comes into existence, differing profusely from the ancestral type. Natural selection acting as a screen leads to differential survival and differential reproduction. In the present outlook about the origin of species, Darwin's struggle for existence may not be in the form of a competition, but the selective value has been found to be more important in differential survival of different variations.

The modern theory explains the evolution of the giraffe in the following way: Every generation of the short-necked ancestral stock must have included a few *mutant* types, with shorter or longer necks than the average neck-length of the population. The longer-necked individuals are in a more advantageous position. In the subsequent generation they will produce more longer-necked forms. This will go on through several generations in which changes in the gene would produce mutants, and natural selection acting as a screen would again and again eliminate the short-necked individuals, until the appearance of the modern giraffe with very long neck.

This modern theory is known as the *synthetic theory*. Several investigators of the synthetic school, such as Haldane, Ford, Waddington, Müller, Dobzhansky, and others have contributed their bit in its shaping. It is nothing but a completely remodelled *natural selection theory* minus its weaknesses.

CHAPTER XXI

ADAPTATION

The study of the preceding chapters has revealed that living bodies—plants and animals may grow in diverse environmental conditions, and their bodies are perfectly suited to live and thus adjust themselves in those situations. The characters of an individual which help it to survive in a particular environment is called *adaptation.*

Plants and animals are not only very large in number, but also they are equally varied in nature. Apart from the infinitely small bacteria and microbes which may grow in all conceivable situations, most of them are either aquatic organisms or terrestrial, i.e. land-dwellers. Aquatic situation is more homogeneous in nature in comparison to terrestrial one where the degree of diversity is much more pronounced. Hence terrestrial living bodies exhibit innumerable diversities as regards adaptation.

Adaptation in Plants. Aquatic plants are those which live in water or situations with a higher percentage of water. They include plants of all groups from the lowest algae to highest angiosperms. Generally they are put into three categories, viz., (1) those floating on the water surface, (2) those submerged in water, and (3) those growing mainly in water often having roots put into the muddy soil. They have simpler usually soft and delicate bodies for adjusting themselves to aquatic habitat. The absorbing organs like the roots are feebly developed as they can absorb water and nutrients by their entire body surface. Occurrence of air spaces for giving buoyancy and storage of gases is another distinctive character. Multiplication by vegetative parts far exceeds normal sexual reproduction.

The terrestrial plants have well-organised bodies to make best use of available water. Common land plants growing around us have distinct roots with root hairs for absorption, mechanical tissues for protection against stresses and strains operating from outside, well-formed expanded leaves suitable for gaseous exchange, particularly for manufacture of food by utilising solar energy, and formation of protective cuticle on the epidermis of exposed organs for preventing excessive loss of water. Maintenance of water balance, i.e. balance between intake and output of water is a significant factor. The xerophytes or plants growing in deserts or situations with great dearth of water are usually short ones with extensive root system, much reduced leaves which are often rolled or folded, modified into hard spines with heavily developed cuticle and waxy coating on the epidermis. It is evident that they are nicely adapted for effective absorption of available water and also for retaining the same by reducing the evaporating surface to the maximum extent. Plants growing in saline marshes or near sea shore have interesting adaptive features. Many roots usually come vertically upwards for utilising oxygen from air. The adaptive characters of the plants stated above reveal that plants maintain a wonderful balance between the intake and output of water.

Plants like orchids growing on the trunks of trees with roots hanging the air are the epiphytes. Those roots possess a special tissue for absorbing moisture from the air. *Cuscuta* or Swarnalata is a well-known parasite which is a leafless angiosperm. It grows at the cost of another living plant referred to as host, by sending out special sucking roots into the body of the host for nourishment. Partial parasites are also noticed. Sandalwood tree of Mysore grows as a partial root parasite. *Rafflesia* yielding the largest flower of the world grows as a root parasite. Normally many fungi grow on dead rotten organic matters saprophytically, but there are saprophytic angiosperms as well. Symbiosis or leading a life of mutual friendship between plants of two groups is another interesting feature, e.g. lichens.

Adaptations in Animals

Like plants the animals may also be aquatic or terrestrial. The aquatic forms include animals from all major groups—from Protozoa to mammals, but due to homogeneous nature of the environment they exhibit a tendency towards convergence of structural organisation. But the land-dwellers on the contrary have to survive in diverse environmental conditions and thus possess more pronounced adaptive features.

As already stated aquatic medium, whether fresh or saline, presents a homogeneous condition and is much denser than air. The animals living there must possess bodies capable of tolerating the pressure of the medium, offer least resistance during locomotion, keen sense organ, mechanism to keep osmotic balance and adequate respiratory devices, these being the basic needs of aquatic life.

Lower aquatic animals may be fixed or may move with water current. Respiration may be carried on by direct diffusion through body surface or specialised structures for the purpose may be present. Appendages in some cases, as in prawn, may be built for locomotion and other processes. Many insects have appendages modified for swimming and to act as respiratory gills.

The fishes exhibit the triumph in aquatic life. They have laterally compressed body capable of offering least resistance during swimming, forwardly directed mouth for procuring food and producing water current to bathe the gills, occurrence of different fins to help in swimming and the specialised organ—air bladder, for maintaining desired plane of floating. The mucus-laden scaly covering prevents the body from becoming wet. Blood vessels are so arranged that oxygenated blood may reach the different parts of the body from the gills. Some salt-water fishes retain their excretory product in the blood and maintain osmotic balance. Accessory respiratory structures for breathing air are present in ones inhabiting water with poor oxygen concentration, as in Koi, Magur and Singhi.

The amphibians exhibit distinct features of transition from water to land life. As found in toads and frogs, structures like webbed feet and partly naked skin are helpful in staying in water. In spite of their ability to live on land all amphibians return to water for egg-laying and their larval forms have perforce almost fish-like bodies.

Mammals like whales are completely aquatic with distinct fish-like bodies, and they cannot survive on land.

Life on land where the animals had migrated for some reasons in the past, presents a much more complicated picture from the point of view of adaptation. Here the medium is air, with lighter density than water. The basic requirements for survival on land are—suitable body for living in a medium of lighter density, adequate mechanism for using air as respiratory medium, arrangement for preventing desiccation, efficient sense organs for meeting aerial stimuli, suitable anatomical adjustments for counteracting the force of gravity and proper mechanism for protecting the progeny from the hazards of life.

Early land animals were crawlers. The anterior and posterior limbs gradually became long to give support to the body. A neck then appeared between the head and trunk to make the former more flexible and suitable for procuring food. Tail in certain forms is used for support. Arrest of growth after a certain period is a notable change as the increase in weight makes life on land more difficult, particularly in case of those adapted to flight. Special organs, lungs, appeared for facilitating respiration in air. In animals like reptiles the skin is covered by dry scales for preventing loss of water. Special receptors for chemical stimuli are restricted to the mouth cavity and nostrils. The most important changes occurred in the eyes and ears. Various types of eyelids with glands to maintain a watery film in front of cornea and attainment of binocular vision are specific devices against the problems of desiccation, mechanical injury and alteration of refractive index of the eyes. The construction of ear—by way of addition of a new part, middle ear, is conductive to effective sound reception and balancing. The amphibians lay their eggs in water, the reptiles and birds have impervious shell round the egg, the mammals excelled others in having a system of internal development—a wonderful device for protection of the progeny.

Animals like horses, deers, and dogs have to move very fast for safety and for procuring food. Their bodies possess distinctive adaptive features for the purpose, viz., (1) streamlined and spindle-shaped body to offer least resistance during movement, (2) specialisation of limbs, particularly hindlimbs for running, (3) changes in the posture of foot, so that body weight is carried on the tips of the fingers and toes, (4) reduction of number of digits in the limbs, and of ulna of the forelimb and fibula of the hindlimb, (5) limbs capable of movement in antero-posterior plane, distal segments being more elongated than proximal ones, and (6) young ones either kept in seclusion or able to follow the parents immediately after birth.

Burrowing animals like moles and others are able to dig actively or live in the burrows thus prepared. Often burrows made by other animals are used as abode. They have elongate bodies with pointed head and weak jaws. Eyes and ears are either undeveloped or vestigial. The limbs are the digging organs and so are strongly built having sharply clawed digits. The incisor teeth may also be used as digging organ.

Climbing animals are usually of three types, viz., (1) those like house lizard which climb on the wall; (2) bears and jaguars which normally live on land but are good climbers, and (3) the permanent tree climbers like the monkeys. They have strong chest, ribs and girdles; the clavical bone of pectoral girdle is well-developed for climbing and hanging on tree trunks. They either grasp the branches of the trees and prehensile feet or they have curved claws on the feet for the purpose. Presence of prehensile tail is another distinctive feature.

The flying animals both passive and active fliers, have usually two sets of adaptive features—one set for flying and the other for staying on the land. The birds are the most successful flying animals. The act of flying requires excess energy and for this reason the metabolic rates of birds are high. They have streamlined bodies to offer minimum resistance during passage through air and well-developed wings with feathers. Some bones are hollow and air filled which make the body light. The presence of air-sacs with lungs makes the body light and at the same time supply excess of oxygenated air. The evaporation is prevented by feathery covering, formation of semi-solid urine and frequent intake of water. In flying fishes the pectoral fins are extended to act as wings. The birds have powerful sense organs and efficient nervous system. Eyes of the birds are specially mechanised to adjudge the distance. The nocturnal flying animals like bats have keen hearing mechanism.

The cave-dwellers have usually slender bodies with long thinned out appendages. Loss of pigmentation and degeneration of eyes are common features, but they have powerful sense organs to receive smell and sound vibrations.

The desert animals have adaptive features for meeting three necessities, viz., getting moisture, conservation of moisture and defence. Their bodies are usually covered by thick skin often scales for preventing evaporation. Some of them, e.g. camel, usually have special chambers called water cells in their stomach for storing water. These animals usually take up the colouration of the background. Spines, poison glands and sharp claws are present for self-defence. Eyes, ears and nostrils have special protective arrangements against sand particles.

CHAPTER XXII

HEREDITY AND GENETICS

The offspring of any particular species resembles its parents in a general way. The young of cats are always kittens and never puppies. With respect to details, however, there is a lot of variation. No two kittens are exactly alike. Some may be black, others white, and still others may have a mottled fur. In some instances the young may resemble one of their parents or they may have characteristics similar to those of their grand-parents rather than their own parents. This passage of characters from one generation to another is called inheritance or heredity. It is resemblance based upon descent. **Heredity** may be defined as the occurrence in the living organisms of qualities that are derived from their ancestors. Heredity includes not only physical but also physiological qualities. These qualities may be either *expressed* or they may be *latent* in the offspring.

Differences among members of any particular species are spoken of as **variations.** These may be *fluctuating* or *phenotypic* variations which depend upon environmenttal factors or these may be *germinal* or *genotypic* variations which are not influenced by the environment. The phenotypic variations are not heritable, whereas the genotypic variations are handed down to the progeny from one generation to another. The science of **Genetics** seeks to account for the resemblance or difference between the young and its ancestors. It is concerned with the heritable or genotypic variations of an individual and with laws governing the process of inheritance.

Mechanism of Heredity

Since higher organisms reproduce sexually and since the sperm and the egg are the only materials which pass from the parents to the offspring, the mechanism of heredity must be located in the gametes. Since only the head of the sperm enters the egg and since the head is the nucleus of the sperm it is reasonable to state that the nucleus of the male gamete is the bearer of paternal characters. The egg contains not only a nucleus but also some amount of cytoplasm. So the matter is not clear in this case, but it may be assumed that the maternal and paternal characteristics are handed on to the new individual in the same way. It has been proved by experiments that (1) mature eggs caused to develop by *parthenogenesis* produce individuals with only maternal characteristics; (2) an egg which is first deprived of its nucleus and then fertilized by a sperm produces an individual with only paternal characters. The results of these two experiments confirm the view that the transmission of hereditary characters depends upon the nucleus.

Modern geneticists have observed certain ultra microscopic particles called **genes** which are arranged in a linear series upon the chromosomes of the gametes. The genes are composed of special kinds of nucleoproteins whose chemical and physical properties are obscure. They appear as beads or cross bands upon certain chromosomes. A gene is regarded as a *factor* for a single heritable character. It is the unit of inheritance that is transmitted from the parents to the offspring, and is responsible for a particular character, such as colour of the eye or texture of the hair. A gene always acts by interaction with other genes, the cytoplasm, and the environment. Recently Müller has been able to produce gene mutation in the fruitfly *Drosophila* by exposing them to the bombarding action of the X-rays. Normally gene mutations occur during meiosis by random sorting of chromosomes —any single gamete receiving either one or the other of each homologous pair.

Thus gene mutations are the real cause of heritable variations. Phenotypic variations are not inherited.

Laws of Heredity : Mendelism

Experimental studies concerning the way in which characters are inherited was first undertaken by an Austrian monk, Gregor Johann Mendel (1822–1884). He chose the garden pea, *Pisum sativum* for his observations and spent 2 years in selecting **pure stocks** having distinctive and **contrasted characters.** He performed numerous experimental crosses by artificial pollination, each of which was carried through three or four generations, and kept a careful record of all plants and seeds of each kind that was produced in the course of his studies. From these data Mendel deduced three most important laws of heredity. His research was published in an obscure journal in 1866 and was soon forgotten.

It was brought to world-wide fame in 1900 when De Vries and two other workers arrived at the same conclusions independently of one another. So Mendel's laborious works, which formed the very basis of genetics, received recognition long after his death.

Mendel believed that there were unit carriers of heredity which occurred in pairs. One of each pair came from the female parent and the other from the male parent. These unit carriers are now called *genes*. They are situated in pairs on the *homologous chromosomes*, one of which is derived from each parent at the time of fertilization.

Fig. 198. GREGOR JOHANN MENDEL (1822–1884).

Mendel selected pure plants having contrasted pairs of characters, for example, tall or dwarf stems, yellow or green cotyledons, smooth or wrinkled seeds. When he crossed a green with a yellow pea, he found that all the offspring, in the first generation or F_1 generation, were yellow. Similarly the offspring were all tall when a tall variety was crossed with a dwarf one. This led him to the **law of dominance** which states that amongst the contrasting characters (genes) one is *dominant* and the other is *recessive*. In other words, if an individual has one dominant and one recessive gene, it always exhibits the dominant trait. Such contrasting pairs of genes are designated as *allelomorphs* or simply *allels*. It must be noted that intermediate forms such as yellow-green were never met with in pure crosses.

Mendel's original pure stocks were all *homozygous;* this means that each individual had like units or genes in both the chromosomes of a homologous pair. To be more explicit, genes for a particular character, say yellowness, were present in both the chromosomes. The product of crosses were however **hybrids** or *heterozygous*. This means that in each pair of homologous chromosomes in the hybrid, the genes are unlike, one of them containing the dominant gene and the other containing the recessive gene. The hybrid, however, exhibits the dominant trait. When Mendel crossed his hybrid yellow peas obtained by crossing pure yellow with pure green, he found that in the second generation or F_2 generation the proportion of yellow to green was 3 : 1. Of the three yellow peas in the F_2 generation one was pure yellow but the other two were hybrids with dominant trait. The only green pea obtained in this experiment was pure, having two genes for greenness. This led Mendel to postulate his **law of segregation,** which states that the unit characters or genes preserve their individuality in the hybrid state; they separate from each other, in each succeeding generation, to be redistributed in the offspring.

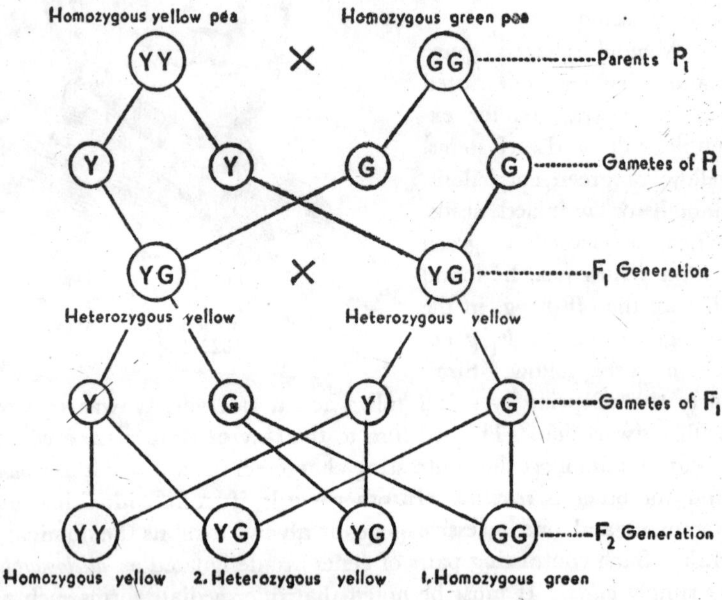

Fig. 199. Diagrammatic representation of cross between a pure yellow and a pure green pea, showing 3 : 1 ratio in the F_2 generation.

The ratio obtained by taking into account one set of contrasting characters or *allels* is referred to as the **monohybrid ratio.** (Fig. 199).

In crosses in which more than one characters were observed (dihybrid crosses) Mendel found that each pair of unit-characters behaved independently of the other pairs. Thus when a yellow smooth pea was crossed with a green-wrinkled variety, all of the first generation were yellow and smooth. If two hybrid plants of this first generation were crossed, all possible combinations appeared. Some of the offspring were yellow and smooth, some yellow and wrinkled, others were green and smooth, and still others were green and wrinkled. This led Mendel to the **law of independent assortment,** because one gene has apparently no influence on another gene for a different character.

The discoveries of Mendel form the fundamental basis of genetics and are often referred to as **Mendel's laws of heredity.** It asserts that (i) characters are determined by paired genes, one of which is dominant over the other; (ii) the genes segregate and redistribute themselves independently of each other from generation to generation; (iii) during redistribution one gene has no influence on another gene for a different character.

CHAPTER XXIII

INTERDEPENDENCE OF PLANTS AND ANIMALS

The living things, plants and animals, exist in all conceivable types of situations in the world. They are present on the snow-clad tops of the mountains, in the fathomless depth of the ocean, in barren desert with scorching heat and extremely adverse climatic conditions, in water, on land and even in air. The minutest bacteria, which cannot be seen without very high magnification of a microscope, are also plants; and they occur everywhere—in food we take, in water we drink, in air we breathe, in clothes we wear and even inside our own bodies. The vastness and variety of plants and animals have not only impressed inquisitive men but have also presented very interesting problems for scientific investigations.

An insight into the subject of organic evolution would reveal that living things were similar in the dim past. At one stage during that very long process bifurcation occurred and there had been progressive elaboration in the two lines culminating in modern plants and animals. These plants and animals are living in the world side by side deriving mutual benefit.

Plants are *autotrophic* or self-supporting bodies. Due to presence of the green pigment, chlorophyll, they can synthesise complex organic food matters like carbohydrates—sugar and starch, out of water and carbon dioxide gas utilising the energy from sunlight. This unique process is known as photosynthesis. Sugar forms the ground substance for the synthesis of other organic food matters. The whole animal kingdom including ourselves has to depend on plants directly or indirectly for the supply of food. The herbivorous animals feed purely on plant parts. The carnivores use animals as their food, which in their turn had their food from the plants.

During the process of photosynthesis plants take in injurious carbon dioxide gas from the air and give out almost equal volume of oxygen. All plants and animals respire day and night when oxygen is absorbed and carbon dioxide is given out. The amount of carbon dioxide given out by plants alone during respiration is not enough for photosynthesis; they need some more than what is supplied by animals. At any rate if the plants had not absorbed carbon dioxide gas for photosynthesis, the air would have been vitiated, so much so that life would have been impossible on earth. Thus besides providing us with the most essential thing, food, plants are purifying the air. For proper nutrition plants require a number of elements which are absorbed from the soil and air. Animal bodies and excreta are broken down to simpler elements by bacteria and other micro-organisms. They are added to the soil increasing its fertility when plants can have healthy growth by utilising those elements. Nitrogen is a very important element for the plants. Though enough of nitrogen is present in the air, plants can hardly utilise it. It mainly comes as nitrates and ammonia salts from the soil. The protein molecules of animal bodies and excreta are broken down making nitrogen available to the plants. Some insects have the power of fixing free nitrogen, and some micro-organisms living in symbiotic association in the lumen of some herbivorous animals can do the same. The use of animal excreta, cowdung, etc., and bone dusts as manures is a very common practice with farmers. The insectivorous plants like *Nepenthes* (pitcher plant), bladder wort, etc., get part of their nitrogen requirement from the bodies of the insects they catch. Many fungi grow at the expense of decomposed animal bodies or they lead parasitic existence on living animals. All these illustrations go to show that even for nutrition the plants and animals are interdependent.

Flowers are specialised organs of plants for securing reproduction or multiplication of race. For the formation of fruits and seeds union of two opposite gametes or reproductive cells present in the male and female part of the flower is essential. The pollen grains containing the male gametes must be carried to the carpel, the female organ of the flower. This transference, called pollination, is brought about by animals in many cases. The brightly coloured or sweetly scented flowers have many insect visitors like bees, butterflies, wasps, etc. Even flowers with very disagreeable odours have their own visitors. These insects in course of their visits in search of honey or pollen carry the pollen grains from one flower to another. Slow-moving animals like snails are found to crawl on the inflorescence axis of aroids when they accomplish pollination. Birds are the pollinating agents in plants like silk cotton with large flowers.

Every seed has the potentiality of giving rise to a new plant. In fact, the baby plant or embryo, remains in the seed in sleeping condition. If all the fruits would discharge their seeds in the shade of the mother plant the result would be disastrous. The seeds crowded in a small area with limited resources would try to germinate and grow. A keen struggle for existence would ensue and only the fittest would survive and the rest perish. To prevent this danger of wastage plants disperse or scatter their seeds over far and wide areas. In many cases dispersal is secured through animals. The grazing herbivorous animals very often carry the seeds and fruits during their movement in search of food. The ants and other insects carry small seeds and thus cause dispersal. The seeds of many fleshy fruits remain uninjured even when they pass through the alimentary canals of birds, squirrels, bats, etc. Those animals discharge the seeds with excreta on far off places—trunks of big trees, roofs and cornices of houses, where the seeds get very suitable medium—the excreta, for germination. Dispersal through human agency is also a noticeable feature. Many plants of economic and ornamental value have been carried by human beings to different countries where the plants in course of time got naturalised. The water hyacinth, a veritable pest of Bengal today, was brought to this country for the beautiful lilac flowers.

Ants and other stinging insects often take shelter on big trees, particularly fruit trees like mango and litchi. These insects act as an army and protect the plants from enemies whenever necessary. In some American acacias certain part of the leaf becomes hollow to be used as an abode of the insects and even some food materials are kept ready for them. The insects in return work as a troop of body guards and render defence service in time of necessity. This interesting device is known as ***myrmecophily***, literally meaning love for insects.

INDEX

INDEX

BOTANY

A

B

C

D

E

N

O

P

Q

R

S

T

U

V

W

X

Y

Z

ZOOLOGY

A

D

E

H

M

N

O

P

Q

R